In Memory of My Father

Shri Hira Nand

Modern Control System Theory

M. GOPAL
Department of Electrical Engineering
Indian Institute of Technology, Delhi
On deputation from
Indian Institute of Technology, Bombay

WILEY EASTERN LIMITED
New Delhi Bangalore Bombay Calcutta

WILEY EASTERN LIMITED
4835/24 Ansari Road, Daryaganj, New Delhi 110 002
4654/21 Daryaganj, New Delhi 110 002
6 Shri B.P. Wadia Road, Basavangudi, Bangalore 560 004
Abid House, Dr. Bhadkamkar Marg, Bombay 400 007
40/8 Ballygunge Circular Road, Calcutta 700 019

ISBN 0 85226 321 X

Published by Mohinder Singh Sejwal for Wiley Eastern Limited, 4835/24 Ansari Road, Daryaganj, New Delhi 110 002 and printed by Abhav Rastogi at Prabhat Press, 20/1 Nauchandi Grounds, Meerut 250 002. Printed in India.

PREFACE

Over the past two decades, Modern Control System Theory has been gaining great importance, for being potentially applicable to an increasing number of widely different disciplines of human activity. Basically, the Modern Control System Theory has involved the study of analysis and control of any dynamical system—whether engineering, economic, managerial, medical, social or even political. This theory first gained considerable maturity in the discipline of engineering and has been successfully applied in a variety of branches of engineering, particularly receiving great impetus from aerospace engineering. Recently, it has been applied in economics and other disciplines as well, and has proved very promising. A fascinating fact is that all these widely different disciplines of application depend upon a common core of mathematical techniques of the Modern Control System Theory. It is these techniques that I have exposed in this book, emphasizing their application in the engineering discipline.

Modern control theory has no doubt been presented in varying depths by many prominent authors. Among many excellent presentations, a few stand out in my mind because of the repeated use I have made of them over the years: *Linear System Theory*—C. T. Chen; *Linear Optimal Control Systems*—H. Kwakernaak and R. Sivan; *An Introduction to Linear Control Systems*—T. E. Fortmann and K. L. Hitz; *Optimal Control Theory: An Introduction*—D. E. Kirk; *Optimum Systems Control*—A. P. Sage and C. C. White, III However, despite having these and many other good books on modern control theory, while teaching undergraduate and postgraduate students of various engineering branches and while guiding doctoral students, I have experienced a strong need for a book that meets the following requirements:

1. A thorough exposure of modern control theory through the application of its potential concepts consistently to a variety of practical system examples drawn from various engineering disciplines.
2. A significant provision of the necessary topics that enables a research student to comprehend various technical papers in which modern control theory techniques are employed to solve many systems and control problems.
3. A flexibility which enables a teacher to add recent potential topics of Linear Multivariable System Theory.

I have therefore made an attempt to meet these requirements which are eagerly sought after, but have not been met in any of the existing works. The present book is the result of such an attempt.

This book presents both continuous-time and discrete-time systems, and brings out, in particular, the similarities which reinforce many ideas. Also, an attempt has been made to point out many exceptions that occur which warrant a careful study in each case. Special attention has been given to important control problems such as deadbeat control, non-zero set points, and external disturbances and sensitivity problems in optimal linear regulators. Further, at appropriate places, the recent proposals for system design have been emphasized, such as modal control, state observers and estimators, suboptimal control, etc.

Specifically, the book provides lucid illustrations of modern control system theory concepts by applying them repeatedly to five practical control problems from various engineering disciplines. It was not possible to avoid simplification of the real-life problems for the sake of tractability; yet, particular care has been taken to retain the essence of the real-life problems in their simplified versions. In fact, the same chosen five real-life problems have been thoroughly developed over several chapters to reinforce the concepts.

The book is organised as follows:
Chapter 2 presents the basic core background, namely: linear spaces and linear operators. Chapters 3 and 4 deal with the issues of modelling of systems. Chapters 5 through 8 primarily present methods of analysis; also, they project some of the potential design techniques that evolve from analysis. Finally, Chapters 9 through 12 completely address themselves to the design of controllers for several classes of plants.

Most of the theoretical results have been presented in a manner suitable for digital computer programming, along with the necessary algorithms for numerical computations. However, detailed discussion of these algorithms has been deliberately avoided; instead, suitable references for further study have been suggested. Exercise problems, tailored particularly to help the reader understand and apply the results presented in the text, have been given at the end of each chapter. In fact, some of these problems also serve the purpose of extending the subject matter of the text.

A basic working understanding in the following areas has been assumed on the part of the reader: calculus, linear differential and difference equations, transform theory, matrix theory and probability theory. It would be additionally facilitating, though not necessary, for the reader to have taken a course on classical control theory.

For teaching the material covered in this book, I suggest—based on the successful class-testing of the material in the courses EE 658 and EE 660 at the Indian Institute of Technology, Bombay—that two courses may be offered:

1. A one-semester course at senior undergraduate or postgraduate level covering Chapters 1 through 8.

2. A one-semester course at postgraduate level covering Chapters 9 through 12.

I have great pleasure in expressing the acknowledgements which I owe to many persons in writing this book. I warmly recognize the continuing debt to my 'mentor' in Control System Theory Prof. I. J. Nagrath of the Birla Institute of Technology and Science, Pilani. It is in him that I have found my teacher, friend, and source of inspiration. At the Indian Institute of Technology, Bombay, I have been influenced and assisted by a great many people while preparing this book. I acknowledge pleasant association with Dr. M. C. Srisailam, Dr. S. D. Agashe, Dr. H. Narayanan, Prof. V. V. Athani, Dr. (Mrs.) Y. S. Apte and Dr. M. P. R. Vittal Rao. I must record, separately, my appreciation of the help given by my doctoral students, Mr. J. G. Ghodekar, Mr. P. Pratapchandran Nair and Dr. S. I. Mehta during the crucial period of the growth of this book. Finally, I wish to express my gratitude to my wife Lakshmi, my son Ashwani and daughter Anshu for their interest, encouragement, and understanding and for bearing with me through the project.

I am grateful to the authorities of the Indian Institute of Technology, Bombay for aiding this book writing project through Curriculum Development Cell.

I warmly welcome suggestions and criticism from the users of this book. I shall consider it a pleasure to respond to specific questions concerning the use of the material in this book.

M. GOPAL

CONTENTS

Modern Control System Theory

1. INTRODUCTION

1.1 SYSTEMS: MODELLING, ANALYSIS AND CONTROL

The word 'system' implies essentially two concepts: (1) interaction within a set of given or chosen entities, and (2) a boundary (real or imaginary), separating the entities inside the system from its outside entities. The interaction is among entities inside the system that influence or get influenced by those outside the system. The boundary is. however, completely flexible. For instance, one may choose to confine oneself to only a constituent part of an original system as a system itself; or, on the contrary, one may choose to strech the boundary of the original system to include new entities as well.

In this book, we confine ourselves to only *physical systems*. In the study of physical systems, the entities of interest are certain physical quantities; these quantities being interrelated in accordance with some principles based on fundamental physical laws. Under the influence of external *inputs* (entities outside the boundary), the interactions arise in the system in a manner entirely attributable to the character of the inputs and the bonds of interaction. The inputs, while affecting the system behaviour, are usually not reciprocally affected and therefore are arbitrary in their time-behaviour.

In dealing with control systems, we will invariably be concerned with the dynamic characteristics of the system. Our main interest will be focussed upon only some of the entities of *dynamic* systems, namely those whose behaviour we wish to control. These entities—the *outputs* of the system, are normally accessible for purposes of measurement.

The study of physical systems often broadly consists of the following stages: (1) Modelling (2) Analysis (3) Design and synthesis.

Zadeh and Desoer (1963) have defined a 'system' as a collection of all of its *input-output pairs*. For a quantitative analysis of a system, we determine mathematical relations which can be used to generate all input-output pairs belonging to the system. In our terminology, the mathematical relations used for generating all possible input-output pairs of a system will be referred to as the *mathematical model* of the system. The mathematical model that we develop for a system should account for the fact that to each input to a dynamical system, there are, in general, a number of possible outputs. As we shall see in later chapters, the nonuniqueness of response to a given input reflects the dependence of output not merely on the input but also on the *initial status* (*initial conditions*) of the dynamical system. A set of differential/difference equations is a well-known form of mathematical description of a dynamical system which accounts for initial conditions of the system and gives rise to set of all possible input-output pairs.

Another useful form of mathematical description of a dynamical system is *state variable formulation*. As we shall see in later chapters of this book, the *state* of a system is a mathematical entity that mediates between the inputs and outputs. For given inputs and initial status of the dynamic system, the change in state variables with time is first determined and therefrom the values of outputs are obtained. The reader will observe the fact that the set of state variables, in general, is not a quantity that is directly measurable; it is introduced merely as a mathematical convenience. The only variables that have physical meaning are those that we can generate or observe, namely the inputs and outputs.

It is important to note that for the purpose of mathematical modelling, certain idealizing assumptions are always made, since a system, generally, cannot be represented in its full intricacies. An idealized physical system is often called *physical model* which is a conceptual physical system resembling the actual system in certain salient features but which is simpler and therefore more amenable to analytical studies. In many cases, the idealizing assumptions involve neglecting effects which are clearly negligible. Many of these effects are in fact neglected as a matter of course without a clear statement of implied assumptions. For example, the effect of mechanical vibrations on the performance of an electronic circuit is ordinarily not considered. Similarly, in electric power equipment the induced voltage due to surrounding time-varying electromagnetic fields (radio waves etc.) is usually neglected.

The above mentioned examples concern factors which are completely negligible in situations of interest. However, in many other situations there are more important effects which will still be neglected frequently to define a problem so that it can be handled mathematically without much complexity. For example, one early approximation which is usually made is to consider a system with *distributed parameters* as an equivalent system with *lumped parameters*. (If it is required to control the temperature of a room, it will first be assumed that the room is isothermal; to do otherwise would lead to very complicated partial differential equations representing the heat flow, both conductive and convective, between any two points in the room). Similarly, in physical systems we are uncertain to varying degrees about the values of parameters, measurements, expected inputs and disturbances (*stochastic systems*). In many practical applications, the uncertainties can be neglected and we proceed as though all quantities have definite values that are known precisely. This assumption gives us a *deterministic model* of the system.

Generally, crude approximations are made in a first attack on the problem so as to get a quick feel of the predominant effects. These assumptions are then gradually given up to obtain a more accurate physical model. A point of diminishing return is reached when the gain in accuracy of representation is not commensurate with the increased complexity of the computations required. In fact, beyond a certain point, there may be an undetermined loss in accuracy of representation due to flow of errors in complex computations.

The principal characteristic of a physical model is that some but not all of the features of the real system are reflected in the model. The model contains only those aspects of the real system that are considered to be important to the attributes of the real system under study. A physical system can be modelled in a number of ways depending upon the specific problem to be dealt with and the desired accuracy. For example, a communication satellite may be modelled as a point, a rigid body or a flexible body depending upon the type of study to be carried out. An automobile may be modelled as a point if we are studying traffic flow, but may be modelled as a spring-mass-damper system (Fig. 1.1) if we are interested in the study of vertical

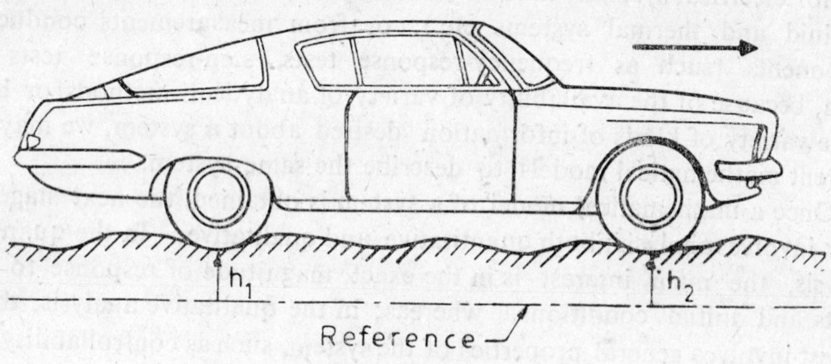

(a) An automobile on road

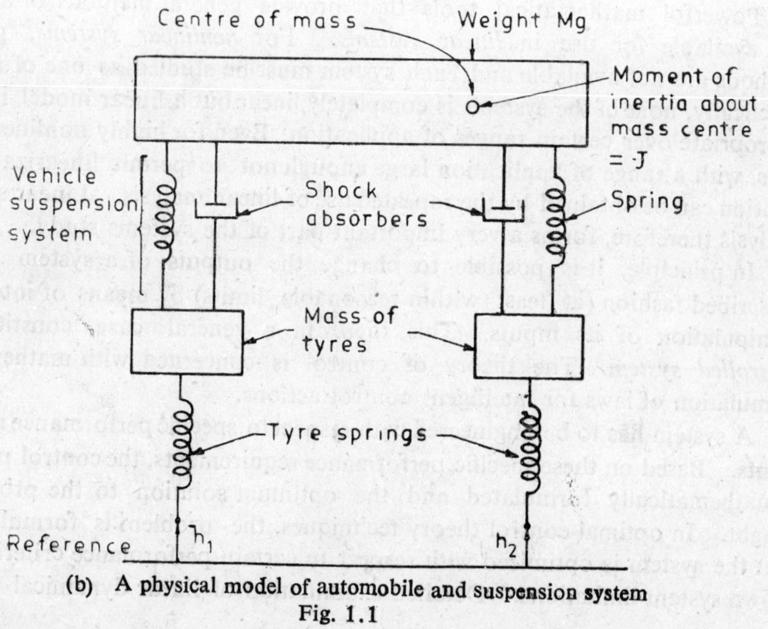

(b) A physical model of automobile and suspension system

Fig. 1.1

translational motion of the automobile (i.e., translational motion of the centre of mass) and rotational motion (rocking) of the automobile about the centre of mass.

In fact, the development of physical models of physical systems requires knowledge of each specific field. We have not to concern ourselves with this development, but direct the reader to specialised books on the subject, e.g., Cannon (1967). Instead, we consider the physical models of real physical systems as systems themselves. And therefore, in our terminology, *a physical system is a real object or collection of such objects in the real world; and a 'system' is a physical model of the real-world system resembling it in certain salient features.*

In practice, mathematical models for a system can be derived from fundamental laws (such as Newton's laws for mechanical systems, Kirchoff's laws for electrical systems, laws of thermodynamics and transport phenomena for fluid and thermal systems etc.); or from measurements conducted on components (such as frequency-response tests, step-response tests etc.). Often, because of the availability of variety of analytical methods or because of the variety of kinds of information desired about a system, we may set up different mathematical models to describe the same system.

Once a mathematical model of a system is obtained, the next stage in the study involves analysis; both quantitative and qualitative. In the quantitative analysis, the main interest is in the exact magnitude of response to certain inputs and initial conditions, whereas; in the qualitative analysis, the main interest involves general properties of the system, such as controllability, observability, stability and sensitivity. Often, synthesis and design techniques evolve from this study.

Powerful mathematical tools that provide general methods of analysis are available for use in *linear systems*. For *nonlinear systems*, general methods are not available and each system must be studied as one of a kind. Practically, none of the systems is completely linear but a linear model is often appropriate over certain ranges of application. Even for highly nonlinear systems, with a range of application large enough not to permit linearization, a solution can be obtained by the repeated use of linear analysis. Linear systems analysis therefore, forms a very important part of the systems study.

In principle, it is possible to change the outputs of a system in any prescribed fashion (at least within reasonable limits) by means of intelligent manipulation of its inputs. This then, in a general case, constitutes a *controlled system*. The theory of control is concerned with mathematical formulation of laws for intelligent control actions.

A system has to be engineered with respect to specific performance requirements. Based on these specific performance requirements, the control problem is mathematically formulated and the optimum solution to the problem is sought. In optimal control theory techniques, the problem is formulated so that the system is optimized with respect to certain performance criterion and known system limitations. Detailed examination of linear dynamical systems by number of investigators has led to some new proposals for system design—pole shifting controllers, noninteracting control etc.

It is always useful to determine various structural solutions of carrying out the system requirements and objectives at the initial stage itself. The more solutions initially considered, the greater is the probability of success in the final system. The optimum solution is selected on the basis of the functional and hardware requirements of the system.

The final stage involves: (1) physical implementation of the solution; (2) testing the overall system; and (3) correlating the test data with the system requirements and refining the process of modelling, analysis and design of control action, if necessary.

1.2 CONTROL SYSTEM: COTINUOUS-TIME AND DISCRETE-TIME

Figure 1.2 shows the general structure of a *multiple-input/multiple-output* (*multivariable*) control system. The *plant* is that part of the system which is to be controlled. It is generally a fixed component (*continuous-time subsystem*) of the system whose physical characteristics are beyond control. The output of the plant is measured by q variables $y_1(t), y_2(t), \ldots, y_q(t)$ whose values give an indication of plant performance. Direct control of the plant is exerted by means of p control forces $u_1(t), \ldots, u_p(t)$. These forces are applied by some controlling device, called *controller* which determines proper control action based upon the *reference commands* $r_1(t), \ldots, r_q(t)$ and information obtained via output sensors concerning the actual output. The feedback of output information results in a closed-loop signal flow and the term *closed-loop control* is often used for such a control action. In *open-loop control systems*, the controller operates on some pre-set pattern without taking account of the outputs. Optimal control systems, to be discussed in later chapters of this book, are often of the open-loop type.

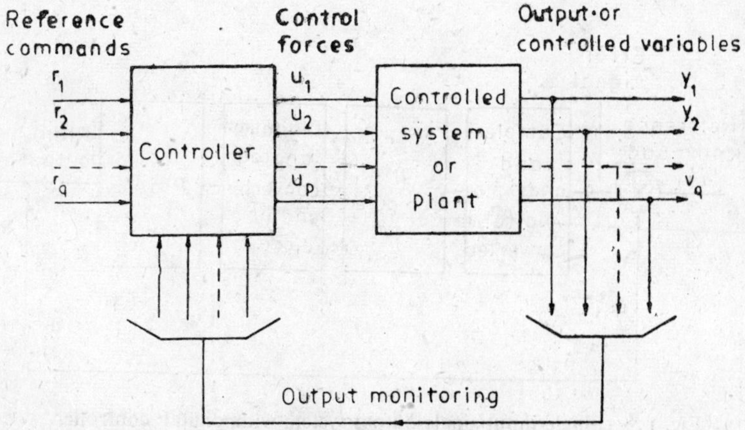

Fig. 1.2 General structure of control system

The plant accepts continuous-time signals as inputs and gives out continuous-time signals as outputs. If the controller elements are such that the controller produces continuous-time control signals from continuous-time input signals (*analog controller*), then the overall control system is a *continuous-time system*, where in the signal at every point is a continuous function of time.

The complexity of the controller needed to implement a control law is a function of the plant and the stringency of the control requirements. The cost of an analog controller rises steeply with increasing control function complexity. In fact, implementing a complex control function may even become technically infeasible if one is restricted to use only analog elements. A *digital controller*, in which either a special purpose or a general purpose computer forms the heart is usually an ideal choice for complex control systems. A general purpose computer, if used, lends itself to time-shared use for other control functions in the plant or process. A digital controller also has versatility in the sense that its control function can be easily modified by changing a few program instructions.

Digital controllers have the inherent characteristic of accepting the data in the form of short duration pulses (i.e., *sampled* or *discrete-time signals*), and producing a similar kind of output as control signal. Figure 1.3 shows a simple control scheme employing a digital controller for a *single-input/single-output* (*scalar*) system. The sampler and analog-to-digital converter are needed at the computer input; the sampler converts the continuous-time error signal into a sequence of pulses which are then expressed in a numerical code. Numerically coded output of the digital computer are decoded into a continuous-time signal by digital-to-analog converter and hold circuit. This continuous-time signal then controls the plant. The overall system is *hybrid*, in which the signal is in a sampled form in the digital controller and in a continuous form in the rest of the system. A system of this kind is referred to as a *sampled-data control system.*

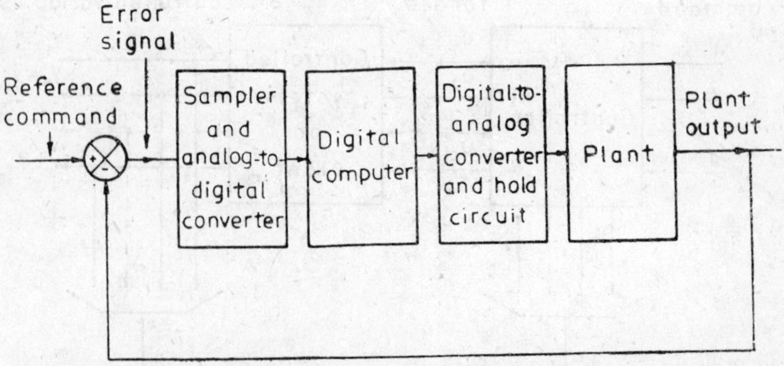

Fig. 1.3 Single-input/single-output system with digital controller (Sampled-data control system)

1.3 SCOPE AND ORGANIZATION OF THE BOOK

It would be appropriate to review in brief the history of the development of control theory to appreciate the emergence of Modern Control System Theory as a new discipline.

The control systems built until about the 1940s were *regulating systems*: the speed of an engine or of a hydraulic turbine had to be maintained accurately at a constant value. Steady-state accuracy and avoidance of instability were the major design criteria.

During the second World War, another important class of control systems emerged where the transient performance was more important than the steady-state behaviour; this is the class of *servomechanisms*, *follow-up systems* or *tracking systems*. An example would be the control system for a gun which is required to track a moving target with the help of a radar. It was discovered that much of the theory needed for designing such systems had already been developed in the field of communication engineering. The frequency-response theory developed by communication engineers from Nyquist's key result provided the analytical tools needed by control engineers for the design of military control systems having a high quality transient performance as well as static accuracy.

Another important contribution to the theory was made in 1948 by W.R. Evens. His root locus technique formed the basis for time-domain design methods, complementary to the frequency-domain methods developed from Nyquist's theorem. With this contribution, the first stage in the development of control was essentially complete; this theory is now commonly known as *Classical Control Theory*. The theory was refined subsequently and numerous text books were published; among the classics are Chestnut and Mayer (1951, 1955), Truxal (1955) and Newton et al. (1957).

The classical control theory had quite serious limitations which prevented its immediate application to more complex systems that became increasingly important in aerospace and process applications in late 1950s. This theory was restricted to linear time-invariant scalar systems. Further, in many control problems, typically those associated with space efforts such as launching and steering of a rocket, a satellite or a space ship, it is the transient behaviour which is important, because a steady-state may never be reached at all; the control action may be terminated after a finite duration. Also, there may be no notion of error as such, whether steady-state error or transient error. Such problems could not be cast in the format for which classical control theory had a solution. One was forced to consider a different kind of performance index. The maximization or minimization of the performance index became the basis of design. Optimization methods developed by Bellman, Pontryagin and their students (Bellman 1957, Pontryagin et al. 1962) provided powerful working tools to this 'modern' school of thought. With the contributions of Bellman and Pontryagin, the second stage in the development of control theory began from about 1960. This stage is commonly known as the *Modern Control Theory*. Considerable further developments and refinements have taken place in this theory which relies heavily on state-space descriptions of systems (Athans and Falb 1966, Sage and White 1977).

These developments in control theory, arising from the serious limitations of the classical theory, are by no means complete; each area abounds with unsolved problems and much research is still going on. Since late 1960s, considerable research work has been reported on new proposals of the design of linear time-invariant control systems—pole-shifting controllers, state observers, noninteracting control etc. The techniques of classical control theory are being generalized for linear time-invariant multivariable systems. These contributions may be termed as the third stage of the development in control theory; the theory is being reported under the title *Linear Multivariable Control Theory* (Chen 1970, Rosenbrock 1970, Wolovich 1974, Wonham 1979, Kailath 1980).

This brief survey would convince the reader that control theory has become a very large field. Obviously no single book can hope to give an adequate introduction to all branches of the subject. In this text on Modern Control System Theory, we discuss state-space techniques for modelling, analysis and control problems. Important results of Linear Multivariable Control Theory have also been covered.

The chapter-wise organization of the book is as follows:

Chapter 2, 'Linear Spaces and Linear Operators', reviews the basic concepts of linear algebra. The main objective of the chapter is to enable the reader to carry out similarity transformations for equivalent system representations, to solve linear algebraic equations, to transform any square matrix into a Jordan canonical form and to compute functions of matrices.

In Chapter 3, 'State Variable Descriptions', we introduce the state variable descriptions of systems from a very general setting. These descriptions are developed from the concepts of relaxedness, causality, time-invariance and linearity.

Chapter 4, 'Physical Systems and State Assignment', answers the following question. Given a real-world system—be it electrical, mechanical, chemical, ..., how do we obtain a state variable description of the system? In addition, plant models of some illustrative control systems are derived in this chapter, which are extensively used in later chapters of the book.

The objective of Chapter 5, 'Solution of State Equations', is to present standard techniques for solving state variable equations. The fundamental questions of the existence and uniqueness of solutions are also dealt with in this chapter.

In Chapter 6, 'Controllability and Observability', we discuss qualitative aspects of controllability and observability properties of systems. Various necessary and sufficient conditions for a dynamical system to be controllable and/or observable are derived. Some of the design techniques which evolve from controllability and observability analysis are also discussed in this chapter.

Chapter 7, 'Relationship Between State Variable and Input-Output

Descriptions', is a bridge between input-output analysis and state variable analysis of linear time-invariant systems. It deals with the ways of going from a state variable description to an input-output description and vice versa, using the standard controllable and standard observable representations.

The purpose of Chapter 8, 'Stability', is to introduce the reader to rigorous definitions of stability, Lyapunov's stability methods and the role that the state variable approach plays.

In Chapter 9, 'Modal Control', we study the practical implications of the concepts of controllability, observability and stability. Recent proposals for system design—pole shifting controllers, state observers etc., are discussed in this chapter.

In Chapter 10, 'Optimal Control:General Mathematical Procedures', we present various mathematical procedures which can be used to construct optimal control for engineering systems. Simple illustrations at this stage acquaint the reader with these methods, presenting both the computational difficulties and the advantages of certain methods over others.

In Chapter 11, 'Optimal Feedback Control', the methods developed in Chapter 10 are applied to certain important classes of optimization problems, namely, linear state regulator problem, time-optimal problem, etc.

In Chapter 12, 'Stochastic Optimal Linear Estimation and Control', we touch upon the enormous body of knowledge that is perhaps best described by the term 'filtering theory'.

REFERENCES

1. Athans, M., and P.L. Falb, *Optimal Control*, New York: McGraw-Hill, 1966.
2. Bellman, R.E., *Dynamic Programming*, Princeton, N.J.: Princeton Uni. Press, 1957.
3. Cannon, R.H., *Dynamics of Physical Systems*, New York: McGraw-Hill, 1967.
4. Chen, C.T., *Introduction to Linear System Theory*, New York: Holt, Rinehart and Winston, 1970.
5. Chestnut, H., and R.W. Mayer, *Servomechanisms and Regulating System Design*, Vol. 1., New York: Wiley, 1951.
6. Chestnut, H., and R.W. Mayer, *Servomechanisms nnd Regulating System Design*, Vol. 2, New York: Wiley, 1955.
7. Kailath, T., *Linear Systems*, Englewood Cliffs, N.J.: Prentice-Hall, 1980.
8. Newton, G., L. Gould, and J. Kaiser, *Analytical Design of Linear Feedback Controls*, New York: Wiley, 1957.
9. Pontryagin, L.S., V.G. Boltyanskii, R.V. Gamkrelidze, and E.F. Mischenko, *The Mathematical Theory of Optimal Processes*, New York: Wiley, 1962.
10. Rosenbrock, H.H., *State Space and Multivariable Theory*, New York: Wiley. 1970.
11. Sage, A.P., and C.C. White III, *Optimum Systems Control*, 2nd edition, Englewood Cliffs, N.J.: Prentice-Hall, 1977.
12. Truxal, J.G., *Control System Synthesis*, New York: McGraw-Hill, 1955.
13. Wolovich, W.A., *Linear Multivariable Systems*, Berlin: Springer-Verlag, 1974.
14. Wonham, W.M., *Linear Multivariable Control: A Geometric Approach*, 2nd Edition, Berlin: Springer-Verlag, 1979.
15. Zadeh, L.A., and C.A. Desoer, *Linear System Theory*, New York: McGraw-Hill, 1963.

2. LINEAR SPACES AND LINEAR OPERATORS

2.1 INTRODUCTION

We shall use a number of mathematical results and techniques throughout this book. In an attempt to make the book reasonably self contained, we review in this chapter the basic results of linear algebra. Our treatment will be fairly rapid, as we shall assume that the reader has been exposed to some of the mathematical notions we need. Also, our treatment will be incomplete in the sense that we shall present only that material which is used in the sequel. Basic knowledge of matrix theory, e.g., determinants, matrix addition, matrix multiplication and matrix inversion, is assumed on the part of the reader. Some useful results of matrix algebra and matrix calculus are summarized in Appendix III.

Readers well conversent with linear algebra may skip this chapter, or read it quickly to be conversent with the notations we use. For those who did not have a previous exposure to these topics, this chapter should serve as a working reference. A detailed account of the material presented in this chapter can be found in Gantmacher (1959), Hadley (1961), Zadeh and Desoer (1963), DeRusso et al. (1965), Bellman (1968), Nering (1970), Padulo and Arbib (1974), Noble and Daniel (1977), Goult (1978) and Friedberg et al. (1979).

Preliminary Notions

Some mathematical notations used in this and later chapters are given below:

$\mathcal{R}$ for set of all real numbers

$\mathcal{C}$ for set of all complex numbers

$\triangleq$ for 'equal to by definition'

$\in$ for 'belongs to'

$\subset$ for 'is a subset of'

$\forall$ for 'for all'

$\Rightarrow$ for 'implies that'

$\{x \mid P(x)\}$ for 'the set of x for which $P(x)$ is true'

$\mathscr{F}$ for set of all rational functions;

$$\mathscr{F} = \left\{ \frac{p(x)}{q(x)} \,\middle|\, p(x) \text{ and } q(x) \text{ are polynomials of arbitrary but finite degree with real coefficients; } q(x) \neq 0 \right\}$$

(a, b) for ordered pair of variables

$A \times B$ for $\{(a, b) \mid a \in A, b \in B\}$; the cartesian product of sets A and B.

When dealing with intervals of real numbers, we use the following notations:

$$(a, b) = \{x \in \mathcal{R} \mid a < x < b\}$$

$$[a, b) = \{x \in \mathcal{R} \mid a \leqslant x < b\}$$

$$(a, b] = \{x \in \mathcal{R} \mid a < x \leqslant b\}$$

$$[a, b] = \{x \in \mathcal{R} \mid a \leqslant x \leqslant b\}$$

Note that the symbol (a, b) is used in two different senses:

(i) as an ordered pair of variables a and b,

(ii) as an open interval defined above.

Next we consider the notion of *mapping*. A mapping from set A to set B is an assignment to every element of A, of a unique element of B. We use the symbol

$$f: A \to B$$

to mean that to every $a \in A$, the *map/function/operator/transformation* f assigns one and only one element $f(a)$ and that $f(a) \in B$. The set A is called *domain* of f, the set B is *co-domain* of f and the set of all elements of the form $f(a)$ for $a \in A$: $\{f(a) \mid a \in A\}$, is called the *range* of f.

We shall encounter functions which operate on n different kinds of variables, say, $a_1, a_2, \ldots, a_n$, with $a_1 \in A_1, \ldots, a_n \in A_n$, and which associate with each *ordered n-tuple* $(a_1, a_2, \ldots, a_n)$ an element in some set B; the domain of such a map is $A_1 \times A_2 \times \ldots \times A_n$ and co-domain is B and we write

$$f: A_1 \times A_2 \times \ldots \times A_n \to B$$

Throughout our discussion, t stands for time. The range of this variable will be denoted by $\mathscr{T}$ with the understanding that unless otherwise indicated, $\mathscr{T}$ is the real line $(-\infty, \infty)$. In the case of discrete-time systems, $\mathscr{T}$ will be assumed to be a set of integers.

A time function v is understood to be a set of pairs $\{(t, v(t)) \mid t \in \mathscr{T}\}$ where $v(t)$ denotes the value of v at time t. The range of $v(t)$ is the set $\mathcal{R}$. If v is a time function defined over $\mathscr{T} = (-\infty, \infty)$, then the set of pairs: $\{(t, v(t)) \mid t \in \mathscr{I}\}$ where $\mathscr{I}$ is an interval of $\mathscr{T}$, is called a segment of v. Such a segment will be denoted by $v_{[t_0, t_1]}$, if $\mathscr{I} = [t_0, t_1]$. Where no confusion with the time function v can arise, the segment $v_{[t_0, t_1]}$ will be abbreviated to v.

2.2 FIELDS, VECTORS AND VECTOR SPACES

In the study of linear systems, it is necessary to define a vector space over a field.

Fields

A *field* $\mathcal{F}$ is a set of elements, called *scalars*, together with the two operations of addition and multiplication for which the following axioms hold:

1. For any pair of elements $a, b \in \mathcal{F}$, there is always a unique sum $a + b$ in $\mathcal{F}$ and a unique product $a \cdot b$ (or simply ab) in $\mathcal{F}$. Further,

$$\left.\begin{aligned} a + b &= b + a \\ ab &= ba \end{aligned}\right] \text{ commutative laws}$$

2. For any three elements $a, b, c \in \mathcal{F}$,

$$\left.\begin{aligned} a + (b + c) &= (a + b) + c \\ a(bc) &= (ab)\, c \end{aligned}\right] \text{ associative laws}$$

$$a(b + c) = ab + ac] \text{ distributive law}$$

3. $\mathcal{F}$ contains a zero element, denoted by 0, and a unity element, denoted by 1, such that $a + 0 = a$ and $1 \cdot a = a$ for every $a \in \mathcal{F}$.
4. $\mathcal{F}$ contains an additive inverse; for every $a \in \mathcal{F}$, there exists an element $b \in \mathcal{F}$ such that $a + b = 0$.
5. $\mathcal{F}$ contains a multiplicative inverse; for every $a \in \mathcal{F}$ which is not the element 0, there exists an element $b \in \mathcal{F}$ such that $a \cdot b = 1$.

Example 2.1: We give below some examples to illustrate the concept of field.

(i) The set $\mathcal{R}$ of all real numbers with usual rules of addition and multiplication forms a field. This field is called *real field*.

(ii) The set C of all complex numbers with usual rules of addition and multiplication forms a field. This set is called the *complex field*. Note that $\mathcal{R}$ is a subfield of C, i.e., a field whose elements also belong to C.

(iii) The set of all positive (negative) real numbers with usual rules of addition and multiplication does not form a field; it violates axiom 4.

(iv) The set of all integers with usual rules of addition and multiplication does not form a field; it violates axiom 5.

(v) The set $\mathscr{F}$ of all rational functions forms a field with usual rules of polynomial addition and multiplication.

□

In view of the example above, it should be clear that a field may consist of either a finite (see Problem 2.2) or infinite number of scalars and, in general, is dependent upon a defined set of arithmetical rules for addition and multiplication. In this book, the fields that concern us are the *number fields* $\mathcal{R}$ and $\mathcal{C}$ and the *function field* $\mathcal{F}$.

Vectors

The reader will already be familiar with geometric vectors in two and three dimensions. A three dimensional vector **x** is a set of three real numbers x_1, x_2, x_3 which represent the coordinates of the vector relative to some defined set of coordinate axes.

Let us look at the vector from a slightly different point of view. If the state of a system can be characterized by the result of say 3 measurements, then it is mathematically convenient to treat such collections of 3-measurements as a set of 'all measurements with 3 components'. Thus a 3-dimensional vector may refer to 3 degrees of freedom in carrying out measurements and not to 3 spatial dimensions in the space.

A natural extension of these ideas is to consider the set of n-tuples of real numbers,

$$\mathbf{x} = \begin{bmatrix} x_1 \\ x_2 \\ \vdots \\ x_n \end{bmatrix}, \ \mathbf{y} = \begin{bmatrix} y_1 \\ y_2 \\ \vdots \\ y_n \end{bmatrix}, \cdots \tag{2.1}$$

with the addition and multiplication by a real number defined in the usual way:

$$\mathbf{z} = \mathbf{x} + \mathbf{y} = \begin{bmatrix} x_1 + y_1 \\ x_2 + y_2 \\ \vdots \\ x_n + y_n \end{bmatrix}; \ \mathbf{w} = \alpha\mathbf{x} = \begin{bmatrix} \alpha x_1 \\ \alpha x_2 \\ \vdots \\ \alpha x_n \end{bmatrix}, \alpha \in \mathcal{R} \tag{2.2}$$

This gives us, for different values of positive integer n, a family of vector spaces, the ordinary 3-dimensional vector space being a special case for $n = 3$. It will be a simple matter to verify that vector spaces thus defined have the basic algebraic properties listed below. These properties are the axiomatic properties which define a more general vector-space, also called *linear space* or *linear vector space.*

Vector Spaces

A *vector space* over the field $\mathcal{F}$ is a set **V**, denoted as $\mathbf{V}(\mathcal{F})$, the elements of $\mathcal{F}$ being called scalars and the elements of **V** being called vectors; together with two operations of vector addition and scalar multiplication for which the following axioms hold:

1. For any pair of vectors $\mathbf{x}, \mathbf{y} \in \mathbf{V}$, there is always a unique sum $\mathbf{x}+\mathbf{y}$ in $\mathbf{V}$. Further

$\mathbf{x}+\mathbf{y}=\mathbf{y}+\mathbf{x}$] commutative law

2. For any vector $\mathbf{x} \in \mathbf{V}$ and any scalar $\alpha \in \mathcal{F}$, there is always a unique product $\alpha\mathbf{x}$ in $\mathbf{V}$.
3. For any vectors $\mathbf{x}, \mathbf{y}, \mathbf{z} \in \mathbf{V}$,

$(\mathbf{x}+\mathbf{y})+\mathbf{z}=\mathbf{x}+(\mathbf{y}+\mathbf{z})$] associative law

4. For any vectors $\mathbf{x}, \mathbf{y} \in \mathbf{V}$ and any scalar $\alpha \in \mathcal{F}$,

$\alpha(\mathbf{x}+\mathbf{y})=\alpha\mathbf{x}+\alpha\mathbf{y}$] distributive law.

5. For any scalars $\alpha, \beta \in \mathcal{F}$ and any vector $\mathbf{x} \in \mathbf{V}$,

$\alpha(\beta\mathbf{x})=\alpha\beta\mathbf{x}$] associative law

$(\alpha+\beta)\mathbf{x}=\alpha\mathbf{x}+\beta\mathbf{x}$] distributive law

6. $\mathbf{V}$ contains a *zero (null) vector*, denoted by $\mathbf{0}$, such that $\mathbf{x}+\mathbf{0}=\mathbf{x}$ for every $\mathbf{x} \in \mathbf{V}$.
7. For every $\mathbf{x} \in \mathbf{V}$, there exists an element $-\mathbf{x} \in \mathbf{V}$, such that $\mathbf{x}+(-\mathbf{x})=\mathbf{0}$.
8. For every $\mathbf{x} \in \mathbf{V}$, $1 \cdot \mathbf{x}=\mathbf{x}$ where 1 denotes the unity of $\mathcal{F}$.

Example 2.2: Some examples to illustrate the concept of vector space are given below.

(i) An important linear vector space is $\mathcal{F}^n(\mathcal{F})$: the linear space of n-tuples in $\mathcal{F}$ over the field $\mathcal{F}$. The elements of this space are of the form (2.1) with vector addition and scalar multiplication defined by (2.2). Using the axioms of field $\mathcal{F}$, we can show that $\mathcal{F}^n(\mathcal{F})$ is a linear space.

The most common examples of the linear space $\mathcal{F}^n(\mathcal{F})$ are $\mathcal{R}^n(\mathcal{R})$ and $\mathcal{C}^n(\mathcal{C})$. $\mathcal{R}^n(\mathcal{R})$ is called the n-dimensional *real vectar space* and $\mathcal{C}^n(\mathcal{C})$ is called the n-dimensional *complex vector space.*

(ii) Let $\mathbf{\Omega}(\mathcal{R}, \mathcal{R}^n)=\{\mathbf{f} \mid \mathbf{f}: \mathcal{R} \to \mathcal{R}^n\}$ denote the set of real-valued vector functions $\mathbf{f}, \mathbf{g}, \ldots$ mapping $\mathcal{R}$ into $\mathcal{R}^n$.

Note that if $\mathbf{f}$ is a function from $\mathcal{R}$ into $\mathcal{R}^n$, then we have for $t \in \mathcal{R}$,

$$\mathbf{f}(t)=\begin{bmatrix} f_1(t) \\ f_2(t) \\ \vdots \\ f_n(t) \end{bmatrix}$$

where $f_1, f_2, \ldots, f_n$ are real-valued functions, called the components of $\mathbf{f}$. On $\mathbf{\Omega}$ $(\mathcal{R}, \mathcal{R}^n)$, we define addition by

$$[\mathbf{f}+\mathbf{g}](t)=\mathbf{f}(t)+\mathbf{g}(t) \;\forall\, t \in \mathcal{R}, \;\forall\, \mathbf{f}, \mathbf{g} \in \mathbf{\Omega}$$

and scalar multiplication by

$$(\alpha\,\mathbf{f})(t) = \alpha\,\mathbf{f}(t),\ \forall\ \alpha \in \mathcal{R},\ \forall t \in \mathcal{R},\ \forall \mathbf{f} \in \boldsymbol{\Omega}$$

$\boldsymbol{\Omega}$ ($\mathcal{R}$, $\mathcal{R}^n$) with the prescription above for addition and scalar multiplication is a linear space, which is sometimes called the *function space* (refer Athans and Falb (1966)). The zero vector in this linear space is the *null function* Ⓗ defined by

$$Ⓗ(t) = \mathbf{0},\ \forall t \in \mathcal{R},\ \mathbf{0} \in \mathcal{R}^n$$

Subspaces: Let **V** ($\mathcal{F}$) be a linear space and let $\mathbf{W} \subset \mathbf{V}$. **W** ($\mathcal{F}$) is said to be a *subspace* of **V** ($\mathcal{F}$) if under the operations of **V** ($\mathcal{F}$), **W** itself forms a vector space over $\mathcal{F}$.

From this definition it follows that

$$\mathbf{w}_1, \mathbf{w}_2 \in \mathbf{W} \Rightarrow \mathbf{w}_1 + \mathbf{w}_2 \in \mathbf{W}$$

$$\alpha \in \mathcal{F}, \mathbf{w} \in \mathbf{W} \Rightarrow \alpha\,\mathbf{w} \in \mathbf{W}$$

Example 2.3: We give some examples to illustrate this concept.

(i) Consider the three-dimensional space wherein a vector **x** is characterized by three real numbers x_1, x_2, x_3 which represent the coordinates of the vector relative to coordinate axes X_1, X_2, X_3 perpendicular to each other. The planes $X_1\,X_2$, $X_2\,X_3$ and $X_3\,X_1$, containing the origin are two-dimensional subspaces.

(ii) Consider a function space $\boldsymbol{\Omega}$ ($\mathcal{R}$, $\mathcal{R}^n$). Let there be a set $\mho$ of functions ($\mho \subset \boldsymbol{\Omega}$) which are equal to null function for t belonging to some fixed subset of $\mathcal{R}$, e.g.,

$$\mho\,([t_0, t_1], \mathcal{R}^n) = \{\mathbf{f} \mid \mathbf{f}: [t_0, t_1] \rightarrow \mathcal{R}^n\}$$

$\mho$ ($[t_0, t_1]$, $\mathcal{R}^n$) is a subspace of the function space $\boldsymbol{\Omega}$ ($\mathcal{R}$, $\mathcal{R}^n$).

Product Spaces: Linear spaces can be combined to form product speces. Let **V** and **W** be linear spaces over the same field. For any $\mathbf{v} \in \mathbf{V}$ and $\mathbf{w} \in \mathbf{W}$, we consider the ordered pair (**v**, **w**) as a 'vector' in a new space in which addition is defined by

$$(\mathbf{v}_1, \mathbf{w}_1) + (\mathbf{v}_2, \mathbf{w}_2) = (\mathbf{v}_1 + \mathbf{v}_2, \mathbf{w}_1 + \mathbf{w}_2);\quad \forall\, \mathbf{v}_1, \mathbf{v}_2 \in \mathbf{V},\ \forall\, \mathbf{w}_1, \mathbf{w}_2 \in \mathbf{W}$$

and scalar multiplication is defined by

$$\alpha\,(\mathbf{v}, \mathbf{w}) = (\alpha\,\mathbf{v}, \alpha\,\mathbf{w});\ \forall\, \mathbf{v} \in \mathbf{V},\ \forall\, \mathbf{w} \in \mathbf{W},\ \forall\, \alpha \in \mathcal{F}$$

The collection of all such vectors $(\mathbf{v}, \mathbf{w})$ with prescription above for addition and scalar multiplication constitutes a new linear space over $\mathcal{F}$. It is called the product space of $\mathbf{V}(\mathcal{F})$ and $\mathbf{W}(\mathcal{F})$ and is denoted as $\mathbf{V}\times\mathbf{W}(\mathcal{F})$.

□

In this book the linear vector spaces that concern us the most are $\mathcal{R}^n(\mathcal{R})$, $C^n(C)$ and $\mathbf{\Omega}\,(\mathcal{R}^m, \mathcal{R}^n)$.

2.3 LINEAR COMBINATIONS AND BASES

Linear Independence

Suppose that $\mathbf{V}(\mathcal{F})$ is a vector space and that $\mathbf{v}_1, \mathbf{v}_2, \ldots, \mathbf{v}_n$ are elements of $\mathbf{V}$. We say that a vector $\mathbf{v}\in\mathbf{V}$ is a *linear combination* of $\mathbf{v}_i$; $i = 1, 2, .., n$, if there are scalars $\alpha_1, \alpha_2, .., \alpha_n \in \mathcal{F}$ such that

$$\mathbf{v} = \alpha_1\mathbf{v}_1 + \alpha_2\mathbf{v}_2 + \ldots + \alpha_n\mathbf{v}_n = \sum_{i=1}^{n} \alpha_i\mathbf{v}_i$$

We say that the set $\{\mathbf{v}_1, \mathbf{v}_2, .., \mathbf{v}_n\}$ of elements of $\mathbf{V}$ is a *linearly dependent* set if $\mathbf{0}$ is a linear combination of the $\mathbf{v}_i$; $i = 1, 2, .., n$, in which not all of the α_i; $i = 1, 2, .., n$, are 0. In other words, $\mathbf{v}_1, \mathbf{v}_2, .., \mathbf{v}_n$ are linearly dependent if and only if

$$\mathbf{0} = \sum_{i=1}^{n} \alpha_i\mathbf{v}_i$$

and there is some $\alpha_i \neq 0$.

On the other hand, if

$$\mathbf{0} = \sum_{i=1}^{n} \alpha_i\mathbf{v}_i$$

implies that $\alpha_i = 0 \;\forall\; i$, then the set $\{\mathbf{v}_1, \mathbf{v}_2, .., \mathbf{v}_n\}$ is called *linearly independent* set.

Example 2.4: We give some examples to illustrate the concept.

(i) The vectors $\begin{bmatrix} 1 \\ 2 \\ 3 \end{bmatrix}$, $\begin{bmatrix} -1 \\ 0 \\ 2 \end{bmatrix}$ and $\begin{bmatrix} 0 \\ 2 \\ 5 \end{bmatrix}$ are linearly dependent,

since

$$\sum_{i=1}^{3} \alpha_i\mathbf{v}_i = \alpha_1\begin{bmatrix} 1 \\ 2 \\ 3 \end{bmatrix} + \alpha_2\begin{bmatrix} -1 \\ 0 \\ 2 \end{bmatrix} + \alpha_3\begin{bmatrix} 0 \\ 2 \\ 5 \end{bmatrix} = \begin{bmatrix} 0 \\ 0 \\ 0 \end{bmatrix}$$

can be satisfied by, say

$$\boldsymbol{\alpha} = \begin{bmatrix} \alpha_1 \\ \alpha_2 \\ \alpha_3 \end{bmatrix} = \begin{bmatrix} 1 \\ 1 \\ -1 \end{bmatrix}$$

(ii) The vectors $\begin{bmatrix} 1 \\ 0 \\ 0 \end{bmatrix}$, $\begin{bmatrix} 0 \\ 1 \\ 0 \end{bmatrix}$ and $\begin{bmatrix} 0 \\ 0 \\ 1 \end{bmatrix}$ are linearly independent,

since

$$\sum_{i=1}^{3} \alpha_i \mathbf{v}_i = \alpha_1 \begin{bmatrix} 1 \\ 0 \\ 0 \end{bmatrix} + \alpha_2 \begin{bmatrix} 0 \\ 1 \\ 0 \end{bmatrix} + \alpha_3 \begin{bmatrix} 0 \\ 0 \\ 1 \end{bmatrix} = \begin{bmatrix} 0 \\ 0 \\ 0 \end{bmatrix}$$

reduces to

$$\boldsymbol{\alpha} = \begin{bmatrix} \alpha_1 \\ \alpha_2 \\ \alpha_3 \end{bmatrix} = \begin{bmatrix} 0 \\ 0 \\ 0 \end{bmatrix}$$

(iii) Consider a set of vectors $\{\mathbf{v}_1, \mathbf{v}_2, .., \mathbf{v}_n\}$ in which $\mathbf{v}_1 = \mathbf{0}$. This set of vectors is always linearly dependent because we can choose $\alpha_1 = 1$, $\alpha_2 = \alpha_3 = .. = \alpha_n = 0$ to satisfy the condition

$$\mathbf{0} = \sum_{i=1}^{n} \alpha_i \mathbf{v}_i$$

Thus, any set of vectors that contains a zero vector is a linearly dependent set.

□

Let us now consider linear independence of vector functions. We shall be concerned with real-valued functions. Let $\mathbf{f}_i(t)$; $i = 1, 2, .., n$, be $1 \times m$ real-valued functions of t. These functions are linearly independent on $[t_1, t_2]$ if and only if

$$\alpha_1 \mathbf{f}_1(t) + \alpha_2 \mathbf{f}_2(t) + .. + \alpha_n \mathbf{f}_n(t) = \mathbf{0}_{1\times m};\ \alpha_i \in \mathcal{R} \tag{2.3}$$

implies that $\alpha_1 = \alpha_2 = .. = \alpha_n = 0$ for all t in $[t_1, t_2]$. If any of the α_i's are nonzero, then the $\mathbf{f}_i(t)$'s are linearly dependent on $[t_1, t_2]$.

We can rewrite eqn. (2.3) in matrix notation on the interval $[t_1, t_2]$ as

$$[\alpha_1\alpha_2\ldots\alpha_n]\begin{bmatrix}\mathbf{f}_1(t)\\ \mathbf{f}_2(t)\\ \vdots\\ \mathbf{f}_n(t)\end{bmatrix} \stackrel{\Delta}{=} \boldsymbol{\alpha}\mathbf{F}(t)=\mathbf{0} \tag{2.4}$$

where

$$\boldsymbol{\alpha}_{1\times n} \stackrel{\Delta}{=} [\alpha_1\alpha_2\ldots\alpha_n]$$

$$\mathbf{F}_{n\times m}(t) \stackrel{\Delta}{=} \begin{bmatrix}\mathbf{f}_1(t)\\ \mathbf{f}_2(t)\\ \vdots\\ \mathbf{f}_n(t)\end{bmatrix}$$

The linear independence is achieved if and only if

$$\boldsymbol{\alpha}=\mathbf{0}\,\forall t \in [t_1, t_2].$$

In Section 2.5, we shall give efficient tests to check linear independence of vectors and vector functions.

Dimension of Linear Space

The maximum number of linearly independent vectors in a linear space is called the dimension of the linear space.

Example 2.5:

(i) The significance of n in the linear space $\mathcal{R}^n$ $(\mathcal{R})$ is now clear; there are n linearly independent vectors in this linear space. One such set is

$$\mathbf{e}_1=\begin{bmatrix}1\\0\\ \vdots\\0\end{bmatrix},\quad \mathbf{e}_2=\begin{bmatrix}0\\1\\ \vdots\\0\end{bmatrix},\quad \ldots,\ \mathbf{e}_n=\begin{bmatrix}0\\0\\ \vdots\\1\end{bmatrix}$$

(ii) Consider the function space

$$\Omega\,(\mathcal{R}, \mathcal{R}) = \{f \mid f \text{ maps all } t \in (-\infty, \infty) \text{ into } \mathcal{R}\}$$

Let a set of functions $f_1, f_2, \ldots, f_i, \ldots$ in Ω be $t, t^2, \ldots, t^i, \ldots$. Since

$$\sum_i \alpha_i t^i = 0 \Rightarrow \alpha_i = 0,$$

the set of functions $\{t, t^2, \ldots, t^i, \ldots\}$ is linearly independent. There are infinitely many of these functions, therefore the dimension of function space Ω is infinity.

The Notion of Bases

In a classic geometric sense, we think of a 2-dimensional vector

$$\mathbf{v}=\begin{bmatrix}x_1\\x_2\end{bmatrix}$$

as the vectorial sum of two components, each represented by a linear combination of unit vectors $\mathbf{e}_1$ and $\mathbf{e}_2$ in the directions of abscissa and ordinate respectively for two coordinate axes which are at right angles to each other (Fig. 2.1a). The unit vectors $\mathbf{e}_1$ and $\mathbf{e}_2$, which are linearly independent, define an *orthogonal basis.*

However, there is nothing in the notion of a vector space which tells us about lengths or angles of basis vectors. From Fig. 2.1a, we observe that in terms of a pair of vectors $\mathbf{u}_1$, $\mathbf{u}_2$, the vector $\mathbf{v}$ could be uniquely expressed as the sum $y_1 \mathbf{u}_1 + y_2 \mathbf{u}_2$. Thus the basis vectors $\mathbf{u}_1$ and $\mathbf{u}_2$ allow us to assign y_1 and y_2 as the components of the vector $\mathbf{v}$ and we say that

$$\begin{bmatrix} y_1 \\ y_2 \end{bmatrix}$$

is the *representation* of $\mathbf{v}$ with respect to the basis $\{\mathbf{u}_1, \mathbf{u}_2\}$. The vectors $\mathbf{u}_1$, $\mathbf{u}_2$ define a *nonorthogonal basis.*

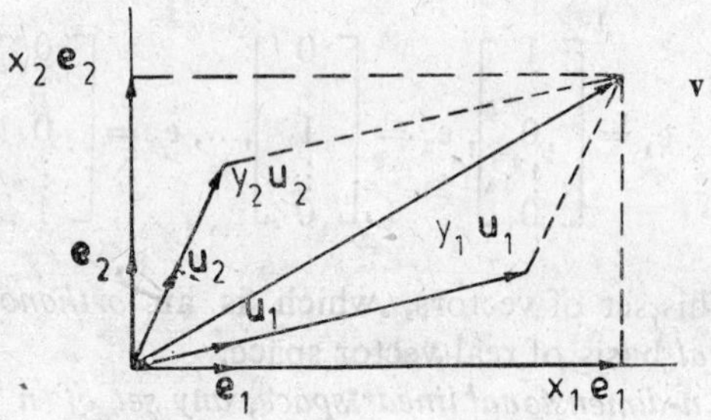

Fig. 2.1(a) $\mathbf{v}=x_1\mathbf{e}_1+x_2\mathbf{e}_2$
$=y_1\mathbf{u}_1+y_2\mathbf{u}_2$

From Figs. 2.1a, b and c, the following observations can be easily established:

(i) Consider the set of vectors $\mathbf{u}_1$ which consists of only one vector. This set is linearly independent if and only if $\mathbf{u}_1 \neq \mathbf{0}$. If $\mathbf{u}_1 \neq \mathbf{0}$, the only way to have $\alpha_1 \mathbf{u}_1=\mathbf{0}$ is $\alpha_1=0$.
This set of linearly independent vectors does not form a basis for two-dimensional space.

(ii) The set of linearly dependent vectors cannot form a basis. Consider for example

(a) $$\mathbf{u}_1=\begin{bmatrix} 1 \\ 0 \end{bmatrix}, \quad \mathbf{u}_2=\begin{bmatrix} -1 \\ 0 \end{bmatrix}$$

All vectors of the plane cannot be represented uniquely in terms of $\mathbf{u}_1$ and $\mathbf{u}_2$ (Fig. 2.1b).

(b) Any vector in the plane will not have unique representation in terms of three vectors $\mathbf{u}_1$, $\mathbf{u}_2$, $\mathbf{u}_3$ which are linearly dependent (Fig. 2.1c).

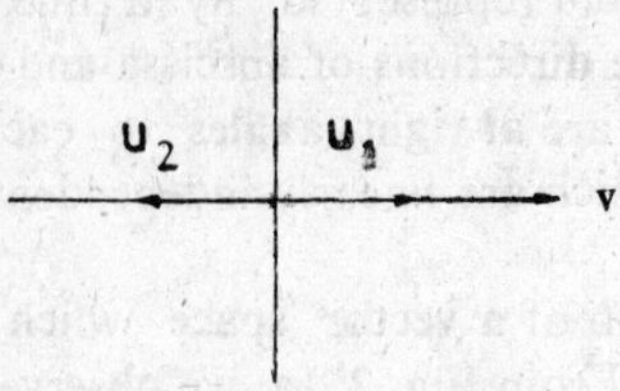

Fig. 2.1(b) $\mathbf{v} = 2\mathbf{u}_1 + 0\mathbf{u}_2$
$= 3\mathbf{u}_1 + 1\mathbf{u}_2$
$= \ldots\ldots$

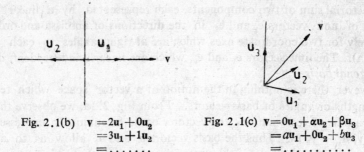

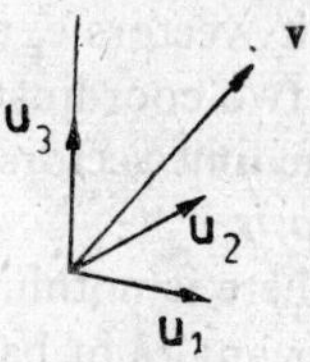

Fig. 2.1(c) $\mathbf{v} = 0\mathbf{u}_1 + \alpha\mathbf{u}_2 + \beta\mathbf{u}_3$
$= a\mathbf{u}_1 + 0\mathbf{u}_2 + b\mathbf{u}_3$
$= \ldots\ldots\ldots\ldots$

These observations lead us to the definition of basis.

A set of linearly independent vectors of a vector space $\mathbf{V}(\mathcal{F})$ forms a basis of $\mathbf{V}$ if every vector in $\mathbf{V}$ can be expressed as a unique linear combination of these vectors.

For example, for the linear space $\mathcal{R}^n(\mathcal{R})$, the set of vectors

$$\mathbf{e}_1 = \begin{bmatrix} 1 \\ 0 \\ \vdots \\ 0 \end{bmatrix}, \mathbf{e}_2 = \begin{bmatrix} 0 \\ 1 \\ \vdots \\ 0 \end{bmatrix}, \ldots, \mathbf{e}_n = \begin{bmatrix} 0 \\ 0 \\ \vdots \\ 1 \end{bmatrix}$$

forms a basis. This set of vectors, which is an *orthonormal* set, is called *natural* or *canonical* basis of real vector space.

In fact, *in an n-dimensional linear space, any set of n linearly independent vectors forms a basis.*

To prove this, let $\mathbf{b}_1, \mathbf{b}_2, \ldots, \mathbf{b}_n \in \mathbf{V}(\mathcal{F})$—an n-dimensional space, be any linearly independent set of vectors. When an element $\mathbf{v} \in \mathbf{V}$ is added to this set, the augmented set is linearly dependent. Therefore,

$$\alpha_0 \mathbf{v} + \alpha_1 \mathbf{b}_1 + \ldots + \alpha_n \mathbf{b}_n = \mathbf{0}$$

where all $\alpha_i \neq 0$. Clearly $\alpha_0 \neq 0$, since otherwise $\mathbf{b}_1, \ldots, \mathbf{b}_n$ become linearly dependent which contradicts the given assumption. Hence

$$\mathbf{v} = -\frac{\alpha_1}{\alpha_0}\mathbf{b}_1 - \frac{\alpha_2}{\alpha_0}\mathbf{b}_2 - \ldots - \frac{\alpha_n}{\alpha_0}\mathbf{b}_n$$
$$= x_1\mathbf{b}_1 + x_2\mathbf{b}_2 + \ldots + x_n\mathbf{b}_n$$

Thus $\mathbf{v}$ can be expressed as a linear combination of the linearly independent set $\{\mathbf{b}_i\}$. This property qualifies this set as a basis.

The uniqueness of the representation can be proved as follows. Suppose

$$\mathbf{v} = \sum_{i=1}^{n} y_i\mathbf{b}_i = \sum_{i=1}^{n} x_i\mathbf{b}_i$$

Therefore,

$$\sum_{i=1}^{n} (y_i - x_i)\,\mathbf{b}_i = \mathbf{0}$$

Since $\mathbf{b}_i$, $i = 1, 2, \ldots, n$ are linearly independent,

$$y_i - x_i = 0$$

This proves the uniqueness of the expression for $\mathbf{v}$.

Representation of Vectors in an n-dimensional Space: In an n-dimensional vector space $\mathbf{V}(\mathcal{F})$, if a basis is chosen, then every vector in $\mathbf{V}$ can be uniquely represented by a set of n scalars $x_1, x_2, \ldots, x_n$ in $\mathcal{F}$, i.e.,

$$\mathbf{v} = x_1\mathbf{b}_1 + x_2\mathbf{b}_2 + \cdots + x_n\mathbf{b}_n$$

$$= \sum_{i=1}^{n} x_i\mathbf{b}_i$$

$$= [\mathbf{b}_1\mathbf{b}_2\cdots\mathbf{b}_n]\,\mathbf{x}$$

where

$$\mathbf{x} = [x_1\ x_2 \cdots x_n]^T$$

The $n \times 1$ vector $\mathbf{x}$ can be considered as a vector in $\mathcal{F}^n(\mathcal{F})$. $\mathbf{x}$ is called the *representation* of $\mathbf{v}$ with respect to the basis $\{\mathbf{b}_1, \mathbf{b}_2, \ldots, \mathbf{b}_n\}$.

The representation described above, of the vectors in the n-dimensional space $\mathbf{V}(\mathcal{F})$ by vectors $\mathcal{F}^n(\mathcal{F})$, has the added virtue of preserving the algebraic properties of the space. Hence if $\mathbf{v}$ is represented by $\mathbf{x}$ and $\mathbf{w}$ is represented by $\mathbf{y}$, it is a simple matter to show that $\mathbf{v} + \mathbf{w}$ is represented by $\mathbf{x} + \mathbf{y}$. Similarly for any α in $\mathcal{F}$, $\alpha\mathbf{v}$ is represented by $\alpha\mathbf{x}$. The earlier remark that the vector space $\mathcal{F}^n(\mathcal{F})$ is important, now should be clear. The argument given above shows that every n-dimensional space $\mathbf{V}(\mathcal{F})$ is precisely modelled by $\mathcal{F}^n(\mathcal{F})$.

We shall be mostly concerned with real field $\mathcal{R}$ and the complex field $\mathcal{C}$; any n-dimensional vector space over $\mathcal{R}$ or $\mathcal{C}$ can be modelled by $\mathcal{R}^n(\mathcal{R})$ or $\mathcal{C}^n(\mathcal{C})$ respectively.

Given an n-tuple of real or complex numbers

$$\mathbf{x} = \begin{bmatrix} x_1 \\ x_2 \\ \vdots \\ x_n \end{bmatrix};$$

we can interpret it in two ways:

(i) It is a representation of a vector with respect to some fixed unknown basis.

(ii) It itself is a vector or the representation of a vector with respect to the basis $\{\mathbf{e}_1, \mathbf{e}_2, \ldots, \mathbf{e}_n\}$ because with respect to this particular basis, the representation and the vector itself are identical, i.e.,

$$\begin{bmatrix} x_1 \\ x_2 \\ \vdots \\ x_n \end{bmatrix} = [\mathbf{e}_1\ \mathbf{e}_2 \ldots \mathbf{e}_n] \begin{bmatrix} x_1 \\ x_2 \\ \vdots \\ x_n \end{bmatrix}$$

Given an array of numbers, unless it is tied up with some basis, we shall always consider it as a vector.

Change of Basis in an n-dimensional Space: We have shown that a vector $\mathbf{v}$ in $\mathbf{V}(\mathcal{F})$ has different representations with respect to different bases. It is natural to ask what the relationships are between these different representations. In this subsection, we study this problem.

If $\mathbf{V}(\mathcal{F})$ is an n-dimensional vector space with a fixed basis $\{\mathbf{b}_1, \mathbf{b}_2, \ldots, \mathbf{b}_n\}$, each vector $\mathbf{v}$ in $\mathbf{V}$ is associated with a unique vector $\mathbf{x}$ in $\mathcal{F}^n$. The components of $\mathbf{x}$ are defined by the equation

$$\mathbf{v} = \sum_{i=1}^{n} x_i \mathbf{b}_i = [\mathbf{b}_1\ \mathbf{b}_2 \ldots \mathbf{b}_n]\ \mathbf{x} \tag{2.5}$$

If instead of choosing $\{\mathbf{b}_1, \mathbf{b}_2, \ldots, \mathbf{b}_n\}$ as basis, we refer $\mathbf{v}$ to an alternative basis $\{\mathbf{c}_1, \mathbf{c}_2, \ldots, \mathbf{c}_n\}$, we will obtain the expression

$$\mathbf{v} = \sum_{i=1}^{n} y_i \mathbf{c}_i = [\mathbf{c}_1\ \mathbf{c}_2 \ldots\ \mathbf{c}_n]\ \mathbf{y} \tag{2.6}$$

Thus, with our new choice of basis, $\mathbf{v}$ is represented by the vector

$$\mathbf{y} = \begin{bmatrix} y_1 \\ y_2 \\ \vdots \\ y_n \end{bmatrix}$$

in $\mathcal{F}^n$.

In geometric terms, the choice of basis defines the reference axes and if we change the axes, we will obtain new coordinates for each point in space. Our problem is how to compute directly the new representation $\mathbf{y}$ from the original representation $\mathbf{x}$.

In order to derive the relationship between $\mathbf{x}$ and $\mathbf{y}$, we need either the information of the representations of $\mathbf{b}_i$; $i = 1, 2, \ldots, n$ with respect to basis $\mathbf{c}_i$; $i = 1, 2, \ldots, n$ or the information of the representation of $\mathbf{c}_i$; $i = 1, 2, \ldots, n$ with respect to the basis $\mathbf{b}_i$; $i = 1, 2, \ldots, n$. Let the representation of $\mathbf{b}_i$; $i = 1, 2, \ldots, n$ with respect to basis $\mathbf{c}_i$; $i = 1, 2, \ldots, n$ be $[p_{1i}\ p_{2i} \ldots p_{ni}]^T$, i.e.,

$$\mathbf{b}_i = \sum_{j=1}^{n} p_{ji}\mathbf{c}_j = [\mathbf{c}_1\ \mathbf{c}_2\ \ldots\ \mathbf{c}_n] \begin{bmatrix} p_{1i} \\ p_{2i} \\ \vdots \\ p_{ni} \end{bmatrix}$$

Therefore

$$[\mathbf{b}_1 \mathbf{b}_2 \ldots \mathbf{b}_n] = [\mathbf{c}_1\ \mathbf{c}_2\ \ldots \mathbf{c}_n] \begin{bmatrix} p_{11} & p_{12} \cdots p_{1n} \\ p_{21} & p_{22} \cdots p_{2n} \\ \vdots & \vdots \quad\ \ \vdots \\ p_{n1} & p_{n2} \cdots p_{nn} \end{bmatrix}$$

$$\triangleq [\mathbf{c}_1\mathbf{c}_2 \ldots \mathbf{c}_n]\,\mathbf{P} \tag{2.7}$$

Substituting (2.7) in (2.5) we get

$$\mathbf{v} = [\mathbf{b}_1\mathbf{b}_2 \ \ldots \ \mathbf{b}_n]\,\mathbf{x}$$
$$= [\mathbf{c}_1\mathbf{c}_2 \ \ldots \ \mathbf{c}_n]\,\mathbf{Px} \tag{2.8}$$

Since (2.6) and (2.8) are two equivalent expressions for $\mathbf{v}$, we have

$$\mathbf{y} = \mathbf{Px} \tag{2.9}$$

where

$$\mathbf{P} = \begin{bmatrix} i\text{th column: the representation} \\ \text{of } \mathbf{b}_i \text{ with respect to basis} \\ \{\mathbf{c}_1, \mathbf{c}_2, \ldots, \mathbf{c}_n\} \end{bmatrix}$$

This establishes the relation between $\mathbf{y}$ and $\mathbf{x}$.

If the representation of $\mathbf{c}_i$; $i = 1, 2, \ldots, n$ with respect to $\mathbf{b}_i$; $i = 1, 2 \ldots, n$ is used, then we shall obtain

$$\mathbf{x} = \mathbf{Qy} \tag{2.10}$$

where

$$\mathbf{Q} = \begin{bmatrix} i\text{th column: the representation} \\ \text{of } \mathbf{c}_i \text{ with respect to basis} \\ \{\mathbf{b}_1, \mathbf{b}_2, \ldots, \mathbf{b}_n\} \end{bmatrix}$$

From (2.9) and (2.10),

$$\mathbf{y} = \mathbf{Px} = \mathbf{PQy}$$

Hence

$$\mathbf{PQ} = \mathbf{I}, \text{ the identity matrix}$$

or

$$\mathbf{P} = \mathbf{Q}^{-1} \tag{2.11}$$

Because any set of n linearly independent vectors can be taken as a basis of $\mathbf{V}$, we can obtain a new basis of $\mathbf{V}$ by taking the column vectors of any $n \times n$ matrix with linearly independent columns as representation of a new set of basis vectors in terms of the original ones. This establishes a one to one correspondence between bases of $\mathbf{V}$ and $n \times n$ matrices with linearly independent column vectors. Later, we shall prove that such matrices are nonsingular.

Example 2.6:

(i) The matrix $\mathbf{P}$ which is associated with the change from basis $\{\mathbf{e}_1, \mathbf{e}_2, \mathbf{e}_3\}$ in $\mathcal{R}^3$ to basis $\{\mathbf{u}_1, \mathbf{u}_2, \mathbf{u}_3\}$;

$$\mathbf{u}_1 = \begin{bmatrix} 1 \\ 1 \\ 0 \end{bmatrix}, \mathbf{u}_2 = \begin{bmatrix} 0 \\ 1 \\ 1 \end{bmatrix}, \mathbf{u}_3 = \begin{bmatrix} 1 \\ 0 \\ 1 \end{bmatrix}$$

is

$$\mathbf{P} = \begin{bmatrix} 1/2 & 1/2 & -1/2 \\ -1/2 & 1/2 & 1/2 \\ 1/2 & -1/2 & 1/2 \end{bmatrix}$$

(ii) Consider a vector **v** whose representation **x** with respect to the basis $\{\mathbf{b}_1, \mathbf{b}_2\}$;

$$\mathbf{b}_1 = \begin{bmatrix} -1 \\ 1 \end{bmatrix}, \ \mathbf{b}_2 = \begin{bmatrix} 1 \\ 0 \end{bmatrix}, \text{ is}$$

$$\mathbf{x} = \begin{bmatrix} 1 \\ 2 \end{bmatrix}.$$ This is illustrated in Fig. 2.2a.

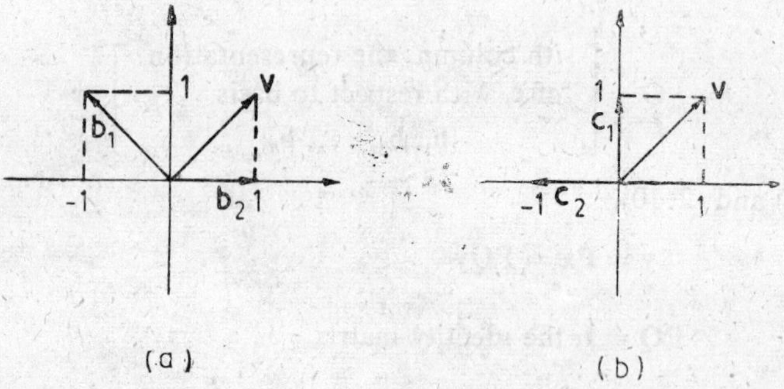

Fig. 2.2

Consider now any 2×2 matrix **P** with linearly independent vectors, say

$$\mathbf{P} = \begin{bmatrix} 1 & 0 \\ 1 & -1 \end{bmatrix}$$

Then

$$\mathbf{y} = \mathbf{P}\mathbf{x} = \begin{bmatrix} 1 & 0 \\ 1 & -1 \end{bmatrix}\begin{bmatrix} 1 \\ 2 \end{bmatrix} = \begin{bmatrix} 1 \\ -1 \end{bmatrix}$$

is the representation of **v** with respect to some new basis. The new basis $\{\mathbf{c}_1, \mathbf{c}_2\}$ may be obtained from relation (2.8)

$$\begin{bmatrix} -1 & 1 \\ 1 & 0 \end{bmatrix} = [\mathbf{c}_1 \ \mathbf{c}_2]\begin{bmatrix} 1 & 0 \\ 1 & -1 \end{bmatrix}$$

This gives

$$\mathbf{c}_1 = \begin{bmatrix} 0 \\ 1 \end{bmatrix}, \mathbf{c}_2 = \begin{bmatrix} -1 \\ 0 \end{bmatrix}$$

The representation of vector **v** with respect to new basis is shown in Fig. 2.2b.

2.4 LINEAR TRANSFORMATIONS AND MATRICES

This section is devoted to the theory of linear transformations. These are mappings of one vector space into another which preserve linear properties of spaces. In the case of finite dimensional spaces, the transformations can be represented by matrices and a number of useful results in matrix algebra can be obtained from their properties.

A function **L** that maps the vector space $\mathbf{V}(\mathcal{F})$ into the vector space $\mathbf{W}(\mathcal{F})$ is said to be a *linear map/linear operatar/linear transformation* if and only if **L** is both additive and homogeneous.

We say that $\mathbf{L}: \mathbf{V} \rightarrow \mathbf{W}$ is additive if and only if

$$\mathbf{L}(\mathbf{v}_1 + \mathbf{v}_2) = \mathbf{L}(\mathbf{v}_1) + \mathbf{L}(\mathbf{v}_2) \ \forall\ \mathbf{v}_1, \mathbf{v}_2 \in \mathbf{V}$$

Note that the vectors $\mathbf{L}(\mathbf{v}_1)$ and $\mathbf{L}(\mathbf{v}_2)$ are in **W**.

We say that $\mathbf{L}: \mathbf{V} \rightarrow \mathbf{W}$ is homogeneous if and only if

$$\mathbf{L}(\alpha \mathbf{v}) = \alpha \mathbf{L}(\mathbf{v}) \ \forall\ \mathbf{v} \in \mathbf{V}, \alpha \in \mathcal{F}.$$

It is clear that for homogeneity, both **V** and **W** must be vector spaces over the same field $\mathcal{F}$; for all $\alpha \in \mathcal{F}$, $\alpha\mathbf{v} \in \mathbf{V}$ and $\alpha\mathbf{L}(\mathbf{v}) \in \mathbf{W}$.

If $\mathbf{L}: \mathbf{V} \rightarrow \mathbf{W}$ is both additive and homogeneous, we say that **L** is linear. Both additivity and homogeneity are taken into account in the following definition of *linearity*.

$\mathbf{L}: \mathbf{V} \rightarrow \mathbf{W}$ is linear if and only if

$$\mathbf{L}(\alpha \mathbf{v}_1 + \beta \mathbf{v}_2) = \alpha \mathbf{L}(\mathbf{v}_1) + \beta \mathbf{L}(\mathbf{v}_2) \ \forall\ \mathbf{v}_1, \mathbf{v}_2 \in \mathbf{V}, \forall\ \alpha, \beta \in \mathcal{F} \tag{2.12}$$

Example 2.7: Consider the transformation that rotates a point in a geometric plane counter-clockwise ϕ degrees about the origin as shown in Fig. 2.3a.

Given any two vectors $\mathbf{x}_1$ and $\mathbf{x}_2$ in the plane, it is easy to varify that for any real numbers α, β, the vector that is sum of the two vectors $\alpha\mathbf{x}_1$ and $\beta\mathbf{x}_2$ after rotation by ϕ degrees is equal to the rotation of the vector $\alpha\mathbf{x}_1 + \beta\mathbf{x}_2$ by ϕ degrees (Fig. 2.3b). Hence the rotation is a linear transformation. The spaces $\mathrm{V}(\mathcal{F})$ and $\mathrm{W}(\mathcal{F})$ of this example are $\mathcal{R}^2(\mathcal{R})$.

Matrix Representation of Linear Operator

In the following, we shall see that every linear operator that maps finite dimensional space $\mathbf{V}(\mathcal{F})$ to finite dimensional space $\mathbf{W}(\mathcal{F})$ has matrlx representation with coefficients in the field $\mathcal{F}$.

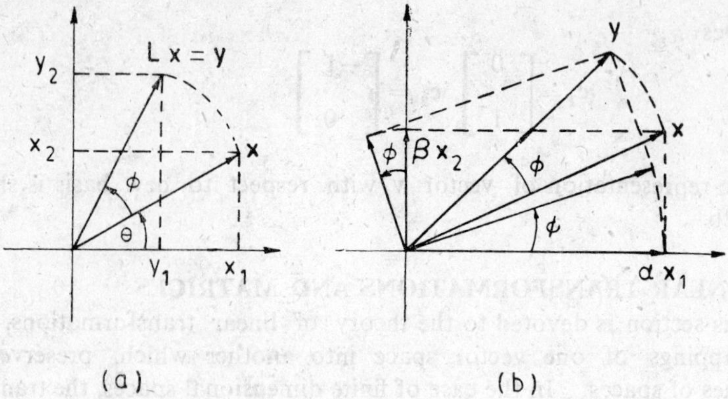

Fig. 2.3

Suppose that $\mathbf{V}$ is n-dimensional space with $\{\mathbf{v}_1, \mathbf{v}_2, \ldots, \mathbf{v}_n\}$ as basis and $\mathbf{W}$ is m-dimensional space with $\{\mathbf{w}_1, \mathbf{w}_2, \ldots, \mathbf{w}_m\}$ as basis and that $\mathbf{L}$ is a linear transformation of $\mathbf{V}$ into $\mathbf{W}$. For any $\mathbf{v} \in \mathbf{V}$, we have a unique expression

$$\mathbf{v} = \sum_{j=1}^{n} x_j \mathbf{v}_j = [\mathbf{v}_1 \mathbf{v}_2 \ldots \mathbf{v}_n] \begin{bmatrix} x_1 \\ x_2 \\ \vdots \\ x_n \end{bmatrix} \tag{2.13}$$

where $x_1, x_2, \ldots, x_n$ are scalars in $\mathcal{F}$.

Therefore

$$\mathbf{L}(\mathbf{v}) = \mathbf{L}\Big(\sum_{j=1}^{n} x_j \mathbf{v}_j\Big) = \sum_{j=1}^{n} x_j \mathbf{L}(\mathbf{v}_j); \text{ using the linearity property of } \mathbf{L}$$

$$= [\mathbf{L}(\mathbf{v}_1)\ \mathbf{L}(\mathbf{v}_2) \ldots \mathbf{L}(\mathbf{v}_n)] \begin{bmatrix} x_1 \\ x_2 \\ \vdots \\ x_n \end{bmatrix} \tag{2.14}$$

$\mathbf{L}(\mathbf{v})$ is thus completely expressed in terms of the transformations of the basis vectors of $\mathbf{V}$.

Now each of the transformed vectors $\mathbf{L}(\mathbf{v}_j)$ can be expressed in terms of the chosen basis $\{\mathbf{w}_1, \mathbf{w}_2, \ldots, \mathbf{w}_m\}$ of $\mathbf{W}$. Suppose

$$\mathbf{L}(\mathbf{v}_j) = \sum_{i=1}^{m} a_{ij} \mathbf{w}_i\ ; \quad j = 1, 2, \ldots, n$$

$$= [\mathbf{w}_1\ \mathbf{w}_2 \ldots \mathbf{w}_m] \begin{bmatrix} a_{1j} \\ a_{2j} \\ \vdots \\ a_{mj} \end{bmatrix} \tag{2.15}$$

Substituting from (2.15) into (2.14) for $\mathbf{L}(\mathbf{v}_j)$ we get

$$\mathbf{L}(\mathbf{v}) = [\mathbf{w}_1\ \mathbf{w}_2 \ldots \mathbf{w}_m] \begin{bmatrix} a_{11} & a_{12} \ldots a_{1n} \\ a_{21} & a_{22} \ldots a_{2n} \\ \vdots & \vdots \quad \vdots \\ a_{m1} & a_{m2} \ldots a_{mn} \end{bmatrix} \begin{bmatrix} x_1 \\ x_2 \\ \vdots \\ x_n \end{bmatrix} \tag{2.16}$$

$$\triangleq [\mathbf{w}_1\ \mathbf{w}_2 \ldots \mathbf{w}_m]\mathbf{Ax}$$

If

$$\mathbf{y} = \mathbf{Ax} \tag{2.17}$$

where

$$\mathbf{A} = \begin{bmatrix} j\text{th column : representation of} \\ \mathbf{L}(\mathbf{v}_j) \text{ with respect to basis} \\ \{\mathbf{w}_1, \mathbf{w}_2, \ldots, \mathbf{w}_m\} \end{bmatrix}$$

then (2.16) reduces to

$$\mathbf{L}(\mathbf{v}) = [\mathbf{w}_1\ \mathbf{w}_2 \ldots \mathbf{w}_m]\mathbf{y} = \sum_{i=1}^{m} y_i \mathbf{w}_i = \mathbf{w}$$

Relative to the chosen basis $\{\mathbf{v}_1, \mathbf{v}_2, \ldots, \mathbf{v}_n\}$ of V and $\{\mathbf{w}_1, \mathbf{w}_2, \ldots, \mathbf{w}_m\}$ of W, v is represented by

$$\mathbf{x} = \begin{bmatrix} x_1 \\ x_2 \\ \vdots \\ x_n \end{bmatrix} \text{ and } \mathbf{w} = \mathbf{y}\mathbf{b}\mathbf{v}) \text{ is represented by } \mathbf{y} = \begin{bmatrix} y_1 \\ y_2 \\ \vdots \\ y_m \end{bmatrix}$$

The relationship between x and y is given by eqn. (2.17). The $m \times n$ matrix **A** is said to represent the linear transformation **L** with respect to the chosen basis.

It may be noted that matrix **A** gives the relation between representations **x** and **y** and not the vectors **v** and **w**. We also see that **A** depends upon the basis chosen; for different bases, we have different representations of the same operator.

Example 2.8: Consider the linear operator **L** of Example 2.7 shown in Fig. 2.3. From Fig. 2.3a,

$$\sin(\theta + \phi) = y_2/(x_1^2 + x_2^2)^{1/2}$$

and

$$\cos(\theta + \phi) = y_1/(x_1^2 + x_2^2)^{1/2}$$

These equations give

$$y_2 = x_1 \sin\phi + x_2 \cos\phi$$

$$y_1 = x_1 \cos\phi - x_2 \sin\phi$$

or

$$\begin{bmatrix} y_1 \\ y_2 \end{bmatrix} = [\mathbf{A}] \begin{bmatrix} x_1 \\ x_2 \end{bmatrix} = \begin{bmatrix} \cos\phi & -\sin\phi \\ \sin\phi & \cos\phi \end{bmatrix} \begin{bmatrix} x_1 \\ x_2 \end{bmatrix}$$

Matrix **A** is the representation of the linear operator **L** with respect to orthogonal basis $\{\mathbf{e}_1, \mathbf{e}_2\}$.

□

A class of linear transformations of practical significance is that in which the vector space $\mathbf{V}(\mathcal{F})$ is transformed into itself.

Consider $$\mathbf{L} : \mathbf{V}(\mathcal{F}) \to \mathbf{V}(\mathcal{F})$$

Let $\{\mathbf{v}_1, \mathbf{v}_2, \ldots, \mathbf{v}_n\}$ be the chosen basis of **V**. Suppose we change the basis of **V** to $\{\bar{\mathbf{v}}_1, \bar{\mathbf{v}}_2, \ldots, \bar{\mathbf{v}}_n\}$. Relative to the old and new basis, the representation of $\mathbf{v} \in \mathbf{V}$ is say **x** and $\bar{\mathbf{x}}$ respectively. There will be a nonsingular matrix **P** such that (refer eqn. (2.9))

$$\bar{\mathbf{x}} = \mathbf{P}\mathbf{x}$$

Similarly, the representation of **L**(**v**) with respect to old and new basis is say **y** and $\bar{\mathbf{y}}$ respectively. Then

$$\bar{\mathbf{y}} = \mathbf{P}\mathbf{y}$$

In terms of the original basis, **L** is represented **L** $n \times n$ square matrix, say **A**. Then we have (eqn. (2.17))

$$\mathbf{y} = \mathbf{A}\mathbf{x}$$

In terms of the new basis, **L** is represented by the equation

$$\bar{\mathbf{y}} = \mathbf{P}\mathbf{A}\mathbf{P}^{-1}\bar{\mathbf{x}} = \bar{\mathbf{A}}\,\bar{\mathbf{x}}$$

This equation shows that **L** is now represented by the matrix

$$\bar{\mathbf{A}} = \mathbf{P}\mathbf{A}\mathbf{P}^{-1} \tag{2.18}$$

Two matrices **A** and $\bar{\mathbf{A}}$ are said to be equivalent or similar if there exists a nonsingular matrix satisfying the transformation (2.18). This transformation is called *similarity transformation or equivalence transformation.*

Given an $n \times n$ matrix **A** with coefficients in field $\mathcal{F}$;

$$\mathbf{A} = \begin{bmatrix} a_{11} & a_{12} \cdots a_{1n} \\ a_{21} & a_{22} \cdots a_{2n} \\ \vdots & \vdots \quad\;\; \vdots \\ a_{n1} & a_{n2} \cdots a_{nn} \end{bmatrix} = [\,\mathbf{a}_1\ \mathbf{a}_2 \ldots \mathbf{a}_n\,],$$

we can interpret it in two ways:

(i) It is a representation of some unknown linear operator **L**.

(ii) It itself is a linear operator or the representation of a linear operator $\mathbf{A} : \mathcal{F}^n(\mathcal{F}) \to \mathcal{F}^n(\mathcal{F})$ with $\{\mathbf{e}_1, \mathbf{e}_2, \ldots, \mathbf{e}_n\}$ as the basis of $\mathcal{F}^n(\mathcal{F})$, because in this case the representation is identical with linear operator. This can be checked by using the fact that the ith column of representation of operator **A** is equal to the representation of $\mathbf{Ae}_i$ with respect to the basis $\{\mathbf{e}_1, \mathbf{e}_2, \ldots, \mathbf{e}_n\}$. Now $\mathbf{Ae}_i = \mathbf{a}_i$ and the representation of $\mathbf{a}_i$ with respect to the basis $\{\mathbf{e}_1, \mathbf{e}_2, \ldots, \mathbf{e}_n\}$ is identical with itself. Thus the representation of a matrix (or linear operator) with respect to basis $\{\mathbf{e}_1, \mathbf{e}_2, \ldots, \mathbf{e}_n\}$ is identical with itself.

Given a matrix **A**, unless it is tied up with some linear operator **L**, we shall always consider it itself as a linear operator.

In matrix theory, a matrix is introduced as an array of numbers. With the concepts of linear operator and representation, we have given a new interpretation of a matrix. Athans and Falb (1966) have shown how matrix algebra carries over to operations with linear transformations.[1]

2.5 SCALAR PRODUCT AND NORMS

Scalar (Inner) Product of Vectors

The concepts introduced in this section are applicable to any linear space over the field of real numbers or over the field of complex numbers. However, for convenience, we restrict our discussion to the real vector space $\mathcal{R}^n(\mathcal{R})$.

The scalar (inner) product of two vectors **x** and **y** in $\mathcal{R}^n(\mathcal{R})$ is a real number denoted by $\langle \mathbf{x}, \mathbf{y} \rangle$ and computed as

$$\langle \mathbf{x}, \mathbf{y} \rangle = \mathbf{x}^T\mathbf{y} = \mathbf{y}^T\mathbf{x} = \sum_{i=1}^{n} x_i y_i \tag{2.19}$$

The scalar product has the following properties:

$$\begin{aligned} &\text{(i)} \quad \langle \mathbf{x}, \mathbf{y} \rangle = \langle \mathbf{y}, \mathbf{x} \rangle \\ &\text{(ii)} \quad \langle \alpha_1\mathbf{x}_1 + \alpha_2\mathbf{x}_2, \mathbf{y} \rangle = \alpha_1\langle \mathbf{x}_1, \mathbf{y} \rangle + \alpha_2\langle \mathbf{x}_2, \mathbf{y} \rangle; \ \alpha_1, \alpha_2 \in \mathcal{R} \\ &\text{(iii)} \quad \langle \mathbf{x}, \mathbf{x} \rangle > 0 \ \forall \ \mathbf{x} \neq \mathbf{0} \end{aligned} \tag{2.20}$$

For orthogonal vectors **x** and **y**,

$$\langle \mathbf{x}, \mathbf{y} \rangle = 0 \tag{2.21}$$

Quadratic Functions and Definite Matrices: One important property of the scalar product involving column vectors and matrices is as follows. Suppose that $\mathbf{x} \in \mathcal{R}^n$ with components $x_1, x_2, \ldots, x_n$ and **A** is a symmetric $n \times n$ matrix with elements $a_{ij} \in \mathcal{R}$. Then

[1] The material in Sections 2.5–2.7 is not used until Chapter 5.

$$\langle \mathbf{x}, \mathbf{A}\mathbf{x} \rangle = \sum_{i=1}^{n} \sum_{j=1}^{n} a_{ij} x_i x_j = \mathbf{x}^T \mathbf{A}\mathbf{x}$$

The scalar valued function $f(\mathbf{x}) = \langle \mathbf{x}, \mathbf{A}\mathbf{x} \rangle$ involves multiplication by pairs of the elements x_i of $\mathbf{x}$ and is called *quadratic function.*

If for all $\mathbf{x} \neq \mathbf{0}$,

(i) $f(\mathbf{x}) = \langle \mathbf{x}, \mathbf{A}\mathbf{x} \rangle \geqslant 0$,

$f(\mathbf{x})$ is called a positive semidefinite form and $\mathbf{A}$ is called a *positive semidefinite matrix*;

(ii) $f(\mathbf{x}) = \langle \mathbf{x}, \mathbf{A}\mathbf{x} \rangle > 0$,

$f(\mathbf{x})$ is called a positive definite form and $\mathbf{A}$ is called a *positive definite matrix*;

(iii) $f(\mathbf{x}) = \langle \mathbf{x}, \mathbf{A}\mathbf{x} \rangle \leqslant 0$,

$f(\mathbf{x})$ is called a negative semidefinite form and $\mathbf{A}$ is called a *negative semidefinite matrix*;

(iv) $f(\mathbf{x}) = \langle \mathbf{x}, \mathbf{A}\mathbf{x} \rangle < 0$,

$f(\mathbf{x})$ is called a negative definite form and $\mathbf{A}$ is called a *negative definite matrix.*

Sign definiteness of quadratic functions can be determined by *Sylvester's theorem* which states that the necessary and sufficient conditions for

$$f(\mathbf{x}) = \mathbf{x}^T \mathbf{A}\mathbf{x} = [x_1 \; x_2 \ldots x_n] \begin{bmatrix} a_{11} & a_{12} & \ldots & a_{1n} \\ a_{12} & a_{22} & \ldots & a_{2n} \\ \vdots & \vdots & & \vdots \\ a_{1n} & a_{2n} & \ldots & a_{nn} \end{bmatrix} \begin{bmatrix} x_1 \\ x_2 \\ \vdots \\ x_n \end{bmatrix} \tag{2.22a}$$

to be positive definite are that all the successive principal determinants of $\mathbf{A}$ be positive, i.e.,

$$a_{11} > 0; \quad \begin{vmatrix} a_{11} & a_{12} \\ a_{12} & a_{22} \end{vmatrix} > 0; \ldots; \det \mathbf{A} > 0 \tag{2.22b}$$

The function is semidefinite if any of the above determinants is zero.

Gram Determinant: Consider a set of m vectors $\mathbf{x}_1$, $\mathbf{x}_2, \ldots, \mathbf{x}_m$ each having n components. These vectors are linearly independent provided no set of constants $\alpha_1, \alpha_2, \ldots, \alpha_m$ exists (at least one α_i must be nonzero) such that

$$\alpha_1 \mathbf{x}_1 + \alpha_2 \mathbf{x}_2 + \ldots + \alpha_m \mathbf{x}_m = \mathbf{0} \tag{2.23}$$

By successively taking the scalar products of $\mathbf{x}_i$ on both sides of (2.23), the set of equations

$$\begin{aligned}
\alpha_1 \langle \mathbf{x}_1, \mathbf{x}_1 \rangle + \alpha_2 \langle \mathbf{x}_1, \mathbf{x}_2 \rangle + \ldots + \alpha_m \langle \mathbf{x}_1, \mathbf{x}_m \rangle &= 0 \\
\alpha_1 \langle \mathbf{x}_2, \mathbf{x}_1 \rangle + \alpha_2 \langle \mathbf{x}_2, \mathbf{x}_2 \rangle + \ldots + \alpha_m \langle \mathbf{x}_2, \mathbf{x}_m \rangle &= 0 \\
&\vdots \\
\alpha_1 \langle \mathbf{x}_m, \mathbf{x}_1 \rangle + \alpha_2 \langle \mathbf{x}_m, \mathbf{x}_2 \rangle + \ldots + \alpha_m \langle \mathbf{x}_m, \mathbf{x}_m \rangle &= 0
\end{aligned}$$

is obtained. As shown in the next section, this set of homogeneous equations possesses a nontrivial solution for α_i only if the determinant of the coefficient matrix vanishes. This determinant is called the *Gramian* or *Gram determinant* and is

$$G = \begin{vmatrix} \langle \mathbf{x}_1, \mathbf{x}_1 \rangle & \langle \mathbf{x}_1, \mathbf{x}_2 \rangle & \ldots & \langle \mathbf{x}_1, \mathbf{x}_m \rangle \\ \langle \mathbf{x}_2, \mathbf{x}_1 \rangle & \langle \mathbf{x}_2, \mathbf{x}_2 \rangle & \ldots & \langle \mathbf{x}_2, \mathbf{x}_m \rangle \\ \cdots & \cdots & \cdots & \cdots \\ \langle \mathbf{x}_m, \mathbf{x}_1 \rangle & \langle \mathbf{x}_m, \mathbf{x}_2 \rangle & \ldots & \langle \mathbf{x}_m, \mathbf{x}_m \rangle \end{vmatrix} \tag{2.24}$$

Therefore *a set of vectors is linearly dependent if and only if the Gramian of the set of vectors is zero.*

Vector and Matrix Norms

The concept of the norm of a vector is a generalization of the idea of length.

For the vector

$$\mathbf{x} = \begin{bmatrix} x_1 \\ x_2 \\ \vdots \\ x_m \end{bmatrix}$$

the euclidean norm, denoted by $\| \mathbf{x} \|_2$ is defined by

$$\| \mathbf{x} \|_2 = (x_1^2 + x_2^2 + \ldots + x_n^2)^{1/2} = \sqrt{\langle \mathbf{x}, \mathbf{x} \rangle} \tag{2.25}$$

It should be clear that the value of $\| \mathbf{x} \|_2$ provides us with an idea of how big $\mathbf{x}$ is. The euclidean norm defined by (2.25) satisfies the following conditions:

(i) $\| \mathbf{x} \|_2 \geqslant 0$; $\| \mathbf{x} \|_2 = 0$ if and only if $\mathbf{x} = \mathbf{0}$

(ii) $\| \alpha \mathbf{x} \|_2 = | \alpha | \, \| \mathbf{x} \|_2$ for all $\alpha \in \mathcal{R}$

(iii) $\| \mathbf{x} + \mathbf{y} \|_2 \leqslant \| \mathbf{x} \|_2 + \| \mathbf{y} \|_2$ for all $\mathbf{x}, \mathbf{y} \in \mathcal{R}^n$

For many applications, the euclidean norm is not the most convenient to use in algebraic manipulations. For this reason, we generalize the notion of the norm of a vector in the following way.

A scalar-valued function of $\mathbf{x} \in \mathcal{R}^n$ qualifies as a norm, $\| \mathbf{x} \|$ of $\mathbf{x}$, provided that the following properties are satisfied:

$$\begin{aligned}
&\text{(i)} \quad \| \mathbf{x} \| > 0 \;\forall\; \mathbf{x} \neq \mathbf{0} \\
&\text{(ii)} \quad \| \alpha \mathbf{x} \| = | \alpha | \, \| \mathbf{x} \| \;\forall\; \alpha \in \mathcal{R} \\
&\text{(iii)} \quad \| \mathbf{x} + \mathbf{y} \| \leqslant \| \mathbf{x} \| + \| \mathbf{y} \| \;\forall\; \mathbf{x}, \mathbf{y} \in \mathcal{R}^n
\end{aligned} \tag{2.26}$$

Some valid norms are

$$\| \mathbf{x} \|_1 \triangleq \sum_{i=1}^{n} | x_i | \tag{2.27a}$$

$$\| \mathbf{x} \|_2 \triangleq [\sum_{i=1}^{n} x_i^2]^{1/2} \tag{2.27b}$$

$$\| \mathbf{x} \|_p \triangleq [\sum_{i=1}^{n} | x_i |^p]^{1/p} \tag{2.27c}$$

$$\| \mathbf{x} \|_\infty \triangleq \max_i | x_i | \tag{2.27d}$$

$$\| \mathbf{x} \| \triangleq (\mathbf{x}^T \mathbf{A} \mathbf{x})^{1/2}, \ \mathbf{A} \text{ is positive definite} \tag{2.27e}$$

Next we turn our attention to the concept of a norm of a matrix. To motivate the definition, we note that a vector $\mathbf{x} \in \mathcal{R}^n$ may be viewed as $n \times 1$ matrix. Thus the properties of vector norms can be extended to the matrix norms in the following way.

Let **A** and **B** be real $n \times m$ matrices with elements a_{ij} and b_{ij} ($i = 1, 2, \ldots, n$; $j = 1, 2, \ldots, m$) respectively. A scalar valued function $\| \mathbf{A} \|$ of **A** qualifies as the norm of **A** provided that the following properties are satisfied:

(i) $\| \mathbf{A} \| > 0$ provided not all $a_{ij} = 0$

(ii) $\| \alpha \mathbf{A} \| = | \alpha | \, \| \mathbf{A} \| \ \forall \ \alpha \in \mathcal{R}$ (2.28)

(iii) $\| \mathbf{A} + \mathbf{B} \| \leqslant \| \mathbf{A} \| + \| \mathbf{B} \|$

Some valid matrix norms are

$$\| \mathbf{A} \|_1 = \sum_{i=1}^{n} \sum_{j=1}^{m} | a_{ij} | \tag{2.29a}$$

$$\| \mathbf{A} \|_2 = (\sum_{i=1}^{n} \sum_{j=1}^{m} a_{ij}^2)^{1/2} \tag{2.29b}$$

$$\| \mathbf{A} \|_\infty = \max_i \sum_{j=1}^{m} | a_{ij} | \tag{2.29c}$$

The following two important properties of norms can be easily proved by using the norm definitions given above.

(i) $\| \mathbf{A}\mathbf{x} \| \leqslant \| \mathbf{A} \| \, \| \mathbf{x} \|$ (2.30)

(ii) $\| \mathbf{A}\mathbf{B} \| \leqslant \| \mathbf{A} \| \, \| \mathbf{B} \|$ (2.31)

Scalar Product and Norm of Vector Functions

Let us now consider the set of continuous real functions $\mathbf{\Omega}([t_0, t_1)], \mathcal{R}^n)$. If **f** and **g** are elements of $\mathbf{\Omega}$, then the scalar product **f** and **g**, denoted by $\langle \mathbf{f}, \mathbf{g} \rangle$ is given by

$$\langle \mathbf{f}, \mathbf{g} \rangle = \int_{t_0}^{t_1} \mathbf{f}^T(t)\mathbf{g}(t)dt \tag{2.32}$$

In view of this definition,

$$\| \mathbf{f} \|_2 = \sqrt{\langle \mathbf{f}, \mathbf{f} \rangle} = \left[\int_{t_0}^{t_1} \mathbf{f}^T(t)\mathbf{f}(t)dt \right]^{1/2} \tag{2.33}$$

The Gramian matrix of functions $\mathbf{f}_i$; $i = 1, 2, \ldots, m$, is

$$\mathbf{W}(t_0, t_1) = \int_{t_0}^{t_1} \mathbf{F}(t)\mathbf{F}^T(t)dt \tag{2.34}$$

where $\mathbf{F}$ is $m \times n$ matrix with $\mathbf{f}_i$ as the ith row. The functions $\mathbf{f}_1, \mathbf{f}_2, \ldots, \mathbf{f}_m$ are linearly independent on $[t_0, t_1]$ if and only if the $m \times m$ constant matrix $\mathbf{W}(t_0, t_1)$ is nonsingular.

2.6 SOLUTION OF LINEAR ALGEBRAIC EQUATIONS

A set of linear algebraic equations is a set of equations of the form

$$\begin{aligned} a_{11}x_1 + a_{12}x_2 + \ldots + a_{1n}x_n &= b_1 \\ a_{21}x_1 + a_{22}x_2 + \ldots + a_{2n}x_n &= b_2 \\ \vdots \qquad \vdots \qquad & \vdots \\ a_{m1}x_1 + a_{m2}x_2 + \ldots + a_{mn}x_n &= b_m \end{aligned} \tag{2.35}$$

where the given a_{ij}'s and b_i's are assumed to be known elements of a field $\mathcal{F}$, the unknown x_i's are also required to be in the same field $\mathcal{F}$. (2.35) is a set of m equations for n unknowns. This set of equations can be written in the vector-matrix form as

$$\mathbf{Ax} = \mathbf{b} \tag{2.36}$$

where

$$\mathbf{A} \triangleq \begin{bmatrix} a_{11} & a_{12} & \ldots & a_{1n} \\ a_{21} & a_{22} & \ldots & a_{2n} \\ \vdots & \vdots & & \vdots \\ a_{m1} & a_{m2} & \ldots & a_{mn} \end{bmatrix}; \quad \mathbf{x} \triangleq \begin{bmatrix} x_1 \\ x_2 \\ \vdots \\ x_n \end{bmatrix}; \quad \mathbf{b} \triangleq \begin{bmatrix} b_1 \\ b_2 \\ \vdots \\ b_m \end{bmatrix}$$

No restriction is placed on the integer m in (2.35); it may be larger than, equal to or less than the integer n. There are generally two questions of interest with regard to (2.35):

(i) Whether a solution to (2.35) exists, i.e., whether there is any $\mathbf{x}$ satisfying $\mathbf{Ax} = \mathbf{b}$.

(ii) If the answer to the first question is yes, how many linearly independent solutions occur for the equation $\mathbf{Ax} = \mathbf{b}$?

In order to answer these questions, we introduce the terms range space, rank, null space and nullity of matrix $\mathbf{A}$.

Range Space, Rank, Null Space and Nullity of a Matrix

Note that matrix $\mathbf{A}$ may be considered as a linear operator which maps the vector space $\mathcal{F}^n(\mathcal{F})$ into the vector space $\mathcal{F}^m(\mathcal{F})$. The vector space $\mathcal{F}^n(\mathcal{F})$ is the domain of the linear operator and $\mathcal{F}^m(\mathcal{F})$ is the co-domain. Associated with linear transformation **A** of $\mathcal{F}^n(\mathcal{F})$ into $\mathcal{F}^m(\mathcal{F})$ are two particular subspaces, called the *null space* of **A**(**N**(**A**)) and the *range space* of **A**(**R**(**A**)).

The *range* of the linear operator **A** is the set **R**(**A**) consisting of all elements **b** of $\mathcal{F}^m(\mathcal{F})$ for which there exists at least one vector in $\mathcal{F}^n(\mathcal{F})$ such that $\mathbf{b} = \mathbf{A}\mathbf{x}$.

The range of linear operator **A** is a subspace of $\mathcal{F}^m(\mathcal{F})$. This can be easily established. Suppose $\mathbf{b}_1$ and $\mathbf{b}_2$ are elements of **R** (**A**); then by definition there exist $\mathbf{x}_1$ and $\mathbf{x}_2$ in $\mathcal{F}^n(\mathcal{F})$ such that

$$\mathbf{b}_1 = \mathbf{A}\mathbf{x}_1$$

and

$$\mathbf{b}_2 = \mathbf{A}\mathbf{x}_2$$

By the linearity of **A** we have,

$$\mathbf{A}(\alpha_1\mathbf{x}_1 + \alpha_2\mathbf{x}_2) = \alpha_1\mathbf{b}_1 + \alpha_2\mathbf{b}_2 \quad \forall \alpha_1, \alpha_2 \in \mathcal{F}$$

Thus $\alpha_1\mathbf{b}_1 + \alpha_2\mathbf{b}_2$ is also an element of **R**(**A**). Hence **R**(**A**) is a subspace of $\mathcal{F}^m(\mathcal{F})$.

Equation (2.36) may be expressed as

$$\mathbf{b} = \mathbf{A}\mathbf{x}$$

$$= [\mathbf{a}_1 \ \mathbf{a}_2 \ldots \mathbf{a}_n] \begin{bmatrix} x_1 \\ x_2 \\ \vdots \\ x_n \end{bmatrix}$$

$$= x_1\mathbf{a}_1 + x_2\mathbf{a}_2 + \ldots + x_n\mathbf{a}_n \tag{2.37}$$

where $\mathbf{a}_1, \mathbf{a}_2, \ldots, \mathbf{a}_n$ are the column vectors of **A**. In equation (2.37), all the vectors are known but the coefficients $x_1, x_2, \ldots, x_n$ have to be determined. The problem can be viewed as that of expressing the vector **b** as a linear combination of vectors $\mathbf{a}_1, \mathbf{a}_2, \ldots, \mathbf{a}_n$. With this interpretation, we may say that **R**(**A**) *is the set of all possible linear combinations of the columns of* **A**. Since **R**(**A**) is a linear space, we may define its dimensionality which obviously is the *maximum number of linearly independent columns in* **A**.

We can now define the rank of a matrix **A**. *The rank of matrix* **A**, *denoted as* ρ(**A**) *is the maximum number of linearly independent columns in* **A** *or equivalently the dimension of range space* **R**(**A**).

The *null space* of a linear operator **A** is the set **N**(**A**) consisting of all the elements **x** of $\mathcal{F}^n(\mathcal{F})$ for which $\mathbf{A}\mathbf{x} = \mathbf{0}$. The dimension of **N**(**A**) is called the *nullity* of **A** and is denoted by γ(**A**). In other words, the null space **N**(**A**) is a set of all solutions of the *homogeneous equation* $\mathbf{A}\mathbf{x} = \mathbf{0}$.

It is easy to show that **N**(**A**) is a subspace of domain $\mathcal{F}^n(\mathcal{F})$. If $\gamma(\mathbf{A})=0$, then N(A) consists of only the zero vector and the only solution of $\mathbf{Ax} = \mathbf{0}$ is $\mathbf{x} = \mathbf{0}$. If $\gamma(\mathbf{A}) = \gamma$, then the equation $\mathbf{Ax} = \mathbf{0}$ has γ linearly independent vector solutions.

□

In matrix theory, the rank of a matrix is defined as follows:

Any square array of an $m \times n$ rectangular matrix A will have a determinant. The rank of A is defined as the order of the largest square array whose determinant does not vanish. The square array is formed by removing appropriate rows and columns of A.

This definition and the definition given earlier are in fact equivalent (for proof, refer Hadley (1961)). Using this result, we can prove the following very easily:

(i) An $n \times n$ square matrix A has full rank if and only if the determinant of A is nonzero. Correspondingly, a square matrix A is nonsingular (det A is nonzero) if and only if all the columns of the matrix are linearly independent.

(ii) The maximum number of linearly independent columns of a matrix is equal to the maximum number of linearly independent rows. Therefore, column rank of A = row rank of A = determinantal rank of A.

Homogeneous Equations

Example 2.9: Consider the linear homogeneous equations:

$$4x_1 + 2x_2 + x_3 + 3x_4 = 0$$

$$6x_1 + 3x_2 + x_3 + 4x_4 = 0$$

$$2x_1 + x_2 \qquad + x_4 = 0$$

or

$$\mathbf{Ax} = \mathbf{0}$$

where

$$\mathbf{A} = \begin{bmatrix} 4 & 2 & 1 & 3 \\ 6 & 3 & 1 & 4 \\ 2 & 1 & 0 & 1 \end{bmatrix}$$

It is easy to check that $\rho(\mathbf{A}) = 2$; a highest order array having nonvanishing determinant is

$$\begin{bmatrix} 2 & 3 \\ 1 & 1 \end{bmatrix}$$

which is obtained from A by omitting the first and third columns and the second row. Consequently, a set of linearly independent equations is

$$2x_2 + 3x_4 = -4x_1 - x_3$$
$$x_2 + x_4 = -2x_1$$

or

$$\begin{bmatrix} 2 & 3 \\ 1 & 1 \end{bmatrix}\begin{bmatrix} x_2 \\ x_4 \end{bmatrix} = \begin{bmatrix} -4x_1 - x_3 \\ -2x_1 \end{bmatrix}$$

Therefore

$$\begin{bmatrix} x_2 \\ x_4 \end{bmatrix} = \begin{bmatrix} 2 & 3 \\ 1 & 1 \end{bmatrix}^{-1}\begin{bmatrix} -4x_1 - x_3 \\ -2x_1 \end{bmatrix}$$
$$= \begin{bmatrix} x_3 - 2x_1 \\ -x_3 \end{bmatrix}$$

Note that the number of equations which **x** has to satisfy is equal to ρ(A). There are four components in **x** and two equations governing them; therefore two of the four components can be arbitrarily chosen. For example, let $x_1 = x_3 = 1$.

Then $x_2 = x_4 = -1$. Thus a solution vector is

$$\mathbf{x}_1 = \begin{bmatrix} 1 \\ -1 \\ 1 \\ -1 \end{bmatrix}$$

Similarly $x_1 = 1$, $x_3 = 2$ leads to a second solution vector

$$\mathbf{x}_2 = \begin{bmatrix} 1 \\ 0 \\ 2 \\ -2 \end{bmatrix}$$

Since the Gramian

$$G = \begin{vmatrix} 4 & 5 \\ 5 & 9 \end{vmatrix} \neq 0,$$

$\mathbf{x}_1$ and $\mathbf{x}_2$ are linearly independent. A different choice for x_1 and x_3 leads to a vector **x** which is not linearly independent on $\mathbf{x}_1$ and $\mathbf{x}_2$. Thus any solution $\mathbf{Ax} = \mathbf{0}$ must be a linear combination of $\mathbf{x}_1$ and $\mathbf{x}_2$. Therefore, this set of vectors forms a basis for N(A) and γ(A) = 2.

□

We see from this example that the number of equations that the vectors of N(A) should obey is equal to ρ(A) and that there are n-components (n = number of columns of A) in every vector of N(A); therefore (n-ρ(A)) components of the vectors of N(A) can be arbitrarily chosen. Consequently there are (n-ρ(**A**)) linearly independent vectors in N(**A**) (refer Hadley (1961)), i.e.,

$$\gamma(\mathbf{A}) = n - \rho(\mathbf{A})$$

Thus if $\rho(\mathbf{A}) = n$, then the only solution of $\mathbf{Ax} = \mathbf{0}$ is $\mathbf{x} = \mathbf{0}$ which is the trivial solution. If $\rho(\mathbf{A}) < n$, then we can always find a nonzero vector **x** such that $\mathbf{Ax} = \mathbf{0}$.

Therefore, the set $\mathbf{Ax} = \mathbf{0}$ *of m homogeneous equations in n unknowns has a nontrivial solution if* $m < n$. *For a square matrix* **A**, $\mathbf{Ax} = \mathbf{0}$ *has a nontrivial solution if* $\rho(\mathbf{A}) < n$ *or equivalently det* $\mathbf{A} = 0$.

Nonhomogeneous Equations

Consider nonhomogeneous linear algebraic equations given by (2.36). For the solution of these equations, four cases can be distinguished (refer Noble's paper in the book Nashed (1976)):

1. If $\rho(\mathbf{A}) = \rho(\mathbf{A} : \mathbf{b})$, the equations are consistent and have one or more solutions. We further distinguish two cases.
   (i) If $\rho(\mathbf{A}) = n \leqslant m$, there is a unique solution.
   (ii) If $\rho(\mathbf{A}) < n$, the equations have infinite number of solutions which can be written in the form $\mathbf{x} = \mathbf{x}_0 + \mathbf{z}$ where $\mathbf{x}_0$ is a particular solution and $\mathbf{Az} = \mathbf{0}$, i.e., $\mathbf{z} \in \mathrm{N}(\mathbf{A})$ of dimension $(n - \rho(\mathbf{A}))$.
2. If $\rho(\mathbf{A}) < \rho(\mathbf{A} : \mathbf{b})$, the equations are inconsistent. We may however obtain a *least-squares solution*, i.e., the solution that minimizes $\| \mathbf{Ax} - \mathbf{b} \|_2$. We again distinguish two cases.
   (i) If $\rho(\mathbf{A}) = n$, there exists a unique least-squares solution.
   (ii) If $\rho(\mathbf{A}) < n$, there are infinite least-squares solutions of the form $\mathbf{x} = \mathbf{x}_0 + \mathbf{z}$ where $\mathbf{x}_0$ is a particular least-squares solution and $\mathbf{z} \in \mathrm{N}(\mathbf{A})$.

(**A**: **b**) denotes the augmented $m \times (n + 1)$ matrix

$$\begin{bmatrix} a_{11} & \cdots & a_{1n} & b_1 \\ a_{21} & \cdots & a_{2n} & b_2 \\ \vdots & & \vdots & \vdots \\ a_{m1} & \cdots & a_{mn} & b_m \end{bmatrix} \overset{\Delta}{=} [\mathbf{a}_1\ \mathbf{a}_2\ \ldots\ \mathbf{a}_n\ \mathbf{b}]$$

Now

$$\mathbf{b} = x_1\mathbf{a}_1 + x_2\mathbf{a}_2 + \ldots + x_n\mathbf{a}_n \tag{2.38}$$

Equation (2.36) is soluble if and only if **b** is linearly dependent on vectors $\mathbf{a}_1, \mathbf{a}_2, \ldots, \mathbf{a}_n$. If $\rho(\mathbf{A} : \mathbf{b}) = \rho(\mathbf{A})$, **b** is linearly dependent upon some or all of the vectors $\mathbf{a}_1, \mathbf{a}_2, \ldots, \mathbf{a}_n$. On the other hand, if $\rho(\mathbf{A} : \mathbf{b}) > \rho(\mathbf{A})$, the set $\{\mathbf{a}_1, \mathbf{a}_2, \ldots, \mathbf{a}_n, \mathbf{b}\}$ contains more linearly independent vectors than $\{\mathbf{a}_1, \mathbf{a}_2, \ldots, \mathbf{a}_n\}$ and in this case the equations are insoluble.

Inconsistent Equations: As pointed out earlier, if $\rho(\mathbf{A}) < \rho(\mathbf{A} : \mathbf{b})$, the equations $\mathbf{Ax} = \mathbf{b}$ are inconsistent. In this case it makes sense in many applications to look for a least-squares solution.

(i) If $\rho(\mathbf{A}) = n$, then the solution of the equations $\mathbf{Ax} = \mathbf{b}$ that minimizes the sum of the squares of the residuals $S = (\mathbf{b} - \mathbf{Ax})^T (\mathbf{b} - \mathbf{Ax})$ is obtained as follows (refer Appendix III).

$$S = (\mathbf{b} - \mathbf{Ax})^T (\mathbf{b} - \mathbf{Ax})$$
$$= \mathbf{b}^T\mathbf{b} - \mathbf{x}^T\mathbf{A}^T\mathbf{b} - \mathbf{b}^T\mathbf{Ax} + \mathbf{x}^T\mathbf{A}^T\mathbf{Ax}$$

Setting

$$\frac{\partial S}{\partial x_j} = 0;\ j = 1, 2, \ldots, n$$

we get

$$\mathbf{A}^T\mathbf{Ax} = \mathbf{A}^T\mathbf{b} \tag{2.39}$$

The matrix $\mathbf{A}^T\mathbf{A}$ is nonsingular. Therefore,

$$\mathbf{x} = (\mathbf{A}^T\mathbf{A})^{-1}\,\mathbf{A}^T\mathbf{b} \tag{2.40}$$

In case of $\mathbf{Ax} = \mathbf{b}$ where $\mathbf{A}$ is nonsingular square matrix, we have the unique solution

$$\mathbf{x} = \mathbf{A}^{-1}\mathbf{b} \tag{2.41}$$

(ii) If $\rho(\mathbf{A}) < n$, then there are infinite least squares solutions of the form $\mathbf{x} = \mathbf{x}_0 + \mathbf{z}$ where $\mathbf{x}_0$ is a particular least-squares solution and $\mathbf{z} \in N(\mathbf{A})$. A unique solution is obtained if we wish that

(a) the sum of the squares of the residuals

$S = (\mathbf{b} - \mathbf{Ax})^T (\mathbf{b} - \mathbf{Ax})$ is minimized; and

(b) the sum of the squares of the unknowns $\mathbf{x}^T\mathbf{x}$ is minimized.

This solution will be called *minimal least-squares solution.*

In order to minimize the sum of the squares of the residuals, $\mathbf{x}$ satisfies the relation (eqn. (2.39)),

$$\mathbf{A}^T\mathbf{Ax} = \mathbf{A}^T\mathbf{b} \tag{2.42}$$

We can no longer invert $\mathbf{A}^T\mathbf{A}$ since the matrx may be singular. Assume that $\mathbf{A} = \mathbf{BC}$ where $\mathbf{A}$, $\mathbf{B}$ and $\mathbf{C}$ are respectively $m \times n$, $m \times k$ and $k \times n$ matrices and all the three matrices are of rank $k (k < n)$ (refer Noble and Daniel (1977) for methods of decomposition of $\mathbf{A}$ into the form $\mathbf{BC}$). On introducing $\mathbf{A} = \mathbf{BC}$, eqn. (2.42) becomes

$$\mathbf{C}^T(\mathbf{B}^T\mathbf{B})\mathbf{Cx} = \mathbf{C}^T\mathbf{B}^T\mathbf{b}.$$

We multiply this equation by $\mathbf{C}$. Since $\mathbf{CC}^T$ and $\mathbf{B}^T\mathbf{B}$ are nonsingular matrices, we can multiply both sides of the above equation by $(\mathbf{B}^T\mathbf{B})^{-1}(\mathbf{CC}^T)^{-1}$ to obtain

$$\mathbf{Cx} = (\mathbf{B}^T\mathbf{B})^{-1}\mathbf{B}^T\mathbf{b} \stackrel{\Delta}{=} \mathbf{d} \tag{2.43}$$

The quantity $\mathbf{x}$ that minimizes $\mathbf{x}^T\mathbf{x}$ subject to (2.43) can be found by minimizing

$$M = \mathbf{x}^T\mathbf{x} + 2\boldsymbol{\lambda}^T(\mathbf{d} - \mathbf{Cx})$$

where $\boldsymbol{\lambda}$ is $k \times 1$ column vector of Lagrange multipliers and the factor of 2 is inserted for convenience. (The reader will appreciate this after reading Chapter 10.) Setting

$$\frac{\partial M}{\partial x_j} = 0; \quad j = 1, 2, \ldots, n$$

and

$$\frac{\partial M}{\partial \lambda_i} = 0; \quad i = 1, 2, \ldots, k$$

we get (refer Appendix III)

$$\mathbf{x} = \mathbf{C}^T\boldsymbol{\lambda}$$

$$\mathbf{C}\mathbf{x} = \mathbf{d}$$

Therefore

$$\mathbf{C}\mathbf{C}^T\boldsymbol{\lambda} = \mathbf{d} \quad \text{or} \quad \boldsymbol{\lambda} = (\mathbf{C}\mathbf{C}^T)^{-1}\mathbf{d}$$

Now

$$\mathbf{x} = \mathbf{C}^T\boldsymbol{\lambda}$$

$$= \mathbf{C}^T(\mathbf{C}\mathbf{C}^T)^{-1}\mathbf{d} \tag{2.44}$$

From eqns. (2.43) and (2.44) we obtain

$$\mathbf{x} = \mathbf{C}^T(\mathbf{C}\mathbf{C}^T)^{-1}(\mathbf{B}^T\mathbf{B})^{-1}\mathbf{B}^T\mathbf{b} \tag{2.45}$$

In the special case of full rank matrix **A**, i.e., $\rho(\mathbf{A}) = m$, we get

$$\mathbf{B} = \mathbf{I}, \ \mathbf{C} = \mathbf{A}$$

$$\mathbf{x} = \mathbf{A}^T(\mathbf{A}\mathbf{A}^T)^{-1}\mathbf{b} \tag{2.46}$$

Consistent Equations: As pointed out earlier, if $\rho(\mathbf{A}) = \rho(\mathbf{A} : \mathbf{b})$, the equations $\mathbf{A}\mathbf{x} = \mathbf{b}$ are consistent. The exact solution is therefore possible in this case.

(i) If $\rho(\mathbf{A}) = n$, then the vectors $\mathbf{a}_1, \mathbf{a}_2, \ldots, \mathbf{a}_n$ are linearly independent. Equation (2.38) then determines the coefficients x_i uniquely.
For $n \times n$ square matrix **A**, the exact solution is given by

$$\mathbf{x} = \mathbf{A}^{-1}\mathbf{b} \tag{2.47}$$

For $m \times n$ matrix **A** of rank n, the exact solution is

$$\mathbf{x} = (\mathbf{A}^T\mathbf{A})^{-1}\mathbf{A}^T\mathbf{b} \tag{2.48}$$

(ii) If $\rho(\mathbf{A}) < n$, then there are $(n - \rho(\mathbf{A})) = \gamma(\mathbf{A})$ linearly independent solutions of the equation $\mathbf{A}\mathbf{x} = \mathbf{0}$. If $\mathbf{x} = \mathbf{x}_0$ is any solution of $\mathbf{A}\mathbf{x} = \mathbf{b}$ and if $\mathbf{x} = \mathbf{z}$ is a solution of homogeneous equation $\mathbf{A}\mathbf{x} = \mathbf{0}$, then for any constant k, $\mathbf{x} = \mathbf{x}_0 + k\mathbf{z}$ is a solution of $\mathbf{A}\mathbf{x} = \mathbf{b}$. This can be easily established.

A unique solution is obtained if we desire to minimize the sum of the squares of the unknowns $\mathbf{x}^T\mathbf{x}$. For the case of full rank matrix **A**, i.e., $\rho(\mathbf{A}) = m$, it can be easily shown that

$$\mathbf{x} = \mathbf{A}^T(\mathbf{A}\mathbf{A}^T)^{-1}\mathbf{b} \tag{2.49}$$

□

The solutions given by eqns. (2.40), (2.45), (2.46), (2.48) and (2.49) may be expressed as

$$\mathbf{x} = \mathbf{A}^{\#}\mathbf{b} \tag{2.50}$$

where $\mathbf{A}^{\#}$ is called the *generalized inverse* (*pseudo inverse*) of **A**.

2.7 EIGENVALUES, EIGENVECTORS AND A CANONICAL-FORM REPRESENTATION OF LINEAR OPERATORS

In Section 2.4 we observed that a linear operator has many matrix representations. A representation **A** depends upon the basis chosen; for different bases we have different representations of the same operator. It is possible to select bases vectors such that the resulting representations are simple and useful. One such simple and useful form of a linear operator, mapping $\mathcal{C}^n(\mathcal{C})$ into itself, is discussed in this section.

Consider the equation

$$\mathbf{A}\mathbf{x} = \mathbf{y}$$

and view it as a transformation of vector $\mathbf{x} \in \mathcal{C}^n$ into vector $\mathbf{y} \in \mathcal{C}^n$ by matrix operator **A**. We pose ourselves a question. Does there exist a nonzero vector **x** in $\mathcal{C}^n$ such that linear operator **A** transforms it into vector $\lambda\mathbf{x}$ in $\mathcal{C}^n$ (λ is a scalar $\in \mathcal{C}$), i.e., a vector which is collinear to **x**?

Such a vector **x** is the solution of the equations

$$\mathbf{A}\mathbf{x} = \lambda\mathbf{x} \tag{2.51}$$

or

$$(\lambda\mathbf{I} - \mathbf{A})\mathbf{x} = \mathbf{0} \tag{2.52}$$

The set of homogeneous equations (2.52) has a solution if and only if (refer Section 2.6)

$$|\lambda\mathbf{I} - \mathbf{A}| = 0 \tag{2.53}$$

This equation may be written in the expanded form as

$$\Delta(\lambda) = \lambda^n + \alpha_1\lambda^{n-1} + \alpha_2\lambda^{n-2} + \ldots + \alpha_n = 0 \tag{2.54}$$

The values of λ for which eqn. (2.54) is satisfied are called the *eigenvalues* of matrix **A**, eqn. (2.53) (or eqn. (2.54)) is called the *characteristic equation* corresponding to matrix **A** and $\Delta(\lambda)$ is called the *characteristic polynomial.* Since $\Delta(\lambda)$ is of order n, the $n \times n$ matrix **A** has n eigenvalues (not necessarily all distinct). The coefficients α_i of the polynomial $\Delta(\lambda)$ are all real for the real-valued constant matrices we shall encounter in this book; however the roots of the polynomial are not necessarily all real. For this reason, we shall consider matrix **A** (real-valued) as a linear operator mapping $\mathcal{C}^n(\mathcal{C})$ into itself.

Now for $\lambda = \lambda_i$ satisfying eqn. (2.51), we have from eqn. (2.52)

$$(\lambda_i\mathbf{I} - \mathbf{A})\mathbf{x} = \mathbf{0} \tag{2.55}$$

The solution $\mathbf{x} = \mathbf{m}_i$ of eqn. (2.55) is called the *eigenvector* of $\mathbf{A}$ associated with eigenvalue λ_i.

Example 2.10: Consider the matrix

$$\mathbf{A} = \begin{bmatrix} -4 & 1 & 0 \\ 0 & -3 & 1 \\ 0 & 0 & -2 \end{bmatrix}$$

Corresponding to this matrix, the characteristic equation is

$$|\lambda\mathbf{I} - \mathbf{A}| = \begin{vmatrix} \lambda+4 & -1 & 0 \\ 0 & \lambda+3 & -1 \\ 0 & 0 & \lambda+2 \end{vmatrix}$$

$$= (\lambda+4)(\lambda+3)(\lambda+2) = 0$$

Therefore, the eigenvalues are $\lambda_1 = -2$, $\lambda_2 = -3$, $\lambda_3 = -4$.

The eigenvector $\mathbf{m}_1$ associated with $\lambda_1 = -2$ is obtained from the solution of homogeneous equations

$$(\lambda_1\mathbf{I} - \mathbf{A})\,\mathbf{x} = \mathbf{0}$$

or

$$\begin{bmatrix} 2 & -1 & 0 \\ 0 & 1 & -1 \\ 0 & 0 & 0 \end{bmatrix} \begin{bmatrix} x_1 \\ x_2 \\ x_3 \end{bmatrix} = \begin{bmatrix} 0 \\ 0 \\ 0 \end{bmatrix}$$

$\rho(\lambda_1\mathbf{I} - \mathbf{A})$ is 2; a highest order array having nonvanishing determinant is

$$\begin{bmatrix} 2 & -1 \\ 0 & 1 \end{bmatrix}$$

Consequently, a set of linearly independent equations is

$$2x_1 - x_2 = 0$$

$$x_2 = x_3$$

or

$$\begin{bmatrix} 2 & -1 \\ 0 & 1 \end{bmatrix} \begin{bmatrix} x_1 \\ x_2 \end{bmatrix} = \begin{bmatrix} 0 \\ x_3 \end{bmatrix}$$

or

$$\begin{bmatrix} x_1 \\ x_2 \end{bmatrix} = \begin{bmatrix} x_3/2 \\ x_3 \end{bmatrix}$$

Choosing $x_3 = 2$, we have $x_1 = 1$ and $x_2 = 2$. Therefore

$$\mathbf{m}_1 = \begin{bmatrix} 1 \\ 2 \\ 2 \end{bmatrix}$$

The eigenvectors $\mathbf{m}_2$ and $\mathbf{m}_3$ obtained from the solution of the equations $(\lambda_2\mathbf{I} - \mathbf{A})\,\mathbf{x} = \mathbf{0}$ and $(\lambda_3\mathbf{I} - \mathbf{A})\,\mathbf{x} = \mathbf{0}$ respectively, are

$$\mathbf{m}_2 = \begin{bmatrix} 1 \\ 1 \\ 0 \end{bmatrix}, \quad \mathbf{m}_3 = \begin{bmatrix} 2 \\ 0 \\ 0 \end{bmatrix}$$

It may be pointed out here that since the solution $\mathbf{x}$ of the homogeneous equations $(\lambda\mathbf{I} - \mathbf{A})\,\mathbf{x} = \mathbf{0}$ is nonunique, the eigenvectors of a matrix are also nonunique. However, the eigenvectors corresponding to a particular eigenvalue have a unique direction and they differ only in terms of a scalar multiplier.

□

With these preliminaries, we are ready to introduce a set of vectors such that a linear operator has a canonical form. We study first the case in which the canonical form of the linear operator is a diagonal matrix.

Diagonal-Form Matrix Representation of a Linear Operator

Let $\lambda_1, \lambda_2, .., \lambda_n$ be the eigenvalues (assumed distinct) of matrix $\mathbf{A}$ and let $\mathbf{m}_1, \mathbf{m}_2, .., \mathbf{m}_n$ be their corresponding eigenvectors. Then the set $\{\mathbf{m}_1, \mathbf{m}_2, .., \mathbf{m}_n\}$ is linearly independent (over $\mathcal{C}$) and therefore qualifies as a basis.

This can easily be proved by contradiction. Suppose that $\mathbf{m}_1, \mathbf{m}_2, .., \mathbf{m}_n$ are linearly dependent, then there exist $\alpha_1, \alpha_2, .., \alpha_n$ (not all zero) in $\mathcal{C}$ such that

$$\alpha_1\mathbf{m}_1 + \alpha_2\mathbf{m}_2 + .. + \alpha_n\mathbf{m}_n = \mathbf{0} \tag{2.56}$$

Premultiplying eqn. (2.56) by $\mathbf{A}$ yields

$$\alpha_1\mathbf{A}\mathbf{m}_1 + \alpha_2\mathbf{A}\mathbf{m}_2 + .. + \alpha_n\mathbf{A}\mathbf{m}_n = \mathbf{0}$$

But $\quad \mathbf{A}\mathbf{m}_i = \lambda_i\mathbf{m}_i$. Therefore,

$$\alpha_1\lambda_1\mathbf{m}_1 + \alpha_2\lambda_2\mathbf{m}_2 + .. + \alpha_n\lambda_n\mathbf{m}_n = \mathbf{0} \tag{2.57}$$

Repeating the process on eqn. (2.57) we obtain

$$\begin{aligned} \alpha_1\lambda_1^2\mathbf{m}_1 + \alpha_2\lambda_2^2\mathbf{m}_2 + .. + \alpha_n\lambda_n^2\,\mathbf{m}_n &= \mathbf{0} \\ &\vdots \\ \alpha_1\lambda_1^{n-1}\mathbf{m}_1 + \alpha_2\lambda_2^{n-1}\mathbf{m}_2 + .. + \alpha_n\lambda_n^{n-1}\,\mathbf{m}_n &= \mathbf{0} \end{aligned} \tag{2.58}$$

The n equations in (2.56)–(2.58) can be written in the form

$$[\alpha_1\mathbf{m}_1\ \alpha_2\mathbf{m}_2 \ldots \alpha_n\mathbf{m}_n]\begin{bmatrix} 1 & \lambda_1 & \lambda_1^2 \ldots \lambda_1^{n-1} \\ 1 & \lambda_2 & \lambda_2^2 \ldots \lambda_2^{n-1} \\ \vdots & \vdots & \vdots \quad \vdots \\ 1 & \lambda_n & \lambda_n^2 \ldots \lambda_n^{n-1} \end{bmatrix} = \mathbf{0}$$

The matrix

$$\mathbf{V} = \begin{bmatrix} 1 & \lambda_1 & \lambda_1^2 \ldots \lambda_1^{n-1} \\ 1 & \lambda_2 & \lambda_2^2 \ldots \lambda_2^{n-1} \\ \vdots & \vdots & \vdots \quad \vdots \\ 1 & \lambda_n & \lambda_n^2 \ldots \lambda_n^{n-1} \end{bmatrix} \tag{2.59}$$

is the *Vandermonde matrix* and is nonsingular if $\lambda_i \neq \lambda_j$ for all $i \neq j$ (Problem 2.18). Thus

$$[\alpha_1\mathbf{m}_1\ \alpha_2\mathbf{m}_2 \ldots \alpha_n\mathbf{m}_n] = \mathbf{0}$$

which implies that all α_i's are zero since no $\mathbf{m}_i$ can be a zero vector. This is a contradiction to our earlier assumption. Thus the set of vectors $\{\mathbf{m}_1, \mathbf{m}_2, \ldots, \mathbf{m}_n\}$ is linearly independent and qualifies as a basis.

Let $\mathbf{J}$ be the representation of $\mathbf{A}$ with respect to the basis $\{\mathbf{m}_1, \mathbf{m}_2, \ldots, \mathbf{m}_n\}$. Then the ith column of $\mathbf{J}$ is the representation of $\mathbf{A}\mathbf{m}_i = \lambda_i\mathbf{m}_i$ with respect to $\{\mathbf{m}_1, \mathbf{m}_2, \ldots, \mathbf{m}_n\}$, i.e., $[0\ 0 \ldots \lambda_i \ldots 0\ 0]^T$, where λ_i is the ith component (refer eqn. (2.17)). Therefore, the representation of $\mathbf{A}$ with respect to the basis $\{\mathbf{m}_1, \mathbf{m}_2, \ldots, \mathbf{m}_n\}$ is

$$\mathbf{J} = \begin{bmatrix} \lambda_1 & 0 & 0 \ldots 0 \\ 0 & \lambda_2 & 0 \ldots 0 \\ \vdots & \vdots & \vdots \quad \vdots \\ 0 & 0 & 0 \ldots \lambda_n \end{bmatrix} \tag{2.60}$$

Thus, if the eigenvalues of a linear operator $\mathbf{A}$ that maps $\mathcal{C}^n$ ($\mathcal{C}$) into itself are all distinct, then by choosing the set of eigenvectors as basis, the operator $\mathbf{A}$ has a diagonal matrix representation with the eigenvalues as diagonal elements. (As we shall see later in this section, this is a special case of general *Jordan-form representation.*)

This fact can also be verified using a similarity transformation. If $\mathbf{m}_1, \mathbf{m}_2, \ldots, \mathbf{m}_n$ are the eigenvectors of $\mathbf{A}$ corresponding to the eigenvalues $\lambda_1, \lambda_2, \ldots, \lambda_n$ respectively, then we have

$$\begin{aligned} \mathbf{AM} &= \mathbf{A}[\mathbf{m}_1\ \mathbf{m}_2 \ldots \mathbf{m}_n] \\ &= [\mathbf{A}\mathbf{m}_1\ \mathbf{A}\mathbf{m}_2 \ldots \mathbf{A}\mathbf{m}_n] \\ &= [\lambda_1\mathbf{m}_1\ \lambda_2\mathbf{m}_2 \ldots \lambda_n\mathbf{m}_n] \\ &= \mathbf{MJ} \end{aligned}$$

Thus

$$\mathbf{J} = \mathbf{M}^{-1}\mathbf{AM} \tag{2.61}$$

The matrix constructed by placing the eigenvectors (columns) together is therefore a *diagonalizing matrix* (also called *Modal matrix*) **M** of **A**, i.e.,

$$\mathbf{M} = [\mathbf{m}_1\ \mathbf{m}_2 \ldots \mathbf{m}_n] \tag{2.62}$$

Note that **A** and **J** have the same characteristic equation; therefore the eigenvalues are invariant under this transformation (refer Problem 2.17).

Example 2.11: Consider the matrix

$$\mathbf{A} = \begin{bmatrix} -4 & 1 & 0 \\ 0 & -3 & 1 \\ 0 & 0 & -2 \end{bmatrix}$$

for which we found in Example 2.10, the eigenvalues and eigenvectors to be

$$\lambda_1 = -2,\ \mathbf{m}_1 = \begin{bmatrix} 1 \\ 2 \\ 2 \end{bmatrix}$$

$$\lambda_2 = -3,\ \mathbf{m}_2 = \begin{bmatrix} 1 \\ 1 \\ 0 \end{bmatrix}$$

$$\lambda_3 = -4,\ \mathbf{m}_3 = \begin{bmatrix} 2 \\ 0 \\ 0 \end{bmatrix}$$

The modal matrix

$$\mathbf{M} = \begin{bmatrix} 1 & 1 & 2 \\ 2 & 1 & 0 \\ 2 & 0 & 0 \end{bmatrix}$$

from which

$$\mathbf{M}^{-1} = \frac{1}{4}\begin{bmatrix} 0 & 0 & 2 \\ 0 & 4 & -4 \\ 2 & -2 & 1 \end{bmatrix}$$

$$\mathbf{J} = \mathbf{M}^{-1}\mathbf{A}\,\mathbf{M} = \frac{1}{4}\begin{bmatrix} 0 & 0 & 2 \\ 0 & 4 & -4 \\ 2 & -2 & 1 \end{bmatrix}\begin{bmatrix} -4 & 1 & 0 \\ 0 & -3 & 1 \\ 0 & 0 & -2 \end{bmatrix}\begin{bmatrix} 1 & 1 & 2 \\ 2 & 1 & 0 \\ 2 & 0 & 0 \end{bmatrix}$$

$$= \begin{bmatrix} -2 & 0 & 0 \\ 0 & -3 & 0 \\ 0 & 0 & -4 \end{bmatrix}$$

which is the diagonal matrix with eigenvalues of **A** as diagonal elements. In fact **J** could be written down directly without computing $\mathbf{M}^{-1}\mathbf{A}\mathbf{M}$.

Example 2.12: It is possible, in some rare cases, to have nondistinct eigenvalues yet still be able to find *n* linearly independent eigenvectors to qualify as a basis. Consider for example

$$\mathbf{A} = \begin{bmatrix} 1 & 0 & 0 \\ 0 & 1 & 0 \\ -1 & 0 & 2 \end{bmatrix}$$

The eigenvalues of **A** are $\lambda_1 = 1$, $\lambda_2 = 1$, and $\lambda_3 = 2$. The eigenvector associated with λ_1 can be obtained by solving the following homogeneous equations:

$$(\lambda_1 \mathbf{I} - \mathbf{A})\,\mathbf{x} = \begin{bmatrix} 0 & 0 & 0 \\ 0 & 0 & 0 \\ 1 & 0 & -1 \end{bmatrix} \begin{bmatrix} x_1 \\ x_2 \\ x_3 \end{bmatrix} = \begin{bmatrix} 0 \\ 0 \\ 0 \end{bmatrix} \tag{2.63}$$

Sinnce $\rho(\lambda_1 \mathbf{I} - \mathbf{A}) = 1$, two linearly independent solutions can be found for (2.63) (refer Section 2.6). Clearly

$$\mathbf{m}_1 = \begin{bmatrix} 1 \\ 0 \\ 1 \end{bmatrix}; \mathbf{m}_2 = \begin{bmatrix} 0 \\ 1 \\ 0 \end{bmatrix}$$

are two linearly independent eigenvectors associated with $\lambda_1 = \lambda_2 = 1$.

An eigenvector associated with $\lambda_3 = 2$ can be found as

$$\mathbf{m}_3 = \begin{bmatrix} 0 \\ 0 \\ 1 \end{bmatrix}$$

The modal matrix

$$\mathbf{M} = \begin{bmatrix} 1 & 0 & 0 \\ 0 & 1 & 0 \\ 1 & 0 & 1 \end{bmatrix}$$

and

$$\mathbf{J} = \mathbf{M}^{-1}\mathbf{A}\mathbf{M} = \begin{bmatrix} 1 & 0 & 0 \\ 0 & 1 & 0 \\ 0 & 0 & 2 \end{bmatrix}$$ □

The example above shows that a general $n \times n$ matrix **A** can be reduced to a diagonal form by a similarity transformation if and only if a set of n linearly independent eigenvectors can be found. Corresponding to a distinct eigenvalue, one and only one linearly independent eigenvector is possible. However, for a repeated eigenvalue λ_i, one should check the nullity of $(\lambda_i\mathbf{I} - \mathbf{A})$. If the nullity $\gamma(\lambda_i\mathbf{I} - \mathbf{A})$ is equal to the multiplicity of the eigenvalue λ_i, then a diagonal form for the matrix **A** is possible. If this is not the case, then a diagonal form is not possible as is seen in the following example.

Example 2.13: Consider the matrix

$$\mathbf{A} = \begin{bmatrix} 1 & 1 & 2 \\ 0 & 1 & 3 \\ 0 & 0 & 2 \end{bmatrix}$$

The eigenvalues of **A** are $\lambda_1 = 1$, $\lambda_2 = 1$ and $\lambda_3 = 2$. The eigenvalue $\lambda = 1$ has a multiplicity of 2. The rank of $(\lambda_1\mathbf{I} - \mathbf{A})$ is 2; therefore $\gamma\,(\lambda_1\mathbf{I} - \mathbf{A})$ is 1 and there is only one linearly independent eigenvector for $\lambda = 1$. Therefore, the matrix **A** cannot be diagonalized by a similarity transformation. We can find only one linearly independent eigenvector associated with $\lambda_1 = \lambda_2 = 1$. The second eigenvector can be found corresponding to $\lambda_3 = 2$. One more vector is needed to form a basis. The vector we shall use is called a *generalised eigenvector.*

Generalized Eigenvectors

A nonzero vector **m** for which $(\lambda\mathbf{I} - \mathbf{A})^{k-1}\,\mathbf{m} \neq \mathbf{0}$ but $(\lambda\mathbf{I} - \mathbf{A})^{k}\,\mathbf{m} = \mathbf{0}$ is called a generalized eigenvector of rank k of **A** associated with the eigenvalue λ.

Note that if $k = 1$, we have

$$(\lambda\mathbf{I} - \mathbf{A})^{0}\,\mathbf{m} = \mathbf{I}\mathbf{m} = \mathbf{m} \neq \mathbf{0} \text{ and } (\lambda\mathbf{I} - \mathbf{A})\,\mathbf{m} = \mathbf{0},$$

which is our definition of an eigenvector. Therefore the term 'generalised eigenvector' is well justified (refer eqns. (2.71) for the definition of the power of a matrix).

Let **m** be a generalized eigenvector of rank k associated with the eigenvalue λ. Define

$$\begin{aligned} \mathbf{m}_k &= \mathbf{m} \\ \mathbf{m}_{k-1} &= (\lambda\mathbf{I} - \mathbf{A})\,\mathbf{m} = (\lambda\mathbf{I} - \mathbf{A})\,\mathbf{m}_k \\ \mathbf{m}_{k-2} &= (\lambda\mathbf{I} - \mathbf{A})^2\,\mathbf{m} = (\lambda\mathbf{I} - \mathbf{A})\,\mathbf{m}_{k-1} \\ &\vdots \\ \mathbf{m}_1 &= (\lambda\mathbf{I} - \mathbf{A})^{k-1}\,\mathbf{m} = (\lambda\mathbf{I} - \mathbf{A})\,\mathbf{m}_2 \end{aligned} \tag{2.64}$$

For each i in $1 \leqslant i \leqslant k$, $\mathbf{m}_i$ is a generalized eigenvector of rank i. Thus eqns. (2.64) give us a technique for calculating the k generalized eigenvectors after $\mathbf{m}$ has been found. We call the set of vectors $\{\mathbf{m}_1, \mathbf{m}_2, \ldots, \mathbf{m}_k\}$ a chain of generalized eigenvectors.

Some important results used in the sequel regarding generalized eigenvectors are stated and proved below.

1. The set of generalized eigenvectors $\{\mathbf{m}_1, \mathbf{m}_2, \ldots, \mathbf{m}_k\}$ for eigenvalue λ of matrix $\mathbf{A}$, defined in (2.64), is linearly independent.

   Let us prove this statement by contradiction. Assume that the vectors $\mathbf{m}_1, \mathbf{m}_2, \ldots, \mathbf{m}_k$ are linearly dependent. Then there exist $\alpha_1, \alpha_2, \ldots, \alpha_k$ (not all zero) in $\mathcal{C}$ such that

$$\alpha_1\mathbf{m}_1 + \alpha_2\mathbf{m}_2 + \ldots + \alpha_k\mathbf{m}_k = \mathbf{0} \tag{2.65}$$

   Applying the operator $(\lambda\mathbf{I} - \mathbf{A})^{k-1}$ to (2.65) we obtain

$$\alpha_1(\lambda\mathbf{I} - \mathbf{A})^{k-1}\mathbf{m}_1 + \alpha_2(\lambda\mathbf{I} - \mathbf{A})^{k-1}\mathbf{m}_2 + \ldots + \alpha_k(\lambda\mathbf{I} - \mathbf{A})^{k-1}\mathbf{m}_k = \mathbf{0} \tag{2.66}$$

   We know that for $k \neq i$

$$\begin{aligned} (\lambda\mathbf{I} - \mathbf{A})^{k-1}\,\mathbf{m}_i &= (\lambda\mathbf{I} - \mathbf{A})^{k-1}(\lambda\mathbf{I} - \mathbf{A})^{k-i}\,\mathbf{m} \\ &= (\lambda\mathbf{I} - \mathbf{A})^{2k-i-1}\,\mathbf{m} \\ &= (\lambda\mathbf{I} - \mathbf{A})^{k-i-1}\,(\lambda\mathbf{I} - \mathbf{A})^{k}\,\mathbf{m} \\ &= \mathbf{0}, \text{ since } (\lambda\mathbf{I} - \mathbf{A})^{k}\,\mathbf{m} = \mathbf{0} \text{ by definition.} \end{aligned}$$

   Therefore from (2.66) we have

$$\alpha_k(\lambda\mathbf{I} - \mathbf{A})^{k-1}\,\mathbf{m}_k = \mathbf{0}$$

   which means that $\alpha_k = 0$, since $(\lambda\mathbf{I} - \mathbf{A})^{k-1}\,\mathbf{m} \neq \mathbf{0}$ by definition.

   Next by applying the operator $(\lambda\mathbf{I} - \mathbf{A})^{k-2}$ to (2.65) we can show that $\alpha_{k-1} = 0$. Repeated application of

$$(\lambda\mathbf{I} - \mathbf{A})^{k-j};\ j = 1, 2, 3, \ldots, \text{ gives } \alpha_1 = \alpha_2 = \alpha_3 = \ldots = \alpha_k = 0.$$

   This contradicts the assumption. Hence the vectors $\mathbf{m}_1, \mathbf{m}_2, \ldots, \mathbf{m}_k$ defined in (2.64) are linearly independent.

2. Let us now consider generalized eigenvectors of $\mathbf{A}$ associated with different eigenvalues. Assume that $\mathbf{m}$ is a generalized eigenvector of rank k associated with eigenvalue λ_1. Then, from (2.64)

$$\begin{aligned} \mathbf{m}_k &\overset{\Delta}{=} \mathbf{m} \\ \mathbf{m}_i &\overset{\Delta}{=} (\lambda_1\mathbf{I} - \mathbf{A})\,\mathbf{m}_{i+1} = (\lambda_1\mathbf{I} - \mathbf{A})^{k-i}\,\mathbf{m}; \\ &\qquad i = k-1, k-2, \ldots, 1 \end{aligned} \tag{2.67}$$

Let $\mathbf{n}$ be the generalized eigenvector of rank l associated with eigenvalue λ_2. Analogous to (2.67), we define

$$\mathbf{n}_l \triangleq \mathbf{n}$$

$$\mathbf{n}_j \triangleq (\lambda_2\mathbf{I} - \mathbf{A})\,\mathbf{n}_{j+1} = (\lambda_2\mathbf{I} - \mathbf{A})^{l-j}\,\mathbf{n}; \qquad (2.68)$$

$$j = l-1, l-2, \ldots, 1$$

The generalized eigenvectors $\{\mathbf{m}_1, \mathbf{m}_2, \ldots, \mathbf{m}_k\}$ and $(\mathbf{n}_1, \mathbf{n}_2, \ldots, \mathbf{n}_l\}$ associated with different eigenvalues λ_1 and λ_2 respectively of $\mathbf{A}$ are linearly independent.

The proof of this result is analogous to the previous one (Chen 1970).

3. Let $\mathbf{m}$ and $\mathbf{n}$ be the generalized eigenvectors of ranks k and l respectively associated with the same eigenvalue λ. Define

$$\mathbf{m}_k = \mathbf{m}$$

$$\mathbf{m}_i = (\lambda\mathbf{I} - \mathbf{A})\mathbf{m}_{i+1} = (\lambda\mathbf{I} - \mathbf{A})^{k-i}\mathbf{m};$$

$$i = k-1, \quad k-2, \ldots, 1$$

$$\mathbf{n}_l = \mathbf{n} \qquad (2.69)$$

$$\mathbf{n}_j = (\lambda\mathbf{I} - \mathbf{A})\mathbf{n}_{j+1} = (\lambda\mathbf{I} - \mathbf{A})^{l-j}\mathbf{n};$$

$$j = l-1, \quad l-2, \ldots, 1$$

If the two vectors $\mathbf{m}_1$ and $\mathbf{n}_1$ are linearly independent, then the generalized eigenvectors $\mathbf{m}_1, \mathbf{m}_2, \ldots, \mathbf{m}_k, \mathbf{n}_1, \mathbf{n}_2, \ldots, \mathbf{n}_l$ are linearly independent. Again this result may be proved by the arguments given earlier.

Example 2.14: Reconsider the matrix

$$\mathbf{A} = \begin{bmatrix} 1 & 1 & 2 \\ 0 & 1 & 3 \\ 0 & 0 & 2 \end{bmatrix}$$

of Example 2.13. The eigenvalues of $\mathbf{A}$ are $\lambda_1 = 1$, $\lambda_2 = 1$ and $\lambda_3 = 2$. Since $\gamma(\lambda_1\mathbf{I} - \mathbf{A}) = 1$, there is only one linearly independent eigenvector associated with λ_1 of multiplicity 2. In the following, we determine generalized eigenvectors associated with λ_1.

Here $k = 2$ and the generalized eigenvector $\mathbf{m} = \mathbf{m}_2$ is computed from

$$(\lambda_1\mathbf{I} - \mathbf{A})^2\mathbf{m}_2 = \mathbf{0}$$

if

$$(\lambda_1\mathbf{I} - \mathbf{A})\mathbf{m}_2 \neq \mathbf{0}$$

Thus we are looking for $\mathbf{m}_2$ such that

$$\begin{bmatrix} 0 & -1 & -2 \\ 0 & 0 & -3 \\ 0 & 0 & -1 \end{bmatrix}\begin{bmatrix} 0 & -1 & -2 \\ 0 & 0 & -3 \\ 0 & 0 & -1 \end{bmatrix}\mathbf{m}_2 = \begin{bmatrix} 0 & 0 & 5 \\ 0 & 0 & 3 \\ 0 & 0 & 1 \end{bmatrix}\mathbf{m}_2 = \mathbf{0}$$

and

$$\begin{bmatrix} 0 & -1 & -2 \\ 0 & 0 & -3 \\ 0 & 0 & -1 \end{bmatrix}\mathbf{m}_2 \neq \mathbf{0}$$

$\mathbf{m}_2 = \begin{bmatrix} 0 \\ 1 \\ 0 \end{bmatrix}$ is such a vector. We now calculate $\mathbf{m}_1$ from (2.64.)

$$\mathbf{m}_1 = (\lambda_1 \mathbf{I} - \mathbf{A})\mathbf{m}_2$$

$$= \begin{bmatrix} 0 & -1 & -2 \\ 0 & 0 & -3 \\ 0 & 0 & -1 \end{bmatrix}\begin{bmatrix} 0 \\ 1 \\ 0 \end{bmatrix} = \begin{bmatrix} -1 \\ 0 \\ 0 \end{bmatrix}$$

An eigenvector associated with $\lambda_3 = 2$ is

$$\mathbf{m}_3 = \begin{bmatrix} 5 \\ 3 \\ 1 \end{bmatrix}$$

The vectors $\mathbf{m}_1$, $\mathbf{m}_2$ and $\mathbf{m}_3$ are linearly independent and therefore qualify as basis vectors. The representation $\mathbf{J}$ of $\mathbf{A}$ with respect to new basis is given by $\mathbf{M}^{-1}\mathbf{A}\mathbf{M}$ where

$$\mathbf{M} = [\,\mathbf{m}_1 \quad \mathbf{m}_2 \quad \mathbf{m}_3\,]$$

$$= \begin{bmatrix} -1 & 0 & 5 \\ 0 & 1 & 3 \\ 0 & 0 & 1 \end{bmatrix}$$

This gives

$$\mathbf{J} = \mathbf{M}^{-1}\mathbf{A}\mathbf{M}$$

$$= \begin{bmatrix} 1 & 1 & 0 \\ 0 & 1 & 0 \\ 0 & 0 & 2 \end{bmatrix}$$

The reader might have noted that **J** is not in diagonal form. A matrix of this form is said to be in *Jordan Canonical form.*

Jordan-Form Matrix Representation of a Linear Operator

A *Jordan-form representation* **J** of a matrix **A** is an upper triangular matrix whose principal diagonal entries are of the matrix form

$$\mathbf{J}_{ij}(\lambda_i) = \begin{bmatrix} \lambda_i & 1 & 0\ldots0 & 0 \\ 0 & \lambda_i & 1\ldots0 & 0 \\ \vdots & \vdots & \vdots & \vdots \\ 0 & 0 & 0\ldots\lambda_i & 1 \\ 0 & 0 & 0\ldots0 & \lambda_i \end{bmatrix}$$

These matrices are called *Jordan blocks.* A Jordan block associated with eigenvalue λ_i has λ_i's on the main diagonal and 1's on the diagonal just above the main diagonal. The number of Jordan blocks associated with an eigenvalue λ_i is equal to the number $r(i)$ of linearly independent eigenvectors associated with λ_i. For an $n \times n$ matrix **A** with distinct eigenvalues λ_1, $\lambda_2, \ldots, \lambda_m$ of multiplicities $n_1, n_2, \ldots, n_m$ respectively, the general form of the Jordan matrix will be

$$\underset{(n \times n)}{\mathbf{J}} = \begin{bmatrix} \mathbf{J}_1(\lambda_1) & & & \mathbf{0} \\ & \mathbf{J}_2(\lambda_2) & & \\ & & \ddots & \\ \mathbf{0} & & & \mathbf{J}_m(\lambda_m) \end{bmatrix}$$

$$\underset{(n_i \times n_i)}{\mathbf{J}_i(\lambda_i)} = \begin{bmatrix} \mathbf{J}_{i1}(\lambda_i) & & & \mathbf{0} \\ & \mathbf{J}_{i2}(\lambda_i) & & \\ & & \ddots & \\ \mathbf{0} & & & \mathbf{J}_{ir(i)}(\lambda_i) \end{bmatrix} \qquad (2.70)$$

$$\underset{(n_{ij} \times n_{ij})}{\mathbf{J}_{ij}(\lambda_i)} = \begin{bmatrix} \lambda_i & 1 & 0\ldots0 & 0 \\ 0 & \lambda_i & 1\ldots0 & 0 \\ \vdots & \vdots & \vdots & \vdots \\ 0 & 0 & 0\ldots\lambda_i & 1 \\ 0 & 0 & 0\ldots0 & \lambda_i \end{bmatrix}$$

$$n = \sum_{i=1}^{m} n_i = \sum_{i=1}^{m} \sum_{j=1}^{r(i)} n_{ij}$$

A diagonal matrix is clearly a special case of the Jordan form; all of its Jordan blocks are of the order 1.

Every linear transformation that maps $C^n(C)$ into itself has a Jordan-form representation by a proper choice of basis or equivalently every matrix A can be transformed by a similarity transformation into a Jordan form. A general procedure for computing the similarity transformation and the resulting Jordan canonical form is as follows:

1. Compute the eigenvalues of **A** by solving $\det(\lambda\mathbf{I} - \mathbf{A}) = 0$. Let $\lambda_1, \lambda_2, .., \lambda_m$ be the distinct eigenvalues of **A** with multiplicities $n_1, n_2, .., n_m$ respectively.
2. Compute the number $r(1)$ of linearly independent eigenvectors associated with λ_1. This is also the number of Jordan blocks for λ_1.
   If $r(1) = 1$, then compute n_1 linearly independent generalized eigenvectors associated with λ_1 as follows. Find generalized eigenvector $\mathbf{m} \triangleq \mathbf{m}_{n_1}$ of rank n_1. Define $\mathbf{m}_i = (\lambda_1\mathbf{I} - \mathbf{A})^{n_1-i}\mathbf{m}$ for $i = 1, 2, .., n_1 - 1$. Proceed to step 3.
   If $r(1) > 1$, then compute n_1 linearly independent generalized eigenvectors associated with λ_1 as follows. Compute $(\lambda_1\mathbf{I} - \mathbf{A})^i$ for $i = 1, 2, ..$ until the rank of $(\lambda_1\mathbf{I} - \mathbf{A})^k$ is equal to the rank of $(\lambda_1\mathbf{I} - \mathbf{A})^{k+1}$. Find generalized eigenvector $\mathbf{m} \triangleq \mathbf{m}_k$ of rank k. Define $\mathbf{m}_i = (\lambda_1\mathbf{I} - \mathbf{A})^{k-i}$ for $i = 1, 2, .., k - 1$. This gives us one chain of generalized eigenvectors. Now since $k < n_1$, we have to compute other chains of generalised eigenvectors. If $\rho(\lambda_1\mathbf{I} - \mathbf{A}) = q$, then there are $(n - q)$ chains of generalized eigenvectors associated with λ_1. Find another generalized eigenvector of rank k, if not possible, of rank $(k - 1)$ and so forth and therefrom obtain the second chain of generalized eigenvectors. The procedure is repeated to get a set of n_1 generalized eigenvectors associated with λ_1 (refer Chen (1970) for an example).
3. Repeat step 2 for eigenvalues $\lambda_2, \lambda_3, .., \lambda_m$.
4. The n linearly independent generalized eigenvectors obtained in step 2 and 3 give us the modal matrix **M** and the Jordan matrix $\mathbf{J} = \mathbf{M}^{-1}\mathbf{A}\mathbf{M}$. In fact the Jordan matrix **J** can be written down without actually computing $\mathbf{M}^{-1}\mathbf{A}\mathbf{M}$.

For a Jordan-form representation of a linear operator **A**, the number of Jordan blocks and the order of each Jordan block are uniquely determined by **A**. However, because of different orderings of the basis vectors, we may have different Jordan-form representations of the same matrix. In this book, we use mostly the Jordan form given in (2.70).

A square matrix **A** is said to have a simple structure if for each eigenvalue λ_i of **A**, there exists one and only one eigenvector. In other words, **A** has a simple structure if no two uncoupled Jordan blocks in the canonical form have the same eigenvalue. For example, the matrix **A** of Example 2.14 has a simple structure while that of Example 2.12 does not.

A square matrix **A** with a simple structure is called a *cyclic matrix*. A matrix with distinct eigenvalues is obviously a cyclic matrix.

Example 2.15: Consider the matrix

$$\mathbf{A} = \begin{bmatrix} 0 & 1 & 0 & 0 \\ 0 & 0 & 1 & 0 \\ 0 & 0 & 0 & 1 \\ -2 & -7 & -9 & -5 \end{bmatrix}$$

1. The eigenvalues of matrix **A** are given by the characteristic equation

$$\det(\lambda\mathbf{I} - \mathbf{A}) = \lambda^4 + 5\lambda^3 + 9\lambda^2 + 7\lambda + 2 = 0$$

This gives $\lambda_1 = -1$ of multiplicity $n_1 = 3$

$\lambda_2 = -2$ of multiplicity $n_2 = 1$

2. $$\rho(\lambda_1\mathbf{I} - \mathbf{A}) = \rho \begin{bmatrix} -1 & -1 & 0 & 0 \\ 0 & -1 & -1 & 0 \\ 0 & 0 & -1 & -1 \\ 2 & 7 & 9 & 4 \end{bmatrix} = 3$$

Therefore there is only one linearly independent eigenvector associated with λ_1, i.e., $r(1) = 1$. There will be only one Jordan block

$$\mathbf{J}_1(\lambda_1) = \begin{bmatrix} -1 & 1 & 0 \\ 0 & -1 & 1 \\ 0 & 0 & -1 \end{bmatrix}$$

in the Jordan matrix for λ_1.

Let us compute $n_1 = 3$ linearly independent generalized eigenvectors associated with λ_1. We determine $\mathbf{m} \triangleq \mathbf{m}_3$ from the equations

$$(\lambda_1\mathbf{I} - \mathbf{A})^3 \mathbf{m}_3 = \mathbf{0}$$

and

$$(\lambda_1\mathbf{I} - \mathbf{A})^2 \mathbf{m}_3 \neq \mathbf{0}$$

It can easily be verified that

$$\mathbf{m}_3 = \begin{bmatrix} -1 \\ 0 \\ 0 \\ 1 \end{bmatrix}$$

satisfies these conditions.

The other vectors in the chain of generalized eigenvectors are

$$\mathbf{m}_2 = (\lambda_1\mathbf{I} - \mathbf{A})\mathbf{m}_3 = \begin{bmatrix} -1 \\ 0 \\ 1 \\ -2 \end{bmatrix}$$

$$\mathbf{m}_1 = (\lambda_1\mathbf{I} - \mathbf{A})\mathbf{m}_2 = \begin{bmatrix} -1 \\ 1 \\ -1 \\ 1 \end{bmatrix}$$

3. $\lambda_2 = -2$ is a distinct eigenvalue. The eigenvector $\mathbf{m}_4$ associated with this eigenvalue is obtained from $(\lambda_2\mathbf{I} - \mathbf{A})\mathbf{m}_4 = \mathbf{0}$
This gives

$$\mathbf{m}_4 = \begin{bmatrix} -1 \\ 2 \\ -4 \\ 8 \end{bmatrix}$$

4. The modal matrix

$$\mathbf{M} = \begin{bmatrix} -1 & -1 & -1 & -1 \\ 1 & 0 & 0 & 2 \\ -1 & 1 & 0 & -4 \\ 1 & -2 & 1 & 8 \end{bmatrix}$$

and the Jordan matrix

$$= \begin{bmatrix} -1 & 1 & 0 & 0 \\ 0 & -1 & 1 & 0 \\ 0 & 0 & -1 & 0 \\ 0 & 0 & 0 & -2 \end{bmatrix}$$

2.8 FUNCTIONS OF A SQUARE MATRIX

The simplest of the functions of a square matrix A are its powers. If k is a positive integer, we define

$$\mathbf{A}^k \triangleq \mathbf{A}.\mathbf{A}.\mathbf{A}\ldots\mathbf{A} \quad (k \text{ terms}) \tag{2.71a}$$

$$\mathbf{A}^{\circ} \triangleq \mathbf{I} \qquad \text{(unity matrix)} \tag{2.71b}$$

One of the properties of the square matrix $\mathbf{A}^k$ is that it commutes with another square matrix $\mathbf{A}^j$, i.e.,

$$\mathbf{A}^k . \mathbf{A}^j = \mathbf{A}^j . \mathbf{A}^k \tag{2.72}$$

In terms of the definition (2.71), a matrix polynomial of a square matrix may be written as

$$f(\mathbf{A}) = \alpha_0 \mathbf{I} + \alpha_1 \mathbf{A} + \alpha_2 \mathbf{A}^2 + .. + \alpha_m \mathbf{A}^m \tag{2.73}$$

Two polynomials of the same square matrix **A** commute with each other, i.e.,

$$f(\mathbf{A}) g(\mathbf{A}) = g(\mathbf{A}) f(\mathbf{A}) \tag{2.74}$$

Two polynomials of square matrices **A** and **B** respectively, commute if **A** commutes with **B**.

A matrix polynomial $f(\mathbf{A})$ can be factorized in a manner analogous to the factorization of scalar polynomials. Consider for example, the scalar polynomial

$$f(x) = 2 + 3x + x^2$$

This polynomial may be factorized as

$$f(x) = (2 + x)(1 + x)$$

The corresponding matrix polynomial is given by

$$\begin{aligned} f(\mathbf{A}) &= 2\mathbf{I} + 3\mathbf{A} + \mathbf{A}^2 \\ &= (2\mathbf{I} + \mathbf{A})(\mathbf{I} + \mathbf{A}) \end{aligned}$$

Consider now the infinite series in a scalar variable x,

$$\begin{aligned} f(x) &= \alpha_0 + \alpha_1 x + \alpha_2 x^2 + \dots \\ &= \sum_{i=0}^{\infty} \alpha_i x^i \end{aligned} \tag{2.75}$$

with the radius of convergence r.

We can define infinite series in matrix variable **A** as

$$\begin{aligned} f(\mathbf{A}) &= \alpha_0 \mathbf{I} + \alpha_1 \mathbf{A} + \alpha_2 \mathbf{A}^2 + \dots \\ &= \sum_{i=0}^{\infty} \alpha_i \mathbf{A}^i \end{aligned} \tag{2.76}$$

Note that the two series $f(\mathbf{A})$ and $g(\mathbf{A})$ of the form (2.76) will commute. However, $f(\mathbf{A})$ and $g(\mathbf{B})$ commute if **A** and **B** commute.

An important relation between the scalar power series (2.75) and matrix power series (2.76) is that if the absolute values of the eigenvalues of **A** are smaller than r, then the matrix power series (2.76) converges (for proof, refer Lefschetz (1957)).

Consider, in particular, the scalar power series

$$f(x) = 1 + x + \frac{1}{2!} x^2 + \ldots + \frac{1}{k!} x^k + \ldots$$

$$= \sum_{i=0}^{\infty} \frac{1}{i!} x^i \tag{2.77a}$$

It is well known that this power series converges on to the exponential e^x for all finite x, so that

$$f(x) = e^x \tag{2.77b}$$

It follows from this result that the matrix power series

$$f(\mathbf{A}) = \mathbf{I} + \mathbf{A} + \frac{1}{2!} \mathbf{A}^2 + \ldots + \frac{1}{k!} \mathbf{A}^k + \ldots$$

converges for all $\mathbf{A}$. By analogy with the power series in (2.77) for the ordinary exponential function, we adopt the following nomenclature:

If $\mathbf{A}$ is an $n \times n$ matrix, the *matrix exponential* of $\mathbf{A}$ is

$$e^{\mathbf{A}} \triangleq \mathbf{I} + \mathbf{A} + \frac{1}{2!} \mathbf{A}^2 + \ldots + \frac{1}{k!} \mathbf{A}^k + \ldots$$

$$\triangleq \sum_{i=0}^{\infty} \frac{1}{i!} \mathbf{A}^i \tag{2.78}$$

Cayley-Hamilton Theorem

This theorem is one of the most important theorems in matrix analysis which is extremely versatile and useful. It states that *every square matrix* $\mathbf{A}$ *satisfies its own characteristic equation.*

Thus if we have, for an $n \times n$ matrix $\mathbf{A}$, the characteristic equation

$$\Delta(\lambda) = \lambda^n + \alpha_1 \lambda^{n-1} + \ldots + \alpha_{n-1}\lambda + \alpha_n = 0 \tag{2.79}$$

then according to this theorem,

$$\Delta(\mathbf{A}) = \mathbf{A}^n + \alpha_1 \mathbf{A}^{n-1} + \ldots + \alpha_{n-1}\mathbf{A} + \alpha_n \mathbf{I} = \mathbf{0} \tag{2.80}$$

where $\mathbf{I}$ is an identity matrix and $\mathbf{0}$ is a null matrix. (Cayley-Hamilton theorem will be proved in Chapter 7. Refer Problem 7.3).

The degree of a matrix polynomial can be reduced in the light of Cayley-Hamilton theorem. Given an $n \times n$ matrix with eigenvalues $\lambda_1, \lambda_2, \ldots, \lambda_n$, the characteristic polynomial of $\mathbf{A}$ is

$$\Delta(\lambda) = \prod_{i=1}^{n} (\lambda - \lambda_i) \tag{2.81}$$

Consider the matrix polynomial

$$f(\mathbf{A}) = a_0 \mathbf{I} + a_1 \mathbf{A} + a_2 \mathbf{A}^2 + \ldots + a_n \mathbf{A}^n + a_{n+1} \mathbf{A}^{n+1} + \ldots \tag{2.82}$$

which is of degree higher than the order of $\mathbf{A}$. The corresponding scalar polynomial is

$$f(\lambda) = a_0 + a_1\lambda + a_2\lambda^2 + \ldots + a_n\lambda^n + a_{n+1}\lambda^{n+1} + \ldots \tag{2.83}$$

Dividing $f(\lambda)$ by the characteristic polynomial $\Delta(\lambda)$ we get

$$\frac{f(\lambda)}{\Delta(\lambda)} = q(\lambda) + \frac{g(\lambda)}{\Delta(\lambda)} \tag{2.84}$$

where $g(\lambda)$ is the remainder polynomial of the following form

$$g(\lambda) = \alpha_0 + \alpha_1\lambda + \ldots + \alpha_{n-1}\lambda^{n-1} \tag{2.85}$$

Equation (2.84) may be written as

$$f(\lambda) = q(\lambda)\,\Delta(\lambda) + g(\lambda) \tag{2.86}$$

Since $\Delta(\lambda_i)$, $i = 1, 2, \ldots, n$ is zero, we have from (2.86),

$$f(\lambda_i) = g(\lambda_i);\ i = 1, 2, \ldots, n \tag{2.87}$$

The coefficients $\alpha_0, \alpha_1, \ldots, \alpha_{n-1}$ in (2.85) can be obtained by successively substituting $\lambda_1, \lambda_2, \ldots, \lambda_n$ into eqn. (2.87).

Substituting **A** for λ in eqn. (2.86) we get

$$f(\mathbf{A}) = q(\mathbf{A})\,\Delta(\mathbf{A}) + g(\mathbf{A})$$

Since $\Delta(\mathbf{A})$ is identically zero, it follows that

$$\begin{aligned} f(\mathbf{A}) &= g(\mathbf{A}) \\ &= \alpha_0\mathbf{I} + \alpha_1\mathbf{A} + \ldots + \alpha_{n-1}\mathbf{A}^{n-1} \end{aligned} \tag{2.88}$$

If **A** possesses an eigenvalue λ_k of multiplicity n_k, then only one independent equation can be obtained by substituting λ_k into eqn. (2.87). The remaining $(n_k - 1)$ linear equations, which must be obtained in order to solve for α_i's, can be found by differentiating both sides of eqn. (2.86). Since

$$\frac{d^j}{d\lambda^j}\Delta(\lambda)\bigg|_{\lambda=\lambda_k} = 0;\ j = 0, 1, \ldots, (n_k - 1),$$

it follows that

$$\frac{d^j}{d\lambda^j} f(\lambda)\bigg|_{\lambda=\lambda_k} = \frac{d^j}{d\lambda^j} g(\lambda)\bigg|_{\lambda=\lambda_k};\ j = 1, 2, \ldots, (n_k - 1) \tag{2.89}$$

These results can be stated in a compact form as follows:

Given the $n \times n$ matrix **A** with eigenvalues $\lambda_1, \lambda_2, .., \lambda_m$ of multiplicities $n_1, n_2, .., n_m$ respectively, the characteristic polynomial of **A** is

$$\Delta(\lambda) = \prod_{i=1}^{m} (\lambda - \lambda_i)^{n_i} \tag{2.90}$$

Let f and g be two arbitrary polynomials. If

$$\frac{d^j}{d\lambda^j} f(\lambda)\bigg|_{\lambda=\lambda_i} = \frac{d^j}{d\lambda^j} g(\lambda)\bigg|_{\lambda=\lambda_i}; \tag{2.91}$$

$$j = 0, 1, 2, .., n_i - 1$$

$$i = 1, 2, .., m$$

then

$$f(\mathbf{A}) = g(\mathbf{A})$$

The set of numbers $\left.\frac{d^j}{d\lambda^j} f(\lambda)\right|_{\lambda=\lambda_i}$ for $i = 1, 2, .., m$ and $j = 0, 1, 2, ..,$ $n_i - 1$ are called the *values of f on the spectrum of* **A**.

If **A** is of order n, for any polynomial $f(\lambda)$, we can construct a polynomial $g(\lambda)$ of degree $(n-1)$ of the form (2.85) such that $f(\lambda) = g(\lambda)$ on the spectrum of **A**. Hence any polynomial of **A** can be expressed as

$$f(\mathbf{A}) = g(\mathbf{A}) = \alpha_0 \mathbf{I} + \alpha_1 \mathbf{A} + .. + \alpha_{n-1} \mathbf{A}^{n-1} \tag{2.92}$$

The use of Cayley-Hamilton theorem may be extended to include polynomials as well as functions. If $f(\lambda)$ is a function (not necessarily a polynomial) that is defind on the spectrum of **A** and $g(\lambda)$ is a polynomial that has the same values as $f(\lambda)$ on the spectrum of **A**, then the matrix-valued function $f(\mathbf{A})$ is equal to $g(\mathbf{A})$. We shall use this result in Chapter 5 to compute matrix exponentials.

2.9 CONCLUDING COMMENTS

In this chapter, we have reviewed a number of concepts and results in linear algebra that will be used in the later chapters of the book. We covered the following main topics:

1. Similarity transformations.
2. Solution of linear algebraic equations.
3. Jordan-form representation of a matrix.
4. Functions of a square matrix.

The computational problems such as finding eigenvalues, eigenvectors and Jordan-form representation of a matrix, computing functions of a square matrix, solving a set of linear algebraic equations, etc., can be solved on a digital computer. Any good book on numerical analysis may be referred to for this purpose. Some useful references are Householder (1964), Forsythe and Moler (1967), Acton (1970), Hamming (1971), Stewart (1973) and Wait (1979).

PROBLEMS

2.1 With the usual definition of addition and multiplication, which of the following sets form a field?
(a) The set of purely imaginary numbers.
(b) The set of all rational numbers.
(c) The set of polynomials of degree less than n with real coefficients.
(d) The set of all $n \times n$ matrices.

2.2 Show that 'addition' and 'multiplication' defined below, make the set $\{0, 1\}$ a field (called the binary field).

$$0 + 0 = 0 \ ; \ 0.1 = 0$$

$$1 + 1 = 0 \ ; \ 0.0 = 0$$

$$1 + 0 = 1 \ ; \ 1.1 = 1.$$

2.3 Show that the set of all 2×2 matrices of the form

$$\begin{bmatrix} a & -b \\ b & a \end{bmatrix}; \quad a, b \in \mathcal{R}$$

forms a field with $\begin{bmatrix} 0 & 0 \\ 0 & 0 \end{bmatrix}$ as zero element and $\begin{bmatrix} 1 & 0 \\ 0 & 1 \end{bmatrix}$ as unity element of the field and with usual definition of addition and multiplication of matrices.

2.4 Show that $\mathcal{C}^n(\mathcal{R})$ is a vector space but $\mathcal{R}^n(\mathcal{C})$ is not a vector space.

2.5 Show that $\mathcal{R}^n(\mathcal{R})$ is a subspace of $\mathcal{C}^n(\mathcal{C})$.

2.6 Let X denote the set of all solutions of the homogeneous differential equation

$$\frac{d^2x}{dt^2} + 3\frac{dx}{dt} + 2x = 0$$

Show that $X(\mathcal{R})$ is a vector space with the vector addition and scalar multiplication defined in the usual way.

Is $X(\mathcal{R})$ a vector space if the differential equation is nonhomogeneous?

2.7 Show that the set of all real polynomials in x of degree $\leqslant 2$ forms a linear space with the usual definition of addition and scalar multiplication.

Is the set of all real polynomials in x of degree 2 a vector space?

2.8 Let $\mathcal{F}$ denote the set of all rational functions with real coefficients. Show that $\mathcal{F}(\mathcal{F})$ and $\mathcal{F}(\mathcal{R})$ are linear spaces. What are the dimensions of these linear spaces?

2.9 Which of the following sets of vectors in $\mathcal{R}^4$ are linearly dependent?

(a) (1, 2, 1, 2); (2, 3, 2, 3); (1, 2, 3, 4); (1, 1, 1, 1)

(b) (1, 0, 1, 0); (0, 1, 0, 1); (1, 0, 0, 0); (0, 1, 0, 0)

(c) (2, 3, 1, 2); (1, 2, 3, 1); (4, 7, 7, 4)

2.10 Is the following set of vectors linearly dependent over the field of (i) rational functions (ii) real numbers?

$$\mathbf{x}_1 = \begin{bmatrix} \dfrac{1}{s+1} \\ \\ \dfrac{1}{s+2} \end{bmatrix}; \quad \mathbf{x}_2 = \begin{bmatrix} \dfrac{s+2}{(s+1)(s+3)} \\ \\ \dfrac{1}{s+3} \end{bmatrix}$$

2.11 Prove that the set of all $n \times n$ matrices with real elements forms a linear space of dimension n^2.

2.12 Find a matrix $\mathbf{P}$ which represents the change from basis $\{\mathbf{e}_1, \mathbf{e}_2, \mathbf{e}_3, \mathbf{e}_4\}$ of $\mathcal{R}^4$ to the basis

$$\{\mathbf{w}_1 = (1, 0, 0, 0); \quad \mathbf{w}_2 = (1, -1, 0, 0),$$
$$\mathbf{w}_3 = (0, 1, -1, 0); \quad \mathbf{w}_4 = (0, 0, 1, -1)\}$$

2.13 Given

$$\mathbf{A} = \begin{bmatrix} 3 & 2 & -1 \\ -2 & 1 & 0 \\ 4 & 3 & 1 \end{bmatrix}; \quad \mathbf{b} = \begin{bmatrix} 0 \\ 0 \\ 1 \end{bmatrix}$$

Find the representation of $\mathbf{A}$ with respect to the basis

$$\{\mathbf{b}, \mathbf{Ab}, \mathbf{A}^2\mathbf{b}\}.$$

2.14 Find rank and nullity of the following matrices.

$$\mathbf{A} = \begin{bmatrix} 3 & 4 & 1 & 2 \\ 6 & 8 & 3 & 4 \\ 9 & 12 & 4 & 6 \end{bmatrix}; \quad \mathbf{B} = \begin{bmatrix} 2 & 1 & 9 \\ 4 & 1 & 18 \\ -6 & -3 & -27 \end{bmatrix}$$

Also find the bases of the range space and the null space of the matrices $\mathbf{A}$ and $\mathbf{B}$.

2.15 Consider the homogeneous equation

$$\mathbf{Ax} = \mathbf{0}$$

where

$$\mathbf{A} = \begin{bmatrix} 0 & 1 & 1 & 2 & -1 \\ 1 & 2 & 3 & 4 & -1 \\ 2 & 0 & 2 & 0 & 2 \end{bmatrix}$$

Find linearly independent solutions of this equation.

2.16 Solve, if possible, the equations

(i) $$\begin{bmatrix} 1 & 1 & 2 \\ 2 & 3 & 2 \\ -1 & 1 & 1 \end{bmatrix} \begin{bmatrix} x_1 \\ x_2 \\ x_3 \end{bmatrix} = \begin{bmatrix} 5 \\ 10 \\ 2 \end{bmatrix}$$

(ii) $$\begin{bmatrix} 1 & 1 & 2 \\ 2 & 1 & -1 \\ 3 & 2 & 1 \end{bmatrix} \begin{bmatrix} x_1 \\ x_2 \\ x_3 \end{bmatrix} = \begin{bmatrix} 2 \\ 3 \\ 7 \end{bmatrix}$$

$$\text{(iii)} \begin{bmatrix} 1 & 1 & 2 \\ 2 & 1 & -1 \\ 3 & 2 & 1 \end{bmatrix} \begin{bmatrix} x_1 \\ x_2 \\ x_3 \end{bmatrix} = \begin{bmatrix} 1 \\ 4 \\ 5 \end{bmatrix}$$

2.17 (a) Prove that similar matrices have the same characteristic polynomial and therefore the same eigenvalues.

(b) Show that

$$\det(\mathbf{A}) = \prod_{i=1}^{n} \lambda_i$$

if $\lambda_1, \lambda_2, \ldots, \lambda_n$ are the eigenvaluses of matrix **A**.

2.18 The Vandermonde matrix

$$\mathbf{V} = \begin{bmatrix} 1 & 1 & \cdots & 1 \\ \lambda_1 & \lambda_2 & \cdots & \lambda_n \\ \lambda_1^2 & \lambda_2^2 & \cdots & \lambda_n^2 \\ \vdots & \vdots & & \vdots \\ \lambda_1^{n-1} & \lambda_2^{n-1} & \cdots & \lambda_n^{n-1} \end{bmatrix}$$

occurs frequenty in linear systems work. Show that

$$\det \mathbf{V} = \prod_{1<i<j<n} (\lambda_i - \lambda_j)$$

2.19 The $n \times n$ real matrix **A** is said to be in *companion form* if it has the structure

$$\mathbf{A} = \begin{bmatrix} 0 & 1 & 0 & \cdots & 0 \\ 0 & 0 & 1 & \cdots & 0 \\ \vdots & \vdots & \vdots & & \vdots \\ 0 & 0 & 0 & \cdots & 1 \\ -\alpha_n & -\alpha_{n-1} & -\alpha_{n-2} & \cdots & -\alpha_1 \end{bmatrix}$$

Show that the characteristic polynomial of **A** is

$$\Delta(\lambda) = \lambda^n + \alpha_1\lambda^{n-1} + \alpha_2\lambda^{n-2} + \ldots \alpha_{n-1}\lambda + \alpha_n$$

If $\lambda_1, \lambda_2, \ldots, \lambda_n$ are distinct eigenvalues of **A**, show that the Vandermonde matrix is a modal matrix that transforms **A** to a diagonal form.

2.20 Consider the matrix given in Problem 2.19. Suppose that λ_1 is an eigenvalue of the matrix with multiplicity k and other $(n-k)$ are distinct eigenvalues $\lambda_{k+1}, \lambda_{k+2}, \ldots, \lambda_n$.

Show that the *modified Vandermonde matrix*

$$\mathbf{V}=\begin{bmatrix} 1 & 0 & 0 & 0 & \cdots & 1 & \cdots & 1 \\ \lambda_1 & 1 & 0 & 0 & \cdots & \lambda_{k+1} & \cdots & \lambda_n \\ \lambda_1^2 & 2\lambda_1 & 1 & 0 & \cdots & \lambda_{k+1}^2 & \cdots & \lambda_n^2 \\ \lambda_1^3 & 3\lambda_1^2 & 3\lambda_1 & 1 & \cdots & \lambda_{k+1}^3 & \cdots & \lambda_n^3 \\ \vdots & \vdots & \vdots & \vdots & & \vdots & & \vdots \\ \lambda_1^{n-1} & \frac{d}{d\lambda_1}(\lambda_1^{n-1}) & \frac{1}{2!}\frac{d^2}{d\lambda_1^2}(\lambda_1^{n-1}) & \frac{1}{3!}\frac{d^3}{d\lambda_1^3}(\lambda_1^{n-1}) & \cdots & \lambda_{k+1}^{n-1} & \cdots & \lambda_n^{n-1} \end{bmatrix}$$

is a modal matrix that transforms **A** into a Jordan form.

2.21 Find the eigenvalues, eigenvectors and Jordan-form representations for the following matrices.

(i) $\begin{bmatrix} 0 & 1 & 0 \\ 3 & 0 & 2 \\ -12 & -7 & -6 \end{bmatrix}$

(ii) $\begin{bmatrix} 4 & 1 & -2 \\ 1 & 0 & 2 \\ 1 & -1 & 3 \end{bmatrix}$

(iii) $\begin{bmatrix} 0 & 1 & 0 \\ 0 & 0 & 1 \\ -2 & -4 & -3 \end{bmatrix}$

(iv) $\begin{bmatrix} 0 & 1 & 0 & 0 \\ 0 & 0 & 1 & 0 \\ 0 & 0 & 0 & 1 \\ 4 & -4 & -3 & 4 \end{bmatrix}$

(Hint: You may use the results of Problems 2.19 and 2.20.)

2.22 Consider the matrix

$$\mathbf{A}=\begin{bmatrix} 0 & 1 & 0 \\ 0 & 0 & 1 \\ -15 & -17 & -7 \end{bmatrix}$$

(a) Suggest a modal matrix **M** such that

$$\mathbf{J} = \mathbf{M}^{-1}\mathbf{A}\mathbf{M} = \begin{bmatrix} -2+j1 & 0 & 0 \\ 0 & -2-j1 & 0 \\ 0 & 0 & -3 \end{bmatrix}$$

(b) Suggest a suitable transformation matrix $\mathbf{Q}$ such that

$$\mathbf{Q}^{-1}\mathbf{J}\mathbf{Q} = \begin{bmatrix} -2 & 1 & 0 \\ -1 & -2 & 0 \\ 0 & 0 & -3 \end{bmatrix}$$

Hint: $\mathbf{Q} = \begin{bmatrix} \frac{1}{2} & -j/2 & 0 \\ \frac{1}{2} & j/2 & 0 \\ 0 & 0 & 1 \end{bmatrix}$

2.23 Show that if λ_i is an eigenvalue of matrix $\mathbf{A}$, then $f(\lambda_i)$ is an eigenvalue of the matrix function $f(\mathbf{A})$.
(Hint: Use Cayley-Hamilton theorem to show that there exists a nonzero vector $\mathbf{x}$ such that $f(\mathbf{A})\mathbf{x} = f(\lambda_i)\mathbf{x}$.)

2.24 A *nilpotent matrix* $\mathbf{A}$ is a matrix for which $\mathbf{A}^k = \mathbf{0}$ if $\mathbf{A}^{k-1} \neq \mathbf{0}$. Show that if all eigenvalues of a matrix $\mathbf{A}$ are zero, then $\mathbf{A}$ is nilpotent and investigate the given $\mathbf{A}$ for the value of k which makes it nilpotent.

$$\mathbf{A} = \begin{bmatrix} 0 & 0 & 0 & 0 \\ 0 & 0 & -1 & 0 \\ -3 & 0 & 0 & 0 \\ 1 & 0 & 0 & 0 \end{bmatrix}$$

2.25 The *outer product* (or *diadic product*) of two real column vectors $\mathbf{x}$ and $\mathbf{y}$ is defined as $\mathbf{x}\mathbf{y}^T$. If both $\mathbf{x}$ and $\mathbf{y}$ are $n \times 1$ vectors, show that

$$\mathbf{x}^T\mathbf{y} = \mathbf{y}^T\mathbf{x} = \text{trace}\,(\mathbf{x}\mathbf{y}^T)$$

2.26 The *minimal polynomial* $\psi(\lambda)$ of a matrix $\mathbf{A}$ is the least order monic polynomial (i.e., with the coefficient of highest power equal to 1) such that $\psi(\mathbf{A}) = \mathbf{0}$.
Let $\lambda_1, \lambda_2, \ldots, \lambda_n$ be the eigenvalues of $\mathbf{A}$ with multiplicities $n_1, n_2, \ldots, n_m$ respectively. In the Jordan-form representation of $\mathbf{A}$, let $\bar{n}_i$ be the largest order of the Jordan blocks associated with λ_i. $\bar{n}_i$ is called the *index* of λ_i.
The minimal polynomial of $\mathbf{A}$ is given by (for proof, refer Chen (1970))

$$\psi(\lambda) = \prod_{i=1}^{m} (\lambda - \lambda_i)^{\bar{n}_i}$$

Find the minimal polynomial for

$$A = \begin{bmatrix} 7 & 4 & -1 \\ 4 & 7 & -1 \\ -4 & -4 & 4 \end{bmatrix}$$

2.27 Let A be an $n \times n$ matrix. Using Cayley-Hamilton theorem, show that any A^k with $k \geqslant n$ can be written as a linear combination of $\{I, A, .., A^{n-1}\}$.

What modification can you make if the degree of minimal polynomial of A is known to be m?

REFERENCES

1. Acton, F. S., *Numerical Methods that Work,* New York: Harper and Row, 1970.
2. Athans, M., and P. L. Falb, *Optimal Control,* New York: McGraw-Hill, 1966.
3. Bellman, R. E., *Matrix Analysis,* 2nd Edition, New York: McGraw-Hill, 1968.
4. Chen, C. T., *Introduction to linear System Theory,* New York: Holt, Rinehart and Winston, 1970.
5. DeRusso, P. M., R. J. Roy, and C. M. Close, *State Variables for Engineers,* New York: Wiley, 1965.
6. Forsythe, G., and C. B. Moler, *Computer Solution of Linear Algebraic Systems,* Englewood Cliffs, N. J.: Prentice-Hall, 1967.
7. Friedberg, S. H., A. J. Insel, and L. E. Spence, *Linear Algebra,* Englewood Cliffs, N. J. : Prentice-Hall, 1979.
8. Gantmacher, F. R., *The Theory of Matrices,* vols. 1 and 2, New York: Chelsea 1959.
9. Goult, R. J., *Applied Linear Algebra,* Chichester (England): Ellis Horwood, 1978.
10. Hadley, G., *Linear Algebra,* Reading, Mass.: Addison-Wesley, 1961.
11. Hamming, R. W., *Introduction to Applied Numerical Analysis.* New York: McGraw-Hill, 1971.
12. Householder, A. S., *The Theory of Matrices in Numerical Analysis,* New York: Blaisdell, 1964
13. Lefschetz, S., *Differential Equations: Geometric Theory,* New York: Interscience. 1957.
14. Nashed, M. Z. (Editor), *Generalized Inverses and Applications,* New York: Academic Press, 1976.
15. Nering, E. D., *Linear Algebra and Matrix Theory,* 2nd Edition, New York: Wiley, 1970.
16. Noble, B., and J. W. Daniel, *Applied Linear Algebra,* 2nd Edition, Englewood Cliffs, N. J.: Prentice-Hall, 1977.
17. Padulo, L., and M. A. Arbib, *System Theory: A Unified State-Space Approach to Discrete and Continuous Systems,* Philadelphia: Saunders, 1974.
18. Stewart, G. W., *Introduction to Matrix Computations,* New York: Academic Press, 1973.
19. Wait, R., *The Numerical Solution of Algebraic Equations,* Chichester (England): Wiley-Interscience, 1979.
20. Zadeh, L. A., and C. A. Desoer, *Linear System Theory: A State-Space Approach,* New York: McGraw-Hill, 1963.

3. STATE VARIABLE DESCRIPTIONS

3.1 INTRODUCTION

The very first step in the analytical study of systems is to set up mathematical equations to describe the systems. Because of different analytical methods used, we may often set up different mathematical models to describe the same system. When, for any reason, the analysis in the time-domain is to be preferred, the use of so-called state-space approach will offer a great deal of convenience conceptually, notationally, and sometimes analytically. The conceptual convenience is derived from the elegant representation of the instantaneous condition of the system by the notion of the system *state*. The notational and analytical conveniences come through the use of vector-matrix representation which allows the system equations and the form of solutions to be written compactly. The adaptation of state-space representation to the numerical solution is an added advantage, particularly when the system to be investigated contains time-varying and nonlinear elements.

In this chapter, we shall introduce the state variable descriptions of systems from a very general setting. These descriptions will be developed from the concepts of relaxedness, causality, time-invariance and linearity. These concepts have not been introduced very rigorously. For a more rigorous exposition, see Zadeh and Desoer (1963), DeRusso et al. (1965), Kalman et al. (1969) and Padulo and Arbib (1974).

3.2 THE CONCEPT OF STATE

We know that the properties of a physical process depend on a number of entities. In general, if $v_1, v_2 \ldots, v_n$ are relevant entities of a physical process, we may vary a subset of these, say $v_1, v_2, \ldots, v_p$ with time in a manner specified by the time functions $u_1, u_2, \ldots, u_p$ belonging to a given set of functions $\mathbf{\Omega}$ and the resulting variations $y_1, y_2, \ldots, y_q$ belonging to a set of functions Γ due to u_i; $i = 1, 2, \ldots, p$, may be observed say in entities $v_{p+1}, \ldots, v_{p+q}$ respectively. We then disregard the entities $v_{p+q+1}, \ldots, v_n$.

For convenience, we identify u_1; $i = 1, 2, \ldots, p$, with the components of a p-dimensional vector (called *input vector function*),

$$\mathbf{u} = \begin{bmatrix} u_1 \\ u_2 \\ \vdots \\ u_p \end{bmatrix}$$

We use $\mathbf{u}$ to denote a vector function defined over a time set $\mathscr{T}$ and $\mathbf{u}(t)$ to denote the value of $\mathbf{u}$ at time t. The input vector function $\mathbf{u} \in$ the set $\mathbf{\Omega}$ of admissible input functions and $\mathbf{u}(t) \in$ the set $\mathbf{U}$ of input values. The time set $\mathscr{T}$ may be an infinite interval $(-\infty, \infty)$, semi-infinite interval $[t_0, \infty)$, finite interval $[t_0, t_1]$ or a set of discrete instants of time.

$$\mathbf{u} : \mathscr{T} \to \mathbf{U}$$

i.e., function $\mathbf{u}$ operates on the time set $\mathscr{T}$ to produce values in the set $\mathbf{U}$.

We identify y_i; $i = 1, 2, \ldots, q$, in the q-dimensional vector (called the *output vector function*)

$$\mathbf{y} = \begin{bmatrix} y_1 \\ y_2 \\ \vdots \\ y_q \end{bmatrix}$$

The output vector function $\mathbf{y} \in \Gamma$, the set of output functions and the outputs or responses $\mathbf{y}(t)$ at time $t \in$ the set $\mathbf{Y}$ of output values.

$$\mathbf{y} : \mathscr{T} \to \mathbf{Y}$$

We shall assume that the sets of input vector functions $\mathbf{\Omega}$ and output vector functions Γ defined over $t \in (-\infty, \infty)$ are *closed under segmentation.* This means that for each t_0 and t_1 in $(-\infty, \infty)$ with $t_1 \geqslant t_0$, the segment $\mathbf{u}_{[t_0, t_1]} \in \mathbf{\Omega}$ and the segment $\mathbf{y}_{[t_0, t_1]} \in \Gamma$, i.e., every segment of a vector function which is a member of $\mathbf{\Omega}$ (or Γ) is also a member of $\mathbf{\Omega}$ (or Γ).

In general, the input segments of a system are not allowed to be arbitrary functions but must belong to a restricted class $\mathbf{\Omega}$. The choice of $\mathbf{\Omega}$ may be inferred from physical considerations but more often it is dictated by mathematical expediency. Only mild restrictions are imposed on the class of allowable output segments Γ.

If the output $\mathbf{y}(t_0)$ of a system depends only on the input applied at t_0, the system is called *static* or *zero-memory system.* A network consisting of resistors only is an example of such systems. Most systems of interest however have memory; the output $\mathbf{y}(t_0)$ at time t_0 depends not only on the input $\mathbf{u}(t_0)$ applied at time t_0 but also on the input applied before t_0. Assume that we set such a system, called *system with memory* or *dynamic system*, going at time t_0 and apply the input thereafter. We use $\mathbf{u}_{(t_0, \infty)}$ to denote such an input segment; the notation corresponds to the input applied at $t = t_0^+$. The output segment $\mathbf{y}_{(t_0, \infty)}$ of this system is not uniquely determinable. In fact for different inputs applied before t_0, we will obtain different outputs $\mathbf{y}_{(t_0, \infty)}$, though the same input $\mathbf{u}_{(t_0, \infty)}$ is applied at t_0 and thereafter.

As an example, we consider the situation shown in Fig 3.1, wherein a particle of mass M is actuated upon by force f; $f(t) \in$ real field $\mathcal{R}$.

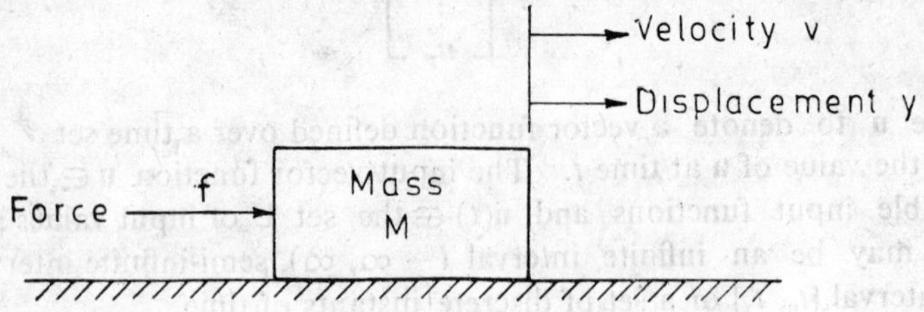

Fig. 3.1

The system is characterized by the relation

$$f = M\frac{dv}{dt} = M\,\dot{v} \tag{3.1}$$

Integrating both sides of this equation, we get

$$v(t) = \frac{1}{M}\int f(\tau)\,d\tau \tag{3.2}$$

The velocity $v(t)$ at any instant t is the result of force f applied to the particle in the entire past. Hence the limits of integration in eqn. (3.2) are from $-\infty$ to t. Thus

$$\begin{aligned} v(t) &= \frac{1}{M}\int_{-\infty}^{t} f(\tau)\,d\tau \\ &= \frac{1}{M}\int_{-\infty}^{t_0} f(\tau)\,d\tau + \frac{1}{M}\int_{t_0}^{t} f(\tau)\,d\tau \\ &= v(t_0) + \frac{1}{M}\int_{t_0}^{t} f(\tau)\,d\tau \end{aligned} \tag{3.3}$$

From this relation it is observed that for the same input $f_{(t_0, t_1]}$, we get different values of output $v(t)$ depending upon our choice of the parameter $v(t_0) \in \mathcal{R}$. $v(t_0)$ is the value of velocity of the particle at $t = t_0$. Equation (3.3) suggests that if we know the initial velocity of the particle and the value of force from t_0 onwards, we obtain a unique value of the output $v(t)$.

If the displacement of the particle in Fig. 3.1 is the desired output, we have

$$\frac{dy}{dt} = \dot{y} = v$$

or

$$\begin{aligned} y(t) &= \int_{-\infty}^{t} v(\theta)\,d\theta = \int_{-\infty}^{t_0} v(\theta)\,d\theta + \int_{t_0}^{t} v(\theta)\,d\theta \\ &= y(t_0) + v(t_0)\,[t - t_0] + \frac{1}{M}\int_{t_0}^{t} d\theta \int_{t_0}^{\theta} f(\tau)\,d\tau \end{aligned} \tag{3.4}$$

Thus to calculate the output uniquely, we require information about velocity at $t = t_0$ and displacement at $t = t_0$ of the particle, in addition to the input applied at t_0 onwards.

From this example, we recognize the necessity of including the initial conditions in system characterization, as these affect the subsequent behaviour of the system. We may therefore conceive of initial conditions as describing the status or the *state of the system* at $t = t_0$. A formal definition of the state will shortly be given.

A system is said to be *relaxed* at time t_0 if the output $\mathbf{y}_{(t_0, \infty)}$ is solely and uniquely excited by the input $\mathbf{u}_{(t_0, \infty)}$. Consider, for example, the system of Fig. 3.1. The initial velocity $v(t_0)$ and the initial displacement $y(t_0)$ have all the relevant information about the entire past, $t \in (-\infty, t_0]$, of the force applied to the particle; if $v(t_0) = y(t_0) = 0$, the system is relaxed[1] at t_0.

We thus observe that if we set a system going at some time t_0 and apply the input $\mathbf{u}_{(t_0, \infty)}$, monitoring the corresponding output $\mathbf{y}_{(t_0, \infty)}$ of the system, the output in general, is not uniquely determinable and one way of associating a unique $\mathbf{y}$ with each $\mathbf{u}$ consists in attaching to each input-output pair $(\mathbf{u}, \mathbf{y})$, a parameter (n-dimensional vector)

$$\mathbf{x}(t_0) \overset{\Delta}{=} \mathbf{x}^\circ \overset{\Delta}{=} \begin{bmatrix} x_1(t_0) \\ x_2(t_0) \\ \vdots \\ x_n(t_0) \end{bmatrix}$$

ranging over a set $\mathbf{X}$ such that $\mathbf{y}$ is uniquely determined by $\mathbf{u}$ and $\mathbf{x}(t_0)$. We shall call $\mathbf{x}(t_0) = \mathbf{x}^0$, the *state of the system at* $t = t_0$.

The initial state $\mathbf{x}(t_0)$ has all the relevant information about the input $\mathbf{u}_{(-\infty, t_0]}$ and characterizes the system completely at $t = t_0$. $\mathbf{x}(t_1)$, the new updated state at $t = t_1 > t_0$ is determined by $\mathbf{x}(t_0)$ and the input $\mathbf{u}$ applied over the interval $(t_0, t_1]$.

The functional dependence of $\mathbf{x}(t)$ on $\mathbf{x}(t_0)$ and $\mathbf{u}_{(t_0, t]}$ may be expressed as

$$\mathbf{x}(t) = \boldsymbol{\Phi}(t, t_0, \mathbf{x}^0, \mathbf{u}) \tag{3.5}$$

where

$$\boldsymbol{\Phi}(.) \overset{\Delta}{=} \begin{bmatrix} \phi_1(.) \\ \phi_2(.) \\ \vdots \\ \phi_n(.) \end{bmatrix}$$

is n-dimensional function vector. Using set-theoretic notation

$$\boldsymbol{\Phi} : \mathcal{T} \times \mathcal{T} \times \mathbf{X} \times \boldsymbol{\Omega} \rightarrow \mathbf{X}$$

1. In a transfer function description of a linear system, the system is always implicitly assumed to be relaxed at $t = 0$ (Chapter 7).

we shall call relation (3.5) the *state transition relation.*

Since the state $\mathbf{x}(t)$ characterizes the system completely at t, it is usual to think of the output of the system as an observation of certain aspects of the state of the system. The output is just an instantaneous 'read out' of those chcaracteristics deemed relevant to particular use of the system. Thus the input affects the output only indirectly through its effect in updating the state. If the system is in state $\mathbf{x}(t_1)$ at time t_1, the output at t_1 will be a function of $\mathbf{x}(t_1)$ only. However in some cases, the output may also depend upon the instantaneous value of the input. The functional dependence of the output may therefore be written as

$$\mathbf{y}(t) = \mathbf{g}(t, \mathbf{x}, \mathbf{u}) \tag{3.6}$$

where

$$\mathbf{g}(.) \triangleq \begin{bmatrix} g_1(.) \\ g_2(.) \\ \vdots \\ g_q(.) \end{bmatrix}$$

is q-dimensional function vector and each $g_k(.)$; $k = 1, 2, \ldots, q$ represents an *algebraic relation* between $x_i(t)$; $i = 1, 2, \ldots, n$ and $u_j(t)$; $j = 1, 2, \ldots, p$.

$$\mathbf{g}: \mathscr{T} \times \mathbf{X} \times \mathbf{U} \rightarrow \mathbf{Y}$$

We shall call relation (3.6) the *output equation.*

Relations (3.5) and (3.6) provide *input-output-state relations* of the system, represented diagrammatically in Fig. 3.2. It should be noted that

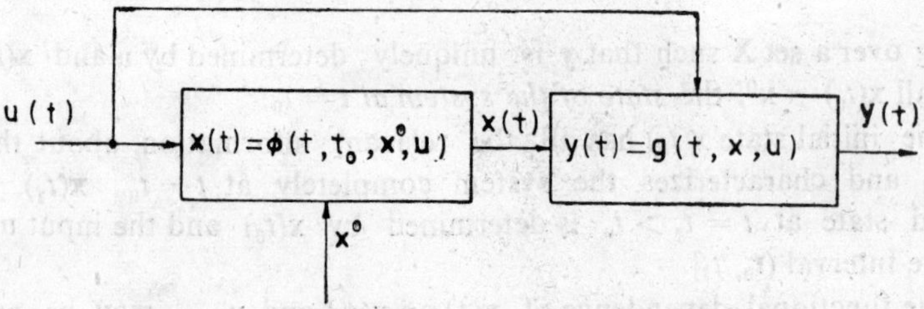

Fig. 3.2 Input-output-state description of a system

these relations are not completely arbitrary. The relations must satisfy the *consistency conditions* given below. If these conditions are satisfied, (3.5)-(3.6) qualify to be called input-output-state relations for the system. Then $\mathbf{x}(t)$ is called the *state of the system at time t*, the components $x_1, x_2, \ldots, x_n$ of the *state vector* $\mathbf{x}$ are called the *state variables* of the system. The set $\mathbf{X}$ of the values of $\mathbf{x}(t)$ is the vector space of the system which we shall assume to be finite dimensional. It means that we shall study systems whose states may be chosen to consist of a finite number of state variables. The state space $\mathbf{X}$ is frequently denoted as Σ.

A plot of state variables in the n-dimensional state space with t as the implicit variable is called the *state trajectory* of the system. The visualization of state trajectory is more practical in two dimensions since then the state space is a plane.

Consistency Conditions

The consistency conditions on input-output-state relations are given below.

1. $\Phi(t_0, t_0, \mathbf{x}^0, \mathbf{u}) = \mathbf{x}^0$ (3.7a)

   for all times t_0, all states $\mathbf{x}^0$ and all admissible input functions $\mathbf{u}$.

2. The output of a system at time t_1 is uniquely determined by its initial state $\mathbf{x}(t_0) \triangleq \mathbf{x}^0$ and input segment $\mathbf{u}_{(t_0,t_1]}$.

   This condition, called the *causality condition*, says two things. First, the input prior to t_0 has no effect on the state $\Phi(t_1, t_0, \mathbf{x}^0, \mathbf{u})$ and hence output $\mathbf{g}(t_1, \Phi(t_1, t_0, \mathbf{x}^0, \mathbf{u}), \mathbf{u})$; the state $\mathbf{x}(t_0)$ contains all relevant information and further information about the input prior to obtaining the state $\mathbf{x}(t_0)$ is irrelevant. Second, the input after t_1 has no effect on the state $\Phi(t_1, t_0, \mathbf{x}^0, \mathbf{u})$ and hence the output $\mathbf{g}(t_1, \Phi(t_1, t_0, \mathbf{x}^0, \mathbf{u}), \mathbf{u})$; the system cannot foresee the future and responds only to inputs it has already received.
   A system whose input-output-state relations satisfy the causality condition is called *causal* or *nonanticipatory* system. If a system is not causal, it is said to be *noncausal* or *anticipatory*. The output of a noncausal system depends not only on the past input but also on the future value of the input. This implies that a noncausal system is able to predict the input that will be applied in future. For example, the input-output relation $y(t)=u(t-1)$, realizable as a 'perfect delay line' represents a causal system while the input-output relation $y(t) = u(t+1)$, a perfect predictor of the future, represents a noncausal system which has no physical realization. In fact *causality is an intrinsic property of every physical system.*

3. For every pair of segments $(\mathbf{u}_{(t_0,t_2]}, \mathbf{y}_{(t_0,t_2]})$ satisfying input-output-state relation, the segment $(\mathbf{u}_{(t_1,t_2]}, \mathbf{y}_{(t_1,t_2]})$ also satisfies it for all $t_0 < t_1 \leqslant t_2$.

   Suppose that the input $\mathbf{u}_{(t_0,t_1]}$ sets a system (in state $\mathbf{x}(t_0) \triangleq \mathbf{x}^0$) going at $t = t_0$. At time t_1, the state of the system will be $\Phi(t_1, t_0, \mathbf{x}^0, \mathbf{u})$ and the output will be $\mathbf{g}(t_1, \Phi(t_1, t_0, \mathbf{x}^0, \mathbf{u}), \mathbf{u})$. If the input segment $\mathbf{u}_{(t_0,t_1]}$ is followed by another segment $\mathbf{u}_{(t_1,t_2]}$, we may compute the new state and output in two ways. We may either look at the updating of the state $\Phi(t_1, t_0, \mathbf{x}^0, \mathbf{u})$ at time t_1 by the input $\mathbf{u}_{(t_1,t_2]}$ to yield the new state $\Phi(t_2, t_1, \Phi(t_1, t_0, \mathbf{x}^0, \mathbf{u}), \mathbf{u})$ and the corresponding output $\mathbf{g}(t_2, \Phi(t_2, t_1, \Phi(t_1, t_0, \mathbf{x}^0, \mathbf{u}), \mathbf{u}), \mathbf{u})$, or we may simply start with the state $\mathbf{x}(t_0)$ at time t_0 and use input $\mathbf{u}_{(t_0,t_2]}$ to obtain the state $\Phi(t_2, t_0, \mathbf{x}^0, \mathbf{u})$ and the corresponding output $\mathbf{g}(t_2, \Phi(t_2, t_0, \mathbf{x}^0, \mathbf{u}), \mathbf{u})$. Both the approaches will yield the same result if

$$\Phi(t_2, t_1, \Phi(t_1, t_0, \mathbf{x}^0, \mathbf{u}), \mathbf{u}) = \Phi(t_2, t_0, \mathbf{x}^0, \mathbf{u}) \tag{3.7b}$$

□

In formulating a system model, we try to choose the state variables so that the state vector gets updated by an appropriate state transition function $\Phi(.)$ and can be read out instantaneously by an appropriate read-out function g(.) . The choice of the state variables, the state transition function and read-out function go hand in hand. These concepts are illustrated in the following examples.

Example 3.1: Consider the simple network shown in Fig. 3.3. The input-output relation for this system is

$$RC\frac{dy}{dt}+y=RC\frac{du}{dt} \tag{3.8}$$

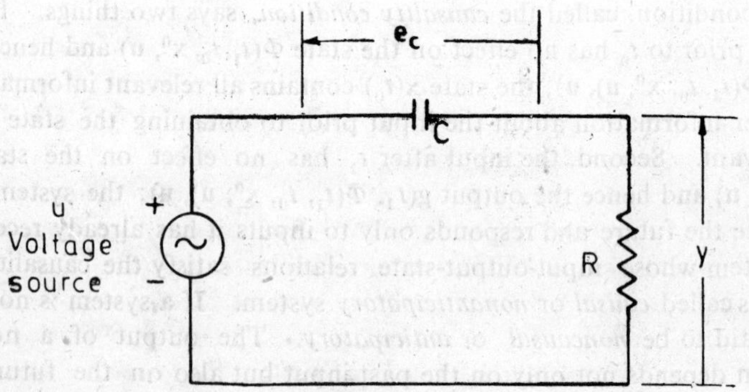

Fig. 3.3

To get the input-output-state relation, we have to choose state variables, function $\Phi(.)$ and function g(.). The capacitor voltage e_C is a candidate for our choice of state variable for this system. Assuming $x=e_C$ to be the state variable for the system (subject to its satisfying the consistency conditions), we have

$$RC\frac{dx}{dt}+x=u$$

or

$$e^{t/RC}\frac{dx}{dt}+e^{t/RC}\frac{x}{RC}=e^{t/RC}\frac{u}{RC}$$

or

$$\frac{d}{dt}(e^{t/RC}.x)=e^{t/RC}\frac{u}{RC}$$

Integrating both sides from t_0 to t we get

$$e^{t/RC}x(t)\Big|_{t_0}^{t}=\frac{1}{RC}\int_{t_0}^{t}e^{\tau/RC}\,u(\tau)d\tau$$

which gives

$$x(t)=\phi(t,\,t_0,\,x^0,\,u)$$

$$=e^{-(t-t_0)/RC}\,x^0+\frac{1}{RC}\int_{t_0}^{t}e^{-(t-\tau)/RC}\,u(\tau)d\tau \tag{3.9a}$$

The output

$$y(t) = g(t, x, u)$$
$$= u(t) - x(t) \tag{3.9b}$$

It can be easily verified that relations (3.9) satisfy the three consistency conditions. Therefore, $x = e_C$ qualifies to be called the state variable of the system, (3.9a) qualifies to be called the state transition relation and (3.9b) qualifies to be called the output equation of the system. Note that the state space for this system is $\Sigma = \mathcal{R}(\mathcal{R})$.

It may here be noted that *the state of a system is not uniquely specified.* For example, for the system in Fig. 3.3, the charge $q(t)$ deposited on the capacitor is given by $q(t) = C\, e_C(t)$. Therefore, instead of $e_C(t)$, the variable $q(t)$ may be taken as state variable to define the state of the system. The functions $\phi(.)$ and $g(.)$ will get modified accordingly.

Can the energy $E = \frac{1}{2} C e_C^2$ stored in the capacitor be taken as a state variable? The answer is 'no', since the corresponding state transition function $\phi(.)$ does not satisfy the consistency condition (3.7b).

Example 3.2: For the network shown in Fig. 3.4, the set of capacitor voltages qualifies as the state of the network.

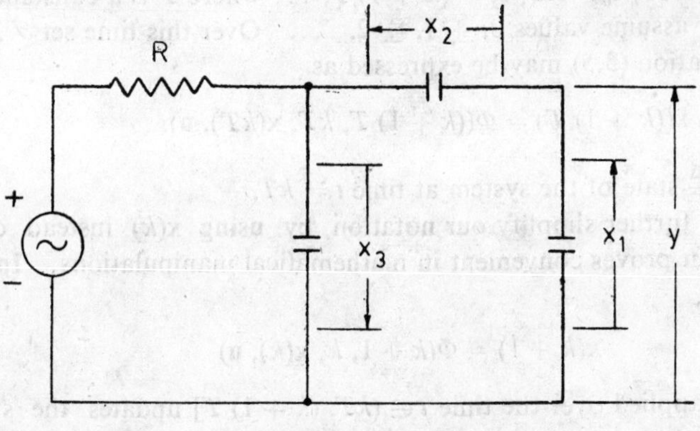

Fig. 3.4

If we apply the Kirchhoff's voltage law to the loop of three capacitors, we have

$$x_1(t) + x_2(t) + x_3(t) = 0 \text{ for all } t.$$

It implies that if any two of x_1, x_2, x_3 are known, the third one is also known. Thus any two of the three capacitor voltages determine the state of the network. If all the three capacitor voltages are chosen as a state, then there is a redundancy.

It is desirable that the state variable set describing a system be a minimal set.

Example 3.3: Let us reconsider the system of Fig. 3.1 described by eqn. (3.4). For this system, displacement y and velocity v of the particle qualify as the state variables. We may define new state variables as

$$x_1 = 2y + v$$
$$x_2 = y + v$$

We can express displacement y and velocity v in terms of the variables x_1 and x_2. Hence the state vector

$$\mathbf{x}(t) = \begin{bmatrix} x_1(t) \\ x_2(t) \end{bmatrix}$$

defines the state of the system at time t.

This example brings out an additional fact that *the state of a system is an auxiliary quantity that may or may not be easily interpretable in physical terms.*

3.3 STATE EQUATIONS FOR DYNAMIC SYSTEMS

The state $\mathbf{x}(t_0)$ at $t_0 \in \mathscr{T}$ of a system is updated to state $\mathbf{x}(t)$ at $t > t_0$ ($t \in \mathscr{T}$) by the state transition relation (3.5).

Assume that $\mathscr{T}$ is an ordered set of discrete instants of time which are uniformly spaced; $t_0 = kT$, $t_1 = (k+1)\,T$, ... where T is a constant interval and k may assume values $0, \pm 1, \pm 2, \ldots$. Over this time set $\mathscr{T}$, the state transition relation (3.5) may be expressed as

$$\mathbf{x}((k+1)\,T) = \Phi((k+1)\,T, kT, \mathbf{x}(kT), \mathbf{u}) \tag{3.10}$$

where $\mathbf{x}(kT) \triangleq$ state of the system at time $t = kT$.

We may further simplify our notation by using $\mathbf{x}(k)$ instead of $\mathbf{x}(kT)$. This notation proves convenient in mathematical manipulations. In terms of this notation,

$$\mathbf{x}(k+1) = \Phi(k+1, k, \mathbf{x}(k), \mathbf{u}) \tag{3.11}$$

The input $\mathbf{u}$ applied over the time $t \in (kT, (k+1)\,T]$ updates the state $\mathbf{x}(k)$ at $t = kT$ to the new state $\mathbf{x}(k+1)$ at $t = (k+1)\,T$. As per our assumption stated earlier (eqn. (3.7a)), the input $\mathbf{u}(k)$ at $t = kT$ arrives simultaneously with the system settling into the state $\mathbf{x}(k)$ and so cannot affect this state. The input $\mathbf{u}(k)$ affects the state $\mathbf{x}(k+1)$. On similar lines, we can argue that the input $\mathbf{u}(k+1)$ does not affect the state $\mathbf{x}(k+1)$ but affects only the later states. Therefore, relation (3.11) may be replaced by another relation given below, which shows dependence of $\mathbf{x}(k+1)$ on $\mathbf{u}(k)$ only.

$$\mathbf{x}(k+1) = \mathbf{f}(k, \mathbf{x}, \mathbf{u}) \tag{3.12a}$$
$$\mathbf{f}: \mathscr{T} \times \boldsymbol{\Sigma} \times \mathbf{U} \rightarrow \boldsymbol{\Sigma}$$

For time set $\mathscr{T}$ defined above, the output equation (3.6) may be written as

$$\mathbf{y}(k) = \mathbf{g}(k, \mathbf{x}, \mathbf{u}) \tag{3.12b}$$

$$\mathbf{g}: \mathscr{T} \times \mathbf{\Sigma} \times \mathbf{U} \rightarrow \mathbf{Y}$$

In the sequel, we shall call eqn. (3.12a) as *state equation* and eqn. (3.12b) as *output equation* of discrete-time systems with uniform sampling interval. Both the equations (3.12a) and (3.12b) together form the *state variable model* of discrete-time systems[2].

State equation (3.12a) is a transition one-step equation. Repeated use of this equation determines the behaviour for all $t \in kT$; $k = 0, \pm 1, \pm 2, \ldots$.

In the case of continuous-time systems, we employ the notion of derivative. Consider the equation

$$\frac{d\mathbf{x}(t)}{dt} \triangleq \dot{\mathbf{x}}(t) = \mathbf{f}(t, \mathbf{x}, \mathbf{u})$$

$$\mathbf{f}: \mathscr{T} \times \mathbf{\Sigma} \times \mathbf{U} \rightarrow \mathbf{\Sigma}$$

At $t = t_0$, the initial state $\mathbf{x}(t_0)$ is known. The knowledge of the input $\mathbf{u}(t_0)$ defines the value of the derivative vector $\dot{\mathbf{x}}(t_0)$; it is given by the function $\mathbf{f}(t_0, \mathbf{x}, \mathbf{u})$. Thus we have a point $\mathbf{x}(t_0)$ and a 'direction' specified by $\dot{\mathbf{x}}(t_0)$. This information can be used to deduce the value of the state at $t = t_0 + \Delta t$ where Δt is small.

$$\mathbf{x}(t_0 + \Delta t) \simeq \mathbf{x}(t_0) + (\mathbf{f}(t_0, \mathbf{x}, \mathbf{u}))\, \Delta t$$

This process can be repeated by starting at $t = t_0 + \Delta t$.

Continuous-time systems may therefore be described by the equations

$$\dot{\mathbf{x}}(t) = \mathbf{f}(t, \mathbf{x}, \mathbf{u});\ \mathbf{x}(t_0) = \mathbf{x}^0 \qquad (3.13a)$$

$$\mathbf{f}: \mathscr{T} \times \mathbf{\Sigma} \times \mathbf{U} \rightarrow \mathbf{\Sigma}$$

$$\mathbf{y}(t) = \mathbf{g}(t, \mathbf{x}, \mathbf{u}) \qquad (3.13b)$$

$$\mathbf{g}: \mathscr{T} \times \mathbf{\Sigma} \times \mathbf{U} \rightarrow \mathbf{Y}$$

with certain restrictions for the existence and uniqueness of the solution of eqn. (3.13a).

In Chapter 5, we shall show that if

(i) $\mathbf{u}$ is a piecewise continuous function of time—segments of $\mathbf{u}$ are allowed to have a finite number of discontinuities over a finite interval of time and that between the discontinuities, the segments of $\mathbf{u}$ are continuous functions of t;

(ii) $\mathbf{f}$ satisfies a *Lipschitz condition*;
then the solutions $\mathbf{x}(t)$ of (3.13a) exist and are unique and continuous for all initial conditions $(t_0, \mathbf{x}^0) \in \mathscr{T} \times \mathbf{\Sigma}$ and $t \in \mathscr{T}$.

We shall further show that $\mathbf{f}$ satisfies a Lipschitz condition if $\mathbf{f}(t, \mathbf{x}, \mathbf{u})$ is differentiable (piecewise differentiable) with respect to $\mathbf{x}$ and the components of the Jacobian matrix $\dfrac{\partial \mathbf{f}}{\partial \mathbf{x}}$ (see Appendix III) are continuous (piecewise conti-

2. Unless otherwise specified, we shall assume that discrete-time systems have uniform sampling interval.

nuous) functions of **x**. Fortunately, these requirements are met by all well-behaved physical systems.

In the sequel, we shall call eqn. (3.13a) as the state equation and eqn. (3.13b) as the output equation of continuous-time systems (or differential systems). Both the equations (3.13a) and (3.13b) together form the state variable model of the continuous-time systems.

We can now define the admissible classes Ω and Γ of input and output functions respectively for continuous-time and discrete-time systems. For continuous-time systems, all piecewise-continuous real functions **u** constitute the class Ω of admissible input functions. Since a necessary condition for realizability of a system is that **y** be a real function of time for all **u**; for an input $\mathbf{u} \in \Omega$, the output **y** belongs to a set Γ of real functions.

For discrete-time systems, all possible real sequences of inputs will be admissible.

3.4 TIME-INVARIANCE AND LINEARITY

Systems are classified in a variety of ways according to their different properties. Two important modes of classification are: linear—nonlinear systems and time-invariant—time-varying systems. These modes of classification depend upon the notions of linearity and time-invariance which we shall discuss in this section.

Time-Invariance

Roughly speaking, a system is *time-invariant* if its characteristics do not change with time. To cast this rough definition into a more precise form, we introduce the notion of a translation operator.

A *translation operator*, denoted as $z^{-\tau}$, is a system characterized by the input-output relation

$$\mathbf{y}(t) = z^{-\tau}(\mathbf{u}(t))$$

$$= \mathbf{u}(t-\tau), \ \forall t, \ \tau \text{ and } \mathbf{u} \qquad (3.14)$$

Essentially, the action of $z^{-\tau}$ on its input **u** consists in translating **u** by a fixed amount τ along the time-axis (Fig. 3.5). A positive τ corresponds to

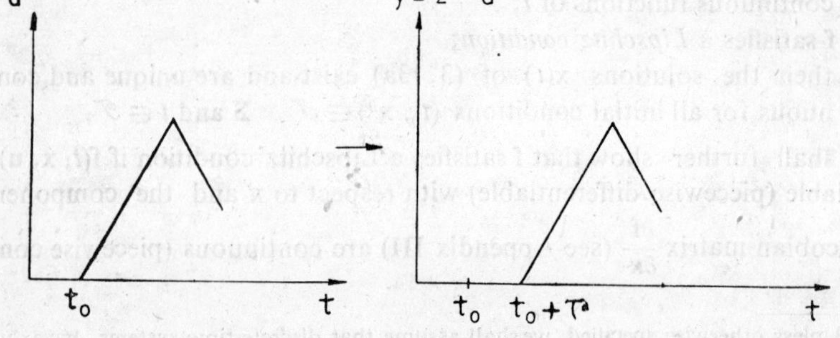

Fig. 3.5

a *delay* of τ units of time whereas a negative τ corresponds to an *advance* of $-\tau$ units of time.

Consider now a system with state transition function $\Phi(t, t_0, \mathbf{x}^\circ, \mathbf{u})$: $\mathscr{T} \times \mathscr{T} \times \mathbf{\Sigma} \times \mathbf{\Omega} \rightarrow \mathbf{\Sigma}$, and the read-out map $\mathbf{g}(t, \mathbf{x}, \mathbf{u})$: $\mathscr{T} \times \mathbf{\Sigma} \times \mathbf{U} \rightarrow \mathbf{Y}$. $\mathbf{\Omega}$ is the set of admissible input functions defined over time set $\mathscr{T}$.

This system is *time-invariant* (or *constant*) if and only if

(i) $\mathscr{T}$ is closed under addition, i.e., $t, t+\tau \in \mathscr{T} \ \forall\ t, \tau$

(ii) $\mathbf{\Omega}$ is closed under the shift operator $z^{-\tau}$, i.e., for all $\mathbf{u} \in \mathbf{\Omega}$

$$z^{-\tau}\mathbf{u} \in \mathbf{\Omega} \qquad \forall\ \tau \in \mathscr{T}$$

(iii) $\Phi(t, t_0, \mathbf{x}^\circ, \mathbf{u}) = \Phi(t+\tau, t_0+\tau, \mathbf{x}^\circ, \mathbf{z}^{-\tau}\mathbf{u}) \ \forall\ \tau \in \mathscr{T}$ (3.15)

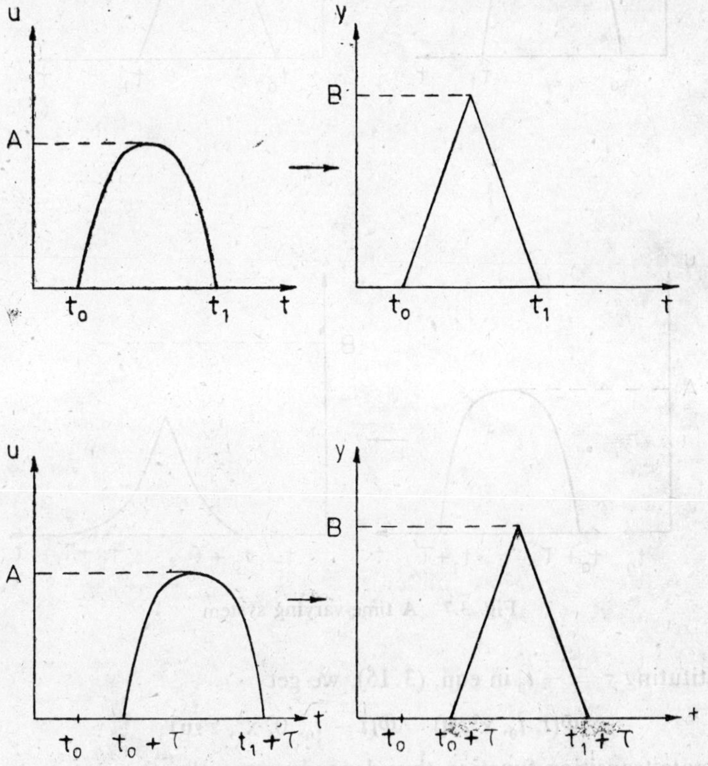

Fig. 3.6 A time-invariant system

This condition stipulates that as long as we start in state $\mathbf{x}^\circ$, any delay in application of input shifts the response correspondingly without affecting the magnitude.

(iv) The map $\mathbf{g}(t, \mathbf{x}, \mathbf{u})$ is independent of time, i.e.,

$$\mathbf{g}(t_1, \hat{\mathbf{x}}, \hat{\mathbf{u}}) = \mathbf{g}(t_2, \hat{\mathbf{x}}, \hat{\mathbf{u}}) \ \forall\ t_1, t_2 \in \mathscr{T} \qquad (3.16)$$

Figure 3.6 gives an example of a time-invariant system while Fig. 3.7 shows a time-dependent (time-varying) system.

We observe that the initial moment of time t_0 is insignificant for time-invariant systems. For convenience, we often shift this initial moment of time to zero as seen below.

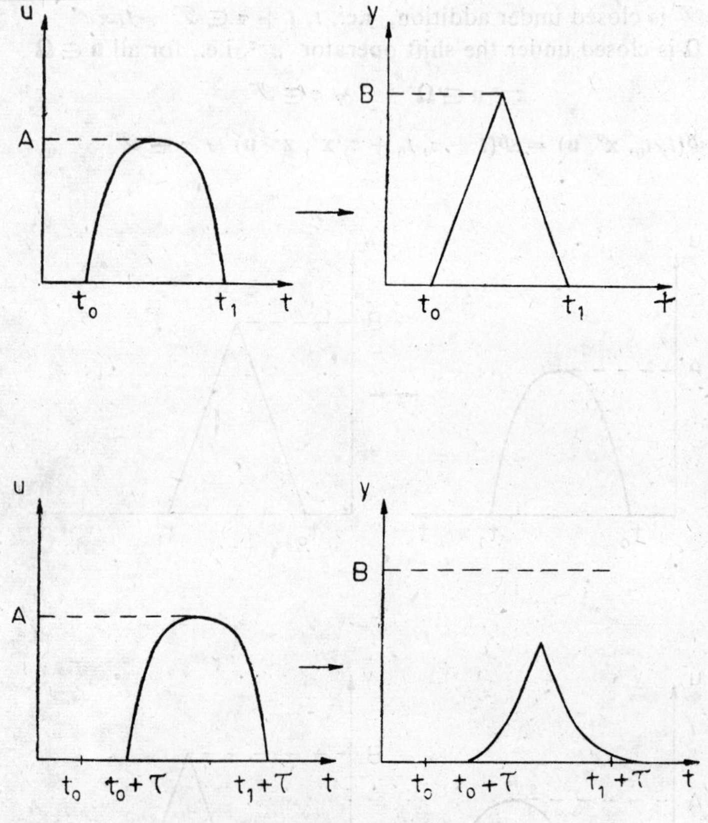

Fig. 3.7 A time-varying system

Substituting $\tau = -t_0$ in eqn. (3.15), we get

$$\Phi(t, t_0, \mathbf{x}^\circ, \mathbf{u}) = \Phi(t - t_0, 0, \mathbf{x}^\circ, z^{t_0}\mathbf{u})$$

The state transition function thus depends upon the time interval $(t - t_0)$ and not on the initial time t_0. For time-invariant systems, we may omit mentioning initial time variable and simply make use of

$$\mathbf{x}(t) = \Phi(t, \mathbf{x}^\circ, \mathbf{u}): \mathscr{T} \times \mathbf{\Sigma} \times \mathbf{\Omega} \to \mathbf{\Sigma} \qquad (3.17a)$$

where $\Phi(.)$ is the state transition function of the system started in state $\mathbf{x}^\circ$ at time $t = 0$.

The output equation for time-invariant systems may be written as

$$\mathbf{y}(t) = \mathbf{g}(\mathbf{x}, \mathbf{u}): \mathbf{\Sigma} \times \mathbf{U} \to \mathbf{Y} \qquad (3.17b)$$

Example 3.4: Let us consider the familiar example of low-pass filter shown in Fig. 3.8. We shall assume that the capacitor is charged at $t = t_0$ and $e_i = u(t)$ for $t > t_0$. The initial voltage across the capacitor $e_0(t_0) = x°$.

From Fig. 3.8 we have

$$RC\frac{de_0}{dt} + e_0 = u \tag{3.18}$$

which gives (refer Example 3.1),

$$e_0(t) \stackrel{\Delta}{=} x(t) = \phi(t,\ t_0,\ x^0,\ u)$$

$$= e^{-(t-t_0)/RC}x^0 + \frac{1}{RC}\int_{t_0}^{t} e^{-(t-\theta)/RC}u(\theta)d\theta \tag{3.19}$$

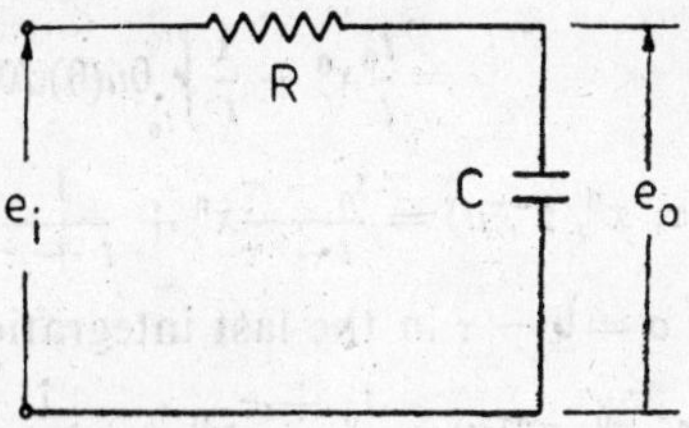

Fig. 3.8 Low-pass filter

Let us now test this system for time-invariance.

$$\phi(t+\tau,\ t_0+\tau,\ x^0,\ z^{-\tau}u) = e^{-(t-t_0)/RC}x^0 + \frac{1}{RC}\int_{t_0+\tau}^{t+\tau} e^{-(t+\tau-\theta)/RC}u(\theta-\tau)d\theta$$

The change of variable $\sigma = \theta - \tau$ in the last integration gives

$$\phi(t+\tau, t_0+\tau,\ x^0, z^{-\tau}u) = e^{-(t-t_0)/RC}x^0 + \frac{1}{RC}\int_{t_0}^{t} e^{-(t-\sigma)/RC}u(\sigma)d\sigma$$

$$= \phi(t,\ t_0,\ x^0,\ u)$$

Output

$$y(t) = g(t,\ x,\ u) = x$$

The map $g(.)$ is thus independent of time.

The system of Fig. 3.8 described by the first-order differential equation (3.18) is therefore time-invariant.

Example 3.5: Consider the differential equation

$$\dot{y} + \frac{1}{t}y = u,\quad y(t_0) = x^0 \tag{3.20}$$

For a general first-order differential equation of the form

$$\dot{y} + a(t)y = u$$

we have

$$\frac{d}{dt}[e^{\int a(t)dt}\cdot y] = e^{\int a(t)dt}\cdot u$$

For $a(t) = \frac{1}{t}$, $e^{\int a(t)dt} = e^{\ln t} = t$, and therefore

$$\frac{d}{dt}(y \cdot t) = t \cdot u$$

Integrating both sides from t_0 to t, we get

$$\int_{t_0}^{t} \frac{d}{d\theta}(\theta y(\theta))d\theta = \int_{t_0}^{t} \theta u(\theta)d\theta$$

or

$$ty(t) - t_0 y(t_0) = \int_{t_0}^{t} \theta u(\theta)d\theta$$

or

$$\begin{aligned} y(t) &= \phi(t, t_0, x^0, u) \\ &= \frac{t_0}{t}x^0 + \frac{1}{t}\int_{t_0}^{t} \theta u(\theta)d\theta \end{aligned} \tag{3.21}$$

$$\phi(t+\tau, t_0+\tau, x^0, z^{-\tau}u) = \frac{t_0+\tau}{t+\tau}x^0 + \frac{1}{t+\tau}\int_{t_0+\tau}^{t+\tau} \theta u(\theta-\tau)d\theta$$

The change of variable $\sigma = \theta - \tau$ in the last integration gives

$$\begin{aligned} \phi(t+\tau, t_0+\tau, x^0, z^{-\tau}u) &= \frac{t_0+\tau}{t+\tau}x^0 + \frac{1}{t+\tau}\int_{t_0}^{t} (\sigma+\tau)u(\sigma)d\sigma \\ &\neq \phi(t, t_0, x^0, u) \end{aligned}$$

Therefore the differential equation (3.20) is time-varying.

□

The dynamical equations (3.12) and (3.13) reduce to the following functional relations for time-invariant systems.

For discrete-time systems,

$$\mathbf{x}(k+1) = \mathbf{f}(\mathbf{x}(k), \mathbf{u}(k)) : \mathbf{\Sigma} \times \mathbf{U} \rightarrow \mathbf{\Sigma} \tag{3.22a}$$

$$\mathbf{y}(k) = \mathbf{g}(\mathbf{x}(k), \mathbf{u}(k), : \mathbf{\Sigma} \times \mathbf{U} \rightarrow \mathbf{Y} \tag{3.22b}$$

For continuous-time systems,

$$\dot{\mathbf{x}}(t) = \mathbf{f}(\mathbf{x}(t), \mathbf{u}(t)) : \mathbf{\Sigma} \times \mathbf{U} \rightarrow \mathbf{\Sigma} \tag{3.23a}$$

$$\mathbf{y}(t) = \mathbf{g}(\mathbf{x}(t), \mathbf{u}(t)) : \mathbf{\Sigma} \times \mathbf{U} \rightarrow \mathbf{Y} \tag{3.23b}$$

Linearity

A system is said to be a linear system if and only if

(i) its state transition function $\Phi(t, t_0, \mathbf{x}^0, \mathbf{u})$ is a linear map, mapping the product space $\mathbf{\Sigma} \times \mathbf{\Omega}$ into the state space $\mathbf{\Sigma}$,

(ii) its read-out function $\mathbf{g}(t, \mathbf{x}, \mathbf{u})$ is a linear map, mapping the product space $\mathbf{\Sigma} \times \mathbf{U}$ into output space $\mathbf{Y}$.

The product space $\mathbf{\Sigma} \times \mathbf{\Omega}$ is a set $\{(\mathbf{x}^0, \mathbf{u}) \mid \mathbf{x}^0 \in \mathbf{\Sigma}, \mathbf{u} \in \mathbf{\Omega}\}$ for which addition is defined as (refer Section 2.2)

$$(\mathbf{x}_1^0, \mathbf{u}_1) + (\mathbf{x}_2^0, \mathbf{u}_2) = (\mathbf{x}_1^0 + \mathbf{x}_2^0, \mathbf{u}_1 + \mathbf{u}_2) \;\forall\; \mathbf{x}_1^0, \mathbf{x}_2^0 \in \mathbf{\Sigma} \;\forall\; \mathbf{u}_1, \mathbf{u}_2 \in \mathbf{\Omega}$$

and scalar multiplication is defined as

$$\alpha(\mathbf{x}^0, \mathbf{u}) = (\alpha\mathbf{x}^0, \alpha\mathbf{u}) \;\forall\; \alpha \in \mathcal{R}, \mathbf{x}^0 \in \mathbf{\Sigma}, \mathbf{u} \in \mathbf{\Omega}$$

The map Φ is linear if (refer eqn. (2.12))

$$\Phi(t, t_0, \alpha(\mathbf{x}_1^0, \mathbf{u}_1) + \beta(\mathbf{x}_2^0, \mathbf{u}_2)) = \Phi(t, t_0, \alpha\mathbf{x}_1^0 + \beta\mathbf{x}_2^0, \alpha\mathbf{u}_1 + \beta\mathbf{u}_2)$$

$$= \alpha\Phi(t, t_0, (\mathbf{x}_1^0, \mathbf{u}_1)) + \beta\Phi(t, t_0, (\mathbf{x}_2^0, \mathbf{u}_2)) \quad (3.24)$$

$$\forall\ \alpha, \beta \in \mathcal{R}$$

$$\mathbf{x}_1^0, \mathbf{x}_2^0 \in \mathbf{\Sigma}$$

$$\mathbf{u}_1, \mathbf{u}_2 \in \mathbf{\Omega}$$

Similarly, the linearity of map $\mathbf{g}(t, \mathbf{x}, \mathbf{u})$ may also be defined.

From the definition of linearity, the following important observations are made:

1. Linear systems have the *decomposition* property which says that the response of a system to the input $\mathbf{u}$ starting in state $\mathbf{x}^0$ at $t = t_0$ is identical with the *zero-input response* ($\mathbf{u} = \Theta$, the null function belonging to $\mathbf{\Omega}$) of the system starting in state $\mathbf{x}^0$ plus the *zero-state response* ($\mathbf{x}^0 = \mathbf{0}$, the *zero* of state space $\mathbf{\Sigma}$) of the system to input $\mathbf{u}$.

   The zero-input response is also referred to as *free*, *natural* or *unforced* response.

2. *Zero-state linearity* is a necessary condition for a system to be linear but not a sufficient condition, and the same argument is applicable to *zero-input linearity*.

These concepts are illustrated with the help of examples.

Example 3.6: Let us reconsider the system of Example 3.4. The RC network of Fig. 3.8 is described by the differential eqn. (3.18). The response given by eqn. (3.19) is

$$x(t) = \phi(t, t_0, x^0, u)$$

$$= \underbrace{e^{-(t-t_0)/RC}x^0}_{\substack{\text{zero-input}\\ \text{response}\\ \text{component}}} + \underbrace{\frac{1}{RC}\int_{t_0}^{t} e^{-(t-\theta)/RC}\, u(\theta)\, d\theta}_{\substack{\text{zero-state response}\\ \text{component}}} \quad (3.25)$$

$$= x_\Sigma(t) + x_U(t)$$

where

$$x_\Sigma(t) = \phi(t, t_0, x^0, 0)$$

$$= e^{-(t-t_0)/RC}\, x^0$$

and

$$x_U(t) = \phi(t, t_0, 0, u)$$

$$= \frac{1}{RC}\int_{t_0}^{t} e^{-(t-\theta)/RC}\, u(\theta)\, d\theta$$

Since

$$\phi(t, t_0, \alpha x_1^0 + \beta x_2^0, 0) = e^{-(t-t_0)/RC}\,(\alpha x_1^0 + \beta x_2^0);\ \alpha, \beta \in \mathcal{R}$$

$$= \alpha e^{-(t-t_0)/RC} x_1^0 + \beta e^{-(t-t_0)/RC} x_2^0$$

$$= \alpha\phi(t, t_0, x_1^0, 0) + \beta\phi(t, t_0, x_2^0, 0);$$

the system is zero-input linear.

Further, since

$$\phi(t, t_0, 0, \alpha u_1 + \beta u_2) = \frac{1}{RC}\int_{t_0}^{t} e^{-(t-\theta)/RC} (\alpha u_1(\theta) + \beta u_2(\theta))\, d\theta;\ \alpha, \beta \in \mathcal{R}$$

$$= \frac{\alpha}{RC}\int_{t_0}^{t} e^{-(t-\theta)/RC} u_1(\theta)\, d\theta + \frac{\beta}{RC}\int_{t_0}^{t} e^{-(t-\theta)/RC} u_2(\theta)\, d\theta$$

$$= \alpha\phi(t, t_0, 0, u_1) + \beta\phi(t, t_0, 0, u_2);$$

the system is zero-state linear.

From eqn. (3.25), we observe that if we double x^0 and do not change u, the response $x(t)$ is not doubled. Similarly, the input αu without changing the initial state, does not yield the response $\alpha x(t)$. Thus zero-state or zero-input linearity does not imply that the system is linear.

The system of Fig. 3.8 is linear if

$$\phi(t, t_0, \alpha(x_1^0, u_1) + \beta(x_2^0, u_2))$$
$$= \phi(t, t_0, \alpha x_1^0 + \beta x_2^0, \alpha u_1 + \beta u_2)$$
$$= \alpha\phi(t, t_0, x_1^0, u_1) + \beta\phi(t, t_0, x_2^0, u_2);\ \alpha, \beta \in \mathcal{R}$$

From eqn. (3.25), we have

$$\phi(t, t_0, \alpha(x_1^0, u_1) + \beta(x_2^0, u_2))$$
$$= e^{-(t-t_0)/RC} (\alpha x_1^0 + \beta x_2^0) + \frac{1}{RC}\int_{t_0}^{t} e^{-(t-\theta)/RC} (\alpha u_1(\theta) + \beta u_2(\theta))\, d\theta$$
$$= \alpha\left[e^{-(t-t_0)/RC} x_1^0 + \frac{1}{RC}\int_{t_0}^{t} e^{-(t-\theta)/RC} u_1(\theta)\, d\theta\right]$$
$$+ \beta\left[e^{-(t-t_0)/RC} x_2^0 + \frac{1}{RC}\int_{t_0}^{t} e^{-(t-\theta)/RC} u_2(\theta)\, d\theta\right]$$
$$= \alpha\phi(t, t_0, x_1^0, u_1) + \beta\phi(t, t_0, x_2^0, u_2)$$

This proves that the system under consideration is linear.

□

As mentioned earlier, in this book we study only the class of systems whose states may be chosen to consist of finite number of variables n (*finite-dimensional systems*). Since the state variables are usually real-valued, and since we consider cases in which states consist of finite number of state variables, state space we encounter in this book usually is the familiar finite-dimensional real vector space $\mathcal{R}^n(\mathcal{R})$.

We shall also be concerned with other vector spaces $\mathbf{U}$, $\mathbf{\Omega}$, $\mathbf{Y}$, $\mathbf{\Gamma}$ defined over real field $\mathcal{R}$. In particular, we shall assume $\mathbf{U} = \mathcal{R}^p(\mathcal{R})$ and $\mathbf{Y} = \mathcal{R}^q(\mathcal{R})$.

In our notation, we may drop the specification of field with vector space; we shall assume the field to be $\mathcal{R}$ unless otherwise stated. Thus $\mathbf{\Sigma} = \mathcal{R}^n$, $\mathbf{U} = \mathcal{R}^p$ and $\mathbf{Y} = \mathcal{R}^q$.

Consider an n-dimensional continuous-time system described by the state variable model

$$\dot{\mathbf{x}}(t) = \mathbf{f}(t, \mathbf{x}, \mathbf{u}): \mathscr{T} \times \mathscr{R}^n \times \mathscr{R}^p \to \mathscr{R}^n$$

$$\mathbf{y}(t) = \mathbf{g}(t, \mathbf{x}, \mathbf{u}): \mathscr{T} \times \mathscr{R}^n \times \mathscr{R}^p \to \mathscr{R}^q$$

If $\mathbf{f}(\,.\,)$ and $\mathbf{g}(\,.\,)$ are linear functions of $\mathbf{x}$ and $\mathbf{u}$, then it is easy to show that they are of the form

$$\mathbf{f}(t, \mathbf{x}, \mathbf{u}) = \mathbf{A}(t)\,\mathbf{x}(t) + \mathbf{B}(t)\,\mathbf{u}(t)$$

$$\mathbf{g}(t, \mathbf{x}, \mathbf{u}) = \mathbf{C}(t)\,\mathbf{x}(t) + \mathbf{D}(t)\,\mathbf{u}(t)$$

where **A**, **B**, **C** and **D** are respectively $n \times n$, $n \times p$, $n \times q$ and $q \times p$ matrices. Hence an n-dimensional linear dynamical state model is of the form

$$\dot{\mathbf{x}}(t) = \mathbf{A}(t)\,\mathbf{x}(t) + \mathbf{B}(t)\,\mathbf{u}(t) \quad : \textit{State equation} \tag{3.26a}$$

$$\mathbf{y}(t) = \mathbf{C}(t)\,\mathbf{x}(t) + \mathbf{D}(t)\,\mathbf{u}(t) \quad : \textit{Output equation} \tag{3.26b}$$

As we shall see in Chapter 5, a sufficient condition for (3.26a) to have a unique solution is that every entry of $\mathbf{A}(\,.\,)$ is a continuous function of t defined over $(-\infty, \infty)$. For convenience, the entries of **B**, **C** and **D** are also assumed to be continuous in $(-\infty, \infty)$.

An n-dimensional time-invariant linear dynamical state model is of the form

$$\dot{\mathbf{x}}(t) = \mathbf{A}\mathbf{x}(t) + \mathbf{B}\mathbf{u}(t): \quad \textit{State equation} \tag{3.27a}$$

$$\mathbf{y}(t) = \mathbf{C}\mathbf{x}(t) + \mathbf{D}\mathbf{u}(t): \quad \textit{Output equation} \tag{3.27b}$$

where **A**, **B**, **C** and **D** are respectively $n \times n$, $n \times p$, $q \times n$ and $q \times p$ real constant matrices.

We shall refer to matrices **A**, **B**, **C** and **D** as *evolution*, *control*, *observation* and *direct transmission* matrices respectively.

For linear discrete-time systems, the following forms of state model are obtained:

$$\mathbf{x}(k+1) = \mathbf{F}(k)\,\mathbf{x}(k) + \mathbf{G}(k)\,\mathbf{u}(k): \quad \textit{State equation} \tag{3.28a}$$

$$\mathbf{y}(k) = \mathbf{C}(k)\,\mathbf{x}(k) + \mathbf{D}(k)\,\mathbf{u}(k): \quad \textit{Output equation} \tag{3.28b}$$

where $k = k_0, k_0 + 1, \ldots$ ($k_0 =$ integer) and **F**, **G**, **C** and **D** are, respectively, $n \times n$, $n \times p$, $q \times n$ and $q \times p$ time-varying matrices.

For time-invariant systems, equations (3.28) take the form

$$\mathbf{x}(k+1) = \mathbf{F}\mathbf{x}(k) + \mathbf{G}\mathbf{u}(k): \textit{State equation} \tag{3.29a}$$

$$\mathbf{y}(k) = \mathbf{C}\mathbf{x}(k) + \mathbf{D}\mathbf{u}(k): \textit{Output equation} \tag{3.29b}$$

where **F**, **G**, **C** and **D** are, respectively, $n \times n$, $n \times p$, $q \times n$ and $q \times p$ real constant matrices and $k = 0, 1, 2, \ldots$

In the study of linear time-invariant equations, we may also apply the Laplace transform and z-transform techniques. Taking the Laplace transform of eqns. (3.27) and assuming $\mathbf{x}(0) = \mathbf{x}^0$, we obtain (see Appendix I)

$$s\hat{\mathbf{x}}(s) - \mathbf{x}^0 = \mathbf{A}\hat{\mathbf{x}}(s) + \mathbf{B}\hat{\mathbf{u}}(s);\ \hat{\mathbf{x}}(s) \overset{\Delta}{=} \mathscr{L}\,[\mathbf{x}(s)]$$

$$\hat{\mathbf{y}}(s) = \mathbf{C}\hat{\mathbf{x}}(s) + \mathbf{D}\hat{\mathbf{u}}(s)$$

From these equations, we have

$$\hat{\mathbf{x}}(s) = (s\mathbf{I} - \mathbf{A})^{-1}\,\mathbf{x}^0 + (s\mathbf{I} - \mathbf{A})^{-1}\,\mathbf{B}\hat{\mathbf{u}}(s) \tag{3.30a}$$

$$\hat{\mathbf{y}}(s) = \mathbf{C}(s\mathbf{I} - \mathbf{A})^{-1}\,\mathbf{x}^0 + \mathbf{C}(s\mathbf{I} - \mathbf{A})^{-1}\,\mathbf{B}\hat{\mathbf{u}}(s) + \mathbf{D}\hat{\mathbf{u}}(s) \tag{3.30b}$$

Equations (3.30) are algebraic equations. If $\mathbf{x}^0$ and $\hat{\mathbf{u}}(s)$ are known, $\hat{\mathbf{x}}(s)$ and $\hat{\mathbf{y}}(s)$ can be computed from these equations. Alternatively, we may use signal flow graph techniques to solve for $\hat{\mathbf{x}}(s)$ and $\hat{\mathbf{y}}(s)$ for given $\mathbf{x}^0$ and $\hat{\mathbf{u}}(s)$. This will be discussed in Section 3.6.

Application of z-transform techniques gives relations similar to (3.30) for linear time-invariant discrete-time equations. Taking z-transform of eqns. (3.29) and assuming $\mathbf{x}(0) = \mathbf{x}^0$, we obtain (see Appendix II)

$$z\hat{\mathbf{x}}(z) - z\mathbf{x}^0 = \mathbf{F}\hat{\mathbf{x}}(z) + \mathbf{G}\hat{\mathbf{u}}(z);\ \hat{\mathbf{x}}(z) \overset{\Delta}{=} \mathscr{Z}\,[\mathbf{x}(z)]$$

$$\hat{\mathbf{y}}(z) = \mathbf{C}\hat{\mathbf{x}}(z) + \mathbf{D}\hat{\mathbf{u}}(z)$$

From these equations, we have

$$\hat{\mathbf{x}}(z) = (z\mathbf{I} - \mathbf{F})^{-1}\,z\mathbf{x}^0 + (z\mathbf{I} - \mathbf{F})^{-1}\,\mathbf{G}\hat{\mathbf{u}}(z) \tag{3.31a}$$

$$\hat{\mathbf{y}}(z) = \mathbf{C}(z\mathbf{I} - \mathbf{F})^{-1}\,z\mathbf{x}^0 + \mathbf{C}(z\mathbf{I} - \mathbf{F})^{-1}\,\mathbf{G}\hat{\mathbf{u}}(z) + \mathbf{D}\hat{\mathbf{u}}(z) \tag{3.31b}$$

3.5 NONUNIQUENESS OF STATE MODEL

The reader will recall the comment made in Example 3.1 that state variable representation of a system is not unique. In fact, one can find an infinite number of state variable representations for the same system. To prove this, let us suppose that we have found one state model of the form

$$\dot{\mathbf{x}}(t) = \mathbf{A}\mathbf{x}(t) + \mathbf{B}\mathbf{u}(t) \tag{3.32a}$$
$$\mathbf{y}(t) = \mathbf{C}\mathbf{x}(t) + \mathbf{D}\mathbf{u}(t) \tag{3.32b}$$

for a given linear time-invariant system where **A**, **B**, **C** and **D** are, respectively, $n \times n$, $n \times p$, $q \times n$ and $q \times p$ real constant matrices, **u** is the $p \times 1$ input vector, **y** is the $q \times 1$ output vector and **x** is the $n \times 1$ state vector. The state space $\mathbf{\Sigma}$ of the dynamical equation is the n-dimensional real vector space and the matrix A maps $\mathbf{\Sigma}$ into itself. It is very convenient to introduce the set of orthonormal vectors $\{\mathbf{e}_1, \mathbf{e}_2, \ldots, \mathbf{e}_n\}$ where $\mathbf{e}_i$ is $n \times 1$ column vector with 1 at its ith component and zero elsewhere, as the basis of the state space. In doing so, we may also think of the matrix **A** as representing a linear operator with respect to this orthonormal basis.

By changing the basis of Σ, we can get dynamical equations equivalent to (3.32). In order to allow a broader class of equivalent dynamical equations, we may extend the field of real numbers to the field of complex numbers and consider the state space as an n-dimensional complex vector space. We shall need this generalization in later chapters to convert (3.32) into the equivalent Jordan-form state model.

Let $\mathbf{P}$ be an $n \times n$ nonsingular matrix with coefficients in the field of complex numbers C. Let $\bar{\mathbf{x}}(t)$ be a column n-vector defined by

$$\bar{\mathbf{x}}(t) = \mathbf{P}\mathbf{x}(t) \tag{3.33}$$

It is easy to show that $\bar{\mathbf{x}}(t)$ satisfies the equations

$$\dot{\bar{\mathbf{x}}}(t) = \bar{\mathbf{A}}\,\bar{\mathbf{x}}(t) + \bar{\mathbf{B}}\mathbf{u}(t) \tag{3.34a}$$

$$\mathbf{y}(t) = \bar{\mathbf{C}}\,\bar{\mathbf{x}}(t) + \mathbf{D}\mathbf{u}(t) \tag{3.34b}$$

where

$$\bar{\mathbf{A}} = \mathbf{P}\mathbf{A}\mathbf{P}^{-1}, \bar{\mathbf{B}} = \mathbf{P}\mathbf{B}, \bar{\mathbf{C}} = \mathbf{C}\mathbf{P}^{-1}.$$

From eqns. (3.34) we conclude that $\bar{\mathbf{x}}(t)$ also qualifies as a state vector for the given system. State model (3.34) is said to be *equivalent* to the model (3.32) and $\mathbf{P}$ is an *equivalence* or *similarity transformation* (refer Section 2.4).

Note that in the substitution $\bar{\mathbf{x}} = \mathbf{P}\mathbf{x}$, we have changed the basis vectors of state space from the orthonormal vectors to the columns of $\mathbf{P}^{-1}$. The matrices $\mathbf{A}$ and $\bar{\mathbf{A}}$ are similar, they are different representations of the same operator. Let

$$\mathbf{P}^{-1} = \mathbf{Q} = [\mathbf{q}_1\,\mathbf{q}_2 \ldots \mathbf{q}_n]$$

Then

$$\mathbf{A}\mathbf{Q} = \mathbf{Q}\bar{\mathbf{A}} \text{ or } \mathbf{A}[\mathbf{q}_1\,\mathbf{q}_2 \ldots \mathbf{q}_n] = [\mathbf{q}_1\,\mathbf{q}_2 \ldots \mathbf{q}_n]\,\bar{\mathbf{A}};$$

or the ith column of $\bar{\mathbf{A}}$ is the representation of $\mathbf{A}\mathbf{q}_i$ with respect to basis $\{\mathbf{q}_1, \mathbf{q}_2, \ldots \mathbf{q}_n\}$. Similarly, from equation $\mathbf{B} = \mathbf{Q}\bar{\mathbf{B}} = [\mathbf{q}_1\,\mathbf{q}_2 \ldots \mathbf{q}_n]\bar{\mathbf{B}}$ we see that the ith column of $\bar{\mathbf{B}}$ is the representation of the ith column of $\mathbf{B}$ with respect to basis $\{\mathbf{q}_1, \mathbf{q}_2, \ldots, \mathbf{q}_n\}$. The matrix $\bar{\mathbf{C}}$ is to be computed from $\mathbf{C}\mathbf{P}^{-1}$. The matrix $\mathbf{D}$ is the direct transmission part between the input and output; it has nothing to do with state space and hence is not affected by any similarity transformation.

The discussion above applies to linear time-varying state models as well. The only conceptual difference is that the basis vectors in the time-invariant case are fixed whereas the basis vectors in time-varying case may change with time (refer Chen (1970)).

3.6 STATE DIAGRAMS

A state diagram is a pictorial representation of a state model. The important significance of the state diagram is that it forms a close relationship[3]

3. Relationship between state models and differential/difference equations will be discussed in Chapter 7.

among the state models, differential/difference equations and computer simulation. As we shall see shortly, a state diagram is essentially a block diagram for the programming of a computer. In Chapter 5 we shall see that state diagrams may also be used for solving linear constant systems analytically.

State Diagrams for Continuous-Time State Models

The state diagram for a linear time-invariant continuous-time state model is a proper interconnection of three basic units:

(i) Scalars

(ii) Adders

(iii) Integrators

These basic units represent the fundamental linear operations that can be performed on an analog computer (Differentiators are not used because these amplify the inevitable noise). The operation of the basic units is obvious from Fig. 3.9. In this figure, both the block diagrams and the signal flow graphs have been used for basic operations.

In the following, we shall consider state diagrams for state and output

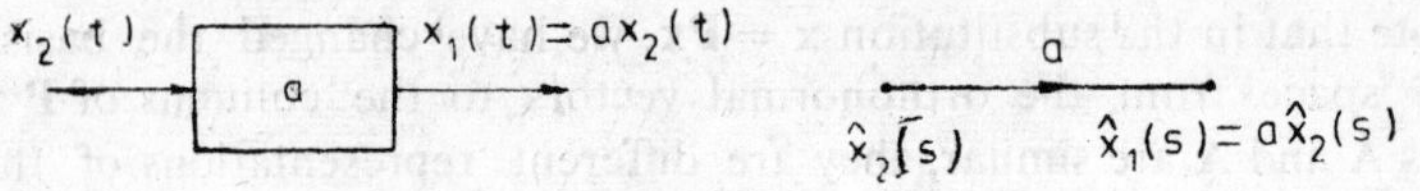

(a) Scalar

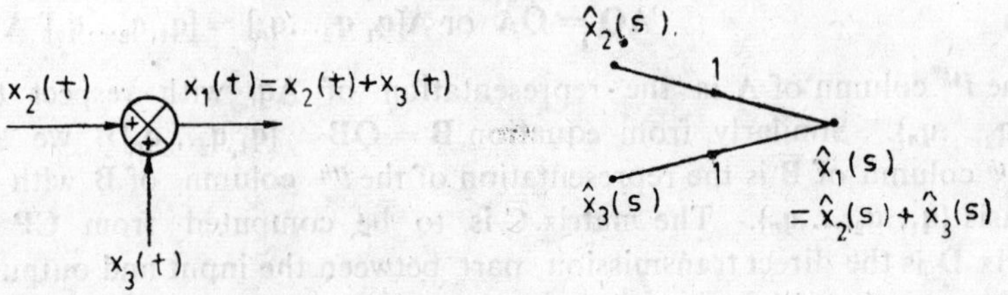

(b) Adder

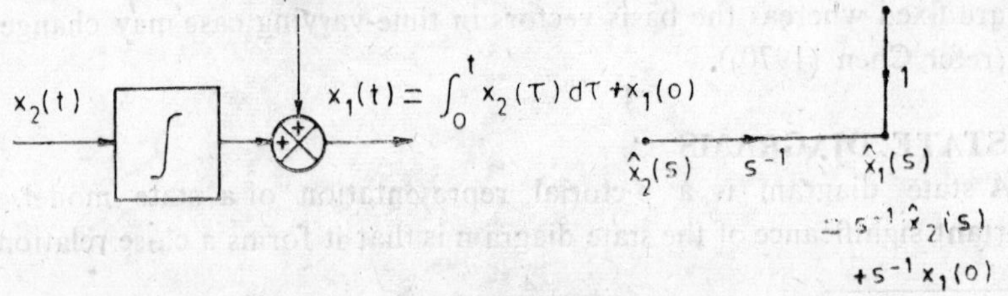

(c) Integrator

Fig. 3.9

equations of differential systems. To begin with, let us consider a single first-order differential equation[4]

$$\dot{x}(t) = ax(t) + bu(t);\ x(0) = x^0 \tag{3.35}$$

The state diagram for this differential equation is shown in Fig. 3.10. We start with signals $x(t)$ and $u(t)$ and form the right hand side of eqn. (3.35) to generate $\dot{x}(t)$. The initial condition $x(0) = x^0$ is added to the output of the integrator.

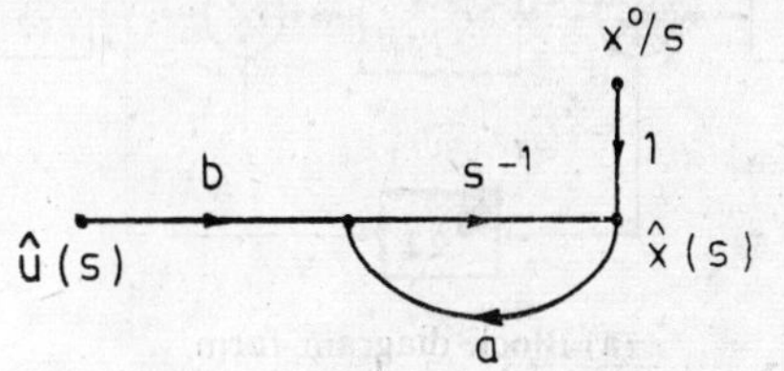

(a) Signal flow graph form

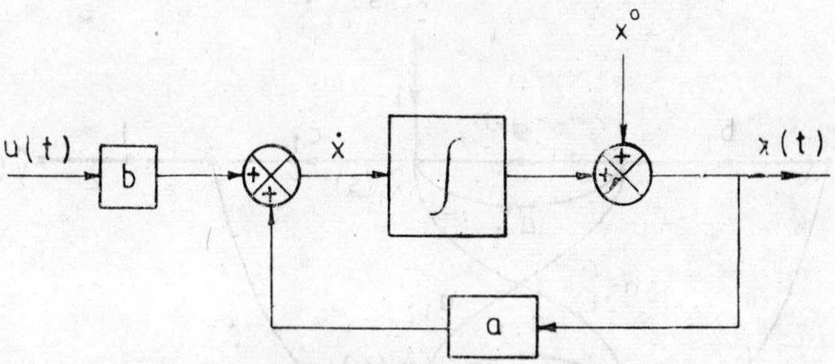

(b) Block diagram form

Fig. 3.10 State diagram for equation (3.35)

Note that if a state diagram of a system is given, then it is simple to obtain the state equation by labelling the integrator outputs as state variables.

We shall next consider a system with a single input $u(t)$ and a single output $y(t)$ (scalar system) and two state variables x_1 and x_2. The system equations are

$$\begin{aligned} \dot{x}_1(t) &= a_{11}x_1(t) + a_{12}x_2(t) + b_1u(t);\ x_1(0) = x_1^0 \\ \dot{x}_2(t) &= a_{21}x_1(t) + a_{22}x_2(t) + b_2u(t);\ x_2(0) = x_2^0 \\ y(t) &= c_1x_1(t) + c_2x_2(t) \end{aligned} \tag{3.36}$$

The state diagram for this set of equations utilizing two integrators is shown in Fig. 3.11.

This procedure can be easily extended for a multi-input, multi-output system (multivariable system) with n state variables. The general philosophy

4. The intial time $t = 0$ can be generalized to $t = t_0$.

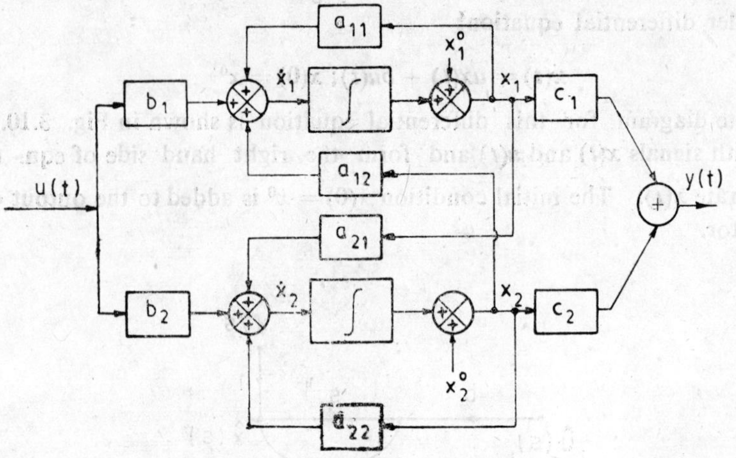

(a) Block diagram form

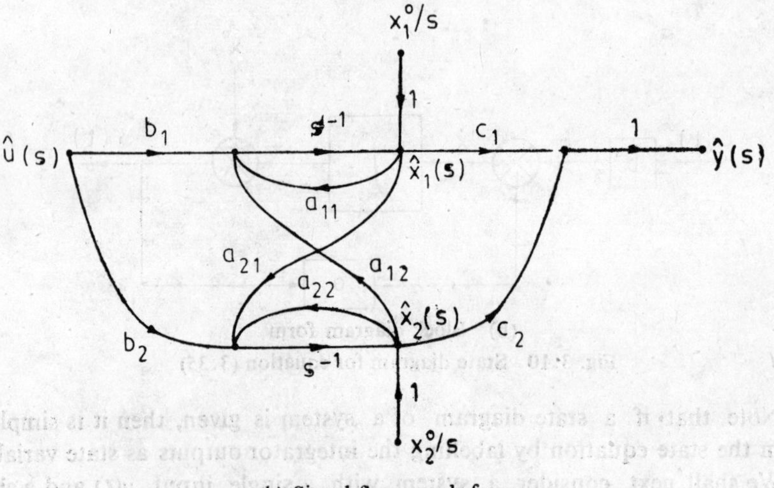

(b) Signal flow graph form

Fig. 3.11 State diagram for equations (3.36)

of the state diagram for the multivariable system (3.27) can be expressed graphically as in Fig. 3.12.

For nonlinear/time-varying state models, the state diagram cannot be obtained in terms of the three basis units—scalars, adders, integrators, described earlier. In addition to these, we require nonlinear function generators and time-dependent devices. Consider for example, the state model

$$\dot{x}_1(t) = x_2(t);\ x_1(0) = x_1^0$$
$$\dot{x}_2(t) = -\,x_1(t) - f(x_2(t));\ x_2(0) = x_2^0 \qquad (3.37)$$

where $f(x_2)$ is a nonlinear function of x_2. (We assume that $f(x_2)$ is a function

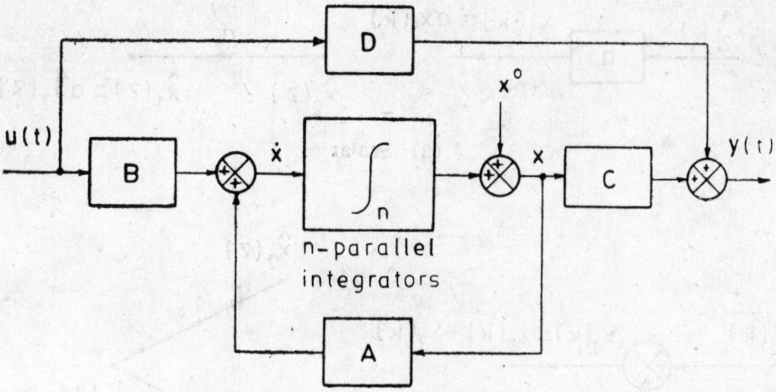

Fig. 3.12 State diagram for linear time-invariant multivariable system

that can be simulated by means of an analog computer.) Figure 3.13 shows the state diagram for this system.

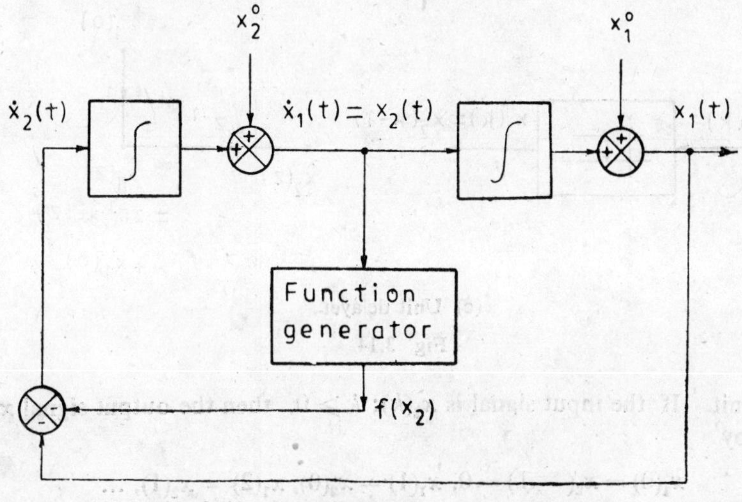

Fig, 3.13 State diagram for state model (3.37)

State Diagrams for Discrete-Time State Models

Similar to the relations between the analog computer simulation and the state diagram for a differential system, the elements of a state diagram for discrete-time systems resemble the computing elements of a digital computer. Some of the operations of a digital computer are:

(i) Multiplication by a constant;
(ii) Addition of several machine variables;
(iii) Time delay or shifting.

These basic operations are diagrammatically represented in Fig. 3.14. The output signal of a unit delayer is the input signal, 'shifted to the right'

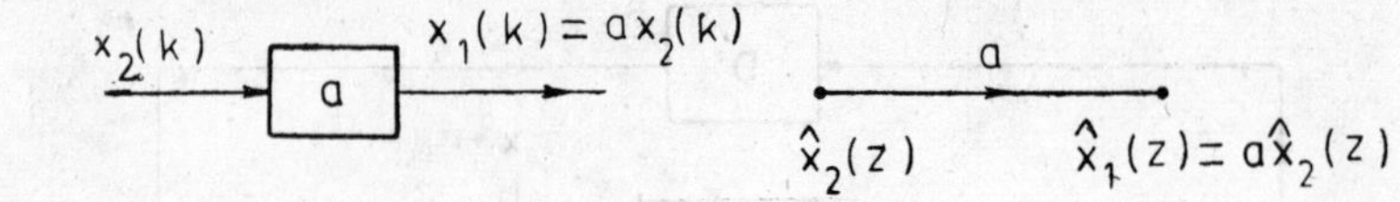

(a) Scalar

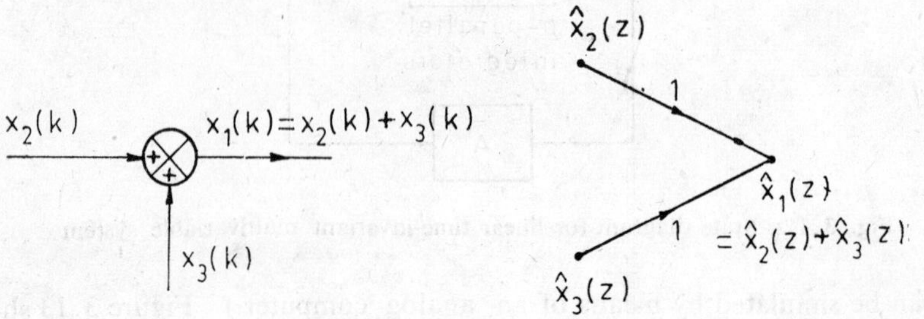

(b) Adder

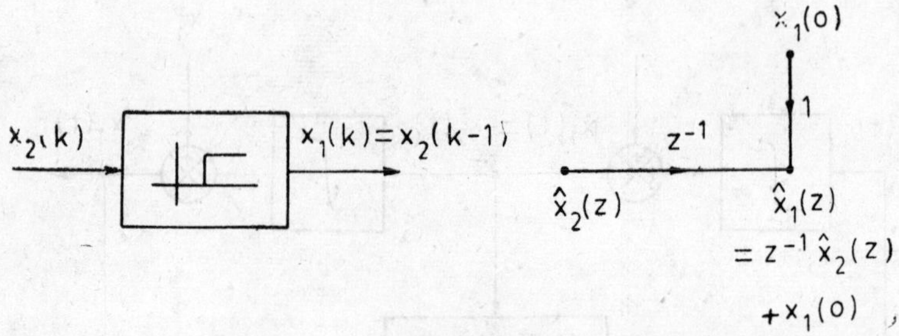

(c) Unit delayer

Fig. 3.14

by one unit. If the input signal is $x_2(k)$; $k \geqslant 0$, then the output signal $x_1(k)$ is given by

$$x_1(0) = x_2(-1) = 0,\ x_1(1) = x_2(0),\ x_1(2) = x_2(1),\ \ldots$$

This one-sided signal convention implies that the output of a unit delayer always has the value 0 at $k = 0$. However, a specified initial condition can always be stored before the commencement of the algorithm in the appropriate register of the digital computer containing $x_1(k)$.

To demonstrate the method for obtaining the state diagram for a given state model, consider the scalar model

$$\begin{aligned} x_1(k+1) &= a_{11}x_1(k) + a_{12}x_2(k) + b_1u(k); \quad x_1(0) = x_1^0 \\ x_2(k+1) &= a_{21}x_1(k) + a_{22}x_2(k) + b_2u(k); \quad x_2(0) = x_2^0 \qquad (3.38) \\ y(k) &= c_1x_1(k) + c_2x_2(k) + du(k) \end{aligned}$$

The state diagram for this set of equations utilizing two delayers is shown in Fig. 3.15.

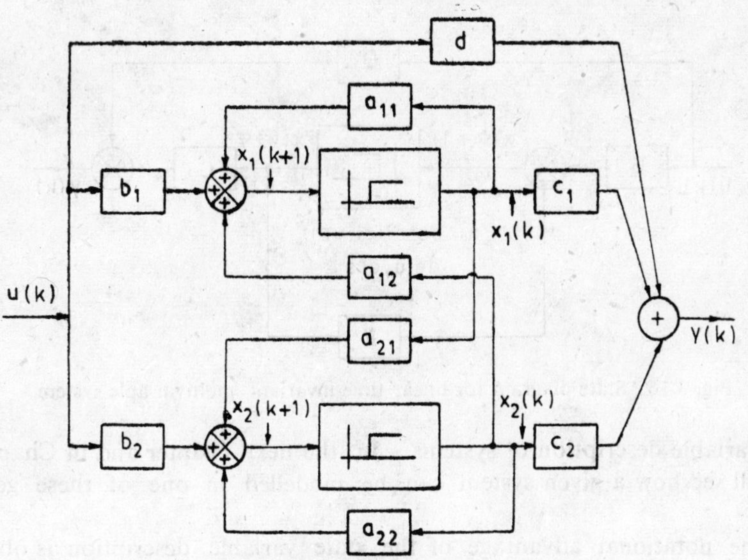

(a) Block diagram form

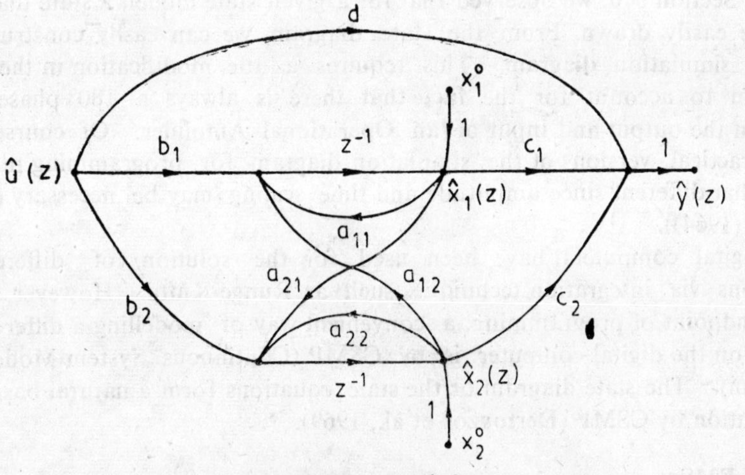

(b) Signal flow graph form

Fig. 3.15 State diagram for equations (3.38)

The procedure can be easily extended to a general multivariable system with n state variables. The general philosophy of the state diagram for the state model (3.29) can be expressed graphically as shown in Fig. 3.16.

Note that if a state diagram of a discrete-time system is given, then it is straightforward to obtain state equation by labelling the unit-delayer outputs as state variables.

3.7 CONCLUDING COMMENTS

In this chapter, we have developed systematically the general forms of

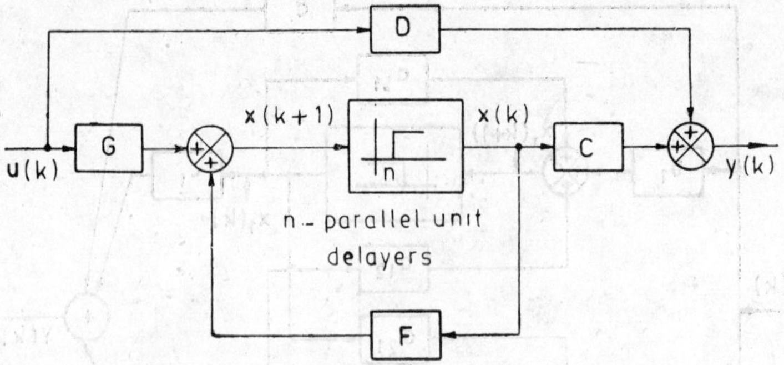

Fig, 3.16 State diagram for linear time-invariant multivariable system

state variable description of systems. In the next chapter and in Chapter 7, we shall see how a given system can be modelled in one of these general forms.

The notational advantage of the state variable description is obvious from the discussion in this chapter. Conceptual and analytical advantages will be clear from the discussion that follows in later chapters.

In Section 3.6, we observed that for a given state model, a state diagram can be easily drawn. From the state diagram, we can easily construct an analog simulation diagram. This requires a little modification in the state diagram to account for the fact that there is always a 180° phase shift between the output and input of an Operational Amplifier. Of course, the final practical version of the simulation diagram for programming may be somewhat different since amplitude and time scaling may be necessary (refer Levine (1964)).

Digital computers have been used for the solution of differential equations via integration techniques such as Runge-Kutta. However, from the standpoint of programming, a convenient way of modelling a differential system on the digital computer is by CSMP (Continuous System Modelling Program). The state diagram or the state equations form a natural basis for the solution by CSMP (Dertouzos et al., 1969).

PROBLEMS

3.1 For the ideal integrator, convince yourself that if you know the value of output y at time $t = t_0$ and input u from t_0 onwards, then you can tell what the output is at any time $t \geqslant t_0$. This entitles the output of the integrator to be the state of the integrator. Could the input u of the ideal integrator be taken as the state?

3.2 (a) The input-output relation of a unit delayer is $y(t) = u(t-1) \;\forall\; t$. Convince yourself that to determine $y_{[t_0,\,\infty)}$ uniquely from $u_{(t_0,\,\infty)}$, you need the information $u_{[t_0-1,\,t_0]}$. This entitles the information $u_{[t_0-1,\,t_0]}$ to be called the state of the delayer at t_0. What is the dimension of the state space?

(b) Assume that in the delayer system described above, the output is observed only at discrete instants of time $t = kT$, $k = \ldots -1, 0, 1, 2, \ldots$ and $T =$ constant. The system equation is

$$y(k) = u(k-1)$$

Convince yourself that if you know the value of y at $k = 0$ and input u from $k = 0$ onwards, then you can tell what the output is at any $k \geqslant 0$. This entitles the output of the delayer to be the state of the delayer.

3.3 For a differentiator, convince yourself that to determine output $y_{[t_0, \infty)}$ uniquely from input $u_{(t_0, \infty)}$, you need the information $u_{[t_0-\epsilon, t_0]}$ where ϵ is a small positive constant. This entitles the information $u_{[t_0-\epsilon, t_0]}$ to be called the state of the differentiator at t_0.

3.4 For the electrical circuit of Fig. P-3.4, show that the current through the inductor is an acceptable state variable. Write the state and output equations.

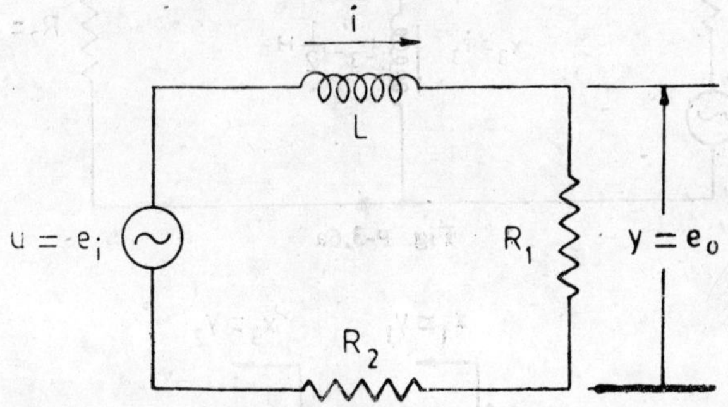

Fig. P-3.4

What do you think about the suitability of
(i) inductor flux linkage
(ii) energy stored in the inductor
as a state variable for the system?

3.5 For the mechanical system of Fig. P-3.5, the problem is to deter-

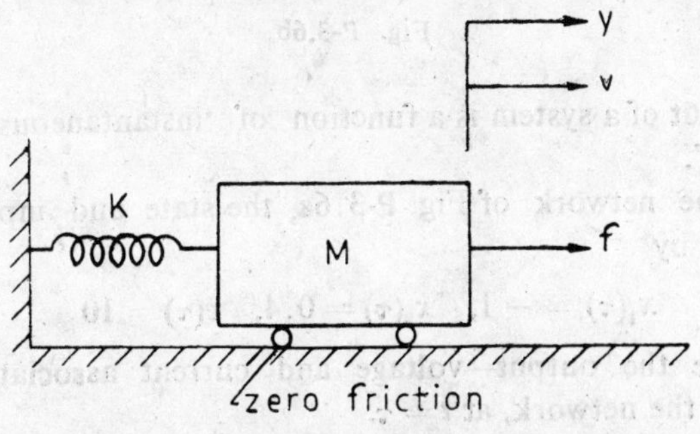

Fig. P-3.5

mine acceleration of mass in response to force f. Show that displacement y and velocity v constitute an acceptable set of state variables. Write the state and output equations.

What do you think about the suitability of spring force (Ky) and velocity v as a set of state variables for the system?

3.6 Rules of thumb, often used in writing state equations for electrical and mechanical networks, are 'choose voltages across capacitors and currents through inductors as the state variables', 'choose positions and velocities of all junctions (points where two or more elements are connected as the state variables'.

Convince yourself that these thumb rules give redundant state variable for systems of Fig. P-3.6a and Fig. P-3.6b.

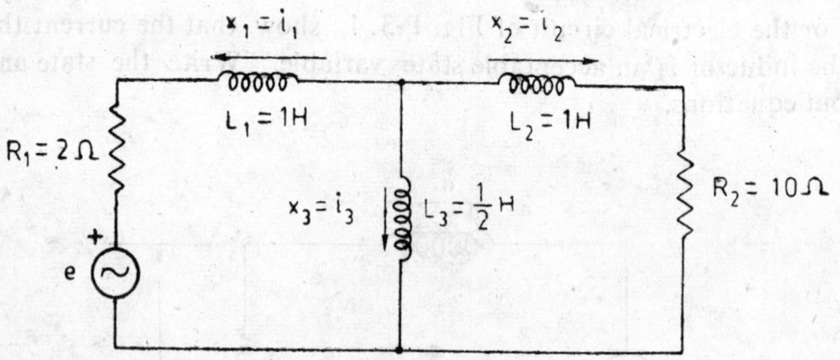

Fig. P-3.6a

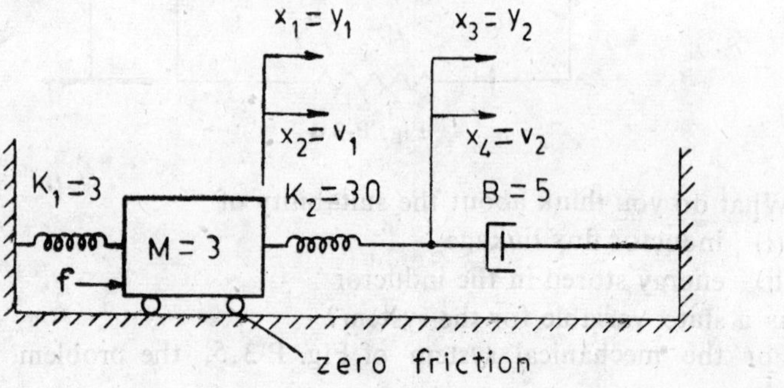

Fig. P-3.6b

3.7 'The output of a system is a function of 'instantaneous values of state and input'.

(a) In the network of Fig. P-3.6a, the state and input at $t = \tau$ are given by

$$x_1(\tau) = -1, \quad x_2(\tau) = 0.4, \quad e(\tau) = 10$$

Determine the output—voltage and current associated with every branch of the network, at $t = \tau$.

(b) In the network of Fig. P-3.6b, the state and input at $t = \tau$ are

$$x_1(\tau) = 0.5, \quad x_2(\tau) = 2.5, \quad x_3(\tau) = 0.2, \quad f(\tau) = 2$$

Determine the output—force, displacement and velocity associated with every element of the system, at $t = \tau$.

3.8 Verify that the systems of Fig. P-3.4 and Fig. P-3.5 are linear and time-invariant.

3.9 Write the state equation for the system of Fig. P-3.9 and show that it is a linear, time-invariant system.

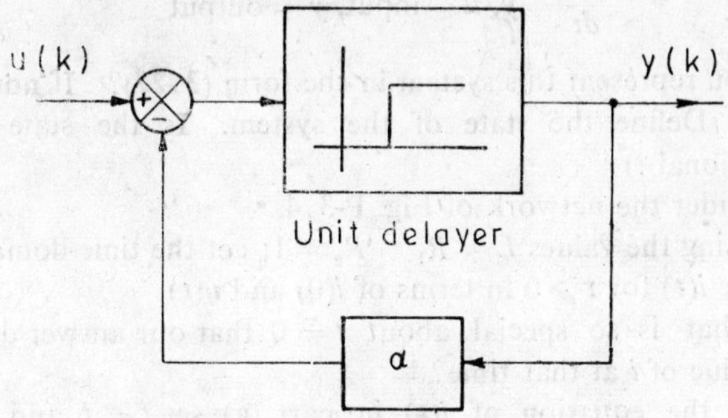

Fig. P-3.9

3.10 Consider a simple pendulum of mass M suspended by a string of length l. Let θ be the angular displacement as shown in Fig. P-3.10. Define the state variables as $x_1 = \theta$ and $x_2 = \dot{\theta}$ and write the state equation in the form

$$\dot{\mathbf{x}}(t) = \mathbf{f}(t, \mathbf{x}, \mathbf{u})$$

Is **f** a linear function ?

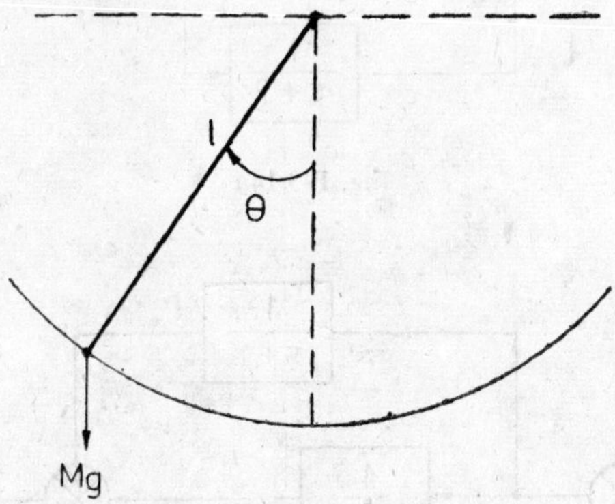

Fig. P-3.10

3.11 Prove that a system described by state model of the form

$$\dot{\mathbf{x}}(t) = \mathbf{A}\mathbf{x}(t) + \mathbf{B}\mathbf{u}(t)$$

$$\mathbf{y}(t) = \mathbf{C}\mathbf{x}(t) + \mathbf{D}\mathbf{u}(t)$$

where **A**, **B**, **C** and **D** are constant matrices, is linear and time-invariant. (Hint: The state transition function Φ is given as (Chapter 5)

$$\mathbf{x}(t) = \Phi(t, t_0, \mathbf{x}^0, \mathbf{u})$$
$$= e^{\mathbf{A}(t-t_0)}\mathbf{x}^0 + \int_{t_0}^{t} e^{\mathbf{A}(t-\tau)}\mathbf{B}\mathbf{u}(\tau)\, d\tau)$$

3.12 A system is described by the equation

$$\frac{du}{dt} = y,\ u = \text{input},\ y = \text{output}$$

Can you represent this system in the form (3.27)? If not, why? (Hint: Define the state of the system. Is the state space Σ finite dimensional?)

3.13 Reconsider the network of Fig. P-3.4.

(a) Using the values $L = R_1 = R_2 = 1$, get the time-domain expression for $i(t)$ for $t \geqslant 0$ in terms of $i(0)$ and $u(t)$.

(b) What is so special about $t = 0$ that our anรwer depends on the value of i at that time?

In the equation of $i(t)$ in part (a), set $t = t_0$ and solve for $i(t_0)$. Obtain a time domain expression for $i(t)$ for $t \geqslant t_0$ in terms of $i(t_0)$ and $u(t)$.

3.14 Find state models for the systems shown in Fig. P-3.14a, and Fig. P-3.14b.

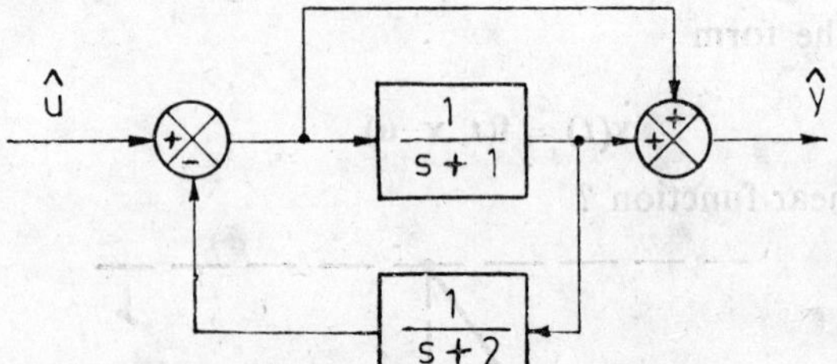

Fig. P-3.14a

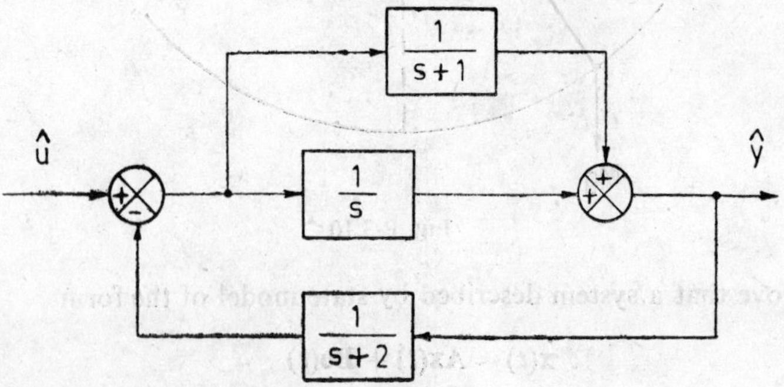

Fig. P-3.14b

(Hint: Assign a component of state vector **x** to the output of each integrator).

3.15 A system is described by the state equation

$$\dot{\mathbf{x}}(t) = \begin{bmatrix} 0 & 1 & 0 \\ 0 & 0 & 1 \\ -1 & 0 & -3 \end{bmatrix} \mathbf{x}(t) + \begin{bmatrix} 0 \\ 0 \\ 1 \end{bmatrix} u(t);\ \mathbf{x}(0) = \mathbf{x}^0$$

Using the Laplace transform technique, transform the state equation into a set of linear algebraic equations in the form

$$\hat{\mathbf{x}}(s) = \hat{\mathbf{G}}(s)\, \mathbf{x}^0 + \hat{\mathbf{H}}(s)\, \hat{u}(s)$$

3.16 A system is described by the state equation

$$\mathbf{x}(k+1) = \begin{bmatrix} -3 & 1 & 0 \\ -4 & 0 & 1 \\ -1 & 0 & 0 \end{bmatrix} \mathbf{x}(k) + \begin{bmatrix} -3 \\ -7 \\ 0 \end{bmatrix} u(k);\ \mathbf{x}(0) = \mathbf{x}^0$$

Using z-transform technique, transform the state equation into a set of linear algebraic equations in the form

$$\hat{\mathbf{x}}(z) = \hat{\mathbf{G}}(z)\, \mathbf{x}^0 + \hat{\mathbf{H}}(z)\, \hat{u}(z)$$

3.17 Give a block diagram for the programming of the system of Problem 3.15 on an analog computer.

3.18 Give a block diagram for the digital network realization of the state equation of Problem 3.16.

3.19 A system is described by the state equation

$$\dot{\mathbf{x}}(t) = \mathbf{A}\mathbf{x}(t) + \mathbf{B}\mathbf{u}(t);\ \mathbf{x}(0) = \mathbf{x}^0$$

$$\mathbf{y}(t) = \mathbf{c}\mathbf{x}(t)$$

where

$$\mathbf{A} = \begin{bmatrix} -5 & -4 & 2 \\ 3 & 3 & -2 \\ 0 & 2 & -2 \end{bmatrix},\ \mathbf{B} = \begin{bmatrix} -1 & 0 \\ 1 & 1 \\ 0 & 2 \end{bmatrix},$$

$$\mathbf{c} = [1 \quad 1 \quad 0]$$

Draw state diagram and therefrom obtain the state equation in the form

$$\hat{\mathbf{x}}(s) = \hat{\mathbf{G}}(s)\, \mathbf{x}^0 + \hat{\mathbf{H}}(s)\, \hat{\mathbf{u}}(s)$$

(Hint: You may draw the state diagram in signal flow graph form and then use Mason's gain formula (refer Nagrath and Gopal (1982)) to obtain $\hat{\mathbf{G}}(s)$ and $\hat{\mathbf{H}}(s)$. Mason's gain formula for the determination of system gain is given by

$$T = \frac{1}{\Delta} \sum_K P_K \Delta_K$$

where

P_K = path gain of K-th forward path

Δ = determinant of the graph

= 1 − (sum of loop gains of all individual loops) + (sum of gain products of all possible combinations of two non-touching loops) − (sum of gain products of all possible combinations of three non-touching loops) + ...

$$= 1 - \sum_m P_{m1} + \sum_m P_{m2} - \sum_m P_{m3} + \ldots$$

P_{mr} = gain product of m-th possible combination of r non-touching loops

Δ_K = the value of Δ for that part of the graph not touching the K-th forward path

T = overall gain of the system).

3.20 The state diagrams of two linear systems are shown in Fig. P-3.20a and Fig. P-3.20b. Assign the state variables and write the dynamic equations of the systems.

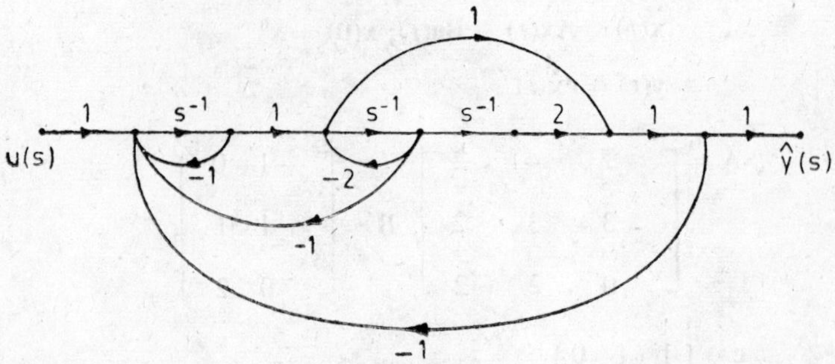

Fig. P-3.20a

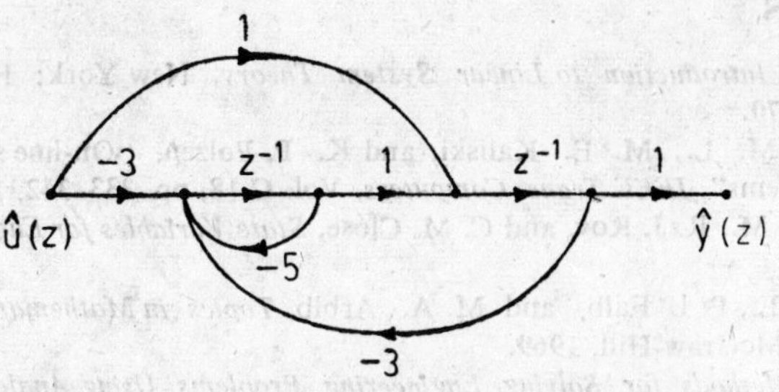

Fig. P-3.20b

3.21 Set up state models for systems of Figs. P-3.21a and P-3.21b.

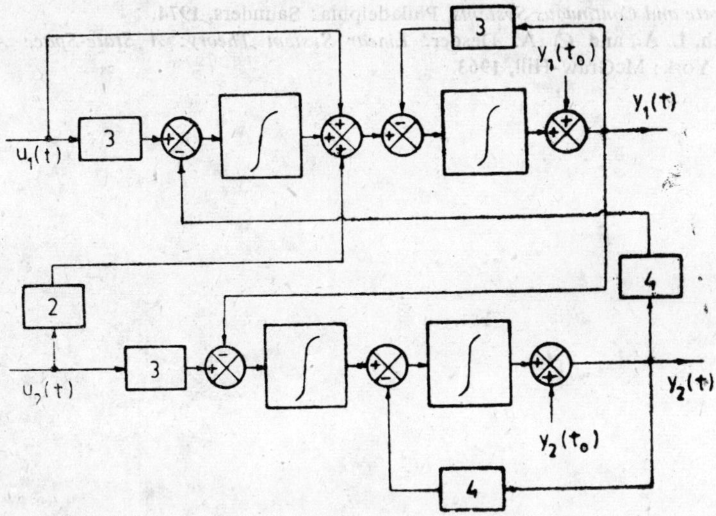

Fig. P-3.21a

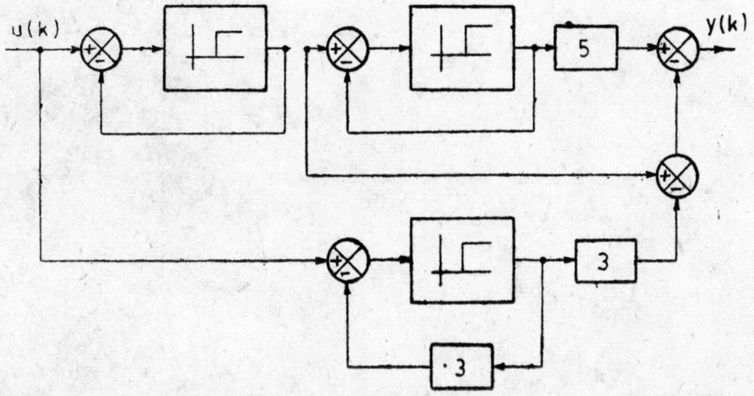

Fig. P-3.21b

REFERENCES

1. Chen, C. T., *Introduction to Linear System Theory*, New York: Holt, Rinehart and Winston, 1970.
2. Dertouzos, M. L., M. E. Kaliski, and K. P. Polzen, "On-line simulation of block-diagram systems", *IEEE Trans. Computers*, Vol. C-18, pp. 333–342, 1969.
3. DeRusso, P. M., R. J. Roy, and C. M. Close, *State Variables for Engineers*, New York: Wiley, 1965.
4. Kalman, R. E., P. L Falb, and M. A. Arbib, *Topics in Mathematical System Theory*, New York: McGraw-Hill, 1969.
5. Levine, L., *Methods for Solving Engineering Problems Using Analog Computers*, New York: McGraw-Hill, 1964.
6. Nagrath, I. J., and M. Gopal, *Control Systems Engineering*, 2nd Edition, New Delhi: Wiley Eastern, 1982.
7. Padulo, L., and M. A. Arbib, *System Theory: A Unified State-Space Approach to Discrete and Continuous Systems*, Philadelphia: Saunders, 1974.
8. Zadeh, L. A., and C. A. Desoer, *Linear System Theory: A State-Space Approach*, New York: McGraw–Hill, 1963.

4. PHYSICAL SYSTEMS AND STATE ASSIGNMENT

4.1 INTRODUCTION

In the present chapter, we shall deal with the question, "Given a real-world system—be it electrical, mechanical, chemical, ..., how do we obtain the state variable description of the system ?"

The selection of state variables depends upon the system structure and the goals of study. Even after the goals of the project have been well defined and the structure of the model has been established, we shall see that many equivalent sets of state variables describe that same model. The engineer's intuition and 'feel' for the situation must provide the major guidance. The guidelines suggested below and elsewhere are no more than indications of where to start (Auslander et al. 1974).

Some systems, by their actual structure, make the choice of state variables easy. Such is the case when a system's primary energy (or mass) storage modes are associated with parts of the system that are physically separated. It is then customary to assign one state variable to each independent storage element. Once the state variables of an energetic system are assigned for each energy storage element, they can be related to each other by applying the appropriate fundamental laws[1].

Consider a mass-spring-damper system shown in Fig. 4.1. In this system, the energy storing elements are the mass and the spring. The kinetic energy stored in mass M at time t is $\frac{1}{2}Mv^2(t)$, where $v(t)$ is the relative velocity of the mass at t; the potential energy stored in the spring of stiffness K at time t is $\frac{1}{2}Ky^2(t)$ where $y(t)$ is the displacement (relative) of spring at t.

The principle of conservation of energy states that

Rate of accumulation of energy in a system = rate of flow of energy into the system − rate of flow of energy out of the system.

For the system of Fig. 4.1,

$$\frac{d}{dt}\left[\tfrac{1}{2}Mv^2(t)+\tfrac{1}{2}Ky^2(t)\right]=f(t)v(t)-Bv^2(t)$$

[1]It is beyond the scope of this book to try to cover in a comprehensive fashion, mathematical model building as it would apply to all phases of engineering. Indeed, in most undergraduate curricula, considerable time is spent in describing various engineering processes. Our aim is to discuss a few examples drawn from various engineering disciplines.

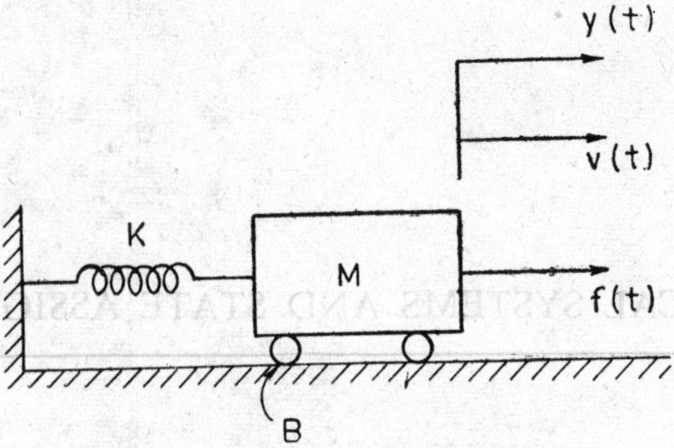

Fig. 4.1 A mass-spring-damper system

where $f(t)v(t)$ is the power (rate of energy) received by the system and $Bv^2(t)$ (B=coefficient of viscous friction) is the power dissipated in the damper.

From the energy equation we have

$$M\dot{v}(t)+Bv(t)+Ky(t)=f(t)$$

This equation is nothing but Newton's law (conservation of force) applied to mass M in Fig. 4.1.

From this example, we observe that the dynamical changes in a mass-spring-damper system are caused basically by a redistribution of energy within the system which takes place with finite flow rates. The total energy at time t defines the *energy state* of the system at that time. This suggests that energy terms associated with various energy storing elements may be taken as state variables. However, we have seen earlier (Chapter 3) that energy itself is not a convenient state variable from the computational point of view. We therefore look for characterizing variables whose value at any time t gives the energy state of the system at that time. Velocities associated with mass elements and displacements associated with spring elements in mass-spring-damper systems are the characterizing variables and therefore may be taken as state variables. Thus the number of dynamically independent energy storage possibilities within a system equals the number of dependent state variables, and the complexity or dimensionality of the system is directly related to this number.

We may remind the reader at this stage that we have to select the state vector so that consistent 'state transition' and 'read-out' functions are obtained. For example, in the mechanical system of Fig. 3.1, there is one energy storing element; therefore mass velocity $v(t)$ defines the energy state of the system. If, however, in addition to energy and velocity at time t, we are interested in displacement also, we require two state variables to get consistent 'state transition' and 'read-out' functions.

In mass-spring-damper systems, we are usually interested in the displacement, velocity and force associated with various elements of the systems.

Normally, an independent set of displacements and velocities of masses and displacements of springs constitutes a state vector for the system whose value at time t defines the energy state of the system at that time and, in addition, the consistency conditions on state transition and read-out functions are met.

In mechanical rotational systems with inertia, torsional spring and damper elements, normally an independent set of angular velocities and displacements associated with inertia elements and angular displacements associated with torsional spring elements constitutes a state vector.

In the study of inductance-capacitance-resistance networks, the desired output information is usually the voltages and currents associated with various branches of the network. It is well known that this information at time t can be obtained if voltages across capacitors and currents through inductors of the network at that time are known, in addition to tha values of the inputs. The dynamical equations are set up in terms of rates of change of capacitor voltages and inductor currents. This selection of state variables is linked with the energy concept. At time t, the energy stored in a capacitor of capacitance C is $\frac{1}{2}Ce^2(t)$ and the energy stored in an inductor of inductance L is $\frac{1}{2}Li^2(t)$ where $e(t)$ is the voltage across the capacitor and $i(t)$ is the current through the inductor. Thus, the capacitor voltages and inductor currents are the characterizing variables which give the energy state of the system. The number of independent energy storage possibilities in a system equals the number of dependent state variables.

In the study of fluid systems, mass (or volume) flow rates of fluids and associated pressures are of interest to us. The energy stored in a fluid storage tank is given by $\frac{1}{2}CP^2(t)$ where C is fluid capacitance of the tank which depends upon cross-sectional area A of the tank and density ρ of the fluid in the tank; P is the pressure (relative) of the fluid in the tank. The storage of energy due to inertia of the fluid flowing through a pipe is given by $\frac{1}{2}LQ^2(t)$ where L is the fluid inertance which depends on dimensions of pipe and density of fluid; Q is fluid flow rate. We observe that an independent set of variables—fluid flow rates Q associated with inertia elements and pressures P or heads $h(P=\rho gh)$ associated with capacitance elements, are the characterizing variables which define the energy state of the fluid system. This gives us a convenient choice of state variables.

In the study of thermal systems, heat flow rates and associated temperatures are of interest to us. Heat energy stored in a substance of mass M is given by McT where c is the heat capacity of the substance and T is the temperature (relative) of the substance. Thus an independent set of temperatures associated with various energy storing elements of a thermal system constitutes a state vector for the system.

In isothermal chemical processes (temperature assumed constant), redistribution of masses of various chemical species takes place. The mass of a substance A(usually measured in *moles*) in a solution $V(m^3)$ is given by VC_A where C_A(moles/m^3) is the concentration of A in that solution. An independent set of mass variables (or concentrations) of various reactants constitutes a state vector for the system.

In nonisothermal chemical systems, in addition to redistribution of mass, redistribution of thermal energy also takes place. Mass variables of reactants and temperatures of energy storing elements give us the state vector for the system.

These ideas are illustrated through numerous examples in this chapter. Athans et al. (1974), Cannon (1967), Crandall et al. (1968), Luyben (1973), MacFarlane (1970), Perkins and Cruz (1969), Russell and Denn (1972) and Smith et al. (1970) are some of the books which deal with the basic concepts of modelling.

4.2 LINEAR CONTINUOUS-TIME MODELS

In this section, we study the problem of state assignment in linear continuous-time systems. The general form of the state model for linear time-invariant, continuous-time systems is (eqns. (3.27))

$$\dot{\mathbf{x}}(t) = \mathbf{A}\mathbf{x}(t) + \mathbf{B}\mathbf{u}(t) \tag{4.1a}$$

$$\mathbf{y}(t) = \mathbf{C}\mathbf{x}(t) + \mathbf{D}\mathbf{u}(t) \tag{4.1b}$$

where **A**, **B**, **C** and **D** are respectively, $n \times n$, $n \times p$, $q \times n$ and $q \times p$ real constant matrices, **u** is the $p \times 1$ input vector, **y** is the $q \times 1$ output vector and **x** is the $n \times 1$ state vector. As pointed out earlier, a unique solution of the state equation (4.1a) exists for any $\mathbf{u} \in$ admissible class $\mathbf{\Omega}$ of input functions.

The general form of the state model of linear time-varying continuous-time systems is (eqns. (3.26))

$$\dot{\mathbf{x}}(t) = \mathbf{A}(t)\,\mathbf{x}(t) + \mathbf{B}(t)\,\mathbf{u}(t) \tag{4.2a}$$

$$\mathbf{y}(t) = \mathbf{C}(t)\,\mathbf{x}(t) + \mathbf{D}(t)\,\mathbf{u}(t) \tag{4.2b}$$

Here, the coefficients of matrices **A**, **B**, **C** and **D** are functions of time t. As pointed out earlier, a unique solution of the state eqns. (4.2a) exists for any $\mathbf{u} \in$ admissible class $\mathbf{\Omega}$ of input functions if coefficients $a_{ij}(t)$ and $b_{ij}(t)$ of matrices **A** and **B**, respectively, are continuous functions of time.

We know that almost all physical systems are time-dependent to some extent. However, if the parameters of a system are changing very slowly with time, it is reasonably valid to design or analyze the system assuming constant parameters and hope that the performance of the system will not change very much with the passage of time. With this assumption, the state model of a linear system will be of the form (4.1).

However, if the parameters change appreciably over the time constants of the system, the state model of the system will be in the form of eqns. (4.2). On the other hand, there are systems wherein the parameters are made time-dependent intentionally. We shall see in Chapters 10 and 11 that the implementation of optimal control law, in many cases, requires amplifiers with time-varying gains.

In the following we give linear time-invariant models of some physical systems. The reader will find examples of linear time-varying models in Chapters 10 and 11.

Before taking up examples of linear models, one comment is in order. Almost all physical systems are nonlinear in nature and therefore linear models only approximately represent the actual systems. However, in the examples that follow, this approximation gives satisfactory results over wide ranges of parameters and variables of interest. In the cases where the range is not wide, the powerful tools af linear system analysis can be used only in a limited sense by local linearization of the nonlinear state model around the equilibrium point. This subject will be taken up in a later section of this chapter.

Example 4.1: Consider the mass-spring-damper system shown in Fig. 4.2. The system consists of two platforms coupled to each other and to a fixed

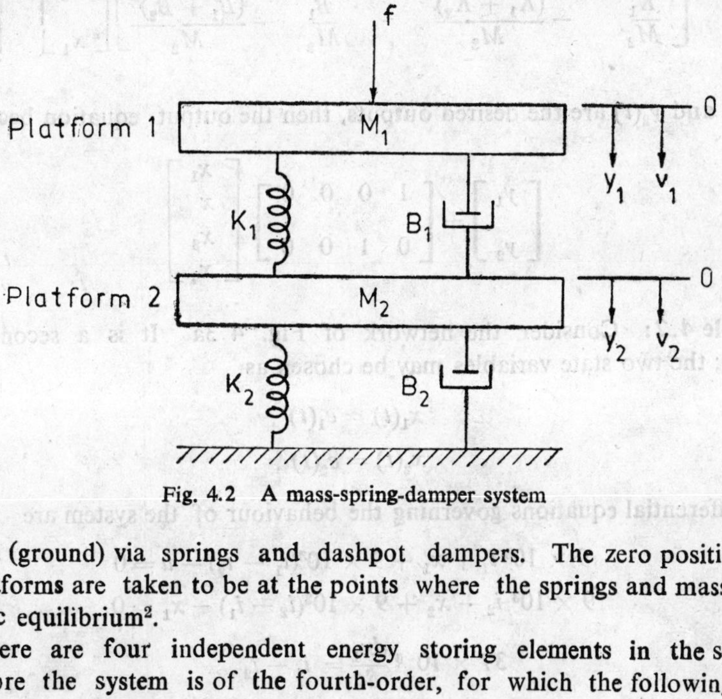

Fig. 4.2 A mass-spring-damper system

support (ground) via springs and dashpot dampers. The zero positions of the platforms are taken to be at the points where the springs and masses are in static equilibrium[2].

There are four independent energy storing elements in the system. Therefore the system is of the fourth-order, for which the following state variables may be chosen:

$$x_1(t) = y_1(t), \text{ the displacement of platform 1}$$
$$x_2(t) = y_2(t), \text{ the displacement of platform 2}$$
$$x_3(t) = v_1(t), \text{ the velocity of platform 1}$$
$$x_4(t) = v_2(t), \text{ the velocity of platform 2}$$

The input variable

$$u(t) = f(t), \text{ the applied force.}$$

[2]Note that the gravitational effect is eliminated by this choice of zero position.

The differential equations describing the dynamics of the system are given below:

$$\dot{x}_1 = x_3$$
$$\dot{x}_2 = x_4$$
$$M_1\dot{x}_3 + B_1(x_3 - x_4) + K_1(x_1 - x_2) = u$$
$$M_2\dot{x}_4 + B_1(x_4 - x_3) + K_1(x_2 - x_1) + B_2x_4 + K_2x_2 = 0$$

These equations yield the following state equation for the system.

$$\begin{bmatrix} \dot{x}_1 \\ \dot{x}_2 \\ \dot{x}_3 \\ \dot{x}_4 \end{bmatrix} = \begin{bmatrix} 0 & 0 & 1 & 0 \\ 0 & 0 & 0 & 1 \\ -\dfrac{K_1}{M_1} & \dfrac{K_1}{M_1} & -\dfrac{B_1}{M_1} & \dfrac{B_1}{M_1} \\ \dfrac{K_1}{M_2} & -\dfrac{(K_1+K_2)}{M_2} & \dfrac{B_1}{M_2} & -\dfrac{(B_1+B_2)}{M_2} \end{bmatrix} \begin{bmatrix} x_1 \\ x_2 \\ x_3 \\ x_4 \end{bmatrix} + \begin{bmatrix} 0 \\ 0 \\ \dfrac{1}{M_1} \\ 0 \end{bmatrix} u \tag{4.3a}$$

If $y_1(t)$ and $y_2(t)$ are the desired outputs, then the output equation becomes

$$\begin{bmatrix} y_1 \\ y_2 \end{bmatrix} + \begin{bmatrix} 1 & 0 & 0 & 0 \\ 0 & 1 & 0 & 0 \end{bmatrix} \begin{bmatrix} x_1 \\ x \\ x_3 \\ x_4 \end{bmatrix} \tag{4.3b}$$

Example 4.2: Consider the network of Fig. 4.3a. It is a second-order system; the two state variables may be chosen as

$$x_1(t) = e_1(t)$$
$$x_2(t) = e_2(t)$$

The differential equations governing the behaviour of the system are

$$9 \times 10^3\, i_1 + x_1 + 9 \times 10^3(i_1 - i_2) - u = 0$$
$$9 \times 10^3\, i_2 + x_2 + 9 \times 10^3(i_2 - i_1) - x_1 = 0$$
$$37 \times 10^{-6} \frac{dx_1}{dt} = i_1 - i_2$$
$$37 \times 10^{-6} \frac{dx_2}{dt} = i_2$$

These equations give us the following state model:

$$\begin{bmatrix} \dot{x}_1 \\ \dot{x}_2 \end{bmatrix} = \begin{bmatrix} -2 & 1 \\ 1 & -2 \end{bmatrix} \begin{bmatrix} x_1 \\ x_2 \end{bmatrix} + \begin{bmatrix} 1 \\ 1 \end{bmatrix} u \tag{4.4a}$$

$$y = [0 \quad 1] \begin{bmatrix} x_1 \\ x_2 \end{bmatrix} \tag{4.4b}$$

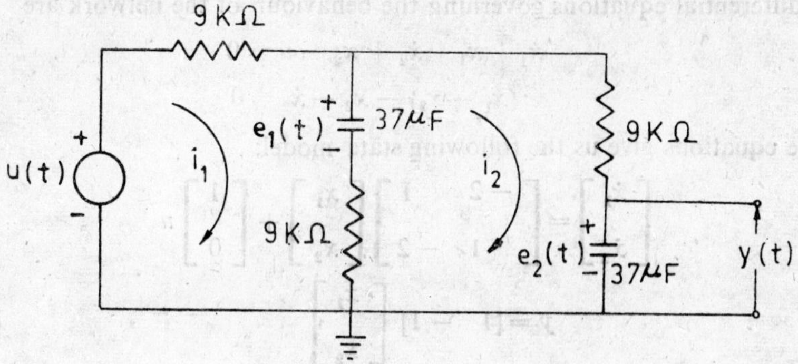

Fig. 4.3a

The state diagram for this state model is shown in Fig. 4.3b.

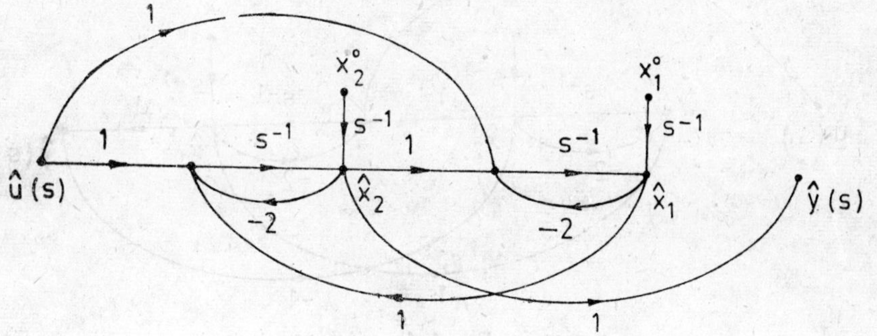

Fig. 4.3b

Consider now the network of Fig. 4.4a. For this network, let

$$x_1(t) = i_1(t)$$
$$x_2(t) = i_2(t)$$

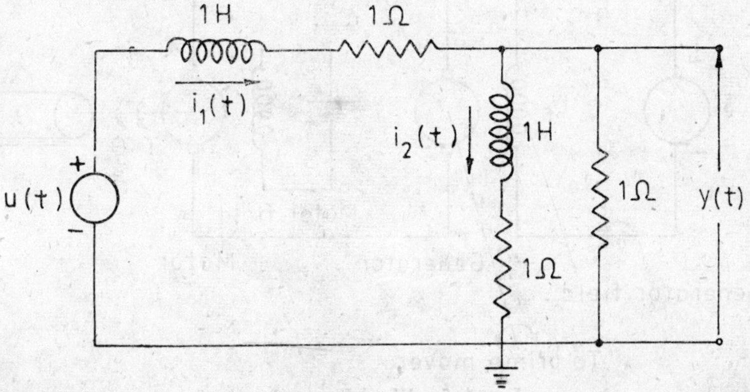

Fig. 4.4a

The differential equations governing the behaviour of the network are

$$\dot{x}_1 + x_1 + \dot{x}_2 + x_2 - u = 0$$

$$(x_1 - x_2) - x_2 - \dot{x}_2 = 0$$

These equations give us the following state model:

$$\begin{bmatrix} \dot{x}_1 \\ \dot{x}_2 \end{bmatrix} = \begin{bmatrix} -2 & 1 \\ 1 & -2 \end{bmatrix} \begin{bmatrix} x_1 \\ x_2 \end{bmatrix} + \begin{bmatrix} 1 \\ 0 \end{bmatrix} u \tag{4.5a}$$

$$y = [1 \quad -1] \begin{bmatrix} x_1 \\ x_2 \end{bmatrix} \tag{4.5b}$$

The state diagram for this state model is shown in Fig. 4.4b.

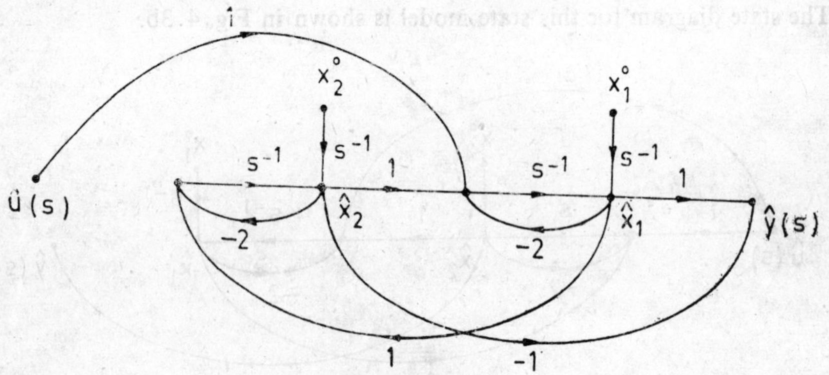

Fig. 4.4b

Example 4.3: In the Ward Leonard system of Fig. 4.5, a prime mover drives a d.c. generator. The output voltage of the generator is controlled by varying the generator field current. The output of the generator is

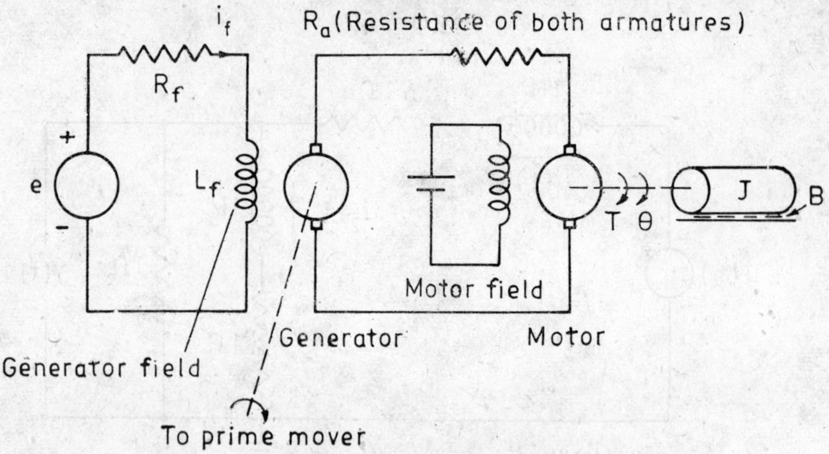

Fig. 4.5 Ward Leonard system

connected to the armature of a d.c. motor with a constant field, thereby controlling the motion of the motor and the associated load.

Consider first the generator field circuit with resistance R_f and inductance L_f. The field is saparately excited by a source of voltage $e(t)$. For the field circuit,

$$L_f \frac{di_f}{dt} + R_f i_f = e$$

where i_f is the field current.

If we assume the prime mover to be a source of constant speed, and further assume the field to be operating in the linear magnetization range, then the generator emf is given by

$$e_g = K_g i_f$$

where K_g is the generator gain constant.

Consider now the motor action. The torque T developed by the motor is proportional to the product of the armature current and air-gap flux. The air-gap flux being constant by virtue of the constant field current, the motor torque is given by

$$T = K_T i_a$$

where K_T is motor torque constant.

The motor back emf is proportional to its speed and is therefore given by

$$e_b = K_b \frac{d\theta}{dt}$$

where K_b = motor back emf constant. The back emf has a polarity opposite to its cause i_a

The differential equation for the armature circuit is

$$R_a i_a + e_b = e_g$$

where armature inductance has been neglected.
The torque equation is

$$J\frac{d^2\theta}{dt^2} + B\frac{d\theta}{dt} = T = K_T i_a$$

Figure 4.6 gives the block diagram for the electromechanical system of Fig. 4.5. Taking outputs of integrating blocks as state variables we have

$$x_1 = \theta(t)$$

$$x_2 = \dot{\theta}(t)$$

$$x_3 = i_f(t)$$

The input variable $u = e(t)$.

The state equations for this selection of state variables are

$$\dot{x}_1 = x_2$$

$$\dot{x}_2 = -\left(\frac{B}{J} - \frac{K_T K_b}{J R_a}\right) x_2 + \frac{K_T K_g}{J R_a} x_3$$

$$\dot{x}_3 = -\frac{R_f}{L_f} x_3 + \frac{1}{L_f} u$$

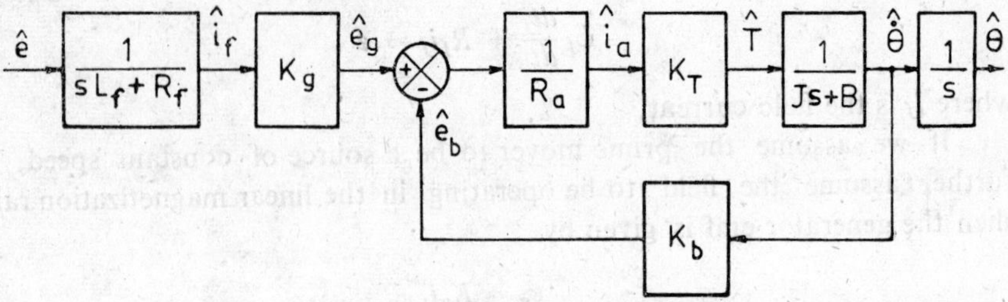

Fig. 4.6 Block diagram for system of Fig. 4.5

In the vector-matrix notation, the state equation is

$$\dot{\mathbf{x}} = \begin{bmatrix} 0 & 1 & 0 \\ 0 & -\left(\frac{B}{J} + \frac{K_T K_b}{J R_a}\right) & \frac{K_T K_g}{J R_a} \\ 0 & 0 & -\frac{R_f}{L_f} \end{bmatrix} \mathbf{x} + \begin{bmatrix} 0 \\ 0 \\ \frac{1}{L_f} \end{bmatrix} u \qquad (4.6)$$

(Refer Nagrath and Gopal (1982a) for further details of electromechanical systems).

Example 4.4: Consider the liquid-level system shown in Fig. 4.7. In this

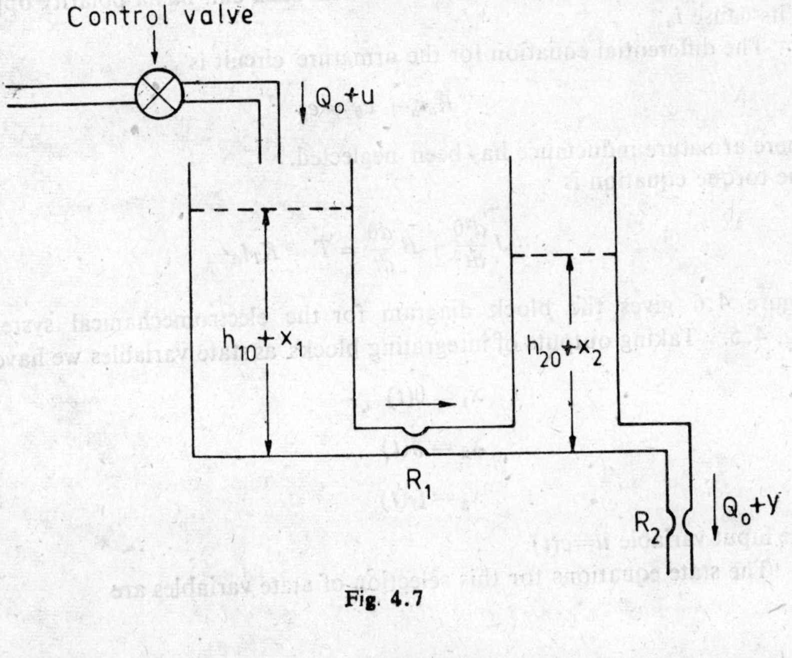

Fig. 4.7

system, a tank having cross-sectional area A_1 is supplying liquid through a pipe of resistance R_1 to another tank of a cross-sectional area A_2, which delivers the liquid through a pipe of resistance R_2. The steady-state flow rate is Q_0 and steady-state heads are h_{10} and h_{20}, respectively.

There are two energy storing elements in this system. The system is thus a second-order system for which the two state variables may be chosen as the heads of liquid level in tanks 1 and 2, respectively.

Let u be a small deviation of the inflow rate from steady-state value. This results in

x_1 = small deviation of the head of tank 1 from its steady-state value,

x_2 = small deviation of the head of tank 2 from its steady-state value,

y_1 = small deviation of the outflow rate of tank 1 from its steady-state value

$= (x_1 - x_2)/R_1$

y = small deviation of the outflow rate of tank 2 from its steady-state value

$= x_2/R_2$

The flow balance equation for tank 1 is (Nagrath and Gopal 1982b)

Rate of storage in tank 1 = rate of inflow in tank 1 − rate of outflow from tank 1

or

$$A_1 \frac{dx_1}{dt} = u(t) - \frac{x_1 - x_2}{R_1}$$

Similarly, for tank 2 we have

$$A_2 \frac{dx_2}{dt} = \frac{x_1 - x_2}{R_1} - \frac{x_2}{R_2}$$

The state equation for the system is

$$\begin{bmatrix} \dot{x}_1 \\ \dot{x}_2 \end{bmatrix} = \begin{bmatrix} -\dfrac{1}{R_1A_1} & \dfrac{1}{R_1A_1} \\ \dfrac{1}{R_1A_2} & -\dfrac{1}{R_1A_2} - \dfrac{1}{R_2A_2} \end{bmatrix} \begin{bmatrix} x_1 \\ x_2 \end{bmatrix} + \begin{bmatrix} \dfrac{1}{A_1} \\ 0 \end{bmatrix} u \tag{4.7a}$$

If the outflow rate of tank 2 is the desired output, the output equation becomes

$$y = \begin{bmatrix} 0 & \dfrac{1}{R_2} \end{bmatrix} \begin{bmatrix} x_1 \\ x_2 \end{bmatrix} \tag{4.7b}$$

(Note that the range of linear operation is quite wide if the liquid flow is laminar).

Example 4.5: Consider the thermal system shown in Fig. 4.8. The input voltage would be varied to control the temperature of the fluid in the vessel.

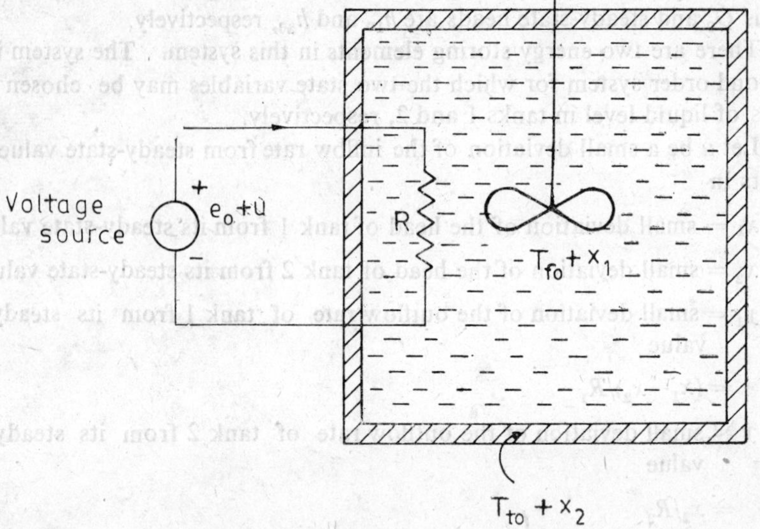

Fig. 4.8

To reduce the complexity of the problem, simplifying assumptions given below are made.

(i) The liquid in the tank is well stirred so that its state can be described by a single temperature.
(ii) The tank is well insulated so that heat loss through its walls is negligible.
(iii) The resistance of the heater does not vary significantly with a rise in temperature.
(iv) The heat storage capacity of the heater element is negligible.

With the assumption of zero heat storage in the heater element, there are two energy storing elements: the fluid in the tank and the tank body. The system can therefore be described by a second-order state model. The two state variables may be chosen as the deviations of fluid temperature and tank-body temperature from their steady-state values.

The heat balance equation for the fluid is (Nagrath and Gopal (1982b))

$$C_f \frac{d}{dt}(T_{f0}+x_1) = \frac{(e_0+u)^2}{RJ} - \frac{(T_{f0}+x_1)-(T_{t0}+x_2)}{R_{CV}+R_{CD}} \tag{4.8}$$

where

C_f = thermal capacitance of fluid in the tank

T_{f0} = steady-state temperature of fluid in the tank

x_1 = deviation of fluid temperature from its steady-state value

e_0 = input voltage corresponding to steady-state of the system

u = deviation in input voltage

R = resistance of heater element

J = Joules constant

T_{t0} = steady-state temperature of tank body

x_2 = deviation of tank-body temperature from its steady-state value

R_{CV} = convective thermal resistance (for convective heat transfer at liquid-solid interface between fluid and tank wall)

R_{CD} = conductive thermal resistance (conduction through tank wall to tank body)

From eqn. (4.8), we obtain the following describing equation in terms of incremental values around the operating point:

$$C_f \frac{dx_1}{dt} = \frac{2e_0}{RJ} u - \frac{x_1 - x_2}{R_{CV} + R_{CD}}$$

If has been assumed that $u^2 \simeq 0$ for all incremental values. Similarly, we obtain the following heat balance equation for the tank body:

$$C_t \frac{dx_2}{dt} = \frac{x_1 - x_2}{R_{CV} + R_{CD}}$$

where C_t = thermal capacitance of tank body.

The state equation in vector-matrix notation is

$$\begin{bmatrix} \dot{x}_1 \\ \dot{x}_2 \end{bmatrix} = \begin{bmatrix} \dfrac{-1}{C_f(R_{CV} + R_{CD})} & \dfrac{1}{C_f(R_{CV} + R_{CD})} \\ \dfrac{1}{C_t(R_{CV} + R_{CD})} & \dfrac{-1}{C_t(R_{CV} + R_{CD})} \end{bmatrix} \begin{bmatrix} x_1 \\ x_2 \end{bmatrix} + \begin{bmatrix} \dfrac{2e_0}{C_f RJ} \\ 0 \end{bmatrix} u \tag{4.9}$$

Example 4.6: Consider a continuously stirred tank reactor (CSTR) shown in Fig. 4.9. A stream of a chemically active species A flows into the tank at rate Q(volume/time). The concentration of species A in the feed stream is C_f (moles/vol), the density of the feed stream is ρ_f (mass/vol) and the temperature is T_f(deg). The volumetric hold up of the liquid in the tank is V(vol), its density is ρ and temperature T. Let it be assumed that stirring causes perfect mixing so that the density, composition and temperature of the liquid in the tank are uniform throughout.

Assume that irreversible reaction taking place in the tank is

$$A \rightarrow B$$

i.e., one mole of A decomposes to form one mole of B. In terms of reaction rate, let

r_A = rate in moles/vol-time at which A decomposes
$\quad = rC$ (first-order reaction)

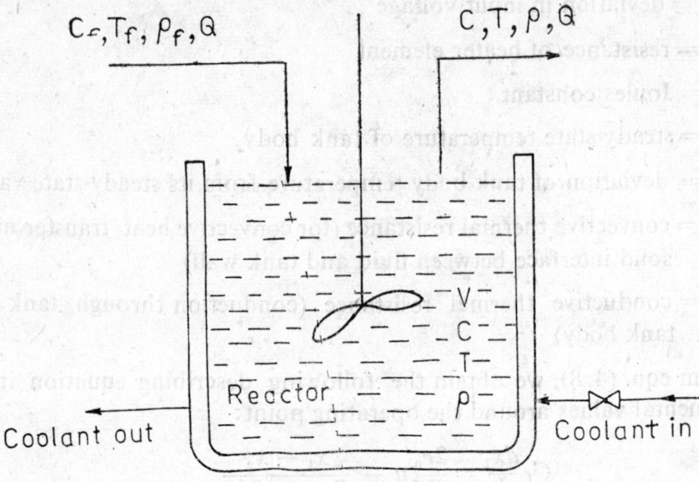

Fig. 4.9 CSTR system

where

$r =$ reaction constant (time^{-1})

$C =$ concentration of A inside the tank.

The reactant mass balance is given by (Nagrath and Gopal 1982b):

Time rate of change of moles of A inside the system

= rate of flow of moles of A into the system

−rate of flow of moles of A out of the system (4.10)

If the inflow rate $Q =$ outflow rate, the volume hold-up V of the tank is constant. Further, we assume that the density remains constant ($\rho = \rho_f$). Then the mass balance equation (4.10) gives

$$V\frac{dC}{dt} = QC_f - QC - rVC \tag{4.11}$$

Let us now write the energy balance equation. Assume that the reaction in the tank is exothermic and ΔH (energy/mole) = heat liberated by decomposition of one mole of A.

To remove the heat of reaction, a cooling jacket surrounds the reactor as shown in Fig. 4.9. Let q(assumed constant) be the heat removal rate.

The energy balance equation for the reactor is

$$V\rho c_p \frac{dT}{dt} = Q\rho c_p T_f - Q\rho c_p T + (\Delta H)rVC - q \tag{4.12}$$

where

c_p (energy/mass-deg) = specific heat of the liquid (assumed constant).

Taking C and $(T-T_f)$; $T_f =$ constant, as the state variables x_1 and x_2 respectively, we have from eqns. (4.11) and (4.12),

$$\dot{\mathbf{x}} = \begin{bmatrix} -\left(r+\dfrac{Q}{V}\right) & 0 \\ \dfrac{(\Delta H)r}{\rho c_p} & -\dfrac{Q}{V} \end{bmatrix} \mathbf{x} + \begin{bmatrix} \dfrac{Q}{V} & 0 \\ 0 & -\dfrac{1}{V\rho c_p} \end{bmatrix} \begin{bmatrix} u_1 \\ u_2 \end{bmatrix} \quad (4.13)$$

where

$$u_1 = C_f$$

$$u_2 = q$$

4.3 LINEAR DISCRETE-TIME MODELS

In this section, we study the problem of state assignment in linear discrete-time systems. The general form of the state model for linear time-invariant discrete-time systems is (eqns. (3.29))

$$\mathbf{x}(k+1) = \mathbf{F}\mathbf{x}(k) + \mathbf{G}\mathbf{u}(k) \quad (4.14a)$$

$$\mathbf{y}(k) \quad = \mathbf{C}\mathbf{x}(k) + \mathbf{D}\mathbf{u}(k) \quad (4.14b)$$

where **F**, **G**, **C** and **D** are, respectively $n \times n$, $n \times p$, $q \times n$ and $q \times p$ real constant matrices, **u** is the $p \times 1$ input vector, **y** is the $q \times 1$ output vector and **x** is the $n \times 1$ state vector.

The general form of the state model for linear time-dependent discrete-time systems is (eqns. (3.28))

$$\mathbf{x}(k+1) = \mathbf{F}(k)\,\mathbf{x}(k) + \mathbf{G}(k)\,\mathbf{u}(k) \quad (4.15a)$$

$$\mathbf{y}(k) \quad = \mathbf{C}(k)\,\mathbf{x}(k) + \mathbf{D}(k)\,\mathbf{u}(k) \quad (4.15b)$$

Here the coefficients of matrices **F**, **G**, **C** and **D** are functions of time k.

A unique solution for state equations (4.14a) and (4.15a) exists for any finite discrete input sequence $\mathbf{u}(k)$.

Discrete-time dynamical systems can be classified into two types:

1. Inherently discrete-time systems, such as digital computers, digital filters, monetary and inventory systems. In such systems, it makes sense to consider the system at discrete instants of time only.
2. Discrete-time systems that result from considering continuous-time systems at discrete instants of time. This may be done for various reasons, e.g., for analyzing a continuous-time system on a digital computer or controlling continuous-time systems using digital controllers. Such systems have been referred to as sampled-data control systems in Chapter 1. We shall study such systems in Section 5.7.

Discrete state variable formulation of the form (4.14) and (4.15) has also been used to describe steady-state behaviour of stage-by-stage operation in chemical processes. Let us discuss an example.

Example 4.7: Consider the sequence of N perfectly stirred chemical reactors shown in Fig. 4.10. In each reactor, the reactant A decomposes irreversibly with first-order kinetics and the reaction liberates heat. The steady-state mass and energy balance on a typical reactor, say reactor k, are (refer eqns. (4.11) and (4.12)):

$$\begin{aligned} &Q[C(k+1) - C(k)] - rV(k)\, C(k) = 0 \\ &Q\rho c_p[T(k+1) - T(k)] + (\Delta H)\, rV(k)\, C(k) - q(k) = 0 \end{aligned} \tag{4.16}$$

where

$C(k)$ = concentration of A in k-th reactor

$V(k)$ = hold up volume in k-th reactor

$T(k)$ = temperature of the liquid in k-th reactor

$q(k)$ = heat removal rate from k-th reactor

$k = 1, 2, \ldots, N.$

(Time delay associated with transport of liquid between two tanks has been neglected).

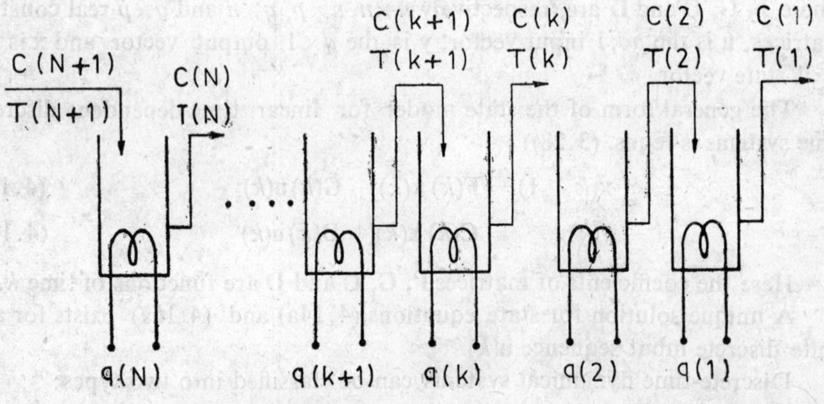

Fig. 4.10 Reactor train for Example 4.7

In an N-stage process, $C(N+1) = C_f$, $T(N+1) = T_f$. Note that the time variable k in the usual sense has been identified with the number k of the reactor stage.

Define the state variables

$$x_1(k) = C(k)$$

$$x_2(k) = T(k)$$

In terms of these state variables, we have from eqns. (4.16)

$$\begin{bmatrix} x_1(k+1) \\ x_2(k+1) \end{bmatrix} = \begin{bmatrix} 1 + \dfrac{rV(k)}{Q} & 0 \\ -\dfrac{(\Delta H)\, rV(k)}{Q\rho c_p} & 1 \end{bmatrix} \begin{bmatrix} x_1(k) \\ x_2(k) \end{bmatrix} + \begin{bmatrix} 0 \\ \dfrac{1}{Q\rho c_p} \end{bmatrix} q(k) \tag{4.17}$$

State equation (4.17) is of the form (4.15). For the special case in which all reactors have identical hold-up, $V(k) = V$ and eqn. (4.17) becomes

$$\mathbf{x}(k+1)\begin{bmatrix} 1+\dfrac{rV}{Q} & 0 \\ -\dfrac{(\Delta H)}{Q\rho c_p} & 1 \end{bmatrix}\mathbf{x}(k) + \begin{bmatrix} 0 \\ \dfrac{1}{Q\rho c_p} \end{bmatrix} q(k) \tag{4.18}$$

State equation (4.18) is of the form (4.14).

4.4 NONLINEAR MODELS

In this section, we study the problem of state assignment in nonlinear systems. The general form of the state model for nonlinear continuous-time systems is (eqn. (3.13))

$$\dot{\mathbf{x}}(t) = \mathbf{f}(t, \mathbf{x}, \mathbf{u}) \tag{4.19a}$$

$$\mathbf{y}(t) = \mathbf{g}(t, \mathbf{x}, \mathbf{u}) \tag{4.19b}$$

where $\mathbf{f}(.)$ and $\mathbf{g}(.)$ are, respectively, $n \times 1$ and $q \times 1$ function vectors, $\mathbf{u}$ is the $p \times 1$ input vector, $\mathbf{y}$ is the $q \times 1$ output vector and $\mathbf{x}$ is $n \times 1$ state vector. As pointed out earlier, a unique solution of state equation (4.19a) for any $\mathbf{u} \in$ admissible class $\mathbf{\Omega}$ of input functions exists if the function $\mathbf{f}$ is continuous (piecewise continuous) and has continuous (piecewise continuous) derivative with respect to $\mathbf{x}$.

The general form of the state model for nonlinear discrete-time systems is (eqns. (3.12))

$$\mathbf{x}(k+1) = \mathbf{f}(k, \mathbf{x}, \mathbf{u}) \tag{4.20a}$$

$$\mathbf{y}(k) = \mathbf{g}(k, \mathbf{x}, \mathbf{u}) \tag{4.20b}$$

A unique solution of state equation (4.20a) exists for any finite discrete sequence $\mathbf{u}(k)$.

Almost all physical systems are nonlinear to some extent. However, as we have seen in earlier sections, many systems can be approximated with sufficient accuracy by linear models.

There are situations wherein a linearized model displaying all the important characteristics of the system is not possible. In some systems, we deliberately introduce nonlinear controllers to fulfil certain specific objectives. In the following, we give an example of nonlinear state models.

Example 4.8: Consider the CSTR system of Fig. 4.9 (Example 4.6). If the reaction $A \rightarrow B$ is of second-order, we have

$$r_A = RC^2$$

where R = reaction constant.

The mass balance and energy balance equations become (refer eqns. (4.11) and (4.12))

$$V\frac{dC}{dt} = Q(C_f - C) - RVC^2$$

$$V\rho c_p \frac{dT}{dt} = Q\rho c_p(T_f - T) + (\Delta H)\,RVC^2 - q \tag{4.21}$$

Define

$$x_1 = C$$

$$x_2 = (T - T_f) \text{ for a constant } T_f$$

$$u_1 = C_f$$

$$u_2 = q$$

In terms of these variables, eqns. (4.21) are

$$\dot{x}_1 = -\frac{Q}{V}x_1 - Rx_1^2 + \frac{Q}{V}u_1 = f_1(x_1, u_1)$$

$$\dot{x}_2 = \frac{Q}{V}x_2 + \frac{(\Delta H)\,R}{\rho c_p}x_1^2 - \frac{1}{V\rho c_p}u_2 = f_2(x_1, x_2, u_2) \tag{4.22}$$

The functions f_1 and f_2 are continuous in **x** and have continuous derivatives with respect to **x**. Therefore x_1 and x_2 are the state variables of the system and (4.22) is the state model.

4.5 LOCAL LINEARIZATION OF NONLINEAR MODELS

In Sections 4.2 and 4.3, we discussed examples of systems which are accurately represented by linear differential/difference equations over wide ranges of system variables. In most of these examples, the equations describing the systems are, in fact, quite nonlinear. Linear relationships could be obtained only by restricting the system variables to sufficiently small deviations about a normal 'operating point' or equilibrium value. If such assumptions cannot validly be made, one has no choice but to deal with the nonlinear equations directly.

However, the use of linear models which are accurate only for relatively small ranges of variables has a great deal of practical value. One reason is that many automatic control systems are in fact designed to maintain themselves as close to a desired equilibrium as possible. Another reason is that even in highly nonlinear systems small perturbations from the known solution of the nonlinear equations of the system usually can be described by a linear model.

Consider a nonlinear system described by the state and output equations of the form (4.19). We assume that all the partial derivatives

$$\frac{\partial f_i}{\partial x_i}, \frac{\partial f_i}{\partial u_k}; \; i, j = 1, 2, \ldots, n$$

$$k = 1, 2, \ldots, p$$

$$\frac{\partial g_l}{\partial x_i}, \frac{\partial g_l}{\partial u_k}; \; l = 1, 2, \ldots, q$$

exist and are continuous for all **x** and **u**. Now suppose that the time functions **x***, **u*** and **y*** constitute a nominal solution of nonlinear equations (4.19). We refer to **u*** as the nominal input and **x*** and **y*** as the nominal state and output trajectories respectively. These variables will satisfy eqns. (4.19):

$$\left.\begin{aligned} \dot{\mathbf{x}}^*(t) &= \mathbf{f}(t, \mathbf{x}^*, \mathbf{u}^*) \\ \mathbf{y}^*(t) &= \mathbf{g}(t, \mathbf{x}^*, \mathbf{u}^*) \end{aligned}\right\} ; \; t_0 \leqslant t \leqslant t_1 \tag{4.23}$$

We now assume that the system is operated close to nominal conditions, i.e., **u**, and **x** deviate only slightly from **u*** and **x*** respectively. We may therefore write

$$\mathbf{u}(t) = \mathbf{u}^*(t) + \tilde{\mathbf{u}}(t); \; t_0 \leqslant t \leqslant t_1$$

$$\mathbf{x}(t_0) = \mathbf{x}^*(t_0) + \tilde{\mathbf{x}}(t_0)$$

where $\tilde{\mathbf{u}}(t)$ and $\tilde{\mathbf{x}}(t_0)$ are small perturbations. Correspondingly,

$$\left.\begin{aligned} \mathbf{x}(t) &= \mathbf{x}^*(t) + \tilde{\mathbf{x}}(t) \\ \mathbf{y}(t) &= \mathbf{y}^*(t) + \tilde{\mathbf{y}}(t) \end{aligned}\right\} ; \; t_0 \leqslant t \leqslant t_1$$

Substituting **x** and **u** into the state equation (4.19a) and making a Taylor series expansion, we get the following linear approximation (See Appendix III).

$$\dot{\mathbf{x}}^*(t) + \dot{\tilde{\mathbf{x}}}(t) = \mathbf{f}(t, \mathbf{x}^*, \mathbf{u}^*) + \left.\frac{\partial \mathbf{f}}{\partial \mathbf{x}}\right|_* \tilde{\mathbf{x}}(t) + \left.\frac{\partial \mathbf{f}}{\partial \mathbf{u}}\right|_* \tilde{\mathbf{u}}(t); \; t_0 \leqslant t \leqslant t_1 \tag{4.24}$$

where

$$\frac{\partial \mathbf{f}}{\partial \mathbf{x}} \triangleq \begin{bmatrix} \frac{\partial f_1}{\partial x_1} & \frac{\partial f_1}{\partial x_2} & \cdots & \frac{\partial f_1}{\partial x_n} \\ \frac{\partial f_2}{\partial x_1} & \frac{\partial f_2}{\partial x_2} & \cdots & \frac{\partial f_2}{\partial x_n} \\ \vdots & \vdots & & \vdots \\ \frac{\partial f_n}{\partial x_1} & \frac{\partial f_n}{\partial x_2} & \cdots & \frac{\partial f_n}{\partial x_n} \end{bmatrix} = \text{Jacobian matrix of } \mathbf{f} \text{ with respect to } \mathbf{x}$$

and

$$\frac{\partial \mathbf{f}}{\partial \mathbf{u}} \triangleq \begin{bmatrix} \frac{\partial f_1}{\partial u_1} & \cdots & \frac{\partial f_1}{\partial u_p} \\ \vdots & & \vdots \\ \frac{\partial f_n}{\partial u_1} & \cdots & \frac{\partial f_n}{\partial u_p} \end{bmatrix} = \text{Jacobian matrix of } \mathbf{f} \text{ with respect to } \mathbf{u}$$

Therefore, the locally linearized state equation is

$$\dot{\tilde{\mathbf{x}}}(t) = \mathbf{A}(t)\,\tilde{\mathbf{x}}(t) + \mathbf{B}(t)\,\tilde{\mathbf{u}}(t);\ t_0 \leqslant t \leqslant t_1 \tag{4.25a}$$

where

$$\mathbf{A}(t) = \frac{\partial \mathbf{f}}{\partial \mathbf{x}}\bigg|_* \text{ and } \mathbf{B}(t) = \frac{\partial \mathbf{f}}{\partial \mathbf{u}}\bigg|_*$$

Similarly, locally linearized output equation is given by

$$\tilde{\mathbf{y}}(t) = \mathbf{C}(t)\,\tilde{\mathbf{x}}(t) + \mathbf{D}(t)\,\tilde{\mathbf{u}}(t);\ t_0 \leqslant t \leqslant t_1 \tag{4.25b}$$

where

$$\mathbf{C}(t) = \frac{\partial \mathbf{g}}{\partial \mathbf{x}}\bigg|_* \text{ and } \mathbf{D}(t) = \frac{\partial \mathbf{g}}{\partial \mathbf{u}}\bigg|_*$$

It is important to bear in mind the fundamental assumption here, that the perturbation remains sufficiently small to validate the linear analysis. The question of how small this should be is difficult to answer with any generality, and we usually verify the validity of the assumption in various specific cases by quantitative analysis (solution of the two models).

Locally linearized state and output equations from a discrete-time nonlinear state model can be written down in an analogous way.

Examples of local linearization are given in the next section.

4.6 PLANT MODELS OF SOME ILLUSTRATIVE CONTROL SYSTEMS

Many existing textbooks on control systems have been criticized for not including adequate practical control problems. A reason for this seems to be that most real-world problems are highly complex and therefore are to be simplified, and in the process much of the realism may be lost.

In this book, we attempt to illustrate the control theory with the help of some practical control problems from various engineering disciplines. Simplification of the problems could not be avoided but due care has been taken to see that essence of real-world problem remains in its simplified version. Plant models of illustrative control systems are given in this section.

Position Servo

In a radar system, an electromagnetic pulse is radiated from an antenna into space. An echo pulse is received back if a conducting surface such as an airplane appears in the path of the signal. When the radar is in search of a target, the antenna is continuously rotated. When the target is located, the antenna is stopped and pointed towards the target by varying its angular direction until a maximum echo is heard.

The position coordinate $\theta(t)$ of the antenna can be controlled via a gear train by a d.c. servomotor. The torque of the motor can be varied in both magnitude and direction by means of a control voltage obtained from the output of an amplifier. The control problem is to command the motor such that

$$\theta(t) \simeq \theta_r(t), \text{ a reference angular position.}$$

A possible method of forcing the antenna to point toward the reference position is shown in Fig. 4.11. $\theta(t)$ and $\theta_r(t)$ are converted into electrical signals, the difference is amplified which serves as the input voltage to the

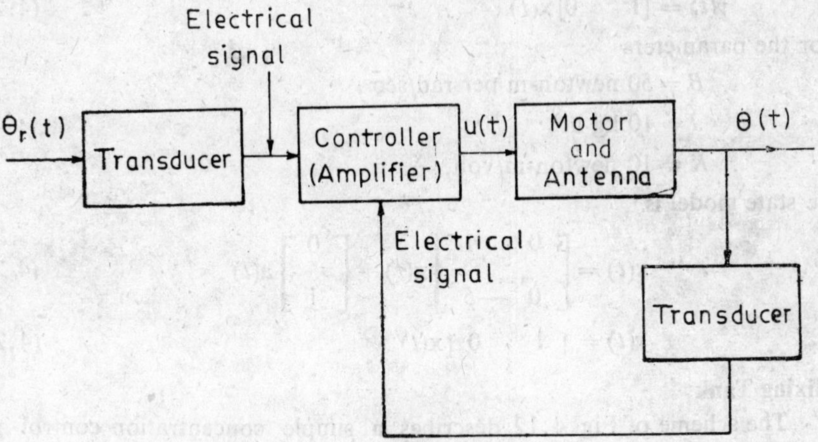

Fig. 4.11 A position servo system

motor. As a result, when $\theta_r(t) - \theta(t)$ is positive, positive input voltage is applied to the motor which makes the antenna rotate in a direction so that the difference between $\theta_r(t)$ and $\theta(t)$ is reduced. A negative input voltage makes the antenna rotate in the opposite direction. This scheme obviously represents a closed-loop controller.

Antenna and motor constitute the plant of the system. The motion of the antenna (in absence of disturbance) can be described by the differential equation

$$J\ddot{\theta}(t) + B\dot{\theta}(t) = T(t)$$

where

J = moment of inertia of motor and load

B = coefficient of viscous friction

$T(t)$ = torque developed by the motor.

The torque developed by the motor is assumed to be proportional to $u(t)$, the input voltage to the motor, so that

$$T(t) = Ku(t)$$

Defining the state variables as

$$x_1(t) = \theta(t)$$

$$x_2(t) = \dot{\theta}(t),$$

the state model of the system becomes

$$\dot{\mathbf{x}}(t) = \begin{bmatrix} 0 & 1 \\ 0 & -\dfrac{B}{J} \end{bmatrix} \mathbf{x}(t) + \begin{bmatrix} 0 \\ \dfrac{K}{J} \end{bmatrix} u(t) \tag{4.26a}$$

$$y(t) = [1 \quad 0]\,\mathbf{x}(t) \tag{4.26b}$$

For the parameters

$B = 50$ newton-m per rad/sec

$J = 10$ kg-m^2

$K = 10$ newton-m/volt,

the state model is

$$\dot{\mathbf{x}}(t) = \begin{bmatrix} 0 & 1 \\ 0 & -5 \end{bmatrix} \mathbf{x}(t) + \begin{bmatrix} 0 \\ 1 \end{bmatrix} u(t) \tag{4.27a}$$

$$y(t) = [\,1 \quad 0\,]\,\mathbf{x}(t) \tag{4.27b}$$

Mixing Tank

The scheme of Fig. 4.12 describes a simple concentration control process. Two concentrated solutions of some chemical with constant concentrations C_1 and C_2 are fed with flow rates Q_1 rnd Q_2 respectively and are continuously mixed in the tank. The outflow from the mixing tank is at a rate $Q(t)$ with concentration $C(t)$. Let it be assumed that stirring causes perfect mixing so that the concentration of the solution in the tank is uniform throughout and equals that of the outflow. We shall also assume that the density remains constant.

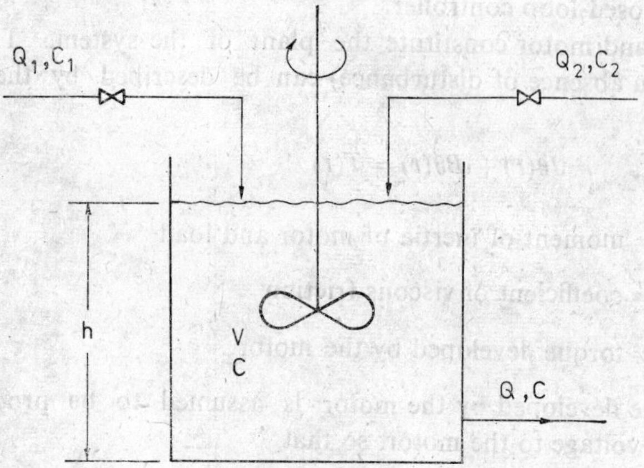

Fig. 4.12 Process in a mixing tank

Let $V(t)$ be the volume of the fluid in the tank. The mass balance equations are

$$\frac{d}{dt} V(t) = Q_1(t) + Q_2(t) - Q(t) \tag{4.28}$$

$$\frac{d}{dt}[C(t)\,V(t)] = C_1 Q_1(t) + C_2 Q_2(t) - C(t) Q(t) \tag{4.29}$$

The flow $Q(t)$ is characterized by the turbulent flow relation

$$Q(t) = k\sqrt{h(t)} = k\sqrt{\frac{V(t)}{A}} \tag{4.30}$$

where $h(t)$ is the head of the liquid in the tank, A is the cross-sectional area of the tank and k is a constant. Assume that under steady-state

$$Q_1(t) = Q_{10},\ Q_2(t) = Q_{20},\ Q(t) = Q_0,\ V(t) = V_0 \text{ and } C(t) = C_0$$

The steady-state operation is described by the equations (from eqns. (4.28) (4.30))

$$0 = Q_{10} + Q_{20} - Q_0 \tag{4.31}$$

$$0 = C_1Q_{10} + C_2Q_{20} - C_0Q_0 \tag{4.32}$$

$$Q_0 = k\sqrt{\frac{V_0}{A}} \tag{4.33}$$

For small perturbations about the steady-state, eqn. (4.30) can be linearized using the technique discussed in Section 4.5. The linear approximation of the Taylor series expansion of eqn. (4.30) about the operating point (Q_0, V_0) gives

$$Q(t) - Q_0 = \frac{k}{\sqrt{A}} \left.\frac{\partial\sqrt{V(t)}}{\partial V(t)}\right|_{V=V_0} (V(t) - V_0)$$

or

$$\tilde{Q}(t) = \frac{k}{2V_0}\sqrt{\frac{V_0}{A}}\,\tilde{V}(t) \tag{4.34}$$

where

$$\tilde{Q}(t) = Q(t) - Q_0$$

$$\tilde{V}(t) = V(t) - V_0$$

Let

$$Q_1(t) = Q_{10} + \tilde{Q}_1(t)$$

$$Q_2(t) = Q_{20} + \tilde{Q}_2(t)$$

$$C(t) = C_0 + \tilde{C}(t)$$

From eqns. (4.28) to (4.34), we obtain the following equations describing perturbations about steady-state:

$$\dot{\tilde{V}}(t) = \tilde{Q}_1(t) + \tilde{Q}_2(t) - \frac{1}{2}\frac{Q_0}{V_0}\tilde{V}(t) \tag{4.35}$$

$$\dot{\tilde{C}}(t)V_0 + C_0\dot{\tilde{V}}(t) = C_1\tilde{Q}_1(t) + C_2\tilde{Q}_2(t) - \frac{1}{2}\frac{C_0Q_0}{V_0}\tilde{V}(t) - Q_0\tilde{C}(t) \tag{4.36}$$

The hold up time of the tank is

$$\tau = \frac{V_0}{Q_0}$$

$$x_1(t)=\tilde{V}(t),\ x_2(t)=\tilde{C}(t),\ u_1(t)=\tilde{Q}_1(t),\ u_2(t)=\tilde{Q}_2(t)$$

$$y_1(t)=\tilde{Q}(t) \text{ and } y_2(t)=\tilde{C}(t)$$

In terms of these variables, we get the following state model from eqns. (4.35)—(4.36).

$$\dot{\mathbf{x}}(t)=\begin{bmatrix} -\dfrac{1}{2\tau} & 0 \\ 0 & -\dfrac{1}{\tau} \end{bmatrix}\mathbf{x}(t)+\begin{bmatrix} 1 & 1 \\ \dfrac{C_1-C_0}{V_0} & \dfrac{C_2-C_0}{V_0} \end{bmatrix}\mathbf{u}(t) \tag{4.37a}$$

$$\mathbf{y}(t)=\begin{bmatrix} \dfrac{1}{2\tau} & 0 \\ 0 & 1 \end{bmatrix}\mathbf{x}(t) \tag{4.37b}$$

For the parameters

$$Q_{10}=10 \text{ litres/sec}$$
$$Q_{20}=20 \text{ litres/sec}$$
$$C_1=9 \text{ g-moles/litre}$$
$$C_2=18 \text{ g-moles/litre}$$
$$V_0=1500 \text{ litres},$$

the state model becomes

$$\dot{\mathbf{x}}(t)=\begin{bmatrix} -0.01 & 0 \\ 0 & -0.02 \end{bmatrix}\mathbf{x}(t)+\begin{bmatrix} 1 & 1 \\ -0.004 & 0.002 \end{bmatrix}\mathbf{u}(t) \tag{4.38a}$$

$$\mathbf{y}(t)=\begin{bmatrix} 0.01 & 0 \\ 0 & 1 \end{bmatrix}\mathbf{x}(t) \tag{4.38b}$$

Inverted Pendulum

Consider the inverted pendulum shown in Fig. 4.13. The pivot of the pendulum is mounted on a carriage which can move in a horizontal direction. The carriage is driven by a small motor that at time t exerts a force $u(t)$ on the carriage. This force is the input variable for the system. This somewhat artificial system example represents a dynamic model of a space booster on take off—the booster is balanced on top of the rocket engine thrust vector (Refer Cannon (1967) and Elgerd (1967) for this system example).

The pendulum is obviously unstable as it cannot remain in the desired position or state without the assistance of the control force. Through this example, we shall be able to emphasize in later chapters that even unstable

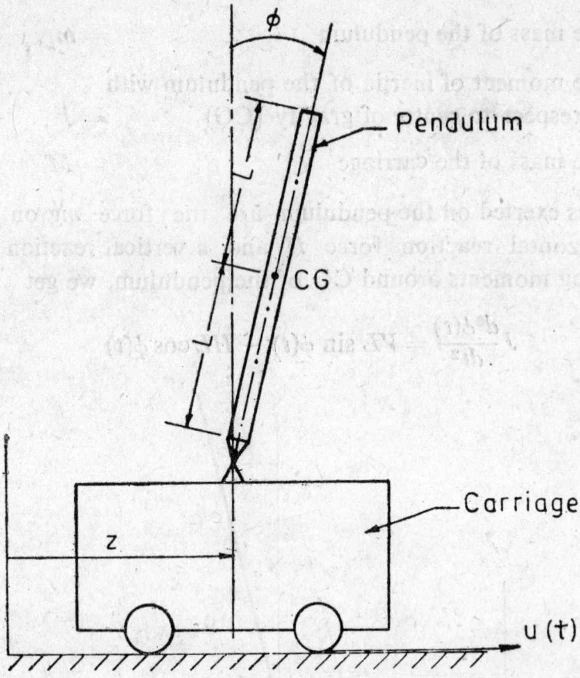

Fig. 4.13

systems can be properly controlled. Here in this chapter, our aim is to construct a simple mathematical model of the system.

The centres of gravity of each body (the carriage and the pendulum) have the following space coordinates in relation to an arbitrarily chosen fixed origin:

Carriage: Horizontal position $= z$

Pendulum: Horizontal position $= z + L \sin \phi$

Vertical position $= L \cos \phi$

The dynamic behaviour of the system is completely described in terms of position and velocity of the carriage and angular position and angular velocity of the pendulum; so we may define the state vector to be

$$\mathbf{x}(t) = \begin{bmatrix} z(t) \\ \dot{z}(t) \\ \phi(t) \\ \dot{\phi}(t) \end{bmatrix}$$

Let

the mass of the pendulum $= m$,

the moment of inertia of the pendulum with respect to centre of gravity (CG) $= J$,

the mass of the carriage $= M$

The forces exerted on the pendulum are the force mg on the centre of gravity, a horizontal reaction force H and a vertical reaction force V (Fig. 4.14a). Taking moments around CG of the pendulum, we get

$$J\frac{d^2\phi(t)}{dt^2} = VL\sin\phi(t) - HL\cos\phi(t) \tag{4.39}$$

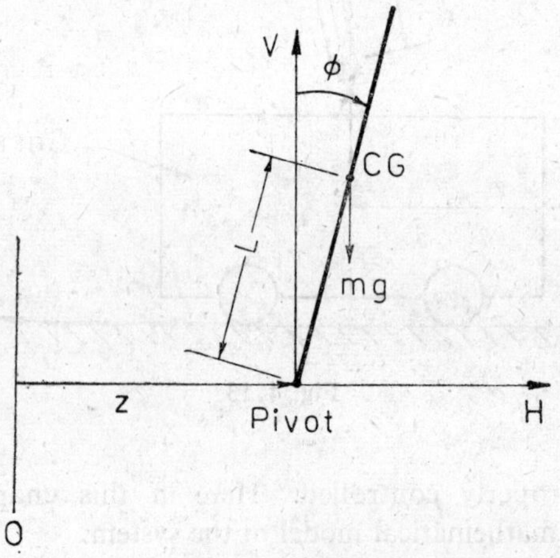

Fig. 4.14a

Summing up all forces on the pendulum in vertical and horizontal directions, we obtain

$$V - mg = m\frac{d^2}{dt^2}(L\cos\phi(t)) \tag{4.40}$$

$$H = m\frac{d^2}{dt^2}(z(t) + L\sin\phi(t)) \tag{4.41}$$

Summing up all forces on the carriage in the horizontal direction (Fig. 4.14b), we get

$$u - H = M\frac{d^2z(t)}{dt^2} \tag{4.42}$$

In our problem, since the object is to keep the pendulum upright, it seems reasonable to assume that ϕ and $\dot{\phi}$ will remain close to zero. In view of this, we can set with sufficient accuracy $\sin\phi \simeq \phi$; $\cos\phi \simeq 1$. With this approximation, we get from eqns. (4.39)–(4.42),

$$mL\ddot{\phi} + (m+M)\ddot{z} = u \tag{4.43}$$

$$(J + mL^2)\ddot{\phi} + mL\ddot{z} - mgL\phi = 0 \tag{4.44}$$

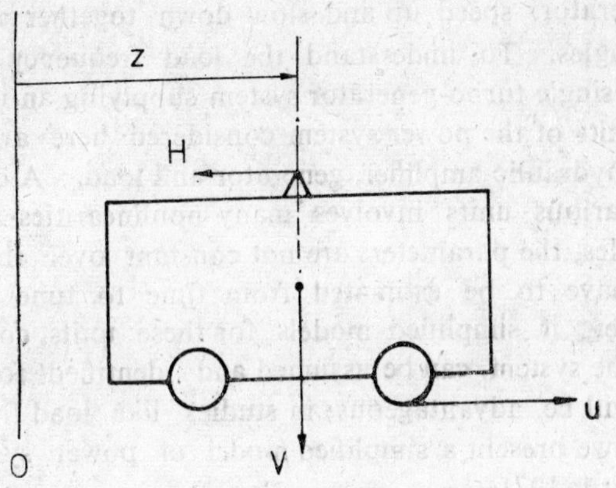

Fig. 4.14b

Suppose that the system parameters are

$$M = 1 \text{ kg}, \quad m = 0.15 \text{ kg}, \quad L = 1 \text{ m}$$

Recall that

$$g = 9.81 \text{ m/sec}^2$$

$$J = \frac{4}{3} mL^2 = 0.2 \text{ kg-m}^2$$

For these parameters, we have from eqns. (4.43)–(4.44),

$$0.15\ddot{\phi} + 1.15\ddot{z} = u$$

$$0.35\ddot{\phi} + 0.15\ddot{z} - 0.15 \times 9.81\phi = 0$$

These equations give the following state model:

$$\dot{\mathbf{x}} = \begin{bmatrix} 0 & 1 & 0 & 0 \\ 0 & 0 & -0.5809 & 0 \\ 0 & 0 & 0 & 1 \\ 0 & 0 & 4.4537 & 0 \end{bmatrix} \mathbf{x} + \begin{bmatrix} 0 \\ 0.9211 \\ 0 \\ -0.3947 \end{bmatrix} u$$

$$= \mathbf{A}\mathbf{x} + \mathbf{b}u \tag{4.45a}$$

$$y = [0 \;\; 0 \;\; 1 \;\; 0]\mathbf{x} = \mathbf{c}\mathbf{x} \tag{4.45b}$$

In the model for inverted pendulum, a large number of zero entries appear in matrix A. In matrix theory, such a matrix is called a *sparse* matrix.

Power System

Let us consider the problem of controlling the power output of the generators of a closely knit electric area so as to maintain the scheduled frequency. All the generators in such an area constitute a *coherent* group so that all the generators speed up and slow down together maintaining their relative power angles. To understand the load frequency control problem, let us consider a single turbo-generator system supplying an isolated load.

The basic units of the power system considered here are boiler, turbine, speed governor, hydraulic amplifier, generator and load. A detailed representation of the various units involves many nonlinearities and several time-constants. Besides, the parameters are not constant over the entire range of operation and have to be estimated from time to time for any realistic control. However, if simplified models for these units, consistent with the performance of the system, can be assumed and identified for each operating condition, they will be advantageous in studies like load frequency control. In the following, we present a simplified model of power system (for further details, refer Elgerd (1971)).

Figure 4.15 shows schematically, the speed governing system of a steam turbine. By controlling the position measured by coordinate x_D of the steam valve, we can exert control over the flow of high-pressure steam through the turbine and thus the turbine power.

Assume that the system is initially operating under steady state—steam valve opened by a definite magnitude and the turbine running at the

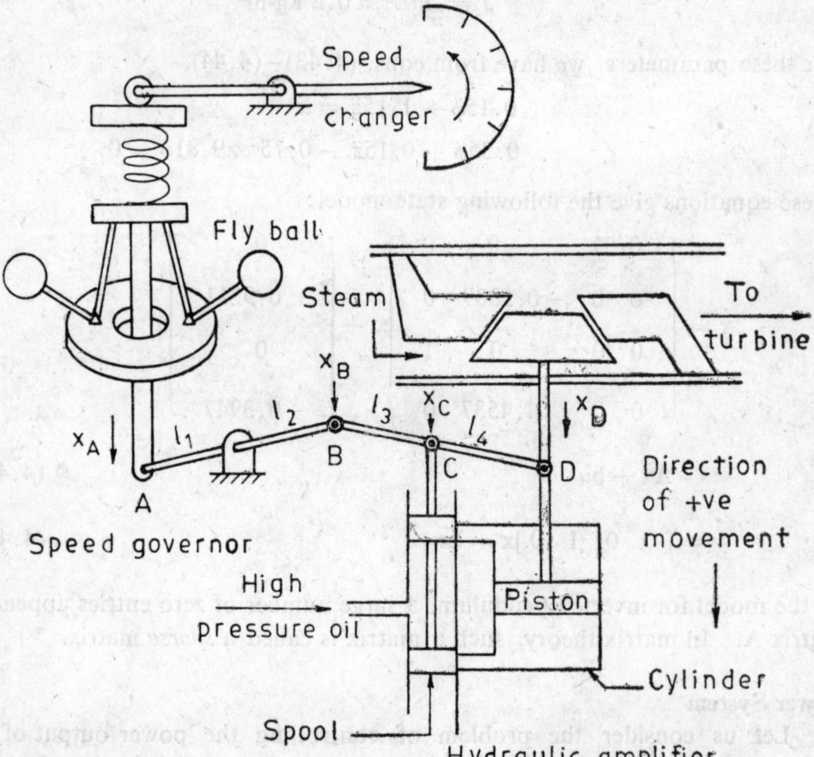

Fig. 4.15

constant speed with the turbine power output balancing the generator load. Let the operating conditions be characterized by

$x_D^* =$ steam valve setting

$f^* =$ system frequency (speed)

$P_G^* =$ generator output = turbine output (neglecting generator loss).

We shall obtain a linear incremental model around these operating conditions. Let the speed changer be moved upwards, moving point A downwards by amount Δx_A. As we shall see shortly, this command causes the turbine output power to increase; therefore we can write

$$\Delta x_A = k_1 \Delta P_C \tag{4.46}$$

where ΔP_C is commanded increase in power and k_1 is a positive constant.

With the movement of point A downwards, the spool moves up admitting high pressure oil into the cylinder. This causes the piston to move down; the steam valve opening consequently increases and the frequency (speed) goes up by Δf. The increase in frequency causes the centrifugal force of the flyballs to increase; thus increasing the force exerted on the bottom of the spring, causing point A to move upwards by $k_2 \Delta f$ where k_2 is a positive constant.

The net movement of point B is

$$\begin{aligned}\Delta x_B &= -\left(\frac{l_2}{l_1}\right) k_1 \Delta P_C + \left(\frac{l_2}{l_1}\right) k_2 \Delta f \\ &= -k_3 \Delta P_C + k_4 \Delta f\end{aligned} \tag{4.47}$$

The net movement of point C is

$$\begin{aligned}\Delta x_C &= \left(\frac{l_4}{l_3 + l_4}\right) \Delta x_B + \left(\frac{l_3}{l_3 + l_4}\right) \Delta x_D \\ &= k_5 \Delta x_B + k_6 \Delta x_D\end{aligned} \tag{4.48}$$

Assuming leakage and compressibility flows to be negligible, the rate q of oil flow into the cylinder is proportional to the rate at which the piston moves, i.e.,

$$q = a \frac{dx_D}{dt}$$

where a is the area of the piston.

Further, if oil flow into the cylinder is proportional to spool displacement, we may write

$$\Delta x_D = k_7 \int (-\Delta x_C) dt \tag{4.49}$$

where k_7 is a positive constant.

Manipulation of eqns. (4.46)–(4.49) gives the block diagram of Fig. 4.16. K_G, the gain of the speed governor; T_G, the time constant of the speed governor and R, the speed regulation coefficient of the governor can be obtained in terms of constants k_i, $i=1$ to 7 in eqns. (4.46)–(4.49).

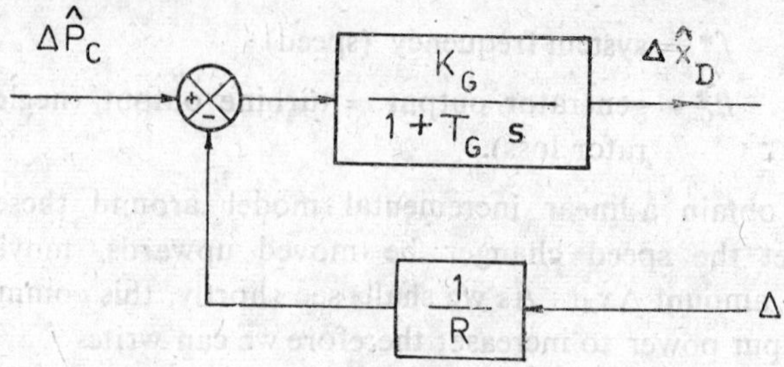

Fig. 4.16

Let us now relate the dynamic response of the steam turbine in terms of changes in power output to changes in the steam valve opening Δx_D. In a simplified representation, the turbine is described by a time-constant T_T (which is required to be identified) and a gain K_T with $K_G K_T = 1$. Figure 4.17 represents a simplified linear model of a turbine controller with a speed governing mechanism.

We now turn to generator-load representation. The increment in power input to generator-load system is $(\Delta P_G - \Delta P_D)$ where $\Delta P_G = \Delta P_T$ is incremental turbine power output (assuming the generator incremental loss to be negligible) and ΔP_D is the load increment. The increment in power input to the system is accounted for in two ways:

(i) By increasing the stored kinetic energy in the generator rotor.
(ii) By an increased load consumption.

The kinetic energy varies as the square of speed or frequency; therefore kinetic energy

$$W = \left(\frac{f^* + \Delta f}{f^*}\right)^2 W^* \simeq \left(1 + 2\frac{\Delta f}{f^*}\right) W^*$$

where W^* is the kinetic energy at scheduled frequeney f.*

The rate of change of kinetic energy is therefore

$$\frac{dW}{dt} = \frac{2W^*}{f^*}\frac{d}{dt}(\Delta f)$$

The rate of change of load with respect to frequency, i.e., $\frac{\partial P_D}{\partial f}$ can be regarded as *merely* constant for small changes in frequency. Therefore, the increase in load consumption due to increase Δf in the frequency is given by

$$\left(\frac{\partial P_D}{\partial f}\right)\Delta f = D\Delta f$$

where the constant D can be determined empirically (D is positive for a predominantly motor load).

Writing the power balance equation we have

$$\Delta P_G - \Delta P_D = \frac{2W^*}{f^*}\frac{d}{dt}(\Delta f) + D(\Delta f) \tag{4.50}$$

Equation (4.50) may be represented by the block diagram of Fig. 4.18. K_P, the power system gain and T_P, the power system time-constant can be obtained in terms of coefficients W^*, f^* and D in eqn. (4.50).

A complete block diagram representation of an isolated power system comprising a turbine, generator, governor and load is easily obtained by combining the block diagrams of Figs. 4.16 to 4.18. This is shown in Fig. 4.19.

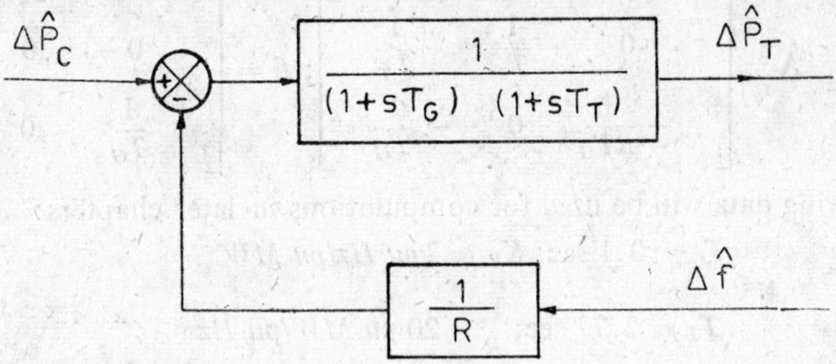

Fig. 4.17

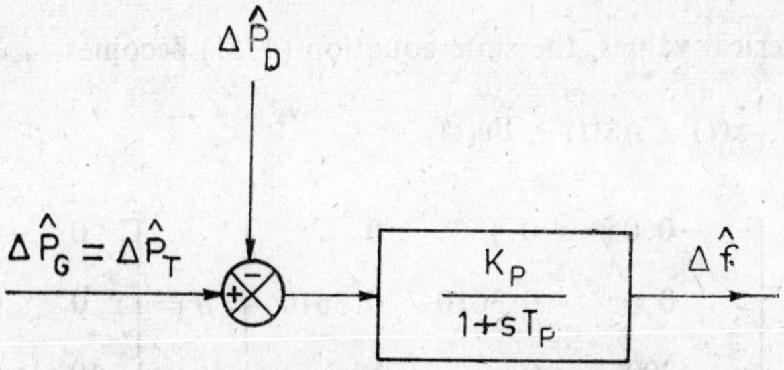

Fig. 4.18

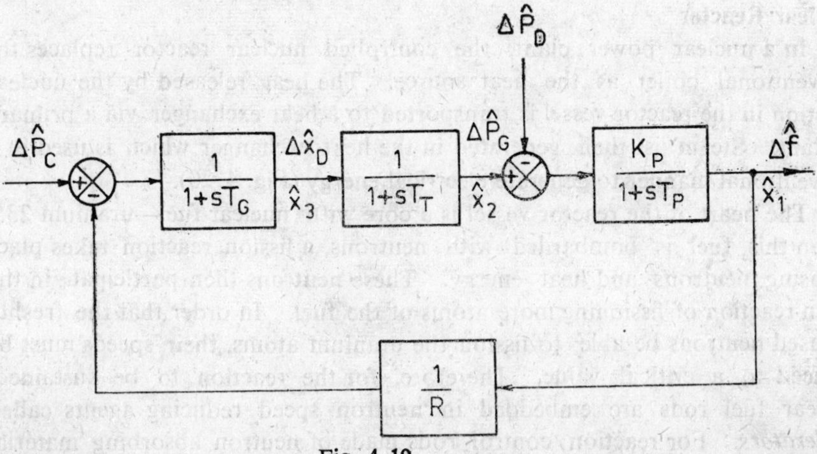

Fig. 4.19

The system of Fig. 4.19 can be represented as

$$\dot{\mathbf{x}}(t) = \mathbf{A}\mathbf{x}(t) + \mathbf{B}\mathbf{u}(t) \tag{4.51}$$

where

$$\mathbf{x} = \begin{bmatrix} x_1 \\ x_2 \\ x_3 \end{bmatrix} = \begin{bmatrix} \Delta f \\ \Delta P_G \\ \Delta x_D \end{bmatrix}; \quad \mathbf{u} = \begin{bmatrix} u_1 \\ u_2 \end{bmatrix} = \begin{bmatrix} \Delta P_C \\ \Delta P_D \end{bmatrix}$$

$$\mathbf{A} = \begin{bmatrix} -\frac{1}{T_P} & \frac{K_P}{T_P} & 0 \\ 0 & -\frac{1}{T_T} & \frac{1}{T_T} \\ -\frac{1}{RT_G} & 0 & -\frac{1}{T_G} \end{bmatrix}; \quad B = \begin{bmatrix} 0 & -\frac{K_P}{T_P} \\ 0 & 0 \\ \frac{1}{T_G} & 0 \end{bmatrix}$$

The following data will be used for computations in later chapters:

$$T_G = 0.1 \text{ sec}; \; K_P = 2 \, pu \; Hz/pu \; MW$$

$$T_T = 2.77 \text{ sec}; \; \frac{1}{R} = 20 \, pu \; MW/pu \; Hz$$

$$T_P = 20 \text{ sec}$$

For these numerical values, the state equation (4.51) becomes

$$\dot{\mathbf{x}}(t) = \mathbf{A}\mathbf{x}(t) + \mathbf{B}\mathbf{u}(t) \tag{4.52}$$

where

$$\mathbf{A} = \begin{bmatrix} -0.05 & 0.1 & 0 \\ 0 & -0.3610 & 0.3610 \\ -200 & 0 & -10 \end{bmatrix}; \quad B = \begin{bmatrix} 0 & -0.1 \\ 0 & 0 \\ 10 & 0 \end{bmatrix}$$

Nuclear Reactor

In a nuclear power plant, the controlled nuclear reactor replaces the conventional boiler as the heat source. The heat released by the nuclear reaction in the reactor vessel is transported to a heat exchanger via a primary coolant. Steam is then generated in the heat exchanger which is used in a conventional manner to generate electrical energy (Fig. 4.20).

The heart of the reactor vessel is a core with nuclear fuel—uranium 235. When this fuel is bombarded with neutrons, a fission reaction takes place releasing neutrons and heat energy. These neutrons then participate in the chain reaction of fissioning more atoms of the fuel. In order that the freshly released neutrons be able to fission the uranium atoms, their speeds must be reduced to a critical value. Therefore, for the reaction to be sustained, nuclear fuel rods are embedded in neutron speed reducing agents called *moderators*. For reaction control, rods made of neutron absorbing material

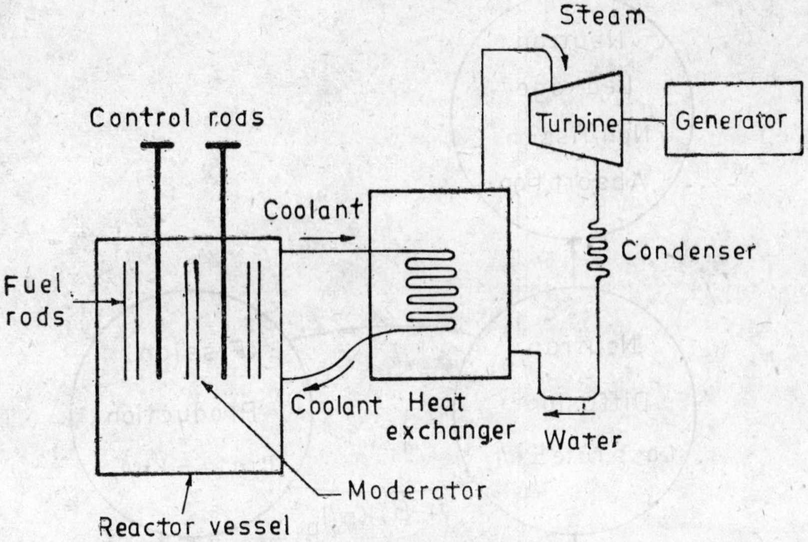

Fig. 4.20 Schematic diagram of a nuclear power plant

are used which when inserted into the reactor vessel control the amount of neutron flux, thereby controlling the rate of reaction.

In the fission process, more than 99 percent of neutrons generated in the process are given off instantly. A small fraction of neutrons are emitted from certain of the fission fragments at discrete amounts of time after the actual fission process occurs. These are called *delayed neutrons* and the fission fragments emitting delayed neutrons are called *delayed neutron precursors.* In practice it is found that an adequate representation of delayed neutron precursors at time t can be given if six 'effective' precursor groups emitting different quantities of neutrons at different times are used. The symbol β is used to denote the total fraction of delayed neutrons with β_i being the fraction of delayed neutrons in the i-th group. Similarly λ_i (sec^{-1}) represents the decay constant of the i-th group of precursors. For certain problems, it is convenient to treat all of the delayed neutrons as a single delayed group having a total fraction β and an average decay constant λ.

Ignoring spatial distribution and considering the reactor vessel as a 'point' generator, we derive the kinetic equations (Hetrick 1971). The derivation is based on the concept of neutron life time. The neutron life cycle is shown in Fig. 4.21. Let

$n(t)$ = the number of neutrons in the system at time t

l_0 = neutron life time—average time between successive neutron generations

Then $$\frac{n(t)}{l_0} = \text{total loss rate}$$

Let K be the multiplication factor, which is defined as the total neutrons, both prompt and delayed, produced per neutron loss.

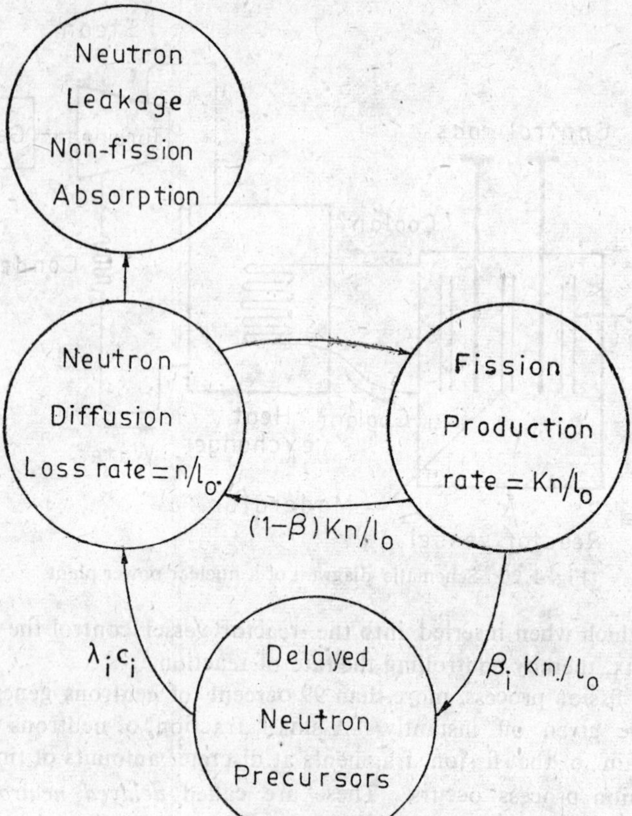

Fig. 4.21 Simplified neutron life cycle

Therefore, the production rate $= K\frac{n(t)}{l_0}$

Prompt neutrons are back in circulation immediately and delayed neutrons are in the bank for various mean times $1/\lambda_i$ sec. If β_i is the delayed neutron fraction belonging to i-th group, then

i-th group delayed neutron precursors are produced at the rate $= \beta_i \frac{Kn(t)}{l_0}$

Therefore prompt neutron production rate $= \frac{(1-\beta)Kn(t)}{l_0}$; $\beta = \sum_i \beta_i$

Let

$c_i(t)$ = number of precursors of i-th group in the system at time t

The delayed neutron production rate is therefore $= \sum_i \lambda_i c_i(t)$

where λ_i = decay constant of i-th group precursor.

The neutron balance equation is

$$\frac{dn(t)}{dt} = (1-\beta)\, K \frac{n(t)}{l_0} + \sum_i \lambda_i c_i(t) - \frac{n(t)}{l_0}$$

Rearranging this equation we get

$$\frac{dn(t)}{dt} = \frac{K-1-\beta K}{l_0}\, n(t) + \sum_i \lambda_i c_i(t)$$

$$= \frac{\dfrac{K-1}{K} - \beta}{l_0/K}\, n(t) + \sum_i \lambda_i c_i(t) \tag{4.53}$$

The term $(K-1)/K$ is known as *reactivity* and is represented by the symbol ρ. Reactivity, which is dimensionless, is defined as the relative deviation of the multiplication factor from unity. Sometimes, reactivity is expressed in multiples of β, in which case the units of 'dollars' or 'cents' are employed; $\rho = \beta$ implies a reactivity of one dollar. The term l_0/K is known as neutron generation time and is denoted by the symbol l.

The balance equation for the delayed neutron emitters is given by

$$\frac{dc_i(t)}{dt} = \frac{\beta_i K}{l_0}\, n(t) - \lambda_i c_i(t) \tag{4.54}$$

Equations (4.53)-(4.54) may be expressed as

$$\frac{dn(t)}{dt} = \frac{\rho - \beta}{l}\, n(t) + \sum_{i=1}^{6} \lambda_i c_i(t) \tag{4.55}$$

$$\frac{dc_i(t)}{dt} = \frac{\beta_i}{l}\, n(t) - \lambda_i c_i(t);\ i = 1, 2, \ldots, 6 \tag{4.56}$$

These are the point kinetic equations of a nuclear reactor (six delayed neutron group model).

In one delayed neutron group model of nuclear reactors, all the delayed neutrons are assumed to be released together in contrast to the six delayed neutron group model wherein the delayed neutrons are assumed to appear in six groups. Taking corresponding effective values of parameters, we have the following model:

$$\frac{dn(t)}{dt} = \frac{\rho(t) - \beta}{l}\, n(t) + \lambda c(t) \tag{4.57}$$

$$\frac{dc(t)}{dt} = \frac{\beta}{l}\, n(t) - \lambda c(t) \tag{4.58}$$

Note that the equations are nonlinear because of the occurrence of the product of $\rho(t)$ and $n(t)$.

The number of neutrons in the core is proportional to the number of fissions occurring and for 3×10^{10} fissions per sec, approximately 1 watt of power is produced. The power output of a reactor is thus proportional to the number of neutrons in the core in any given time interval and the symbol n is used to designate the neutron level or power level.

The discussion above provides a cursory knowledge of reactor dynamics. Familiarity with nuclear physics is essential to obtain a mature understanding of the subject. However, such a maturity is not necessary to follow the optimal control examples discussed in later chapters.

We shall use the following parameters of a research reactor in our subsequent discussion:

$$\beta = 0.0064$$

$$\lambda = 0.1 \text{ sec}^{-1}$$

$$l = 10^{-3} \text{ sec}$$

Power level at steady-state $(n^0) = 10$ kw

In terms of these parameters, we have the following model from eqns. (4.57)-(4.58):

$$\dot{\mathbf{x}}(t) = \mathbf{f}(\mathbf{x}, u);\ \mathbf{x}(t) = \begin{bmatrix} n(t) \\ c(t) \end{bmatrix},\ u(t) = \rho(t) \tag{4.59a}$$

$$\begin{aligned} \dot{x}_1(t) &= f_1(\mathbf{x}, u) \\ &= -6.4x_1(t) + 0.1x_2(t) + 10^3 x_1(t)\, u(t) \end{aligned} \tag{4.59b}$$

$$\begin{aligned} \dot{x}_2(t) &= f_2(\mathbf{x}, u) \\ &= 6.4x_1(t) - 0.1x_2(t) \end{aligned} \tag{4.59c}$$

Linearization of these equations about the steady-state gives (refer eqn. (4.25a))

$$\dot{\tilde{\mathbf{x}}}(t) = \mathbf{A}\tilde{\mathbf{x}}(t) + \mathbf{b}\tilde{u}(t)$$

where $\tilde{\mathbf{x}}$ and $\tilde{\mathbf{u}}$ are deviations from the steady-state $(x_1^0 = n^0,\ x_2^0 = c^0,\ u_{ss} = 0)$ and

$$\mathbf{A} = \begin{bmatrix} \dfrac{\partial f_1}{\partial x_1} & \dfrac{\partial f_1}{\partial x_2} \\ \dfrac{\partial f_2}{\partial x_1} & \dfrac{\partial f_2}{\partial x_2} \end{bmatrix}_{(n^0,\, c^0,\, 0)} = \begin{bmatrix} -\beta/l & \lambda \\ \beta/l & -\lambda \end{bmatrix}$$

$$\mathbf{b} = \begin{bmatrix} \dfrac{df_1}{du_1} \\ \dfrac{df_2}{du_2} \end{bmatrix}_{(n^0,\, c^0,\, 0)} = \begin{bmatrix} n^0/l \\ 0 \end{bmatrix}$$

Note that the steady-state value c^0 is obtained by setting $\rho = 0$ and $\dfrac{dn}{dt} = 0$ in eqn. (4.57). This gives

$$c^0 = \frac{\beta}{\lambda l} n^0 = 640$$

In terms of the given parameters, the linearized model is as follows:

$$\dot{\tilde{\mathbf{x}}}(t) = \mathbf{A}\tilde{\mathbf{x}}(t) + \mathbf{b}\,\tilde{u}(t) \tag{4.60}$$

$$\mathbf{A} = \begin{bmatrix} -6.4 & 0.1 \\ 6.4 & -0.1 \end{bmatrix}; \quad \mathbf{b} = \begin{bmatrix} 10^4 \\ 0 \end{bmatrix}$$

4.7 CONCLUDING COMMENTS

This chapter has been concerned with the mathematical modelling of physical systems in a state variable format. In Sections 4.2–4.4 we illustrated the procedure of modelling through a series of examples. We observed that in all but the simplest electrical and mechanical networks, linear state models are only an approximate description of real systems, valid under appropriate assumptions. The most important assumption is that the behaviour of the system is studied only for small changes of variables from the steady-state. This assumption allows nonlinear relations to be replaced by linear approximations; a formal procedure of doing this was given in Section 4.5.

In Section 4.6 we derived plant models of 5 illustrative control systems. These examples will be used to clarify a significant portion of the text and to emphasize the practical applicability of the results. These design examples will be developed over several chapters of the book.

PROBLEMS

4.1 Write the state and output equations for systems of Figs. P-4.1a and P-4.1b.

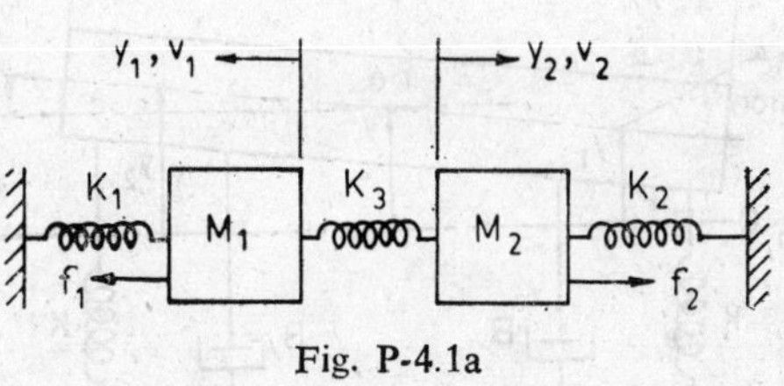

Fig. P-4.1a

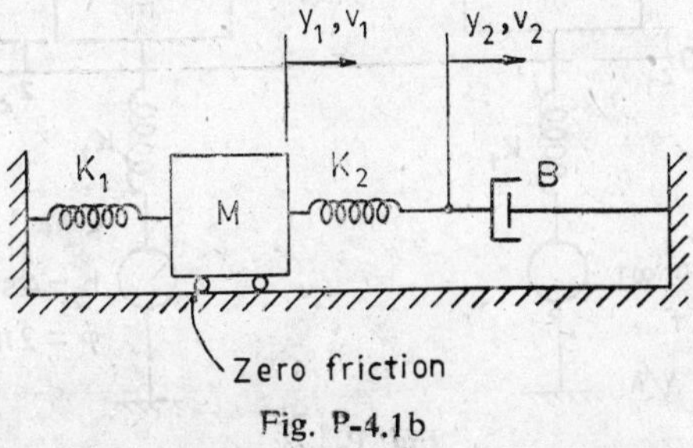

Fig. P-4.1b

4.2 A rail-road car approaches at a constant speed, a massless bumper (see Fig. P-4.2) and couples with the bumper at $t=0$ without backlash and energy loss. Determine a state model describing the system for $t>0$.

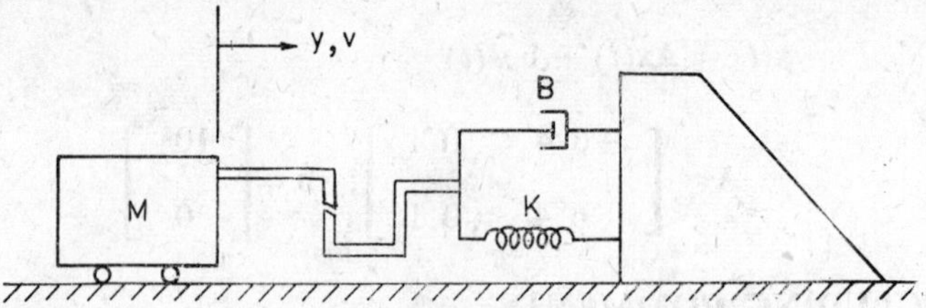

Fig. P-4.2

4.3 The study of the vertical motion and rocking motion of a four-wheeled vehicle travelling over a rough road is a somewhat complicated problem. A less complicated version does provide some insight into this.

Excitation to the system, the road elevation $h(t)$ at any time t, can be determined from the shape of the road and the speed of the vehicle. The shape of the road is so arbitrary that it is very difficult to model it by an analytical function. The worst excitation to the system from viewpoint of passenger discomfort will probably come from a sinusoidally varying road elevation. The passenger in the vehicle will be subjected to sustained oscillations and dizziness (sickness) will result at approximately 10 Hz (Nagrath and Gopal 1982b).

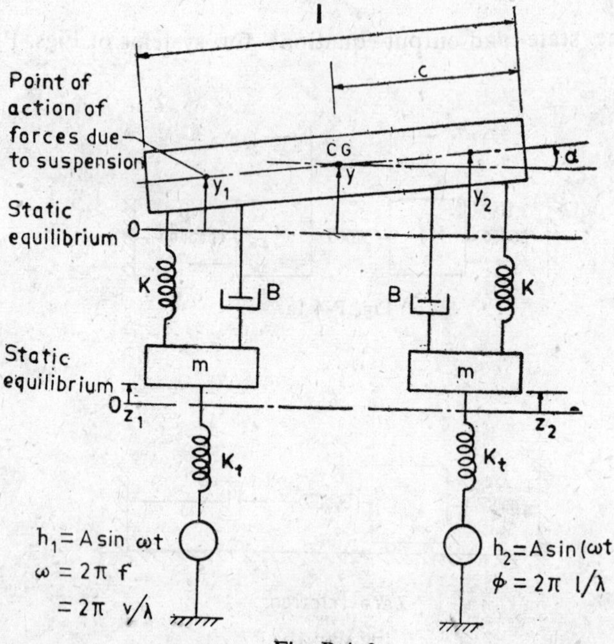

Fig. P-4.3

Let the velocity of the vehicle be v m/sec and the wavelength of the hypothetical sinusoidal road be λ m. The frequency of the input is then $f = v/\lambda$. If the length of the vehicle is much less than λ, then we can treat the body of the vehicle as a point mass. However, we shall assume that length l is comparable to λ and so the body mass must be treated as a rigid body (see Fig. 1.1).

With these assumptions, we model the whole situation as in Fig. P-4.3. The sinusoidal vertical motions of the front and rear wheels of the vehicle have a phase displacement of $\phi = 2\pi l/\lambda$ as shown in the figure.

Construct a suitable state model to determine y and α in response to h_1 and h_2. The mass of vehicle body is M and its moment of inertia about the centre of gravity (CG) is J (refer Nagrath and Gopal (1982b)).

4.4 Consider the oscillations of the spring-constrained inverted pendulum shown in Fig. P-4.4. Construct a suitable state equation of this system.

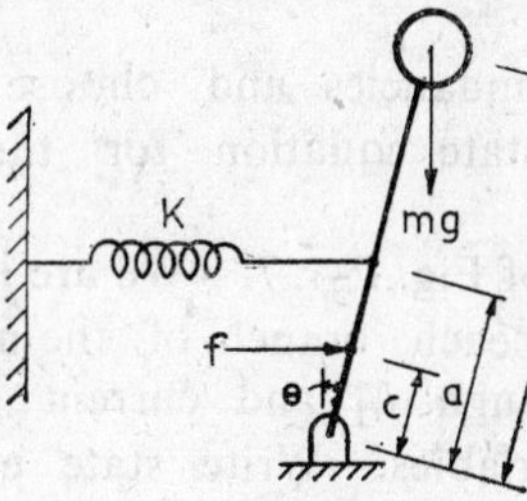

Fig. P-4.4

4.5 Consider the electromechanical system shown in Fig. P-4.5. Assume that the coil has back emf $e_b = k_1 \dfrac{dx}{dt}$ and the coil current i produces a force $f_c = k_2 i$ on the mass M. Construct a state model to obtain the displacement $x(t)$ of mass M in response to voltage $u(t)$.

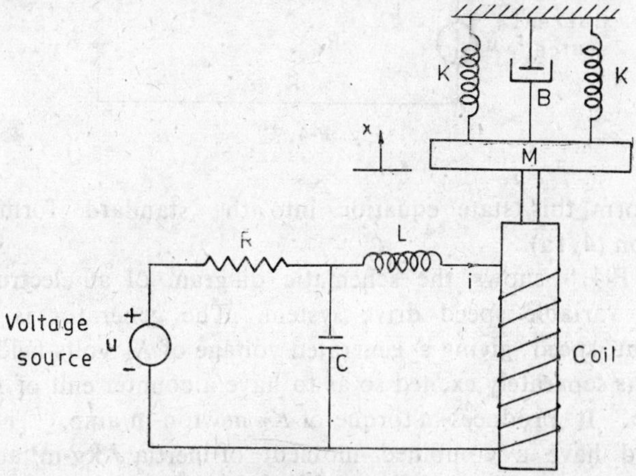

Fig. P-4.5

4.6 The problem illustrates the notion of redundant state variables.

For the network of Fig. P-4.6, choose all capacitor voltages as state variables and get a three-dimensional state equation.

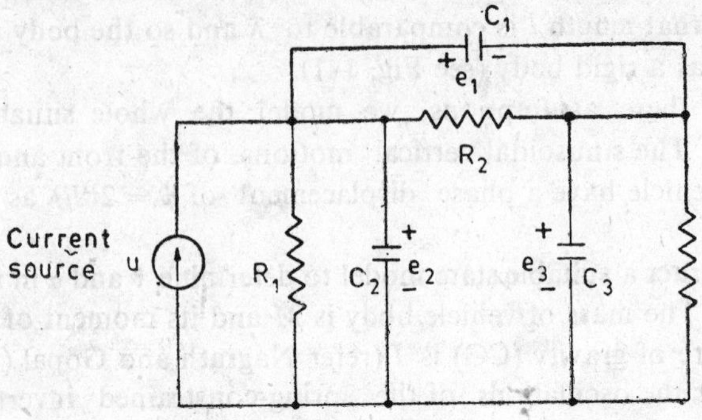

Fig. P-4.6

Now eliminate redundancies and choose state variables to give minimal-dimensional state equation for the network. Obtain the state equation.

4.7 Consider the network of Fig. P-4.7. We are interested in current and voltage associated with each branch of the network. Convince yourself that current through L_1 and current through L_2 constitute a minimal set of state variables. Write state equation for the network. Note that it takes the form

$$\dot{\mathbf{x}}(t) = \mathbf{A}\mathbf{x}(t) + \mathbf{B}\mathbf{u}(t) + \mathbf{K}\dot{\mathbf{u}}(t)$$

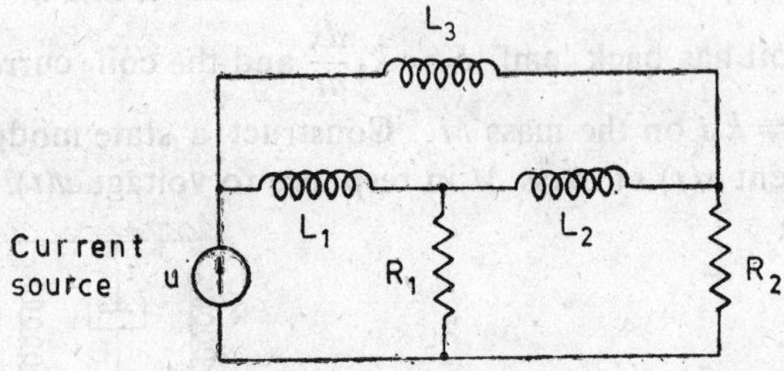

Fig. P-4.7

Transform this state equation into the standard form given by equation (4.1a).

4.8 Figure P-4.8 shows the schematic diagram of an electronically controlled variable speed drive system. The generator is driven at a constant speed giving a generated voltage of K_g volts/field amp. The motor is separately excited so as to have a counter emf of K_b volts per rad/sec. It produces a torque of K_T newton-m/amp. The motor and its load have a combined moment of inertia J kg-m² and friction B newton-m per rad/sec.

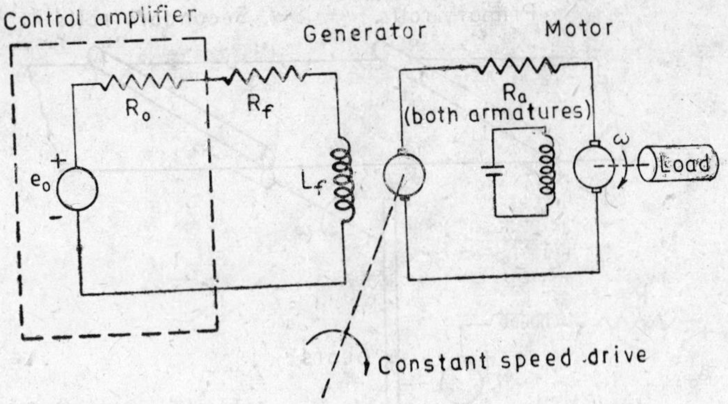

Fig. P-4.8

Given

$R_0 = 2000\ \Omega$ (effective source resistance)

$R_f = 500\ \Omega$

$L_f = 100H$

$K_g = 50$ volts/amp

$R_a = 2\ \Omega$

$K_T = 1.5$ newton-m/amp

$J = 6$ kg-m²

$B = 12$ newton-m per rad/sec.

Construct a suitable state model for the system.
(Hint: $K_b = K_T$ in MKS units (refer Nagrath and Gopal (1982a))

4.9 A servomechanism system is used to position the angle θ of an output load to some desired value θ_r. This system employs the speed control system of Problem 4.8 and is represented in Fig. P-4.9. Construct a suitable state model for the system.

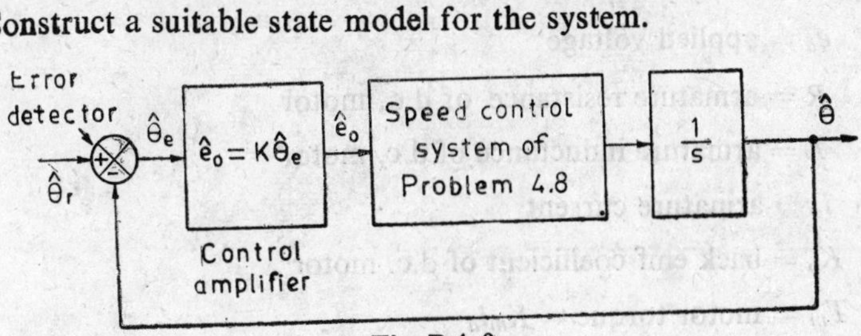

Fig. P-4.9

4.10 Consider the system of Fig. P-4.10. The tension T in the web is the variable to be controlled. The primary pinch rolls are driven at constant speed ω_d. As the speed ω of the secondary pinch rolls varies, the positional synchronism of the rolls changes, thereby changing the tension in the web. The load torque on the motor is

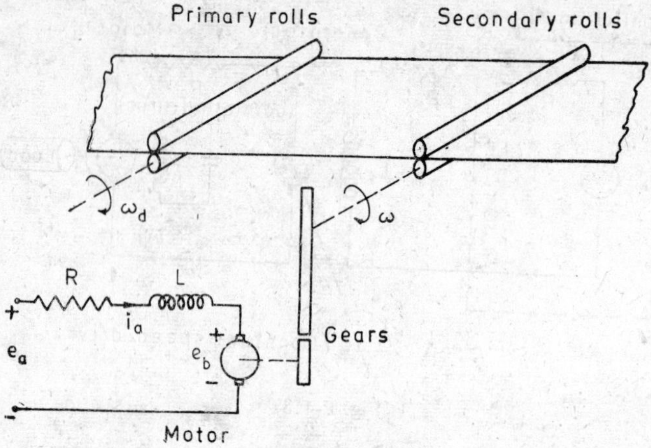

Fig. P-4.10

$$T_L = KT$$

where

K = constant, which depends on roll geometry, gear ratio and width of web;

T = web tension, given by the relation

$$\frac{dT}{dt} = rK_\omega(\omega - \omega_d)$$

r = radius of rolls

K_ω = coefficient of elasticity of web material.

Construct a suitable state model to obtain speed $\omega(t)$ and tension $T(t)$ as a function of motor armature voltage $e_a(t)$.

The system parameters and variables are defined as follows:

e_a = applied voltage

R = armature resistance of d.c. motor

L = armature inductance of d.c. motor

i_a = armature current

K_b = back emf coefficient of d.c. motor

T_m = motor torque = $K_m i_a$

J = motor inertia

B = motor friction coefficient.

4.11 Figure P-4.11 illustrates the situation where Q_0 litres/min of liquid is to be poured into the first of the three tanks connected in cascade. Each tank has a bottom outlet furnished with a valve with a flow

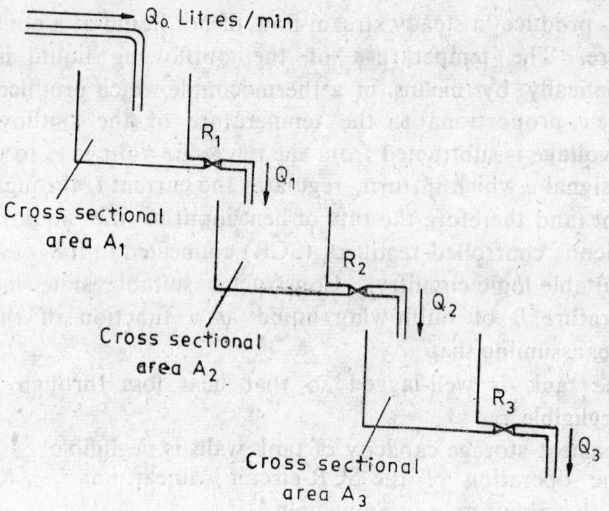

Fig. P-4.11

resistance R. The relation between the head and the flow from the tank is

$$h = RQ^2 \text{ cm}; \ Q \text{ in litres/min}$$

It is required to establish the heads in each tank at time t with the tanks initially empty. Construct a suitable state model for the system.

4.12 Consider the temperature control system of Fig. P-4.12, which is set

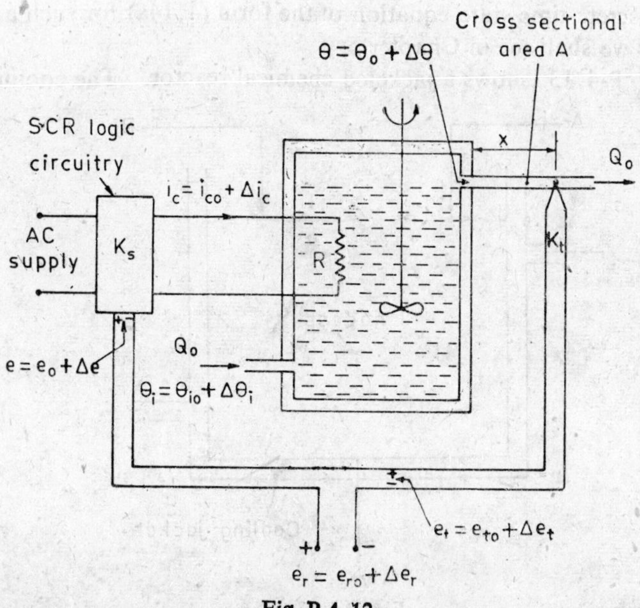

Fig. P-4.12

up to produce a steady stream flow of hot liquid at a controlled temperature. The temperature of the outflowing liquid is regulated automatically by means of a thermocouple which produces an output voltage e_t proportional to the temperature of the outflowing liquid. This voltage is subtracted from the reference voltage e_r to generate the error signal e which in turn, regulates the current i_c through the heater element (and therefore the rate of heat input to the liquid) by means of silicon controlled rectifiers (SCR) connected in full-wave operation and suitable logic circuitry. Construct a suitable state model giving temperature θ of outflowing liquid as a function of the reference voltage, assuming that

(i) the tank is well-lagged so that heat loss through its walls is negligible;

(ii) the heat storage capacity of tank walls is negligible;

(iii) the operation of the SCR circuit is linear, i.e., $i_c = K_s e$ where K_s is the circuit gain in amps/volt.

(iv) In order to allow the fluctuations in temperature due to mixing to smooth out, the temperature is measured not at the mixing point but a certain distance downstream. The measurement of the temperature of the process is thus delayed by $\tau = \dfrac{x}{(Q_0/A)}$. We shall assume that no significant heat is transferred into, out of, or within the fluid as it moves from the mixing point to the measurement point.

Note that a system with pure time delay cannot be described by the state equation of the form (4.1a). We can, however, obtain discrete-time state equation of the form (4.14a) for such a system, as we shall see in Chapter 5.

4.13 Figure P-4.13 shows a jacketed chemical reactor. The cooling jacket

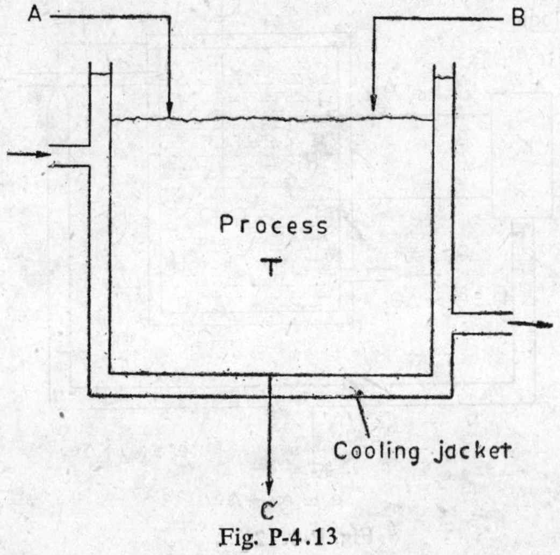

Fig. P-4.13

surrounds the reactor to remove the heat of reaction.
Two streams A and B of dissolved chemical flow continuously into the stirred reactor and react almost immediately producing 225 kcal of heat per kg of solution A. The result is drawn off as stream C. The following numerical values are given:

Rate of flow of $A = 225$ kg/hr

Rate of flow of $B = 360$ kg/hr

Rate of flow of $C = 585$ kg/hr

Initial storage of solution in the reactor $= 1360$ kg

Temperature of $A = 24°C$

Temperature of $B = 30°C$

Specific heats of A, B and C are $c_p = 1$

Heat transfer between the process at temperature T and cooling water at temperature T_w is given by

$$q = G(T - T_w)$$

$$G = 4535 \text{ kcal/hr}$$

Initial value of $T = 16°C$

Initial value of $T_w = 16°C$

Rate of flow of coolant $= 900$ kg/hr

Specific heat of coolant $= 1$

Coolant in jacket (constant) $= 450$ kg

Set up a suitable state model for the system.

4.14 Consider the six-plate absorber of Fig. P-4.14 controlled by the inlet feed streams.

L = rate of flow of liquid, mass/time

G = rate of flow of vapour, mass/time

H = liquid hold up on each plate, mass

h = vapour hold up on each plate, mass

x_n = mass fraction of solute in liquid leaving n^{th} plate

x_{f1} = mass fraction of solute in liquid feed stream

y_n = mass fraction of solute in vapour leaving n^{th} plate

y_{f2} = mass fraction of solute in vapour feed-stream.

The solute mass balance at n^{th} plate gives:
Rate of change of solute mass on n^{th} plate

= rate of flow of solute mass into n^{th} plate

− rate of flow of solute mass out of nth plate.

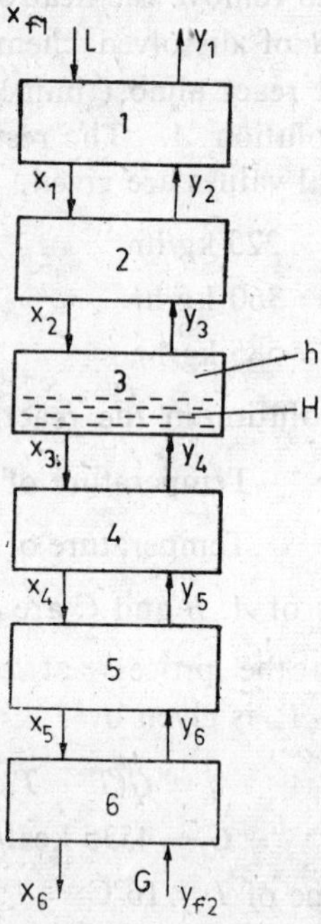

Fig. P-4.14

Determine a suitable state model of the system assuming a linear equilibrium relationship between liquid and vapour at each plate (refer Treybel (1969)):

$$y_n = ax_n$$

Given: $a = 0.70$, $G = 30$ kg/min,

$H = 0.45$ kg, $h = 34$ kg

$L = 18.5$ kg/min

4.15 Present a discrete-time state variable description of the steady-state behaviour of the plate-type gas absorber of Problem 4.14.

4.16 Consider the reactor train of Example 4.7 with second-order kinetics. Obtain a state variable formulation of the system.

REFERENCES

1. Athans, M., M. L. Dertouzos, R. N. Spann, and S. J. Mason, *Systems, Networks and Computation: Multivariable Methods*, New York: McGraw-Hill, 1974.
2. Auslander, D. M., Y. Takahashi, and M. J. Rabins, *Introducing Systems and Control*, Tokyo: McGraw-Hill, 1974.
3. Cannon, R. H., *Dynamics of Physical Systems*, New York: McGraw-Hill, 1967.
4. Crandall, S. H., D. C. Karnopp, E. F. Kurtz, Jr., and D. C. Pridmore-Brown, *Dynamics of Mechanical and Electromechanical Systems*, New York; McGraw-Hill, 1968.
5. Elgerd, O. I., *Control systems Theory*, New York; McGraw-Hill, 1967.

6. Elgerd, O.I., *Electric Energy Systems Theory: An Introduction*, New York; McGraw-Hill, **1971.**
7. Hetrick, D.L., *Dynamics of Nuclear Reactors*, Chicago: University of Chicago Press, **1971.**
8. Luyben, W. L., *Process Modelling, Simulation and Control for Chemical Engineers*, Tokyo: McGraw-Hill, **1973.**
9. MacFarlane, A. G. F., *Dynamical System Models*, London: George G. Harrap, **1970.**
10. Nagrath, I. J., and M. Gopal, *Control Systems Engineering*, 2nd edition, New Delhi: Wiley Eastern, **1982*a*.**
11. Nagrath, I. J., and M. Gopal, *Systems: Modelling and Analysis*, New Delhi: Tata McGraw-Hill, **1982*b*.**
12. Perkins, W. R., and J. B. Cruz, Jr., *Engineering of Dynamic Systems*, New York: Wiley, **1969.**
13. Russell, T. W. F., and M. M. Denn, *Introduction to Chemical Engineering Analysis*, New York: Wiley, **1972.**
14. Smith, C. L., R. W. Pike, and P. W. Murrill, *Formulation and Optimization of Mathematical Models*, Pennsylvania: Intext Publisher, **1970.**
15. Treybel, R. E., *Mass-Transfer Operation*, New York: McGraw-Hill, **1969.**

5. SOLUTION OF STATE EQUATIONS

5.1 INTRODUCTION

We observed in the previous chapter that state-space representations of dynamic physical systems are very frequently in the form of vector-matrix differential equations. These differential equations provide us with the cause-and-effect relationship between several interrelated physical quantities of the system. Our experience indicates that physical processes do not admit nonunique solutions, nor do they fail to have solutions. However, because of the idealization made in arriving at a differential model of a physical process, the resulting differential equations may not have a solution or may have nonunique solutions. Therefore, only a set of differential equations having a unique solution can faithfully represent a well-behaved physical dynamical process.

This chapter discusses the fundamental questions of the existence and uniqueness of solutions of vector-matrix differential equations and presents standard techniques for solving these equations. We shall also deal with the discretization of vector-matrix differential equations to vector-matrix difference equations and the solution of these difference equations.

5.2 EXISTENCE AND UNIQUENESS OF SOLUTIONS TO CONTINUOUS-TIME STATE EQUATIONS

We first consider the n-th order nonlinear state equation

$$\dot{\mathbf{x}}(t) = \mathbf{f}(\mathbf{x}(t));\ \mathbf{x}(t_0) = \mathbf{x}^0,\ t \in [t_0, t_1] \tag{5.1}$$

Note that eqn. (5.1) has been obtained from eqn. (3.23a) by setting $\mathbf{u}(t) = \mathbf{0}$; (5.1) thus corresponds to a *free or unforced system*. Further, t does not appear in the argument of $\mathbf{f}$; (5.1) is thus time-invariant. A system which is both free and time-invariant is called an *autonomous system.*

For the state equation (5.1), a vector-valued function $\Phi(t)$ such that $\Phi(t) \in \mathcal{R}^n$ for each $t \in [t_0, t_1]$ is a solution if

(i) $$\Phi(t_0) = \mathbf{x}^0 \tag{5.2}$$

(ii) $$\dot{\Phi}(t) = \mathbf{f}(\Phi(t))\ \forall\, t \in [t_0, t_1] \tag{5.3}$$

The fundamental questions to be investigated before solving the set of equations (5.1) are:

1. Under what conditions can we guarantee the existence of a solution?
2. If a solution exists, is it unique?

These fundamental questions of existence and uniqueness are dealt with in this section.

We shall first make certain general assumptions about the mathematical properties of the function **f**. Fortunately, these mathematical assumptions are satisfied by all well-behaved physical processes. The assumptions are (Athans et al. 1974):

(i) $\mathbf{f}(\mathbf{x})$ is a continuous function of $\mathbf{x}$ for all $\mathbf{x} \in \mathcal{R}^n$.

(ii) $\mathbf{f}(\mathbf{x})$ is a bounded function, i.e., there exists an F such that for all finite values of $\mathbf{x}$ (refer Section 2.5)

$$\|\mathbf{f}(\mathbf{x})\| \leqslant F \tag{5.4}$$

(iii) $\mathbf{f}(\mathbf{x})$ satisfies a *Lipschitz condition.* To explain this, suppose we change the vector $\mathbf{x}$ from say $\mathbf{x}_1$ to $\mathbf{x}_2$.
We say that $\mathbf{f}(\mathbf{x})$ satisfies a Lipschitz condition if the relation

$$\|\mathbf{f}(\mathbf{x}_1) - \mathbf{f}(\mathbf{x}_2)\| \leqslant L\|\mathbf{x}_1 - \mathbf{x}_2\| \tag{5.5}$$

holds for any two vectors $\mathbf{x}_1$ and $\mathbf{x}_2$ in $\mathcal{R}^n$, where L (the Lipschitz constant) is constant and finite. We shall assume with no loss of generality that $L = F$.

A special class of functions satisfying the Lipschitz condition (5.5) is the class of functions $\mathbf{f}(\mathbf{x})$ which are differentiable with respect to $\mathbf{x}$, i.e., the partial derivatives

$$\frac{\partial f_i}{\partial x_j}(x_1, x_2, \ldots, x_n);\; i, j = 1, 2, \ldots, n$$

exist and are continuous functions of x_i's. To prove that this class of functions automatically satisfies the Lipschitz condition, let us expand $\mathbf{f}(\mathbf{x})$ about $\mathbf{x} = \mathbf{x}_2$ using a Taylor series (refer Appendix III)

$$\mathbf{f}(\mathbf{x}) \simeq \mathbf{f}(\mathbf{x}_2) + \left.\frac{\partial \mathbf{f}(\mathbf{x})}{\partial \mathbf{x}}\right|_{\mathbf{x}=\mathbf{x}_2} (\mathbf{x} - \mathbf{x}_2);\; \|\mathbf{x} - \mathbf{x}_2\| \leqslant K \tag{5.6}$$

where $\dfrac{\partial \mathbf{f}(\mathbf{x})}{\partial \mathbf{x}}$ is a Jacobian matrix.

For $\mathbf{x} = \mathbf{x}_1$ we obtain

$$\mathbf{f}(\mathbf{x}_1) - \mathbf{f}(\mathbf{x}_2) \simeq \left.\frac{\partial \mathbf{f}(\mathbf{x})}{\partial \mathbf{x}}\right|_{\mathbf{x}=\mathbf{x}_2} (\mathbf{x}_1 - \mathbf{x}_2)$$

Hence

$$\|\mathbf{f}(\mathbf{x}_1) - \mathbf{f}(\mathbf{x}_2)\| \leqslant \left\| \left.\frac{\partial \mathbf{f}(\mathbf{x})}{\partial \mathbf{x}}\right|_{\mathbf{x}=\mathbf{x}_2} \right\| \|\mathbf{x}_1 - \mathbf{x}_2\| \tag{5.7}$$

If the Jacobian matrix exists and is continuous, its elements are bounded and therefore

$$\left\| \left.\frac{\partial \mathbf{f}(\mathbf{x})}{\partial \mathbf{x}}\right|_{\mathbf{x}=\mathbf{x}_2} \right\| \leqslant L \tag{5.8}$$

From eqns. (5.6)-(5.8) we conclude that the class of functions $\mathbf{f}(\mathbf{x})$ which are differentiable with respect to $\mathbf{x}$ satisfies the Lipschitz condition.

Existence of Solutions

Integrating both sides of eqn. (5.1) from t_0 to t yields

$$\int_{t_0}^{t} \dot{\mathbf{x}}(\tau)d\tau = \int_{t_0}^{t} \mathbf{f}(\mathbf{x}(\tau))d\tau$$

or

$$\mathbf{x}(t) = \mathbf{x}^0 + \int_{t_0}^{t} \mathbf{f}(\mathbf{x}(\tau))d\tau \tag{5.9}$$

We want to establish the existence of a function $\Phi(t)$ such that

$$\Phi(t) = \mathbf{x}^0 + \int_{t_0}^{t} \mathbf{f}(\Phi(\tau))d\tau \tag{5.10}$$

in some finite time interval $[t_0, t_1]$ of interest. To establish this, we define a sequence of time functions $\Phi_0(t)$, $\Phi_1(t)$...in the following manner:

$$\begin{aligned}
\Phi_0(t) &\triangleq \mathbf{x}^0 \\
\Phi_1(t) &\triangleq \mathbf{x}^0 + \int_{t_0}^{t} \mathbf{f}(\Phi_0(\tau))d\tau \\
\Phi_2(t) &\triangleq \mathbf{x}^0 + \int_{t_0}^{t} \mathbf{f}(\Phi_1(\tau))d\tau \\
&\vdots \\
\Phi_{j+1}(t) &\triangleq \mathbf{x}^0 + \int_{t_0}^{t} \mathbf{f}(\Phi_j(\tau))d\tau
\end{aligned} \tag{5.11}$$

and show that the limit,

$$\lim_{j\to\infty} \Phi_j(t) \ \forall \text{ finite } t$$

exists and that the limit function is $\Phi(t)$.

Note that because of the assumption that $\Phi_0(t)$ is bounded and $\mathbf{f}(\mathbf{x})$ is continuous and bounded, the integrals are well defined for finite values of t. Thus for finite values of index j, the functions $\{\Phi_j(t)\}$ are well behaved and each $\Phi_j(t)$ is continuous in t.

We can write

$$\begin{aligned}
\Phi_j(t) &= \Phi_0(t) - \Phi_0(t) + \Phi_1(t) - \Phi_1(t) + \ldots + \Phi_{j-1}(t) - \Phi_{j-1}(t) + \Phi_j(t) \\
&= \Phi_0(t) + [\Phi_1(t) - \Phi_0(t)] + [\Phi_2(t) - \Phi_1(t)] + \ldots + [\Phi_j(t) - \Phi_{j-1}(t)] \\
&= \Phi_0(t) + \sum_{i=0}^{j-1} [\Phi_{i+1}(t) - \Phi_i(t)]
\end{aligned} \tag{5.12}$$

Thus $\Phi_j(t)$ converges to a limit if we can prove that the series $\lim_{j\to\infty} \sum_{i=0}^{j-1} [\Phi_{i+1} - \Phi_i]$ converges. It is sufficient to prove that the series

$$\lim_{j\to\infty} \sum_{i=0}^{j-1} \| \Phi_{i+1} - \Phi_i \| \tag{5.13}$$

converges. Let us examine the integral equations (5.11).

$$\Phi_1(t)-\Phi_0(t)=\int_{t_0}^{t} \mathbf{f}(\Phi_0(\tau))d\tau$$

Therefore

$$\begin{aligned}\| \Phi_1(t)-\Phi_0(t) \| &= \| \int_{t_0}^{t} \mathbf{f}(\Phi_0(\tau))d\tau \| \\ &\leqslant \int_{t_0}^{t} \| \mathbf{f}(\Phi_0(\tau)) \| \, d\tau \\ &\leqslant \int_{t_0}^{t} F d\tau \\ &\leqslant F(t-t_0) \end{aligned} \tag{5.14}$$

Next we compute

$$\Phi_2(t)-\Phi_1(t)=\int_{t_0}^{t} [\mathbf{f}(\Phi_1(\tau))-\mathbf{f}(\Phi_0(\tau))\, d\tau$$

Using the Lipschitz condition, we may write

$$\begin{aligned}\| \Phi_2(t)-\Phi_1(t) \| &\leqslant \int_{t_0}^{t} F^2(\tau-t_0)d\tau \\ &\leqslant \frac{1}{2!}F^2(t-t_0)^2\end{aligned} \tag{5.15}$$

The form of inequalities (5.14) and (5.15) suggests the following result by induction:

$$\| \Phi_j(t)-\Phi_{j-1}(t) \| \leqslant \frac{1}{j!}F^j(t-t_0)^j \tag{5.16}$$

$$\| \Phi_{j+1}(t)-\Phi_j(t) \| \leqslant \frac{1}{(j+1)!}F^{j+1}(t-t_0)^{j+1} \tag{5.17}$$

Therefore

$$\lim_{j\to\infty} \sum_{i=0}^{j-1} \| \Phi_{i+1}(t)-\Phi_i(t) \| \leqslant \lim_{j\to\infty} \sum_{i=0}^{j-1} \frac{1}{(i+1)!}F^{i+1}(t-t_0)^{i+1} \tag{5.18}$$

The series on the right hand side of (5.18) converges uniformly to the function

$$e^{F(t-t_0)}-1$$

for all finite values of $(t-t_0)$. Hence we conclude that the functions $\Phi_j(t)$ converge uniformly to a limit function $\Phi(t)$ in finite time interval.

Now from (5.11), we get by letting $j\to\infty$;

$$\lim_{j\to\infty} \Phi_{j+1}(t)=\lim_{j\to\infty}\left[\mathbf{x}^0+\int_{t_0}^{t} \mathbf{f}(\Phi_j(\tau))\, d\tau \right]$$

or

$$\Phi(t)=\mathbf{x}^0+\lim_{j\to\infty}\int_{t_0}^{t} \mathbf{f}(\Phi_j(\tau))\, d\tau$$

By virtue of the assumption that f is continuous in its arguments and the fact that $\Phi_j(t)$ converges uniformly to $\Phi(t)$, we can take the limit inside the integral and the function, i.e.,

$$\Phi(t) = \mathbf{x}^0 + \int_{t_0}^{t} \lim_{j\to\infty} \mathbf{f}(\Phi_j(\tau))\, d\tau$$

$$= \mathbf{x}^0 + \int_{t_0}^{t} \mathbf{f}(\lim_{j\to\infty} \Phi_j(\tau))\, d\tau$$

$$= \mathbf{x}^0 + \int_{t_0}^{t} \mathbf{f}(\Phi(\tau))\, d\tau$$

Thus the existence of a solution $\Phi(t)$ has been established for all finite values of time.

Uniqueness of Solutions

We shall now prove (by contradiction) that under our assumptions, the solution $\Phi(t)$ is unique.

Suppose that in addition to the solution $\Phi(t)$, whose existence has been demonstrated above, there is another solution $\psi(t)$, different from $\Phi(t)$. If $\psi(t)$ is a solution, it must satisfy the equation

$$\psi(t) = \mathbf{x}^0 + \int_{t_0}^{t} \mathbf{f}(\psi(\tau))\, d\tau \tag{5.19}$$

From eqns. (5.11) and (5.19), we get

$$\psi(t) - \Phi_0(t) = \int_{t_0}^{t} \mathbf{f}(\psi(\tau))\, d\tau$$

Therefore,

$$\|\psi(t) - \Phi_0(t)\| = \left\| \int_{t_0}^{t} \mathbf{f}(\psi(\tau))\, d\tau \right\|$$

$$\leqslant \int_{t_0}^{t} \|\mathbf{f}(\psi(\tau))\|\, d\tau$$

$$\leqslant \int_{t_0}^{t} F d\tau$$

$$\leqslant F(t-t_0)$$

Similarly

$$\psi(t) - \Phi_1(t) = \int_{t_0}^{t} [\mathbf{f}(\psi(\tau)) - \mathbf{f}(\Phi_0(\tau))]\, d\tau$$

Therefore

$$\|\psi(t) - \Phi_1(t)\| \leqslant \int_{t_0}^{t} F\, \|\psi(\tau) - \Phi_0(\tau)\|\, d\tau$$

$$\leqslant \int_{t_0}^{t} F^2(\tau - t_0)\, d\tau$$

$$\leqslant \frac{1}{2!} F^2(t-t_0)^2$$

We may write, by induction

$$\| \psi(t) - \Phi_j(t) \| \leqslant \frac{1}{(j+1)!} F^{j+1}(t-t_0)^{j+1}$$

or

$$\lim_{j\to\infty} \| \psi(t) - \Phi_j(t) \| \leqslant \lim_{j\to\infty} \frac{1}{(j+1)!} F^{j+1}(t-t_0)^{j+1} \tag{5.20}$$

Since the powers of any fixed number grow more slowly than the factorial grows, the right hand side of inequality (5.20) is equal to zero in the limit for finite values of F and $(t-t_0)$. Further, since $\Phi_j(t)$ converges to $\Phi(t)$, we may write from (5.20),

$$\| \psi(t) - \Phi(t) \| \leqslant 0$$

which implies that

$$\psi(t) = \Phi(t),$$

a contradiction to our earlier assumption that the solution $\psi(t)$ was different from $\Phi(t)$. Therefore, the solution is unique.

□

We now consider the extension of these results to the state equation

$$\dot{\mathbf{x}}(t) = \mathbf{f}(t, \mathbf{x}, \mathbf{u});\ \mathbf{x}(t_0) = \mathbf{x}^0 \tag{5.21}$$

Obviously for a given $\mathbf{u}(t)$, eqn. (5.21) is equivalent to

$$\dot{\mathbf{x}}(t) = \bar{\mathbf{f}}(t, \mathbf{x}) \tag{5.22}$$

and the answer to the existence and uniqueness of solution for eqn. (5.21) may be obtained from the answer to eqn. (5.22) for a given $\mathbf{u}(t)$. Note that input functions need not necessarily be differentiable—piecewise differentiable or piecewise continuous functions are still satisfactory so long as plugging them into the differential equation still yields an integrable differential equation and this will normally be the case. Therefore, we shall enforce the following restrictions on input functions:

(i) $\|\mathbf{u}(t)\| \leqslant K$ for all finite t

(ii) $\mathbf{u}(t)$ is piecewise continuous function of time.

Equation (5.22) may be treated exactly as eqn. (5.1) by introduction of an extended state vector $\mathbf{y}(t)$, where

$$\mathbf{y}(t) = \begin{bmatrix} \mathbf{x}(t) \\ t-t_0 \end{bmatrix}$$

and an extended function vector

$$\mathbf{g}(t) = \begin{bmatrix} \bar{\mathbf{f}}(t) \\ 1 \end{bmatrix}$$

Then eqn. (5.22) becomes

$$\dot{\mathbf{y}}(t) = \mathbf{g}(\mathbf{y}) \; ; \; \mathbf{y}(t_0) = \begin{bmatrix} \mathbf{x}^0 \\ 0 \end{bmatrix} = \mathbf{y}^0$$

The existence and uniqueness theorem proved earlier now applies.

5.3 SOLUTION OF NONLINEAR CONTINUOUS–TIME STATE EQUATIONS

It is usually not possible to write an explicit solution of the general nonlinear state equation of the form (5.21). For the case where the state vector $\mathbf{x}(t)$ has two or at most three components, graphical methods involving the state-plane or state-space may be used to generate pictorial representation of the state behaviour (Gibson 1963). The graphical methods become awkward when the dimension of the state vector is greater than 2. In such cases, one must resort to numerical solution techniques.

Many special numerical methods for solving nonlinear differential equations have been described in the literature. They differ in the amount of programming complexity required and the amount of computer time needed to solve a given equation for a given accuracy. We shall not enter into the discussion of these techniques; we simply present below the algorithm of a very popular solution method—the fourth-order Runge-Kutta method (for proof, refer any book on Numerical Analysis, e.g., Hamming (1971)).

Fourth-Order Runge-Kutta Algorithm

Given the differential equation

$$\dot{\mathbf{x}}(t) = \mathbf{f}(t, \mathbf{x}, \mathbf{u}) \; ; \; \mathbf{x}(t_0) = \mathbf{x}^0$$

Let $\mathbf{x}_i$ be the value of $\mathbf{x}$ at $t = t_i$ and h be the increment (step-size) of the time variable t. The fourth-order Runge-Kutta method uses the formulas given below:

$$\begin{aligned}
\mathbf{S}_1 &= \mathbf{f}(t_i, \mathbf{x}_i, \mathbf{u}(t_i)) \\
\mathbf{S}_2 &= \mathbf{f}\left(t_i + \frac{h}{2}, \mathbf{x}_i + \frac{h}{2}.\mathbf{S}_1, \mathbf{u}\left(t_i + \frac{h}{2}\right)\right) \\
\mathbf{S}_3 &= \mathbf{f}\left(t_i + \frac{h}{2}, \mathbf{x}_i + \frac{h}{2}.\mathbf{S}_2, \mathbf{u}\left(t_i + \frac{h}{2}\right)\right) \qquad (5.23) \\
\mathbf{S}_4 &= \mathbf{f}(t_i + h, \mathbf{x}_i + h.\mathbf{S}_3, \mathbf{u}(t_i + h)) \\
\mathbf{x}_{i+1} &\overset{\Delta}{=} \mathbf{x}(t_i + h) \\
&\simeq \mathbf{x}_i + \frac{h}{6}(\mathbf{S}_1 + 2\mathbf{S}_2 + 2\mathbf{S}_3 + \mathbf{S}_4)
\end{aligned}$$

Example 5.1: Consider the Nuclear Reactor model given by (eqns. (4.59))

$$\dot{x}_1 = f_1(\mathbf{x}, u)$$

$$= -6.4x_1(t) + 0.1x_2(t) + 10^3 x_1(t)\, u(t)$$

$$\dot{x}_2 = f_2(\mathbf{x}, u)$$

$$= 6.4x_1(t) - 0.1x_2(t)$$

The initial value of neutron power $= x_1^0 = 10$ kw
The initial value of delayed neutron precursors (refer Section 4.6) $= x_2^0 = 640$.

The variation of neutron power (x_1) for a reactivity input of $u(t) = 0.009 - 0.1\,(t-0.3)^2$ is shown in Fig. 5.1. The nonlinear state equations were solved on a digital computer using Runge-Kutta method described earlier.

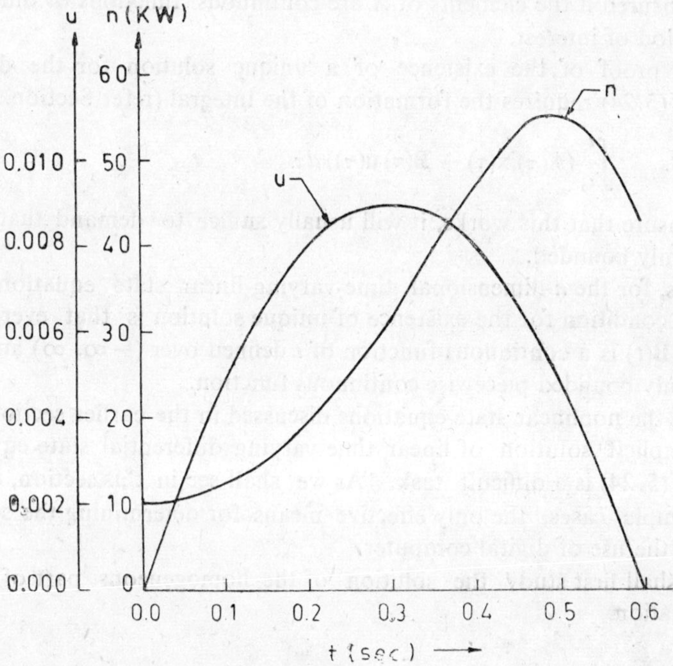

Fig. 5.1

5.4 SOLUTION OF LINEAR TIME-VARYING CONTINUOUS-TIME STATE EQUATIONS

Consider the nth-order linear state equation (eqn. (3.26a))

$$\dot{\mathbf{x}}(t) = \mathbf{A}(t)\,\mathbf{x}(t) + \mathbf{B}(t)\,\mathbf{u}(t);\ \mathbf{x}(t_0) = \mathbf{x}^0 \tag{5.24}$$

where $\mathbf{A}(t)$ and $\mathbf{B}(t)$ are, respectively, $n \times n$ and $n \times p$ real time-varying matrices.

Having fixed upon a particular input function $\mathbf{u}(t)$, we are then considering the case in which

$$\mathbf{f}(t, \mathbf{x}) = \mathbf{A}(t)\,\mathbf{x}(t) + \mathbf{B}(t)\,\mathbf{u}(t)$$

This $\mathbf{f}$ satisfies the Lipschitz condition if there exists a constant L for which

$$\|\mathbf{f}(t, \mathbf{x}_1) - \mathbf{f}(t, \mathbf{x}_2)\| \leqslant L \|\mathbf{x}_1 - \mathbf{x}_2\|; \mathbf{x}_1, \mathbf{x}_2 \in \mathcal{R}^n$$

or

$$\|\mathbf{A}(t)(\mathbf{x}_1 - \mathbf{x}_2)\| \leqslant L \|\mathbf{x}_1 - \mathbf{x}_2\|$$

holds true for all values of finite t.

Using eqn. (2.30), we may write

$$\|\mathbf{A}(t)(\mathbf{x}_1 - \mathbf{x}_2)\| \leqslant \|\mathbf{A}(t)\| \, \|\mathbf{x}_1 - \mathbf{x}_2\|$$

Thus the Lipschitz condition is satisfied if $\mathbf{A}(t)$ is uniformly bounded:

$$\|\mathbf{A}(t)\| \leqslant L \text{ for all finite } t.$$

This is ensured if the elements of $\mathbf{A}$ are continuous functions of time over the finite period of interest.

Our proof of the existence of a unique solution for the differential equation (5.24) requires the formation of the integral (refer Section 5.2)

$$\int_{t_0}^{t} (\mathbf{A}(\tau)\,\mathbf{x}(\tau) + \mathbf{B}(\tau)\,\mathbf{u}(\tau))\, d\tau$$

and to ensure that this works, it will usually suffice to demand that $\mathbf{B}(t)$ also is uniformly bounded.

Thus, for the n-dimensional time-varying linear state equation (5.24), a sufficient condition for the existence of unique solution is that every entry of $\mathbf{A}(t)$ and $\mathbf{B}(t)$ is a continuous function of t defined over $(-\infty, \infty)$ and $\mathbf{u}(t)$ is a uniformly bounded piecewise continuous function.

Like the nonlinear state equations discussed in the earlier section, obtaining the explicit solution of linear time-varying differential state equations of the form (5.24) is a difficult task. As we shall see in this section, except for a few simple cases, the only effective means for determining the solution of (5.24) is the use of digital computer.

We shall first study the solution of the homogeneous part of the linear state equation.

The Homogeneous Solution

Consider the homogeneous part of the state equation (5.24):

$$\dot{\mathbf{x}}(t) = \mathbf{A}(t)\mathbf{x}(t); \; \mathbf{x}(t_0) = \mathbf{x}^0 \tag{5.25}$$

Theorem 5.1: The set of all solutions of (5.25) forms an n-dimensional vector space over the field of real numbers.

Proof: Let ψ_1 and ψ_2 be two arbitrary solutions of (5.25). For any real α_1 and α_2, we have

$$\frac{d}{dt}(\alpha_1\psi_1 + \alpha_2\psi_2) = \alpha_1 \frac{d\psi_1}{dt} + \alpha_2 \frac{d\psi_2}{dt}$$

$$= \alpha_1\mathbf{A}(t)\psi_1 + \alpha_2\mathbf{A}(t)\psi_2$$

$$= \mathbf{A}(t)\,(\alpha_1\psi_1 + \alpha_2\psi_2)$$

Thus $\alpha_1\psi_1 + \alpha_2\psi_2$ is also a solution of (5.25) for any $\alpha_1, \alpha_2 \in \mathcal{R}$. The set of solutions of (5.25), therefore, forms a linear space, called the *solution space*.

To prove that the solution space of (5.25) has the dimension n, we assume that ψ_i are the solutions of (5.25) with the initial conditions $\mathbf{x}_i^0$; $i = 1, 2,\ldots,n$, where $\mathbf{x}_1^0, \mathbf{x}_2^0\ldots, \mathbf{x}_n^0$ are any linearly independent vectors in $\mathcal{R}^n(\mathcal{R})$. Suppose that ψ_i; $i = 1, 2,\ldots,n$ are linearly dependent; then there exist $\alpha_1, \alpha_2,\ldots,\alpha_n$ (not all zero) such that

$$\alpha_1\psi_1(t) + \alpha_2\psi_2(t) + \ldots + \alpha_n\psi_n(t) = \mathbf{0} \,\forall\, t \in (-\infty, \infty)$$

or

$$[\psi_1(t)\ \ \psi_2(t)\ldots\psi_n(t)]\ \boldsymbol{\alpha} = \mathbf{0}\ \ \forall t \in (-\infty, \infty)$$

In particular,

$$[\psi_1(t_0)\ \ \psi_2(t_0)\ldots\psi_n(t_0)]\ \boldsymbol{\alpha} = [\mathbf{x}_1^0\ \ \mathbf{x}_2^0\ldots\mathbf{x}_n^0]\ \boldsymbol{\alpha} = \mathbf{0}$$

which implies that $\mathbf{x}_i^0$; $i = 1, 2,\ldots,n$ are linearly dedendent. This contradicts the hypothesis; hence ψ_i; $i = 1, 2\ldots,n$ are linearly independent over $(-\infty, \infty)$.

Now let us prove that any solution of (5.25) can be expressed as a linear combination of ψ_i; $i = 1, 2,\ldots,n$. Let ψ be any solution of (5.25) and let $\psi(t_0) = \mathbf{x}^0$. Since $\mathbf{x}_1^0, \mathbf{x}_2^0,\ldots, \mathbf{x}_n^0$ are linearly independent vectors in the n-dimensional real vector space, $\mathbf{x}^0$ can be expressed as

$$\mathbf{x}^0 = \sum_{i=1}^{n} \alpha_i\, \mathbf{x}_i^0$$

Since $\sum_{i=1}^{n} \alpha_i\psi_i(t)$ is a solution of (5.25) with the initial condition $\sum_{i=1}^{n} \alpha_i\psi_i(t_0) = \mathbf{x}^0$, from the uniqueness of the solution we conclude that

$$\psi(t) = \sum_{i=1}^{n} \alpha_i\psi_i(t)$$

This completes the proof of the theorem.

□

Let Ψ be a matrix whose n columns consist of n linearly independent solutions of (5.25). The $n \times n$ matrix Ψ is called a *fundamental matrix* (or a *Wronskian matrix*) of (5.25).

Example 5.2: Consider the system

$$\dot{\mathbf{x}}(t) = \begin{bmatrix} 1 & 0 \\ 0 & 2t \end{bmatrix} \mathbf{x}(t)$$

It consists of two equations

$$\dot{x}_1(t) = x_1(t)$$

$$\dot{x}_2(t) = 2tx_2(t)$$

The solutions of these equations are (refer Example 3.5)

$$x_1(t) = x_1(t_0)\, e^{(t - t_0)}$$

$$x_2(t) = x_2(t_0)\, e^{(t^2 - t_0^2)}$$

Thus

$$\mathbf{x}(t) = \begin{bmatrix} x_1(t_0)\, e^{(t - t_0)} \\ x_2(t_0)\, e^{(t^2 - t_0^2)} \end{bmatrix}$$

is the general solution of the given equation.

Now two linearly independent solutions

$$\psi_1 = \begin{bmatrix} e^{(t - t_0)} \\ 0 \end{bmatrix} \text{ and } \psi_2 = \begin{bmatrix} 0 \\ e^{(t^2 - t_0^2)} \end{bmatrix}$$

can be easily obtained by setting $x_1(t_0) = 1, x_2(t_0) = 0$ and $x_1(t_0) = 0, x_2(t_0) = 1$. Hence the matrix

$$\Psi = \begin{bmatrix} e^{(t - t_0)} & 0 \\ 0 & e^{(t^2 - t_0^2)} \end{bmatrix}$$

is a fundamental matrix.

□

It is obvious from the example above that the fundamental matrix of (5.25) is not unique.

Since each column of Ψ satisfies the differential equation (5.25), matrix Ψ satisfies the matrix equation

$$\dot{\Psi} = \mathbf{A}(t)\, \Psi;\ \Psi(t_0) = \mathbf{X}^0, \quad \text{some nonsingular real constant matrix} \tag{5.26}$$

Because the fundamental matrix is nonsingular at some t_0, it will be nonsingular for all t. This follows from Theorem 5.1 wherein we proved that ψ_i; $i = 1, 2, \ldots, n$ are linearly independent over $(-\infty, \infty)$.

Since $\Psi(t)$ is a solution of eqn. (5.26) and since $\Psi(t_0)$ is a nonsingular constant matrix for any t_0, the matrix $\Phi(t, t_0)$ defined by

$$\Phi(t, t_0) = \Psi(t)\, \Psi^{-1}(t_0) \ \forall\ t, t_0 \text{ in } (-\infty, \infty) \tag{5.27}$$

also satisfies (5.26). Thus

$$\frac{d}{dt}\Phi(t, t_0) = \mathbf{A}(t)\, \Phi(t, t_0) \tag{5.28}$$

$$\Phi(t_0, t_0) = \mathbf{I}$$

The matrix $\Phi(t, t_0)$ is called the *state transition matrix* of (5.25). From the definition (5.27), we have the following important properties of state transition matrix.

$$\begin{aligned} \Phi(t, t) &= \mathbf{I} \\ \Phi^{-1}(t, t_0) &= \Psi(t_0)\,\Psi^{-1}(t) = \Phi(t_0, t) \\ \Phi(t_2, t_0) &= \Phi(t_2, t_1)\,\Phi(t_1, t_0) \end{aligned} \tag{5.29}$$

for any t, t_0, t_1 and t_2 in $(-\infty, \infty)$.

The state transition matrix $\Phi(t, t_0)$ is uniquely determined by $\mathbf{A}(t)$ and is independent of particular Ψ chosen. To establish this, let us assume that Ψ and $\overline{\Psi}$ are two different fundamental matrices of (5.25). The columns of Ψ, as well as the columns of $\overline{\Psi}$ qualify as basis vectors; therefore there exists a nonsingular real constant matrix $\mathbf{P}$ such that $\overline{\Psi} = \Psi\mathbf{P}$. Thus

$$\begin{aligned} \Phi(t, t_0) &= \overline{\Psi}(t)\,\overline{\Psi}^{-1}(t_0) \\ &= \Psi(t)\,\mathbf{P}\mathbf{P}^{-1}\Psi^{-1}(t_0) \\ &= \Psi(t)\,\Psi^{-1}(t_0) \end{aligned}$$

Example 5.3: Let us compute state transition matrix for the system of Example 5.2.

In Example 5.2, we found that

$$\Psi(t) = \begin{bmatrix} e^{(t-t_0)} & 0 \\ 0 & e^{(t^2 - t_0^2)} \end{bmatrix}$$

Thus

$$\begin{aligned} \Phi(t, t_0) &= \Psi(t)\,\Psi^{-1}(t_0) \\ &= \begin{bmatrix} e^{(t-t_0)} & 0 \\ 0 & e^{(t^2 - t_0^2)} \end{bmatrix} \begin{bmatrix} 1 & 0 \\ 0 & 1 \end{bmatrix} \\ &= \begin{bmatrix} e^{(t-t_0)} & 0 \\ 0 & e^{(t^2 - t_0^2)} \end{bmatrix} \end{aligned}$$

□

From the concept of state transition matrix, the solution of (5.25) follows immediately. The solution of $\dot{\mathbf{x}}(t) = \mathbf{A}(t)\,\mathbf{x}(t)$ with $\mathbf{x}(t_0) = \mathbf{x}^0$ is given by

$$\mathbf{x}(t) = \Phi(t, t_0)\,\mathbf{x}^0 \tag{5.30}$$

which can be verified by direct substitution. From (5.30) it is obvious that the state transition matrix is a linear transformation that maps the state $\mathbf{x}^0$ at time t_0 into the state $\mathbf{x}$ at time t. It, thus, governs the motion of the state vector in the time interval in which input is identically zero.

Evaluation of State Transition Matrix

It is very difficult to obtain the state transition matrix of a linear time-varying system, except numerically on a digital computer. A simple analytical solution is however possible when $\mathbf{A}(t)$ and $\int_{t_0}^{t} \mathbf{A}(\tau)\, d\tau$ commute for all t, i.e.,

$$\mathbf{A}(t)\left(\int_{t_0}^{t} \mathbf{A}(\tau)\, d\tau\right) = \left(\int_{t_0}^{t} \mathbf{A}(\tau)\, d\tau\right)\mathbf{A}(t) \tag{5.31}$$

Under this condition, the state transition matrix

$$\Phi(t, t_0) = \exp\left[\int_{t_0}^{t} \mathbf{A}(\tau)\, d\tau\right] \tag{5.32}$$

To establish this result, let us assume that the solution to (5.25) is

$$\mathbf{x}(t) = \Phi(t, t_0)\, \mathbf{x}(t_0)$$

where $\Phi(t, t_0)$ is given by (5.32).

Substituting the series form of $\exp\left[\int_{t_0}^{t} \mathbf{A}(\tau)\, d\tau\right]$;

which is

$$\exp\left[\int_{t_0}^{t} \mathbf{A}(\tau)\, d\tau\right] = \mathbf{I} + \int_{t_0}^{t} \mathbf{A}(\tau)\, d\tau + \frac{1}{2!}\int_{t_0}^{t} \mathbf{A}(\tau)\, d\tau \int_{t_0}^{t} \mathbf{A}(\theta)\, d\theta + \ldots$$

and the series form of $\frac{d}{dt}\exp\left[\int_{t_0}^{t} \mathbf{A}(\tau)\, d\tau\right]$;

which is

$$\frac{d}{dt}\exp\left[\int_{t_0}^{t} \mathbf{A}(\tau)\, d\tau\right] = \mathbf{A}(t) + \tfrac{1}{2}\mathbf{A}(t)\int_{t_0}^{t} \mathbf{A}(\theta)\, d\theta + \tfrac{1}{2}\left(\int_{t_0}^{t} \mathbf{A}(\tau)\, d\tau\right)\mathbf{A}(t) + \ldots$$

into eqn. (5.28), we obtain

$$\mathbf{A}(t) + \tfrac{1}{2}\mathbf{A}(t)\int_{t_0}^{t} \mathbf{A}(\tau)d\tau + \tfrac{1}{2}\left(\int_{t_0}^{t} \mathbf{A}(\tau)d\tau\right)\mathbf{A}(t) + \ldots$$
$$= \mathbf{A}(t) + \mathbf{A}(t)\int_{t_0}^{t} \mathbf{A}(\tau)d\tau + \frac{1}{2!}\mathbf{A}(t)\int_{t_0}^{t} \mathbf{A}(\tau)d\tau \int_{t_0}^{t} \mathbf{A}(\theta)d\theta + \ldots$$

This equation is true if and only if,

$$\tfrac{1}{2}\mathbf{A}(t)\left[\int_{t_0}^{t} \mathbf{A}(\tau)d\tau\right] + \tfrac{1}{2}\left[\int_{t_0}^{t} \mathbf{A}(\tau)d\tau\right]\mathbf{A}(t) = \mathbf{A}(t)\left[\int_{t_0}^{t} \mathbf{A}(\tau)d\tau\right]$$

which is true if and only if

$$\mathbf{A}(t)\left[\int_{t_0}^{t} \mathbf{A}(\tau)d\tau\right] = \left[\int_{t_0}^{t} \mathbf{A}(\tau)d\tau\right]\mathbf{A}(t)$$

i.e., if and only if $\mathbf{A}(t)$ and $\int_{t_0}^{t} \mathbf{A}(\tau)d\tau$ commute for all t.

It is clear that this commutativity requirement is satisfied if $\mathbf{A}(t)$ is a diagonal matrix or a constant matrix. However, most of the time-varying matrices do not have this commutativity property and (5.31) does not hold. In these cases, there is no simple relation between $\mathbf{A}(t)$ and $\Phi(t, t_0)$ and solving for Φ is generally a very difficult task. We frequently have to resort to numerical integration for computing $\Phi(t, t_0)$ (Refer Wu (1974a, 1974b, 1975) for a method of computing state transition matrix for a class of linear time-varying systems).

Example 5.4: The condition given by (5.31) is satisfied by

$$\mathbf{A}(t) = \begin{bmatrix} 1 & 0 \\ 0 & 2t \end{bmatrix}$$

Therefore, from (5.32) we have

$$\Phi(t, t_0) = \exp\left[\int_{t_0}^{t} \mathbf{A}(\tau)d\tau\right]$$

$$= \exp\begin{bmatrix} (t-t_0) & 0 \\ 0 & (t^2-t_0^2) \end{bmatrix}$$

$$= \begin{bmatrix} e^{(t-t_0)} & 0 \\ 0 & e^{(t^2-t_0^2)} \end{bmatrix}$$

The Nonhomogeneous Solution

Theorem 5.2: The solution of the state equation (5.24) is given by

$$\mathbf{x}(t) = \Phi(t, t_0)\mathbf{x}^0 + \int_{t_0}^{t} \Phi(t, \tau)\mathbf{B}(\tau)\mathbf{u}(\tau)d\tau \tag{5.33a}$$

$$= \Phi(t, t_0)\left[\mathbf{x}^0 + \int_{t_0}^{t} \Phi(t_0, \tau)\mathbf{B}(\tau)\mathbf{u}(\tau)d\tau\right] \tag{5.33b}$$

where $\Phi(t, \tau)$ is the unique solution of

$$\frac{d}{dt}\Phi(t, \tau) = \mathbf{A}(t)\Phi(t, \tau)$$

$$\Phi(\tau, \tau) = \mathbf{I} \tag{5.34}$$

Proof:

By direct substitution, we can show that (5.33a) is a solution of (5.24).

$$\frac{d}{dt}\mathbf{x}(t) = \frac{d}{dt}\Phi(t, t_0)\mathbf{x}^0 + \frac{d}{dt}\int_{t_0}^{t} \Phi(t, \tau)\mathbf{B}(\tau)\mathbf{u}(\tau)d\tau$$

$$= \mathbf{A}(t)\Phi(t, t_0)\mathbf{x}^0 + \Phi(t, \tau)\mathbf{B}(\tau)\mathbf{u}(\tau)\Big|_{\tau=t} + \int_{t_0}^{t} \frac{d}{dt}\Phi(t, \tau)\mathbf{B}(\tau)\mathbf{u}(\tau)d\tau$$

$$= \mathbf{A}(t)\Phi(t, t_0)\mathbf{x}^0 + \Phi(t, t)\mathbf{B}(t)\mathbf{u}(t) + \int_{t_0}^{t} \mathbf{A}(t)\Phi(t, \tau)\mathbf{B}(\tau)\mathbf{u}(\tau)d\tau$$

$$= \mathbf{A}(t)\left[\Phi(t, t_0)\mathbf{x}^0 + \int_{t_0}^{t} \Phi(t, \tau)\mathbf{B}(\tau)\mathbf{u}(\tau)d\tau\right] + \mathbf{B}(t)\mathbf{u}(t)$$

$$= \mathbf{A}(t)\mathbf{x}(t) + \mathbf{B}(t)\mathbf{u}(t)$$

Equation (5.33b) is obtained from (5.33a) by using the property

$$\Phi(t, \tau) = \Phi(t, t_0)\Phi(t_0, \tau)$$

□

To be more informative, we use $\boldsymbol{\Phi}(t, t_0, \mathbf{x}^0, \mathbf{u})$ to denote the solution of (5.24) at time t due to initial condition $\mathbf{x}(t_0) = \mathbf{x}^0$ and input $\mathbf{u}(t)$.

If $\mathbf{u} \equiv \mathbf{0}$, then eqn. (5.33a) reduces to

$$\boldsymbol{\Phi}(t, t_0, \mathbf{x}^0, \mathbf{0}) = \Phi(t, t_0)\mathbf{x}^0 \tag{5.35}$$

If $\mathbf{x}^0 = \mathbf{0}$, eqn. (5.33a) gives

$$\boldsymbol{\Phi}(t, t_0, \mathbf{0}, \mathbf{u}) = \int_{t_0}^{t} \Phi(t, \tau)\mathbf{B}(\tau)\mathbf{u}(\tau)d\tau \tag{5.36}$$

$\boldsymbol{\Phi}(t, t_0, \mathbf{x}^0, \mathbf{0})$ given by (5.35) is the *zero-input response* and $\boldsymbol{\Phi}(t, t_0, \mathbf{0}, \mathbf{u})$ given by (5.36) is the *zero-state response* of state equation (5.24). Note that

$$\boldsymbol{\Phi}(t, t_0, \mathbf{x}^0, \mathbf{u}) = \boldsymbol{\Phi}(t, t_0, \mathbf{x}^0, \mathbf{0}) + \boldsymbol{\Phi}(t, t_0, \mathbf{0}, \mathbf{u}) \tag{5.37}$$

Example 5.5: Consider the system described by the state equation

$$\dot{\mathbf{x}}(t) = \mathbf{A}(t)\mathbf{x}(t) + \mathbf{b}u(t)$$

$$= \begin{bmatrix} 1 & e^{-t} \\ 0 & -1 \end{bmatrix}\mathbf{x}(t) + \begin{bmatrix} 0 \\ 1 \end{bmatrix}u(t); \ \mathbf{x}(0) = \mathbf{x}^0 = \mathbf{0}$$

The homogeneous part of the state equation is

$$\dot{\mathbf{x}}(t) = \begin{bmatrix} 1 & e^{-t} \\ 0 & -1 \end{bmatrix}\mathbf{x}(t)$$

or

$$\dot{x}_1(t) = x_1(t) + e^{-t}x_2(t)$$
$$\dot{x}_2(t) = -x_2(t)$$

The solution (refer Example 3.5)

$$\mathbf{x}(t) = \begin{bmatrix} e^{(t-t_0)}x_1^0 + (\frac{1}{3}e^{(t-2t_0)} - \frac{1}{3}e^{(-2t+t_0)})x_2^0 \\ e^{-(t-t_0)}x_2^0 \end{bmatrix}$$

$$= \begin{bmatrix} e^{(t-t_0)} & \frac{1}{3}e^{(t-2t_0)} - \frac{1}{3}e^{(-2t+t_0)} \\ 0 & e^{-(t-t_0)} \end{bmatrix}\begin{bmatrix} x_1^0 \\ x_2^0 \end{bmatrix}$$

$$= \Phi(t, t_0)\mathbf{x}^0$$

Therefore, the state transition matrix

$$\Phi(t, \tau) = \begin{bmatrix} e^{(t-\tau)} & \frac{1}{3}[e^{(t-2\tau)} - e^{(-2t+\tau)}] \\ 0 & e^{-(t-\tau)} \end{bmatrix}$$

The solution of the nonhomogeneous equation is

$$\mathbf{x}(t) = \Phi(t, 0)\left[\mathbf{x}^0 + \int_0^t \Phi(0, \tau)\, \mathbf{b}u(\tau)d\tau\right]$$

$$= \Phi(t, 0)\int_0^t \Phi(0, \tau)\, \mathbf{b}u(\tau)d\tau$$

$$= \begin{bmatrix} e^t & \frac{1}{3}(e^t - e^{-2t}) \\ 0 & e^{-t} \end{bmatrix} \int_0^t \begin{bmatrix} e^{-\tau} & \frac{1}{3}(e^{-2\tau} - e^{\tau}) \\ 0 & e^{\tau} \end{bmatrix} \begin{bmatrix} 0 \\ 1 \end{bmatrix} u(\tau)d\tau$$

$$= \begin{bmatrix} e^t & \frac{1}{3}(e^t - e^{-2t}) \\ 0 & e^{-t} \end{bmatrix} \int_0^t \begin{bmatrix} \frac{1}{3}(e^{-2\tau} - e^{\tau}) \\ e^{\tau} \end{bmatrix} u(\tau)\, d\tau$$

For a unit step input applied at $t = 0$,

$$\mathbf{x}(t) = \begin{bmatrix} e^t & \frac{1}{3}(e^t - e^{-2t}) \\ 0 & e^{-t} \end{bmatrix} \begin{bmatrix} \frac{1}{3}(\frac{3}{2} - e^t - \frac{1}{2}e^{-2t}) \\ e^t - 1 \end{bmatrix}$$

$$= \begin{bmatrix} \frac{1}{6}e^t - \frac{1}{2}e^{-t} + \frac{1}{3}e^{-2t} \\ 1 - e^{-t} \end{bmatrix}$$

Adjoint Equations

Suppose that the $n \times n$ matrix $\Phi^a(t, t_0)$ satisfies the equation

$$\frac{d}{dt}\Phi^a(t, t_0) = -\Phi^a(t, t_0)\,\mathbf{A}(t) \tag{5.38}$$

$$\Phi^a(t_0, t_0) = \mathbf{I}$$

Equation (5.38) is called the *adjoint* of eqn. (5.28). To show how the adjoint systems of equations are related, we multiply eqn. (5.28) by $\Phi^a(t, t_0)$ on the left, eqn. (5.38) by $\Phi\,(t, t_0)$ on the right and then add to get

$$\frac{d}{dt}\,[\Phi^a(t, t_0)\,\Phi(t, t_0)] = 0$$

which has the solution

$$\Phi^a(t, t_0)\,\Phi(t, t_0) = \text{constant.}$$

From the relations $\Phi^a(t_0, t_0) = \Phi(t_0, t_0) = \mathbf{I}$, it follows that the constant is a unit matrix **I**. Therefore,

$$\Phi^a(t, t_0) = \Phi^{-1}(t, t_0) = \Phi(t_0, t) \tag{5.39}$$

The matrix equation (5.38) can be transformed into a vector equation as follows. We take the transpose of both sides in (5.38) to obtain

$$\frac{d}{dt}\,\Phi^{a^T}(t, t_0) = -\,\mathbf{A}^T(t)\,\Phi^{a^T}(t, t_0) \tag{5.40}$$

$\Phi^{a^T}(t, t_0)$ is thus the transition matrix for dynamic system with the homogeneous state equation

$$\frac{d\mathbf{x}}{dt} = -\,\mathbf{A}^T(t)\,\mathbf{x}(t) \tag{5.41}$$

Equation (5.41) is called the adjoint of eqn. (5.25).

The significance of adjoint equations may be seen by multiplying eqn. (5.33a) on the left by $\Phi^{-1}(t, t_0)$. This gives

$$\mathbf{x}(t_0) = \Phi^{-1}(t, t_0)\,\mathbf{x}(t) - \int_{t_0}^{t} \Phi(t_0, \tau)\,\mathbf{B}(\tau)\,\mathbf{u}(\tau)\,d\tau$$

$$= \Phi^a(t, t_0)\,\mathbf{x}(t) - \int_{t_0}^{t} \Phi^a(\tau, t_0)\,\mathbf{B}(\tau)\,\mathbf{u}(\tau)\,d\tau \tag{5.42}$$

Equation (5.42) enables us to compute the initial condition $\mathbf{x}(t_0)$ from a given final condition, i.e., to compute backwards. Therefore to solve (5.33a) in the backward direction, we can solve the adjoint equation (5.41) in the forward direction for the transition matrix $\Phi^a(t, t_0)$ and then use eqn. (5.42) in a direct manner.

5.5 SOLUTION OF LINEAR TIME-INVARIANT CONTINUOUS-TIME STATE EQUATIONS

In this section we study the n-th order linear state equation (eqn. (3.27a))

$$\dot{\mathbf{x}}(t) = \mathbf{A}\mathbf{x}(t) + \mathbf{B}\mathbf{u}(t);\ \mathbf{x}(t_0) = \mathbf{x}^0 \tag{5.43}$$

where $\mathbf{A}$ and $\mathbf{B}$ are, respectively, $n \times n$ and $n \times p$ real constant matrices.

Since eqn. (5.43) is a special case of linear time-varying equation (5.24), all the results derived in the previous section can be applied here. Consider first the homogeneous part of eqn. (5.43), i.e.,

$$\dot{\mathbf{x}}(t) = \mathbf{A}\mathbf{x}(t);\ \mathbf{x}(t_0) = \mathbf{x}^0 \tag{5.44}$$

The solution of this equation is given by (eqn. (5.30))

$$\mathbf{x}(t) = \Phi(t, t_0)\mathbf{x}^0 \tag{5.45}$$

where $\Phi(t, t_0)$ is the state transition matrix of (5.44) satisfying the differential equation (eqn. (5.28))

$$\frac{d}{dt}\,\Phi(t, t_0) = \mathbf{A}\Phi(t, t_0) \tag{5.46a}$$

and the boundary condition

$$\Phi(t_0, t_0) = \mathbf{I} \tag{5.46b}$$

From eqns. (5.31) and (5.32) we find that the state transition matrix of (5.44) is given by

$$\Phi(t, t_0) = e^{\mathbf{A}(t-t_0)} = \Phi(t-t_0) \tag{5.47}$$

If $t_0=0$, as is usually assumed in time-invariant systems, then the state transition matrix becomes

$$\Phi(t) = e^{\mathbf{A}t} \tag{5.48}$$

A matrix exponential has been defined earlier by a convergent series in eqn. (2.78). This gives

$$e^{\mathbf{A}t} = \mathbf{I} + \mathbf{A}t + \frac{1}{2!}\mathbf{A}^2t^2 + \ldots + \frac{1}{k!}\mathbf{A}^kt^k + \ldots$$

$$= \sum_{i=0}^{\infty} \frac{1}{i!}\mathbf{A}^i t^i \tag{5.49}$$

which converges for all **A** and all finite t. By differentiation, term by term, of (5.48) we have

$$\frac{d}{dt}e^{\mathbf{A}t} = \sum_{i=1}^{\infty} \frac{1}{(i-1)!}\mathbf{A}^{i-1}t^{i-1}$$

$$= \mathbf{A}\left(\sum_{i=0}^{\infty} \frac{1}{i!}\mathbf{A}^i t^i\right)$$

$$= \mathbf{A}e^{\mathbf{A}t} = e^{\mathbf{A}t}\,\mathbf{A} \tag{5.50}$$

Here we have used the fact given by eqn. (2.74) that functions of the same matrix commute.

Substituting $t=0$ in (5.49), we get

$$e^{\mathbf{A}0} = e^{\mathbf{0}} = \mathbf{I} \tag{5.51}$$

Thus matrix exponential $e^{\mathbf{A}t}$ satisfies eqns. (5.46).

In the following we state some of the properties of matrix exponential, which follow from the definition in (5.49).

$$e^{\mathbf{A}t}.e^{\mathbf{A}\tau} = e^{\mathbf{A}(t+\tau)} \tag{5.52}$$

In (5.52), if we choose $\tau = -t$, then from the fact that $e^{\mathbf{0}} = \mathbf{I}$, we have

$$(e^{\mathbf{A}t})^{-1} = e^{-\mathbf{A}t} \tag{5.53}$$

$$e^{(\mathbf{A}+\mathbf{B})t} = e^{\mathbf{A}t}.e^{\mathbf{B}t} \text{ if and only if } \mathbf{AB}=\mathbf{BA} \tag{5.54}$$

It follows from eqn. (5.33) that the solution of (5.43) is

$$\mathbf{x}(t) = \Phi(t, 0, \mathbf{x}^0, \mathbf{u})$$

$$= \Phi(t)\mathbf{x}^0 + \int_0^t \Phi(t-\tau)\mathbf{B}\mathbf{u}(\tau)d\tau \tag{5.55a}$$

$$= e^{\mathbf{A}t}\,\mathbf{x}^0 + \int_0^t e^{\mathbf{A}(t-\tau)}\mathbf{B}\mathbf{u}(\tau)d\tau \tag{5.55b}$$

$$=e^{\mathbf{A}t}\left[\mathbf{x}^0+\int_0^t e^{-\mathbf{A}\tau}\,\mathbf{Bu}(\tau)d\tau\right] \tag{5.55c}$$

If the initial state is known at $t=t_0$ rather than $t=0$, eqns. (5.55) become

$$\mathbf{x}(t)=\Phi(t,\,t_0,\,\mathbf{x}^0,\mathbf{u})$$

$$=\Phi(t-t_0)\mathbf{x}^0+\int_{t_0}^{t}\Phi(t-\tau)\,\mathbf{Bu}(\tau)d\tau \tag{5.56a}$$

$$=e^{\mathbf{A}(t-t_0)}\mathbf{x}^0+\int_{t_0}^{t}e^{\mathbf{A}(t-\tau)}\,\mathbf{Bu}(\tau)d\tau \tag{5.56b}$$

$$=e^{\mathbf{A}(t-t_0)}\left[\mathbf{x}^0+\int_{t_0}^{t}e^{\mathbf{A}(t_0-\tau)}\,\mathbf{Bu}(\tau)d\tau\right] \tag{5.56c}$$

The entire discussion of adjoints given in earlier section also applies to the special case when $\mathbf{A}(t)=\mathbf{A}$, a constant matrix.

Evaluation of Matrix Exponential

The effective computation of the matrix $e^{\mathbf{A}t}$, defined by the convergent series (5.49) is the topic of this subsection. Dozens of methods for computing $e^{\mathbf{A}t}$ have been reported in the literature on the subject. Some of the methods have been proposed as specific algorithms, while others are based on less constructive characterizations. Moler and VanLoan (1978) give a bibliography which concentrates on recent papers with strong algorithmic content. Some commonly used methods are discussed below.

Series Evaluation : This is a straightforward way to evaluate the matrix exponential using a digital computer. This method simply approximates $e^{\mathbf{A}t}$ by evaluating only the first, say, N terms in the series expansion (5.49). An algorithm for evaluation of matrix series is given in Section 5.7.

Evaluation Using Similarity Transformation : Suppose that **A** is a square matrix with distinct eigenvalues $\lambda_1, \lambda_2, \ldots, \lambda_n$. We define the diagonal matrix **J** as

$$\mathbf{J}=\begin{bmatrix}\lambda_1 & 0 & 0\ldots 0\\ 0 & \lambda_2 & 0\ldots 0\\ \vdots & \vdots & \vdots\\ 0 & 0 & 0\ldots\lambda_n\end{bmatrix}$$

A and **J** are similar matrices; there exists a nonsingular modal matrix **M** such that

$$\mathbf{J}=\mathbf{M}^{-1}\mathbf{AM}$$

Now

$$\mathbf{M}^{-1}e^{\mathbf{A}}\mathbf{M}=\mathbf{M}^{-1}\left[\mathbf{I}+\mathbf{A}+\frac{1}{2!}\mathbf{A}^2+\ldots\right]\mathbf{M}$$

$$=\mathbf{I}+\mathbf{M}^{-1}\mathbf{AM}+\frac{1}{2!}\mathbf{M}^{-1}\mathbf{A}^2\mathbf{M}+\ldots$$

$$= \mathbf{I} + \mathbf{M}^{-1}\mathbf{A}\mathbf{M} + \frac{1}{2!}\mathbf{M}^{-1}\mathbf{A}\mathbf{M}\mathbf{M}^{-1}\mathbf{A}\mathbf{M} + \dots$$

$$= \mathbf{I} + \mathbf{J} + \frac{1}{2!}\mathbf{J}^2 + \dots$$

$$= e^{\mathbf{J}}$$

Thus the matrices $e^{\mathbf{A}}$ and $e^{\mathbf{J}}$ are similar. Since $\mathbf{J}$ is a diagonal matrix, $e^{\mathbf{J}}$ is given by

$$e^{\mathbf{J}} = \begin{bmatrix} \sum_{k=0}^{\infty} \frac{1}{k!}\lambda_1^k & 0 & \dots & 0 \\ 0 & \sum_{k=0}^{\infty} \frac{1}{k!}\lambda_2^k & \dots & 0 \\ \vdots & \vdots & & \vdots \\ 0 & 0 & \dots & \sum_{k=0}^{\infty} \frac{1}{k!}\lambda_n^k \end{bmatrix}$$

$$= \begin{bmatrix} e^{\lambda_1} & 0 \dots 0 \\ 0 & e^{\lambda_2} \dots 0 \\ \vdots & \vdots \quad \vdots \\ 0 & 0 \dots e^{\lambda_n} \end{bmatrix}$$

The matrix exponential $e^{\mathbf{A}t}$ for a matrix $\mathbf{A}$ with eigenvalues $\lambda_1, \lambda_2, \dots, \lambda_n$ may therefore be evaluated using the following relation:

$$e^{\mathbf{A}t} = \mathbf{M}e^{\mathbf{J}t}\mathbf{M}^{-1}$$

$$= \mathbf{M}\begin{bmatrix} e^{\lambda_1 t} & 0 \dots 0 \\ 0 & e^{\lambda_2 t} \dots 0 \\ \vdots & \vdots \quad \vdots \\ 0 & 0 \dots e^{\lambda_n t} \end{bmatrix}\mathbf{M}^{-1} \tag{5.57}$$

where $\mathbf{M}$ is a modal matrix that transforms $\mathbf{A}$ into the diagonal form.

For the general case wherein matrix $\mathbf{A}$ has eigenvalues $\lambda_1, \lambda_2, \dots, \lambda_m$ with multiplicities $n_1, n_2, \dots, n_m$ respectively, we have (refer eqns. (2.70))

$$e \;= \mathbf{M}e^{\mathbf{J}t}\mathbf{M}^{-1}$$

$$= \mathbf{M}\begin{bmatrix} e^{\mathbf{J}_1 t} & & \mathbf{0} \\ & e^{\mathbf{J}_2 t} \ddots & \\ \mathbf{0} & & e^{\mathbf{J}_m t} \end{bmatrix}\mathbf{M}^{-1} \tag{5.58a}$$

where $\mathbf{M}$ is a modal matrix that transforms $\mathbf{A}$ into the Jordan form

$$\mathbf{J} = \begin{bmatrix} \mathbf{J}_1(\lambda_1) & & \mathbf{0} \\ & \mathbf{J}_2(\lambda_2) \ddots & \\ \mathbf{0} & & \mathbf{J}_m(\lambda_m) \end{bmatrix} \tag{5.58b}$$

The exponentials of Jordan blocks $\mathbf{J}_i$ can be given in the closed form. For example, if

$$\underset{(n_i \times n_i)}{\mathbf{J}_i} = \begin{bmatrix} \lambda_i & 1 & 0 \ldots 0 \\ 0 & \lambda_i & 1 \ldots 0 \\ \vdots & \vdots & \vdots \\ 0 & 0 & 0 \ldots \lambda_i \end{bmatrix} \tag{5.58c}$$

then

$$e^{\mathbf{J}_i t} = \begin{bmatrix} e^{\lambda_i t} & te^{\lambda_i t} & t^2 e^{\lambda_i t}/2! \ldots t^{n_i-1} e^{\lambda_i t}/(n_i-1)! \\ 0 & e^{\lambda_i t} & te^{\lambda_i t} \quad \ldots t^{n_i-2} e^{\lambda_i t}/(n_i-2)! \\ \vdots & \vdots & \vdots \quad\quad \vdots \\ 0 & 0 & 0 \quad \ldots \quad e^{\lambda_i t} \end{bmatrix} \tag{5.58d}$$

This result can be easily proved by induction or by using the Cayley-Hamilton theorem (Problem 5.12).

Note that every element of $e^{\mathbf{J}_i t}$ is of the form $t^k e^{\lambda_i t}$ for $k = 0, 1, \ldots, n_i - 1$, $i = i, 2, \ldots, m$. Hence every element of $e^{\mathbf{A}t}$ is a linear combination of these factors. Since the function $t^k e^{\lambda_i t}$ is an analytic function over $(-\infty, \infty)$, we conclude that $e^{\mathbf{A}t}$ is analytic over $(-\infty, \infty)$.

Evaluation Using Cayley-Hamilton Technique: Consider the matrix **A** with characteristic polynomial

$$\Delta(\lambda) = \det(\lambda\mathbf{I} - \mathbf{A}) = \lambda^n - \sum_{i=0}^{n-1} a_i \lambda^i$$

From the Cayley-Hamilton theorem (Section 2.8), we have

$$\Delta(\mathbf{A}) = \mathbf{A}^n - \sum_{i=0}^{n-1} a_i \mathbf{A}^i$$

or

$$\mathbf{A}^n = a_0 \mathbf{I} + a_1 \mathbf{A} + \ldots + a_{n-1} \mathbf{A}^{n-1}$$

This shows that any power of **A** can be expressed in terms of $\mathbf{I}, \mathbf{A}, \ldots, \mathbf{A}^{n-1}$, i.e.,

$$\mathbf{A}^k = \sum_{i=0}^{n-1} b_{ki} \mathbf{A}^i$$

The matrix $f(\mathbf{A}) = e^{\mathbf{A}t}$ is a polynomial in **A** with analytic coefficients in t:

$$e^{\mathbf{A}t} = \mathbf{I} + \mathbf{A}t + \frac{\mathbf{A}^2 t^2}{2!} + \ldots$$

This polynomial can be expressed as a polynomial in **A** of degree $(n-1)$.

$$e^{\mathbf{A}t} = \sum_{k=0}^{\infty} \frac{t^k}{k!} \mathbf{A}^k = \sum_{k=0}^{\infty} \frac{t^k}{k!} \left[\sum_{i=0}^{n-1} b_{ki} \mathbf{A}^i \right]$$

$$\sum_{i=0}^{n-1} \left[\sum_{k=0}^{\infty} b_{ki} \frac{t^k}{k!} \right] \mathbf{A}^i$$

$$= \sum_{i=0}^{n-1} \alpha_i(t)\mathbf{A}^i \tag{5.59}$$

The coefficients α_i are obtained by equating the values of

$$f(\lambda)=e^{\lambda t} \text{ with } g(\lambda)=\sum_{i=0}^{n-1} \alpha_i\lambda^i$$

on the spectrum of A (refer equation (2.91)).

Example 5.6: Compute $f(\mathbf{A})=e^{\mathbf{A}t}$ for

$$\mathbf{A}=\begin{bmatrix} 0 & 2 \\ -2 & -4 \end{bmatrix}$$

Solution:

Matrix A has the eigenvalues $\lambda_1=\lambda_2=-2$. Since A is of second-order, the polynomial $g(\lambda)$ will be of the form (refer eqn. (5.59))

$$g(\lambda)=\alpha_0+\alpha_1\lambda$$

The coefficients α_0 and α_1 are evaluated from the following equations.

$$f(-2)=e^{-2t}=\alpha_0-2\alpha_1$$

$$\frac{d}{d\lambda}f(\lambda)\Big|_{\lambda=-2}=t\,e^{-2t}=\frac{d}{d\lambda}\,g(\lambda)\Big|_{\lambda=-2}=\alpha_1$$

The result is

$$\alpha_0=(1+2t)e^{-2t}$$

$$\alpha_1=t\,e^{-2t}$$

Hence

$$f(\mathbf{A})=e^{\mathbf{A}t}=\alpha_0\mathbf{I}+\alpha_1\mathbf{A}$$

$$=(1+2t)e^{-2t}\begin{bmatrix} 1 & 0 \\ 0 & 1 \end{bmatrix}+te^{-2t}\begin{bmatrix} 0 & 2 \\ -2 & -4 \end{bmatrix}$$

$$=\begin{bmatrix} (1+2t)e^{-2t} & 2t\,e^{-2t} \\ -2t\,e^{-2t} & (1-2t)e^{-2t} \end{bmatrix}$$

Evaluation Using Inverse Laplace Transforms: The Laplace transform of $e^{\mathbf{A}t}$ is given by

$$\mathscr{L}[e^{\mathbf{A}t}]=\mathscr{L}\left[\sum_{k=0}^{\infty}\frac{t^k}{k!}\mathbf{A}^k\right]$$

From the Table of Laplace transforms in Appendix I, we find

$$\mathscr{L}\left[\frac{t^k}{k!}\right]=\frac{1}{s^{k+1}}$$

Therefore

$$\mathscr{L}[e^{\mathbf{A}t}]=\sum_{k=0}^{\infty} s^{-(k+1)}\,\mathbf{A}^k$$

$$= \sum_{k=0}^{\infty} (s^{-1}\mathbf{A})^k \tag{5.60}$$

It is well known that the scalar power series

$$f(x) = \frac{1}{1-x} = 1 + x + x^2 + \ldots = \sum_{k=0}^{\infty} x^k$$

converges for $|x| < 1$. From the relationship of the convergence conditions of scalar power series and matrix power series (eqns. (2.75) and (2.76)), it follows that the matrix power series

$$\sum_{k=0}^{\infty} (s^{-1}\mathbf{A})^k$$

converges to $(\mathbf{I} - s^{-1}\mathbf{A})^{-1}$ if the absolute values of all the eigenvalues of $(s^{-1}\mathbf{A})$ are smaller than 1, which can be assured if s is chosen sufficiently large. Therefore, from (5.60) we write

$$\mathscr{L}[e^{\mathbf{A}t}] = s^{-1}(\mathbf{I} - s^{-1}\mathbf{A})^{-1}$$

$$= (s\mathbf{I} - \mathbf{A})^{-1} \tag{5.61a}$$

$$= \hat{\Phi}(s) \tag{5.61b}$$

The matrix $\hat{\Phi}(s)$ is known as *resolvent matrix*. The entries of the resolvent matrix are rational functions of s. In Chapter 7, we shall see that

$$(s\mathbf{I} - \mathbf{A})^{-1} = \frac{\mathbf{Q}(s)}{\Delta(s)} = \frac{1}{\Delta(s)} [\mathbf{Q}_1 s^{n-1} + \mathbf{Q}_2 s^{n-2} + \ldots + \mathbf{Q}_{n-1} s + \mathbf{Q}_n] \tag{5.62}$$

where $\mathbf{Q}_i$ are constant matrices and $\Delta(s)$ is the characteristic polynomial of $\mathbf{A}$. The coefficients of the characteristic polynomial $\Delta(s)$ and matrix polynomial $\mathbf{Q}(s)$ may be determined sequentially by resolvent algorithm (convenient for digital computer). We shall give this algorithm in Chapter 7.

The inverse transform

$\mathscr{L}^{-1}[\mathbf{Q}(s)/\Delta(s)] = e^{\mathbf{A}t}$ can be expressed as a power series in t.

□

In the classroom, the cases of matrices of the order $n \leqslant 3$ are often discussed. In the following we give explicit formulas covering all matrices of order $\leqslant 3$ (Apostol (1969)).

1. $\mathbf{A}$ has eigenvalues λ, λ.

$$e^{\mathbf{A}t} = e^{\lambda t}[\mathbf{I} + t(\mathbf{A} - \lambda\mathbf{I})] \tag{5.63}$$

2. $\mathbf{A}$ has eigenvalues λ_1, λ_2.

$$e^{\mathbf{A}t} = e^{\lambda_1 t}\left(\frac{\mathbf{A} - \lambda_2\mathbf{I}}{\lambda_1 - \lambda_2}\right) + e^{\lambda_2 t}\left(\frac{\mathbf{A} - \lambda_1\mathbf{I}}{\lambda_2 - \lambda_1}\right) \tag{5.64}$$

3. $\mathbf{A}$ has eigenvalues $\lambda, \lambda, \lambda$.

$$e^{\mathbf{A}t} = e^{\lambda t}[\mathbf{I} + t(\mathbf{A} - \lambda\mathbf{I}) + \tfrac{1}{2}t^2(\mathbf{A} - \lambda\mathbf{I})^2] \tag{5.65}$$

4. A has eigenvalues $\lambda_1, \lambda_2, \lambda_3$.

$$e^{\mathbf{A}t} = e^{\lambda_1 t}\left\{\frac{(\mathbf{A}-\lambda_2\mathbf{I})(\mathbf{A}-\lambda_3\mathbf{I})}{(\lambda_1-\lambda_2)(\lambda_1-\lambda_3)}\right\} + e^{\lambda_2 t}\left\{\frac{(\mathbf{A}-\lambda_1\mathbf{I})(\mathbf{A}-\lambda_3\mathbf{I})}{(\lambda_2-\lambda_1)(\lambda_2-\lambda_3)}\right\}$$
$$+ e^{\lambda_3 t}\left\{\frac{(\mathbf{A}-\lambda_1\mathbf{I})(\mathbf{A}-\lambda_2\mathbf{I})}{(\lambda_3-\lambda_1)(\lambda_3-\lambda_2)}\right\} \tag{5.66}$$

5. A has eigenvalues $\lambda_1, \lambda_1, \lambda_2$.

$$e^{\mathbf{A}t} = e^{\lambda_1 t}\,[\mathbf{I} + t(\mathbf{A}-\lambda_1\mathbf{I})]$$
$$+ \frac{e^{\lambda_2 t}-e^{\lambda_1 t}}{(\lambda_2-\lambda_1)^2}(\mathbf{A}-\lambda_1\mathbf{I})^2 - \frac{te^{\lambda_1 t}}{(\lambda_2-\lambda_1)}(\mathbf{A}-\lambda_1\mathbf{I})^2 \tag{5.67}$$

Example 5.7: Consider the network of Fig. 4.3a. The state equation governing the behaviour of the network is (eqn. (4.4a))

$$\dot{\mathbf{x}}(t) = \mathbf{A}\mathbf{x}(t) + \mathbf{b}u(t);\ \mathbf{x}(0) = \mathbf{x}^0 \tag{5.68}$$

where

$$\mathbf{A} = \begin{bmatrix} -2 & 1 \\ 1 & -2 \end{bmatrix};\ \mathbf{b} = \begin{bmatrix} 1 \\ 1 \end{bmatrix}$$

It follows from eqn. (3.30a) that transform-domain representation of eqn. (5.68) is given by

$$\hat{\mathbf{x}}(s) = (s\mathbf{I}-\mathbf{A})^{-1}\mathbf{x}^0 + (s\mathbf{I}-\mathbf{A})^{-1}\mathbf{b}\hat{u}(s) \tag{5.69a}$$
$$= \hat{\mathbf{G}}(s)\,\mathbf{x}^0 + \hat{\mathbf{H}}(s)\,\hat{u}(s) \tag{5.69b}$$

In the following, we obtain $\hat{\mathbf{G}}(s)$ and $\hat{\mathbf{H}}(s)$ from the state diagram of (5.68) shown in Fig. 5.2, using Mason's gain formula (refer Problem 3.19).

$$\begin{bmatrix} \hat{x}_1 \\ \hat{x}_2 \end{bmatrix} = \begin{bmatrix} \hat{G}_{11} & \hat{G}_{12} \\ \hat{G}_{21} & \hat{G}_{22} \end{bmatrix}\begin{bmatrix} x_1^0 \\ x_2^0 \end{bmatrix} + \begin{bmatrix} \hat{H}_1 \\ \hat{H}_2 \end{bmatrix}\hat{u}$$

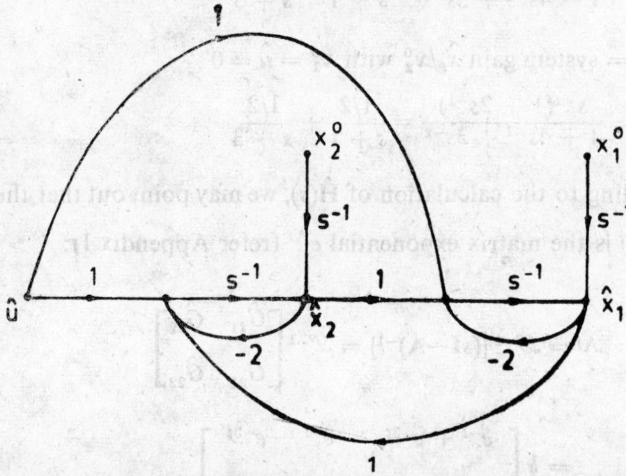

Fig. 5.2 State diagram for Example 5.7

$$\hat{G}_{11} = \text{system gain } \hat{x}_1/x_1^0 \text{ with } x_2^0 = \hat{u} = 0$$

$$= \frac{P_1 \Delta_1}{\Delta}$$

where P_1 = path gain of forward path $= s^{-1}$

Δ = determinant of the graph

$= 1 -$ (sum of loop gains of all individual loops) $+$ (sum of gain products of all possible combinations of two non-touching loops) $- \ldots$

$$= 1 - (-2s^{-1} - 2s^{-1} + s^{-2}) + (-2s^{-1})(-2s^{-1})$$

$$= 1 + 4s^{-1} + 3s^{-2}$$

Δ_1 = the value of Δ for that part of the graph not touching the forward path

$$= 1 - (-2s^{-1}) = 1 + 2s^{-1}$$

Therefore

$$\hat{G}_{11} = \frac{s^{-1}(1 + 2s^{-1})}{1 + 4s^{-1} + 3s^{-2}}$$

$$= \frac{s+2}{(s+3)(s+1)} = \frac{1/2}{s+1} + \frac{1/2}{s+3}$$

$$\hat{G}_{12} = \text{system gain } \hat{x}_1/x_2^0 \text{ with } x_1^0 = \hat{u} = 0$$

$$= \frac{s^{-2}(1)}{1 + 4s^{-1} + 3s^{-2}} = \frac{1/2}{s+1} + \frac{-1/2}{s+3}$$

$$\hat{G}_{21} = \text{system gain } \hat{x}_2/x_1^0 \text{ with } x_2^0 = \hat{u} = 0$$

$$= \frac{s^{-2}(1)}{1 + 4s^{-1} + 3s^{-2}} = \frac{1/2}{s+1} + \frac{-1/2}{s+3}$$

$$\hat{G}_{22} = \text{system gain } \hat{x}_2/x_2^0 \text{ with } x_1^0 = \hat{u} = 0$$

$$= \frac{s^{-1}(1 + 2s^{-1})}{1 + 4s^{-1} + 3s^{-2}} = \frac{1/2}{s+1} + \frac{1/2}{s+3}$$

Before proceeding to the calculation of $\hat{\mathbf{H}}(s)$, we may point out that the Laplace inverse of $\hat{\mathbf{G}}(s)$ is the matrix exponential $e^{\mathbf{A}t}$ (refer Appendix I):

$$e^{\mathbf{A}t} = \mathscr{L}^{-1}[(s\mathbf{I}-\mathbf{A})^{-1}] = \mathscr{L}^{-1}\begin{bmatrix} \hat{G}_{11} & \hat{G}_{12} \\ \hat{G}_{21} & \hat{G}_{22} \end{bmatrix}$$

$$= \frac{1}{2}\begin{bmatrix} e^{-t} + e^{-3t} & e^{-t} - e^{-3t} \\ e^{-t} - e^{-3t} & e^{-t} + e^{-3t} \end{bmatrix}$$

The homogeneous solution (zero-input response) is therefore

$$\Phi(t, 0, \mathbf{x}^0, 0) = e^{\mathbf{A}t}\mathbf{x}^0 = \tfrac{1}{2}\begin{bmatrix} e^{-t}+e^{-3t} & e^{-t}-e^{-3t} \\ e^{-t}-e^{-3t} & e^{-t}+e^{-3t} \end{bmatrix}\begin{bmatrix} x_1^0 \\ x_2^0 \end{bmatrix}$$

To obtain the zero-state response, we require $\hat{\mathbf{H}}(s)$.

$$\hat{H}_1 = \text{system gain } \hat{x}_1/\hat{u} \text{ with } x_1^0 = x_2^0 = 0$$

$$= \frac{P_1\Delta_1 + P_2\Delta_2}{\Delta}$$

$$= \frac{s^{-2}(1) + s^{-1}(1 + 2s^{-1})}{1 + 4s^{-1} + 3s^{-2}} = \frac{1}{s+1}$$

$$\hat{H}_2 = \text{system gain } \hat{x}_2/\hat{u} \text{ with } x_1^0 = x_2^0 = 0$$

$$= \frac{1}{s+1}$$

For a unit step input, $\hat{u}(s) = 1/s$. This gives

$$\Phi(t, 0, \mathbf{0}, u) = \begin{bmatrix} 1 - e^{-t} \\ 1 - e^{-t} \end{bmatrix}$$

This example illustrates the use of state diagrams for solving linear time-invariant systems analytically. We could have obtained the matrix exponential $e^{\mathbf{A}t}$ by any of the methods discussed earlier and then used eqn. (5.55) to obtain the solution.

Example 5.8: Consider the position servo system shown in Fig. 5.3. Find the response to a step input $r(t) = 10$. Assume that the output position and velocity are both zero initially.

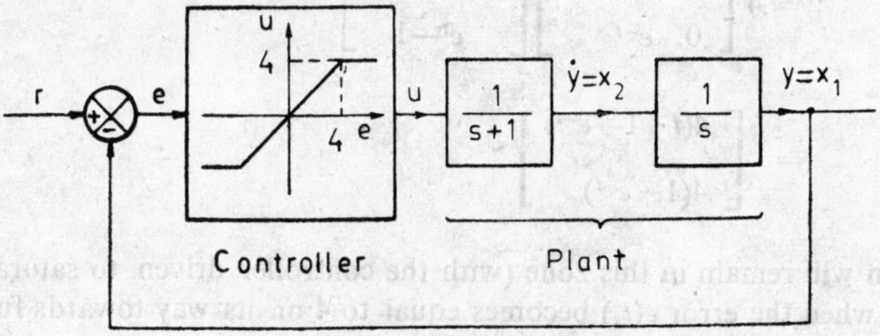

Fig. 5.3

Solution: The state variable model of the plant is given by

$$\dot{\mathbf{x}} = \mathbf{A}\mathbf{x} + \mathbf{b}u$$

or

$$\begin{bmatrix} \dot{x}_1 \\ \dot{x}_2 \end{bmatrix} = \begin{bmatrix} 0 & 1 \\ 0 & -1 \end{bmatrix}\begin{bmatrix} x_1 \\ x_2 \end{bmatrix} + \begin{bmatrix} 0 \\ 1 \end{bmatrix}u;\ \mathbf{x}(0) = \mathbf{x}^0 = \mathbf{0} \qquad (5.70)$$

The eigenvalues of **A** are $\lambda_1 = 0, \lambda_2 = -1$

From eqn. (5.64), we obtain

$$e^{\mathbf{A}t} = \begin{bmatrix} 1 & 1-e^{-t} \\ 0 & e^{-t} \end{bmatrix}$$

The zero-state response

$$\mathbf{x}(t) = \int_0^t e^{\mathbf{A}(t-\tau)}\,\mathbf{b}u(\tau)d\tau \tag{5.71}$$

At $t = 0$, $e(0) = r(0) - c(0)$

$$= 10$$

Therefore the controller will operate in its positive saturation zone and the plant will have an input $u = 4$. Equation (5.71) then gives

$$\mathbf{x}(t) = 4e^{\mathbf{A}t} \int_0^t e^{-\mathbf{A}\tau}\,\mathbf{b}\,d\tau$$

$$= 4e^{\mathbf{A}t} \int_0^t \begin{bmatrix} 1 & 1-e^{\tau} \\ 0 & e^{\tau} \end{bmatrix} \begin{bmatrix} 0 \\ 1 \end{bmatrix} d\tau$$

$$= 4e^{\mathbf{A}t} \begin{pmatrix} \int_0^t (1-e^{\tau})d\tau \\ \int_0^t e^{\tau}d\tau \end{pmatrix}$$

$$= 4 \begin{bmatrix} 1 & 1-e^{-t} \\ 0 & e^{-t} \end{bmatrix} \begin{bmatrix} 1+t-e^{t} \\ e^{t}-1 \end{bmatrix}$$

$$= \begin{bmatrix} 4(t-1+e^{-t}) \\ 4(1-e^{-t}) \end{bmatrix}$$

The system will remain in this zone (with the controller driven to saturation) till $t = t_1$, when the error $e(t_1)$ becomes equal to 4 on its way towards further reduction, i.e., when

$$4 = r - x_1(t_1)$$

$$= 10 - 4(t_1 - 1 + e^{-t_1})$$

This is a transcendental equation which may be solved for t_1 by numerical methods. Approximate calculation gives

$$t_1 = 2.4, \quad x_1(t_1) = 6, \quad x_2(t_1) = 3.6.$$

Immediately after $t = t_1$, the controller operates in the linear range. Consequently, the system may be looked upon as a linear system with the initial conditions matching the boundary conditions at $t = t_1$. The input $u(t) = 10 - x_1(t);\ t \geqslant t_1$. The linear state model becomes

$$\dot{\mathbf{x}}(t) = \begin{bmatrix} 0 & 1 \\ -1 & -1 \end{bmatrix}\mathbf{x}(t) + \begin{bmatrix} 0 \\ 10 \end{bmatrix};\quad \mathbf{x}(t_1) = \begin{bmatrix} 6 \\ 3.6 \end{bmatrix};\quad t \geqslant t_1$$

The solution

$$\mathbf{x}(t) = e^{\overline{\mathbf{A}}(t-t_1)}\mathbf{x}(t_1) + \int_{t_1}^{t} e^{\overline{\mathbf{A}}(t-\tau)}\begin{bmatrix} 0 \\ 10 \end{bmatrix}d\tau;\quad t \geqslant t_1$$

where $\overline{\mathbf{A}} = \begin{bmatrix} 0 & 1 \\ -1 & -1 \end{bmatrix}$.

The calculations to solve for $\mathbf{x}(t);\ t \geqslant t_1$ are left as an exercise for the reader.

From this example, it is clear that controllers with nonlinear input-output characteristics may be approximated by suitable piecewise linear function and analyzed in a manner outlined above.

System Modes

Although eqn. (5.58) shows the explicit dependence of the solution of time-invariant linear state equation (5.43) on the eigenvalues of matrix **A**, it masks some of the interesting geometric issues of the solution. In this sub-section, we give some insight into the qualitative nature of solutions of (5.43).

Let us first assume that $\lambda_1, \lambda_2, \ldots, \lambda_n$ are the distinct eigenvalues of matrix **A**. Let $\mathbf{m}_1, \mathbf{m}_2, \ldots, \mathbf{m}_n$ be the corresponding eigenvectors. Since this set of vectors is linearly independent, it can be used as a basis for the solution $\mathbf{x}(t)$. We can therefore write $\mathbf{x}(t)$ as

$$\mathbf{x}(t) = z_1(t)\mathbf{m}_1 + z_2(t)\mathbf{m}_2 + \ldots + z_n(t)\mathbf{m}_n \tag{5.72}$$

Since the eigenvectors are constant, the time variation of $\mathbf{x}(t)$ must be found in variable coefficients $z_i(t);\ i = 1, 2, \ldots, n$. Substituting (5.72) in (5.43) we obtain

$$\begin{aligned}&\dot{z}_1(t)\mathbf{m}_1 + \dot{z}_2(t)\mathbf{m}_2 + \ldots + \dot{z}_n(t)\mathbf{m}_n \\ &\qquad = \mathbf{A}z_1(t)\mathbf{m}_1 + \mathbf{A}z_2(t)\mathbf{m}_2 + \ldots + \mathbf{A}z_n(t)\mathbf{m}_n + \mathbf{B}\mathbf{u}(t)\end{aligned} \tag{5.73}$$

At any time t, $\mathbf{Bu}(t)$ belongs to the vector space of $\mathbf{x}(t)$ and therefore can be expressed as

$$\mathbf{Bu}(t) = \beta_1(t)\mathbf{m}_1 + \beta_2(t)\mathbf{m}_2 + \ldots + \beta_n(t)\mathbf{m}_n \tag{5.74}$$

From eqns. (5.73) and (5.74) we get

$$[\dot{z}_1(t)-\mathbf{A}z_1(t)-\beta_1(t)]\mathbf{m}_1+[\dot{z}_2(t)-\mathbf{A}z_2(t)-\beta_2(t)]\mathbf{m}_2(t)+\ldots=\mathbf{0} \tag{5.75}$$

Since $\mathbf{A}\mathbf{m}_i=\lambda_i\mathbf{m}_i$, and the eigenvectors are linearly independent, we can write from (5.75)

$$\dot{z}_i(t)=\lambda_i z_i(t)+\beta_i(t);\quad i=1,2,\ldots,n \tag{5.76}$$

Thus we can solve for each $z_i(t)$ independently. The $z_i(t)$; $i=1,2,\ldots,n$ decompose the state trajectory into its components along the eigenvectors of **A**. We call these fundamental motions along the eigenvectors of **A**, the *fundamental modes* (also called the *normal modes* or *natural modes*) of the system and we call the expansion in (5.72) the *eigenvector expansion* of $\mathbf{x}(t)$.

The zero-input solution is

$$z_i(t)=e^{\lambda_i t}z_i(0)$$

It is also customary to speak of λ_i itself or $e^{\lambda_i t}$ as being a mode.

Example 5.9: Consider a linear system described by the state equation

$$\dot{\mathbf{x}}=\mathbf{A}\mathbf{x}+\mathbf{b}u$$

$$=\begin{bmatrix}0 & 1\\ -2 & -3\end{bmatrix}\mathbf{x}+\begin{bmatrix}b_1\\ b_2\end{bmatrix}u;\quad \mathbf{x}(0)=\mathbf{x}^0$$

The eigenvalues of matrix **A** are $\lambda_1=-1$, $\lambda_2=-2$. The corresponding eigenvectors are

$$\mathbf{m}_1=\begin{bmatrix}1\\ -1\end{bmatrix},\quad \mathbf{m}_2=\begin{bmatrix}1\\ -2\end{bmatrix}$$

The Homogeneous Solution

Let us write initial state $\mathbf{x}^\circ$ as a linear combination of $\mathbf{m}_1$ and $\mathbf{m}_2$, i.e.,

$$\mathbf{x}^0=z_1(0)\mathbf{m}_1+z_2(0)\mathbf{m}_2$$

This gives

$$\begin{bmatrix}z_1(0)\\ z_2(0)\end{bmatrix}=\begin{bmatrix}2 & 1\\ -1 & -1\end{bmatrix}\begin{bmatrix}x_1^0\\ x_2^0\end{bmatrix}$$

and the homogeneous solution is

$$\mathbf{x}(t)=(2x_1^0+x_2^0)e^{-t}+(-x_1^0-x_2^0)e^{-2t}$$

Upon further inspection of these equations, we see that if $\mathbf{x}^0=k\mathbf{m}_1$; k is any constant, i.e., if initial state coincides with one of the eigenvectors, the solution

$$\mathbf{x}(t)=ke^{\lambda_1 t}\mathbf{m}_1$$

remains for all time along the straight line defined by this eigenvector; also the time variation is determined by the associated eigenvalue. Mode $z_2(t)$ never appears in the response. This is called *mode-suppression*.

Nonhomogeneous Solution

The input function $\mathbf{b}u(t)$ as a linear combination of $\mathbf{m}_1$ and $\mathbf{m}_2$ is given as

$$\mathbf{b}u(t) = \beta_1(t)\mathbf{m}_1 + \beta_2(t)\mathbf{m}_2$$

For a step input,

$$\begin{bmatrix} \beta_1(t) \\ \beta_2(t) \end{bmatrix} = \begin{bmatrix} 2 & 1 \\ -1 & -1 \end{bmatrix} \begin{bmatrix} b_1 \\ b_2 \end{bmatrix}$$

$$\dot{z}_1(t) = -z_1(t) + \beta_1(t)$$
$$\dot{z}_2(t) = -2z_2(t) + \beta_2(t)$$

The inspection of these equations reveals that if $2b_1 = -b_2$, then $\beta_1(t)=0$ and system trajectory $\mathbf{x}(t)$ starting from zero initial state will always travel along the vector $\mathbf{m}_2$. The mode $e^{\lambda_1 t}$ will be suppressed. Alternatively, if $b_1=-b_2$, the mode $e^{\lambda_2 t}$ will be suppressed.

Example 5.10 : Consider the system described by the state equation

$$\dot{\mathbf{x}}(t) = \mathbf{A}\mathbf{x} + \mathbf{b}u$$
$$= \begin{bmatrix} 0 & 1 \\ -1 & -2 \end{bmatrix} \mathbf{x} + \begin{bmatrix} b_1 \\ b_2 \end{bmatrix} u; \ \mathbf{x}(0) = \mathbf{x}^0$$

The eigenvalues of matrix $\mathbf{A}$ are $\lambda_1 = \lambda_2 = \lambda = -1$. The corresponding generalized eigenvectors are (refer Problem 2.20),

$$\mathbf{m}_1 = \begin{bmatrix} 1 \\ -1 \end{bmatrix}, \ \mathbf{m}_2 = \begin{bmatrix} 0 \\ 1 \end{bmatrix}$$

The eigenvector expansion of the state $\mathbf{x}(t)$ is

$$\mathbf{x}(t) = z_1(t)\,\mathbf{m}_1 + z_2(t)\mathbf{m}_2$$
$$= [\,\mathbf{m}_1 \ \ \mathbf{m}_2\,] \begin{bmatrix} z_1(t) \\ z_2(t) \end{bmatrix}$$
$$= \mathbf{M}\mathbf{z}(t)$$

Substitution into the system equation gives

$$\mathbf{M}\dot{\mathbf{z}}(t) = \mathbf{A}\mathbf{M}\mathbf{z} + \mathbf{b}u$$

or

$$\dot{\mathbf{z}}(t) = (\mathbf{M}^{-1}\mathbf{A}\mathbf{M})\mathbf{z} + \mathbf{M}^{-1}\mathbf{b}u$$
$$= \mathbf{J}\mathbf{z} + \mathbf{M}^{-1}\mathbf{b}u; \ \mathbf{z}(0) = \mathbf{M}^{-1}\mathbf{x}^0$$

where

$$\mathbf{J} = \begin{bmatrix} \lambda & 1 \\ 0 & \lambda \end{bmatrix}$$

From this equation, we may write

$$\dot{z}_1(t) = \lambda z_1(t) + z_2(t) + \beta_1(t) \tag{5.77a}$$

$$\dot{z}_2(t) = \lambda z_2(t) + \beta_2(t) \tag{5.77b}$$

Note that eqn. (5.77a) is not completely uncoupled.

Equation (5.77b) is uncoupled which can be solved for $z_2(t)$ and its substitution in eqn. (5.77a) will make it a function of $z_1(t)$ only.

The zero-input solution of (5.77b) is

$$z_2(t) = z_2(0)\, e^{\lambda t} = z_2(0)\, e^{-t}$$

The zero-input solution of (5.77a) then becomes

$$\begin{aligned} z_1(t) &= e^{\lambda t} z_1(0) + te^{\lambda t} z_2(0) \\ &= e^{-t} z_1(0) + te^{-t} z_2(0) \end{aligned}$$

$e^{\lambda t}$ and $te^{\lambda t}$ are the modes of the system.

5.6 SOLUTION OF LINEAR DISCRETE-TIME STATE EQUATIONS

Mathematical representations of discrete-time systems are given in the form of vector-matrix difference equations. In general, a nonlinear difference equation is expressed as (eqn. (3.12a))

$$\mathbf{x}(k+1) = \mathbf{f}(k, \mathbf{x}, \mathbf{u}) \tag{5.78}$$

A unique solution of this equation exists for any bounded sequence $\mathbf{u}(0)$, $\mathbf{u}(1)$, ...and the solution may be easily obtained by iteration. However, for a general nonlinear equation it may not always be possible to get the solution in the closed form.

In this section, we discuss the methods of obtaining closed-form solutions of linear discrete-time equations. As will be observed, the techniques to be presented here are similar to those discussed in the earlier section.

Time-Varying Case

Consider the n-th order linear state equation (eqn. (3.28a))

$$\mathbf{x}(k+1) = \mathbf{F}(k)\mathbf{x}(k) + \mathbf{G}(k)\mathbf{u}(k); \quad \mathbf{x}(k_0) = \mathbf{x}^0 \tag{5.79}$$

where $\mathbf{F}(k)$ and $\mathbf{G}(k)$ are, respectively, $n \times n$ and $n \times p$ real time-varying matrices and $k = k_0, k_0 + 1, \dots (k_0 = \text{integer})$.

The Homogeneous Solution: Consider first the homogeneous part of (5.79) :

$$\mathbf{x}(k+1) = \mathbf{F}(k)\mathbf{x}(k); \quad \mathbf{x}(k_0) = \mathbf{x}^0 \tag{5.80}$$

Similar to the case of linear vector-matrix differential equations, an $n \times n$ matrix $\Phi(k, k_0)$ such that

$$\begin{aligned} \Phi(k+1, k_0) &= \mathbf{F}(k)\Phi(k, k_0) \\ \Phi(k_0, k_0) &= \mathbf{I} \end{aligned} \tag{5.81}$$

is called the *state transition matrix* of (5.80). Notice that $\Phi(k, k_0)$ has the property that

$$\begin{aligned}
\Phi(k_0, k_0) &= \mathbf{I} \\
\Phi(k_0+1, k_0) &= \mathbf{F}(k_0) \\
\Phi(k_0+2, k_0) &= \mathbf{F}(k_0+1)\mathbf{F}(k_0) \\
\Phi(k_0+3, k_0) &= \mathbf{F}(k_0+2)\,\mathbf{F}(k_0+1)\,\mathbf{F}(k_0) \\
&\vdots \\
\Phi(k, k_0) &= \mathbf{F}(k-1)\mathbf{F}(k-2)\ldots\mathbf{F}(k_0+2)\mathbf{F}(k_0+1)\mathbf{F}(k_0) \\
&= \prod_{i=k0}^{k-1} \mathbf{F}(i) \qquad (5.82)
\end{aligned}$$

The solution of homogeneous equation (5.80) is

$$\mathbf{x}(k) = \Phi(k, k_0)\mathbf{x}^0 \qquad (5.83)$$

The following properties of state transition matrix can easily be established from (5.82).

$$\begin{aligned}
\Phi(k, k) &= \mathbf{I} \\
\Phi^{-1}(k, k_0) &= \Phi(k_0, k) \text{ if } \Phi^{-1}(k, k_0) \text{ exists for all } k \\
\Phi(k_2, k_0) &= \Phi(k_2, k_1)\Phi(k_1, k_0) \qquad (5.84)
\end{aligned}$$

for any k, k_0, k_1, k_2 in $(-\infty, \infty)$.

Note that if $\mathbf{F}(k)$ is singular, then $\Phi^{-1}(k, k_0)$ will not exist for all k. We shall henceforth assume $\mathbf{F}(k)$ to be nonsingular.

The matrix in (5.82) is not in closed form but is a product of matrices. This usually forces a computer solution for time-varying discrete-time equations. However, for some simple cases, it is possible to obtain the state transition matrix in closed form.

Example 5.11: Consider the homogeneous state equation

$$\mathbf{x}(k+1) = \mathbf{F}(k)\mathbf{x}(k)$$

$$= \begin{bmatrix} 1 & k \\ 0 & -1 \end{bmatrix} \mathbf{x}(k),\ \mathbf{x}(0) = \mathbf{x}^0 \qquad (5.85)$$

The state equation may be written as

$$x_1(k+1) = x_1(k) + kx_2(k) \qquad (5.86a)$$

$$x_2(k+1) = -\,x_2(k) \qquad (5.86b)$$

Taking the z-transform of eqn. (5.86b), we get (refer Appendix II)

$$z\hat{x}_2(z) - zx_2^0 = -\,\hat{x}_2(z)$$

or

$$\hat{x}_2(z) = \frac{z}{z+1}\,x_2^0$$

Inverse z-transform of this equation gives

$$x_2(k) = (-1)^k x_2^0;\ k \geqslant 0$$

Substituting $x_2(k)$ in (5.86a), we obtain

$$x_1(k+1) = x_1(k) + k(-1)^k x_2^0$$

Therefore

$$z\hat{x}_1(z) - zx_1^0 = \hat{x}_1(z) + \frac{(-1)z}{(z+1)^2} x_2^0$$

or

$$\hat{x}_1(z) = \frac{z}{z-1} x_1^0 + \frac{(-1)z}{(z-1)(z+1)^2} x_2^0$$

This gives

$$x_1(k) = x_1^0 + [-\tfrac{1}{4} + \tfrac{1}{4}(-1)^k - \tfrac{1}{2}k(-1)^k] x_2^0;\ k \geqslant 0$$

Written in the vector-matrix form, the solution becomes

$$\begin{bmatrix} x_1(k) \\ x_2(k) \end{bmatrix} = \begin{bmatrix} 1 & -\tfrac{1}{4} + \tfrac{1}{4}(-1)^k - \tfrac{1}{2}k(-1)^k \\ 0 & (-1)^k \end{bmatrix} \begin{bmatrix} x_1^0 \\ x_2^0 \end{bmatrix};\ k \geqslant 0 \tag{5.87}$$

or

$$\mathbf{x}(k) = \Phi(k, 0)\mathbf{x}^0$$

The Nonhomogeneous Solution:

Theorem 5.3: The solution of the state equation (5.79) is given by

$$\mathbf{x}(k) = \Phi(k, k_0)\mathbf{x}^0 + \Phi(k, k_0) \sum_{i=k_0+1}^{k} \Phi^{-1}(i, k_0)\mathbf{G}(i-1)\mathbf{u}(i-1);\ k \geqslant k_0 \tag{5.88}$$

where $\Phi(k, k_0)$ is the unique solution of (5.81).

Proof: We can prove the theorem by direct substitution. For $k > k_0$, we obtain from (5.88)

$$\mathbf{x}(k+1) = \Phi(k+1, k_0)\mathbf{x}^0 + \Phi(k+1, k_0) \sum_{i=k_0+1}^{k+1} \Phi^{-1}(i, k_0)\mathbf{G}(i-1)\mathbf{u}(i-1) \tag{5.89}$$

Noting that

$\Phi(k, k_0) = \mathbf{F}(k-1)\mathbf{F}(k-2)\ldots\mathbf{F}(k_0+1)\mathbf{F}(k_0)$, eqn (5.89) can be expressed as follows:

$$\begin{aligned} \mathbf{x}(k+1) &= \mathbf{F}(k)\Phi(k, k_0)\mathbf{x}^0 + \mathbf{F}(k)\Phi(k, k_0) \sum_{i=k_0+1}^{k} \Phi^{-1}(i, k_0)\mathbf{G}(i-1)\mathbf{u}(i-1) \\ &\quad + \Phi(k+1, k_0)\Phi^{-1}(k+1, k_0)\,\mathbf{G}(k)\mathbf{u}(k) \\ &= \mathbf{F}(k)\mathbf{x}(k) + \mathbf{G}(k)\mathbf{u}(k) \end{aligned}$$

This completes the proof.

□

The first term on the right hand side in eqn. (5.88) is the zero-input component of the state vector $\mathbf{x}(k)$ while the second term represents the zero-state component of $\mathbf{x}(k)$.

Adjoint Equations: Suppose that the $n \times n$ matrix $\Phi^a(k, k_0)$ satisfies the equation

$$\Phi^a(k+1, k_0) = \Phi^a(k, k_0)\mathbf{F}^{-1}(k) \tag{5.90}$$
$$\Phi^a(k_0, k_0) = \mathbf{I}$$

Equation (5.90) is called the *adjoint* of eqn. (5.81). Multiplying eqn. (5.81) by $\Phi^a(k+1, k_0)$ on the left, we obtain

$$\begin{aligned}\Phi^a(k+1, k_0)\Phi(k+1, k_0) &= \Phi^a(k+1, k_0)\mathbf{F}(k)\Phi(k, k_0)\\ &= \Phi^a(k, k_0)\mathbf{F}^{-1}(k)\mathbf{F}(k)\Phi(k, k_0)\\ &= \Phi^a(k, k_0)\Phi(k, k_0)\end{aligned}$$

This implies that the product $\Phi^a\Phi$ is constant. From the relations $\Phi^a(k_0, k_0) = \Phi(k_0, k_0) = \mathbf{I}$, it follows that the constant is a unit matrix $\mathbf{I}$. Therefore,

$$\Phi^a(k, k_0) = \Phi^{-1}(k, k_0) = \Phi(k_0, k) \tag{5.91}$$

From eqn. (5.90), we obtain

$$\Phi^{a^T}(k+1, k_0) = [\mathbf{F}^{-1}(k)]^T\Phi^{a^T}(k, k_0)$$

which shows that $\Phi^{a^T}(k, k_0)$ is the transition matrix for the state vector $\mathbf{x}(k)$ satisfying the equation

$$\mathbf{x}(k+1) = [\mathbf{F}^{-1}(k)]^T\mathbf{x}(k) \tag{5.92}$$

Equation (5.92) is called the adjoint of eqn. (5.80).

Multiplying eqn. (5.88) on left by $\Phi^{-1}(k, k_0)$ we get

$$\begin{aligned}\mathbf{x}(k_0) &= \Phi^{-1}(k, k_0)\,\mathbf{x}(k) - \sum_{i=k_0+1}^{k} \Phi^{-1}(i, k_0)\,\mathbf{G}(i-1)\,\mathbf{u}(i-1)\\ &= \Phi^a(k, k_0)\,\mathbf{x}(k) - \sum_{i=k_0+1}^{k} \Phi^a(i, k_0)\,\mathbf{G}(i-1)\,\mathbf{u}(i-1)\end{aligned} \tag{5.93}$$

As was the case in continuous-time systems, the equation enables the computation backwards.

Example 5.12: Reconsider the system of Example 4.7 (Koppel 1968). The state equation describing the system is (eqn. (4.17))

$$\mathbf{x}(k+1) = \mathbf{F}(k)\mathbf{x}(k) + \mathbf{g}(k)\,u(k);\ \mathbf{x}(N+1) = \mathbf{x}^f;\ k \geqslant 1$$

where

$$\mathbf{F}(k) = \begin{bmatrix} 1+\dfrac{rV(k)}{Q} & 0\\ -\dfrac{(\Delta H)\,rV(k)}{Q\rho c_p} & 1\end{bmatrix};\ \mathbf{g}(k) = \begin{bmatrix} 0\\ \dfrac{1}{Q\rho c_p}\end{bmatrix};\ u(k) = q(k)$$

Let

$$f_{11}(k) = 1 + \frac{rV(k)}{Q}$$

and

$$f_{21}(k) = -\frac{(\Delta H)\,rV(k)}{Q\rho c_p}$$

Using eqn. (5.82), we get

$$\Phi(k, 1) = \begin{bmatrix} \prod_{i=1}^{k-1} f_{11}(i) & 0 \\ \sum_{i=1}^{k-1} f_{21}(i) \prod_{j=1}^{i-1} f_{11}(j) & 1 \end{bmatrix} \tag{5.94a}$$

where

$$\prod_{j=1}^{0} f_{11}(j) \triangleq 1$$

Using eqn. (5.91), we get

$$\Phi^a(k, 1) = \Phi^{-1}(k, 1)$$

$$= \frac{1}{\prod_{i=1}^{k-1} f_{11}(i)} \begin{bmatrix} 1 & 0 \\ -\sum_{i=1}^{k-1} f_{21}(i) \prod_{j=1}^{i-1} f_{11}(j) & \prod_{i=1}^{k-1} f_{11}(i) \end{bmatrix} \tag{5.94b}$$

We are interested in the backward calculation; computation of $\mathbf{x}(1)$ from the feed conditions $\mathbf{x}(N+1)$. Using eqn. (5.93), we obtain

$$\mathbf{x}(1) = \Phi^a(N+1, 1)\, \mathbf{x}(N+1) - \sum_{i=2}^{N+1} \Phi^a(i, 1)\, \mathbf{g}\, u(i-1)$$

$$= \Phi^a(N+1, 1)\, \mathbf{x}^f - \begin{bmatrix} 0 \\ \dfrac{1}{Q\rho c_p} \end{bmatrix} \sum_{i=2}^{N+1} u(i-1) \tag{5.95}$$

Time-Invariant Case

Consider now the state equation (eqn. (3.29a))

$$\mathbf{x}(k+1) = \mathbf{F}\mathbf{x}(k) + \mathbf{G}\mathbf{u}(k);\ \mathbf{x}(k_0) = \mathbf{x}^0 \tag{5.96}$$

where $\mathbf{F}$ and $\mathbf{G}$ are, respectively, $n \times n$ and $n \times p$ real constant matrices. Since eqn. (5.96) is a special case of linear time-varying equation (5.79), all the results derived in the earlier subsection are applicable here. From eqn. (5.82), we have

$$\Phi(k, k_0) = \Phi(k - k_0) = \mathbf{F}^{(k-k_0)} \tag{5.97}$$

The solution of eqn. (5.96) which follows from eqns. (5.88) and (5.97) is

$$\mathbf{x}(k) = \Phi(k - k_0)\mathbf{x}^0 + \Phi(k - k_0) \sum_{i=k_0+1}^{k} \Phi^{-1}(i - k_0)\mathbf{G}\mathbf{u}(i-1) \tag{5.98a}$$

$$= \mathbf{F}^{(k-k_0)}\, \mathbf{x}^0 + \mathbf{F}^{(k-k_0)} \sum_{i=k_0+1}^{k} \mathbf{F}^{-(i-k_0)}\, \mathbf{G}\mathbf{u}(i-1) \tag{5.98b}$$

If $k_0 = 0$, as is usually assumed in the time-invariant case, we have

$$\mathbf{x}(k) = \mathbf{F}^k\mathbf{x}^0 + \sum_{i=1}^{k} \mathbf{F}^{(k-i)}\, \mathbf{G}\mathbf{u}(i-1) \tag{5.99}$$

Example 5.13: Let us reconsider Example 5.12 with the assumption that

$$V(1) = V(2) \dots = V(k) = V$$

Under this assumption, eqn. (5.94b) gives

$$\Phi^a(k-1) = \begin{bmatrix} f_{11}^{-(k-1)} & 0 \\ -f_{21}\left(\dfrac{1-f_{11}^{-(k-1)}}{f_{11}-1}\right) & 1 \end{bmatrix}$$

Substituting into eqn. (5.95), we obtain

$$\begin{bmatrix} x_1(1) \\ x_2(1) \end{bmatrix} = \begin{bmatrix} f_{11}^{-N} x_1^f \\ -f_{21}\left(\dfrac{1-f_{11}^{-N}}{f_{11}-1}\right) x_1^f + x_2^f \end{bmatrix} - \begin{bmatrix} 0 \\ \dfrac{1}{Q\rho c_p} \end{bmatrix} \sum_{i=2}^{N+1} u(i-1)$$

Therefore

$$x_1(1) = \left(1+\frac{rV}{Q}\right)^{-N} x_1^f$$

$$x_2(1) = x_2^f + x_1^f \left[\frac{(\Delta H) rV}{Q\rho c_p}\right] \left[\frac{1-\left(1+\dfrac{rV}{Q}\right)^{-N}}{\dfrac{rV}{Q}}\right] - \frac{1}{Q\rho c_p}\sum_{i=2}^{N+1} u(i-1)$$

$$= x_2^f + \frac{(\Delta H)}{\rho c_p} x_1^f \left[1-\left(1+\frac{rV}{Q}\right)^{-N}\right] - \frac{1}{Q\rho c_p}\sum_{i=2}^{N+1} u(i-1)$$

Evaluation of State Transition Matrix: Various methods of evaluation of matrix exponential discussed in earlier section can be easily used to evaluate $\Phi(k) = F^k$ as demonstrated below.

Evaluation Using Similarity Transformation

Consider an $n \times n$ matrix **F** with eigenvalues $\lambda_1, \lambda_2, \dots, \lambda_m$ with multiplicities $n_1, n_2, \dots, n_m$ respectively. A modal matrix **M** transforms **F** into Jordan canonical form given by eqns. (2.70), i.e.,

$$\mathbf{M}^{-1}\mathbf{F}\mathbf{M} = \mathbf{J}$$

or

$$\mathbf{F} = \mathbf{M}\mathbf{J}\mathbf{M}^{-1}$$

Now

$$\begin{aligned} \mathbf{M}^{-1}\mathbf{F}^k\mathbf{M} &= \mathbf{M}^{-1}[\mathbf{F}.\mathbf{F}\dots\mathbf{F}]\mathbf{M} \\ &= \mathbf{M}^{-1}[(\mathbf{M}\mathbf{J}\mathbf{M}^{-1})(\mathbf{M}\mathbf{J}\mathbf{M}^{-1})\dots(\mathbf{M}\mathbf{J}\mathbf{M}^{-1})]\mathbf{M} \\ &= \mathbf{J}^k \end{aligned}$$

Thus the matrices $\mathbf{F}^k$ and $\mathbf{J}^k$ are similar. The matrix $\mathbf{F}^k$ may therefore be expressed as (refer eqns. (2.70))

$$\mathbf{F}^k = \mathbf{M}\mathbf{J}^k\mathbf{M}^{-1}$$

$$= \mathbf{M}\begin{bmatrix} \mathbf{J}_1^k & & & \mathbf{0} \\ & \mathbf{J}_2^k . & & \\ & & . & \\ \mathbf{0} & & & \mathbf{J}_m^k \end{bmatrix}\mathbf{M}^{-1} \tag{5.100a}$$

The powers of the Jordan blocks $\mathbf{J}_i$ can be given in closed form. For example, if

$$\underset{(n_i \times n_i)}{\mathbf{J}_i} = \begin{bmatrix} \lambda_i & 1 & 0 \ldots & 0 \\ 0 & \lambda_i & 1 \ldots & 0 \\ \vdots & \vdots & \vdots & \vdots \\ 0 & 0 & 0 \ldots & \lambda_i \end{bmatrix} \tag{5.100b}$$

then

$$\mathbf{J}_i^k = \begin{bmatrix} \lambda_i^k & k\lambda_i^{k-1} & \frac{1}{2!}k(k-1)\,\lambda_i^{k-2} \ldots & \frac{k!\,\lambda_i^{k-n_i+1}}{(k-n_i+1)!\,(n_i-1)!} \\ 0 & \lambda_i^k & k\lambda_i^{k-1} \quad \ldots & . \\ 0 & 0 & \lambda_i^k \quad \ldots & . \\ \vdots & \vdots & \vdots & \vdots \\ 0 & 0 & 0 \quad \ldots & \lambda_i^k \end{bmatrix} \tag{5.100c}$$

This result can easily be proved by induction or by using Cayley-Hamilton theorem (Problem 5.12).

Example 5.14: Consider the matrix

$$\mathbf{F} = \begin{bmatrix} 0 & 1 \\ -1 & -2 \end{bmatrix}$$

The eigenvalues of $\mathbf{F}$ are $\lambda_1 = \lambda_2 = -1$.
The modified Vandermonde matrix (refer Problem 2.20)

$$\mathbf{V} = \begin{bmatrix} 1 & 0 \\ \lambda_1 & 1 \end{bmatrix} = \begin{bmatrix} 1 & 0 \\ -1 & 1 \end{bmatrix}$$

is a modal matrix $\mathbf{M}$ and

$$\mathbf{M}^{-1}\mathbf{F}\mathbf{M} = \mathbf{J} = \begin{bmatrix} -1 & 1 \\ 0 & -1 \end{bmatrix}$$

From eqn. (5.100c), we may write

$$\mathbf{F}^k = \mathbf{M}\begin{bmatrix} (-1)^k & k(-1)^{k-1} \\ 0 & (-1)^k \end{bmatrix}\mathbf{M}^{-1}$$

$$=\begin{bmatrix} 1 & 0 \\ -1 & 1 \end{bmatrix}\begin{bmatrix} (-1)^k & k(-1)^{k-1} \\ 0 & (-1)^k \end{bmatrix}\begin{bmatrix} 1 & 0 \\ 1 & 1 \end{bmatrix}$$

$$=(-1)^k\begin{bmatrix} (1-k) & -k \\ k & (1+k) \end{bmatrix}$$

Evaluation Using Cayley-Hamilton Technique

The Cayley-Hamilton Technique has already been explained in the earlier section. We illustrate the use of this technique for evaluation of $\mathbf{F}^k$ by an example.

Example 5.15 : Reconsider matrix **F** of Example 5.14.

$$\mathbf{F}=\begin{bmatrix} 0 & 1 \\ -1 & -2 \end{bmatrix}$$

Let us evaluate $f(\mathbf{F})=\mathbf{F}^k$.

The eigenvalues are $\lambda_1=\lambda_2=-1$.

Since **F** is of second-order, the polynomial $g(\lambda)$ will be of the form (refer eqn. (2.85))

$$g(\lambda)=\alpha_0+\alpha_1\lambda$$

The coefficients α_0 and α_1 are evaluated from the following equations (refer eqn. (2.91)).

$$f(-1)=(-1)^k=\alpha_0-\alpha_1$$

$$\frac{d}{d\lambda}f(\lambda)\bigg|_{\lambda=-1}=k(-1)^{k-1}=\frac{d}{d\lambda}g(\lambda)\bigg|_{\lambda=-1}=\alpha_1$$

The result is

$$\alpha_0=(1-k)\,(-1)^k$$

$$\alpha_1==-k\,(-1)^k$$

Hence

$$f(\mathbf{F})=\mathbf{F}^k=\alpha_0\mathbf{I}+\alpha_1\mathbf{F}$$

$$=(1-k)\,(-1)^k\begin{bmatrix} 1 & 0 \\ 0 & 1 \end{bmatrix}-k(-1)^k\begin{bmatrix} 0 & 1 \\ -1 & -2 \end{bmatrix}$$

$$=(-1)^k\begin{bmatrix} (1-k) & -k \\ k & (1+k) \end{bmatrix}$$

Evaluation Using Inverse z-Transforms

Comparing eqns. (5.99) and (3.31a), we observe that

$$\mathbf{F}^k=\mathscr{Z}^{-1}[(z\mathbf{I}-\mathbf{F})^{-1}z] \tag{5.101}$$

where $\mathscr{Z}^{-1}$ denotes inverse z-transform (Appendix II).

Thus the state transition matrix is the inverse z-transform of $n \times n$ matrix $(z\mathbf{I}-\mathbf{F})^{-1}z$. Now

$$(z\mathbf{I}-\mathbf{F})^{-1}=\frac{\mathbf{Q}(z)}{\Delta(z)}=\frac{1}{\Delta(z)}[\mathbf{Q}_1 z^{n-1}+\mathbf{Q}_2 z^{n-2}+\ldots+\mathbf{Q}_{n+1}z+\mathbf{Q}_n] \tag{5.102}$$

were $\mathbf{Q}_i$ are constant matrices and $\Delta(z)$ is the characteristic polynomial of **F**. The coefficients of the characteristic polynomial $\Delta(z)$ and the matrix polynomial $\mathbf{Q}(z)$ may be determined sequentially by resolvent algorithm (convenient for digital computer). We shall give this algorithm in Chapter 7. The inverse transform

$$\mathscr{Z}^{-1}\left[\frac{\mathbf{Q}(z)z}{\Delta(z)}\right]=\mathbf{F}^k$$

If the dimension of the matrix **F** is 2 or 3, we may use state diagrams to obtain the state transition matrix (refer Example 5.7).

System Modes

This subsection parallels the last subsection of Section 5.5. Assume that a matrix **F** has distinct eigenvalues $\lambda_1, \lambda_2, \ldots, \lambda_n$ and $\mathbf{m}_1, \mathbf{m}_2, \ldots, \mathbf{m}_n$ are the corresponding eigenvectors. The eigenvector expansion for the solution $\mathbf{x}(k)$ of eqn. (5.96) is given by

$$\mathbf{x}(k)=z_1(k)\mathbf{m}_1+z_2(k)\mathbf{m}_2+\ldots+z_n(k)\mathbf{m}_n \tag{5.103}$$

We can also express $\mathbf{Gu}(k)$ as

$$\mathbf{Gu}(k)=\beta_1(k)\mathbf{m}_1+\beta_2(k)\mathbf{m}_2+\ldots+\beta_n(k)\mathbf{m}_n \tag{5.104}$$

From eqns. (5.96), (5.103) and (5.104) we can easily deduce the following

$$z_i(k+1)=\lambda_i z_i(k)+\beta_i(k); \quad i=1, 2, \ldots, n \tag{5.105}$$

Thus, we can solve for each $z_i(k)$ independently. The zero-input solution is

$$z_i(k)=(\lambda_i)^k z_i(0)$$

The $z_i(k)$; $i=1, 2, \ldots, n$ are the fundamental modes of the system (5.96). λ_i itself or $(\lambda_i)^k$ is also referred to as being a mode.

Similar results for the case of repeated eigenvalues can be easily obtained.

Example 5.16: Consider the system described by the state equation

$$\mathbf{x}(k+1)=\mathbf{Fx}(k)$$

$$=\begin{bmatrix} 0 & 1 \\ -1 & -2 \end{bmatrix}\mathbf{x}(k); \quad \mathbf{x}(0)=\mathbf{x}^0$$

In Example 5.14, we found that the eigenvalues of **F** are $\lambda_1=\lambda_2=-1=\lambda$ and the corresponding generalized eigenvectors are

$$\mathbf{m}_1 = \begin{bmatrix} 1 \\ -1 \end{bmatrix}; \quad \mathbf{m}_2 = \begin{bmatrix} 0 \\ 1 \end{bmatrix}$$

The eigenvector expansion of $\mathbf{x}(k)$ is

$$\mathbf{x}(k) = z_1(k)\mathbf{m}_1 + z_2(k)\mathbf{m}_2$$

$$= [\,\mathbf{m}_1 \quad \mathbf{m}_2\,] \begin{bmatrix} z_1(k) \\ z_2(k) \end{bmatrix} = \mathbf{Mz}(k)$$

Therefore

$$\mathbf{Mz}(k+1) = \mathbf{FMz}(k)$$

or

$$\mathbf{z}(k+1) = (\mathbf{M}^{-1}\mathbf{FM})\mathbf{z}(k)$$

$$= \mathbf{Jz}(k)$$

$$= \begin{bmatrix} \lambda & 1 \\ 0 & \lambda \end{bmatrix} \mathbf{z}(k); \quad \mathbf{z}(0) = \mathbf{M}^{-1}\mathbf{x}^0$$

This gives (refer Example 5.14)

$$\mathbf{z}(k) = \mathbf{J}^k\mathbf{z}(0)$$

$$= \begin{bmatrix} (\lambda)^k & k(\lambda)^{k-1} \\ 0 & (\lambda)^k \end{bmatrix} \mathbf{z}(0)$$

$(\lambda)^k$ and $k(\lambda)^{k-1}$ are the modes of the given system.

Let us consider some typical initial conditions $\mathbf{x}(0)$. Suppose that $z_2(0) = 0$, which could occur if $x_1^0 = -x_2^0$. (This can be easily verified from the equation $\mathbf{z}(0) = \mathbf{M}^{-1}\mathbf{x}^0$.) This choice of initial conditions suppresses the mode $k(\lambda)^{k-1}$ in the response $\mathbf{x}(k)$.

5.7 STATE EQUATIONS OF SAMPLED-DATA SYSTEMS

Sampled-data systems that consist of an interconnection of a discrete-time system and a continuous-time system are frequently encountered. An example of particular interest is the control of continuous-time plants using a digital computer. Whenever such interconnections exist, there must be some type of interface system that takes care of the communication between the discrete-time and continuous-time systems. In the system of Fig. 1.3, the interface system is represented by the blocks 'Sampler and Analog-to-Digital Converter' and 'Digital-to-Analog Converter and Hold Circuit'.

To obtain a simple model of interface action, we represent the sampling operation by a sampler shown in Fig. 5.4 which gives a mathematical representation of the operation of taking periodic samples from $f(t)$ to produce $f(kT)$; $k = 0, 1, 2, \ldots$ and T is the time interval between samples. The holding operation, performed by the zero-order hold, is represented by

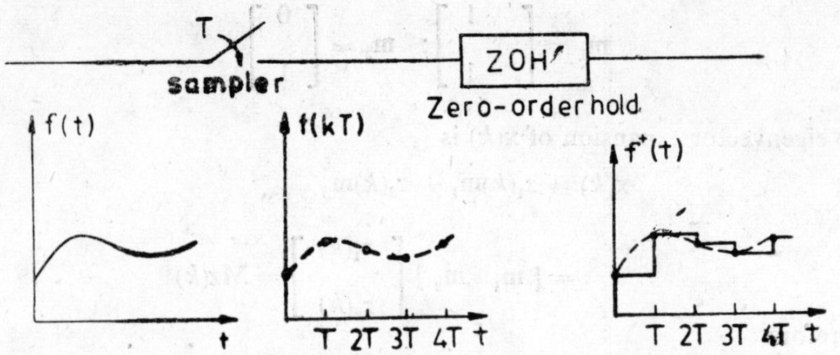

Fig. 5.4 Simple model of converter action of an interface system.

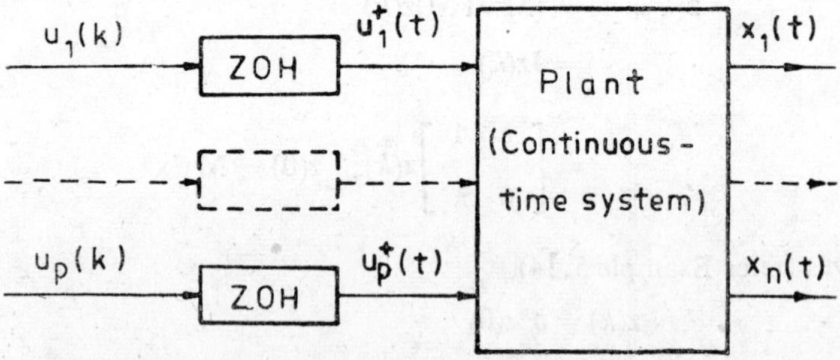

Fig. 5.5 Linear continuous-time system with sampled inputs.

storing the sample as a constant until the next sampling instant. We must emphasize that the signal $f(kT)$ in Fig. 5.4 is not expected to represent a physical signal in the interface system, but is rather introduced to allow us to obtain a simple model of the converter action of interface system (refer Franklin and Powell (1980), Nagrath and Gopal (1982)).

For a sampled-data system such as that shown in Fig. 1.3, we wish to establish a general method for obtaining the difference equations which represent the behaviour of the continuous-time plant. Figure 5.5 shows a portion of the sampled-data system; the digital controller puts out a sequence of numbers $\mathbf{u}(k)$, which are the inputs to the plant.

Because of the hold operation,

$$u_i^+(kT+\tau) = u_i(kT);\ 0 \leqslant \tau < T \tag{5.106a}$$

$$k = 0, 1, 2, \ldots$$

$$i = 1, 2, \ldots, p$$

Therefore

$$\mathbf{u}^+(t) = \mathbf{u}(kT);\ kT \leqslant t < (k+1)T \tag{5.106b}$$

The plant may be described by a state equation of the form

$$\dot{\mathbf{x}}(t) = \mathbf{A}\mathbf{x}(t) + \mathbf{B}\mathbf{u}^+(t) \tag{5.107}$$

where $\mathbf{A}$ and $\mathbf{B}$ are, respectively, $n \times n$ and $n \times p$ real constant matrices.

The solution of eqn. (5.107) with t_0 as initial time is

$$\mathbf{x}(t) = e^{\mathbf{A}(t-t_0)}\mathbf{x}(t_0) + \int_{t_0}^{t} e^{\mathbf{A}(t-\tau)}\mathbf{Bu}^{+}(\tau)d\tau \tag{5.108}$$

In our case the input is sampled, so we shall establish the solution going from one sampling instant $t_0 = kT$ to the next sampling instant $t = (k+1)T$. Thus from (5.108), we may write

$$\mathbf{x}(t) = e^{\mathbf{A}(t-kT)}\mathbf{x}(kT) + \int_{kT}^{t} e^{\mathbf{A}(t-\tau)}\mathbf{Bu}^{+}(\tau)d\tau; kT \leqslant t < (k+1)T \tag{5.109}$$

If we are interested in response at the sampling instants only, we set $t = (k+1)T$. In response to $\mathbf{u}(k)$, the state settles to the value $\mathbf{x}(k+1)$ prior to the application of input $\mathbf{u}(k+1)$.

$$\mathbf{x}(k+1) = \mathbf{Fx}(k) + \mathbf{Gu}(k) \tag{5.110a}$$

where

$$\mathbf{F} = e^{\mathbf{A}T} \tag{5.110b}$$

and

$$\mathbf{G} = \int_{kT}^{(k+1)T} e^{\mathbf{A}[(k+1)T-\tau]}\mathbf{B}d\tau \tag{5.110c}$$

Letting $\sigma = (\tau - kT)$ in (5.110c), we have

$$\mathbf{G} = \int_{0}^{T} e^{\mathbf{A}(T-\sigma)}\mathbf{B}d\sigma$$

With $\theta = T - \sigma$, we get

$$\mathbf{G} = \int_{0}^{T} e^{\mathbf{A}\theta}\,\mathbf{B}d\theta \tag{5.110d}$$

If we are interested in the value of $\mathbf{x}(t)$ between sampling instants, we first solve for $\mathbf{x}(kT)$ for any k using eqn. (5.110) and then use (5.109) to determine $\mathbf{x}(t)$ for $kT \leqslant t < (k+1)T$.

Equation (5.110) offers a discrete simulation of a continuous-time system which is almost completely exact. There are only two sources of error. One is introduced by sampling the input and the other is caused by the iterative procedure for evaluating matrices **F** and **G**. In the following, we discuss the methods of reducing these errors.

Algorithm for Evaluation of Matrix Series

The infinite series expansion for **F** is

$$\mathbf{F} = e^{\mathbf{A}T} = \sum_{i=0}^{\infty} \frac{\mathbf{A}^i T^i}{i!}; \ \mathbf{A}^0 = \mathbf{I} \tag{5.111}$$

For a finite T, this series is uniformly convergent (Section 2.8). It is therefore, possible to evaluate **F** within prescribed accuracy. If the series is truncated at $i = N$, then we may write the finite series sum as

$$\overline{\mathbf{F}} = \sum_{i=0}^{N} \frac{\mathbf{A}^i T^i}{i!} \tag{5.112}$$

which represents the infinite series approximation. The larger the N, the better is the approximation. We evaluate $\overline{\mathbf{F}}$ by a series in the form (Cadzow and Martens 1970)

$$\overline{\mathbf{F}} = \mathbf{I} + \mathbf{A}T\left(\mathbf{I} + \frac{\mathbf{A}T}{2}\left\{\mathbf{I} + \frac{\mathbf{A}T}{3}\left[\mathbf{I} + \ldots + \frac{\mathbf{A}T}{N-1}\left(\mathbf{I} + \frac{\mathbf{A}T}{N}\right)\right]\ldots\right\}\right) \tag{5.113}$$

which has better numerical properties than the direct series of powers. Starting with the innermost factor, this nested product expansion lends itself easily to digital programming. The empirical relation giving the number of terms, N, is

$$N = \min\{3\,\|\mathbf{A}T\| + 6,\ 100\} \tag{5.114}$$

This relation assures that no more than 100 terms are included. The series $e^{\mathbf{A}T}$ will be accurate to at least six significant figures.

The $\mathbf{G}$ integral in (5.110d) can be evaluated term by term to give

$$\mathbf{G} = \sum_{i=0}^{\infty} \frac{\mathbf{A}^i T^{i+1}}{(i+1)!}\mathbf{B} \tag{5.115a}$$

$$= \sum_{i=0}^{\infty} \frac{\mathbf{A}^i T^i}{(i+1)!} T\mathbf{B}$$

$$= \left(\mathbf{I} + \frac{\mathbf{A}T}{2!} + \frac{\mathbf{A}^2T^2}{3!} + \ldots\right) T\mathbf{B}$$

$$= (e^{\mathbf{A}T} - \mathbf{I})\mathbf{A}^{-1}\mathbf{B} \tag{5.115b}$$

The transition from (5.115a) to (5.115b) is possible only for a nonsingular matrix $\mathbf{A}$. For a singular $\mathbf{A}$, we may evaluate $\mathbf{G}$ from (5.115a) by the approximation technique described above. Since the series expansion for $\mathbf{G}$ converges faster than that for $\mathbf{F}$, it suffices to determine N for $\mathbf{F}$ from (5.114) and apply the same value for $\mathbf{G}$. Cadzow and Martens (1970) have given a listing of FORTRAN program using this algorithm. Melsa and Jones (1973) give a listing using the Cayley-Hamilton technique. Review of various methods, for evaluation of $e^{\mathbf{A}T}$ is available in Moler and VanLoan (1978); some of these methods have already been discussed in this chapter.

Selection of Sampling Interval

We have a continuous-time plant which is to be controlled. The control action may be either continuous or discrete and must make the plant behave in a desired manner. If discrete control action is thought of, then the problem of selection of sampling interval arises.

The selection of best sampling interval for a digital control system is a compromise among many factors. The basic motivation to lower the sampling rate $1/T$ (or sampling frequency $\omega_s = 2\pi/T$) is the cost. A decrease in sampling rate means more time is available for control calculations, hence slower computers are possible for a given control function or more control capacity is available for a given computer. Thus economically, the best choice is the slowest possible sampling rate that meets all the performance specifications.

On the other hand, if the sampling rate is too low, the sampler discards part of the information present in a continuous-time signal. The minimum sampling rate or frequency has a definite relationship with the highest significant signal frequency (i.e., signal bandwidth). This relationship is given by the *Sampling Theorem* (Nagrath and Gopal (1982)) according to which the information contained in a signal is fully preserved in its sampled version so long as the sampling frequency is at least twice the highest significant frequency contained in the signal. This sets an absolute lower bound to the sample rate selection.

We are usually satisfied with the trial and error method of selection of sampling interval. We compare the response of the continuous-time plant with models discretized for various sampling rates. Then the model with the slowest sampling rate which gives a response within tolerable limits is selected for future work. However, the method is not rigorous in approach. Also a wide variety of inputs must be given to each prospective model to ensure that it is a true representative of the plants.

There are a number of rules of thumb available in literature. A 'practical' rule mentioned by Gustovsson (1975) is to choose the sampling interval equal to the smallest time-constant or the inverse of the largest real eigenvalue (or real part of complex eigenvalue) of interest. However, this may prove to be a very conservative choice and half this sampling interval may suffice. Another rule of thumb mentioned by Gustovsson is to use one-tenth the major time-constant as the sampling interval.

Like the trial and error method, these rules too are without much of a theoretical basis. Some effectiv analysis tools to guide the designer in making the sampling rate choice are given by Franklin and Powell (1980).

In Chapter 6 we shall study the effect of sampling rate on the controllability and observability properties of a system and in Chapter 8 we shall study the effect on stability.

Example 5.17: Consider a system described by the state equation

$$\dot{\mathbf{x}} = \mathbf{A}\mathbf{x} + \mathbf{b}u \tag{5.116}$$

$$y = \mathbf{c}\mathbf{x}$$

where

$$\mathbf{A} = \begin{bmatrix} 0 & 1 \\ 0 & -1 \end{bmatrix}; \quad \mathbf{b} = \begin{bmatrix} 0 \\ 1 \end{bmatrix}; \quad \mathbf{c} = [1 \quad 0]$$

Let us examine this system when the input signal is sampled and held. The discretized state equation is obtained as (eqns. (5.110))

$$\mathbf{x}(k+1) = \mathbf{F}\mathbf{x}(k) + \mathbf{g}u(k) \tag{5.117}$$

where

$$\mathbf{F} = e^{\mathbf{A}T} = \begin{bmatrix} 1 & 1-e^{-T} \\ 0 & e^{-T} \end{bmatrix}$$

$$= \int_0^T e^{\mathbf{A}\theta}\mathbf{b}\,d\theta = \begin{bmatrix} \int_0^T (1-e^{-\theta})d\theta \\ \int_0^T e^{-\theta}d\theta \end{bmatrix} = \begin{bmatrix} T-1+e^{-T} \\ 1-e^{-T} \end{bmatrix}$$

$T =$ sampling interval

Equations of the form (5.117) can easily be programmed on a digital computer. The programming of eqn. (5.117) allows the analyst to change the value of T and evaluate the effect of sampling rate on the system response. Also the system response may be determined for several inputs by changing $u(kT)$.

Returning to the example, for a specific case of $T = 1$ sec, we obtain

$$\mathbf{F} = \begin{bmatrix} 1 & 0.632 \\ 0 & 0.368 \end{bmatrix} \tag{5.118}$$

$$\mathbf{g} = \begin{bmatrix} 0.368 \\ 0.632 \end{bmatrix}$$

Consider now the feedback system of Fig. 5.6 with the plant described by eqns. (5.116). Equation (5.117) now modifies to the following:

$$\begin{aligned} \mathbf{x}(k+1) &= \mathbf{Fx}(k) + \mathbf{g}e(k) \\ &= \mathbf{Fx}(k) + \mathbf{g}(r(k) - y(k)) \end{aligned}$$

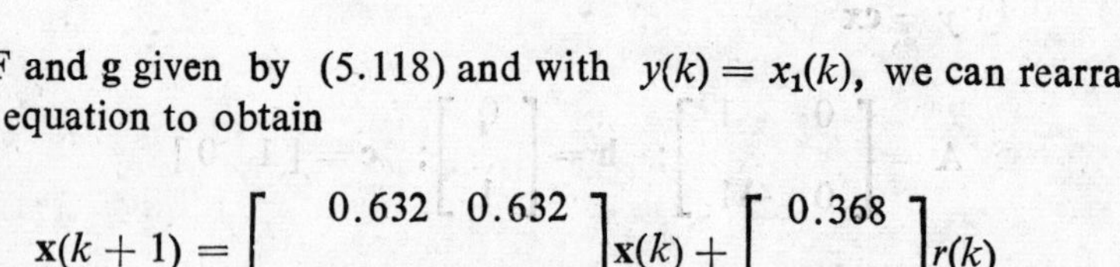

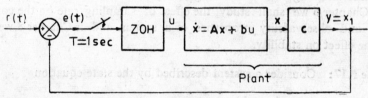

Fig. 5.6

With F and g given by (5.118) and with $y(k) = x_1(k)$, we can rearrange the above equation to obtain

$$\mathbf{x}(k+1) = \begin{bmatrix} 0.632 & 0.632 \\ -0.632 & 0.368 \end{bmatrix}\mathbf{x}(k) + \begin{bmatrix} 0.368 \\ 0.632 \end{bmatrix} r(k) \tag{5.119}$$

Consider now the introduction of digital compensator as shown in Fig. 5.7. Assume that the compensator is described by the input-output relation

$$e_2(k+1) + ae_2(k) = be_1(k)$$

Digital network simulation of the compensator is shown in Fig. 5.8.

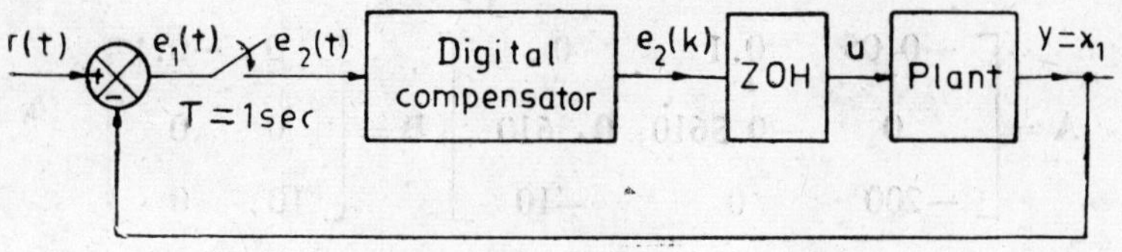

Fig. 5.7

The state vector $\mathbf{x}(k) = \begin{bmatrix} x_1(k) \\ x_2(k) \end{bmatrix}$ of the plant is obtained from the (eqn. (5.117))

$$\begin{bmatrix} x_1(k+1) \\ x_2(k+1) \end{bmatrix} = \begin{bmatrix} 1 & 0.632 \\ 0 & 0.368 \end{bmatrix} \begin{bmatrix} x_1(k) \\ x_2(k) \end{bmatrix} + \begin{bmatrix} 0.368 \\ 0.632 \end{bmatrix} e_2(k) \quad (5.120a)$$

As seen from Fig. 5.8, $e_2(k)$ may be taken as the third state variable $x_3(k)$ whose dynamics are given by

$$\begin{aligned} x_3(k+1) &= -ax_3(k) + be_1(k) \\ &= -ax_3(k) + b(r(k) - x_1(k)) \\ &= -bx_1(k) - ax_3(k) + br(k) \end{aligned} \quad (5.120b)$$

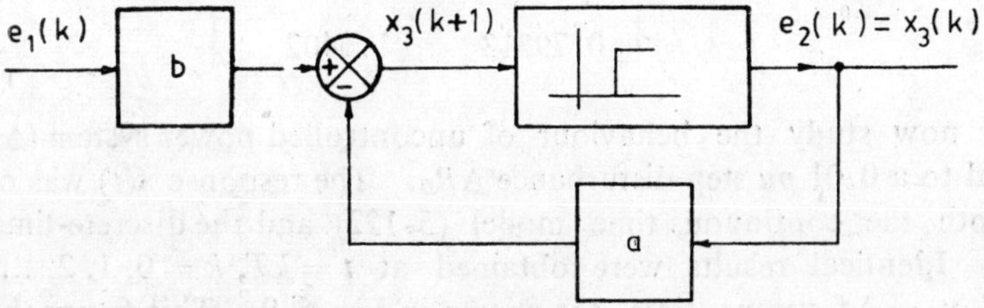

Fig. 5.8

From eqns. (5.120a) and (5.120b), we get the following state model for system of Fig. 5.7.

$$\begin{bmatrix} x_1(k+1) \\ x_2(k+1) \\ x_3(k+1) \end{bmatrix} = \begin{bmatrix} 1 & 0.632 & 0.368 \\ 0 & 0.368 & 0.632 \\ -b & 0 & -a \end{bmatrix} \begin{bmatrix} x_1(k) \\ x_2(k) \\ x_3(k) \end{bmatrix} + \begin{bmatrix} 0 \\ 0 \\ b \end{bmatrix} r(k) \quad (5.121)$$

Example 5.18: Let us consider the Power System of Section 4.6. The state equation governing the behaviour of the system is (eqn. (4.52))

$$\dot{\mathbf{x}}(t) = \mathbf{A}\mathbf{x}(t) + \mathbf{B}\mathbf{u}(t) \quad (5.122)$$

where

$$\mathbf{A}=\begin{bmatrix} -0.05 & 0.1 & 0 \\ 0 & -0.3610 & 0.3610 \\ -200 & 0 & -10 \end{bmatrix}, \mathbf{B}=\begin{bmatrix} 0 & -0.1 \\ 0 & 0 \\ 10 & 0 \end{bmatrix}$$

$$x_1=\Delta f,\ x_2=\Delta P_G,\ x_3=\Delta x_D,\ u_1=\Delta P_C \text{ and } u_2=\Delta P_D.$$

The eigenvalues of matrix **A** are

$$-10.075,\ -0.168\pm j\,0.8437$$

For discrete-time model of the power system, we choose sampling interval $T=1$ sec. With this value of T, eqn. (5.122) was discretized using digital computer. The result obtained is as follows:

$$\mathbf{x}(k+1)=\mathbf{F}\mathbf{x}(k)+\mathbf{G}\mathbf{u}(k) \tag{5.123}$$

where

$$\mathbf{F}=e^{\mathbf{A}T}=\begin{bmatrix} 0.70243 & 0.074384 & 0.0025042 \\ -5.0083 & 0.47109 & 0.019364 \\ -15.043 & -1.3873 & -0.045943 \end{bmatrix}$$

$$\mathbf{G}=\int_0^T e^{\mathbf{A}\theta}\mathbf{B}\,d\theta=\begin{bmatrix} 0.012641 & -0.089529 \\ 0.25674 & 0.25281 \\ 0.79313 & 1.6402 \end{bmatrix}$$

We now study the behaviour of uncontrolled power system ($\Delta P_C=0$) subjected to a 0.01 *pu* step disturbance ΔP_D. The response $\mathbf{x}(t)$ was obtained from both the continuous-time model (5.122) and the discrete-time model (5.123). Identical results were obtained at $t=kT$, $k=0, 1, 2, \ldots$. The plot of $x_1=\Delta f$ versus time t is shown in Fig. 5.9. This figure shows the effect of step disturbance on the deviation in frequency. In Chapter 11, we shall design a controller for this power system to damp out the oscillations in frequency.

5.8 CONCLUDING COMMENTS

The solutions of dynamical state equations for both continuons-time and discrete-time systems were studied in this chapter. If a specific solution to nonlinear or linear state equation is of interest, we may compute it on a digital computer by direct integration. The numerical solution, however, does not allow us to easily discern the qualitative properties of the system such as controllability, observability and stability (Chapters 6 and 8).

For linear systems, we can obtain the solution in an analytical form. The solution hinges on the state transition matrix. For time-varying systems, the state transition matrix is very difficult to compute. For time-invariant

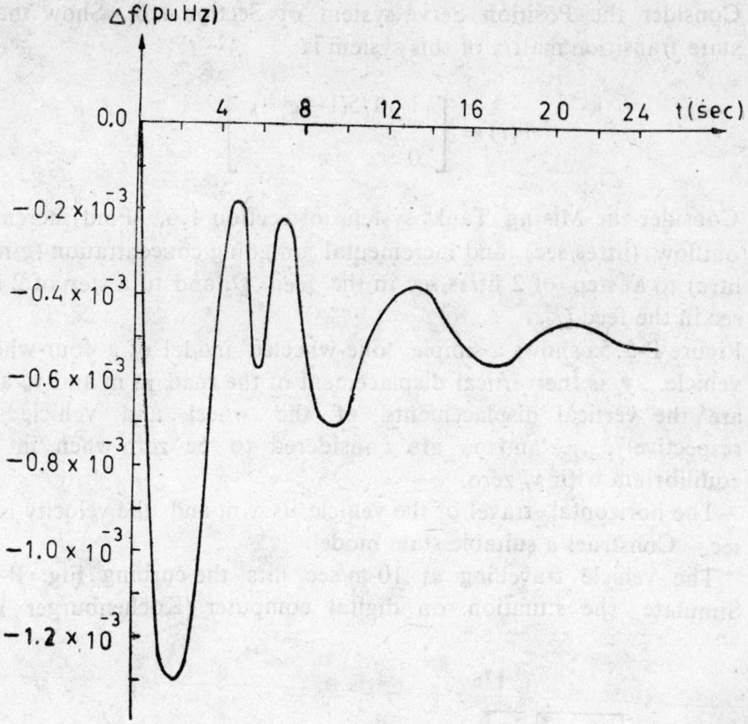

Fig. 5.9

systems, the state transition matrix may be computed using the methods discussed in Sections 5.5 and 5.6.

PROBLEMS

5.1 Find the state transition matrix of the following homogeneous equation.

$$\dot{\mathbf{x}}(t) = \begin{bmatrix} t & 1 \\ 1 & t \end{bmatrix} \mathbf{x}(t)$$

(Hint: A(t) and $\int_{t_0}^{t} \mathbf{A}(\tau)\, d\tau$ commute).

5.2 Consider the system

$$\dot{\mathbf{x}}(t) = \begin{bmatrix} 0 & 1 \\ \frac{1}{t^2} & -\frac{1}{t} \end{bmatrix} \mathbf{x}(t) + \begin{bmatrix} 0 \\ \frac{1}{t^2} \end{bmatrix} u(t); \ \mathbf{x}(1) = 0, \ t \geqslant 1$$

Find the response to the input

$$u(t) = \begin{cases} t-1 \ ; & t \geqslant 1 \\ 0 \ ; & t < 1 \end{cases}$$

5.3 Consider the Position Servo system of Section 4.6. Show that the state transition matrix of this system is

$$\Phi(t) = \begin{bmatrix} 1 & 1/5(1-e^{-5t}) \\ 0 & e^{-5t} \end{bmatrix}$$

5.4 Consider the Mixing Tank system of Section 4.6. Find incremental outflow (litres/sec) and incremental outgoing concentration (g-moles/litre) to a step of 2 litres/sec in the feed Q_1 and to a step of 2 litres/sec in the feed Q_2.

5.5 Figure P-5.5a shows a simple 'one-wheeled' model of a 'four-wheeled' vehicle. y_r is the vertical displacement of the road in m and y_w and y_b are the vertical displacements of the wheel and vehicle body respectively. y_w and y_b are considered to be zero when in static equilibrium with y_r zero.

The horizontal travel of the vehicle is x m and the velocity is v m/sec. Construct a suitable state model.

The vehicle travelling at 10 m/sec hits the curbing Fig. P-5.5b. Simulate the situation on digital computer (Kochenburger 1972).

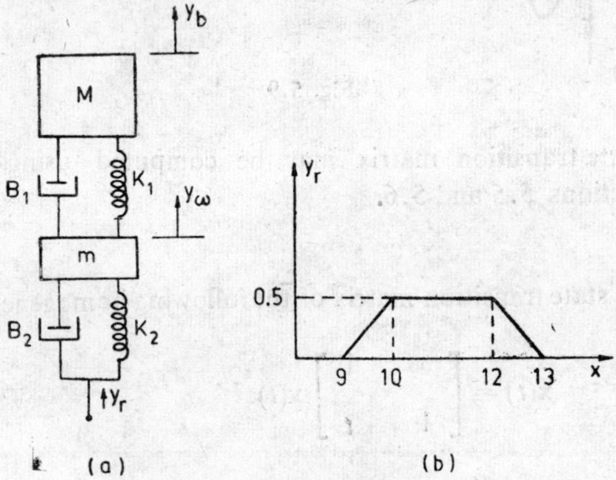

Fig. P-5.5

Given:

$Mg = 1360$ kg; $M =$ mass of vehicle

$K_1 = 8930$ kg/m

$B_1 = 2230$ kg-sec/m

$mg = 45$ kg; $m =$ mass of wheels

$K_2 = 74400$ kg/m

$B_2 = 1190$ kg-sec/m

(Hint: $y_r = 0; \quad 0.0 \leqslant t < 0.9$
$= 5t - 4.5; \quad 0.9 \leqslant t < 1$
$= 0.5; \quad 1 \leqslant t < 1.2$
$= -5t + 6.5; \quad 1.2 \leqslant t < 1.3$).

5.6 Simulate the servomechanism system of Fig. P-4.9 on a digital computer and investigate the effect of adjustment of K on the response of the system to step changes in θ_r. Assume the system to be initially at rest.

5.7 Consider the state equation

$$\dot{\mathbf{x}}(t) = \begin{bmatrix} 0 & 1 \\ -1 & -2 \end{bmatrix} \mathbf{x}(t)$$

Find a set of states $x_1(1)$ and $x_2(1)$ such that $x_1(2) = 2$.
(Hint: Use eqn. (5.42))

5.8 For the state equation

$$\dot{\mathbf{x}}(t) = \begin{bmatrix} 0 & 1 & 0 \\ 3 & 0 & 2 \\ -12 & -7 & -6 \end{bmatrix} \mathbf{x}(t)$$

find the initial condition vector $\mathbf{x}(0)$ which will only excite the mode corresponding to the eigenvalue with the most negative real part.

5.9 Consider the homogenous equation

$$\dot{\mathbf{x}}(t) = \begin{bmatrix} 0 & 1 \\ -2 & 3 \end{bmatrix} \mathbf{x}(t)$$

$$= \mathbf{A}\mathbf{x}(t)$$

(a) Find the eigenvalues λ_1, λ_2 and the corresponding eigenvectors $\mathbf{m}_1$, $\mathbf{m}_2$ of matrix $\mathbf{A}$.

(b) Find the response $\mathbf{x}(t)$ when

(i) $\mathbf{x}(0) = \begin{bmatrix} 1 \\ 1 \end{bmatrix}$ (ii) $\mathbf{x}(0) = \begin{bmatrix} 1 \\ 2 \end{bmatrix}$

(Hint: $\mathbf{x}(t) = e^{\lambda_i t}\mathbf{m}_i$ if $\mathbf{x}(0) = \mathbf{m}_i$)

5.10 Given

$$\mathbf{A} = \begin{bmatrix} \sigma & \omega \\ -\omega & \sigma \end{bmatrix}$$

Compute $e^{\mathbf{A}t}$.

(Hint: $e^{\mathbf{A}+\mathbf{B}} = e^{\mathbf{A}}.e^{\mathbf{B}}$ if $\mathbf{AB} = \mathbf{BA}$.

$$\begin{bmatrix} \sigma & \omega \\ -\omega & \sigma \end{bmatrix} = \begin{bmatrix} \sigma & 0 \\ 0 & \sigma \end{bmatrix} + \begin{bmatrix} 0 & \omega \\ -\omega & 0 \end{bmatrix})$$

5.11 The following facts are known about the linear system

$$\dot{\mathbf{x}}(t) = \mathbf{A}\mathbf{x}(t)$$

If $\mathbf{x}(0) = \begin{bmatrix} 1 \\ -2 \end{bmatrix}$, then $\mathbf{x}(t) = \begin{bmatrix} e^{-2t} \\ -2e^{-2t} \end{bmatrix}$

If $\mathbf{x}(0) = \begin{bmatrix} 1 \\ -1 \end{bmatrix}$, then $\mathbf{x}(t) = \begin{bmatrix} e^{-t} \\ -e^{-t} \end{bmatrix}$

Find $e^{\mathbf{A}t}$ and hence $\mathbf{A}$.

5.12 Given

$$\underset{(n\times n)}{\mathbf{J}} = \begin{bmatrix} \lambda_1 & 1 & 0 \ldots & 0 \\ 0 & \lambda_1 & 1 \ldots & 0 \\ \vdots & \vdots & \vdots & \vdots \\ 0 & 0 & 0 \ldots & 1 \\ 0 & 0 & 0 \ldots & \lambda_1 \end{bmatrix}$$

Compute

(i) $e^{\mathbf{J}t}$ (ii) $(\mathbf{J})^k$

using Cayley-Hamilton technique discussed in Section 2.8.
(Hint: Let $g(\lambda) = \alpha_0 + \alpha_1(\lambda - \lambda_1) + \alpha_2(\lambda - \lambda_1)^2 + \ldots + \alpha_{n-1}(\lambda - \lambda_1)^{n-1}$
Then the conditions (2.91) give

$$\alpha_0 = f(\lambda_1),\ \alpha_1 = \left.\frac{df(\lambda)}{d\lambda}\right|_{\lambda=\lambda_1},\ \ldots\ \alpha_{n-1} = \left.\frac{d^{n-1}}{d\lambda^{n-1}} f(\lambda)\right|_{\lambda=\lambda_1}$$

Take (i) $f(\lambda) = e^{\lambda t}$ (ii) $f(\lambda) = (\lambda)^k$)

5.13 Find the response of the system

$$\mathbf{x}(k+1) = \begin{bmatrix} 1 & k \\ 0 & -1 \end{bmatrix} \mathbf{x}(k) + \begin{bmatrix} 0 \\ 1 \end{bmatrix} u(k);\ \mathbf{x}(0) = \begin{bmatrix} 1 \\ -1 \end{bmatrix}$$

if $u(k) = 1$ for $k = 0, 1, 2, \ldots$

5.14 Set up a state model to find current in any loop of the ladder network shown in Fig. P-5.14. The resistance in each branch is 1 ohm.

(Hint: For the k-th loop

$$(i_k - i_{k-1}) + i_k + (i_k - i_{k+1}) = 0$$

Take $i_k = x_1(k)$, $i_{k+1} = x_2(k)$ and determine the state model

$$\mathbf{x}(k+1) = \mathbf{F}\mathbf{x}(k).$$

Given $i(0) = x_1(0) = 4$. From 10-th loop we obtain $i_9 = 3i_{10}$. This condition can be used to determine $x_2(0)$.)

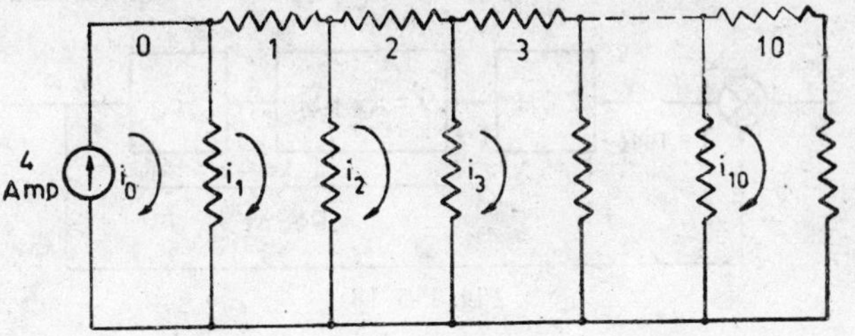

Fig. P-5.14

5.15 Show that the discrete-time model of Position Servo system of Section 4.6 is given by

$$\mathbf{x}(k+1) = \begin{bmatrix} 1 & 0.0787 \\ 0 & 0.6065 \end{bmatrix} \mathbf{x}(k) + \begin{bmatrix} 0.0043 \\ 0.0787 \end{bmatrix} u(k)$$

$$y(k) = [1 \quad 0]\, \mathbf{x}(k)$$

for $T = 0.1$ sec.

5.16 Show that the discrete-time model of Mixing Tank system of Section 4.6 is given by

$$\mathbf{x}(k+1) = \begin{bmatrix} 0.9512 & 0 \\ 0 & 0.9048 \end{bmatrix} \mathbf{x}(k) + \begin{bmatrix} 4.88 & 4.88 \\ -0.019 & 0.0095 \end{bmatrix} \mathbf{u}(k)$$

$$\mathbf{y}(k) = \begin{bmatrix} 0.01 & 0 \\ 0 & 1 \end{bmatrix} \mathbf{x}(k)$$

for $T = 5$ sec. Find the response $\mathbf{y}(k)$ to a step of 2 litres/sec in the feed Q_1 and compare it with the one obtained in Problem 5.4.

5.17 The block diagram of a sampled-data system is shown in Fig. P-5.17.

(a) Obtain discrete-time state model for the system.
(b) Obtain the equation for intersample response of the system.

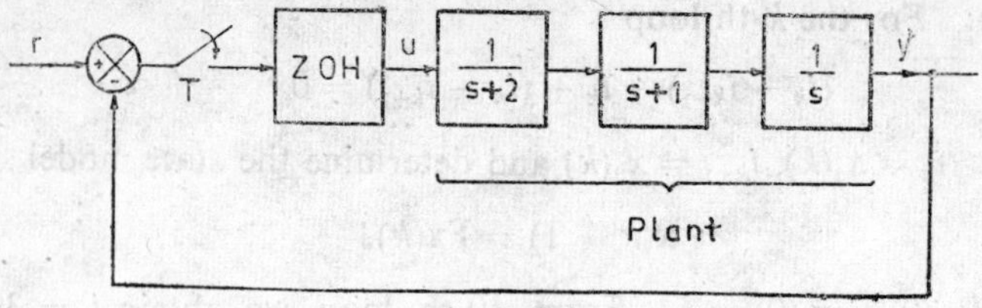

Fig. P-5.17

5.18 The block diagram of a sampled-data system is shown in Fig. P-5.18.

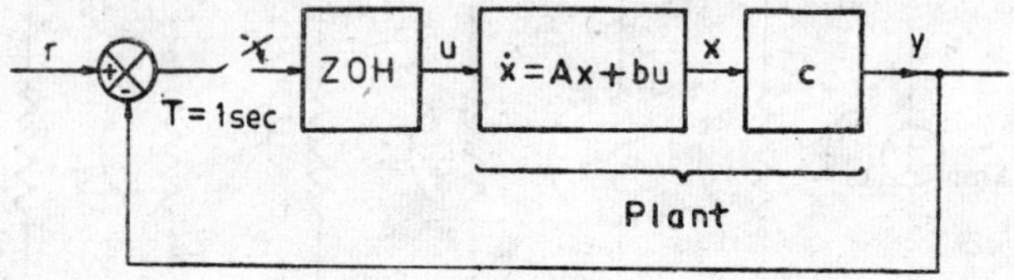

Fig. P-5.18

(a) Obtain discrete-time state model for the system.
(b) Find the response of the system for a unit step input.
(c) What is the effect on system response when

(i) $T=0.5$ sec (ii) $T=1.5$ sec?

Given:

$$\mathbf{A}=\begin{bmatrix} 1 & 1 \\ -2 & -3 \end{bmatrix};\ \mathbf{b}=\begin{bmatrix} 0 \\ 1 \end{bmatrix};\ \mathbf{c}=[1 \quad 0]$$

5.19 A closed-loop computer control system is shown in Fig. P-5.19. The digital compensator is described by the difference equation

$$e_2(k+1)+2e_2(k)=e_1(k)$$

The state model of the plant is as given in Problem 5.18. Obtain discrete-time state model for the system.

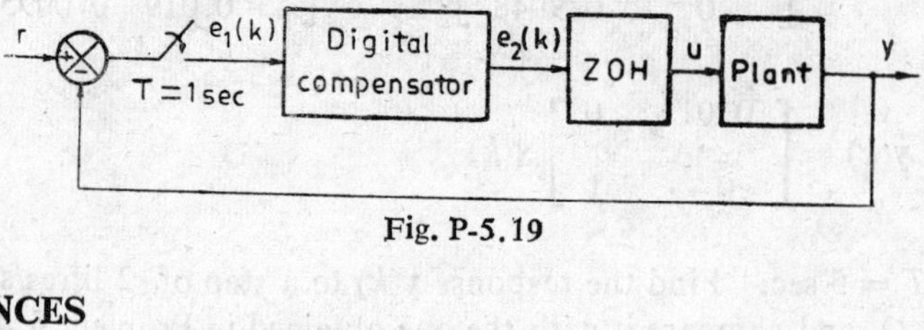

Fig. P-5.19

REFERENCES

1. Apostol, T. M., "Some explicit formulas for the exponential matrix e^{tA}", *The American Mathematical Monthly*, vol. 76, pp 289-292, 1969.
2. Athans, M., M. L. Dertouzos, R. N. Spann, and S. J. Mason, *Systems, Networks and Computation: Multivariable Methods*, New York: McGraw-Hill, 1974.

3. Cadzow, J. A., and H. R. Martens, *Discrete-Time and Computer Control Systems*, Englewood Cliffs, N. J.: Prentice-Hall, 1970.
4. Franklin, G. F., and J. D. Powell, *Digital Control of Dynamic Systems*, Reading, Mass.: Addison-Wesley, 1980.
5. Gibson, J. E., *Nonlinear Automatic Control*, New York: McGraw-Hill, 1963.
6. Gustovsson, I., "Survey of applications of identification in chemical and physical sciences", *Automatica*, vol. II, 1975.
7. Hamming, R. W., *Introduction to Applied Numerical Analysis*, New York: McGraw-Hill, 1971.
8. Kochenburger, R. J., *Computer Simulation of Dynamic Systems*, Englewood Cliffs, N. J.: Prentice-Hall, 1972.
9. Koppel, L. B., *Introduction to Control Theory*, Englewood Cliffs, N. J.: Prentice-Hall, 1968.
10. Melsa, J. L., and S. K. Jones, *Computer Programs for Computational Assistance in the Study of Linear Control Systems*, 2nd Edition, New York: McGraw-Hill, 1973.
11. Moler, C. B., and C. VanLoan, "Nineteen dubious ways to compute the exponential of a matrix," *SIAM Review*, pp. 801-836, 1978.
12. Nagrath, I. J., and M. Gopal, *Control Systems Engineering*, 2nd Edition, New Delhi: Wiley Eastern, 1982.
13. Wu, M-Y, "A new method of computing the state transition matrix of linear time-varying systems", *Proc. IEEE Int. Symp. Circuits and System Theory*, San Francisco, Calif., April 22–24, 1974a.
14. Wu, M-Y, "An extension of a new method of computing the state transition matrix of linear time-varying systems," *IEEE Trans. Automatic Control*, Vol. AC-19, pp. 619-620, 1974b.
15. Wu, M-Y, "Some new results in linear time-varying systems", *IEEE Trans. Automatic Control*, Vol. AC-20, pp. 159–160, 1975.

6. CONTROLLABILITY AND OBSERVABILITY

6.1 INTRODUCTION

Up to this point, we have been primarily concerned with the modelling and quantitative analysis of dynamical systems and nothing has been said regarding the control of these systems. We now discuss in this chapter some of the fundamental notions associated with the control of linear multivariable systems represented by

$$\dot{\mathbf{x}}(t) = \mathbf{A}(t)\mathbf{x}(t) + \mathbf{B}(t)\mathbf{u}(t) \tag{6.1a}$$

$$\mathbf{y}(t) = \mathbf{C}(t)\mathbf{x}(t) + \mathbf{D}(t)\mathbf{u}(t) \tag{6.1b}$$

where $\mathbf{x}$ is the $n\times 1$ state vector, $\mathbf{u}$ is the $p\times 1$ input vector, $\mathbf{y}$ is the $q\times 1$ output vector and $\mathbf{A}, \mathbf{B}, \mathbf{C}$ and $\mathbf{D}$ are, respectively, $n\times n$, $n\times p$, $q\times n$ and $q\times p$ real matrices whose entries are continuous functions of t defined over $(-\infty, \infty)$.

The solution of state equation (6.1a) with $\mathbf{x}(t_0) = \mathbf{x}^0$ is given by (refer eqns. (5.33))

$$\begin{aligned}\mathbf{x}(t) &= \Phi(t, t_0)\mathbf{x}^0 + \int_{t_0}^{t} \Phi(t, \tau)\mathbf{B}(\tau)\mathbf{u}(\tau)\, d\tau \\ &= \Phi(t, t_0)[\mathbf{x}^0 + \int_{t_0}^{t} \Phi(t_0, \tau)\mathbf{B}(\tau)\mathbf{u}(\tau)\, d\tau]\end{aligned} \tag{6.2}$$

where $\Phi(t, \tau)$ is the state transition matrix of $\dot{\mathbf{x}}(t) = \mathbf{A}(t)\mathbf{x}(t)$. From eqns. (6.1b) and (6.2), the output

$$\mathbf{y}(t) = \mathbf{C}(t)\Phi(t, t_0)[\mathbf{x}^0 + \int_{t_0}^{t} \Phi(t_0, \tau)\mathbf{B}(\tau)\mathbf{u}(\tau)\, d\tau] + \mathbf{D}(t)\mathbf{u}(t) \tag{6.3}$$

For linear time-invariant systems, eqns. (6.1) take the form:

$$\dot{\mathbf{x}}(t) = \mathbf{A}\mathbf{x}(t) + \mathbf{B}\mathbf{u}(t) \tag{6.4a}$$

$$\mathbf{y}(t) = \mathbf{C}\mathbf{x}(t) + \mathbf{D}\mathbf{u}(t) \tag{6.4b}$$

where $\mathbf{A}, \mathbf{B}, \mathbf{C}$ and $\mathbf{D}$ are, respectively, $n\times n$, $n\times p$, $q\times n$ and $q\times p$ real constant matrices. The time interval of interest is $[0, \infty)$.

The solution of state equation (6.4a) with $\mathbf{x}(0) = \mathbf{x}^0$ is given by (refer eqns. (5.55))

$$\mathbf{x}(t) = e^{\mathbf{A}t}\mathbf{x}^0 + \int_0^t e^{\mathbf{A}(t-\tau)}\mathbf{B}\mathbf{u}(\tau)d\tau$$

$$= e^{\mathbf{A}t}\,[\mathbf{x}^0 + \int_0^t e^{-\mathbf{A}\tau}\,\mathbf{B}\mathbf{u}(\tau)\,d\tau] \tag{6.5}$$

The output

$$\mathbf{y}(t) = \mathbf{C}e^{\mathbf{A}t}\,[\mathbf{x}^0 + \int_0^t e^{-\mathbf{A}\tau}\,\mathbf{B}\mathbf{u}(\tau)\,d\tau] + \mathbf{D}\mathbf{u}(t) \tag{6.6}$$

In this chapter, we shall also be concerned with linear discrete-time systems represented by

$$\mathbf{x}(k+1) = \mathbf{F}(k)\,\mathbf{x}(k) + \mathbf{G}(k)\,\mathbf{u}(k) \tag{6.7a}$$

$$\mathbf{y}(k) = \mathbf{C}(k)\,\mathbf{x}(k) + \mathbf{D}(k)\,\mathbf{u}(k) \tag{6.7b}$$

where $k = k_0, k_0+1, k_0 = 2,\ldots$ (k_0 = integer) and $\mathbf{F}$, $\mathbf{G}$, $\mathbf{C}$ and $\mathbf{D}$ are, respectively, $n \times n$. $n \times p$, $q \times n$ and $q \times p$ time-varying real matrices.

The solution of state equation (6.7a) with $\mathbf{x}(k_0) = \mathbf{x}^0$ is given by (refer eqn. 5.88)).

$$\mathbf{x}(k) = \Phi(k, k_0)[\mathbf{x}^0 + \sum_{j=k_0+1}^{k} \Phi^{-1}(j, k_0)\mathbf{G}(j-1)\mathbf{u}(j-1)] \tag{6.8}$$

where $\Phi(k, j)$ is the state transition matrix of $\mathbf{x}(k+1) = \mathbf{F}(k)\mathbf{x}(k)$.

The output

$$\mathbf{y}(k) = \mathbf{C}(k)\,\Phi(k, k_0)[\mathbf{x}^0 + \sum_{j=k_0+1}^{k} \Phi^{-1}(j, k_0)\,\mathbf{G}(j-1)\,\mathbf{u}(j-1)] + \mathbf{D}(k)\,\mathbf{u}(k) \tag{6.9}$$

For linear time-invariant discrete-time systems, eqns. (6.7) take the form:

$$\mathbf{x}(k+1) = \mathbf{F}\mathbf{x}(k) + \mathbf{G}\mathbf{u}(k) \tag{6.10a}$$

$$\mathbf{y}(k) = \mathbf{C}\mathbf{x}(k) + \mathbf{D}\mathbf{u}(k) \tag{6.10b}$$

where $\mathbf{F}$, $\mathbf{G}$, $\mathbf{C}$ and $\mathbf{D}$ are, respectively, $n \times n$, $n \times p$, $q \times n$ and $q \times p$ real constant matrices and $k = 0, 1, 2,\ldots$ The solution of state equation (6.10a) with $\mathbf{x}(0) = \mathbf{x}^0$ is given by (refer eqn. (5.99))

$$\mathbf{x}(k) = \mathbf{F}^k\,[\mathbf{x}^0 + \sum_{j=1}^{k} \mathbf{F}^{-j}\,\mathbf{G}\mathbf{u}(j-1)] \tag{6.11}$$

The output

$$\mathbf{y}(k) = \mathbf{C}\mathbf{F}^k\,[\mathbf{x}^0 + \sum_{j=1}^{k} \mathbf{F}^{-j}\,\mathbf{G}\mathbf{u}(j-1)] + \mathbf{D}\mathbf{u}(k) \tag{6.12}$$

We shall first discuss the fundamental notions for the control of linear continuous-time systems and then extend the results to linear discrete-time systems.

The two basic questions that occur in the control of linear dynamic systems are:

1. Is it possible to transfer any state $\mathbf{x}(t_0)$ in the state space $\mathbf{\Sigma}$ to any other state $\mathbf{x}^1$ in $\mathbf{\Sigma}$ in finite time $t_1 - t_0 (t_1 > t_0)$, by the proper choice of the input $\mathbf{u}_{[t_0,\, t_1]}$?

2. Is it possible to identify any state $\mathbf{x}^0$ at time t_0 in the state space Σ by observing the output $\mathbf{y}_{[t_0, t_1]}$ and input $\mathbf{u}_{[t_0, t_1]}$ over a finite time interval $t_1 - t_0 (t_1 > t_0)$?

The answers to these questions were conceptualized by Kalman (1960) into what is known as the *controllability* and *observability* of a system. Further original effort in this area is due to Kalman, Ho and Narendra (Kalman et al. 1962, Kalman 1962, 1963), Kreindler and Sarachik (1964), Lee (1963) and Gilbert (1963). Precise definitions of controllability and observability are given in later sections of this chapter.

The concepts of controllability and observability lead to some very important conclusions regarding behaviour of linear dynamical systems. In this and the next chapter, we shall see that a linear state model may be canonically decomposed into various parts and that input-output characterizations such as the transfer function and impulse response describe only that portion which is both controllable and observable. In Chapters 10 and 11, we shall see that the conditions of controllability and observability often govern the existence of a solution to an optimal control problem. In Chapter 9 we shall see that a controllable system is always stabilizable and further the eigenvalues of a controllable system can be placed arbitrarily using state-variable feedback. In Chapter 9 we shall also see that an observer can be constructed to estimate the state of any observable system and the eigenvalues of the observer can be placed arbitrarily, The design of observers for stochastic systems will be taken up in Chapter 12. The general references for this chapter are Ogata (1967), Cadzow and Martens (1968), Chen (1970), Kuo (1970), Kwakernaak and Sivan (1972), Podulo and Arbib (1974), Fortmann and Hitz (1977) and Kailath (1980).

6.2 GENERAL CONCEPT OF CONTROLLABILITY

Consider the block diagram of Fig. 6.1. The plant is said to be *completely controllable* if every state variable $\mathbf{x}(t)$ can be affected or controlled to reach a desired value in finite time by some unconstrained control $\mathbf{u}(t)$. Consider a case wherein one of the state variables is independent of the control $\mathbf{u}(t)$. There would be no way of driving this particular state variable to a desired state in finite time by means of any control effort. Therefore,

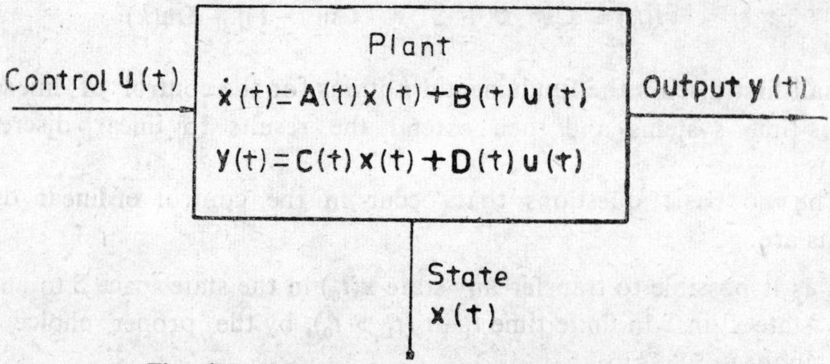

Fig. 6.1 General structure of a control system.

this particular state variable is said to be uncontrollable and as long as there is at least one uncontrollable state the system is said to be *not completely controllable* or *uncontrollable*. The following is a simple example of uncontrollable linear system.

Example 6.1: Consider the Mixing Tank system discussed in Section 4.6. Suppose the feeds Q_1 and Q_2 have equal concentrations, i.e., $C_1 = C_2 = \bar{C}$ (Fig. 4.12). Then the steady-state concentration C_0 in the tank is also $\bar{C}$ and from eqn. (4.37a), we have

$$\dot{\mathbf{x}}(t) = \begin{bmatrix} -\frac{1}{2\tau} & 0 \\ 0 & -\frac{1}{\tau} \end{bmatrix} \mathbf{x}(t) + \begin{bmatrix} 1 & 1 \\ 0 & 0 \end{bmatrix} \mathbf{u}(t)$$

The corresponding state diagram is shown in Fig. 6.2. The input $\mathbf{u}(t)$ affects only the state variable $x_1(t)$, the incremental volume. The variable $x_2(t)$, the incremental concentration, has no connection with the input $\mathbf{u}(t)$.

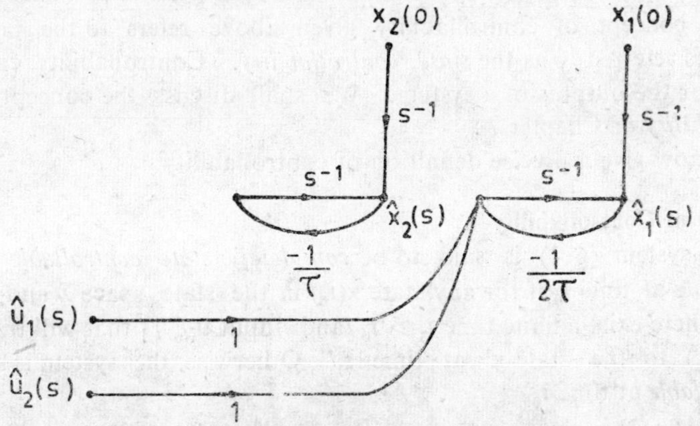

Fig. 6.2 State diagram for Example 6.1.

In other words, it would be impossible to drive $x_2(t)$ from an initial state $x_2(t_0)$ to a desired state $x_2(t_1)$ in a finite time interval $t_1 - t_0$. This is also clear physically. If $C_1 \neq C_2$, the system is completely controllable (Problem 6.4).

Note that complete decoupling of a state variable from the input is a sufficient but not necessary condition for the system to be uncontrollable. To illustrate this point, we consider the following simple example.

Example 6.2: The state equation governing the circuit of Fig. 4.3a is (refer Example 4.2)

$$\dot{\mathbf{x}}(t) = \begin{bmatrix} -2 & 1 \\ 1 & -2 \end{bmatrix} \mathbf{x}(t) + \begin{bmatrix} 1 \\ 1 \end{bmatrix} u(t)$$

$$= \mathbf{A}\mathbf{x}(t) + \mathbf{b}u(t)$$

The corresponding state diagram is shown in Fig. 4.3b. We find that both the state variables x_1 and x_2 have connection with the input $u(t)$. We now study controllability property of this system.

For the initial state $\mathbf{x}(0) = \mathbf{0}$, we have (Example 5.7)

$$\mathbf{x}(t) = \int_0^t e^{\mathbf{A}(t-\tau)}\mathbf{b}u(\tau)d\tau$$

$$= \begin{bmatrix} 1 \\ 1 \end{bmatrix}\left(\int_0^t e^{-(t-\tau)}u(\tau)d\tau\right)$$

Thus $x_1(t) = x_2(t)$ and no input $u(t)$ can be found to allow both $x_1(t_1)$ and $x_2(t_1)$ to be arbitrary, i.e., there is no control which will drive the state $\mathbf{x}(0)$ to a point where $x_1(t_1) \neq x_2(t_1)$. The system is therefore uncontrollable, though the state diagram is completely connected. This conclusion is, of course, obvious from the topology of the circuit.

□

The concept of controllability given above refers to the states and is sometimes referred to as the *state controllability*. Controllability can also be defined for the outputs of a system. We shall discuss the concept of *output controllability in* Chapter 7.

We now give a precise definition of controllability.

Definition of Controllability

The system (6.1) is said to be *completely state controllable* or simply *controllable* at time t_0 if for any state $\mathbf{x}(t_0)$ in the state space Σ and any state $\mathbf{x}^1$ in Σ, there exist a finite time $t_1 > t_0$ and input $\mathbf{u}_{[t_0, t_1]}$ that will transfer the state $\mathbf{x}(t_0)$ to the state $\mathbf{x}^1$ at time t_1. Otherwise, the system is said to be *uncontrollable* at time t_0.

□

It may be noted that according to this definition, there is no constraint imposed on the input or on the trajectory which the state should follow. Further, the system is said to be uncontrollable although it may be 'controllable in part', i.e., it may be possible to transfer certain states to any desired states or any state in Σ to certain regions in the state space.

Controllability is a property of the coupling between the input and the state and, therefore, involves the matrices **A** and **B** of system (6.1). The controllability of system (6.1) is frequently referred to as the controllability of the pair {**A**, **B**}.

From the definition of controllability we observe that by complete controllability of a plant, we mean that we can make the plant do whatever we please. Perhaps this definition is too restrictive in the sense that we are asking too much of the plant. But if we are able to show that system equations satisfy this definition, certainly there can be no intrinsic limitation on the design of the control system for the plant. However, if the system turns out to be uncontrollable, it does not necessarily mean that the plant can never be operated in a satisfactory manner. Provided that a control system will maintain the important variables in an acceptable region, the fact that the plant is not completely controllable is immaterial.

Another important point which the reader must bear in mind is that almost all physical systems are nonlinear in nature to a certain extent and a linear model is obtained after making certain approximations. Small perturbation of the elements of A and **B** may cause an uncontrollable system to become controllable (Lee and Markus 1967). It may also be possible to increase the number of control variables and make the plant completely controllable.

From the simple examples given earlier, we see that it is difficult to guess whether a system is completely controllable or not from its governing state equation. Some simple mathematical tests which answer the question of controllability have been developed and will be discussed in later sections of this chapter.

6.3 GENERAL CONCEPT OF OBSERVABILITY

Consider the block diagram of Fig. 6.1. We observed earlier that the plant is controllable if no variable in state vector $\mathbf{x}(t)$ is beyond the influence of the input $\mathbf{u}(t)$. It is obvious that the fact that we can satisfy the definition of controllability does not provide any information about how to design an acceptable controller. The controller design will be discussed in later chapters. There we shall see that very often the controller is based on the control law

$$\mathbf{u}(t) = \mathbf{K}\mathbf{x}(t)$$

In order to implement this control law, we need to feedback all the state variables. Unfortunately, in practice, not all the state variables are accessible to measurement. We can assume that only the outputs and inputs are measurable. We therefore need a subsystem that performs the observation of the state variables based on the information received from the measurement of the input $\mathbf{u}(t)$ and the output $\mathbf{y}(t)$. This subsystem is called an *observer* (discussed in Chapters 9 and 12) whose design is based on the observability property discussed in this section.

In the problem of observability, we want to find out the possibility of estimating the state $\mathbf{x}(t)$ from the output $\mathbf{y}(t)$. The plant in Fig. 6.1 is *completely observable* if all the variables in $\mathbf{x}(t)$ can be observed from the measurements of the outputs $\mathbf{y}(t)$ and the inputs $\mathbf{u}(t)$. Consider a case wherein one of the state variables has no effect on system output $\mathbf{y}(t)$. There would be no way of observing this particular state variable from the information on $\mathbf{y}(t)$ and $\mathbf{u}(t)$. Therefore this particular state variable is said to be unobservable and as long as there is at least one unobservable state, the system is said to be *not completely observable* or *unobservable*, The state diagram of Fig. 6.3 gives a simple example of unobservable linear system. The state $x_2(t)$ does not affect output $y(t)$ in any way.

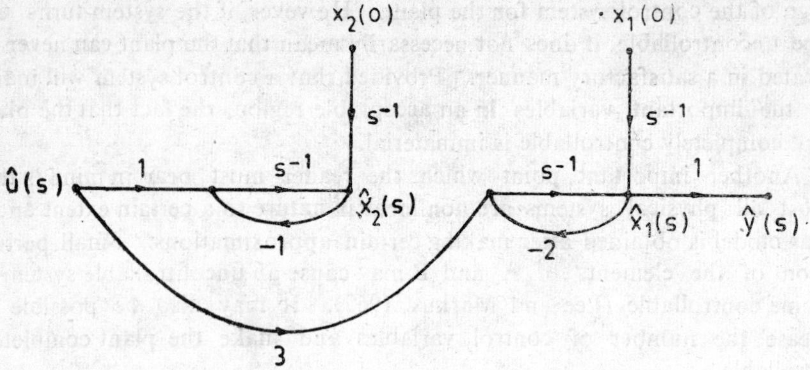

Fig. 6.3 State diagram of a system that is not completely observable.

Complete decoupling of a state variable from the output is sufficient but not necessary condition for the system to be unobservable. To illustrate this point, we consider the following simple example.

Example 6.3: The state model governing the circuit of Fig 4.4a is (refer Example 4.2)

$$\dot{\mathbf{x}}(t) = \begin{bmatrix} -2 & 1 \\ 1 & -2 \end{bmatrix} \mathbf{x}(t) + \begin{bmatrix} 1 \\ 0 \end{bmatrix} u(t)$$

$$= \mathbf{A}\mathbf{x}(t) + \mathbf{b}u(t)$$

$$y(t) = [1 \quad -1]\, \mathbf{x}(t)$$

$$= \mathbf{c}\mathbf{x}(t)$$

The corresponding state diagram is shown in Fig. 4.4b. We find that both the state variables $x_1(t)$ and $x_2(t)$ have a connection with the output $y(t)$. We now study the observability property of this system.

For the input $u(t) = 0$, we have (Example 5.7)

$$y(t) = \mathbf{c}e^{\mathbf{A}t}\,\mathbf{x}^0$$

$$= (x_1^0 - x_2^0)\, e^{-3t}$$

From this expression, it is obvious that the output $y(t)$ depends on the difference $(x_1^0 - x_2^0)$. The initial state with $x_1^0 = x_2^0$ produces no response at the output; therefore the system is unobservable. This conclusion is, of course, obvious from the topology of the circuit.

□

We now give a precise definition of observability.

Definition of Observability

The system (6.1) is said to be *completely observable* or simply *observable* at time t_0 if for any state $\mathbf{x}^0$ at time t_0 in the state space Σ, there exists a finite $t_1 > t_0$ such that the knowledge of the input $\mathbf{u}_{[t_0, t_1]}$ and the output $\mathbf{y}_{[t_0, t_1]}$ suffices to determine the state $\mathbf{x}^0$. Otherwise the system is said to be *unobservable* at t_0. □

Note that the system is said to be unobservable although it may be 'observable in part', i.e., it may be possible to determine some of the variables in $\mathbf{x}(t)$. The notion of observability implies the ability to observe all the variables in $\mathbf{x}(t)$ from a knowledge of $\mathbf{y}(t)$ and $\mathbf{u}(t)$ alone. Plants that are not observable can often be made observable by making more measurements.

Observability is a property of the coupling between state and the output and involves only the matrices **A** and **C**; the observability of system (6.1) is frequently referred to as the observability of the pair {**A**, **C**}.

Some mathematical tests which answer the question of observability will be discussed in later sections of this chapter.

6.4 CONTROLLABILITY TESTS FOR CONTINUOUS-TIME SYSTEMS

Time-Varying Case

We shall develop a basic theorem utilizing state transition matrix $\Phi(t, \tau)$ for the determination of controllability of time-varying system (Kreindler and Sarachik 1964) represented by eqns. (6.1).

Theorem 6.1: The system (6.1) is controllable at time t_0 if and only if there exists a finite $t_1 > t_0$ such that the n rows of the $n \times p$ matrix function $\Phi(t_0, t)$ $\mathbf{B}(t)$ are linearly independent on $[t_0, t_1]$.

Proof:

Let us first show the sufficiency. If the rows of $\Phi(t_0, t)\,\mathbf{B}(t)$ are linearly independent on $[t_0, t_1]$, then we must show that there exists a control $\mathbf{u}_{[t_0, t_1]}$ to drive the system from the arbitrary state $\mathbf{x}(t_0) = \mathbf{x}^0$ to the arbitrary state $\mathbf{x}^1$. If the rows of $\Phi(t_0, t)\,\mathbf{B}(t)$ are linearly independent on $[t_0, t_1]$ then, as per eqn. (2.34), the $n \times n$ constant matrix (the Grammian matrix)

$$\mathbf{W}(t_0, t_1) \overset{\Delta}{=} \int_{t_0}^{t_1} \Phi(t_0, t)\,\mathbf{B}(t)\mathbf{B}^T(t)\Phi^T(t_0, t)\,dt \tag{6.13}$$

is nonsingular and thus $\mathbf{W}^{-1}(t_0, t_1)$ exists. We claim that the input

$$\mathbf{u}(t) = -\,\mathbf{B}^T(t)\,\Phi^T(t_0, t)\,\mathbf{W}^{-1}(t_0, t_1)\,[\mathbf{x}^0 - \Phi(t_0, t_1)\mathbf{x}^1] \tag{6.14}$$

will transfer $\mathbf{x}^0$ to $\mathbf{x}^1$ at time t_1. This can be easily verified by substituting (6.14) into (6.2).

$$\begin{aligned}
\mathbf{x}(t_1) &= \Phi(t_1, t_0)[\mathbf{x}^0 - \{\int_{t_0}^{t_1} \Phi(t_0, t)\,\mathbf{B}(t)\,\mathbf{B}^T(t)\,\Phi^T(t_0, t)\,dt\} \\
&\qquad\qquad \{\mathbf{W}^{-1}(t_0, t_1)[\mathbf{x}^0 - \Phi(t_0, t_1)\mathbf{x}^1]\}] \\
&= \Phi(t_1, t_0)[\mathbf{x}^0 - \mathbf{W}(t_0, t_1)\,\mathbf{W}^{-1}(t_0, t_1)\{\mathbf{x}^0 - \Phi(t_0, t_1)\mathbf{x}^1\}] \\
&= \Phi(t_1, t_0)\Phi(t_0, t_1)\,\mathbf{x}^1 \\
&= \mathbf{x}^1
\end{aligned}$$

We have found a control and therefore the system is controllable.

The necessity requires that we show that if (6.1) is controllable, then the n rows of $\Phi(t_0, t)\mathbf{B}(t)$ are linearly independent. Let us use contradiction and assume that (6.1) is controllable but the rows of $\Phi(t_0, t)\mathbf{B}(t)$ are linearly dependent. Thus there exists a nonzero, constant $1 \times n$ real vector $\boldsymbol{\alpha}$ such that

$$\boldsymbol{\alpha}\Phi(t_0, t)\mathbf{B}(t) = \mathbf{0} \text{ for all } t \text{ in } [t_0, t_1] \tag{6.15}$$

Choose $\mathbf{x}(t_0) = \mathbf{x}^0 = \boldsymbol{\alpha}^T$ as an initial state and $\mathbf{x}(t_1) = \mathbf{x}^1 = \mathbf{0}$ as a terminal state. This a possible since the system is controllable. Then from (6.2) we have

$$\mathbf{0} = \Phi(t_1, t_0)[\boldsymbol{\alpha}^T + \int_{t_0}^{t_1} \Phi(t_0, t)\mathbf{B}(t)\mathbf{u}(t)\,dt]$$

Premultiplying both sides of this equation by $\boldsymbol{\alpha}\Phi(t_0, t_1)$ we get

$$\mathbf{0} = \boldsymbol{\alpha}\boldsymbol{\alpha}^T + \int_{t_0}^{t_1} \boldsymbol{\alpha}\Phi(t_0, t)\mathbf{B}(t)\mathbf{u}(t)\,dt \tag{6.16}$$

Since $\boldsymbol{\alpha}\Phi(t_0, t)\mathbf{B}(t) = \mathbf{0}$ for all t in $[t_0, t_1]$, eqn. (6.16) reduces to

$$\boldsymbol{\alpha}\boldsymbol{\alpha}^T = \mathbf{0}$$

which in turn implies that $\boldsymbol{\alpha} = \mathbf{0}$. This is a contradiction.

□

The controllability condition given above is rather difficult to check since the transition matrix $\Phi(t, t_0)$ must be computed and the integral (6.13) must be solved. If the matrices $\mathbf{A}(t)$ and $\mathbf{B}(t)$ are sufficiently differentiable, then we can develop a controllability criterion based on the matrices $\mathbf{A}$ and $\mathbf{B}$ without solving the state equation. Such a criterion, given in Chen (1970), gives a sufficient but not necessary condition for the controllability of a system.

Minimum Energy Control: The result of Theorem 6.1 is powerful because it gives not only necessary and sufficient conditions for controllability but also a method to compute the control $\mathbf{u}(t)$. As pointed out earlier, for the controllable system (6.1), there are generally many different inputs that can transfer $\mathbf{x}(t_0)$ to $\mathbf{x}^1$ at time t_1 because the trajectory between $\mathbf{x}(t_0)$ and $\mathbf{x}^1$ is not specified. The control function $\mathbf{u}(t)$ given by (6.14) gives *minimum-energy control* in the sense that 'control energy expenditure' measured as (refer Chapter 10)

$$\int_{t_0}^{t_1} \mathbf{u}^T(t)\,\mathbf{u}(t)\,dt \tag{6.17}$$

is minimal when the system is driven by this input (This result is proved below). Therefore if the energy given by (6.17) is used as a criterion, the input (eqn. (6.14))

$$\mathbf{u}^*(t) = -\mathbf{B}^T(t)\Phi^T(t_0, t)\mathbf{W}^{-1}(t_0, t_1)\,[\,\mathbf{x}^0 - \Phi(t_0, t_1)\mathbf{x}^1\,] \tag{6.18}$$

is the optimal control (Kalman 1962).

Theorem 6.2: For controllable system (6.1), let $\mathbf{u}_{[t_0, t_1]}$ be any control that transfers the state $\mathbf{x}(t_0)$ to the state $\mathbf{x}^1$ at time t_1 and let $\mathbf{u}^*(t)$ be the control defined in (6.18) that accomplishes the same transfer; then

$$\int_{t_0}^{t_1} \mathbf{u}^T(t)\mathbf{u}(t)dt \geqslant \int_{t_0}^{t_1} \mathbf{u}^{*T}(t)\mathbf{u}^*(t)dt$$

Proof:

Since $\mathbf{u}_{[t_0, t_1]}$ is a control that transfers the state $\mathbf{x}(t_0) \triangleq \mathbf{x}^0$ to the state $\mathbf{x}^1$ at time t_1, we have from eqn. (6.2)

$$\mathbf{x}^1 = \mathbf{x}(t_1) = \Phi(t_1, t_0)\left[\mathbf{x}^0 + \int_{t_0}^{t_1} \Phi(t_0, t)\mathbf{B}(t)\mathbf{u}(t)dt\right] \quad (6.19)$$

Since $\mathbf{u}^*(t)$ given by (6.18) also accomplishes the same transfer, we have

$$\mathbf{x}^1 = \mathbf{x}(t_1) = \Phi(t_1, t_0)\left[\mathbf{x}^0 + \int_{t_0}^{t_1} \Phi(t_0, t)\mathbf{B}(t)\mathbf{u}^*(t)dt\right] \quad (6.20)$$

From eqns. (6.19) and (6.20) we have

$$\int_{t_0}^{t_1} \Phi(t_0, t)\mathbf{B}(t)(\mathbf{u}(t) - \mathbf{u}^*(t))dt = 0$$

Multiplying on the left by $[\Phi(t_0, t_1)\mathbf{x}^1 - \mathbf{x}^0]^T \mathbf{W}^{-1}(t_0, t_1)$, we get

$$\int_{t_0}^{t_1} [\Phi(t_0, t_1)\mathbf{x}^1 - \mathbf{x}^0]^T \mathbf{W}^{-1}(t_0, t_1)\Phi(t_0, t)\mathbf{B}(t)(\mathbf{u}(t) - \mathbf{u}^*(t))dt = 0 \quad (6.21)$$

From (6.21) and (6.18) we can write

$$\int_{t_0}^{t_1} \mathbf{u}^{*T}(t)(\mathbf{u}(t) - \mathbf{u}^*(t))\, dt = 0$$

or (refer Section 2.5)

$$\int_{t_0}^{t_1} \mathbf{u}^{*T}(t)\mathbf{u}(t)dt = \int_{t_0}^{t_1} \mathbf{u}^{*T}(t)\mathbf{u}^*(t)dt = \int_{t_0}^{t_1} \|\mathbf{u}^*(t)\|^2 dt \quad (6.22a)$$

Next, we consider the obvious inequality

$$0 \leqslant \int_{t_0}^{t_1} \|\mathbf{u}(t) - \mathbf{u}^*(t)\|^2\, dt \quad (6.22b)$$

Now

$$\begin{aligned}\|\mathbf{u}(t) - \mathbf{u}^*(t)\|^2 &= |(\mathbf{u}(t) - \mathbf{u}^*(t))^T (\mathbf{u}(t) - \mathbf{u}^*(t))| \\ &= |\mathbf{u}^T(t)\mathbf{u}(t) - 2\mathbf{u}^{*T}(t)\mathbf{u}(t) + \mathbf{u}^{*T}(t)\mathbf{u}^*(t)|\end{aligned} \quad (6.22c)$$

From (6.22a), (6.22b) and (6.22c) we get

$$0 \leqslant \int_{t_0}^{t_1} (\|\mathbf{u}(t)\|^2 - \|\mathbf{u}^*(t)\|^2)\, dt$$

or

$$\int_{t_0}^{t_1} \|\mathbf{u}(t)\|^2\, dt \geqslant \int_{t_0}^{t_1} \|\mathbf{u}^*(t)\|^2\, dt$$

This proves the result.

The control $\mathbf{u}^*(t)$ given by (6.18) is open-loop: an input function is precomputed and then applied to the system over some time interval with no feedback to adjust for variations and errors. The subject of closed-loop control will be taken up in later chapters.

Example 6.4: Consider a system described by the state equation

$$\dot{\mathbf{x}}(t) = \mathbf{A}(t)\,\mathbf{x}(t) + \mathbf{b}\,u(t)$$

where

$$\mathbf{A}(t) = \begin{bmatrix} 1 & e^{-t} \\ 0 & -1 \end{bmatrix}; \quad \mathbf{b} = \begin{bmatrix} 0 \\ 1 \end{bmatrix}$$

Is this system controllable at $t = 0$? If yes, find the minimum-energy control to drive it from

$$\mathbf{x}(0) = \mathbf{0} \text{ to } \mathbf{x}^1 = \begin{bmatrix} 1 \\ 1 \end{bmatrix} \text{ at } t = 1.$$

Solution:

The state transition matrix $\Phi(t, \tau)$ of $\dot{\mathbf{x}}(t) = \mathbf{A}(t)\mathbf{x}(t)$ is (Example 5.5)

$$\Phi(t, \tau) = \begin{bmatrix} e^{t-\tau} & \frac{1}{3}(e^{t-2\tau} - e^{-2t+\tau}) \\ 0 & e^{-t+\tau} \end{bmatrix}$$

Therefore,

$$\Phi(0, \tau) = \begin{bmatrix} e^{-\tau} & \frac{1}{3}(e^{-2\tau} - e^{\tau}) \\ 0 & e^{\tau} \end{bmatrix}$$

The Grammian matrix (eqn. (6.13))

$$\mathbf{W}(0, 1) = \int_0^1 \Phi(0, \tau)\mathbf{b}\mathbf{b}^T\Phi^T(0, \tau)\, d\tau$$

$$= \int_0^1 \begin{bmatrix} \frac{1}{9}(e^{-4\tau} - 2e^{-\tau} + e^{2\tau}) & \frac{1}{3}(e^{-\tau} - e^{2\tau}) \\ \frac{1}{3}(e^{-\tau} - e^{2\tau}) & e^{2\tau} \end{bmatrix} d\tau$$

$$= \begin{bmatrix} 0.2417 & -0.8541 \\ -0.8541 & 3.1945 \end{bmatrix}$$

$$|\mathbf{W}(0, 1)| = 0.04262 \neq 0.$$

The system is therefore controllable at $t = 0$.

The minimum-energy control is (eqn. (6.18))

$$u^*(t) = -[0 \quad 1]\begin{bmatrix} e^{-t} & 0 \\ \frac{1}{3}(e^{-2t}-e^{t}) & e^{t} \end{bmatrix}\begin{bmatrix} 74.9531 & 20.0399 \\ 20.0399 & 5.6710 \end{bmatrix}$$

$$\left[\begin{pmatrix} 0 \\ 0 \end{pmatrix} - \begin{pmatrix} e^{-1} & \frac{1}{3}(e^{-2}-e^{1}) \\ 0 & e^{1} \end{pmatrix}\begin{pmatrix} 1 \\ 1 \end{pmatrix}\right]$$

$$= -[0 \quad 1]\begin{bmatrix} e^{-t} & 0 \\ \frac{1}{3}(e^{-2t}-e^{t}) & e^{t} \end{bmatrix}\begin{bmatrix} 74.9531 & 20.0399 \\ 20.0399 & 5.6710 \end{bmatrix}\begin{bmatrix} 0.4931 \\ -2.7183 \end{bmatrix}$$

$$= 5.8384e^{-2t} - 0.3026e^{t}$$

The system driven by $\mathbf{u}^*(t)$ follows the trajectory (eqn. (6.2))

$$\mathbf{x}^*(t) = \begin{bmatrix} e^{t} & \frac{1}{3}(e^{t}-e^{-2t}) \\ 0 & e^{-t} \end{bmatrix}\left[\int_0^t \begin{pmatrix} \frac{1}{3}(e^{-2\tau}-e^{\tau}) \\ e^{\tau} \end{pmatrix}(5.8384e^{-2\tau}-0.3026e^{\tau})\,d\tau\right]$$

$$= \begin{bmatrix} 0.3856e^{t} + 0.1513 - 1.9966e^{-2t} + 1.4596e^{-3t} \\ -0.1513e^{t} + 5.9897e^{-t} - 5.8384e^{-2t} \end{bmatrix}$$

Figure 6.4 shows the minimum-energy control and the corresponding system trajectories.

Time-Invariant Case

In this sub-section, we study the controllability property of the n-dimensional linear time-invariant systems described by equations of the form (6.4).

Theorem 6.3: The following statements regarding the linear time-invariant dynamical system (6.4) are equivalent:

a. The system is completely controllable.

b. The rows of $e^{-\mathbf{A}t}\mathbf{B}$ (and hence of $e^{\mathbf{A}t}\mathbf{B}$) are linearly independent over the real field $\mathcal{R}$ for all $t \in [0, \infty)$.

c. $\mathbf{W}(0, t) = \int_0^t e^{-\mathbf{A}\tau}\mathbf{B}\mathbf{B}^T e^{-\mathbf{A}^T\tau}\,d\tau$ is nonsingular for any $t > 0$.

d. The rank of the $(n \times np)$ *controllability matrix* $\mathbf{U}$. $(\mathbf{U} \triangleq [\mathbf{B}\mathbf{A}\mathbf{B}\ldots\mathbf{A}^{n-1}\mathbf{B}])$ is n, i.e., $\rho(\mathbf{U}) = n$.

Proof:

Applying Theorem 6.1 to the system (6.4): the system is controllable at time $t = 0$ if and only if there exists a finite $t > 0$ such that the n rows of $n \times p$ matrix function $\Phi(0, t)\mathbf{B} = e^{-\mathbf{A}t}\mathbf{B}$ are linearly independent on $[0, t]$. In the following we shall see that because of the analyticity property of $e^{-\mathbf{A}t}\mathbf{B}$,

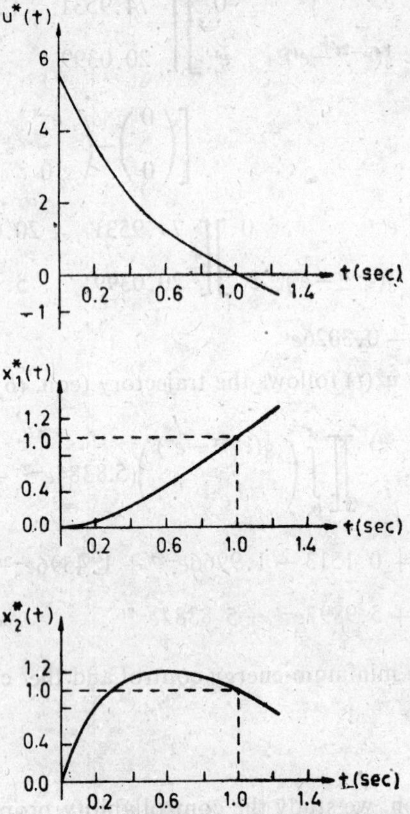

Fig. 6.4 Minimum energy control and trajectories for Example 6.4.

if (6.4) is controllable at some t, then it is controllable at every t in $(0, \infty)$. Statements a and b are therefore equivalent.

The equivalence of statements b and c follows directly from eqn (2.34). The n rows of $e^{-\mathbf{A}t}\mathbf{B}$ are linearly independent over $\mathcal{R}$ if

$$\mathbf{W}(0, t) = \int_0^t e^{-\mathbf{A}\tau}\mathbf{B}\mathbf{B}^T e^{-\mathbf{A}^T\tau}\, d\tau \tag{6.23}$$

is nonsingular for any $t > 0$.

The minimum-energy control that transfers $\mathbf{x}(0) \triangleq \mathbf{x}^0$ to $\mathbf{x}^1$ at t_1 is (eqn. (6.18))

$$\mathbf{u}^*(t) = -\mathbf{B}^T e^{-\mathbf{A}^T t}\mathbf{W}^{-1}(0, t_1)[\mathbf{x}^0 - e^{-\mathbf{A}t_1}\mathbf{x}^1] \tag{6.24}$$

The equivalence of statements b and d can be established as follows We first prove the result:

The n rows of $e^{-\mathbf{A}t}\mathbf{B}$ are linearly independent over $\mathcal{R}$ for all $t \in [0, \infty)$ if and only if the n rows of the $(n \times \infty)$ matrix valued functions of time

$[e^{-\mathbf{A}t}\mathbf{B} \vdots -e^{-\mathbf{A}t}\mathbf{AB} \vdots \ldots \vdots (-1)^{n-1} e^{-\mathbf{A}t}\mathbf{A}^{n-1}\mathbf{B} \vdots \ldots]$ *are linearly independent over* $\mathcal{R}$ *for all* $t \in [0, \infty)$.

We establish the necessity first by contradiction. Assume that the n rows of $e^{-\mathbf{A}t}\mathbf{B}$ are linearly independent but that

$$\rho\,[e^{-\mathbf{A}t}\mathbf{B} \vdots -e^{-\mathbf{A}t}\mathbf{AB} \vdots \ldots \vdots (-1)^{n-1} e^{-\mathbf{A}t}\mathbf{A}^{n-1}\mathbf{B} \vdots \ldots] < n$$

at some time $t_0 \in [0, \infty)$. This implies that for some nonzero real vector $\boldsymbol{\alpha}$,

$$\boldsymbol{\alpha}\,[e^{-\mathbf{A}t_0}\mathbf{B} \vdots -e^{-\mathbf{A}t_0}\mathbf{AB} \vdots \ldots \vdots (-1)^{n-1} e^{-\mathbf{A}t_0}\mathbf{A}^{n-1}\mathbf{B} \vdots \ldots] = \mathbf{0} \tag{6.25}$$

Since the entries of $e^{-\mathbf{A}t}\mathbf{B}$ are analytic (Section 5.5) everywhere for all $t \in [0, \infty)$ including t_0, $e^{-\mathbf{A}t}\mathbf{B}$ can be expanded via a Taylor series about t_0, i.e.,

$$e^{-\mathbf{A}t}\mathbf{B} = \sum_{i=0}^{\infty} \frac{(t-t_0)^i}{i!}(-1)^i e^{-\mathbf{A}t_0}\mathbf{A}^i\mathbf{B} \tag{6.26}$$

Equation (6.26) holds wherever $e^{-\mathbf{A}t}\mathbf{B}$ is analytic, i.e., it holds for all $t \in [0, \infty)$. Premultiplying both sides of (6.26) by $\boldsymbol{\alpha}$, we get

$$\sum_{i=0}^{\infty} \frac{(t-t_0)^i}{i!}(-1)^i \boldsymbol{\alpha}\, e^{-\mathbf{A}t_0}\mathbf{A}^i\mathbf{B} = \boldsymbol{\alpha}\, e^{-\mathbf{A}t}\mathbf{B} \tag{6.27}$$

for all t where (6.26) holds, i.e., for all $t \in [0, \infty)$. In view of (6.25), the right-hand side of (6.27) is equal to $\mathbf{0}$, which gives

$$\boldsymbol{\alpha}\, e^{-\mathbf{A}t}\mathbf{B} = \mathbf{0}$$

for all $t \in [0, \infty)$.

This is contrary to the assumption that the rows of $e^{-\mathbf{A}t}\mathbf{B}$ are linearly independent on $[0, \infty)$. Necessity is therefore established by contradiction. We note in view of the analyticity property of $e^{-\mathbf{A}t}\mathbf{B}$ that if the rows of $e^{-\mathbf{A}t}\mathbf{B}$ are linearly independent (dependent) at any time t_0, then they are linearly independent (dependent) for all $t \in [0, \infty)$.

Sufficiency is also readily established by contradiction. We assume that

$$\rho\,[e^{-\mathbf{A}t}\mathbf{B} \vdots -e^{-\mathbf{A}t}\mathbf{AB} \vdots \ldots \vdots (-1)^{n-1} e^{-\mathbf{A}t}\mathbf{A}^{n-1}\mathbf{B} \vdots \ldots] = n \tag{6.28}$$

but for some nonzero real vector $\boldsymbol{\alpha}$,

$$\boldsymbol{\alpha}\, e^{-\mathbf{A}t}\mathbf{B} = \mathbf{0} \quad \text{for all } t \in [0, \infty) \tag{6.29}$$

By successively differentiating (6.29) with respect to time, we find that

$$-\boldsymbol{\alpha}\, e^{-\mathbf{A}t}\mathbf{AB} = \boldsymbol{\alpha}\, e^{-\mathbf{A}t}\mathbf{A}^2\mathbf{B} = \ldots = (-1)^{n-1}\boldsymbol{\alpha}\, e^{-\mathbf{A}t}\mathbf{A}^{n-1}\mathbf{B} = \ldots = \mathbf{0}$$

or that

$$\boldsymbol{\alpha}\,[e^{-\mathbf{A}t}\mathbf{B} \vdots -e^{-\mathbf{A}t}\mathbf{AB} \vdots \ldots \vdots (-1)^{n-1} e^{-\mathbf{A}t}\mathbf{A}^{n-1}\mathbf{B} \vdots \ldots] = \mathbf{0}$$

This is contrary to the assumption. Thus the rows of $e^{-\mathbf{A}t}\mathbf{B}$ are linearly independent on $[0, \infty)$ if and only if (6.28) is true for all $t \in [0, \infty)$.

Let $t = 0$, then (6.28) reduces to the following:

$$\rho\,[\mathbf{B} \,\vdots\, -\mathbf{AB} \,\vdots\, \ldots \,\vdots\, (-1)^{n-1}\mathbf{A}^{n-1}\mathbf{B} \,\vdots\, (-1)^{n}\mathbf{A}^{n}\mathbf{B} \,\vdots\, \ldots] = n \tag{6.30}$$

This expression can now be simplified if we employ the Cayley-Hamilton theorem (Seetion 2.8) which enables us to express all terms of the form $\mathbf{A}^i\mathbf{B}$ with $i \geqslant n$ as some linear combination of the n terms: $\mathbf{B}, \mathbf{AB}, \ldots, \mathbf{A}^{n-1}\mathbf{B}$. Hence the columns of $\mathbf{A}^i\mathbf{B}$ with $i \geqslant n$ are directly dependent on the columns of $\mathbf{B}, \mathbf{AB}, \ldots, \mathbf{A}^{n-1}\mathbf{B}$. Consequently, (6.30) is equivalent to

$$\rho\,[\mathbf{B} \,\vdots\, -\mathbf{AB} \,\vdots\, \ldots \,\vdots\, (-1)^{n-1}\mathbf{A}^{n-1}\mathbf{B}] = n$$

Furthermore, since the rank of a matrix does not change if we multiply any column by (-1), it follows that rows of $e^{-\mathbf{A}t}\mathbf{B}$ (and hence of $e^{\mathbf{A}t}\mathbf{B}$) are linearly independent if and only if

$$\rho\,[\mathbf{U}] = \rho\,[\mathbf{B} \,\vdots\, \mathbf{AB} \,\vdots\, \ldots \,\vdots\, \mathbf{A}^{n-1}\mathbf{B}] = n$$

This proves the equivalence of statements b and d.

□

Note that the determination of the rank of controllability matrix $\mathbf{U}$ is the test usually employed to ascertain whether or not the pair $\{\mathbf{A}, \mathbf{B}\}$ is controllable.

Example 6.5: Consider the platform system of Example 4.1. Let us neglect platform masses in order to get a simple expression. Also, we take $K_1 = B_1 = K_2 = 1$ and make the coefficient B_2 adjustable. The displacements x_1 and x_2 of the platforms are then governed by the equation

$$\dot{\mathbf{x}}(t) = \mathbf{A}\mathbf{x}(t) + \mathbf{b}u(t)$$

where

$$\mathbf{A} = \begin{bmatrix} -1 & \dfrac{B_2 - 1}{B_2} \\ 0 & -\dfrac{1}{B_2} \end{bmatrix}; \quad \mathbf{b} = \begin{bmatrix} \dfrac{B_2 + 1}{B_2} \\ \dfrac{1}{B_2} \end{bmatrix}$$

The controllability matrix

$$\mathbf{U} = [\mathbf{b} \,\vdots\, \mathbf{Ab}]$$

$$= \begin{bmatrix} \dfrac{B_2 + 1}{B_2} & \dfrac{-1 - B_2^2}{B_2^2} \\ \dfrac{1}{B_2} & -\dfrac{1}{B_2^2} \end{bmatrix}$$

$\rho(\mathbf{U}) = 2$ if $B_2 \neq 1$. Therefore the system is controllable if $B_2 \neq 1$.

As pointed out earlier, merely the satisfaction of the test of controllability does not provide any information about how to design an acceptable controller. Consider the problem of controlling the platform system under consideration. For $B_2 = 0.5$, the state model becomes

$$\dot{\mathbf{x}}(t) = \begin{bmatrix} -1 & -1 \\ 0 & -2 \end{bmatrix} \mathbf{x}(t) + \begin{bmatrix} 3 \\ 2 \end{bmatrix} \mathbf{u}(t)$$

If the initial displacements of the unforced platforms are $x_1(0)$ and $x_2(0)$, the platforms will vibrate. The control problem is to find a suitable $u(t)$ so that the platforms can be forced to rest in t_1 sec.

Taking $\mathbf{x}(0) = \mathbf{x}^0 = [10 \quad 10]^T$ and $t_1 = 2$ sec, we derive the minimum-energy control law.

From eqns. (6.24) and (6.23) we have

$$\mathbf{u}^*(t) = -\mathbf{b}^T e^{-\mathbf{A}^T t} \mathbf{W}^{-1}(0, 2)\, \mathbf{x}^0$$

$$\mathbf{W}(0, 2) = \int_0^2 e^{-\mathbf{A}\tau} \mathbf{b}\mathbf{b}^T e^{-\mathbf{A}^T\tau}\, d\tau$$

$$= \int_0^2 \begin{bmatrix} e^{\tau} & -e^{\tau}+e^{2\tau} \\ 0 & e^{2\tau} \end{bmatrix} \begin{bmatrix} 3 \\ 2 \end{bmatrix} [3 \quad 2] \begin{bmatrix} e^{\tau} & 0 \\ -e^{\tau}+e^{2\tau} & e^{2\tau} \end{bmatrix} d\tau$$

$$= \begin{bmatrix} 3543.331 & 3248.246 \\ 3248.246 & 2979.96 \end{bmatrix}$$

$$\mathbf{W}^{-1}(0, 2) = \begin{bmatrix} 0.3781 & -0.4121 \\ -0.4121 & 0.44955 \end{bmatrix}$$

Therefore

$$u^*(t) = -[3 \quad 2] \begin{bmatrix} e^{t} & 0 \\ -e^{t}+e^{2t} & e^{2t} \end{bmatrix} [\mathbf{W}^{-1}(0, 2)] \begin{bmatrix} 10 \\ 10 \end{bmatrix}$$

$$= 0.34e^{t} - 0.068e^{2t}$$

The system driven by $\mathbf{u}^*(t)$ follows the trajectory

$$\mathbf{x}^*(t) = \begin{bmatrix} 0.3967e^{t} - 0.0567e^{2t} - 0.1473e^{-t} + 9.8073e^{2t} \\ 0.2267e^{t} - 0.034e^{2t} + 9.8073e^{-2t} \end{bmatrix}$$

The behaviour of x_1^*, x_2^* and u^* is shown in Fig. 6.5 by solid lines.

In Fig. 6.5, we have also plotted by using dotted lines the input $u^*(t)$ that transfers $\mathbf{x}(0) = [10 \quad 10]^T$ to zero in 1 sec. We see from the figure that in transferring $\mathbf{x}(0)$ to zero, the smaller the time interval the larger the amplitude of input. Thus, if there is no rest. iction on amplitude of input, we can transfer $\mathbf{x}(0)$ to zero in an arbitrarily small interval of time. However, all practical systems have limitation of the amplitude of the input. This limitation restricts the minimum time over which the transfer of $\mathbf{x}(0)$ to zero can be made possible. For example, if we want to force the platform to rest with $|u(t)| \leqslant 1$, then we might not be able to do so in less than 2 seconds.

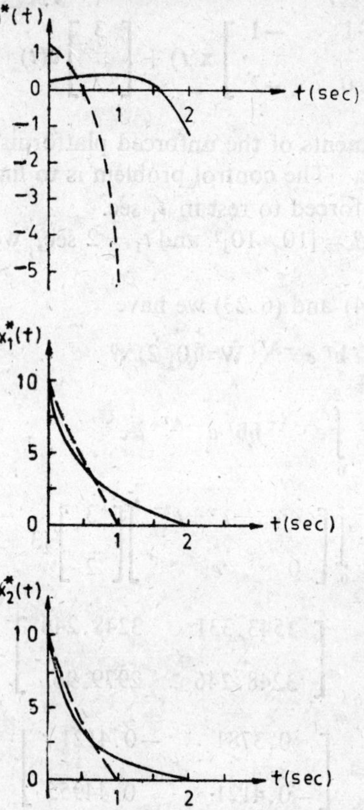

Fig. 6.5 Minimum energy control and trajectories for Example 6.5

Example 6.6: Consider the Inverted Pendulum system discussed in Section 4.6 We found the linearized equations governing the system to be

$$\dot{\mathbf{x}}(t) = \mathbf{A}\mathbf{x}(t) + \mathbf{b}\, u(t)$$

where

$$\mathbf{A} = \begin{bmatrix} 0 & 1 & 0 & 0 \\ 0 & 0 & -0.5809 & 0 \\ 0 & 0 & 0 & 1 \\ 0 & 0 & 4.4537 & 0 \end{bmatrix}; \ \mathbf{b} = \begin{bmatrix} 0 \\ 0.9211 \\ 0 \\ -0.3947 \end{bmatrix}$$

The controllability matrix

$$\mathbf{U} = \begin{bmatrix} 0 & 0.9211 & 0 & 0.2293 \\ 0.9211 & 0 & 0.2293 & 0 \\ 0 & -0.3947 & 0 & -1.7579 \\ -0.3947 & 0 & -1.7579 & 0 \end{bmatrix}$$

$|\mathbf{U}| = 2.3205 \neq 0$. Therefore $\rho(\mathbf{U}) = 4$ and the system is controllable.

6.5 OBSERVABILITY TESTS FOR CONTINUOUS-TIME SYSTEMS

Time Varying Case

The general observability theorems parallel those of controllability. This is not an accident; we shall prove in this section that controllability and observability are duals of each other.

Theorem 6.4: The system (6.1) is observable at time t_0 if and only if there exists a finite $t_1 > t_0$ such that the n columns of $q \times n$ matrix function $\mathbf{C}(t)\Phi(t, t_0)$ are linearly independent on $[t_0, t_1]$.

Proof:

Let us first show the sufficiency. If the columns of $\mathbf{C}(t)\Phi(t, t_0)$ are linearly independent on $[t_0, t_1]$, then we must show that (6.1) is observable.

The output (eqn. (6.3))

$$\mathbf{y}(t) = \mathbf{C}(t)\Phi(t, t_0)\,[\mathbf{x}^0 + \int_{t_0}^{t} \Phi(t_0, \tau)\mathbf{B}(\tau)\mathbf{u}(\tau)d\tau] + \mathbf{D}(t)\mathbf{u}(t) \tag{6.31}$$

In the study of observability, the output $\mathbf{y}(t)$ and the input $\mathbf{u}(t)$ are assumed to be known; $\mathbf{x}(t_0) = \mathbf{x}^0$ is the only unknown. Equation (6.31) may be written as

$$\hat{\mathbf{y}}(t) = \mathbf{C}(t)\Phi(t, t_0)\mathbf{x}^0 \tag{6.32}$$

where

$$\hat{\mathbf{y}}(t) \triangleq \mathbf{y}(t) - \mathbf{C}(t)\Phi(t, t_0)\int_{t_0}^{t} \Phi(t_0, \tau)\mathbf{B}(\tau)\mathbf{u}(\tau)d\tau - \mathbf{D}(t)\mathbf{u}(t)$$

The observability problem is to determine $\mathbf{x}^0$ in (6.32) with the knowledge $\hat{\mathbf{y}}(t)$, $\mathbf{C}(t)$ and $\Phi(t, t_0)$. Once $\mathbf{x}^0$ is known, the state $\mathbf{x}(t)$; $t > t_0$ can be computed from (6.2).

Multiplying both sides of eqn. (6.32) by $\Phi^T(t, t_0)\mathbf{C}^T(t)$ and integrating from t_0 to t_1 we get

$$\int_{t_0}^{t_1} \Phi^T(t, t_0)\mathbf{C}^T(t)\hat{\mathbf{y}}(t)dt = \left(\int_{t_0}^{t_1} \Phi^T(t, t_0)\mathbf{C}^T(t)\mathbf{C}(t)\Phi(t, t_0)dt\right)\mathbf{x}^0 \triangleq \mathbf{V}(t_0, t_1)\mathbf{x}^0 \tag{6.33}$$

where the Gramian matrix

$$\mathbf{V}(t_0, t_1) = \int_{t_0}^{t_1} \Phi^T(t, t_0)\mathbf{C}^T(t)\mathbf{C}(t)\Phi(t, t_0)\,dt \tag{6.34}$$

We have assumed that the columns $\mathbf{C}(t)\Phi(t, t_0)$ are linearly independent on

$[t_0, t_1]$; therefore the rows of $\Phi^T(t, t_0)\mathbf{C}^T(t)$ are linearly independent on $[t_0, t_1]$. Then by eqn. (2.34), the $n \times n$ constant matrix $\mathbf{V}(t_0, t_1)$ defined above is non-singular and $\mathbf{V}^{-1}(t_0, t_1)$ exists. Equation (6.33) gives

$$\mathbf{x}^0 = \mathbf{V}^{-1}(t_0, t_1) \int_{t_0}^{t_1} \Phi^T(t, t_0)\, \mathbf{C}^T(t)\, \hat{\mathbf{y}}(t)\, dt \tag{6.35}$$

We have found the initial state and therefore the system is observable.

The necessity requires that we show that if (6.1) is observable, then the n columns of $\mathbf{C}(t)\,\Phi(t, t_0)$ are linearly independent. Let us use contradiction and assume that (6.1) is observable but the columns of $\mathbf{C}(t)\,\Phi(t, t_0)$ are linearly dependent. Thus there exists a nonzero, constant $n\times 1$ real vector $\boldsymbol{\alpha}$ such that

$$\mathbf{C}(t)\,\Phi(t, t_0)\,\boldsymbol{\alpha} = \mathbf{0} \quad \text{for all } t \text{ in } [t_0, t_1]$$

Choose $\mathbf{x}(t_0) = \mathbf{x}^0 = \boldsymbol{\alpha}$, then

$$\mathbf{C}(t)\,\Phi(t, t_0)\,\mathbf{x}^0 = \hat{\mathbf{y}}(t) = \mathbf{0} \quad \text{for all } t \text{ in } [t_0, t_1]$$

Hence the initial state $\mathbf{x}^0 = \boldsymbol{\alpha}$ cannot be detected. This is a contradiction.

□

Since observability depends on only $\mathbf{C}(t)$ and $\Phi(t, t_0)$ (or equivalently only on $\mathbf{C}(t)$ and $\mathbf{A}(t)$), it is usually more convenient to let $\mathbf{u}(t) = \mathbf{0}$ and calculate observability for the system

$$\dot{\mathbf{x}}(t) = \mathbf{A}(t)\,\mathbf{x}(t) \tag{6.36a}$$

$$\mathbf{y}(t) = \mathbf{C}(t)\,\mathbf{x}(t) \tag{6.36b}$$

The controllability of (6.1) is determined by the linear independence of the rows of $\Phi(t_0, t)\,\mathbf{B}(t)$ whereas the observability is determined by the linear independence of the columns of $\mathbf{C}(t)\,\Phi(t, t_0)$. There seems to be a connecting thread between the two concepts. In fact this link is provided by the adjoint system; recall from Section 5.4 that far a system described by $\dot{\mathbf{x}}(t) = \mathbf{A}(t)\,\mathbf{x}(t)$, the adjoint system is described by $\dot{\mathbf{z}}(t) = -\mathbf{A}^T(t)\,\mathbf{z}(t)$ and the state transition matrix $\Phi(t, t_0)$ of $\dot{\mathbf{x}}(t) = \mathbf{A}(t)\,\mathbf{x}(t)$ is related to state transition matrix $\Phi^{aT}(t, t_0)$ of $\dot{\mathbf{z}}(t) = -\mathbf{A}^T(t)\,\mathbf{z}(t)$ by the equation

$$\Phi^a(t, t_0) = \Phi^{-1}(t, t_0) = \Phi(t_0, t)$$

The relationship between the controllability and observability concepts is established below.

Principle of Duality: Consider the system (6.1) and the system

$$\dot{\mathbf{z}}(t) = -\mathbf{A}^T(t)\,\mathbf{z}(t) + \mathbf{C}^T(t)\,\mathbf{v}(t) \tag{6.37a}$$

$$\mathbf{w}(t) = \mathbf{B}^T(t)\,\mathbf{z}(t) + \mathbf{D}^T(t)\,\mathbf{v}(t) \tag{6.37b}$$

The system (6.1) is controllable (observable) at t_0 if and only if the system (6.37) is observable (controllable) at t_0.

Let us verify this principle. The system (6.1) is controllable if and only if the n rows of $\Phi(t_0, t)\mathbf{B}(t)$ are linearly independent in t on $[t_0, t_1]$ (Theorem 6.1). The system (6.37) is observable if and only the n columns of $\mathbf{B}^T(t)\,\Phi^{aT}(t, t_0)$ are linearly independent in t on $[t_0, t_1]$ (Theorem 6.4) or equivalently if the n rows of $[\mathbf{B}^T(t)\,\Phi^{aT}(t, t_0)]^T$ are linearly independent on $[t_0, t_1]$.

But

$$[\mathbf{B}^T(t)\,\Phi^{aT}(t, t_0)]^T = \Phi^a(t, t_0)\,\mathbf{B}(t)$$
$$= \Phi(t_0, t)\,\mathbf{B}(t)$$

Hence (6.1) is controllable if and only if (6.37) is observable. Similarly we can verify that (6.1) is observable if and only if (6.37) is controllable.

Time-Invariant Case

The following theorem is the dual of Theorem 6.3 which may be proved either directly or by applying the principle of duality to Theorem 6.3.

Theorem 6.5: The following statements regarding the linear time-invariant dynamical system (6.4) are equivalent:

a. The system is completely observable.

b. The columns of $\mathbf{C}\,e^{\mathbf{A}t}$ are linearly independent over the real field $\mathcal{R}$ for all $t \in [0, \infty)$.

c. $\mathbf{V}(0, t) = \int_0^t e^{\mathbf{A}^T\tau}\,\mathbf{C}^T\mathbf{C}\,e^{\mathbf{A}\tau}\,d\tau$ is nonsingular for any $t > 0$.

d. The rank of the $(nq \times n)$ *observability matrix* $\mathbf{V}$

$$\left(\mathbf{V} \triangleq \begin{bmatrix} \mathbf{C} \\ \mathbf{CA} \\ \vdots \\ \mathbf{CA}^{n-1} \end{bmatrix}\right) \text{ is } n;\ \text{i.e., } \rho(\mathbf{V}) = n.$$

□

It should be noted that as in the case of controllability, the determination of the rank of observability matrix $\mathbf{V}$ is the test usually employed to ascertain whether or not the pair $\{\mathbf{A}, \mathbf{C}\}$ is observable.

Example 6.7: Consider the Inverted Pendulum system discussed in Section 4.6. The linearized equations governing the system are

$$\dot{\mathbf{x}}(t) = \mathbf{A}\mathbf{x}(t) + \mathbf{b}u(t)$$

where

$$\mathbf{x}(t) = \begin{bmatrix} z(t) \\ \dot{z}(t) \\ \phi(t) \\ \dot{\phi}(t) \end{bmatrix};\quad \mathbf{A} = \begin{bmatrix} 0 & 1 & 0 & 0 \\ 0 & 0 & -0.5809 & 0 \\ 0 & 0 & 0 & 1 \\ 0 & 0 & 4.4537 & 0 \end{bmatrix};\quad \mathbf{b} = \begin{bmatrix} 0 \\ 0.9211 \\ 0 \\ -0.3947 \end{bmatrix}$$

Taking angular position $\phi(t)$ of the pendulum as the output variable, we have the output equation

$$y(t) = \mathbf{c}\mathbf{x}(t)$$

where

$$\mathbf{c} = [0 \quad 0 \quad 1 \quad 0]$$

The observability matrix

$$\mathbf{V} = \begin{bmatrix} \mathbf{c} \\ \mathbf{cA} \\ \mathbf{cA}^2 \\ \mathbf{cA}^3 \end{bmatrix} = \begin{bmatrix} 0 & 0 & 1 & 0 \\ 0 & 0 & 0 & 1 \\ 0 & 0 & 4.4537 & 0 \\ 0 & 0 & 0 & 4.4537 \end{bmatrix}$$

$|\mathbf{V}| = 0$ and therefore the system is not completely observable.

Consider now the displacement $z(t)$ of the carriage as the output variable. Then

$$\mathbf{c} = [1 \quad 0 \quad 0 \quad 0]$$

and observability matrix

$$\mathbf{V} = \begin{bmatrix} 1 & 0 & 0 & 0 \\ 0 & 1 & 0 & 0 \\ 0 & 0 & -0.5809 & 0 \\ 0 & 0 & 0 & -0.5809 \end{bmatrix}.$$

$|\mathbf{V}| = 0.3374 \neq 0$; the system is therefore observable. The values of $\dot{z}(t)$, $\phi(t)$ and $\dot{\phi}(t)$ can all be determined by observing $z(t)$ over an arbitrary time interval.

6.6 CONTROLLABILITY AND OBSERVABILITY OF DISCRETE-TIME SYSTEMS

Time-Varying Case

The concepts of controllability and observability discussed in Sections 6.2–6.5 are equally relevant for discrete-time systems represented by (6.7) with obvious modifications. The definitions carry over to the discrete-time case if the discrete-time variable k is substituted for the continuous-time variable t.

Controllability: The direct analog to Theorem 6.1 is given below.

Theorem 6.6: System (6.7) is controllable at time k_0 if and only if there exists a finite $k_1 \geqslant k_0 + 1$ such that the n rows of the $n \times p$ matrix $\Phi(k_0, k+1)\,\mathbf{G}(k)$ are linearly independent on $[k_0, k_1]$.

Further, the 'minimum-energy control' to drive a controllable system from $\mathbf{x}(k_0) \triangleq \mathbf{x}^0$ to $\mathbf{x}^1$ at k_1 is given by

$$\mathbf{u}^*(k) = -\,\mathbf{G}^T(k)\Phi^T(k_0, k+1)\,\mathbf{W}^{-1}(k_0, k_1)[\mathbf{x}^0 - \Phi(k_0, k_1)\mathbf{x}^1] \qquad (6.38a)$$

where

$$\mathbf{W}(k_0, k_1) = \sum_{j=k_0}^{k_1-1} \Phi(k_0, j+1)\mathbf{G}(j)\mathbf{G}^T(j)\Phi^T(k_0, j+1) \qquad (6.38b)$$

Here $\Phi(k, k_0)$ is the state transition matrix of the system.

Example 6.8: Consider the system of Example 4.7. The state equation of the system is (eqn. (4.17))

$$\mathbf{x}(k+1) = \begin{bmatrix} 1 + \dfrac{rV(k)}{Q} & 0 \\ -\dfrac{(\Delta H)rV(k)}{Q\rho c_p} & 1 \end{bmatrix}\mathbf{x}(k) + \begin{bmatrix} 0 \\ \dfrac{1}{Q\rho c_p} \end{bmatrix} q(k)$$

The state transition matrix (eqn. (5.94a))

$$\Phi(k, k_0) = \begin{bmatrix} \prod_{i=k_0}^{k-1}\left(1 + \dfrac{rV(i)}{Q}\right) & 0 \\ \sum_{i=k_0}^{k-1} -\dfrac{(\Delta H)rV(i)}{Q\rho c_p} \prod_{j=k_0}^{i-1}\left(1 + \dfrac{rV(j)}{Q}\right) & 1 \end{bmatrix}$$

It can be easily verified using Theorem 6.6 that the system is not completely controllable at any k.

Observability: For observability, we can write a result analogous to Theorem 6.4.

Theorem 6.7: The system (6.7) is observable at time k_0 if and only if there exists a finite $k_1 \geqslant k_0 + 1$ such that the n columns of $q \times n$ matrix function $\mathbf{C}(k)\Phi(k, k_0)$ are linearly independent on $[k_0, k_1]$.

Further, $\mathbf{x}(k_0)$ is given by

$$\mathbf{x}(k_0) = \mathbf{V}^{-1}(k_0, k_1) \sum_{k=k_0}^{k_1} \Phi^T(k, k_0)\mathbf{C}^T(k)\hat{\mathbf{y}}(k) \qquad (6.39a)$$

where

$$\mathbf{V}(k_0, k_1) = \sum_{k=k_0}^{k_1} \Phi^T(k, k_0)\mathbf{C}^T(k)\mathbf{C}(k)\Phi(k, k_0) \qquad (6.39b)$$

and

$$\hat{\mathbf{y}}(k) = \mathbf{y}(k) - \mathbf{D}(k)\mathbf{u}(k) - \mathbf{C}(k)\Phi(k, k_0) \sum_{j=k_0+1}^{k} \Phi^{-1}(j, k_0)\mathbf{G}(j-1)\mathbf{u}(j-1) \qquad (6.39c)$$

Here $\Phi(k, k_0)$ is the state transition matrix of the system.

Example 6.9: Consider the system described by the state model

$$\mathbf{x}(k+1) = \mathbf{F}(k)\mathbf{x}(k) + \mathbf{g}(k)\mathbf{u}(k)$$

$$y(k) = \mathbf{c}(k)\mathbf{x}(k)$$

where

$$\mathbf{F}(k) = \begin{bmatrix} 1 & k \\ 0 & -1 \end{bmatrix}; \mathbf{g}(k) = \begin{bmatrix} 2 \\ k \end{bmatrix}; \mathbf{c}(k) = [1 \quad k]$$

Is this system observable at $k = 1$? If yes, find $\mathbf{x}(1)$ from the following measurements:

$$u(k) = (-1)^k;\ k \geqslant 1$$

$$y(1) = 3;\ y(2) = -5$$

Solution: The Gramian matrix (eqn. (6.39b))

$$\mathbf{V}(1.\ 2) = \Phi^T(1,\ 1)\mathbf{c}^T(1)\Phi(1,\ 1) + \Phi^T(2,\ 1)\mathbf{c}^T(2)\mathbf{c}(2)\Phi(2,\ 1)$$

$$= \begin{bmatrix} 1 & 1 \\ 1 & 1 \end{bmatrix} + \begin{bmatrix} 1 & -1 \\ -1 & 1 \end{bmatrix} = \begin{bmatrix} 2 & 0 \\ 0 & 2 \end{bmatrix}$$

is nonsingular. Therefore the system is observable at $k = 1$.

From eqns. (6.39a) and (6.39c),

$$\mathbf{x}(1) = \mathbf{V}^{-1}(1,\ 2)\ [\Phi^T(1,\ 1)\mathbf{c}^T(1)\hat{y}(1) + \Phi^T(2,\ 1)\mathbf{c}^T(2)\hat{y}(2)]$$

$$= \begin{bmatrix} \frac{1}{2} & 0 \\ 0 & \frac{1}{2} \end{bmatrix}\left[\begin{pmatrix} 3 \\ 3 \end{pmatrix} + \begin{pmatrix} -1 \\ 1 \end{pmatrix}\right]$$

$$= \begin{bmatrix} 1 \\ 2 \end{bmatrix}$$

Time-Invariant Case

In the preceding subsection, we have seen that controllability and observability Gramians for discrete-time systems are defined analogous to the continuous-time case and the results for 'minimum-energy control' and observation of $\mathbf{x}(k_0)$ from $\mathbf{y}_{[k_0,\ k_1]}$ and $\mathbf{u}_{[k_0,\ k_1]}$ are also equivalent.

All these results can be applied for time-invariant systems with obvious modifications. However, as we shall see, it is more convenient in discrete-time case to solve control and observation problems by solving a set of linear equations obtained from state and output equations of the system by iteration. Such is not the case in the continuous-time counterpart.

Controllability: A simple test for controllability is given by the following theorem.

Theorem 6.8: The discrete-time system (6.10) is controllable if and only if the rank of $(n \times np)$ *controllability matrix* $\mathbf{U}(\mathbf{U} \overset{\Delta}{=} [\mathbf{G} \mid \mathbf{FG} \mid \ldots \mid \mathbf{F}^{n-1}\mathbf{G}])$ is n; i.e., $\rho(\mathbf{U}) = n$.

Proof:

Solving (6.10) by successive substitution, we get

$$\mathbf{x}(1) = \mathbf{Fx}(0) + \mathbf{Gu}(0)$$

$$\mathbf{x}(2) = \mathbf{Fx}(1) + \mathbf{Gu}(1)$$

$$= \mathbf{F}^2\mathbf{x}(0) + \mathbf{FGu}(0) + \mathbf{Gu}(1)$$

$$\ldots \quad \ldots \quad \ldots \quad \ldots \quad \ldots \quad \ldots \quad \ldots \quad \ldots \quad \ldots$$

$$\mathbf{x}(N) = \mathbf{F}^N\mathbf{x}(0) + \mathbf{F}^{N-1}\mathbf{Gu}(0) + \ldots + \mathbf{FGu}(N-2) + \mathbf{Gu}(N-1) \tag{6.40}$$

According to the definition of controllability, system (6.10) is controllable if there exists a set of unconstrained input vectors $\mathbf{u}(kT)$; $k = 0, 1, \ldots, (N-1)$ which will transfer an arbitrary state $\mathbf{x}(0)$ to an arbitrary state $\mathbf{x}^1$ in N sampling periods where N is a finite positive integer.

Equation (6.40) can be written as

$$\mathbf{x}^1 - \mathbf{F}^N\mathbf{x}(0) = \mathbf{F}^{N-1}\mathbf{Gu}(0) + \ldots + \mathbf{FGu}(N-2) + \mathbf{Gu}(N-1)$$

or

$$\hat{\mathbf{x}} = [\mathbf{G} \mid \mathbf{FG} \mid \ldots \mid \mathbf{F}^{N-1}\mathbf{G}] \begin{bmatrix} \mathbf{u}(N-1) \\ \mathbf{u}(N-2) \\ \vdots \\ \mathbf{u}(0) \end{bmatrix} \tag{6.41a}$$

or

$$\underset{(n\times 1)}{\hat{\mathbf{x}}} = \underset{(n\times Np)}{[\mathbf{U}_N]} \; \underset{(Np\times 1)}{[\mathbf{u}]} \tag{6.41b}$$

For (6.41) to be satisfied for arbitrary $\mathbf{x}(0)$ and $\mathbf{x}^1$, it is necessary that (refer Section 2.6)

$$\rho(\mathbf{U}_N) = \rho(\mathbf{U}_N \mid \hat{\mathbf{x}}),$$

Now $\rho(\mathbf{U}_N)$ is by definition the dimension of range space of $\mathbf{U}_N$. Thus if $\rho(\mathbf{U}_N) = n$, then range space of $\mathbf{U}_N$ is $\mathcal{R}^n(\mathcal{R})$ and for any arbitrary $\hat{\mathbf{x}}$ in $\mathcal{R}^n(\mathcal{R})$, the necessary condition for existence of solution is satisfied.

A state can thus be transferred to some other state in at most n steps if and only if

$$\rho(\mathbf{U}) = \rho[\mathbf{G} \mid \mathbf{FG} \mid \ldots \mid \mathbf{F}^{n-1}\mathbf{G}] = n \tag{6.42}$$

If a state cannot be transferred to some other state in n sampling periods, no matter how long the input sequence $\mathbf{u}(0), \mathbf{u}(1), \ldots \mathbf{u}(N-1)$; $N > n$ is, it still cannot be achieved. This statement can easily be established using the Cayley-Hamilton theorem.

Deadbeat Control: The result of Theorem 6.8 gives not only necessary and sufficient conditions for controllability but also a method to compute the control.

Consider first a single input system described by the state equation

$$\mathbf{x}(k+1) = \mathbf{F}\mathbf{x}(k) + \mathbf{g}u(k) \tag{6.43}$$

where $\mathbf{x}$ is the $n \times 1$ state vector, u is the scalar input and $\mathbf{F}$ and $\mathbf{g}$ are, respectively, $n \times n$ and $n \times 1$ real constant matrices.

The system is assumed to be controllable; therefore

$$\rho(\mathbf{U}) = \rho(\mathbf{g} \mid \mathbf{F}\mathbf{g} \mid \ldots \mid \mathbf{F}^{n-1}\mathbf{g}) = n \tag{6.44}$$

Suppose that we wish to control the system in such a manner as to transfer the system from some arbitrary initial state $\mathbf{x}(0)$ to a desired state $\mathbf{x}^1$. A control sequence $u(0), u(1), \ldots, u(N-1)$ satisfying the relation (eqn. (6.41))

$$\hat{\mathbf{x}} = \mathbf{x}^1 - \mathbf{F}^N\mathbf{x}(0) = [\mathbf{g} \mid \mathbf{F}\mathbf{g} \mid \ldots \mid \mathbf{F}^{N-1}\mathbf{g}] \begin{bmatrix} u(N-1) \\ u(N-2) \\ \vdots \\ u(0) \end{bmatrix} \tag{6.45}$$

will accomplish the desired transfer in N sampling intervals. We shall consider three cases.

Case 1 ($N < n$):

If N is less than n, i.e., if the number of control periods is less than the order of the system, the solution of (6.45) may not exist for an arbitrary $\hat{\mathbf{x}}$ in the entire state space. We can determine subspace for $N < n$ so that any initial state in this subspace can be transferred to any desired state in this subspace in $N(< n)$ sampling periods (Desoer and Wing 1961a, 1961b).

Some simplification in the thought process, at least, is obtained if we assume that $\mathbf{x}^1 = \mathbf{0}$, i.e., $\mathbf{x}(0)$ is to be transferred to the origin of the state space[1].

---

[1]There is no loss of generality in this assumption. Consider the time-invariant linear system described by state equation (6.4a):

$$\dot{\mathbf{x}}(t) = \mathbf{A}\mathbf{x}(t) + \mathbf{B}\mathbf{u}(t); \ \mathbf{x}(0) \text{ and } \mathbf{x}^1 \text{ are specified} \tag{i}$$

Define a new state vector

$$\mathbf{y}(t) = \mathbf{x}(t) - \mathbf{x}^1 \tag{ii}$$

and let $\mathbf{u}^1$ be the control needed to maintain $\mathbf{x}(t)$ at the set point, i.e.,

$$\mathbf{A}\mathbf{x}^1 + \mathbf{B}\mathbf{u}^1 = \mathbf{0} \quad \text{or} \quad \underset{(n\times p)}{\mathbf{B}} \ \underset{(p\times 1)}{\mathbf{u}^1} = - \underset{(n\times 1)}{\mathbf{A}\mathbf{x}^1} \tag{iii}$$

If $\rho(\mathbf{B}) = p$ — which is usually the case, since the control variables in a system are usually independent, we get least squares solution (refer Section 2.6)

$$\mathbf{u}^1 = -(\mathbf{B}^T\mathbf{B})^{-1}\mathbf{B}^T\mathbf{A}\mathbf{x}^1 \tag{iv}$$

Continued

With this assumption, we get from eqn. (6.45)

$$\mathbf{x}(0) = -\mathbf{F}^{-1}\mathbf{g}u(0) - \mathbf{F}^{-2}\mathbf{g}u(1) - \ldots - \mathbf{F}^{-N}\mathbf{g}u(N-1) \tag{6.46a}$$

$$= \mathbf{f}_1 u(0) + \mathbf{f}_2 u(1) + \ldots + \mathbf{f}_N u(N-1) \tag{6.46b}$$

If this equation is satisfied, then it is guaranteed that

$$\mathbf{x}(N) = \mathbf{x}^1 = \mathbf{0}$$

Taking $N = 1$, we find from eqn. (6.46) that all initial states governed by the relation

$$\mathbf{x}(0) = \mathbf{f}_1 u(0) \tag{6.47}$$

can be forced to the origin in one sampling interval. Thus initial states positioned on the line defined by the vector $\mathbf{f}_1$ can be forced to zero in one sampling period and all other initial states cannot be forced to zero in just one sampling period.

Taking $N = 2$, we find from eqn. (6.46) that all initial states governed by the relation

$$\mathbf{x}(0) = \mathbf{f}_1 u(0) + \mathbf{f}_2 u(1) \tag{6.48}$$

can be forced to origin in two sampling intervals.

$\mathbf{x}(0)$ is (6.48) is simply a linear combination of the two $n \times 1$ vectors $\mathbf{f}_1$ and $\mathbf{f}_2$.

We might think of the set of initial states in (6.47) as forming a one-dimensional subspace of the n-dimensional state space (refer Section 6.9). Similarly the initial states in (6.48) can be thought of as forming a two-dimensional subspace ($\mathbf{f}_1$ and $\mathbf{f}_2$ are linearly independent) of the n-dimensional state space. This notion can be extended to $N = 3, 4, 5, \ldots$. The set of initial states that may be forced to the origin in N sampling periods forms N-dimensional subspace ($\mathbf{f}_1, \mathbf{f}_2, \ldots, \mathbf{f}_N$ are linearly independent) of n-dimensional state space, which is given by

$$\mathscr{R}^N = \{\mathbf{x}(0) : \mathbf{x}(0) = \sum_{i=0}^{N-1} \mathbf{f}_{i+1} u(i);\ \mathbf{f}_i = -\mathbf{F}^{-i}\mathbf{g}\} \tag{6.49}$$

Now, define a control vector $\hat{\mathbf{u}}(t)$ where

$$\hat{\mathbf{u}}(t) = \mathbf{u}(t) - \mathbf{u}^1 \tag{v}$$

Substituting for $\mathbf{x}(t)$ and $\mathbf{u}(t)$ from (ii) and (v) respectively, in (i) we obtain

$$\dot{\mathbf{y}}(t) = \mathbf{A}\mathbf{y}(t) + \mathbf{B}\hat{\mathbf{u}}(t);\ \mathbf{y}(0) = \mathbf{x}(0) - \mathbf{x}^1 \text{ and } \mathbf{y}^1 = \mathbf{0} \tag{vi}$$

Thus the problem of transferring $\mathbf{x}(0)$ to $\mathbf{x}^1$ is equivalent to the problem of transferring $\mathbf{y}(0)$ to $\mathbf{0}$ in the new state space which is obtained merely by shifting of the origin of original state space by $\mathbf{x}^1$.

Therefore, we conclude that any arbitrary initial state cannot be forced to zero in $N \ll n$ sampling periods. If $\mathbf{x}(N)$ can be forced to zero state, the control sequence required for $N < n$ is unique.

If it is not possible to force $\mathbf{x}(N) = \mathbf{0}$, then we may find a control sequence that minimizes the euclidean distance

$$\|\mathbf{x}(N)\|^{1/2} = \left(\sum_{i=1}^{n} \mathbf{x}_i^2\,(N)\right)^{1/2} \tag{6.50}$$

Such a control sequence is the least squares solution of eqn. (6.46) and is given by (Section 2.6)

$$\begin{bmatrix} u(0) \\ u(1) \\ \vdots \\ u(N-1) \end{bmatrix} = (\mathbf{S}^T\,\mathbf{S})^{-1}\,\mathbf{S}^T\mathbf{x}(0) \tag{6.51}$$

where

$$\mathbf{S} = [-\mathbf{F}^{-1}\mathbf{g} \,|\, -\mathbf{F}^{-2}\mathbf{g} \,|\, \ldots \,|\, -\mathbf{F}^{-N}\mathbf{g}]$$

Case 2 ($N = n$):

When $N = n$, the subspace $\mathcal{R}^N$ defined by (6.49) becomes the entire state space $\mathcal{R}^n$ and hence any arbitrary state space can be forced to zero or equivalently any arbitrary initial state $\mathbf{x}(0)$ in state space can be transferred to any other arbitrary state $\mathbf{x}^1$ in state space.

For $N = n$, we get from eqn. (6.45)

$$\begin{bmatrix} u(n-1) \\ u(n-2) \\ \vdots \\ u(0) \end{bmatrix} = [\mathbf{g} \,|\, \mathbf{F}\mathbf{g} \,|\, \ldots \,|\, \mathbf{F}^{n-1}\mathbf{g}]^{-1}\,[\mathbf{x}^1 - \mathbf{F}^n\mathbf{x}(0)] \tag{6.52}$$

The *minimum-time control law* (6.52) is an open-loop control law, since the required control vector depends on the initial state $\mathbf{x}(0)$. Therefore, once the initial state $\mathbf{x}(0)$ is monitored and the desired state $\mathbf{x}^1$ is given, the control sequence $u(0), \ldots, u(n-1)$ is immediately calculated from (6.52). Further, this sequence is unique[2].

Case 3 ($N > n$):

Now consider the case: $N > n$. For this case, eqns. (6.45) have an infinite number of solutions (Section 2.6). The one which minimizes

$$\|\mathbf{u}\|^2 = \sum_{k=0}^{N-1} u^2(k) \tag{6.53}$$

is given by

$$\mathbf{u}^* = \mathbf{U}_N^T\,(\mathbf{U}_N\,\mathbf{U}_N^T)^{-1}(\mathbf{x}^1 - \mathbf{F}^N\mathbf{x}(0)) \tag{6.54}$$

---

[2]If the system is not completely controllable, then pseudoinverse of $\mathbf{U}$ may be used in place of $\mathbf{U}^{-1}$. Refer Section 2.6.

where

$$\mathbf{U}_N \overset{\Delta}{=} [\mathbf{g} \mid \mathbf{F}\mathbf{g} \mid \dots \mid \mathbf{F}^{N-1}\mathbf{g}]$$

and

$$\mathbf{u}^* = \begin{bmatrix} u^*(N-1) \\ u^*(N-2) \\ \vdots \\ u^*(0) \end{bmatrix}$$

If control energy is measured by the quantity given by (6.53), then (6.54) is the *minimum-energy control law*.

The value of N normally depends on the constraints of input energy, input amplitude, time of transfer etc. (Desoer and Wing 1961c). See Example 6.10.

Consider now the multi-input system (6.10). The system is assumed to be controllable, i.e.,

$$\rho(\mathbf{U}) = \rho[\mathbf{G} \mid \mathbf{F}\mathbf{G} \mid \dots \mid \mathbf{F}^{n-1}\mathbf{G}] = n$$

The assumption of controllability guarantees a solution to the equation

$$\mathbf{x}^1 - \mathbf{F}^N\mathbf{x}(0) = [\mathbf{G} \mid \mathbf{F}\mathbf{G} \mid \dots \mid \mathbf{F}^{N-1}\mathbf{G}] \begin{bmatrix} \mathbf{u}(N-1) \\ \mathbf{u}(N-2) \\ \vdots \\ \mathbf{u}(0) \end{bmatrix}; \; N \geqslant n \tag{6.55}$$

i.e., there exists a control vector which will transfer the state $\mathbf{x}(0)$ to the state $\mathbf{x}^1$ ln $N \geqslant n$ steps where n is the order of the system. For a selected N, there are infinite solutions to the equation (6.55) and the one which provides 'minimum-energy control' is given by

$$\begin{bmatrix} \mathbf{u}^*(N-1) \\ \mathbf{u}^*(N-2) \\ \vdots \\ \mathbf{u}^*(0) \end{bmatrix} = \mathbf{U}_N^T (\mathbf{U}_N \mathbf{U}_N^T)^{-1} (\mathbf{x}^1 - \mathbf{F}^N \mathbf{x}(0)) \tag{6.56}$$

where

$$\mathbf{U}_N = [\mathbf{G} \mid \mathbf{F}\mathbf{G} \mid \dots \mid \mathbf{F}^{N-1}\mathbf{G}]$$

Suppose now that we wish to control the system in such a manner so as to transfer the system from initial state $\mathbf{x}(0)$ to a desired state $\mathbf{x}^1$ in the minimum time, i.e., we wish to determine the minimum number of iteration times N (obviously $< n$) that are required to force $\mathbf{x}(0)$ to $\mathbf{x}^1$. We can write from (6.55)

$$[\mathbf{x}^1 - \mathbf{F}^n\mathbf{x}(0)] = [\mathbf{G} \mid \mathbf{F}\mathbf{G} \mid \dots \mid \mathbf{F}^{n-1}\mathbf{G}] \begin{bmatrix} \mathbf{u}(n-1) \\ \mathbf{u}(n-2) \\ \vdots \\ \mathbf{u}(0) \end{bmatrix} \tag{6.57a}$$

or

$$\underset{(n\times 1)}{\hat{\mathbf{x}}} = \underset{(n\times np)}{\mathbf{U}} \quad \underset{(np\times 1)}{\mathbf{u}} \tag{6.57b}$$

Given an arbitrary initial state $\mathbf{x}(0)$ and arbitrary desired state $\mathbf{x}^1$, the minimum-time control strategy is determined from (6.57) by selecting n linearly independent columns of $\mathbf{U}$. It is apparent that there is a certain degree of nonuniqueness in this selection. We shall define a minimum-time control strategy by selecting the first n linearly independent columns of $\mathbf{U}$. Let the m-th column of $\mathbf{U}$ be n-th linearly independent column and $\mathbf{Q}$ be the $n\times m$ matrix formed by the first m columns of $\mathbf{U}$. The minimum-time interval is determined from the last column of matrix $\mathbf{Q}$. If the last column of matrix $\mathbf{Q}$ is from $\mathbf{F}^k\mathbf{G}$ in $\mathbf{U}$, then minimum time required is $N=k+1$.

Let $\mathbf{u}^*$ be the vector containing the last m rows of

$$\mathbf{u} = \begin{bmatrix} \mathbf{u}(n-1) \\ \mathbf{u}(n-2) \\ \vdots \\ \mathbf{u}(0) \end{bmatrix}$$

Then

$$\mathbf{u}^* = \mathbf{Q}^T(\mathbf{Q}\mathbf{Q}^T)^{-1}[\mathbf{x}^1 - \mathbf{F}^N\mathbf{x}(0)] \tag{6.58}$$

is the minimum-time control.

Example 6.10: Consider the Position Servo system discussed in Section 4.6. The continuous-time positioning system is described by the state equation (eqn. (4.27a))

$$\dot{\mathbf{x}}(t) = \begin{bmatrix} 0 & 1 \\ 0 & -5 \end{bmatrix} \mathbf{x}(t) + \begin{bmatrix} 0 \\ 1 \end{bmatrix} u(t) \tag{6.59}$$

Suppose that this system is part of a control system that is commanded by a digital computer (Fig. 6.6). As a result, the input u changes at discrete

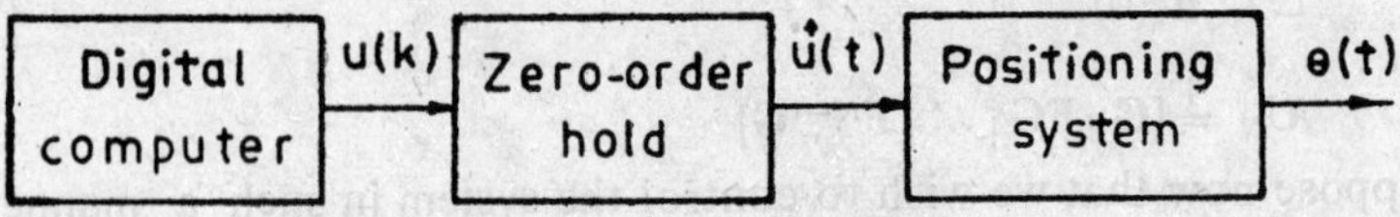

Fig. 6.6 A digital positioning system.

instants only. If these instants are separated by time period $T = 0.1$ sec, then the discrete-time description of the system becomes (Problem 5.15)

$$\mathbf{x}(k+1) = \mathbf{F}\mathbf{x}(k) + \mathbf{g}u(k) \tag{6.60}$$

where

$$\mathbf{F} = \begin{bmatrix} 1 & 0.0787 \\ 0 & 0.6065 \end{bmatrix}; \quad \mathbf{g} = \begin{bmatrix} 0.0043 \\ 0.0787 \end{bmatrix}$$

Since $\det[\mathbf{g} \vdots \mathbf{Fg}] = \det\begin{bmatrix} 0.0043 & 0.0105 \\ 0.0787 & 0.0477 \end{bmatrix} = -0.0006 \neq 0,$

the system (6.60) is controllable.

Suppose now that it is desired to transfer $\theta(t) = x_1(t)$, the position coordinate of the antenna from 0 rad to 0.1 rad. We would like that $x_1 = \theta(t)$ equals 0.1 rad with $x_2 = \dot{\theta}$ equal to zero. This will assure that the antenna will not move from the desired position once it has reached it. It can be easily established that such a transfer is possible in a minimum of 2 sampling periods (0.2 sec). The control sequence for the minimum-time control is given by (eqn. (6.45))

$$\begin{bmatrix} u^*(1) \\ u^*(0) \end{bmatrix} = [\mathbf{g} \vdots \mathbf{Fg}]^{-1}[\mathbf{x}^1 - \mathbf{F}^2\,\mathbf{x}(0)]$$

$$= \begin{bmatrix} 0.0043 & 0.0105 \\ 0.0787 & 0.0477 \end{bmatrix}^{-1} \begin{bmatrix} 0.1 \\ 0 \end{bmatrix}$$

$$\begin{bmatrix} -76.9355 & 16.9355 \\ 126.9355 & -6.9355 \end{bmatrix} \begin{bmatrix} 0.1 \\ 0 \end{bmatrix} = \begin{bmatrix} -7.6936 \\ 12.6936 \end{bmatrix} \tag{6.61}$$

The optimal control sequence $u^*(t)$ and the corresponding response $\mathbf{x}(t)$, derived from eqn. (6.59), are shown in Fig. 6.7.

Let us now determine minimum energy solution for $N = 3, 4$.

$$\mathbf{U}_N = [\mathbf{g} \vdots \mathbf{Fg} \vdots \ldots \vdots \mathbf{F}^{N-1}\mathbf{g}]$$

Therefore

$$\mathbf{U}_3 = [\mathbf{g} \vdots \mathbf{Fg} \vdots \mathbf{F}^2\mathbf{g}]$$

$$= \begin{bmatrix} 0.0043 & 0.0105 & 0.0143 \\ 0.0787 & 0.0477 & 0.0289 \end{bmatrix}$$

$$\mathbf{U}_4 = [\mathbf{g} \vdots \mathbf{Fg} \vdots \mathbf{F}^2\mathbf{g} \vdots \mathbf{F}^3\mathbf{g}]$$

$$= \begin{bmatrix} 0.0043 & 0.0105 & 0.0143 & 0.0166 \\ 0.0787 & 0.0477 & 0.0289 & 0.0175 \end{bmatrix}$$

For $N = 3$ (refer eqn. (6.56))

$$\begin{bmatrix} u^*(2) \\ u^*(1) \\ u^*(0) \end{bmatrix} = \mathbf{U}_3^T(\mathbf{U}_3\mathbf{U}_3^T)^{-1} \begin{bmatrix} 0.1 \\ 0 \end{bmatrix}$$

$$= \begin{bmatrix} -3.8238 \\ 2.4777 \\ 6.3236 \end{bmatrix}.$$

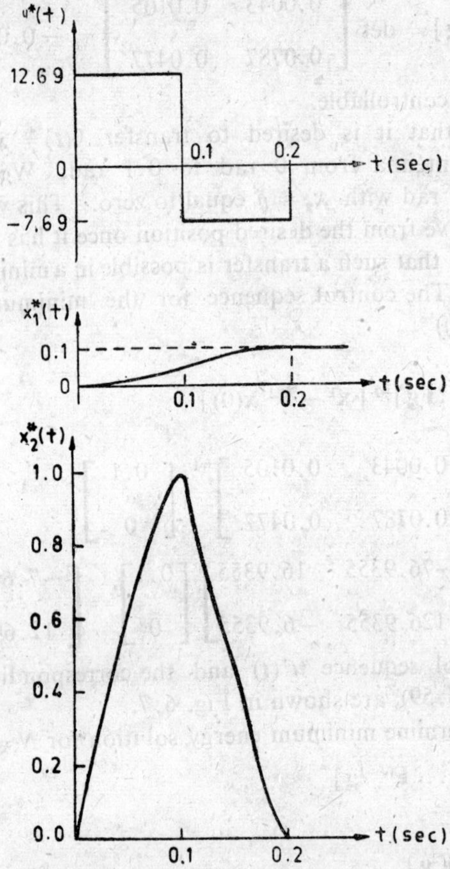

Fig. 6.7 Deadbeat control and trajectories for Example 6.10.

For $N = 4$

$$\begin{bmatrix} u^*(3) \\ u^*(2) \\ u^*(1) \\ u^*(0) \end{bmatrix} = \mathbf{U}_4^T(\mathbf{U}_4\mathbf{U}_4^T)^{-1}\begin{bmatrix} 0.1 \\ 0 \end{bmatrix} = \begin{bmatrix} -2.3088 \\ 0.7870 \\ 2.6755 \\ 3.8220 \end{bmatrix}$$

Summarizing we get

$$N = 2: u^*(0) = 12.6936,\ u^*(1) = -7.6036;\ u^*(k) = 0,\ k \geqslant 2$$

$$N = 3: u^*(0) = 6.3236,\quad u^*(1) = 2.4777,\quad u^*(2) = -3.8238,$$
$$u^*(k) = 0,\ k \geqslant 3$$

$$N = 4: u^*(0) = 3.8200,\quad u^*(1) = 2.6755,\quad u^*(2) = 0.7870$$
$$u^*(3) = -2.3088,\quad u^*(k) = 0,\ k \geqslant 4$$

Although each of the above control sequences will force the antenna from zero position to 0.1 rad position, it will be noted that by making N progressively larger the energy requirements are drastically reduced; i.e.,

$$N = 2 : \sum_{i=0}^{1} u^2(i) = 220.3185$$

$$N = 3 : \sum_{i=0}^{2} u^2(i) = 60.7483$$

$$N = 4 : \sum_{i=0}^{3} u^3(i) = 27.7006$$

The selection of N for a system thus depends on the limitations on the amplitude of the control.

Note that minimum-time control sequence given by (6.61) yields

$$x_1(2) = \theta(2) = 0.1, \; x_2(2) = \dot{\theta}(2) = 0$$

where θ is the position of the antenna. The control sequence thus forces the antenna from zero position to the desired position of 0.1 rad in minimum-time and thereafter makes it stay there. This type of system response is referred to as output *deadbeat response* and (6.61) gives an *output deadbeat control* sequence. It may be noted that because of free integrator of the plant (6.59), it was possible to force θ to the desired value $\dot{\theta}$ to zero at $t = 0.2$ sec. In Section 7.3 we shall obtain the results for the general case; plants with/without a free integrator.

Example 6.11 : Consider the Mixing Tank system discussed in Section 4.6. The continuous-time system is described by the state equation (eqn. (4.38a))

$$\dot{\mathbf{x}}(t) = \begin{bmatrix} -0.01 & 0 \\ 0 & -0.02 \end{bmatrix} \mathbf{x}(t) + \begin{bmatrix} 1 & 1 \\ -0.004 & 0.002 \end{bmatrix} \mathbf{u}(t)$$

Suppose that this system forms part of a process commanded by a process control computer. As a result, the valve settings change at discrete instants only and remain constant in between. If these instants are separated by time period $T = 5$ sec, then the discrete-time description of the system becomes (Problem 5.16)

$$\mathbf{x}(k+1) = \mathbf{F}\mathbf{x}(k) + \mathbf{G}\mathbf{u}(k) \tag{6.62}$$

where

$$\mathbf{F} = \begin{bmatrix} 0.9512 & 0 \\ 0 & 0.9048 \end{bmatrix}; \quad \mathbf{G} = \begin{bmatrix} 4.88 & 4.88 \\ -0.019 & 0.0095 \end{bmatrix}$$

The state equation (6.62) is controllable, this is obvious from this equation and no further calculations are required (the reader will agree with this statement after going through Section 6.7).

Let us design a minimum-time controller for this system which transfers any nonzero initial state $\mathbf{x}^0$ to zero in the minimum number of steps.

The controllability matrix

$$\mathbf{U} = [\mathbf{G} \mid \mathbf{FG}] = \begin{bmatrix} 4.88 & 4.88 & 4.6419 & 4.6419 \\ -0.019 & 0.0095 & -0.0172 & 0.0086 \end{bmatrix}$$

The first two columns of $\mathbf{U}$ are linearly independent. Therefore $N = 1$ and the minimum-time required for the transfer is one step. The minimum-time control sequence is obtained from the equation (eqn. (6.58))

$$\begin{bmatrix} u_1^*(0) \\ u_2^*(0) \end{bmatrix} = \begin{bmatrix} 4.88 & 4.88 \\ -0.019 & 0.0095 \end{bmatrix}^{-1} \left(- \begin{bmatrix} 0.9512 & 0 \\ 0 & 0.9048 \end{bmatrix} \begin{bmatrix} x_1^0 \\ x_2^0 \end{bmatrix} \right)$$

$$= - \begin{bmatrix} 0.065 & -31.7428 \\ 0.1299 & 31.7428 \end{bmatrix} \begin{bmatrix} x_1^0 \\ x_2^0 \end{bmatrix} \tag{6.63a}$$

$$\mathbf{u}^*(k) = \mathbf{0};\ k = 1, 2, \ldots \tag{6.63b}$$

Note that the control sequence given by eqn. (6.63) will transfer any initial state to zero state in minimum-time and keep it there for all time thereafter. This is because $\mathbf{x}(N) = \mathbf{0}$ and $\mathbf{x}(t)$ cannot change from $\mathbf{x}(N)$ to $\mathbf{x}(N+1)$. Equations (6.63) thus define a *state deadbeat control* for the system and the corresponding response from (6.62) is the state *deadbeat response.*

Observability: A simple test for observability is given by the following theorem.

Theorem 6.9: The discrete-time system (6.10) is observable if and only if the rank of the $(nq \times n)$ *observability matrix*

$$\mathbf{V}\left(\mathbf{V} \triangleq \begin{bmatrix} \mathbf{C} \\ \mathbf{CF} \\ \mathbf{CF}^2 \\ \vdots \\ \mathbf{CF}^{n-1} \end{bmatrix}\right) \text{ is } n, \text{ i.e., } \rho(\mathbf{V}) = n.$$

Proof: From eqns. (6.10), we have

$$\begin{aligned} \mathbf{y}(0) &= \mathbf{Cx}(0) + \mathbf{Du}(0) \\ \mathbf{y}(1) &= \mathbf{Cx}(1) + \mathbf{Du}(1) \\ &= \mathbf{CFx}(0) + \mathbf{CGu}(0) + \mathbf{Du}(1) \\ \ldots\ &\ldots\ \ldots\ \ldots\ \ldots\ \ldots\ \ldots\ \ldots \\ \mathbf{y}(N) &= \mathbf{CF}^N\mathbf{x}(0) + \mathbf{CF}^{N-1}\mathbf{Gu}(0) + \mathbf{CF}^{N-2}\mathbf{Gu}(1) \\ &\quad + \ldots + \mathbf{CGu}(N-1) + \mathbf{Du}(N) \end{aligned}$$

From these equations, we may write

$$\begin{bmatrix} \mathbf{y}(0) - \mathbf{Du}(0) \\ \mathbf{y}(1) - \mathbf{CGu}(0) - \mathbf{Du}(1) \\ \vdots \\ \mathbf{y}(n-1) - \mathbf{CF}^{n-2}\mathbf{Gu}(0) - \mathbf{CF}^{n-3}\mathbf{Gu}(1) - \ldots - \mathbf{CGu}(n-2) - \mathbf{Du}(n-1) \end{bmatrix}$$

$$= \begin{bmatrix} \mathbf{C} \\ \mathbf{CF} \\ \vdots \\ \mathbf{CF}^{n-1} \end{bmatrix} \mathbf{x}(0) \tag{6.64}$$

Equation (6.64) uniquely determines $\mathbf{x}(0)$ if and only if the matrix on the right has full rank n. This proves the theorem.

Note that eqn. (6.64) also provides the solution.[3]

Example 6.12: Consider a linear system described by the state model

$$\mathbf{x}(k+1) = \mathbf{Fx}(k) + \mathbf{Gu}(k)$$
$$\mathbf{y}(k) \quad = \mathbf{Cx}(k) + \mathbf{Du}(k)$$

where

$$\mathbf{F} = \begin{bmatrix} 0 & 1 & 0 \\ 0 & 0 & 1 \\ -2 & -1 & -1 \end{bmatrix}; \mathbf{G} = \begin{bmatrix} 0 & 1 \\ 1 & 0 \\ 1 & 1 \end{bmatrix}; \mathbf{C} = \begin{bmatrix} 1 & 0 & 1 \\ 0 & 1 & 0 \end{bmatrix}; \mathbf{D} = \begin{bmatrix} 1 & 0 \\ 0 & 1 \end{bmatrix}$$

Given

$$\mathbf{u}(k) = \begin{bmatrix} 1 \\ 1 \end{bmatrix}; k \geqslant 0$$

$$\mathbf{y}(0) = \begin{bmatrix} 3 \\ 1 \end{bmatrix}; \mathbf{y}(1) = \begin{bmatrix} 1 \\ 3 \end{bmatrix} \text{ and } \mathbf{y}(2) = \begin{bmatrix} 3 \\ 1 \end{bmatrix}$$

We shall determine $\mathbf{x}(0)$ from this information.

Let us first check the observability property of the system. The observability matrix

---

[3] If the system is not completely observable, then pseudoinverse of $\mathbf{V}$ may be used in place of $\mathbf{V}^{-1}$ (Chidambara and Wells 1965). Refer Section 2.6.

$$\begin{bmatrix} \mathbf{C} \\ \mathbf{CA} \\ \mathbf{CA} \end{bmatrix} = \begin{bmatrix} 1 & 0 & 1 \\ 0 & 1 & 0 \\ -2 & 0 & -1 \\ 0 & 0 & 1 \\ 2 & -1 & 1 \\ -2 & -1 & -1 \end{bmatrix}$$

Since $\rho(\mathbf{V}) = 3$, the system is observable. With the given data, eqn. (6.64) becomes

$$\begin{bmatrix} 2 \\ 0 \\ -3 \\ 1 \\ 3 \\ -3 \end{bmatrix} = \begin{bmatrix} 1 & 0 & 1 \\ 0 & 1 & 0 \\ -2 & 0 & -1 \\ 0 & 0 & 1 \\ 2 & -1 & 1 \\ -2 & -1 & -1 \end{bmatrix} \begin{bmatrix} x_1^0 \\ x_2^0 \\ x_3^0 \end{bmatrix}$$

Since $\rho(\mathbf{V}) = 3$, we can solve any three linearly independent equations for $\mathbf{x}^0$. Top three rows of the matrix $\mathbf{V}$ are linearly independent; therefore

$$\begin{bmatrix} x_1^0 \\ x_2^0 \\ x_3^0 \end{bmatrix} = \begin{bmatrix} 1 & 0 & 1 \\ 0 & 1 & 0 \\ -2 & 0 & -1 \end{bmatrix}^{-1} \begin{bmatrix} 2 \\ 0 \\ -3 \end{bmatrix} = \begin{bmatrix} 1 \\ 0 \\ 1 \end{bmatrix}$$

6.7 CONTROLLABILITY AND OBSERVABILITY OF STATE MODEL IN JORDAN CANONICAL FORM

We have been using linear transformations of state variables in earlier chapters to derive equivalent systems. An obvious question that should arise is whether these linear transformations affect controllability and/or observability. The answer in both cases is 'no' and is explained in the following theorem.

Theorem 6.10: The controllability and observability of a linear time-invariant system (6.4) are invariant under any equivalence transformation.

Proof:

Consider the state model (6.4). Let $\mathbf{P}$ be an $n \times n$ constant nonsingular matrix. Defining a state vector

$$\hat{\mathbf{x}} = \mathbf{P}\mathbf{x}$$

we can transform (6.4) to the following equivalent state model.

$$\dot{\hat{\mathbf{x}}}(t) = \hat{\mathbf{A}}\hat{\mathbf{x}}(t) + \hat{\mathbf{B}}\mathbf{u}(t) \tag{6.65a}$$

$$\mathbf{y}(t) = \hat{\mathbf{C}}\hat{\mathbf{x}}(t) + \mathbf{D}\mathbf{u}(t) \tag{6.65b}$$

where

$$\hat{\mathbf{A}} = \mathbf{P}\mathbf{A}\mathbf{P}^{-1}$$

$$\hat{\mathbf{B}} = \mathbf{P}\mathbf{B}$$

$$\hat{\mathbf{C}} = \mathbf{C}\mathbf{P}^{-1}$$

We know that (6.65) is controllable if and only if

$$\rho\,[\hat{\mathbf{B}} \,\vdots\, \hat{\mathbf{A}}\hat{\mathbf{B}} \,\vdots \ldots \vdots\, \hat{\mathbf{A}}^{n-1}\hat{\mathbf{B}}] = n$$

But

$$\begin{aligned}[\hat{\mathbf{B}} \,\vdots\, \hat{\mathbf{A}}\hat{\mathbf{B}} \,\vdots \ldots \vdots\, \hat{\mathbf{A}}^{n-1}\hat{\mathbf{B}}] &= [\mathbf{P}\mathbf{B} \,\vdots\, \mathbf{P}\mathbf{A}\mathbf{B} \,\vdots \ldots \vdots\, \mathbf{P}\mathbf{A}^{n-1}\mathbf{B}] \\ &= \mathbf{P}\,[\mathbf{B} \,\vdots\, \mathbf{A}\mathbf{B} \,\vdots \ldots \vdots\, \mathbf{A}^{n-1}\mathbf{B}]\end{aligned}$$

Since the rank of a matrix does not change by multiplication of a nonsingular matrix, we have

$$\rho\,[\hat{\mathbf{B}} \,\vdots\, \hat{\mathbf{A}}\hat{\mathbf{B}} \,\vdots \ldots \vdots\, \hat{\mathbf{A}}^{n-1}\hat{\mathbf{B}}] = \rho\,[\mathbf{B} \,\vdots\, \mathbf{A}\mathbf{B} \,\vdots \ldots \vdots\, \mathbf{A}^{n-1}\mathbf{B}]$$

Consequently, (6.4) is controllable if and only if (6.65) is controllable. The observability part can be proved on the similar lines.

□

This theorem can be extended to linear time-varying state models (Swisher 1976).

If the system equations are known in Jordan-form, then one need not resort to controllability and observability tests given in the earlier section. These properties can be determined almost by inspection of the system equations as will be shown below (Ho 1962, Gilbert 1963).

Consider the n-dimensional linear time-invariant Jordan-form state model

$$\dot{\mathbf{x}}(t) = \mathbf{J}\mathbf{x}(t) + \mathbf{B}\mathbf{u}(t) \tag{6.66a}$$

$$\mathbf{y}(t) = \mathbf{C}\mathbf{x}(t) + \mathbf{D}\mathbf{u}(t) \tag{6.66b}$$

where

$$\underset{(n\times n)}{\mathbf{J}} = \begin{bmatrix} \mathbf{J}_1 & & \\ & \mathbf{J}_2 & \\ & & \ddots \\ & & & \mathbf{J}_m \end{bmatrix}; \quad \underset{(n\times p)}{\mathbf{B}} = \begin{bmatrix} \mathbf{B}_1 \\ \mathbf{B}_2 \\ \vdots \\ \mathbf{B}_m \end{bmatrix}; \quad \underset{(q\times n)}{\mathbf{C}} = [\mathbf{C}_1 \mathbf{C}_2 \ldots \mathbf{C}_m]$$

$\mathbf{J}_1, \mathbf{J}_2, \ldots, \mathbf{J}_m$ are the m Jordan blocks corresponding to m distinct eigenvalues $\lambda_1, \lambda_2, \ldots, \lambda_m$ of multiplicity $n_1, n_2, \ldots, n_m$ respectively; $n = \sum_{i=1}^{m} n_i$. The $(n_i \times n_i)$ Jordan block $\mathbf{J}_i$ corresponding to eigenvalue λ_i will have on its principal diagonal $r(i)$ blocks of the form (2.70). The number $r(i)$ depends on the structure of n_i eigenvectors/generalized eigenvectors corresponding to eigenvalue λ_i. Partitioning of various matrices in (6.66a) and (6.66b) is shown below:

$$\underset{(n_i \times n_i)}{\mathbf{J}_i} = \begin{bmatrix} \mathbf{J}_{i1} & & \\ & \mathbf{J}_{i2} & \\ & & \ddots & \\ & & & \mathbf{J}_{ir(i)} \end{bmatrix}; \quad \underset{(n_i \times p)}{\mathbf{B}_i} = \begin{bmatrix} \mathbf{B}_{i1} \\ \mathbf{B}_{i2} \\ \vdots \\ \mathbf{B}_{ir(i)} \end{bmatrix}; \quad \underset{(q \times n_i)}{\mathbf{C}_i} = [\mathbf{C}_{i1} \mathbf{C}_{i2} \cdots \mathbf{C}_{ir(i)}] \tag{6.66c}$$

$$i = 1, 2, \ldots, m$$

$$\underset{(k \times k)}{\mathbf{J}_{ij}} = \begin{bmatrix} \lambda_i & 1 & 0 \ldots 0 & 0 \\ 0 & \lambda_i & 1 \ldots 0 & 0 \\ \vdots & & \vdots & \\ 0 & 0 & 0 \ldots \lambda_i & 1 \\ 0 & 0 & 0 \ldots 0 & \lambda_i \end{bmatrix}; \quad \underset{(k \times p)}{\mathbf{B}_{ij}} = \begin{bmatrix} \mathbf{b}_{1ij} \\ \mathbf{b}_{2ij} \\ \vdots \\ \mathbf{b}_{kij} \end{bmatrix}; \quad \underset{(q \times k)}{\mathbf{C}_{ij}} = [\mathbf{c}_{1ij}\, \mathbf{c}_{2ij} \ldots \mathbf{c}_{kij}] \tag{6.66d}$$

$$j = 1, 2, \ldots, r(i).$$

Once system equations are known in the form of eqns. (6.66), the controllability and observability can be determined using the following theorem (Chen and Desoer 1968, Chen 1970).

Theorem 6.11: The system (6.66) is controllable if and only if for each $i = 1, 2, \ldots, m$, the rows of $r(i) \times p$ matrix

$$\underset{(r(i) \times p)}{\mathbf{B}_i^k} \triangleq \begin{bmatrix} \mathbf{b}_{ki1} \\ \mathbf{b}_{ki2} \\ \vdots \\ \mathbf{b}_{kir(i)} \end{bmatrix} \tag{6.67a}$$

are linearly independent over the field of complex numbers. The system (6.66) is observable if and only if for each $i = 1, 2, \ldots, m$, the columns of $q \times r(i)$ matrix

$$\underset{(q \times r(i))}{\mathbf{C}_i^l} \triangleq [\mathbf{c}_{li1}\, \mathbf{c}_{li2} \ldots \mathbf{c}_{lir(i)}] \tag{6.67b}$$

are linearly independent over the field of complex numbers.

Proof:
In Theorem 6.3 we established the result that the linear time-invariant system (6.4) is controllable if and only if the rows of $e^{\mathbf{A}t}\mathbf{B}$ are linearly independent

over the field of real numbers for all $t \in [0, \infty)$.

Taking the Laplace transform of $e^{\mathbf{A}t}\mathbf{B}$, we have

$$\mathscr{L}[e^{\mathbf{A}t}\mathbf{B}] = (s\mathbf{I}-\mathbf{A})^{-1}\mathbf{B}$$

Since Laplace transform is one-to-one linear operator which maps real vector space into complex vector space, the rows of $e^{\mathbf{A}t}\mathbf{B}$ are linearly independent over $\mathcal{R}$ which implies that the rows of $(s\mathbf{I}-\mathbf{A})^{-1}\mathbf{B}$ are linearly independent over $\mathcal{C}$.

Let us examine the rows of $(s\mathbf{I}-\mathbf{J})^{-1}\mathbf{B}$ where $\mathbf{J}$ and $\mathbf{B}$ are as defined in (6.66).

$$(s\mathbf{I}-\mathbf{J})^{-1}\mathbf{B} = \begin{bmatrix} (s\mathbf{I}-\mathbf{J}_1)^{-1}\mathbf{B}_1 \\ (s\mathbf{I}-\mathbf{J}_2)^{-1}\mathbf{B}_2 \\ \vdots \\ (s\mathbf{I}-\mathbf{J}_m)^{-1}\mathbf{B}_m \end{bmatrix}$$

$$(s\mathbf{I}-\mathbf{J}_i)^{-1}\mathbf{B}_i = \begin{bmatrix} (s\mathbf{I}-\mathbf{J}_{i1})^{-1}\mathbf{B}_{i1} \\ (s\mathbf{I}-\mathbf{J}_{i2})^{-1}\mathbf{B}_{i2} \\ \vdots \\ (s\mathbf{I}-\mathbf{J}_{ir(i)})^{-1}\mathbf{B}_{ir(i)} \end{bmatrix}; \; i = 1, 2, \ldots, m$$

$$(s\mathbf{I}-\mathbf{J}_{ij})^{-1}\mathbf{B}_{ij} = \begin{bmatrix} \frac{1}{(s-\lambda_i)} & \frac{1}{(s-\lambda_i)^2} & \frac{1}{(s-\lambda_i)^3} \cdots & \frac{1}{(s-\lambda_i)^k} \\ 0 & \frac{1}{(s-\lambda_i)} & \frac{1}{(s-\lambda_i)^2} \cdots & \frac{1}{(s-\lambda_i)^{k-1}} \\ 0 & 0 & \frac{1}{(s-\lambda_i)} \cdots & \frac{1}{(s-\lambda_i)^{k-2}} \\ \vdots & \vdots & \vdots & \vdots \\ 0 & 0 & 0 \;\cdots & \frac{1}{(s-\lambda_i)} \end{bmatrix} \begin{bmatrix} \mathbf{b}_{1ij} \\ \mathbf{b}_{2ij} \\ \mathbf{b}_{3ij} \\ \vdots \\ \mathbf{b}_{kij} \end{bmatrix}$$

$$j = 1, 2, \ldots, r(i)$$

Let us first show the necessity. If the rows $\mathbf{b}_{kij}$; $j = 1, 2, \ldots, r(i)$ of matrix $\mathbf{B}_i^k$ defined in (6.67a) are not linearly independent for some i, then we must show that the system (6.66) is not completely controllable. In the matrix $(s\mathbf{I}-\mathbf{J}_i)^{-1}\mathbf{B}_i$, there are $r(i)$ rows of the form

$$\begin{matrix} \frac{1}{(s-\lambda_i)}\mathbf{b}_{ki1} \\ \frac{1}{(s-\lambda_i)}\mathbf{b}_{ki2} \\ \vdots \\ \frac{1}{(s-\lambda_i)}\mathbf{b}_{kir(i)} \end{matrix} \qquad (6.68)$$

Therefore, if the rows $\mathbf{b}_{kij}; j = 1, 2. ..., r(i)$ are not linearly independent, so are the rows given by (6.68) and hence the rows of $(s\mathbf{I}-\mathbf{J}_i)^{-1}\mathbf{B}_i$; the system (6.66) is then not completely controllable. Note that if any of the rows of the matrix $\mathbf{B}_i^k$ defined in (6.67a) is a zero row, the set of rows of this matrix becomes linearly dependent.

The sufficiency requires that we show that if the rows of $\mathbf{B}_i^k$ are linearly independent for all i, then the system (6.66) is completely controllable. Look at the p elements of the first row of $(s\mathbf{I}-\mathbf{J}_{ij})^{-1}\mathbf{B}_{ij}$. Each element has an additive term given by the corresponding element of the row vector

$$\frac{1}{(s-\lambda_i)^k}\,\mathbf{b}_{kij}$$

The second row of $(s\mathbf{I}-\mathbf{J}_{ij})^{-1}\mathbf{B}_{ij}$ consists of p elements, wherein each element has an additive term given by the corresponding element of the row vector

$$\frac{1}{(s-\lambda_i)^{k-1}}\,\mathbf{b}_{kij}$$

The last row of $(s\mathbf{I}-\mathbf{J}_{ij})^{-1}\mathbf{B}_{ij}$ consists of p elements given by the row vector

$$\frac{1}{(s-\lambda_i)}\,\mathbf{b}_{kij}$$

Thus if $\mathbf{b}_{kij} \neq \mathbf{0}$, then all the rows of $(s\mathbf{I}-\mathbf{J}_{ij})^{-1}\mathbf{B}_{ij}$ are linearly independent. It can be similarly shown that if $\mathbf{b}_{kij}$, $j = 1, 2, ..., r(i)$ are linearly independent, then all the rows of $(s\mathbf{I}-\mathbf{J}_i)^{-1}\mathbf{B}_i$ are linearly independent.

The observability part of the theorem can be proved on similar lines.

□

The following results can be easily established from Theorem 6.11;

1. For the system (6.66) with distinct eigenvalues $\lambda_1, \lambda_2, ..., \lambda_m$, if there is at most one Jordan block associated with each eigenvalue, then the condition of controllability is that none of the rows of **B** matrix corresponding to last row of each Jordan block should be a zero row.
   The observability condition is that none of the columns of **C** matrix corresponding to the first row of each Jordan block should be a zero column.
2. If all the n eigenvalues $\lambda_1, \lambda_2, ..., \lambda_n$ are distinct, then (6.66) is controllable if and only if none of the rows of **B** matrix is a zero row and (6.66) is observable if and only if none of the columns of **C** matrix is a zero column.

All the results given in this section have obvious extensions for linear time-invariant discrete-time systems.

Example 6.13: Consider the Jordan-form state model

$$\dot{\mathbf{x}}(t) = \mathbf{J}\mathbf{x}(t) + \mathbf{b}u(t);\ \mathbf{x}(0) = \mathbf{0}$$

$$y(t) = \mathbf{c}\mathbf{x}(t)$$

where

$$\mathbf{J} = \begin{bmatrix} \lambda_1 & 1 & 0 & 0 & 0 & 0 \\ 0 & \lambda_1 & 1 & 0 & 0 & 0 \\ 0 & 0 & \lambda_1 & 0 & 0 & 0 \\ 0 & 0 & 0 & \lambda_4 & 1 & 0 \\ 0 & 0 & 0 & 0 & \lambda_4 & 0 \\ 0 & 0 & 0 & 0 & 0 & \lambda_6 \end{bmatrix}; \mathbf{b} = \begin{bmatrix} b_1 \\ b_2 \\ b_3 \\ b_4 \\ b_5 \\ b_6 \end{bmatrix}$$

$$\mathbf{c} = [c_1 \quad c_2 \quad c_3 \quad c_4 \quad c_5 \quad c_6]$$

It is obvious that the state model is controllable if and only if b_3, b_5, b_6 are nonzero and the state model is observable if and only if c_1, c_4 and c_6 are nonzero.

The physical meaning of these conditions can be seen from the block diagram of the given state model shown in Fig. 6.8. In this figure, each

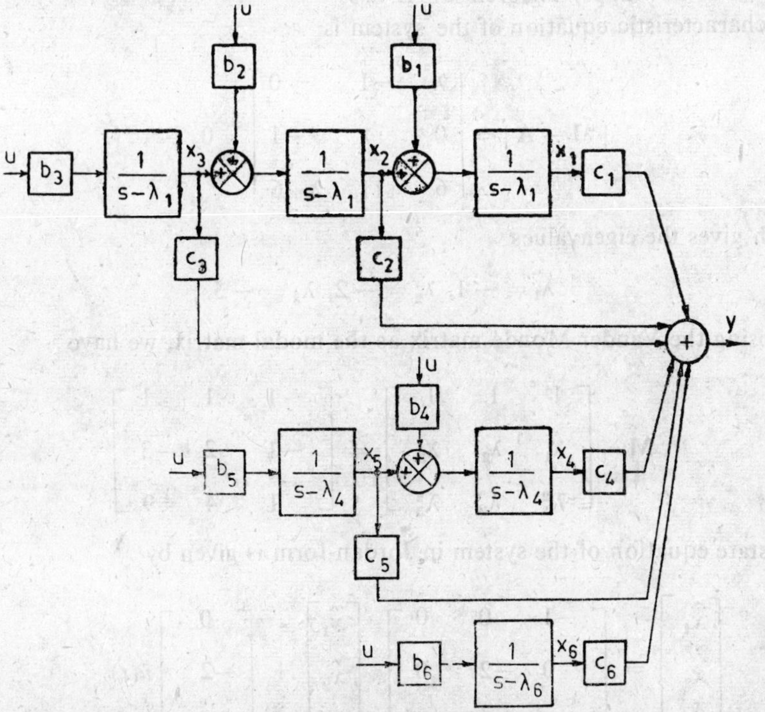

Fig. 6.8 Block diagram for state model of Example 6.13.

chain of blocks corresponds to a Jordan block in the given state model. We see that if $b_3 \neq 0$, then all the state variables in that chain can be controlled; if $c_1 \neq 0$, then all the state variables in that chain can be observed. Similar arguments apply to the other two chains.

Example 6.14: Consider a system described by the state equation

$$\dot{\mathbf{x}}(t) = \mathbf{A}\mathbf{x}(t) + \mathbf{b}u(t)$$

where

$$\mathbf{A} = \begin{bmatrix} 0 & 1 & 0 \\ 0 & 0 & 1 \\ -6 & -11 & -6 \end{bmatrix}; \; \mathbf{b} = \begin{bmatrix} -1 \\ 1 \\ 1 \end{bmatrix}$$

The controllability matrix

$$\mathbf{U} = [\mathbf{b} \mid \mathbf{A}\mathbf{b} \mid \mathbf{A}^2\mathbf{b}] = \begin{bmatrix} -1 & 1 & 1 \\ 1 & 1 & -11 \\ 1 & -11 & 49 \end{bmatrix}$$

$|\mathbf{U}| = 0$; therefore by Theorem 6.3, the system is not completely controllable.

Let us now apply Theorem 6.11.

The characteristic equation of the system is

$$|\lambda\mathbf{I} - \mathbf{A}| = \begin{vmatrix} \lambda & -1 & 0 \\ 0 & \lambda & -1 \\ 6 & 11 & \lambda+6 \end{vmatrix} = 0$$

which gives the eigenvalues

$$\lambda_1 = -1, \; \lambda_2 = -2, \; \lambda_3 = -3.$$

Choosing the Vander Monde matrix as the modal matrix, we have

$$\mathbf{M} = \begin{bmatrix} 1 & 1 & 1 \\ \lambda_1 & \lambda_2 & \lambda_3 \\ \lambda_1^2 & \lambda_2^2 & \lambda_3^2 \end{bmatrix} = \begin{bmatrix} 1 & 1 & 1 \\ -1 & -2 & -3 \\ 1 & 4 & 9 \end{bmatrix}$$

The state equation of the system in Jordan-form is given by

$$\begin{bmatrix} \dot{\hat{x}}_1 \\ \dot{\hat{x}}_2 \\ \dot{\hat{x}}_3 \end{bmatrix} = \begin{bmatrix} -1 & 0 & 0 \\ 0 & -2 & 0 \\ 0 & 0 & -3 \end{bmatrix} \begin{bmatrix} \hat{x}_1 \\ \hat{x}_2 \\ \hat{x}_3 \end{bmatrix} + \begin{bmatrix} 0 \\ -2 \\ 1 \end{bmatrix} u(t)$$

where $$\hat{\mathbf{x}}(t) = \mathbf{M}\hat{\mathbf{x}}(t)$$

From this equation we obtain the same conclusion; the system is not completely controllable. This equation also tells us that the modes e^{-2t} and e^{-3t} of the system under consideration are controllable and the mode e^{-t} is not controllable.

□

The example above makes it clear that the criteria for controllability and observability given by Theorem 6.11 are useful because of the geometric insight provided. The state model in Jordan canonical-form reveals the modes that are uncontrollable (unobservable). However, it is obvious that time-consuming calculation of the modal matrix is necessary.

6.8 LOSS OF CONTROLLABILITY AND OBSERVABILITY DUE TO SAMPLING

Consider the open-loop sampled-data system of Fig. 5·5. The plant is described by the state equation of the form (6.4a) which is reproduced below:

$$\dot{\mathbf{x}}(t) = \mathbf{A}\mathbf{x}(t) + \mathbf{B}\mathbf{u}(t) \tag{6.69}$$

Because of sample and hold operations, the input to the plant

$$\mathbf{u}^+(kT+\tau) = \mathbf{u}(kT)\,; \quad 0 \leqslant \tau < T\,;$$
T is the sampling period

and eqn. (6.69) can be replaced by the following discrete-time equation (eqn. (5.110)):

$$\mathbf{x}(k+1) = \mathbf{F}\mathbf{x}(k) + \mathbf{G}\mathbf{u}(k) \tag{6.70}$$

where

$$\mathbf{F} = e^{\mathbf{A}T}$$

$$\mathbf{G} = \int_0^T e^{\mathbf{A}\lambda}\,\mathbf{B}\,d\lambda$$

If state equation (6.69) is controllable, a natural question arises whether the system remains controllable after the introduction of sampling, i.e., whether state equation (6.70) is controllable. The matrices **F** and **G** in (6.70) depend on the sampling period T and hence the controllability condition of Theorem 6.8 depends on T. We study this dependence in the following Theorem (Kalman et al. 1962, Chen 1970).

Theorem 6.12: The system (6.69) which is completely state controllable in the absence of sampling, remains completely state controllable after the introduction of sampling if, for every eigenvalue of the characteristic equation,

$$Re(\lambda_i) = Re(\lambda_j)$$

implies

$$\operatorname{Im}(\lambda_i - \lambda_j) \neq \frac{2\alpha\pi}{T};\ \alpha = \pm 1, \pm 2, \ldots;\ T = \text{sampling period.}$$

Proof:

We assume without loss of generality that the given continuous-time plant is known in the Jordan-form (6.66). Then

$$\mathbf{F} = e^{\mathbf{J}T} = \begin{bmatrix} e^{\mathbf{J}_1 T} & & & \\ & e^{\mathbf{J}_2 T} & & \\ & & \ddots & \\ & & & e^{\mathbf{J}_m T} \end{bmatrix} \tag{6.71a}$$

where

$$e^{\mathbf{J}_i T} = \begin{bmatrix} e^{\mathbf{J}_{i1} T} & & & \\ & e^{\mathbf{J}_{i2} T} & & \\ & & \ddots & \\ & & & e^{\mathbf{J}_{ir(i)} T} \end{bmatrix};\ i = 1, 2, \ldots, m$$

and

$$e^{\mathbf{J}_{ij} T} = e^{\lambda_i T} \begin{bmatrix} 1 & T & \frac{T^2}{2!} & \cdots & \frac{T^{k-1}}{(k-1)!} \\ 0 & 1 & T & \cdots & \frac{T^{k-2}}{(k-2)!} \\ \vdots & \vdots & \vdots & & \vdots \\ 0 & 0 & 0 & \cdots & 1 \end{bmatrix};\ \begin{aligned} i &= 1, 2, \ldots, m \\ j &= 1, 2, \ldots, r(i). \end{aligned}$$

Since $e^{\mathbf{J}_{ij} T}$ is a triangular matrix, it is easy to verify that $(z\mathbf{I} - e^{\mathbf{J}_{ij} T})^{-1}$

$$= \begin{bmatrix} (z - e^{\lambda_i T})^{-1} & Te^{\lambda_i T}(z - e^{\lambda_i T})^{-2} & \cdots & [(Te^{\lambda_i T})^{k-1}(z - e^{\lambda_i T})^{-k} + \ldots \\ 0 & (z - e^{\lambda_i T})^{-1} & \cdots & [(Te^{\lambda_i T})^{k-2}(z - e^{\lambda_i T})^{-k+1} + \ldots \\ \vdots & \vdots & & \vdots \\ 0 & 0 & \cdots & (z - e^{\lambda_i T})^{-1} \end{bmatrix}$$

$$\mathbf{G} = \int_0^T e^{\mathbf{J}\lambda} \mathbf{B}\, d\lambda \tag{6.71b}$$

From Theorem 6.8 we know that (6.70) is controllable if and only if $\rho(\mathbf{U}) = \rho[\mathbf{G} \mid \mathbf{FG} \mid \ldots \mid \mathbf{F}^{n-1}\mathbf{G}] = n$ or equivalently we say that (6.70) is controllable if and only if the n rows of matrix $\mathbf{U}$ are linearly independent over the complex field $\mathcal{C}$. In the following we show that $\rho[\mathbf{G} \mid \mathbf{FG} \mid \ldots \mid \mathbf{F}^{n-1}\mathbf{G}] = n$ if and only if the n rows of $(z\mathbf{I} - \mathbf{F})^{-1}\mathbf{G}$ are linearly independent over the field $\mathcal{C}$.

$$(z\mathbf{I} - \mathbf{F})^{-1}\mathbf{G} = z^{-1}(\mathbf{I} - z^{-1}\mathbf{F})^{-1}\mathbf{G}$$

From eqns, (5.60) and (5.61)

$$(\mathbf{I}-z^{-1}\mathbf{F})^{-1}=\sum_{i=0}^{\infty}(z^{-1}\mathbf{F})^{i}$$

Therefore

$$(z\mathbf{I}-\mathbf{F})^{-1}\mathbf{G}=z^{-1}\mathbf{G}+z^{-2}\mathbf{F}\mathbf{G}+z^{-3}\mathbf{F}^{2}\mathbf{G}+\dots$$

It is obvious that if all the rows of $\mathbf{U}$ are linearly independent over the field of complex numbers, so are the rows of $(z\mathbf{I}-\mathbf{F})^{-1}\mathbf{G}$. Now if $\rho(\mathbf{U})<n$, by definition there exists a nonzero constant vector $\boldsymbol{\alpha}$ such that

$$\boldsymbol{\alpha}\mathbf{U}=\mathbf{0};\quad \text{i.e., } \boldsymbol{\alpha}\mathbf{G}=\boldsymbol{\alpha}\mathbf{F}\mathbf{G}=\dots=\boldsymbol{\alpha}\mathbf{F}^{n-1}\mathbf{G}=\mathbf{0}$$

This implies that (Cayley-Hamilton theorem) $\boldsymbol{\alpha}\mathbf{F}^{i}\mathbf{G}=\mathbf{0}$ for $i>n$. Hence $\boldsymbol{\alpha}(z\mathbf{I}-\mathbf{F})^{-1}\mathbf{G}=\mathbf{0}$ which means that the rows of $(z\mathbf{I}-\mathbf{F})^{-1}\mathbf{G}$ are not linearly independent if $\rho(\mathbf{U})<n$.

To prove the theorem, we shall show that the rows of $(z\mathbf{I}-\mathbf{F})^{-1}\mathbf{G}$; $\mathbf{F}$ and $\mathbf{G}$ given by (6.71), are linearly independent over the field of complex numbers.

$$(z\mathbf{I}-\mathbf{F})^{-1}\mathbf{G}=(z\mathbf{I}-e^{\mathbf{J}T})^{-1}\mathbf{M}\mathbf{B}$$

where

$$\mathbf{M}=\int_{0}^{T}e^{\mathbf{J}\lambda}\,d\lambda$$

Substituting for $\mathbf{J}$ from eqn. (6.71), we can show that matrix $\mathbf{M}$ is nonsingular for all $T>0$. Since the matrices $(z\mathbf{I}-e^{\mathbf{J}T})^{-1}$ and $\mathbf{M}$ are functions of the same matrix $\mathbf{J}$, they commute and therefore,

$$(z\mathbf{I}-e^{\mathbf{J}T})^{-1}\mathbf{M}=\mathbf{M}(z\mathbf{I}-e^{\mathbf{J}T})^{-1}$$

or

$$(z\mathbf{I}-e^{\mathbf{J}T})^{-1}\mathbf{G}=\mathbf{M}\,(z\mathbf{I}-e^{\mathbf{J}T})^{-1}\mathbf{B}$$

Consequently, if we show that the rows of $(z\mathbf{I}-e^{\mathbf{J}T})^{-1}\mathbf{B}$ are linearly independent over the field of complex numbers, so are the rows of $(z\mathbf{I}-e^{\mathbf{J}T})^{-1}\mathbf{G}$.

Since $\{\mathbf{J},\mathbf{B}\}$ is controllable, $\mathbf{b}_{kij}$ is different from zero vector. Consequently all the rows of $(z\mathbf{I}-e^{J_{ij}T})^{-1}\mathbf{B}_{ij}$ are linearly independent since the first row contains a term

$$(Te^{\lambda_i T})^{k-1}\,(z-e^{\lambda_i T})^{-k}\mathbf{b}_{ki1};$$

the second row contains a term

$$(Te^{\lambda_i T})^{k-2}\,(z-e^{\lambda_i T})^{-k+1}\mathbf{b}_{ki2}$$

and so forth.

The controllability assumption of (6.69) implies that the set $\mathbf{b}_{kij}$ for $j=1, 2, \dots, r(i)$ is a linearly independent set, hence, all the rows of $(z\mathbf{I}-e^{J_{ij}T})^{-1}\mathbf{B}_{i}$ are linearly independent for each i.

Now the assumption that $I_m[\lambda_i - \lambda_j] \neq \frac{2\pi\alpha}{T}$ implies that if λ_i and λ_j are distinct, so are $\exp(\lambda_i T)$ and $\exp(\lambda_j T)$. Consequently we conclude that all the rows of $(z\mathbf{I} - \exp(\mathbf{J}T))^{-1}\mathbf{B}$ are linearly independent.

□

Note that for single-input systems ($p = 1$), the sufficiency condition given by Theorem 6.12 is necessary as well. This can be established as follows.

If $I_m[\lambda_i - \lambda_j] = 2\pi\alpha/T$ for $\text{Re}[\lambda_i - \lambda_j] = 0$, then $e^{\lambda_i T} = e^{\lambda_j T}$; the last row of $(z\mathbf{I} - e^{\mathbf{J}_i T})^{-1}\mathbf{B}_i$ and the last row of $(z\mathbf{I} - e^{\mathbf{J}_j T})^{-1}\mathbf{B}_j$ are linearly dependent. Hence the single input discrete-time system is not controllable.

It can be easily verified that the result of Theorem 6.12 applies for observability as well.

Example 6.15: Consider a linear system described by the equations

$$\dot{\mathbf{x}} = \begin{bmatrix} 0 & 1 \\ -\omega^2 & 0 \end{bmatrix} \mathbf{x} + \begin{bmatrix} 0 \\ \omega \end{bmatrix} u$$

$$y = x_1$$

It can be easily verified that the system is completely controllable and observable.

With sampled data, the control signal to the process is

$$u(t) = u(kT) \text{ for } kT \leqslant t < (k+1)\,T$$

where T is the sampling period.

The state transition matrix

$$\Phi(t) = \begin{bmatrix} \cos \omega t & 1/\omega \sin \omega t \\ -\omega \sin \omega t & \cos \omega t \end{bmatrix}$$

for $T = \frac{2\pi\alpha}{\omega}$, $\Phi(T) = \Phi(2T) = \ldots = \Phi(\alpha T)$ for α = positive integer and the sampled data system is uncontrollable and unobservable.

6.9 CONTROLLABILITY AND OBSERVABILITY CANONICAL FORMS OF STATE MODEL

In this section we analyze the structure of linear time-invariant systems that are not completely controllable and/or observable. We shall study the continuous-time systems. Results for discrete-time systems are identical (Kalman 1962, Gilbert 1963, Weiss 1968).

Controllable Subspace

Consider the n-dimensional linear time-invariant system described by eqns. (6.4). The controllability matrix of this system is

$$\mathbf{U} = [\mathbf{B} \,\vdots\, \mathbf{AB} \,\vdots\, \dots \,\vdots\, \mathbf{A}^{n-1}\mathbf{B}]$$

We know from Theorem 6.3 that if $\rho(\mathbf{U}) = n$, the system (6.4) is completely controllable which implies that any state $\mathbf{x}^0$ in n-dimensional state space Σ spanned by n linearly independent columns of $\mathbf{U}$, can be transferred to any other state $\mathbf{x}^1 \in \Sigma$ in finite time. If $\rho(\mathbf{U}) = m(< n)$, the system (6.4) is uncontrollable, i.e., any arbitrary state $\mathbf{x}^0$ in Σ cannot be transferred to any other arbitrary state $\mathbf{x}^1$ in Σ in finite time. In such a case, it is of interest to know what part of the state space is controllable. The linear subspace consisting of states which are controllable is called *controllable subspace*.

Let the state $\mathbf{x}^0$ of system (6.4) belong to controllable subspace. Then as per the definition of controllability, there exists a control $\mathbf{u}$ that can transfer $\mathbf{x}^0$ to origin of state space in some finite time t_1, i.e., there exists a $\mathbf{u}_{[0,\,t_1]}$ such that

$$\mathbf{0} = e^{\mathbf{A}t_1}\mathbf{x}^0 + \int_0^{t_1} e^{\mathbf{A}(t_1-\tau)}\mathbf{Bu}(\tau)\,d\tau$$

$$= e^{\mathbf{A}t_1}\mathbf{x}^0 + e^{\mathbf{A}t_1}\int_0^{t_1} e^{-\mathbf{A}\tau}\mathbf{Bu}(\tau)\,d\tau \tag{6.72}$$

Let
$$\mathbf{u}(t) = -\,\mathbf{B}^T e^{-\mathbf{A}^T t}\mathbf{z} \tag{6.73a}$$

where $\mathbf{z} \in \mathcal{R}^n$ is the solution (assuming it exists) of equation

$$\mathbf{W}(0, t_1)\,\mathbf{z} = \mathbf{x}^0; \tag{6.73b}$$

$$\mathbf{W}(0, t_1) = \int_0^{t_1} e^{-\mathbf{A}\tau}\mathbf{BB}^T e^{-\mathbf{A}^T\tau}\,d\tau$$

The control $\mathbf{u}(t)$ given by eqn. (6.73) satisfies eqn. (6.72). Therefore $\mathbf{x}^0$ can be transferred to the origin if eqn. (6.73b) can be solved for $\mathbf{z}$. We know from Section 2.6 that eqn. (6.73b) can be solved for $\mathbf{z}$ if $\mathbf{x}^0$ belongs to range space of $\mathbf{W}(0, t_1)$, i.e., if $\mathbf{x}^0$ belongs to the linear space spanned by linearly independent rows of $e^{\mathbf{A}t}\mathbf{B}$ or equivalently if $\mathbf{x}^0$ belongs to the linear space spanned by linearly independent columns of controllability matrix $\mathbf{U}$ (Theorem 6.3).

Therefore, if $\rho(\mathbf{U}) = m < n$, the controllable subspace is m-dimensional and it is the linear space spanned by the m linearly independent column vectors of $\mathbf{U}$.

Note that if $\mathbf{x}$ belongs to controllable subspace, $\mathbf{Ax}$ is also in controllable subspace, i.e., the controllable subspace is invariant under $\mathbf{A}$. This statement can easily be established. The controllable subspace is spanned by the column vectors of $\mathbf{B}, \mathbf{AB}, \dots, \mathbf{A}^{n-1}\mathbf{B}$. Therefore, the vector $\mathbf{Ax}$, where $\mathbf{x}$ is in the controllable subspace, is in the linear space spanned by the column vectors of

$\mathbf{AB}, \mathbf{A}^2\mathbf{B}, \ldots, \mathbf{A}^n\mathbf{B}$. We know that the column vectors of $\mathbf{A}^n\mathbf{B}$ depend linearly on column vectors of $\mathbf{B}, \mathbf{AB}, \ldots, \mathbf{A}^{n-1}\mathbf{B}$. Hence $\mathbf{Ax}$ is in controllable subspace.

Controllability Canonical Form

We now find a state transformation that represents the system (6.4) in a canonical form which very clearly exhibits the controllability properties of the system.

Theorem 6.13: Consider the n-dimensional linear time-invariant system (6.4). If the controllability matrix of this system has rank $m(<n)$, then there exists an equivalence transformation $\hat{\mathbf{x}} = \mathbf{P}^{-1}\mathbf{x}$, where $\mathbf{P}$ is a constant nonsingular matrix, which transforms (6.4) into *controllability canonical form*

$$\begin{bmatrix} \dot{\hat{\mathbf{x}}}_1(t) \\ \dot{\hat{\mathbf{x}}}_2(t) \end{bmatrix} = \begin{bmatrix} \hat{\mathbf{A}}_{11} & \hat{\mathbf{A}}_{12} \\ 0 & \hat{\mathbf{A}}_{21} \end{bmatrix} \begin{bmatrix} \hat{\mathbf{x}}_1(t) \\ \hat{\mathbf{x}}_1(t) \end{bmatrix} + \begin{bmatrix} \hat{\mathbf{B}}_1 \\ 0 \end{bmatrix} \mathbf{u}(t) \tag{6.74a}$$

$$\mathbf{y}(t) = [\hat{\mathbf{C}}_1 \quad \hat{\mathbf{C}}_2] \begin{bmatrix} \hat{\mathbf{x}}_1(t) \\ \hat{\mathbf{x}}_2(t) \end{bmatrix} + \mathbf{Du}(t) \tag{6.74b}$$

and the m-dimensional subsystem

$$\dot{\hat{\mathbf{x}}}_1(t) = \hat{\mathbf{A}}_{11}\hat{\mathbf{x}}_1(t) + \hat{\mathbf{B}}_1\mathbf{u}(t) + \hat{\mathbf{A}}_{12}\hat{\mathbf{x}}_2 \tag{6.75a}$$

$$\hat{\mathbf{y}}(t) = \hat{\mathbf{C}}_1\hat{\mathbf{x}}_1(t) + \mathbf{Du}(t) \tag{6 75b}$$

is controllable from $\mathbf{u}$, i.e., the pair $\{\hat{\mathbf{A}}_{11}, \hat{\mathbf{B}}_1\}$ is completely controllable.

Proof:

The controllability matrix of system (6.4) is

$$\mathbf{U} = [\mathbf{B} \,\vdots\, \mathbf{AB} \,\vdots\, \ldots \,\vdots\, \mathbf{A}^{n-1}\mathbf{B}]$$

$\mathbf{U}$ has rank m, i.e., $\mathbf{U}$ possesses m linearly independent columns which span the controllable subspace of (6.4). Let us choose a basis $\mathbf{p}_1\, \mathbf{p}_2, \ldots, \mathbf{p}_m$ for the controllable subspace. Furthermore, let $\mathbf{p}_{m+1}, \mathbf{p}_{m+2}, \ldots, \mathbf{p}_n$ be $(n-m)$ linearly independent vectors which together with $\mathbf{p}_1, \mathbf{p}_2, \ldots, \mathbf{p}_m$ span the whole n-dimensional space. Define a nonsingular matrix

$$\mathbf{P} = [\mathbf{P}_1 \quad \mathbf{P}_2] \tag{6.76}$$

where

$$\mathbf{P}_1 = [\mathbf{p}_1\ \mathbf{p}_2\ \cdots\ \mathbf{p}_m];\ \mathbf{P}_2 = [\mathbf{p}_{m+1}\ \mathbf{p}_{m+2}\ \cdots\ \mathbf{p}_n]$$

Note that the vectors $\mathbf{p}_1, \mathbf{p}_2, \ldots, \mathbf{p}_m$ may be chosen as the m linearly independent column vectors of $\mathbf{U}$. The vectors $\mathbf{p}_{m+1}, \mathbf{p}_{m+2}, \ldots, \mathbf{p}_n$ are entirely arbitrary so long as the matrix $\mathbf{P}$ is nonsingular.

We now introduce a transformed state vector $\hat{\mathbf{x}}(t)$ defined by

$$\hat{\mathbf{x}}(t) = \mathbf{P}^{-1}\mathbf{x}(t)$$

Substituting this into (6.4), we obtain

$$\mathbf{P}\dot{\hat{\mathbf{x}}}(t) = \mathbf{A}\mathbf{P}\hat{\mathbf{x}}(t) + \mathbf{B}\mathbf{u}(t)$$

or

$$\dot{\hat{\mathbf{x}}}(t) = \mathbf{P}^{-1}\mathbf{A}\mathbf{P}\hat{\mathbf{x}}(t) + \mathbf{P}^{-1}\mathbf{B}\mathbf{u}(t) \qquad (6{\cdot}77\text{a})$$

Partition $\mathbf{P}^{-1}$ as

$$\mathbf{P}^{-1} = \begin{bmatrix} \mathbf{Q}_1 \\ \mathbf{Q}_2 \end{bmatrix} \qquad (6.77\text{b})$$

where $\mathbf{Q}_1$ has m rows and $\mathbf{Q}_2$ has $(n - m)$ rows. In terms of partitionings of $\mathbf{P}$ and $\mathbf{P}^{-1}$, we may write

$$\mathbf{P}^{-1}\mathbf{A}\mathbf{P} = \begin{bmatrix} \mathbf{Q}_1 \\ \mathbf{Q}_2 \end{bmatrix} \mathbf{A}\,[\mathbf{P}_1 \ \ \mathbf{P}_2] = \begin{bmatrix} \mathbf{Q}_1\mathbf{A}\mathbf{P}_1 & \mathbf{Q}_1\mathbf{A}\mathbf{P}_2 \\ \mathbf{Q}_2\mathbf{A}\mathbf{P}_1 & \mathbf{Q}_2\mathbf{A}\mathbf{P}_2 \end{bmatrix} \qquad (6.77\text{c})$$

$$\mathbf{P}^{-1}\mathbf{B} = \begin{bmatrix} \mathbf{Q}_1 \\ \mathbf{Q}_2 \end{bmatrix} \mathbf{B} = \begin{bmatrix} \mathbf{Q}_1\mathbf{B} \\ \mathbf{Q}_2\mathbf{B} \end{bmatrix} \qquad (6.77\text{d})$$

Now

$$\mathbf{P}^{-1}\mathbf{P} = \begin{bmatrix} \mathbf{Q}_1 \\ \mathbf{Q}_2 \end{bmatrix} [\mathbf{P}_1 \ \ \mathbf{P}_2] = \begin{bmatrix} \mathbf{Q}_1\mathbf{P}_1 & \mathbf{Q}_1\mathbf{P}_2 \\ \mathbf{Q}_2\mathbf{P}_1 & \mathbf{Q}_2\mathbf{P}_2 \end{bmatrix} = \begin{bmatrix} \mathbf{I}_m & \mathbf{0} \\ \mathbf{0} & \mathbf{I}_{n-m} \end{bmatrix}$$

which gives

$$\mathbf{Q}_2\mathbf{P}_1 = \mathbf{0} \qquad (6{\cdot}78)$$

$\mathbf{P}_1$ is composed of vectors which span the controllable subspace. This means that (6.78) implies $\mathbf{Q}_2\mathbf{x} = \mathbf{0}$ for any vector $\mathbf{x}$ in the controllable subspace. Columns of $\mathbf{B}$ are part of controllability matrix; these are obviously in the controllable subspace. Therefore,

$$\mathbf{Q}_2\mathbf{B} = \mathbf{0} \qquad (6.79\text{a})$$

Since controllable supspace is invariant under $\mathbf{A}$, the columns of $\mathbf{A}\mathbf{P}_1$ are in controllable subspace. This yields

$$\mathbf{Q}_2\mathbf{A}\mathbf{P}_1 = \mathbf{0} \qquad (6.79\text{b})$$

The transformed eqn. (6.77) is thus of the form of (6.74). This proves the first part of the Theorem. To prove second part, we may compute the controllability matrix of (6.74),

$$\hat{\mathbf{U}} = \begin{bmatrix} \hat{\mathbf{B}}_1 & \hat{\mathbf{A}}_{11}\hat{\mathbf{B}}_1 \ldots \hat{\mathbf{A}}_{11}^{n-1}\hat{\mathbf{B}}_1 \\ \mathbf{0} & \mathbf{0} \quad \ldots \quad \mathbf{0} \end{bmatrix}$$

$$\rho(\hat{\mathbf{U}}) = m$$

Since the zero rows contribute nothing,

$$\rho\,[\hat{\mathbf{B}}_1 \mid \hat{\mathbf{A}}_{11}\hat{\mathbf{B}}_1 \mid \ldots \mid \hat{\mathbf{A}}_{11}^{n-1}\hat{\mathbf{B}}_1] = m$$

It follows from the Cayley-Hamilton theorem that

$$\rho\,[\hat{\mathbf{B}}_1 \mid \hat{\mathbf{A}}_{11}\hat{\mathbf{B}}_1 \mid \ldots \mid \hat{\mathbf{A}}_{11}^{m-1}\hat{\mathbf{B}}_1] = m$$

Therefore the subsystem (6.75) is controllable.

□

In the equivalence transformation $\hat{\mathbf{x}} = \mathbf{P}^{-1}\mathbf{x}$, the n-dimensional state space Σ of system (6.4) is divided into two subspaces. One is the m-dimensional controllable subspace which consists of all vectors $\begin{bmatrix} \hat{\mathbf{x}}_1 \\ \mathbf{0} \end{bmatrix}$. The other in $(n-m)$ dimensional uncontrollable subspace which consists of all vectors $\begin{bmatrix} \mathbf{0} \\ \hat{\mathbf{x}}_2 \end{bmatrix}$.

This state of affairs is shown schematically in Fig. 6.9. We note that $\hat{\mathbf{x}}_2$ behaves completely independently, while $\hat{\mathbf{x}}_1$ is influenced both by $\hat{\mathbf{x}}_2$ and the input $\mathbf{u}$. The additional driving term $\hat{\mathbf{A}}_{12}\hat{\mathbf{x}}_2$ has no effect on controllability.

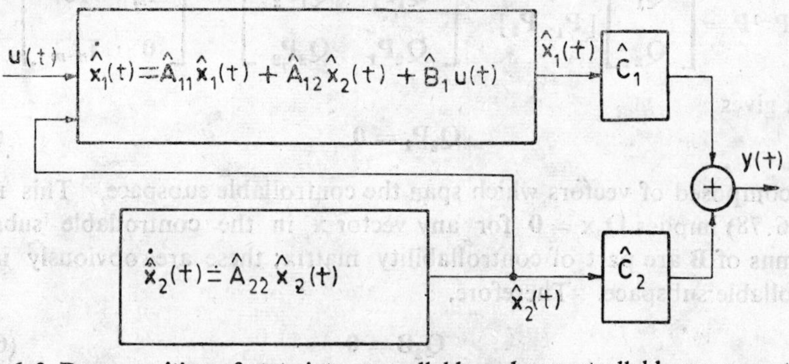

Fig. 6.9 Decomposition of state into controllable and uncontrollable components.

It should also be noted that the controllability canonical form is not unique. However, for any transformation matrix **P**, the eigenvalues of both $\hat{\mathbf{A}}_{11}$ and $\hat{\mathbf{A}}_{22}$ are always the same; eigenvalues of $\hat{\mathbf{A}}_{11}$ are referred to as *controllable eigenvalues* and eigenvalues of $\hat{\mathbf{A}}_{22}$ are referred to as *uncontrollable eigenvalues* of the system.

Example 6.16: Reconsider system of Example 6.2. A state equation of circuit of Fig. 4.3a is

$$\dot{\mathbf{x}}(t) = \mathbf{A}\mathbf{x}(t) + \mathbf{b}u(t)$$

where

$$\mathbf{A} = \begin{bmatrix} -2 & 1 \\ 1 & -2 \end{bmatrix}; \ \mathbf{b} = \begin{bmatrix} 1 \\ 1 \end{bmatrix}$$

The controllability matrix

$$\mathbf{U} = [\mathbf{b} \,\vdots\, \mathbf{Ab}] = \begin{bmatrix} 1 & -1 \\ 1 & -1 \end{bmatrix}$$

$\rho(\mathbf{U}) = 1$; the system is therefore uncontrollable. This fact was earlier proved in Example 6.2.

The equivalence transformation **P** can be chosen as

$$\mathbf{P} = \begin{bmatrix} 1 & 1 \\ 1 & 0 \end{bmatrix}; \ \mathbf{P}^{-1} = \begin{bmatrix} 0 & 1 \\ 1 & -1 \end{bmatrix}$$

The first column of **P** is the first column of **U**; the second column of **P** is chosen arbitrarily to make **P** nonsingular. If we let $\hat{\mathbf{x}}(t) = \mathbf{P}^{-1}\mathbf{x}(t)$, then we get

$$\begin{bmatrix} \dot{\hat{x}}_1(t) \\ \dot{\hat{x}}_2(t) \end{bmatrix} = \begin{bmatrix} -1 & 1 \\ 0 & -3 \end{bmatrix} \begin{bmatrix} \hat{x}_1(t) \\ \hat{x}_2(t) \end{bmatrix} + \begin{bmatrix} 1 \\ 0 \end{bmatrix} u(t)$$

In the transformation $\mathbf{P}\hat{\mathbf{x}} = \mathbf{x}$, the basis of the state space changes from the set of orthonormal vectors $\begin{bmatrix} 1 \\ 0 \end{bmatrix}, \begin{bmatrix} 0 \\ 1 \end{bmatrix}$ to the set of vectors $\begin{bmatrix} 1 \\ 1 \end{bmatrix}, \begin{bmatrix} 1 \\ 0 \end{bmatrix}$ as shown in Fig. 6.10. The state variable $\hat{\mathbf{x}}_1$ is controllable. This fact was established in Example 6.2: state **x** can be controlled only if $\mathbf{x}_1 = \mathbf{x}_2$.

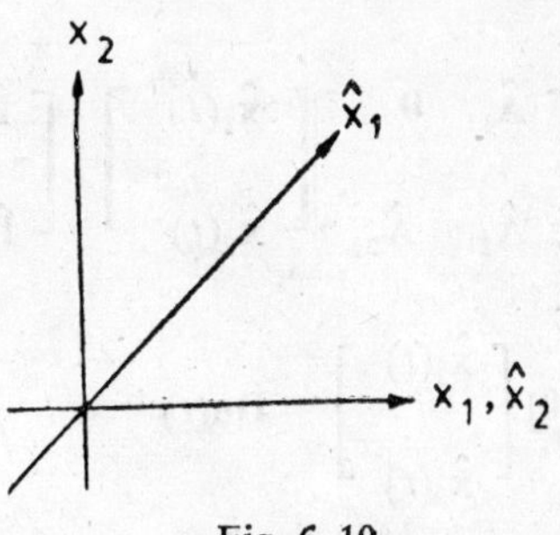

Fig. 6.10

Unobservable Subspace

We know that if a system is not completely observable, it is not possible to establish uniquely from the output what the state of the system is. It is of interest to know exactly what uncertainty remains.

The unobservable subspace of linear time-invariant system (6.4) is the linear subspace consisting of the state $\mathbf{x}^0$ for which $\mathbf{y}(t) = \mathbf{0}$ for $t \geqslant 0$.

For the system (6.4) with $\mathbf{u}(t) = \mathbf{0}$, we have from eqn. (6.6)

$$\mathbf{y}(t) = \mathbf{C}e^{\mathbf{A}t}\,\mathbf{x}^0$$

Multiplying both sides by $e^{\mathbf{A}^T t}\,\mathbf{C}^T$ and integrating between 0 and t_1, we get

$$\int_0^{t_1} e^{\mathbf{A}^T t}\,\mathbf{C}^T\,\mathbf{y}(t)dt = \int_0^{t_1} e^{\mathbf{A}^T t}\,\mathbf{C}^T\mathbf{C}e^{\mathbf{A}t}\,\mathbf{x}^0\,dt$$

$$= \mathbf{V}(0, t_1)\,\mathbf{x}^0$$

$\mathbf{V}(0, t_1)\,\mathbf{x}^0 = \mathbf{0}$ if $\mathbf{x}^0$ belongs to null space of $\mathbf{V}(0, t_1)$ (Section 2.6) or null space of the observability matrix (Theorem 6.5)

$$\mathbf{V} = \begin{bmatrix} \mathbf{C} \\ \mathbf{CA} \\ \vdots \\ \mathbf{CA}^{n-1} \end{bmatrix}$$

Thus the unobservable subspace of system (6.4) is the null space of its observability matrix $\mathbf{V}$ and the observable subspace of this system is the range space of $\mathbf{V}$.

Observability Canonical Form

Theorem 6.14: Consider the n-dimensional linear time-invariant system (6.4). If the observability matrix of this system has rank $m\,(< n)$, then there exists an equivalent transformation $\hat{\mathbf{x}} = \mathbf{P}\mathbf{x}$, where $\mathbf{P}$ is a constant nonsingular matrix, which transforms (6.4) into *observability canonical form*

$$\begin{bmatrix} \dot{\hat{\mathbf{x}}}_1(t) \\ \dot{\hat{\mathbf{x}}}_2(t) \end{bmatrix} = \begin{bmatrix} \hat{\mathbf{A}}_{11} & \mathbf{0} \\ \hat{\mathbf{A}}_{21} & \hat{\mathbf{A}}_{22} \end{bmatrix} \begin{bmatrix} \hat{\mathbf{x}}_1(t) \\ \hat{\mathbf{x}}_2(t) \end{bmatrix} + \begin{bmatrix} \hat{\mathbf{B}}_1 \\ \hat{\mathbf{B}}_2 \end{bmatrix} \mathbf{u}(t) \tag{6.80a}$$

$$\mathbf{y}(t) = [\hat{\mathbf{C}}_1 \quad \mathbf{0}] \begin{bmatrix} \hat{\mathbf{x}}_1(t) \\ \hat{\mathbf{x}}_2(t) \end{bmatrix} + \mathbf{D}\mathbf{u}(t) \tag{6.80b}$$

and the m-dimensional subsystem

$$\dot{\hat{\mathbf{x}}}_1(t) = \hat{\mathbf{A}}_{11}\,\hat{\mathbf{x}}_1(t) + \hat{\mathbf{B}}_1\mathbf{u}(t) \tag{6.81a}$$

$$\mathbf{y}(t) = \hat{\mathbf{C}}_1\hat{\mathbf{x}}_1(t) + \mathbf{D}\mathbf{u}(t) \tag{6.81b}$$

is observable.

Proof:

The observability matrix of (6.4) is

$$\mathbf{V} = \begin{bmatrix} \mathbf{C} \\ \mathbf{CA} \\ \vdots \\ \mathbf{CA}^{n-1} \end{bmatrix}$$

V has rank m, i.e., **V** possesses m linearly independent rows which span the observable subspace of (6.4). Let the row vectors $\mathbf{p}_1, \mathbf{p}_2, \ldots, \mathbf{p}_m$ be a basis for this subspace. Furthermore let $\mathbf{p}_{m+1}, \mathbf{p}_{m+2}, \ldots, \mathbf{p}_n$ be $(n-m)$ linearly independent row vectors which together with $\mathbf{p}_1, \mathbf{p}_2, \ldots, \mathbf{p}_m$ span the whole n-dimensional space. Define a nonsingular matrix

$$\mathbf{P} = \begin{bmatrix} \mathbf{P}_1 \\ \mathbf{P}_2 \end{bmatrix} \tag{6.82}$$

where

$$\mathbf{P} = \begin{bmatrix} \mathbf{p}_1 \\ \mathbf{p}_2 \\ \vdots \\ \mathbf{p}_m \end{bmatrix}; \quad \mathbf{P}_2 = \begin{bmatrix} \mathbf{p}_{m+1} \\ \mathbf{p}_{m+2} \\ \vdots \\ \mathbf{p}_n \end{bmatrix}$$

We now introduce a transformed state vector defined by

$$\hat{\mathbf{x}}(t) = \mathbf{P}\mathbf{x}(t)$$

Substituting this into (6.4) we obtain

$$\dot{\hat{\mathbf{x}}}(t) = \mathbf{PAP}^{-1}\hat{\mathbf{x}}(t) + \mathbf{PB}\mathbf{u}(t) \tag{6.83a}$$

$$\mathbf{y}(t) = \mathbf{CP}^{-1}\hat{\mathbf{x}}(t) + \mathbf{D}\mathbf{u}(t) \tag{6.83b}$$

On the lines of proof of Theorem 6.13, it can easily be established that

$$\mathbf{PAP}^{-1} = \begin{bmatrix} \mathbf{P}_1\mathbf{AQ}_1 & \mathbf{0} \\ \mathbf{P}_2\mathbf{AQ}_1 & \mathbf{P}_2\mathbf{AQ}_2 \end{bmatrix}; \tag{6.84a}$$

$$\mathbf{CP}^{-1} = [\mathbf{CQ}_1 \quad \mathbf{0}] \tag{6.84b}$$

where

$$\mathbf{P}^{-1} = [\mathbf{Q}_1 \quad \mathbf{Q}_2]; \tag{6.84c}$$

$\mathbf{Q}_1$ has m and $\mathbf{Q}_2$ has $(n-m)$ columns.

The second part of the theorem, the observability of subsystem (6.81), can also be proved easily.

□

Figure 6.11 schematically shows the decomposition of state into observable and unobservable components. We note that nothing about $\hat{\mathbf{x}}_2$ can be inferred by observing the output $\mathbf{y}(t)$.

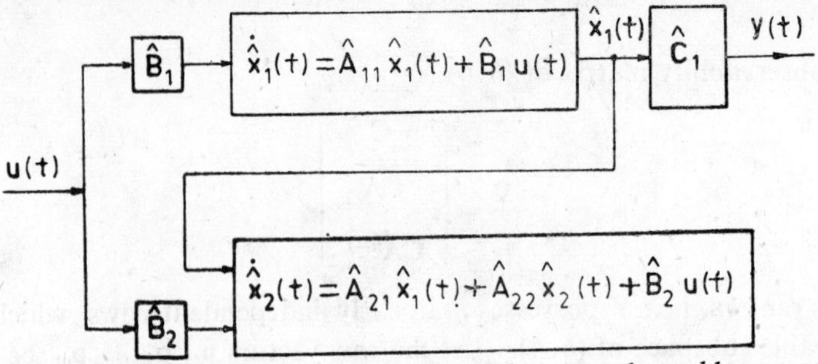

Fig. 6.11 Decomposition of state into observable and unobservable components.

It should also be noted that the observability canonical form is not unique. However, for any **P**, the eigenvalues of $\hat{\mathbf{A}}_{11}$ and $\hat{\mathbf{A}}_{22}$ are always the same, eigenvalues of $\hat{\mathbf{A}}_{11}$ are referred to as *observable eigenvalues* and those of $\hat{\mathbf{A}}_{22}$ as *unobservable eigenvalues.*

Example 6.17: Reconsider system of Example 6.3. A state model for circuit of Fig. 4.4a is

$$\dot{\mathbf{x}}(t)=\begin{bmatrix} -2 & 1 \\ 1 & -2 \end{bmatrix}\mathbf{x}(t)+\begin{bmatrix} 1 \\ 0 \end{bmatrix}u(t)$$

$$=\mathbf{Ax}(t)+\mathbf{b}u(t)$$

$$y(t)=[1 \quad -1]\,\mathbf{x}(t)$$

$$=\mathbf{cx}(t).$$

The observability matrix

$$\mathbf{V}=\begin{bmatrix} 1 & -1 \\ -3 & 3 \end{bmatrix}$$

$\rho(\mathbf{V})=1$; the system is therefore unobservable. This fact was earlier proved in Example 6.3.

The equivalence transformation **P** can be chosen as

$$\mathbf{P}=\begin{bmatrix} 1 & -1 \\ 1 & 0 \end{bmatrix};\ \mathbf{P}^{-1}=\begin{bmatrix} 0 & 1 \\ -1 & 1 \end{bmatrix}$$

If we let $\hat{\mathbf{x}}(t)=\mathbf{Px}(t)$, then we get the state model in observability canonical form

$$\begin{bmatrix} \dot{\hat{x}}_1(t) \\ \dot{\hat{x}}_2(t) \end{bmatrix} = \begin{bmatrix} -3 & 0 \\ -1 & -1 \end{bmatrix} \begin{bmatrix} \hat{x}_1(t) \\ \hat{x}_2(t) \end{bmatrix} + \begin{bmatrix} 1 \\ 1 \end{bmatrix} u(t)$$

$$y(t) = [1 \quad 0] \begin{bmatrix} \hat{x}_1(t) \\ \hat{x}_2(t) \end{bmatrix}$$

In the transformation $\mathbf{P}^{-1}\hat{\mathbf{x}} = \mathbf{x}$, the basis of the state space changes from the set of orthonormal vectors $\begin{bmatrix} 1 \\ 0 \end{bmatrix}, \begin{bmatrix} 0 \\ 1 \end{bmatrix}$ to the set of vectors $\begin{bmatrix} 0 \\ -1 \end{bmatrix}, \begin{bmatrix} 1 \\ 1 \end{bmatrix}$.

The state variable $\hat{x}_2$ is unobservable. This fact was established in Example 6.3: the initial state with $x_1^0 = x_2^0$ produces no response at the output.

Canonical Decomposition Theorem

In the following, we give a simplified version of the canonical decomposition theorem (For general form, see Kalman (1962, 1963)).

Theorem 6.15: Consider the n-dimensional linear time-invariant system (6.4). By equivalence transformation, (6.4) can be transformed into the following canonical form

$$\begin{bmatrix} \dot{\bar{\mathbf{x}}}_1 \\ \dot{\bar{\mathbf{x}}}_2 \\ \dot{\bar{\mathbf{x}}}_3 \end{bmatrix} = \begin{bmatrix} \bar{\mathbf{A}}_{11} & \bar{\mathbf{A}}_{12} & \bar{\mathbf{A}}_{13} \\ \mathbf{0} & \bar{\mathbf{A}}_{22} & \bar{\mathbf{A}}_{23} \\ \mathbf{0} & \mathbf{0} & \bar{\mathbf{A}}_{33} \end{bmatrix} \begin{bmatrix} \bar{\mathbf{x}}_1 \\ \bar{\mathbf{x}}_2 \\ \bar{\mathbf{x}}_3 \end{bmatrix} + \begin{bmatrix} \bar{\mathbf{B}}_1 \\ \bar{\mathbf{B}}_2 \\ \mathbf{0} \end{bmatrix} \mathbf{u} \qquad (6.85a)$$

$$\mathbf{y} = [\mathbf{0} \quad \bar{\mathbf{C}}_2 \quad \bar{\mathbf{C}}_3]\,\bar{\mathbf{x}} + \mathbf{D}\mathbf{u} \qquad (6.85b)$$

where the vector $\bar{\mathbf{x}}_1$ is controllable but not observable, $\bar{\mathbf{x}}_2$ is controllable and observable and $\bar{\mathbf{x}}_3$ is not controllable.

Proof:

Consider the system (6.4). If the system is not controllable, it can be transformed into the form (6.74).

Now consider the controllable subsystem in (6.74), i.e., $(\hat{\mathbf{A}}_{11}, \hat{\mathbf{B}}_1, \hat{\mathbf{C}}_1, \mathbf{D})$. If this subsystem is not observable, it can be transformed into the form (6.80). Combining these two transformations, we immediately obtain (6.85).

□

The canonical decomposition theorem may be illustrated symbolically as in Fig. 6.12, in which the uncontrollable part is further decomposed into observable and unobservable parts, We shall see in Chapter 7 that the input-output description of the system depends solely on the subsystem which is both controllable and observable.

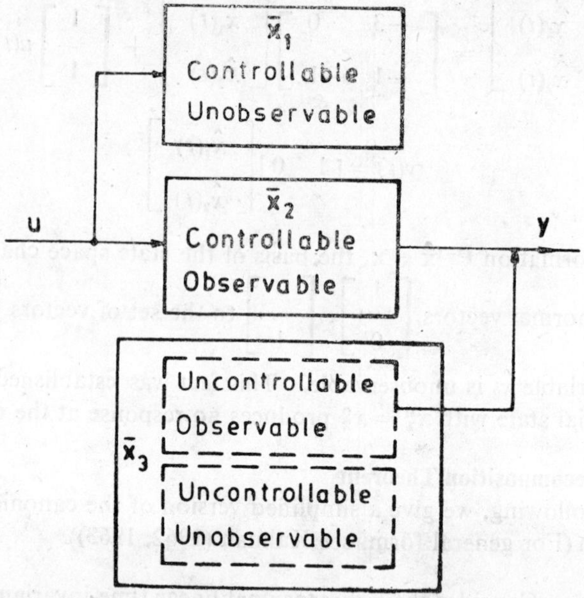

Fig. 6.12 Canonical decomposition of a linear time-invariant system.

6.10 CONCLUDING COMMENTS

In this chapter, we have examined two fundamental aspects of linear systems. The first of these was controllability which concerns the amount of control one can exercise over the state of a system from the input. Theorem 6.1 is an important result on controllability of linear time-varying continuous-time systems. It is derived with the least assumption of continuity. Additional assumptions of continuous differentiability and analyticity yield simpler tests of controllability (Chang 1965, Weiss 1965, Silvermann and Meadows 1967, Chen 1970). Statement (d) in Theorem 6.3 gives an important test for controllability of linear time-invariant differential systems: for the system to be controllable, its controllability matrix must be of full rank. Transformation to the Jordan canonical form gives some additional information about the system in addition to the controllability property but suffers from the limitation of time-consuming computation. All the results on controllability of differential systems were extended to discrete-time systems. An important result that was established was that improper choice of sampling period may destroy the controllability and observability of a system. For further reading, reference may be made to Chen and Desoer (1967) for controllability of composite systems and Lin (1974), Shields and Pearson (1976) and Davison (1977) for structural controllability.

The other fundamental aspect of linear systems discussed in this chapter is observability. It concerns the extent to which the state of a system influences the output. The theorems on observability were derived from those on controllability using the principle of duality.

In Section 6.9, the underlying structure of a linear system with regard to controllability and observability properties was explored, resulting in the canonical decomposition set out in Theorem 6.15.

Several selected optimization problems using the concepts of controllability and observability have also been investigated in this chapter. The deadbeat control problem involved the solution of system of linear equations. Lapidus and Luus (1967) solve the problem using linear programming.

The reader is reminded that only a very limited number of design techniques have been treated here. A more extensive treatment of optimal design techniques will follow in later chapters.

PROBLEMS

6.1 Investigate the controllability and observability of the following systems.

(a) $$\dot{\mathbf{x}}(t) = \begin{bmatrix} 1 & 0 \\ 0 & 2t \end{bmatrix} \mathbf{x}(t) + \begin{bmatrix} 0 \\ 1 \end{bmatrix} u(t)$$

$$y(t) = [0 \quad 1]\,\mathbf{x}(t)$$

(b) $$\mathbf{x}(k+1) = \begin{bmatrix} 1 & k \\ 0 & -1 \end{bmatrix} \mathbf{x}(k) + \begin{bmatrix} 0 \\ 1 \end{bmatrix} u(k)$$

$$y(k) = [1 \quad 1]\,\mathbf{x}(k)$$

6.2 Consider the system

$$\dot{\mathbf{x}}(t) = \begin{bmatrix} 1 & e^{-t} \\ 0 & -1 \end{bmatrix} \mathbf{x}(t)$$

$$y(t) = [1 \quad 1]\,\mathbf{x}(t)$$

Is this system observable at $t = 0$? If yes, find

$\mathbf{x}(0)$ when $y(t) = e^t$.

6.3 Consider the system

$$\mathbf{x}(k+1) = \begin{bmatrix} 1 & k \\ 0 & -1 \end{bmatrix} \mathbf{x}(k) + \begin{bmatrix} 2 \\ k \end{bmatrix} u(k)$$

Is this system controllable at $k = 1$? If yes, find a control sequence to drive the system from

$$\mathbf{x}(0) = \begin{bmatrix} 0 \\ 0 \end{bmatrix} \text{ to } \mathbf{x}(2) = \begin{bmatrix} 1 \\ 1 \end{bmatrix}.$$

6.4 Consider the Mixing Tank system discussed in Section 4.6. Show that the system is completely controllable.

6.5 Show that the pair {A, C} is completely observable for all values of α_i's.

$$\mathbf{A} = \begin{bmatrix} 0 & 0 & 0 \dots 0 & -\alpha_n \\ 1 & 0 & 0 \dots 0 & -\alpha_{n-1} \\ 0 & 1 & 0 \dots 0 & -\alpha_{n-2} \\ \vdots & \vdots & \vdots \quad \vdots & \vdots \\ 0 & 0 & 0 \dots 0 & -\alpha_2 \\ 0 & 0 & 0 \dots 1 & -\alpha_1 \end{bmatrix}$$

$$\mathbf{C} = [0 \quad 0 \dots 0 \quad 1]$$

6.6 Show that the pair {A, b} is completely state controllable for all values of α_i's

$$\mathbf{A} = \begin{bmatrix} 0 & 1 & 0 & \dots & 0 \\ 0 & 0 & 1 & \dots & 0 \\ \vdots & \vdots & \vdots & & \vdots \\ 0 & 0 & 0 & \dots & 1 \\ -\alpha_n & -\alpha_{n-1} & -\alpha_{n-2} & \dots & -\alpha_1 \end{bmatrix}; \quad \mathbf{b} = \begin{bmatrix} 0 \\ 0 \\ \vdots \\ 0 \\ 1 \end{bmatrix}$$

6.7 Given the system

$$\dot{\mathbf{x}}(t) = \begin{bmatrix} -1 & 1 & 0 \\ 0 & -1 & 0 \\ 0 & 0 & -2 \end{bmatrix} \mathbf{x}(t) + \begin{bmatrix} 0 & 1 \\ 2 & 0 \\ 0 & 1 \end{bmatrix} \mathbf{u}(t)$$

$$\mathbf{y}(t) = \begin{bmatrix} 0 & 1 & 2 \\ 0 & 1 & 0 \end{bmatrix} \mathbf{x}(t)$$

What can we say about controllability and observability without making any further calculations?

6.8 Convert the following state model into the Jordan canonical form and therefrom comment on controllability and observability.

$$\dot{\mathbf{x}}(t) = \begin{bmatrix} 0 & 1 & 0 \\ 0 & 0 & 1 \\ -2 & -4 & -3 \end{bmatrix} \mathbf{x}(t) + \begin{bmatrix} 1 & 0 \\ 0 & 1 \\ -1 & 1 \end{bmatrix} \mathbf{u}(t)$$

$$\mathbf{y}(t) = \begin{bmatrix} 0 & 1 & -1 \\ 1 & 2 & 1 \end{bmatrix} \mathbf{x}(t)$$

6.9 The motion of a satellite in the equatorial (r, θ) plane is given by the state equation (Fortmann and Hitz 1977)

$$\begin{bmatrix} \dot{x}_1(t) \\ \dot{x}_2(t) \\ \dot{x}_3(t) \\ \dot{x}_4(t) \end{bmatrix} = \begin{bmatrix} 0 & 1 & 0 & 0 \\ 3\omega^2 & 0 & 0 & 2\omega \\ 0 & 0 & 0 & 1 \\ 0 & -2\omega & 0 & 0 \end{bmatrix} \begin{bmatrix} x_1(t) \\ x_2(t) \\ x_3(t) \\ x_4(t) \end{bmatrix} + \begin{bmatrix} 0 & 0 \\ 1 & 0 \\ 0 & 0 \\ 0 & 1 \end{bmatrix} \begin{bmatrix} u_1 \\ u_2 \end{bmatrix}$$

where ω is the angular frequency of the satellite in circular, equatorial orbit, $x_1(t)$ and $x_3(t)$ are, respectively, the deviations in position variables $r(t)$ and $\theta(t)$ of the satellite, $x_2(t)$ and $x_4(t)$ are, respectively, the deviations in velocity variables $\dot{r}(t)$ and $\dot{\theta}(t)$. The inputs u_1 and u_2 are the thrusts u_r and u_θ in the radial and tangential directions respectively, applied by small rocket engines or gas jets ($\mathbf{u} = \mathbf{0}$ when $\mathbf{x} = \mathbf{0}$).

(a) Prove that the system is completely controllable.

(b) Suppose the tangential thruster becomes inoperable ($u_2 = 0$). Prove that the satellite is not completely controllable with the radial thruster alone.

(c) Suppose the radial thruster becomes inoperable ($u_1 = 0$). Prove that the satellite is completely controllable with tangential thruster alone.

6.10 Reconsider the satellite system of Problem 6.9.

(a) Prove that the system is completely observable from radial ($x_1 = r$) and tangential ($x_3 = \theta$) measurements.

(b) Suppose that the tangential measuring device becomes inoperable. Prove that the state is not completely observable from radial measurements only.

(c) Suppose the radial measurements are lost. Prove that the state is completely observable from tangential measurements only.

6.11 Consider the system

$$\dot{\mathbf{x}}(t) = \begin{bmatrix} -1 & 0 \\ 0 & -2 \end{bmatrix} \mathbf{x}(t) + \begin{bmatrix} 1 \\ 1 \end{bmatrix} u(t)$$

$$y(t) = [1 \quad 1]\,\mathbf{x}(t)$$

(a) Find, if possible, a control law, which will drive the system from

$$\mathbf{x}(0) = \mathbf{0} \text{ to } \mathbf{x}^1 = \begin{bmatrix} 1 \\ 1 \end{bmatrix}$$

(b) Find, if possible, the state $\mathbf{x}(0)$ when $y(t)=\frac{1}{2}e^{-2t}+\frac{3}{2}$ for $u(t)=1$, $t>0$.

6.12 Consider the system

$$\dot{\mathbf{x}}(t)=\begin{bmatrix} 0 & 1 & 0 \\ 0 & 0 & 1 \\ -2 & -5 & -4 \end{bmatrix}\mathbf{x}(t)+\begin{bmatrix} 0 & 1 \\ 1 & 0 \\ 1 & 1 \end{bmatrix}\mathbf{u}(t)$$

Find, if possible, a control law to transfer the state from

$$\mathbf{x}(1)=\begin{bmatrix} 1 \\ 0 \\ 0 \end{bmatrix} \text{ to } \mathbf{x}(4)=\begin{bmatrix} 3 \\ 2 \\ 0 \end{bmatrix}$$

6.13 Consider the system

$$\dot{\mathbf{x}}(t)=\begin{bmatrix} 0 & 1 \\ 2 & 1 \end{bmatrix}\mathbf{x}(t)+\begin{bmatrix} 1 \\ 0 \end{bmatrix}u(t)$$

$$y(t)=[1 \quad 2]\,\mathbf{x}(t)$$

(a) Show that the system modes are e^{-t} and e^{2t}.

(b) Is it possible to find a set of initial conditions at $t=t_0$ such that the mode e^{2t} is suppressed in $y(t)$? If yes, find $\mathbf{x}(t_0)$ to do this $(u=0)$.

(c) Is it possible to choose an input $u_{[0,\ t_0]}$ that transfers $\mathbf{x}(0)$ to $\mathbf{x}(t_0)$? If yes, find such a control.

6.14 Show that the state of an unforced observable n-dimensional linear time-invariant system can be determined instantaneously from the output and its derivatives up to $(n-1)$th order.

6.15 Prove that if the rank of $\mathbf{B}$ is r, then the state equation

$$\dot{\mathbf{x}}(t)=\mathbf{A}\mathbf{x}(t)+\mathbf{B}\mathbf{u}(t);\ \mathbf{x}\in\mathcal{R}^n,\ \mathbf{u}\in\mathcal{R}^p$$

is controllable if and only if (refer Chen (1970))

$$\rho\,[\mathbf{B} \mid \mathbf{A}\mathbf{B} \mid \dots \mid \mathbf{A}^{n-r}\mathbf{B}]=n$$

6.16 Prove that if the rank of $\mathbf{C}$ is r, then the state model

$$\dot{\mathbf{x}}(t)=\mathbf{A}\mathbf{x}(t)+\mathbf{B}\mathbf{u}(t);\ \mathbf{x}\in\mathcal{R}^n,\ \mathbf{u}\in\mathcal{R}^p$$

$$\mathbf{y}=\mathbf{C}\mathbf{x}(t)\qquad ;\ \mathbf{y}\in\mathcal{R}^q$$

is observable if and only if (refer Chen (1970))

$$\rho \begin{bmatrix} \mathbf{C} \\ \mathbf{CA} \\ \vdots \\ \mathbf{CA}^{n-r} \end{bmatrix} = n$$

6.17 Assume that the satellite of Problem 6.9 is a synchronous one with a period of 1 day; thus $\omega = 2\pi$. Obtain the system equations if sampling and control take place twice per orbit ($T = 1/2$ day). Show that the sampled system is controllable.

Design a minimum-time controller for this system which transfers any nonzero initial state $\mathbf{x}^0$ to zero in the minimum number of steps.

6.18 Consider the system

$$\mathbf{x}(k+1) = \begin{bmatrix} -2 & -1 \\ -1 & -2 \end{bmatrix} \mathbf{x}(k) + \begin{bmatrix} 0 \\ 1 \end{bmatrix} u(k)$$

$$y(k) = [1 \quad -1]\ \mathbf{x}(k)$$

(a) Find, if possible, a control sequence $\{u(0), u(1)\}$ to derive the system from

$$\mathbf{x}(0) = \begin{bmatrix} 1 \\ 0 \end{bmatrix} \text{ to } \mathbf{x}(2) = \begin{bmatrix} -1 \\ 0 \end{bmatrix}$$

(b) Find, if possible, the state $\mathbf{x}(0)$ when

$$y(0) = 1,\ y(1) = 0,\ y(2) = -1;\ y(3) = 2 \text{ for } u(k) = (-1)^k;\ k \geqslant 0.$$

6.19 Consider the system

$$\dot{\mathbf{x}}(t) = \begin{bmatrix} -1 & 0 \\ 1 & 0 \end{bmatrix} x(t) + \begin{bmatrix} 1 \\ 0 \end{bmatrix} u(t)$$

Discretize the state equation for $T = 1$ sec. Find initial states that can be transferred to zero in 1 sec.

6.20 For the system of Problem 6.19, find suitable number of steps N to transfer

$$\mathbf{x}(0) = \begin{bmatrix} -4 \\ 4 \end{bmatrix} \text{ to origin under the constraint } |u(k)| < 1.$$

6.21 A linear system is described by the equation

$$\dot{\mathbf{x}}(t) = \begin{bmatrix} 0 & 1 \\ -1 & 0 \end{bmatrix} \mathbf{x}(t) + \begin{bmatrix} 0 \\ 1 \end{bmatrix} u(t)$$

Discretize the equation for $T = \pi/2$ sec. Find optimum strategy for $u(k)$ in terms of $\mathbf{x}(k)$ such that the system can be brought to the equilibrium state **0** from any **x**(0) in a minimum number of sampling periods (refer Kuo (1970)).

6.22 Reconsider the linear system of Problem 6.21. Determine the values of the sampling period T which make the system uncontrollable.

6.23 Reconsider Problem 6.17. Prove that if sampling period $T = 1$ day, the sampled system is uncontrollable and unobservable while the original continuons-time system is both controllable and observable.

6.24 In Problem 6.9 we have proved that the satellite is not completely controllable with radial thruster alone. Find the subspace of controllable states.

6.25 In Problem 6.10 we have proved that the satellite state is not completely observable from radial measurements alone. Find the subspace of unobservable states.

REFERENCES

1. Cadzow, J.A., and H.R. Martens, *Discrete-Time and Computer Control Systems*, Englewood Cliffs, N.J. Prentice-Hall, 1968.
2. Chang, A., "An algebraic characterization of controllability", *IEEE Trans. Automat. Contr.*, Vol. AC-10, pp. 112–113, 1965.
3. Chen, C.T., *Introduction to Linear System Theory*, New York: Holt, Rinehart and Winston, 1970.
4. Chen, C.T., and C.A. Desoer, "Controllability and observability of composite systems," *IEEE Trans. Automat. Contr.*, Vol. AC-12, pp. 402–409, 1967.
5. Chen, C.T., and C.A. Desoer, "A proof of controllability of Jordan form state equations," *IEEE Trans. Automat. Contr.*, Vol. AC-13, pp. 195–196, 1968.
6. Chidambara, M.R., and C.H. Wells, "State variable determination for digital control," *IEEE Trans Automat. Contr.*, Vol. AC-11, p. 326, 1966.
7. Davison, E.J., "Connectability and structural controllability of composite systems," *Automatica*, Vol. 13, pp. 109–123, 1977.
8. Desoer, A., and J. Wing, "An optimal strategy for a saturating sampled-data system," *IRE Trans. Automat. Contr.*, Vol. AC-6, pp. 5–15, 1961a.
9. Desoer, A., and J. Wing, "A minimal time discrete system," *IRE Trans. Automat. Contr.*, Vol. AC-6, pp. 111–125, 1961b.
10. Desoer, A., and J. Wing, "The minimum time regulator problem for linear sampled-data systems: General Theory," *J. Franklin Inst.*, Vol. 272, pp. 208–228, 1961c.
11. Fortmann, T E., and K.L. Hitz, *An Introduction to Linear Control Systems*, New York: Marcel Dekker, 1977.
12. Gilbert, E.G., "Controllability and observability in multivariable control systems," *SIAM J. Control*, Vol. 1, pp. 128–151, 1963.
13. Ho, Y.C., "What constitutes a controllable system?" *IRE Trans. Automat. Contr.*, Vol. AC-7, no. 3, p. 76, 1962.
14. Kailath, T., *Linear Systems*, Englewood Cliffs, N.J.: Prentice-Hall, 1980.
15. Kalman, R.E., "On the general theory of control systems," *Proc. First Intern. Congr. Autom. Control*, Butterworth, London, pp. 481-493, 1960.
16. Kalman, R.E., "Canonical structure of linear dynamical systems," *Proc. Natl. Acad. Sci. U.S* Vol. 48, no. 4, pp. 596–600, 1962.
17. Kalman, R.E., "Mathematical description of linear dynamical systems," *SIAM J. Control*, Vol. 1, pp. 152–192, 1963.
18. Kalman, R.E., Y.C. Ho, and K.S. Narendra, "Controllability of linear dynamical systems," *Contrib. Differential Equations*, Vol. 1, pp. 189–213, 1962.

19. Kreindler, A., and P.E. Sarachik, "On the concepts of controllability of linear systems," *IEEE Trans. Automat. Contr.*, Vol. AC-9, pp. 129–136, **1964.**
20. Kuo, B.C., *Discrete-Data Control Systems*, Englewood Cliffs; N.J.: Prentice-Hall, **1970.**
21. Kwakernaak, H., and R. Sivan, *Linear Optimal Control Systems*, New York: Wiley, **1972.**
22. Lapidus, L., and R. Luus, ***Optimal Control of Engineering Processes***, Walthan, Mass: Blaisdell, 1967.
23. Lee, E.B., "On the domain of controllability of linear systems subject to control amplitude constraints," *IRE Trans. Automat. Contr.*, Vol. AC-8, pp. 172–173, **1963.**
24. Lee, E.B., and L. Markus, *Foundations of Optimal Control Theory*, New York: Wiley, **1967.**
25. Lin C.T., "Structural controllability," *IEEE Trans. Automat. Contr.*, Vol. AC-19, pp. 201–208, **1974.**
26. Ogata, K., *State Space Analysis of Control Systems*, Englewood Cliffs, N.J.: Prentice-Hall, **1967.**
27. Podulo, L., and M.A. Arbib, *System Theory*, Philadelphia: Saunders, **1974.**
28. Shields, R.W., and J.B. Pearson, "Structural controllability of multi-input linear systems," *IEEE Trans. Automat. Contr.*, Vol. AC-21, pp. 203–212, **1976.**
29. Silverman, L.M., and H.E. Meadows. "Controllability and observability in time-variable linear systems," *SIAM J. Control*, Vol. 5, pp. 64–73, **1967.**
30. Swisher, G.M., ***Introduction to Linear Systems Analysis***, Champaign, IL.: Matrix, **1976.**
31. Weiss, L., "The concepts of differential controllability and differential observability" *J. Math. Anal. Appl.*, Vol. 10, pp. 442–449, **1965.**
32. Weiss, L., "On the structure theory of linear differential systems" *SIAM J. Control,* Vol. 6, pp. 659–680, **1968.**

7. RELATIONSHIP BETWEEN STATE VARIABLE AND INPUT-OUTPUT DESCRIPTIONS

7.1 INTRODUCTION

So far we have discussed the state variable approach for the mathematical description of dynamical systems almost exclusively. In this chapter, we shall review the basic ideas behind the so-called *input-output approach.* We assume that the reader has encountered this method earlier. For this reason, we shall not elaborate upon this method extensively; basic ideas and definitions will be presented to the extent needed to establish the correspondence between the input-output method and the state variable method. In particular, we shall answer the following questions:

1. How can we go from a state variable to input-output representation? Is this transformation unique?
2. How can we go from an input-output to state variable representation? Is this transformation unique?

7.2 INPUT-OUTPUT MAPS FROM STATE MODELS

Linear Time-Invariant Continuous-Time Systems

Impulse Response Matrix: Consider first an n-dimensional linear time-invariant single-input/single-output system described by the equations

$$\dot{\mathbf{x}}(t) = \mathbf{A}\mathbf{x}(t) + \mathbf{b}u(t) \tag{7.1a}$$

$$y(t) = \mathbf{c}\mathbf{x}(t) \tag{7.1b}$$

where $\mathbf{x}$ is the $n \times 1$ state vector, $\mathbf{u}$ is the scalar input, y is the scalar output, $\mathbf{A}$ is $n \times n$ real constant matrix and $\mathbf{b}$ and $\mathbf{c}$ are respectively $n \times 1$ and $1 \times n$ real constant vectors. The time interval of interest is $[0, \infty)$.

If the system is initially in zero state, i.e., $\mathbf{x}(0) = \mathbf{0}$, the output (eqn. (6.6))

$$y(t) = \mathbf{c}\, e^{\mathbf{A}t} \int_0^t e^{-\mathbf{A}\tau}\mathbf{b}u(\tau)\, d\tau \tag{7.2}$$

where the initial time t_0 is taken as zero for convenience. Assume that at $t = 0$, the input to the system is a signal in the form of a pulse of short duration Δ (Fig. 7.1). Note that the properties of the pulse $\delta_\Delta(t)$ are that

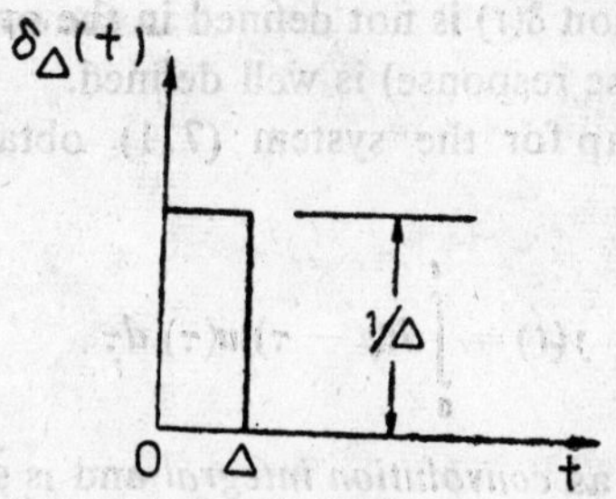

Fig. 7.1

its height increases whenever its duration decreases but in such a way that the 'area' thereby remains constant and is equal to unity, i.e.,

$$\begin{aligned} &\lim_{\Delta\to 0} \delta_\Delta(t) = 0, \quad t \neq 0 \\ &\lim_{\Delta\to 0} \delta_\Delta(t) = \infty, \quad t = 0 \\ &\lim_{\Delta\to 0} \int_{-\infty}^{\infty} \delta_\Delta(t)\, dt = 1 \end{aligned} \tag{7.3}$$

In response to input $\delta_\Delta(t)$, the system output

$$\begin{aligned} y(t) &= \mathbf{c}e^{\mathbf{A}t} \int_0^t e^{-\mathbf{A}\tau}\mathbf{b}\delta_\Delta(\tau)\, d\tau \\ &= \mathbf{c}e^{\mathbf{A}t} \frac{1}{\Delta} \int_0^\Delta e^{-\mathbf{A}\tau}\mathbf{b}\, d\tau \\ &= \mathbf{c}e^{\mathbf{A}t} \frac{1}{\Delta} \int_0^\Delta \left[\mathbf{I} + \mathbf{A}(-\tau) + \frac{\mathbf{A}^2(-\tau)^2}{2!} + \dots\right] \mathbf{b}\, d\tau \end{aligned}$$

Proceeding to the limit we obtain

$$\lim_{\Delta\to 0} y(t) = h(t) = \mathbf{c}\, e^{\mathbf{A}t}\mathbf{b}$$

The 'at the limit' value of function $\delta_\Delta(t)$ in (7.3) is called *unit impulse function* and is denoted as $\delta(t)$. The scalar function

$$\begin{aligned} h(t) &= \mathbf{c}\, e^{\mathbf{A}t}\mathbf{b} \quad ; \quad t \geqslant 0 \\ &= 0 \quad ; \quad t < 0 \end{aligned} \tag{7.4}$$

is called the *impulse response* of the system (7.1).

Note that the function $\delta(t)$ is not defined in the ordinary sense[1]; however the limit in (7.4) (impulse response) is well defined.

An input-output map for the system (7.1) obtained from eqns. (7.2) and (7.4) is

$$y(t) = \int_0^t h(t-\tau)\, u(\tau)\, d\tau \tag{7.5a}$$

This equation is known as *convolution integral* and is sometimes denoted as

$$y = h * u \tag{7.5b}$$

The impulse response $h(t)$ of a system characterizes its input-output behaviour via (7.5) and this behaviour is independent of the particular state vector chosen to represent the system dynamics. Thus $h(t)$ is invariant under a state transformation. To prove this, let us consider a system

$$\dot{\bar{\mathbf{x}}}(t) = \mathbf{PAP}^{-1}\bar{\mathbf{x}}(t) + \mathbf{Pb}u(t) \tag{7.6a}$$

$$y(t) = \mathbf{cP}^{-1}\bar{\mathbf{x}}(t) \tag{7.6b}$$

which is related to system (7.1) by a state transformation

$$\bar{\mathbf{x}} = \mathbf{Px}$$

The impulse response of system (7.6) is given by

$$\bar{h}(t) = \mathbf{cP}^{-1} e^{(\mathbf{PAP}^{-1})t}\,\mathbf{Pb} = \mathbf{cP}^{-1}\mathbf{P}e^{\mathbf{A}t}\mathbf{P}^{-1}\mathbf{Pb}$$

$$= \mathbf{c}\, e^{\mathbf{A}t}\mathbf{b}$$

$$= h(t), \text{ the impulse response of system (7.1).}$$

Let us now add a direct transmission term to the state model (7.1):

$$\dot{\mathbf{x}}(t) = \mathbf{Ax}(t) + \mathbf{b}u(t) \tag{7.7a}$$

$$y(t) = \mathbf{cx}(t) + du(t);\ d \text{ is a scalar const} \tag{7.7b}$$

For $\mathbf{x}(0) = \mathbf{0}$, we have

---

[1]The impulse $\delta(t)$ is a *generalized function* and can be dealt with generalized calculus called the *theory of distributions*. Its definition is based on the properties:

(i) $\delta(t - t_1) = 0$ for $t \neq t_1$

(ii) $\int_{-\infty}^{\infty} f(t)\,\delta(t - t_1) = f(t_1)$ for any function f that is continuous at t_1.

For more details, refer Nagrath and Gopal (1982).

$$y(t) = \int_0^t \mathbf{c}e^{\mathbf{A}(t-\tau)}\,\mathbf{b}u(\tau)d\tau + du(t)$$

$$= \int_0^t \mathbf{c}e^{\mathbf{A}(t-\tau)}\,\mathbf{b}u(\tau)d\tau + \int_0^t d\delta(t-\tau)u(\tau)d\tau$$

$$= \int_0^t \left[\mathbf{c}e^{\mathbf{A}(t-\tau)}\,\mathbf{b} + d\delta(t-\tau)\right]u(\tau)d\tau$$

Comparing this with equation (7.5), we can write

$$\begin{aligned} h(t) &= \mathbf{c}e^{\mathbf{A}t}\,\mathbf{b} + d\delta(t);\ t \geqslant 0 \\ &= 0 \qquad\qquad ;\ t < 0 \end{aligned} \tag{7.8}$$

as impulse response of the system (7.7).

If a system has p input terminals and q output terminals, the input-output map (7.5) becomes

$$\mathbf{y}(t) = \mathbf{H}*\mathbf{u} \tag{7.9a}$$

$$= \int_0^t \mathbf{H}(t-\tau)\,\mathbf{u}(\tau)d\tau \tag{7.9b}$$

where

$$\mathbf{H}(t) = \begin{bmatrix} h_{11}(t) & h_{12}(t) & \dots & h_{1p}(t) \\ h_{21}(t) & h_{22}(t) & \dots & h_{2p}(t) \\ \vdots & \vdots & & \\ h_{q1}(t) & h_{p2}(t) & \dots & h_{qp}(t) \end{bmatrix};$$

$h_{ij}(t)$ is the response at time t at the ith output terminal due to unit impulse applied at $t = 0$ at jth input terminal; the inputs at other terminals being identically zero. Equivalently, $h_{ij}(t)$ is the impulse response between the jth input terminal and ith output terminal, $\mathbf{H}(t)$ is called the *impulse response matrix* of the multi-input/multi-output system.

For a multi-input/multi-output system described by the state model

$$\dot{\mathbf{x}}(t) = \mathbf{A}\mathbf{x}(t) + \mathbf{B}\mathbf{u}(t) \tag{7.10a}$$

$$\mathbf{y}(t) = \mathbf{C}\mathbf{x}(t) + \mathbf{D}\mathbf{u}(t) \tag{7.10b}$$

where $\mathbf{x}$ is the $n\times 1$ state vector, $\mathbf{u}$ is the $p\times 1$ input vector, $\mathbf{y}$ is the $q\times 1$ output vector and $\mathbf{A}$, $\mathbf{B}$, $\mathbf{C}$ and $\mathbf{D}$ are, respectively, $n\times n$, $n\times p$, $q\times n$ and $q\times p$ real constant matrices, the impulse response matrix is

$$\begin{aligned} \mathbf{H}(t) &= \mathbf{C}e^{\mathbf{A}t}\,\mathbf{B} + \mathbf{D}\delta(t);\ t \geqslant 0 \\ &= \mathbf{0} \qquad\qquad ;\ t < 0 \end{aligned} \tag{7.11}$$

Transfer Function Matrix: In the study of linear time-invariant systems, it is of great advantage to use the Laplace transform and z-transform techniques. In Chapter 3, we converted the state differential/difference equations into algebraic equations in the transform domain which could easily be solved using signal flow graph techniques (Chapter 5).

Taking Laplace transform of eqns. (7.10) and assuming $\mathbf{x}(0)=\mathbf{x}^0$, we obtain (eqn. (3.30b))

$$\hat{\mathbf{y}}(s)=\mathbf{C}(s\mathbf{I}-\mathbf{A})^{-1}\mathbf{x}^0+[\mathbf{C}(s\mathbf{I}-\mathbf{A})^{-1}\mathbf{B}+\mathbf{D}]\,\hat{\mathbf{u}}(s) \tag{7.12}$$

The first term on the right hand side is the natural response (zero-input response) which depends on initial state $\mathbf{x}^0$ and the second term is the forced response (zero-state response) which depends on input $\hat{\mathbf{u}}(s)$.

In the case of zero initial state, the input-output behaviour of system (7.10) is determined entirely by the matrix

$$\hat{\mathbf{H}}(s)=\mathbf{C}(s\mathbf{I}-\mathbf{A})^{-1}\mathbf{B}+\mathbf{D} \tag{7.13}$$

This matrix is called the *transfer function matrix* of system (7.10) and it has the property that the input $\hat{\mathbf{u}}(s)$ and output $\hat{y}(s)$ of (7.10) are related by

$$\hat{\mathbf{y}}(s)=\hat{\mathbf{H}}(s)\,\hat{\mathbf{u}}(s)$$

whenever $\mathbf{x}(0)=\mathbf{0}$.

Taking inverse Laplace transform of $\hat{\mathbf{H}}(s)$ in (7.13) (Appendix I), we get

$$\mathbf{H}(t)=\mathbf{C}e^{\mathbf{A}t}\,\mathbf{B}+\mathbf{D}\delta(t) \tag{7.14}$$

which is nothing but the impulse response matrix of system (7.10) as seen from (7.11). Note that Laplace transformation has changed the convolution integral (7.9) in time domain into algebraic equation (7.14) in frequency domain.

For single-input/single-output system (7.7), $\hat{\mathbf{H}}(s)$ reduces to a scalar $\hat{h}(s)$ and is called the *transfer function* of the system.

$$\hat{h}(s)=\mathbf{c}(s\mathbf{I}-\mathbf{A})^{-1}\,\mathbf{b}+d \tag{7.15}$$

Some observations would be in order at this point. In particular, we first recall that the state variable representation for a linear dynamical system specifies the internal (state) as well as external (input-output) behaviour of the system. The transfer function matrix, on the other hand, specifies the external behaviour only. Therefore, certain information which is present in state variable representation is 'lost' when a conversion to transfer function matrix is made. The 'lost' information, as we have seen, involves the initial conditions on state variables. Thus *whenever a transfer function matrix is used, the system is always implicitly assumed to be relaxed at* $t=0$.

Also note that since two equivalent state models have the identical impulse response matrix, they also have the same transfer function matrix.

We further note that the transfer function matrix $\hat{\mathbf{H}}(s)$ given by eqn. (7.13) can be written as

$$\hat{\mathbf{H}}(s) = \frac{\mathbf{C}(s\mathbf{I}-\mathbf{A})^{+}\mathbf{B}}{|s\mathbf{I}-\mathbf{A}|} + \mathbf{D}$$

$$= \frac{\mathbf{C}(s\mathbf{I}-\mathbf{A})^{+}\mathbf{B} + \mathbf{D}\,|s\mathbf{I}-\mathbf{A}|}{|s\mathbf{I}-\mathbf{A}|} \tag{7.16}$$

where $|s\mathbf{I}-\mathbf{A}|$ is the nth order characteristic polynomial of $\mathbf{A}$; $(s\mathbf{I}-\mathbf{A})^{+}$ (i.e., adjoint of $(s\mathbf{I}-\mathbf{A})$) is a matrix of cofactors of $(s\mathbf{I}-\mathbf{A})$. The cofactors of $(s\mathbf{I}-\mathbf{A})$ are determinants of $(n-1)\times(n-1)$ submatrices of $(s\mathbf{I}-\mathbf{A})$ and so they must be polynomials in s of order $(n-1)$ or less. Thus, whenever $\mathbf{D}=\mathbf{0}$, the numerator degree of each entry of $\hat{\mathbf{H}}_{ij}(s)$ of $\hat{\mathbf{H}}(s)$ is strictly less than the denominator degree and $\hat{\mathbf{H}}(s)$ is called a *strictly proper transfer function matrix.* When $\mathbf{D}\neq\mathbf{0}$, the numerator degree of each element of $\hat{\mathbf{H}}(s)$ will be less than or equal to the corresponding denominator degree and $\hat{\mathbf{H}}(s)$ is then called a *proper transfer function matrix.* Further, for any proper transfer function matrix obtained via (7.13), we have

$$\mathbf{D} = \lim_{s\to\infty} \hat{\mathbf{H}}(s) \tag{7.17}$$

□ From the observations given above, we find that $(s\mathbf{I}-\mathbf{A})^{+}$ may always be written as $n\times n$ matrix polynomial:

$$(s\mathbf{I}-\mathbf{A})^{+} = \mathbf{Q}(s) = \mathbf{Q}_1 s^{n-1} + \mathbf{Q}_2 s^{n-2} + \dots + \mathbf{Q}_{n-1}s + \mathbf{Q}_n \tag{7.18a}$$

where $\mathbf{Q}_i$ are constant matrices.

The characteristic polynomial may be expressed as

$$|s\mathbf{I}-\mathbf{A}| = \Delta(s) = s^n + \alpha_1 s^{n-1} + \dots + \alpha_{n-1}s + \alpha_n \tag{7.18b}$$

where α_i are constant scalars.

The $q\times p$ transfer function matrix, therefore, has the form

$$\hat{\mathbf{H}}(s) = \frac{\mathbf{C}\mathbf{Q}(s)\mathbf{B} + \mathbf{D}\Delta(s)}{\Delta(s)} \tag{7.19a}$$

$$= \frac{\mathbf{N}(s)}{\Delta(s)} \tag{7.19b}$$

Each entry $\hat{\mathbf{H}}_{ij}(s)$ of matrix $\hat{\mathbf{H}}(s)$ is thus a rational function of s.

Resolvent Algorithm: The coefficients of the characteristic polynomial $\Delta(s)$ and the matrix polynomial $\mathbf{Q}(s)$ in (7.18) may be determined sequentially by resolvent algorithm described below. In the literature, this algorithm is usually attributed to Leverrier, Souriau and Faddeeva. The algorithm is very convenient for digital computer programming. Melsa and Jones (1973) give a listing of FORTRAN computer program. For proof, refer Faddeeva (1959), Rosenbrock (1970) or Zadeh and Desoer (1963).

Algorithm:

$$\begin{aligned}
\mathbf{Q}_1 &= \mathbf{I} \\
\alpha_1 &= -\operatorname{tr}(\mathbf{Q}_1\mathbf{A}) = -\operatorname{tr}(\mathbf{A}\mathbf{Q}_1) \\
\mathbf{Q}_2 &= \mathbf{Q}_1\mathbf{A} + \alpha_1\mathbf{I} = \mathbf{A}\mathbf{Q}_1 + \alpha_1\mathbf{I} \\
\alpha_2 &= -\tfrac{1}{2}\operatorname{tr}(\mathbf{Q}_2\mathbf{A}) = -\tfrac{1}{2}\operatorname{tr}(\mathbf{A}\mathbf{Q}_2) \\
&\cdots\cdots\cdots\cdots\cdots\cdots\cdots\cdots\cdots\cdots \\
\mathbf{Q}_{n-1} &= \mathbf{Q}_{n-2}\mathbf{A} + \alpha_{n-2}\mathbf{I} = \mathbf{A}\mathbf{Q}_{n-2} + \alpha_2\mathbf{I} \\
\alpha_{n-1} &= -\frac{1}{n-1}\operatorname{tr}(\mathbf{Q}_{n-1}\mathbf{A}) = -\frac{1}{n-1}\operatorname{tr}(\mathbf{A}\mathbf{Q}_{n-1}) \\
\mathbf{Q}_n &= \mathbf{Q}_{n-1}\mathbf{A} + \alpha_{n-1}\mathbf{I} = \mathbf{A}\mathbf{Q}_{n-1} + \alpha_{n-1}\mathbf{I} \\
\alpha_n &= -\frac{1}{n}\operatorname{tr}(\mathbf{Q}_n\mathbf{A}) = -\frac{1}{n}\operatorname{tr}(\mathbf{A}\mathbf{Q}_n) \\
\mathbf{0} &= \mathbf{Q}_n\mathbf{A} + \alpha_n\mathbf{I} = \mathbf{A}\mathbf{Q}_n + \alpha_n\mathbf{I}
\end{aligned} \tag{7.20}$$

where the *trace* of a matrix is defined as the sum of its diagonal elements (Appendix III). □

It is obvious that using resolvent algorithm given above, we can compute the resolvent matrix $(s\mathbf{I}-\mathbf{A})^{-1}$ and hence the state transition matrix $e^{\mathbf{A}t}$. In addition, the determinant of **A** and inverse of **A** can also be computed using this algorithm,

$$\det \mathbf{A} = (-1)^n \alpha_n \tag{7.21a}$$

$$\mathbf{A}^{-1} = -\frac{\mathbf{Q}_n}{\alpha_n} \tag{7.21b}$$

Relations (7.20) can be used to prove the Cayley-Hamilton theorem (Problem 7.3).

Example 7.1: Consider the linear state model

$$\dot{\mathbf{x}}(t) = \mathbf{A}\mathbf{x}(t) + \mathbf{B}\mathbf{u}(t)$$

$$\mathbf{y}(t) = \mathbf{C}\mathbf{x}(t)$$

where

$$\mathbf{A} = \begin{bmatrix} 2 & -1 & 0 \\ 1 & 1 & 2 \\ -1 & 0 & 1 \end{bmatrix}; \ \mathbf{B} = \begin{bmatrix} 1 & 0 \\ 1 & 0 \\ 0 & 2 \end{bmatrix}; \ \mathbf{C} = \begin{bmatrix} 1 & 1 & 0 \\ 1 & 0 & 1 \end{bmatrix}$$

Obtain the transfer function matrix $\hat{\mathbf{H}}(s)$.

Solution:

From sequential equations (7.20), we obtain

$$\alpha_1 = -\operatorname{tr}(\mathbf{A}) = -4$$

$$\mathbf{Q}_2 = \mathbf{A} + \alpha_1 \mathbf{I} = \begin{bmatrix} -2 & -1 & 0 \\ 1 & -3 & 2 \\ -1 & 0 & -3 \end{bmatrix}$$

$$\alpha_2 = -\tfrac{1}{2}\operatorname{tr}(\mathbf{A}\mathbf{Q}_2) = -\tfrac{1}{2}\operatorname{tr}\begin{bmatrix} -5 & 1 & 2 \\ -3 & -4 & -4 \\ 1 & 1 & -3 \end{bmatrix} = 6$$

$$\mathbf{Q}_3 = \mathbf{A}\mathbf{Q}_2 + \alpha_2 \mathbf{I} = \begin{bmatrix} 1 & 1 & -2 \\ -3 & 2 & -4 \\ 1 & 1 & 3 \end{bmatrix}$$

$$\alpha_3 = -\tfrac{1}{3}\operatorname{tr}(\mathbf{A}\mathbf{Q}_3) = -\tfrac{1}{3}\operatorname{tr}\begin{bmatrix} 5 & 0 & 0 \\ 0 & 5 & 0 \\ 0 & 0 & 5 \end{bmatrix} = -5$$

As a numerical check, we see that the relation

$$\mathbf{0} = \mathbf{A}\mathbf{Q}_3 + \alpha_3 \mathbf{I}$$

is satisfied.

Therefore,

$$\Delta(s) = s^3 - 4s^2 + 6s - 5$$

$$(s\mathbf{I} - \mathbf{A})^{+} = \mathbf{Q}_1 s^2 + \mathbf{Q}_2 s + \mathbf{Q}_3 = \mathbf{Q}(s)$$

$$= \begin{bmatrix} s^2 - 2s + 1 & -s + 1 & -2 \\ s - 3 & s^2 - 3s + 2 & 2s - 4 \\ -s + 1 & 1 & s^2 - 3s + 3 \end{bmatrix}$$

$$\hat{\mathbf{H}}(s) = \frac{\mathbf{C}\mathbf{Q}(s)\mathbf{B}}{\Delta(s)}$$

$$= \frac{1}{s^3 - 4s^2 + 6s - 5}\begin{bmatrix} -2(s-3) & 4(s-3) \\ -s^2 + s + 1 & 2(s^2 - 3s + 1) \end{bmatrix}$$

Linear Time-Invariant Discrete-Time Systems

Virtually all of the discussion presented earlier in this section carries over to discrete-time systems with obvious modifications. Some representative results are given below (Rugh 1975).

Pulse Response Matrix: Consider an n-dimensional linear time-invariant single-input/single-output system described by the state model

$$\mathbf{x}(k+1) = \mathbf{F}\mathbf{x}(k) + \mathbf{g}u(k) \tag{7.22a}$$

$$y(k) = \mathbf{c}\mathbf{x}(k) + du(k) \tag{7.22b}$$

where **F** is $n \times n$ real constant matrix, **g** and **c** are, respectively, $n \times 1$ and $1 \times n$ real constant vectors and d is a constant scalar, $k = 0, 1, 2, \ldots$

For a zero initial state ($\mathbf{x}(0) = \mathbf{0}$), the output (eqn. (6.12))

$$y(k) = \sum_{j=0}^{k-1} \mathbf{c}\mathbf{F}^{k-j-1}\mathbf{g}u(j) + du(k) \tag{7.23}$$

Suppose that the input at $k = 0$ is a *unit pulse signal* $\mu(k)$ defined below:

$$\mu(k) \triangleq \begin{bmatrix} 1 \, : \, k = 0 \\ 0 \, ; \, k \neq 0 \end{bmatrix} \tag{7.24}$$

Substituting $u(k) = \mu(k)$ in (7.23), we get

$$y(k) \triangleq h(k) = \begin{bmatrix} \mathbf{c}\mathbf{F}^{k-1}\mathbf{g}; & k \geqslant 1 \\ d \quad ; & k = 0 \\ 0 \quad ; & k < 0 \end{bmatrix} \tag{7.25}$$

The scalar function $h(k)$ is called the *pulse response* of the system (7.22). An input-output map for this system, obtained from eqns. (7.23) and (7.25) is

$$y(k) = h * u \tag{7.26a}$$

$$= \sum_{j=0}^{k} h(k-j)\, u(j); \; k \geqslant 0 \tag{7.26b}$$

For a multi-input/multi-output system described by the state model

$$\mathbf{x}(k+1) = \mathbf{F}\mathbf{x}(k) + \mathbf{G}\mathbf{u}(k) \tag{7.27a}$$

$$\mathbf{y}(k) = \mathbf{C}\mathbf{x}(k) + \mathbf{D}\mathbf{u}(k) \tag{7.27b}$$

where **F**, **G**, **C** and **D** are, respectively, $n \times n$, $n \times p$, $q \times n$ and $q \times p$ real constant matrices, an input-output map is

$$\mathbf{y}(k) = \mathbf{H} * \mathbf{u} \tag{7.28a}$$

$$= \sum_{j=0}^{k} \mathbf{H}(k-j)\,\mathbf{u}(j);\ k \geqslant 0 \tag{7.28b}$$

where

$$\mathbf{H}(k) = \begin{cases} \mathbf{C}\mathbf{F}^{k-1}\mathbf{B}\ ; & k \geqslant 0 \\ \mathbf{D}\ ; & k \geqslant 0 \\ 0\ ; & k < 0 \end{cases} \tag{7.28c}$$

is the *pulse response matrix.*

Pulse Transfer Function Matrix: The *pulse transfer function* for the single-input/single-output system (7.22) is (eqn. (3.31b))

$$\hat{h}(z) = \mathbf{c}(z\mathbf{I}-\mathbf{F})^{-1}\mathbf{g} + d \tag{7.29a}$$

and for multi-input/multi-output system (7.27), the *pulse transfer function matrix*

$$\hat{\mathbf{H}}(z) = \mathbf{C}(z\mathbf{I}-\mathbf{F})^{-1}\,\mathbf{G} + \mathbf{D} \tag{7.29b}$$

Linear Time-Varying Systems

Consider the n-dimensional linear time-varying system described by the state model

$$\dot{\mathbf{x}}(t) = \mathbf{A}(t)\,\mathbf{x}(t) + \mathbf{B}(t)\,\mathbf{u}(t) \tag{7.30a}$$

$$\mathbf{y}(t) = \mathbf{C}(t)\,\mathbf{x}(t) + \mathbf{D}(t)\,\mathbf{u}(t) \tag{7.30b}$$

where **A**, **B**, **C** and **D** are, respectively, $n \times n$, $n \times p$, $q \times n$ and $q \times p$ real matrices whose entries are continuous functions of t defined over $(-\infty, \infty)$.

For zero initial conditions, the output (eqn. (6.3))

$$\mathbf{y}(t) = \int_{t_0}^{t} \mathbf{C}(t)\Phi(t, \tau)\mathbf{B}(\tau)\mathbf{u}(\tau)d\tau + \mathbf{D}(t)\mathbf{u}(t) \tag{7.31}$$

where $\Phi(t, \tau)$ is the state transition matrix of $\dot{\mathbf{x}}(t) = \mathbf{A}(t)\,\mathbf{x}(t)$.

Equation (7.31) may be written as

$$\mathbf{y}(t) = \int_{t_0}^{t} \Big[\mathbf{C}(t)\Phi(t, \tau)\mathbf{D}(\tau) + \mathbf{D}(t)\delta(t-\tau)\Big]\mathbf{u}(\tau)d\tau \tag{7.32a}$$

$$= \int_{t_0}^{t} \mathbf{H}(t, \tau)\mathbf{u}(\tau)\,d\tau \tag{7.32b}$$

where

$$\mathbf{H}(t, \tau) = \begin{cases} \mathbf{C}(t)\,\Phi(t, \tau)\,\mathbf{B}(\tau) + \mathbf{D}(t)\,\delta(t-\tau); & t \geqslant \tau \\ \mathbf{0} & ; \ t < \tau \end{cases}$$

is the $q \times p$ *impulse response matrix* of the system.

Since the transfer function concept does not extend in a useful way to linear time-varying systems, the transfer function matrices do not constitute an effective input-output map for such systems.

Input-output map and its relation with state model for linear discrete-time time-varying systems can be obtained on analogous lines.

7.3 OUTPUT CONTROLLABILITY

We can define output controllability in a manner similar to (state) controllability defined earlier in Chapter 6. The state controllability is a property of the state space description while output controllability is a property of the input-output description of a system; these two concepts are not necessarily related.

Consider the linear time-varying system

$$\dot{\mathbf{x}}(t) = \mathbf{A}(t)\,\mathbf{x}(t) + \mathbf{B}(t)\,\mathbf{u}(t) \tag{7.33a}$$

$$\mathbf{y}(t) = \mathbf{C}(t)\,\mathbf{x}(t) \tag{7.33b}$$

where **A**, **B** and **C** are respectively $n \times n$, $n \times p$ and $q \times n$ real matrices whose entries are continuous function of t. The impulse response matrix of the system is (eqn. (7.32c))

$$\mathbf{H}(t, \tau) = \mathbf{C}(t)\,\Phi(t, \tau)\,\mathbf{B}(\tau) \tag{7.34}$$

where $\Phi(t, \tau)$ is the state transition matrix of $\dot{\mathbf{x}}(t) = \mathbf{A}(t)\mathbf{x}(t)$.

Theorem 7.1: The system (7.33) with impulse response matrix $\mathbf{H}(t, \tau)$ is output controllable at t_0 if and only if there exists a finite $t_1 > t_0$ such that the rows of $\mathbf{H}(t, \tau)$ are linearly independent on $[t_0, t_1]$.

□

The proof of this theorem is analogous to that of Theorem 6.1.

Consider now the linear time-invariant system described by the state model

$$\dot{\mathbf{x}}(t) = \mathbf{A}\mathbf{x}(t) + \mathbf{B}\mathbf{u}(t) \tag{7.35a}$$

$$\mathbf{y}(t) = \mathbf{C}\mathbf{x}(t) \tag{7.35b}$$

where **A**, **B** and **C** are, respectively, $n \times n$, $n \times p$ and $q \times n$ real constant matrices, The impulse response matrix of this system is (eqn. (7.11))

$$\mathbf{H}(t) = \mathbf{C}\,e^{\mathbf{A}t}\mathbf{B} \tag{7.36}$$

Theorem 7.2: The following statements regarding the linear time-invariant dynamical system (7.35) are equivalent:

a. The system is completely output controllable.
b. The rows of $\mathbf{C}\,e^{\mathbf{A}t}\mathbf{B}$ are linearly independent over the real field $\mathcal{R}$ for all $t \in [0, \infty)$.
c. The rank of the $q \times nq$ matrix

$$[\mathbf{CB} \mid \mathbf{CAB} \mid \dots \mid \mathbf{CA}^{n-1}\mathbf{B}]$$

is q. □

The proof of this theorem is analogous to that of Theorem 6.3.

Results of Theorems 7.1 and 7.2 can be easily extended to discrete-time systems.

Note that in state models (7.33) and (7.35) considered in this section, the term corresponding to direct transmission from control input to output is missing. It can be easily shown that the existence of direct transmission always aids output controllability (Problem 7.6). Also note that from the result of Theorem 7.2 (or Theorem 7.1) it can be proved that for the system (7.35), complete state controllability implies complete output controllability if and only if q rows of $\mathbf{C}$ are linearly independent.

Deadbeat Controller

Consider a linear time-invariant system described by the state model (7.35). The corresponding discrete-time state model is (refer eqns. (5.110))

$$\mathbf{x}(k+1) = \mathbf{F}\mathbf{x}(k) + \mathbf{G}\mathbf{u}(k) \tag{7.37a}$$

$$\mathbf{y}(k) = \mathbf{C}\mathbf{x}(k) \tag{7.37b}$$

where $\mathbf{F} = e^{\mathbf{A}t}$, an $n \times n$ constant matrix

$$\mathbf{G} = \int_0^T e^{\mathbf{A}\theta}\mathbf{B}\, d\theta, \text{ an } n \times p \text{ constant matrix}$$

$\mathbf{C}$ is $q \times n$ constant matrix

and T is the sampling interval.

It is desired to control the plant so that output $\mathbf{y}(t)$ responds in a deadbeat manner to step input $\mathbf{r}$. We shall take $\mathbf{x}(0) = \mathbf{0}$.

From eqns. (7.37) we get (refer eqn. (5.99))

$$\mathbf{y}(k) = \sum_{i=0}^{k-1} \mathbf{C}\mathbf{F}^{k-1-i}\mathbf{G}\mathbf{u}(i)$$

The output $\mathbf{y}(k)$ can be forced to the $(q \times 1)$ vector step $\mathbf{r}$ in N sampling intervals if we can solve for $\mathbf{u}(0), \mathbf{u}(1), \dots, \mathbf{u}(N-1)$, the equations

$$\mathbf{r} = [\mathbf{CG} \mid \mathbf{CFG} \mid \dots \mid \mathbf{CF}^{N-2}\mathbf{G} \mid \mathbf{CF}^{N-1}\mathbf{G}] \begin{bmatrix} \mathbf{u}(N-1) \\ \mathbf{u}(N-2) \\ \vdots \\ \mathbf{u}(1) \\ \mathbf{u}(0) \end{bmatrix} \tag{7.38}$$

These equations have a solution if (Section 2.6)

$$\rho[\mathbf{CG} \mid \mathbf{CFG} \mid \dots \mid \mathbf{CF}^{N-2}\mathbf{G} \mid \mathbf{CF}^{N-1}\mathbf{G}] = q$$

It is obvious that output controllability assures a solution to eqns. (7.38).

Example 7.2 : Let us consider the deadbeat control of Position Servo system discussed in Section 4.6. The state model of position servo is (eqns. (4.27))

$$\dot{\mathbf{x}}(t) = \begin{bmatrix} 0 & 1 \\ 0 & -5 \end{bmatrix} \mathbf{x}(t) + \begin{bmatrix} 0 \\ 1 \end{bmatrix} u(t) = \mathbf{A}\mathbf{x} + \mathbf{b}u \tag{7.39a}$$

$$y(t) = [1 \quad 0]\, \mathbf{x}(t) = \mathbf{c}\mathbf{x} \tag{7.39b}$$

The corresponding discrete-time description for $T = 0.1$ sec is (eqn. (6.60))

$$\mathbf{x}(k+1) = \begin{bmatrix} 1 & 0.0787 \\ 0 & 0.6065 \end{bmatrix} \mathbf{x}(k) + \begin{bmatrix} 0.0043 \\ 0.0787 \end{bmatrix} u(k) = \mathbf{F}\mathbf{x}(k) + \mathbf{g}u(k) \tag{7.40a}$$

$$y(k) = \mathbf{c}\mathbf{x}(k) \tag{7.40b}$$

It is desired to transfer the output $y(t) = x_1(t) = \theta(t)$ from 0 to 0.1 rad. From eqn. (7.38), it is observed that such a transfer is possible in one sampling interval ($N = 1$). Setting $r = 0.1$ and **F**, **g** and **c** as given in eqns. (7.39)–(7.40), we obtain

$$u(0) = \frac{0.1}{0.0043} = 23.2558$$

This value of control input will assure that $y(1) = x_1(1) = 0.1$. However, from eqn. (7.40a) we obtain $x_2(1) = \dot{y}(1) = 1.8302$. Thus the system output will be equal to 0.1 at the sampling instant but it will not stay there.

From eqn. (7.40a),

$$x_1(2) = x_1(1) + 0.0787\, x_2(1) + 0.0043\, u(1)$$
$$= 0.2440 + 0.0043\, u(1)$$

To satisfy the requirement: $x_1(2) = 0.1$, the control $u(1) = -33.4969$. For the next period, a similar calculation gives $u(2) = 27.933$.

Figure 7.2 gives the response of the system. Comparing with the deadbeat response of the same system as derived in Example 6.10, we observe the following:

(i) When considering only the response of the angular position at the sampling instants, the system shows output deadbeat response after one sampling interval. However, the response shows a pronounced

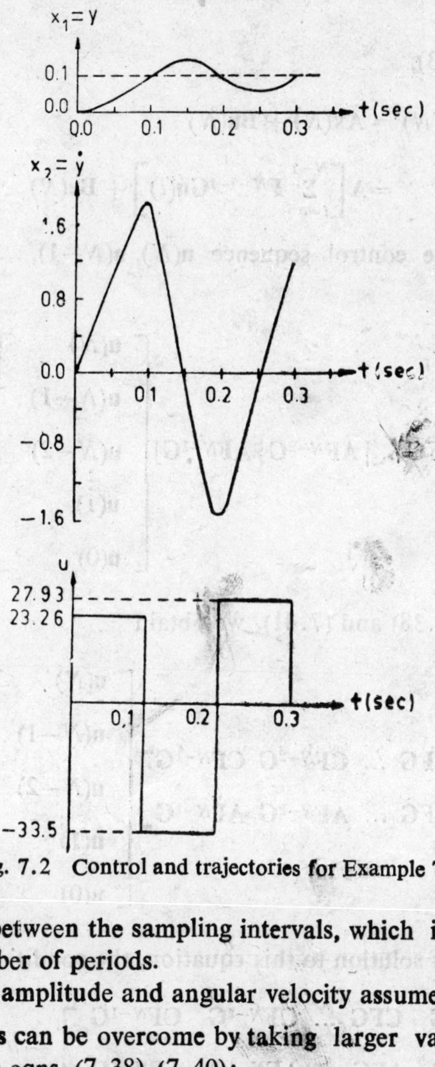

Fig. 7.2 Control and trajectories for Example 7.2

ripple in between the sampling intervals, which is sustained over a large number of periods.

(ii) The input amplitude and angular velocity assume large values.

These problems can be overcome by taking larger value of N. Taking $N=2$, we get from eqns. (7.38)–(7.40):

$$[\mathbf{cg} \mid \mathbf{cFg}] \begin{bmatrix} u(1) \\ u(0) \end{bmatrix} = r$$

The extra degree of freedom in the solution of this equation can be used to force the derivative of the state to zero. Let us first derive the result for the general case of multi-input/multi-output systems (Cadzow and Martens 1970). □

As was illustrated in Example 7.2, the expression (7.38) does not guarantee the deadbeat response; it only forces the output to r. If, however, we are able to force $\dot{\mathbf{x}}(N)$ to zero, then $\mathbf{x}(t)$ cannot change from $t=NT$ to

$t = (N + I)T$.

From eqn. (7.33),

$$\dot{\mathbf{x}}(N) = \mathbf{A}\mathbf{x}(N) + \mathbf{B}\mathbf{u}(N)$$

$$= \mathbf{A}\left[\sum_{i=0}^{N-1} \mathbf{F}^{N-1-i}\mathbf{G}\mathbf{u}(i)\right] + \mathbf{B}\mathbf{u}(N)$$

$\dot{\mathbf{x}}(N) = \mathbf{0}$ if the control sequence $\mathbf{u}(N), \mathbf{u}(N-1), \ldots, \mathbf{u}(0)$ satisfies the relation

$$[\mathbf{B} \,\vdots\, \mathbf{AG} \,\vdots\, \mathbf{AFG} \,\vdots \ldots \vdots\, \mathbf{AF}^{N-2}\mathbf{G} \,\vdots\, \mathbf{AF}^{N-1}\mathbf{G}] \begin{bmatrix} \mathbf{u}(N) \\ \mathbf{u}(N-1) \\ \mathbf{u}(N-2) \\ \vdots \\ \mathbf{u}(1) \\ \mathbf{u}(0) \end{bmatrix} = \mathbf{0} \tag{7.41}$$

Combining eqns. (7.38) and (7.41), we obtain

$$\begin{bmatrix} \mathbf{0} & \mathbf{CG} & \mathbf{CFG} & \ldots & \mathbf{CF}^{N-2}\mathbf{G} & \mathbf{CF}^{N-1}\mathbf{G} \\ \mathbf{B} & \mathbf{AG} & \mathbf{AFG} & \ldots & \mathbf{AF}^{N-2}\mathbf{G} & \mathbf{AF}^{N-1}\mathbf{G} \end{bmatrix} \begin{bmatrix} \mathbf{u}(N) \\ \mathbf{u}(N-1) \\ \mathbf{u}(N-2) \\ \vdots \\ \mathbf{u}(1) \\ \mathbf{u}(0) \end{bmatrix} = \begin{bmatrix} \mathbf{r} \\ \mathbf{0} \end{bmatrix} \tag{7.42}$$

For the existence of solution to this equation, the condition is (Section 2.6)

$$\rho \begin{bmatrix} \mathbf{0} & \mathbf{CG} & \mathbf{CFG} & \ldots & \mathbf{CF}^{N-2}\mathbf{G} & \mathbf{CF}^{N-1}\mathbf{G} \\ \mathbf{B} & \mathbf{AG} & \mathbf{AFG} & \ldots & \mathbf{AF}^{N-2}\mathbf{G} & \mathbf{AF}^{N-1}\mathbf{G} \end{bmatrix} = n + q \tag{7.43}$$

$\mathbf{u}(t) = \mathbf{u}(NT);\ t \geqslant NT$ maintains the condition $\dot{\mathbf{x}}(N) = \mathbf{0}$.

We denote the solution of (7.42) as

$$\mathbf{u}(k) = \mathbf{P}(k)\,\mathbf{r};\ k = 1, 2, \ldots, N \tag{7.44a}$$

Consider the system of Fig. 7.3 for realization of this deadbeat control sequence. Our objective is to determine the transfer function $\hat{\mathbf{D}}(z)$ of the digital controller so that

$$\mathbf{e}_2(k) = \mathbf{P}(k)\,\mathbf{r};\ k = 1, 2, \ldots, N \tag{7.44b}$$

The input to the controller,

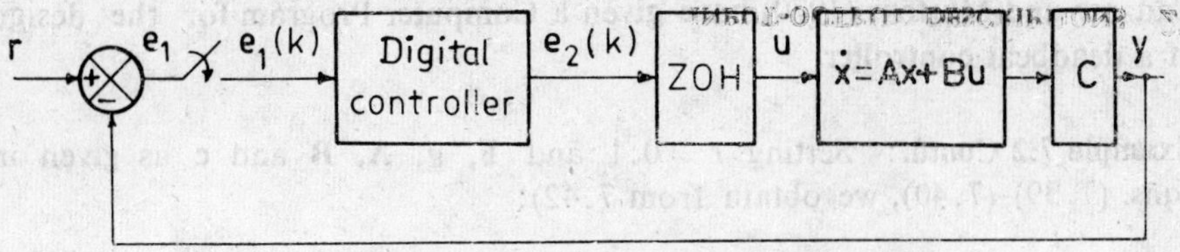

Fig. 7.3 Block diagram of computer control system

$$\begin{aligned}\mathbf{e}_1(k) &= \mathbf{r} - \mathbf{y}(k)\\ &= \mathbf{r} - \sum_{i=0}^{k-1} \mathbf{C}\mathbf{F}^{k-1-i}\mathbf{G}\mathbf{e}_2(i)\\ &= \left[\mathbf{I} - \sum_{i=0}^{k-1} \mathbf{C}\mathbf{F}^{k-1-i}\mathbf{G}\mathbf{P}(i)\right]\mathbf{r} \qquad (7.45)\end{aligned}$$

Taking z-transform of $\mathbf{e}_2(k)$ sequence, we obtain (Appendix II)

$$\begin{aligned}\hat{\mathbf{E}}_2(z) &= \sum_{k=0}^{\infty} \mathbf{P}(k)\mathbf{r}z^{-k}\\ &= \left[\sum_{k=0}^{N-1} \mathbf{P}(k)z^{-k} + \mathbf{P}(N)\sum_{k=N}^{\infty} z^{-k}\right]\mathbf{r}\\ &= \left[\sum_{k=0}^{N-1} \mathbf{P}(k)z^{-k} + \mathbf{P}(N)\frac{z^{-N}}{1-z^{-1}}\right]\mathbf{r} \qquad (7.46)\end{aligned}$$

Taking the z-transform of $\mathbf{e}_1(k)$ sequence, we obtain

$$\begin{aligned}\hat{\mathbf{E}}_1(z) &= \sum_{k=0}^{\infty} \mathbf{e}_1(k)\, z^{-k}\\ &= \sum_{k=0}^{N-1}\left[\mathbf{I} - \sum_{i=0}^{k-1} \mathbf{C}\mathbf{F}^{k-1-i}\mathbf{G}\mathbf{P}(i)\right] z^{-k}\mathbf{r} \qquad (7.47)\end{aligned}$$

The pulse transfer function of the digital controller $\mathbf{D}(z)$ is given by

$$\hat{\mathbf{E}}_2(z) = \hat{\mathbf{D}}(z)\,\hat{\mathbf{E}}_1(z)$$

Substituting $\hat{\mathbf{E}}_2(z)$ and $\hat{\mathbf{E}}_1(z)$ from eqns. (7.46) and (7.47) respectively, we obtain

$$\left[\sum_{k=0}^{N-1} \mathbf{P}(k)z^{-k} + \mathbf{P}(N)\frac{z^{-N}}{1-z^{-1}}\right]\mathbf{r} = \hat{\mathbf{D}}(z)\left[\sum_{k=0}^{N-1}\left[\mathbf{I} - \sum_{i=0}^{k-1} \mathbf{C}\mathbf{F}^{k-1-i}\mathbf{G}\mathbf{P}(i)\right] z^{-k}\right]\mathbf{r}$$

This equation must hold for arbitrary $\mathbf{r}$. Thus

$$\hat{\mathbf{D}}(z) = \left[\sum_{k=0}^{N-1} \mathbf{P}(k)z^{-k} + \mathbf{P}(N)\frac{z^{-N}}{1-z^{-1}}\right]\left[\sum_{k=0}^{N-1}\left(\mathbf{I} - \sum_{i=0}^{k-1} \mathbf{C}\mathbf{F}^{k-1-i}\mathbf{G}\mathbf{P}(i)\right) z^{-k}\right]^{-1} \qquad (7.48)$$

Cadzow and Martens (1970) have given a Computer Program for the design of a deadbeat controller.

Example 7.2 Contd.: Setting $r = 0.1$ and **F**, **g**, **A**, **B** and **c** as given in eqns. (7.39)–(7.40), we obtain from 7.42):

$$\begin{bmatrix} 0 & 0.0043 & 0.0105 \\ 0 & 0.0787 & 0.0477 \\ 1 & -0.3935 & -0.2385 \end{bmatrix} \begin{bmatrix} u(2) \\ u(1) \\ u(0) \end{bmatrix} = \begin{bmatrix} 0.1 \\ 0 \\ 0 \end{bmatrix}$$

This gives

$$\begin{bmatrix} u(2) \\ u(1) \\ u(0) \end{bmatrix} = \begin{bmatrix} 0 \\ -7.69 \\ 12.69 \end{bmatrix} \tag{7.49}$$

The same result was obtained earlier in Example 6.10.

Equation (7.49) may be rearranged as

$$\begin{bmatrix} u(2) \\ u(1) \\ u(0) \end{bmatrix} = \begin{bmatrix} e_2(2) \\ e_2(1) \\ e_2(0) \end{bmatrix} = \begin{bmatrix} 0 \\ -76.9 \\ 126.9 \end{bmatrix} r = \begin{bmatrix} P(2) \\ P(1) \\ P(0) \end{bmatrix} r$$

Thus

$$P(0) = 126.9,\ P(1) = -76.9,\ P(2) = 0$$

The pulse transfer function of the digital controller becomes (eqn. (7.48))

$$\hat{D}(z) = \left[P(0) + P(1)z^{-1} + P(2)\frac{z^{-2}}{1-z^{-1}} \right] \left[1 + (1 - \mathbf{C}GP(0))z^{-1} \right]^{-1}$$

$$= \frac{126.9 - 76.9z^{-1}}{1 + 0.454z^{-1}}$$

The realization of this pulse transfer function is shown in Fig. 7.4. The reader may verify this result after reading Section 7.5 wherein the methods of realization of transfer functions/pulse transfer functions have been presented.

7.4 REDUCIBILITY

In Section 6.9 we studied the Canonical Decomposition Theorem which states that a linear time-invariant system can always be represented, by means of a state transformation, in the form (eqn. (6.85))

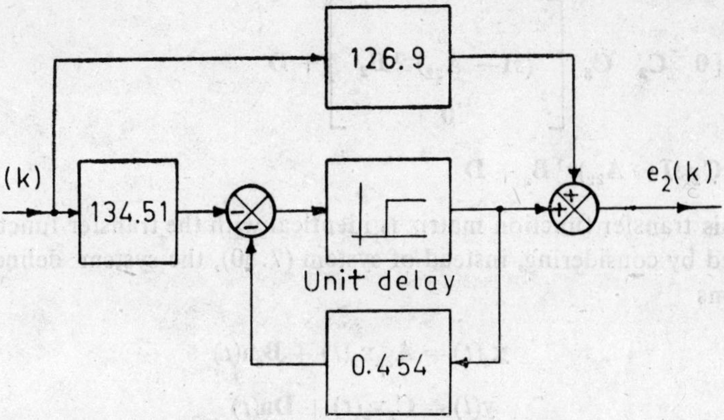

Fig. 7.4 Deadbeat controller for Position Servo

$$\begin{bmatrix} \dot{\mathbf{x}}_1(t) \\ \dot{\mathbf{x}}_2(t) \\ \dot{\mathbf{x}}_3(t) \end{bmatrix} = \begin{bmatrix} \mathbf{A}_{11} & \mathbf{A}_{12} & \mathbf{A}_{13} \\ \mathbf{0} & \mathbf{A}_{22} & \mathbf{A}_{23} \\ \mathbf{0} & \mathbf{0} & \mathbf{A}_{33} \end{bmatrix} \begin{bmatrix} \mathbf{x}_1(t) \\ \mathbf{x}_2(t) \\ \mathbf{x}_3(t) \end{bmatrix} + \begin{bmatrix} \mathbf{B}_1 \\ \mathbf{B}_2 \\ \mathbf{0} \end{bmatrix} \mathbf{u}(t) \qquad (7.50a)$$

$$\mathbf{y}(t) = [\mathbf{0} \quad \mathbf{C}_2 \quad \mathbf{C}_3] \begin{bmatrix} \mathbf{x}_1(t) \\ \mathbf{x}_2(t) \\ \mathbf{x}_3(t) \end{bmatrix} + \mathbf{Du}(t) \qquad (7.50b)$$

where $\mathbf{x}_1$, $\mathbf{x}_2$ and $\mathbf{x}_3$ represent subsets of states as follows:

$\mathbf{x}_1$ — controllable but not observable

$\mathbf{x}_2$ — controllable and observable

$\mathbf{x}_3$ — uncontrollable.

The transfer function matrix associated with state model (7.50) is obtained as follows:

$$\hat{\mathbf{H}}(s) = \mathbf{C}(s\mathbf{I} - \mathbf{A})^{-1}\mathbf{B} + \mathbf{D}$$

$$= [\mathbf{0} \quad \mathbf{C}_2 \quad \mathbf{C}_3] \begin{bmatrix} (s\mathbf{I} - \mathbf{A}_{11}) & -\mathbf{A}_{12} & -\mathbf{A}_{13} \\ \mathbf{0} & (s\mathbf{I} - \mathbf{A}_{22}) & -\mathbf{A}_{23} \\ \mathbf{0} & 3 & (s\mathbf{I} - \mathbf{A}_{33}) \end{bmatrix}^{-1} \begin{bmatrix} \mathbf{B}_1 \\ \mathbf{B}_2 \\ \mathbf{0} \end{bmatrix} + \mathbf{D}$$

$$= [\mathbf{0} \quad \mathbf{C}_2 \quad \mathbf{C}_3] \begin{bmatrix} (s\mathbf{I} - \mathbf{A}_{11})^{-1} & \times & \times \\ \mathbf{0} & (s\mathbf{I} - \mathbf{A}_{22})^{-1} & \times \\ \mathbf{0} & \mathbf{0} & (s\mathbf{I} - \mathbf{A}_{33})^{-1} \end{bmatrix} \begin{bmatrix} \mathbf{B}_1 \\ \mathbf{B}_2 \\ \mathbf{0} \end{bmatrix} + \mathbf{D}$$

$$= [\mathbf{0} \quad \mathbf{C}_2 \quad \mathbf{C}_3] \begin{bmatrix} \times \\ (s\mathbf{I} - \mathbf{A}_{22})^{-1}\mathbf{B}_2 \\ \mathbf{0} \end{bmatrix} + \mathbf{D}$$

$$= \mathbf{C}_2(s\mathbf{I} - \mathbf{A}_{22})^{-1}\mathbf{B}_2 + \mathbf{D}$$

This transfer function matrix is identical with the transfer function matrix obtained by considering, instead of system (7.50), the system defined by the equations

$$\dot{\mathbf{x}}_2(t) = \mathbf{A}_{22}\mathbf{x}_2(t) + \mathbf{B}_2\mathbf{u}(t)$$

$$\mathbf{y}(t) = \mathbf{C}_2\mathbf{x}_2(t) + \mathbf{D}\mathbf{u}(t)$$

This means that the transfer functlon matrix (or the impulse response matrix) represents only the controllable and observable part of a system and loses all information about the uncontrollable and unobservable part. In other words, the input-output description represents only that part of a system which is controllable and observable. This is the most important relation between the input-output description and the state variable description.

From the foregoing discussion we note that if a linear time-invariant model is uncontrollable and/or unobservable, there exists a state model of lesser dimension that has the same transfer function matrix and therefore zero-state response as the original state model. A linear time-invariant state model is said to be *reducible* if there exists a zero-state equivalent linear time-invariant state model of lesser dimension. Otherwise, it is called *irreducible* or *minimal-dimensional* state model.

Theorem 7.3 : A linear time-invariant state model is irreducible if and only if it is both controllable and observable.

Proof :

Consider the linear time-invariant state model

$$\dot{\mathbf{x}}(t) = \mathbf{A}\mathbf{x}(t) + \mathbf{B}\mathbf{u}(t) \tag{7.51a}$$

$$\mathbf{y}(t) = \mathbf{C}\mathbf{x}(t) \tag{7.51b}$$

where **A**, **B** and **C** are respectively $n \times n$, $n \times p$ and $q \times n$ matrices. If this state model is either uncontrollable or unobservable or both uncontrollable and unobservable, then it is reducible as per Theorems 6.13, 6.14 and 6.15. In the following, we prove, by contradiction, that if the given state model (7.51) is controllable and observable, then it is irreducible.

Suppose the given state model (7.51) is controllable and observable and reducible to the following $n_1(< n)$-dimensional state model:

$$\dot{\bar{\mathbf{x}}} = \bar{\mathbf{A}}\bar{\mathbf{x}} + \bar{\mathbf{B}}\mathbf{u} \tag{7.52a}$$

$$\mathbf{y} = \bar{\mathbf{C}}\bar{\mathbf{x}} \tag{7.52b}$$

Since the two models are zero-state equivalent, we have

$$\mathbf{C}e^{\mathbf{A}t}\mathbf{B} = \bar{\mathbf{C}}e^{\bar{\mathbf{A}}t}\bar{\mathbf{B}}$$

This implies that

$$\mathbf{C}\mathbf{A}^i\mathbf{B} = \bar{\mathbf{C}}\,\bar{\mathbf{A}}^i\bar{\mathbf{B}}\ ;\quad i = 0, 1, 2, \ldots$$

Consider the product of observability matrix $\mathbf{V}$ and controllability matrix $\mathbf{U}$ of the given state model (7.51).

$$\underset{(qn\times n)}{\mathbf{V}}\ \underset{(n\times np)}{\mathbf{U}} = \begin{bmatrix} \mathbf{C} \\ \mathbf{CA} \\ \vdots \\ \mathbf{CA}^{n-1} \end{bmatrix} [\mathbf{B}\ \mathbf{AB} \ldots \mathbf{A}^{n-1}\mathbf{B}]$$

$$= \begin{bmatrix} \mathbf{CB} & \mathbf{CAB} & \ldots & \mathbf{CA}^{n-1}\mathbf{B} \\ \mathbf{CAB} & \mathbf{CA}^2\mathbf{B} & \ldots & \mathbf{CA}^n\mathbf{B} \\ \vdots & \vdots & & \vdots \\ \mathbf{CA}^{n-1}\mathbf{B} & \mathbf{CA}^n\mathbf{B} & \ldots & \mathbf{CA}^{2(n-1)}\mathbf{B} \end{bmatrix}$$

$$= \begin{bmatrix} \bar{\mathbf{C}}\bar{\mathbf{B}} & \bar{\mathbf{C}}\bar{\mathbf{A}}\bar{\mathbf{B}} & \ldots & \bar{\mathbf{C}}\bar{\mathbf{A}}^{n-1}\bar{\mathbf{B}} \\ \bar{\mathbf{C}}\bar{\mathbf{A}}\bar{\mathbf{B}} & \bar{\mathbf{C}}\bar{\mathbf{A}}^2\bar{\mathbf{B}} & \ldots & \bar{\mathbf{C}}\bar{\mathbf{A}}^n\bar{\mathbf{B}} \\ \vdots & \vdots & & \vdots \\ \bar{\mathbf{C}}\,\bar{\mathbf{A}}^{n-1} & \bar{\mathbf{C}}\,\bar{\mathbf{A}}^n\bar{\mathbf{B}} & \ldots & \bar{\mathbf{C}}\,\bar{\mathbf{A}}^{2(n-1)}\bar{\mathbf{B}} \end{bmatrix}$$

$$= \begin{bmatrix} \bar{\mathbf{C}} \\ \bar{\mathbf{C}}\bar{\mathbf{A}} \\ \vdots \\ \bar{\mathbf{C}}\bar{\mathbf{A}}^{n-1} \end{bmatrix} [\bar{\mathbf{B}}\ \ \bar{\mathbf{A}}\bar{\mathbf{B}}\ \ldots\ \bar{\mathbf{A}}^{n-1}\bar{\mathbf{B}}]$$

$$= \underset{(qn\times n_1)}{\bar{\mathbf{V}}}\ \underset{(n_1\times np)}{\bar{\mathbf{U}}}$$

Since the given state model is assumed to be controllable and observable, $\rho(\mathbf{V}) = n$, $\rho(\mathbf{U}) = n$ and therefore $\rho(\mathbf{VU}) = n$ (see Appendix III). The rank of $\bar{\mathbf{V}}\bar{\mathbf{U}}$ can be at the most n_1. The equation $\mathbf{V}\,\mathbf{U} = \bar{\mathbf{V}}\,\bar{\mathbf{U}}$, however, implies that $\rho(\bar{\mathbf{V}}\,\bar{\mathbf{U}}) = n$. This is a contradiction. Therefore, if the given state model is controllable and observable, then it is irreducible.

Example 7.3: Consider the state model of Example 6.16:

$$\dot{\mathbf{x}}(t) = \begin{bmatrix} -2 & 1 \\ 1 & -2 \end{bmatrix}\mathbf{x}(t) + \begin{bmatrix} 1 \\ 1 \end{bmatrix}u(t)$$

$$= \mathbf{A}\mathbf{x}(t) + \mathbf{b}u(t)$$

$$y = [0 \quad 1]\,\mathbf{x} = \mathbf{c}\mathbf{x}$$

This model is the representation of network of Fig. 4.3a. The system was found to be uncontrollable.

The transfer function

$$\hat{h}(s) = \mathbf{c}(s\mathbf{I} - \mathbf{A})^{-1}\mathbf{b}$$

$$= [0\ 1]\begin{bmatrix} \dfrac{s+2}{(s+1)(s+3)} & \dfrac{1}{(s+1)(s+3)} \\ \dfrac{1}{(s+1)(s+3)} & \dfrac{s+2}{(s+1)(s+3)} \end{bmatrix}\begin{bmatrix} 1 \\ 1 \end{bmatrix}$$

$$= \frac{1}{s+1}$$

The transfer function thus verifies the fact that the given state model is reducible.

7.5 STATE MODELS FROM INPUT-OUTPUT MAPS

In Section 7.2 we studied the problem—To find the input-output map from the state model of a system. The converse problem—To find a state model from the input-output map of a system, is the subject of discussion in this section. This problem is quite important because there are many design techniques and computational algorithms developed exclusively for state models. In order to apply these techniques and algorithms, experimentally obtained input-output descriptions (impulse response matrices or transfer function matrices) must be realized into state models. We shall discuss the problem of realization of transfer function matrices into state models.

Note the term 'realization' used above. A state model that has a prescribed rational matrix $\hat{\mathbf{H}}(s)$ is the realization of $\hat{\mathbf{H}}(s)$. The term 'realization' is justified by the fact that by using the state model, the system with the transfer function matrix $\hat{\mathbf{H}}(s)$ can be built in the real world by an OP–AMP circuit (refer Section 3.6).

The realization problem, as we shall see, is quite complicated. It actually consists of three problems:

1. Is it possible at all to obtain state variable description from the given transfer function matrix?
2. If yes, is the state variable description unique for a given transfer function matrix?
3. How do we obtain state variable description from the transfer function matrix?

The answer to the first problem has been given in Section 7.2. A rational-function matrix $\hat{\mathbf{H}}(s)$ is realizable by a finite dimensional linear time-invariant state model if and only if $\hat{\mathbf{H}}(s)$ is a proper rational matrix. A proper rational matrix will have state model of the form

$$\dot{\mathbf{x}}(t) = \mathbf{A}\mathbf{x}(t) + \mathbf{B}\mathbf{u}(t) \tag{7.53a}$$

$$\mathbf{y}(t) = \mathbf{C}\mathbf{x}(t) + \mathbf{D}\mathbf{u}(t) \tag{7.53b}$$

where **A**, **B**, **C** and **D** are constant matrices of appropriate dimensions. A strictly proper rational matrix will have a state model of the form

$$\dot{\mathbf{x}}(t) = \mathbf{A}\mathbf{x}(t) + \mathbf{B}\mathbf{u}(t) \tag{7.54a}$$

$$\mathbf{y}(t) = \mathbf{C}\mathbf{x}(t) \tag{7.54b}$$

The scalar transfer function

$$\hat{h}(s) = \frac{\beta_0 s^m + \beta_1 s^{m-1} + \dots + \beta_m}{s^n + \alpha_1 s^{n-1} + \dots + \alpha^n}$$

is realizable if and only if $m \leqslant n$.

Let us now turn to the second problem. Equation (7.16) demands that the number of rows of **C** must be equal to the number of rows of $\hat{\mathbf{H}}(s)$ and the number of columns of **B** must be equal to the number of columns of $\hat{\mathbf{H}}(s)$; the dimension of **D** must be same as $\hat{\mathbf{H}}(s)$. If {**A**, **B**, **C**, **D**} is the realization of a given $\hat{\mathbf{H}}(s)$, we observe that no restriction on the dimension of the square matrix **A** is imposed. Thus the state model corresponding to a given $\hat{\mathbf{H}}(s)$ is not expected to be unique with respect to its order. However, a state model realization with the least possible dimension is a good realization and we know that the *minimal-dimensional* (or *irreducible*) realization must be both controllable and observable (Theorem 7.3).

In Section 3.5 we noted that any state space representation can have an infinite number of equivalent representations of the same order. Thus the irreducible realization corresponding to a given $\hat{\mathbf{H}}(s)$ is not unique with respect to the set of state variables. There can be many irreducible realizations of the same order; however, all these irreducible realizations will be equivalent.

Example 7.4: Let the given transfer function $\hat{h}(s)$ be

$$\hat{h}(s) = \frac{1}{s+1}$$

The following state models realize this transfer function.

(i) $\mathbf{A} = -1$; $\mathbf{b} = 1$; $\mathbf{c} = 1$

The model is controllable and observable

(ii) $\mathbf{A} = \begin{bmatrix} -2 & 1 \\ 1 & -2 \end{bmatrix}$; $\mathbf{b} = \begin{bmatrix} 1 \\ 1 \end{bmatrix}$; $\mathbf{c} = [0 \quad 1]$

The model is observable but not controllable (Example 7.3)

(iii) $\mathbf{A} = \begin{bmatrix} -1 & 0 \\ 0 & -3 \end{bmatrix}$; $\mathbf{b} = \begin{bmatrix} 1 \\ 1 \end{bmatrix}$; $\mathbf{c} = [1 \quad 0]$

The model is controllable but not observable

(iv) $\mathbf{A} = \begin{bmatrix} -2 & 0 \\ 0 & -1 \end{bmatrix}$; $\mathbf{b} = \begin{bmatrix} 0 \\ 1 \end{bmatrix}$; $\mathbf{c} = [0 \quad 1]$

The model is neither controllable nor observable

(v) $\mathbf{A} = \begin{bmatrix} -1 & 0 & 0 \\ 0 & -2 & 0 \\ 0 & 0 & -3 \end{bmatrix}$; $\mathbf{b} = \begin{bmatrix} 1 \\ 0 \\ 0 \end{bmatrix}$; $\mathbf{c} = [1 \quad 1 \quad 1]$

The model is controllable but not observable. □

The statements given above can be easily verified. It may be noted that state models of different orders and with different sets of state variables can realize a given transfer function matrix. Minimal realization is given by the state model which is both controllable and observable.

In the remaining part of this section, we deal with the third problem.

Realization of Scalar Transfer Functions/Differential Equations

A linear time-invariant single-input/single-output system is described by transfer function of the form

$$\hat{h}_1(s) = \frac{\beta_0' s^m + \beta_1' s^{m-1} + \ldots + \beta_m'}{s^n + \alpha_1 s^{n-1} + \ldots + \alpha_n}; \quad m \leqslant n \tag{7.55}$$

where the coefficients α_i and β_i' are real constant scalars. Note that there is no loss in generality to assume the coefficient of s^n to be unity.

In the following, we derive results for $m = n$; these results may be used for the case $m < n$ by setting appropriate β_i' coefficients equal to zero. Therefore, our problem is to obtain a state model corresponding to the transfer function

$$\hat{h}_1(s) = \frac{\beta_0' s^n + \beta_1' s^{n-1} + \ldots + \beta_n'}{s^n + \alpha_1 s^{n-1} + \ldots + \alpha_n} \tag{7.56}$$

By long division, $\hat{h}_1(s)$ can be written as

$$\hat{h}_1(s) = \frac{\beta_1 s^{n-1} + \beta_2 s^{n-2} + \ldots + \beta_n}{s^n + \alpha_1 s^{n-1} + \ldots + \alpha_n} + \beta_0 \tag{7.57}$$

$$\stackrel{\Delta}{=} \hat{h}(s) + d$$

where $\quad d = \hat{h}_1(\infty) = \beta_0'$

Since the constant d immediately gives the direct transmission part of the state model, we need to consider in the following only the strictly proper rational function

$$\hat{h}(s) = \frac{N(s)}{\Delta(s)} = \frac{\beta_1 s^{n-1} + \beta_2 s^{n-2} + \ldots + \beta_n}{s^n + \alpha_1 s^{n-1} + \ldots + \alpha_n} \tag{7.58}$$

The state model for $\hat{h}_1(s)$ may easily be obtained once the state model for $\hat{h}(s)$ is known.

Our problem is equivalent to finding a state variable description of the nth-order differential equation

$$\overset{(n)}{y}(t) + \alpha_1 \overset{(n-1)}{y}(t) + \ldots + \alpha_{n-1}\dot{y}(t) + \alpha_n y(t)$$

$$= \beta_1 \overset{(n-1)}{u}(t) + \beta_2 \overset{(n-2)}{u}(t) + \ldots + \beta_{n-1}\dot{u}(t) + \beta_n u(t) \tag{7.59a}$$

where

$$\overset{(k)}{y}(t) \stackrel{\Delta}{=} \frac{d^k y(t)}{dt^k}$$

In terms of p-operator (Nagrath and Gopal 1982), eqn. (7.59a) may be written as

$$\{\beta^n + \alpha_1 p^{n-1} + \ldots + \alpha_{n-1} p + \alpha_n\}\, y(t)$$

$$= \{\beta_1 p^{n-1} + \beta_2 p^{n-2} + \ldots + \beta_{n-1} p + \beta_n\}\, u(t) \tag{7.59b}$$

where

$$p^k y(t) \stackrel{\Delta}{=} \frac{d^k y(t)}{dt^k}$$

Note that $\hat{h}(s)$ in eqn. (7.58) is the transfer function of (7.59); i.e.,

$$\frac{\hat{y}(s)}{\hat{u}(s)} = \hat{h}(s) = \frac{\beta_1 s^{n-1} + \beta_2 s^{n-2} + \ldots + \beta_n}{s^n + \alpha_1 s^{n-1} + \ldots + \alpha_n} \tag{7.60}$$

It is well known that in nth-order differential eqn. (7.59), in order to have a unique output y for any input u, we need n number of initial conditions: $y(0), \dot{y}(0)\ldots, \overset{(n-1)}{y}(0)$. Hence the state vector will consist of n components.

Phase-Variable Canonical Forms: The transfer function (7.60) of differential equation (7.59) is first assumed to be irreducible, i.e., there are no common factors between numerator and denominator. This will then be followed by the case of differential equation with a reducible transfer function.

Differential equation with irreducible transfer function

Before considering the general case of eqn. (7.59), we first consider the differential equation

$$\overset{(n)}{y}(t) + \alpha_1 \overset{(n-1)}{y}(t) + \ldots + \alpha_{n-1} \dot{y}(t) + \alpha_n y(t) = u(t) \tag{7.61a}$$

The transfer function of this system is

$$\frac{\hat{y}(s)}{\hat{u}(s)} = \hat{h}(s) = \frac{1}{s^n + \alpha_1 s^{n-1} + \ldots + \alpha_{n-1} s + \alpha_n} \tag{7.61b}$$

Taking the Laplace transform of eqn. (7.61a), we have

$$[s^n \hat{y}(s) - s^{n-1} y(0) - s^{n-2} \dot{y}(0) - \ldots - \overset{(n-1)}{y}(0)]$$

$$+ \alpha_1 [s^{n-1} \hat{y}(s) - s^{n-2} y(0) - s^{n-3} \dot{y}(0) - \ldots - \overset{(n-2)}{y}(0)] + \ldots$$

$$+ \alpha_{n-1} [s \hat{y}(s) - y(0)] + \alpha_n \hat{y}(s) = \hat{u}(s)$$

or

$$\hat{y}(s) = \frac{1}{\Delta(s)} \hat{u}(s) + \frac{1}{\Delta(s)} \Big[(s^{n-1} + \alpha_1 s^{n-2} + \ldots + \alpha_{n-1}) y(0)$$

$$+ (s^{n-2} + \alpha_1 s^{n-3} + \ldots + \alpha_{n-2}) \dot{y}(0) + \ldots$$

$$+ (s + \alpha_1) \overset{(n-2)}{y}(0) + \overset{(n-1)}{y}(0) \Big] \tag{7.62}$$

where

$$\Delta(s) = s^n + \alpha_1 s^{n-1} + \ldots + \alpha_{n-1} s + \alpha_n$$

The term $\frac{1}{\Delta(s)} \hat{u}(s)$ gives the response due to input $\hat{u}(s)$; the remaining terms on the right hand side of eqn. (7.62) give the response due to initial conditions. Therefore, if the initial conditions $y(0)$, $\dot{y}(0), \ldots, \overset{(n-2)}{y}(0), \overset{(n-1)}{y}(0)$ are known, then for any u, a unique y can be determined. Consequently, if we choose the state variables as

$$\begin{aligned} x_1(t) &\triangleq y(t) \\ x_2(t) &\triangleq \dot{y}(t) \\ x_3(t) &\triangleq \ddot{y}(t) \\ &\vdots \\ x_n(t) &\triangleq \overset{(n-1)}{y}(t) \end{aligned} \tag{7.63}$$

then $\mathbf{x}(t) = [x_1(t) \quad x_2(t) \ldots x_n(t)]^T$ qualifies as the state vector.

The set of equations in (7.63) yields

$$\dot{x}_1(t) = x_2(t)$$

$$\dot{x}_2(t) = x_3(t)$$

$$\vdots$$

$$\dot{x}_{n-1}(t) = x_n(t)$$

Differentiating $x_n(t) \overset{\Delta}{=} y^{(n-1)}(t)$ once and using (7.61a) we get

$$\dot{x}_n(t) = -\alpha_n x_1(t) - \alpha_{n-1} x_2(t) - \ldots - \alpha_1 x_n(t) + u(t)$$

The foregoing equations can be arranged in the matrix form as

$$\dot{\mathbf{x}}(t) = \mathbf{A}\mathbf{x}(t) + \mathbf{b}u(t) \tag{7.64a}$$

$$y(t) = \mathbf{c}\mathbf{x}(t) \tag{7.64b}$$

where

$$\mathbf{A} = \begin{bmatrix} 0 & 1 & 0 \ldots & 0 \\ 0 & 0 & 1 \ldots & 0 \\ \vdots & \vdots & \vdots & \vdots \\ 0 & 0 & 0 \ldots & 1 \\ -\alpha_n & -\alpha_{n-1} & -\alpha_{n-2} & -\alpha_1 \end{bmatrix}; \quad \mathbf{b} = \begin{bmatrix} 0 \\ 0 \\ \vdots \\ 0 \\ 1 \end{bmatrix}$$

$$\mathbf{c} = [1 \quad 0 \quad 0 \ldots 0]$$

The state diagram for eqns. (7.64) is shown in Fig. 7.5. Applying Mason's gain formula to this state diagram, we find that the transfer function $\hat{y}(s)/\hat{u}(s)$ is equal to the one given by (7.61b). Therefore state model (7.64)

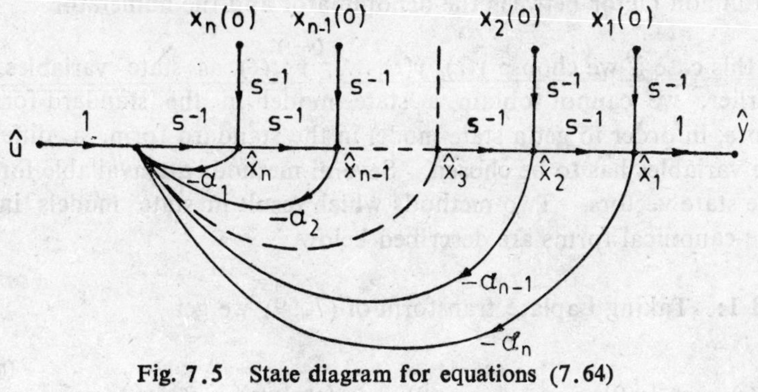

Fig. 7.5 State diagram for equations (7.64)

is the realization of scalar transfer function (7.61b). The realization (7.64) is irreducible or equivalently, (7.64) is controllable and observable. Note that the realization (7.64) can be obtained directly from the coefficients of $\hat{h}(s)$ in (7.61b).

□

When state variables are obtained from one of the system variables and its derivatives, they are referred to as *phase variables.* In (7.63), the state variables have been defined in terms of output and its derivatives.

A single-input time-invariant linear system is said to be in *phase-variable canonical form* if its equations have the form

$$\dot{\mathbf{x}}(t) = \mathbf{A}\mathbf{x}(t) + \mathbf{b}u(t) \tag{7.65a}$$

$$\mathbf{y}(t) = \mathbf{C}\mathbf{x}(t) \tag{7.65b}$$

where

$$\mathbf{A} = \begin{bmatrix} 0 & 1 & 0 & \cdots & 0 \\ 0 & 0 & 1 & \cdots & 0 \\ \vdots & \vdots & \vdots & & \vdots \\ 0 & 0 & 0 & \cdots & 1 \\ -\alpha_n & -\alpha_{n-1} & -\alpha_{n-2} & \cdots & -\alpha_1 \end{bmatrix}; \quad \mathbf{b} = \begin{bmatrix} 0 \\ 0 \\ \vdots \\ 0 \\ 1 \end{bmatrix}$$

No special form is imposed on matrix C in this definition. It is easily verified that system (7.65) is always completely controllable. In fact, as we shall see in Chapter 9, any completely controllable single-input system can be transformed into this phase-variable canonical form.

□

The realization (7.64) of the transfer function (7.61b) is obviously in the phase-variable canonical form defined in (7.65).

We next consider the general differential equation (7.59). The corresponding $\hat{h}(s)$ is given by (7.60) which we assumed to be irreducible, i.e., there is no common factor between the denominator and the numerator.

In this case if we choose $y(t), \dot{y}(t), \ldots, \overset{(n-1)}{y}(t)$ as state variables, as we did earlier, we cannot obtain a state model in the standard form (7.1). Therefore, in order to get a state model in the standard form, a different set of state variables has to be chosen. Several methods are available for finding possible state vectors. Two methods which result in state models in phase-variable canonical forms are described below.

Method 1: Taking Laplace transform of (7.59) we get

$$[s^n\hat{y}(s) - s^{n-1}y(0) - \ldots - \overset{(n-1)}{y}(0)] + \alpha_1[s^{n-1}\hat{y}(s) - s^{n-2}y(0) - \ldots - \overset{(n-2)}{y}(0)] + \ldots + \alpha_{n-1}[s\,\hat{y}(s) - y(0)] + \alpha_n\hat{y}(s)$$

$$= \beta_1[s^{n-1}\hat{u}(s) - s^{n-2}u(0) - \ldots - \overset{(n-2)}{u}(0)] + \ldots + \beta_{n-1}[s\,\hat{u}(s) - u(0)] + \beta_n\hat{u}(s)$$

or

$$\hat{y}(s) = \frac{N(s)}{\Delta(s)}\hat{u}(s) + \frac{1}{\Delta(s)}[s^{n-1}y(0) + s^{n-2}(\dot{y}(0) + \alpha_1 y(0) - \beta_1 u(0)) + \dots$$
$$+ (\overset{(n-1)}{y}(0) + \alpha_1 \overset{(n-2)}{y}(0) + \dots + \alpha_{n-1}y(0) - \beta_1 \overset{(n-2)}{u}(0)$$
$$- \dots - \beta_{n-1}u(0))] \qquad (7.66)$$

where

$$N(s) = \beta_1 s^{n-1} + \dots + \beta_n$$

$$\Delta(s) = s^n + \alpha_1 s^{n-1} + \dots + \alpha_n$$

From eqns. (7.66) we observe that if all the coefficients associated with s^{n-1}, s^{n-2}, …, s^0 are known then for any u, a unique y can be determined. Consequently, we may choose the state variables as

$$\begin{bmatrix} x_n \\ x_{n-1} \\ x_{n-2} \\ \vdots \\ x_1 \end{bmatrix} = \begin{bmatrix} 1 & 0 & 0 & \dots & 0 \\ \alpha_1 & 1 & 0 & \dots & 0 \\ \alpha_2 & \alpha_1 & 1 & \dots & 0 \\ \vdots & \vdots & \vdots & & \vdots \\ \alpha_{n-1} & \alpha_{n-2} & \alpha_{n-3} & \dots & 1 \end{bmatrix} \begin{bmatrix} y \\ \dot{y} \\ \vdots \\ \overset{(n-1)}{y} \end{bmatrix}$$
$$+ \begin{bmatrix} 0 & 0 & 0 & \dots & 0 \\ -\beta_1 & 0 & 0 & \dots & 0 \\ -\beta_2 & -\beta_1 & 0 & \dots & 0 \\ \vdots & \vdots & \vdots & & \vdots \\ -\beta_{n-1} & -\beta_{n-2} & -\beta_{n-3} & \dots & 0 \end{bmatrix} \begin{bmatrix} u \\ \dot{u} \\ \vdots \\ \overset{(n-1)}{u} \end{bmatrix} \qquad (7.67)$$

which is equivalent to

$$x_n = y$$
$$x_{n-1} = \dot{x}_n + \alpha_1 y - \beta_1 u$$
$$x_{n-2} = \dot{x}_{n-1} + \alpha_2 y - \beta_2 u$$
$$\vdots$$
$$x_1 = \dot{x}_2 + \alpha_{n-1} y - \beta_{n-1} u$$

Differentiating $\mathbf{x}$ in (7.67) once and using (7.59) we obtain

$$\dot{x}_1 = -\alpha_n y + \beta_n u$$

The foregoing equations can be arranged in the matrix form as

$$\dot{\mathbf{x}}(t) = \mathbf{A}\mathbf{x}(t) + \mathbf{b}u(t) \qquad (7.68a)$$

$$y(t) = \mathbf{c}\mathbf{x}(t) \qquad (7.68b)$$

where

$$\mathbf{A} = \begin{bmatrix} 0 & 0 & 0 \ldots 0 & -\alpha_n \\ 1 & 0 & 0 \ldots 0 & -\alpha_{n-1} \\ 0 & 1 & 0 \ldots 0 & -\alpha_{n-2} \\ \vdots & \vdots & \vdots \quad \vdots & \vdots \\ 0 & 0 & 0 \ldots 0 & -\alpha_2 \\ 0 & 0 & 0 \ldots 1 & -\alpha_1 \end{bmatrix}; \quad \mathbf{b} = \begin{bmatrix} \beta_n \\ \beta_{n-1} \\ \vdots \\ \beta_2 \\ \beta_1 \end{bmatrix}$$

$$\mathbf{c} = [0 \quad 0 \ldots 0 \quad 1]$$

The state diagram for eqns. (7.68) is shown in Fig. 7.6. Applying Mason's gain formula to this state diagram, we find that the transfer function $\hat{y}(s)/\hat{u}(s)$ is equal to the one given by (7.60). Therefore, state model (7.68)

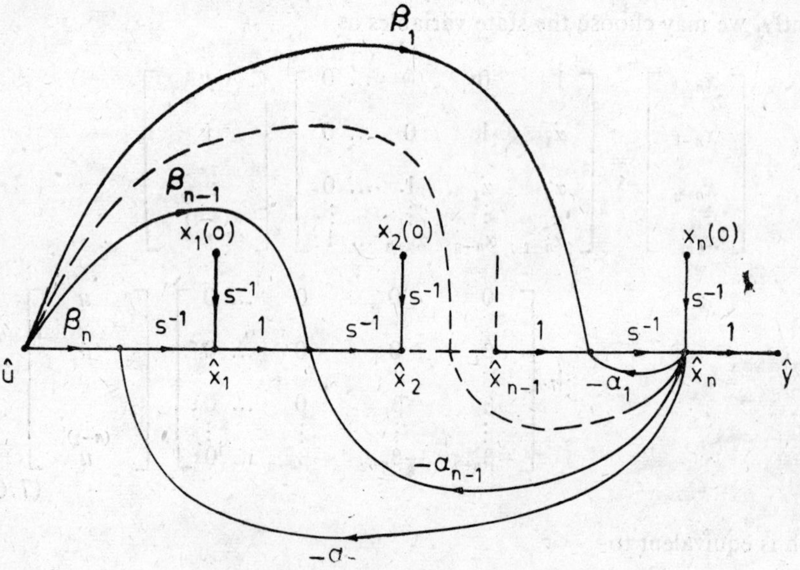

Fig. 7.6 State diagram for equations (7.68)

is the realization of scalar transfer function (7.60). The realization (7.68) is irreducible or equivalently the state model (7.68) is controllable and observable. This realization can be obtained directly from the coefficients of transfer function (7.60).

□

A single-output time-invariant linear system is said to be in the *dual phase-variable canonical form* if its equations have the form

$$\dot{\mathbf{x}}(t) = \mathbf{A}\mathbf{x}(t) + \mathbf{B}\mathbf{u}(t) \tag{7.69a}$$

$$y(t) = \mathbf{c}\mathbf{x}(t) \tag{7.69b}$$

where

$$\mathbf{A} = \begin{bmatrix} 0 & 0 & 0 \dots 0 & -\alpha_n \\ 1 & 0 & 0 \dots 0 & -\alpha_{n-1} \\ 0 & 1 & 0 \dots 0 & -\alpha_{n-2} \\ \vdots & \vdots & \vdots \quad \vdots & \vdots \\ 0 & 0 & 0 \dots 0 & -\alpha_2 \\ 0 & 0 & 0 \dots 1 & -\alpha_1 \end{bmatrix}$$

$$\mathbf{c} = [\,0 \quad 0 \quad 0 \dots 0 \quad 1\,]$$

No special form is imposed on matrix **B** in this definition. It is easily verified that system (7.69) is always completely observable. In fact, as we shall see in Chapter 9, any completely observable single-output system can be transformed into the dual phase-variable canonical form.

The system matrices **A** in the two canonical forms (7.65) and (7.69) are known as *companion matrices*. In the literature, the phase-variable canonical form (7·65) is frequently referred to as *controllable companion form* and the dual phase-variable canonical form (7.69) as *observable companion form.* Related companion forms for multi-input/multi-output systems will be derived in Chapter 9.

When **A** matrix is in companion form, the modal matrix is Vandermonde matrix (Problem 2.18) and therefore the evaluation of $e^{\mathbf{A}t}$ becomes easy. Also the characteristic polynomial $|\lambda\mathbf{I} - \mathbf{A}|$ is immediately apparent by inspection of **A**. Both the companion forms lend themselves easily to interpretation in terms of analog computer models.

There are many other benefits to be derived from companion forms of a state model. These will be discussed in Chapter 9.

□

The realization (7.68) of the transfer function (7.60) is obviously in dual phase-variable canonical form (observable companion form).

Method 2: Consider the system

$$\overset{(n)}{y_1}(t) + \alpha_1 \overset{(n-1)}{y_1}(t) + \dots + \alpha_{n-1}\dot{y}_1(t) + \alpha_n y_1(t) = u(t) \qquad (7.70)$$

and assume that it is initially at rest. Under this assumption, the response of this system to inputs $\dot{u}(t)$, $\ddot{u}(t)$, ... is $\dot{y}_1(t)$, $\ddot{y}_1(t)$, ... respectively. By the superposition principle, the response $y(t)$ to an input $\beta_1 \overset{(n-1)}{u}(t) + \dots + \beta_n u(t)$ is given by

$$y(t) = \beta_1 \overset{(n-1)}{y_1}(t) + \beta_2 \overset{(n-2)}{y_1}(t) + \dots + \beta_n y_1(t) \qquad (7.71)$$

Now let us recall that the state model of eqn. (7.70) is of the from (7.64). Therefore we may define the state variables as

$$\begin{aligned} x_1(t) &= y_1(t) \\ x_2(t) &= \dot{y}_1(t) \\ &\ \vdots \\ x_n(t) &= \overset{(n-1)}{y_1}(t) \end{aligned} \tag{7.72}$$

This set of equations yields

$$\begin{aligned} \dot{x}_1(t) &= x_2(t) \\ \dot{x}_2\{t) &= x_3(t) \\ &\ \vdots \\ \dot{x}_{n-1}(t) &= x_n(t) \end{aligned}$$

Differentiating $x_n(t) \overset{\Delta'}{=} \overset{(n-1)}{y_1}(t)$ once and using eqn. (7.70) we get

$$\dot{x}_n(t) = -\alpha_n x_1(t) - \alpha_{n-1} x_2(t) - \dots - \alpha_1 x_n(t) + u(t)$$

If we substitute (7.72) into (7.71), we readily see that the output

$$y(t) = \beta_n x_1(t) + \beta_{n-1} x_2(t) + \dots + \beta_1 x_n(t)$$

The foregoing equations can be arranged in the matrix form as

$$\dot{\mathbf{x}}(t) = \mathbf{A}\mathbf{x}(t) + \mathbf{b}u(t) \tag{7.73a}$$

$$y(t) = \mathbf{c}\mathbf{x}(t) \tag{7.73b}$$

where

$$\mathbf{A} = \begin{bmatrix} 0 & 1 & 0 \dots 0 \\ 0 & 0 & 1 \dots 0 \\ \vdots & \vdots & \vdots \\ 0 & 0 & 0 \dots 1 \\ -\alpha_n & -\alpha_{n-1} & -\alpha_{n-2} \ -\alpha_1 \end{bmatrix}; \ \mathbf{b} = \begin{bmatrix} 0 \\ 0 \\ \vdots \\ 0 \\ 1 \end{bmatrix}$$

$$\mathbf{c} = [\beta_n \ \ \beta_{n-1} \ \cdots \ \beta_1]$$

Obviously, state model (7.73) is in controllable companion form defined in (7.65). Note that in Method 1 resulting in an observable companion form, we had simple relations between x_i, y and u (eqn. (7.67)). In Method 2, which resulted in a controllable companion form, there are no simple relations between x_i, y and u. From (7.73) we can write

$$\begin{aligned} y(t) &= \mathbf{c}\mathbf{x}(t) \\ \dot{y}(t) &= \mathbf{c}\mathbf{A}\mathbf{x}(t) + \mathbf{c}\mathbf{b}u(t) \\ \ddot{y}(t) &= \mathbf{c}\mathbf{A}^2\mathbf{x}(t) + \mathbf{c}\mathbf{A}\mathbf{b}u(t) + \mathbf{c}\mathbf{b}\dot{u}(t) \\ &\ \vdots \\ \overset{(n-1)}{y}(t) &= \mathbf{c}\mathbf{A}^{n-1}\mathbf{x}(t) + \mathbf{c}\mathbf{A}^{n-2}\mathbf{b}u(t) + \dots + \mathbf{c}\mathbf{A}\mathbf{b}\overset{(n-3)}{u}(t) + \mathbf{c}\mathbf{b}\overset{(n-2)}{u}(t) \end{aligned}$$

Therefore

$$\mathbf{x}(t) = \begin{bmatrix} \mathbf{c} \\ \mathbf{cA} \\ \mathbf{cA}^2 \\ \vdots \\ \mathbf{cA}^{n-1} \end{bmatrix}^{-1} \begin{bmatrix} y(t) \\ \dot{y}(t) - \mathbf{cb}u(t) \\ \ddot{y}(t) - \mathbf{cb}\dot{u}(t) - \mathbf{cAb}u(t) \\ \vdots \\ \overset{(n-1)}{y}(t) - \mathbf{cb}\overset{(n-2)}{u}(t) - \mathbf{cAb}\overset{(n-3)}{u}(t) - \dots - \mathbf{cA}^{n-2}\mathbf{b}u(t) \end{bmatrix} \quad (7.74)$$

The inverse of the matrix $\begin{bmatrix} \mathbf{c} \\ \mathbf{cA} \\ \mathbf{cA}^2 \\ \vdots \\ \mathbf{cA}^{n-1} \end{bmatrix}$ exists because the pair $\{\mathbf{A}, \mathbf{c}\}$ of state model (7.73) is observable, as we shall see shortly.

We can now relax the assumption of zero initial conditions. This is permitted because the state model for a linear differential system with zero initial conditions is also its state model with nonzero initial conditions. Equations (7.73) therefore represent the state model for the differential equation (7.59) with state-input-output relation given by eqn. (7.74).

The state diagram for (7.73) is shown in Fig. 7.7. Applying Mason's gain formula, we find that the transfer function $\hat{y}(s)/\hat{u}(s)$ is same as the one given by (7.60). Therefore, the state model (7.73) is a realization of transfer function (7.60). Further, the state model (7.73) is irreducible (controllable and observable).

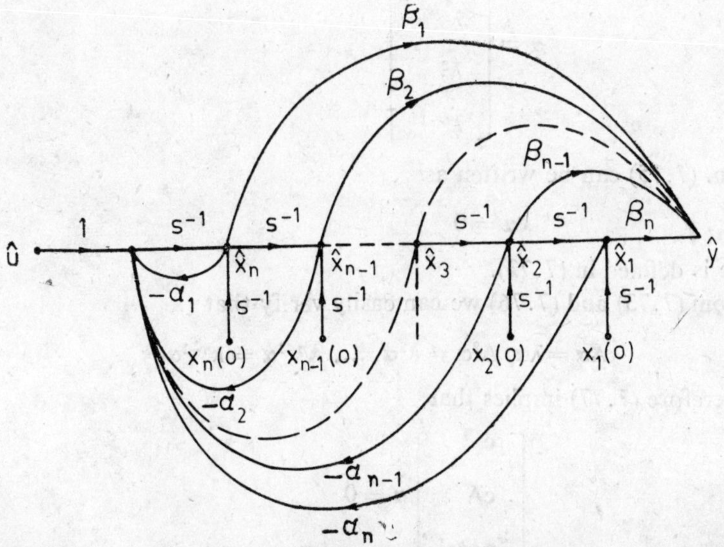

Fig. 7.7 State diagram for equations (7.73)

Differential equation with reducible transfer function

In our discussion so far, we have assumed that the transfer function (7.60) corresponding to differential equation (7.59) is irreducible. Under this assumption, we have obtained controllable and observable realizations of equation (7.59).

Consider now the case wherein the transfer function (7.60) corresponding to differential equation (7.59) has one or more common factors. Applying Method 2 to such a differential equation, we get the state model in the controllable companion form (7.73). The initial condition vector $\mathbf{x}(0)$ is to be obtained from eqn. (7.74). However, if there are common factors in the numerator and denominator of transfer function corresponding to differential equation (7.59), then the pair $\{\mathbf{A}, \mathbf{c}\}$ of (7.73) is not observable and hence the matrix

$$[\mathbf{c} \quad \mathbf{cA} \ \dots \ \mathbf{cA}^{n-1}]^T$$

is singular. This can be established as follows:

Let the transfer function of (7.59) be

$$\hat{h}(s) = \frac{N(s)}{\Delta(s)} = \frac{\beta_1 s^{n-1} + \beta_2 s^{n-2} + \dots + \beta_n}{s^n + \alpha_1 s^{n-1} + \dots + \alpha_n}$$

and $(s-\lambda)$ be a factor of $N(s)$ and $\Delta(s)$.

Then

$$N(\lambda) = \beta_1 \lambda^{n-1} + \beta_2 \lambda^{n-2} + \dots + \beta_n \tag{7.75}$$

$$\Delta(\lambda) = \lambda^n + \alpha_1 \lambda^{n-1} + \dots + \alpha_n \tag{7.76}$$

Define the $n \times 1$ constant vector as

$$\boldsymbol{\alpha} \stackrel{\Delta}{=} \begin{bmatrix} 1 \\ \lambda \\ \lambda^2 \\ \vdots \\ \lambda^{n-1} \end{bmatrix}$$

The eqn. (7.75) can be written as

$$\mathbf{c}\boldsymbol{\alpha} = 0 \tag{7.77}$$

where $\mathbf{c}$ is defined in (7.73).

From (7.73) and (7.76) we can easily verify that

$$\mathbf{A}\boldsymbol{\alpha} = \lambda\boldsymbol{\alpha}, \ \mathbf{A}^2\boldsymbol{\alpha} = \lambda^2\boldsymbol{\alpha}, \ \dots, \ \mathbf{A}^{n-1}\boldsymbol{\alpha} = \lambda^{n-1}\boldsymbol{\alpha}$$

Therefore (7.77) implies that

$$\begin{bmatrix} \mathbf{c} \\ \mathbf{cA} \\ \vdots \\ \mathbf{cA}^{n-1} \end{bmatrix} \boldsymbol{\alpha} = \mathbf{0} \tag{7.78}$$

Since α is a nonzero vector, (7.78) implies that

$$\rho\begin{bmatrix} \mathbf{c} \\ \mathbf{cA} \\ \vdots \\ \mathbf{cA}^{n-1} \end{bmatrix} < n$$

Thus the initial condition vector $\mathbf{x}(0)$ cannot be obtained from the knowledge of the output, input and their derivatives at $t = 0$; thereby the differential equation (7.59) with a reducible transfer function cannot be realized into a state model of the form (7.73).

Applying Method 1 to differential equation (7.59) with reducible transfer function, we get the state model in observable canonical form (7.68). The initial condition vector $\mathbf{x}(0)$ is obtained from eqn. (7.67). However, it may be noted that the realization (7.68) in this case will not be controllable.

Example 7.5: Let us obtain a state model of the form (7.73) for the differential equation

$$\dddot{y} + 6\ddot{y} + 11\dot{y} + 6y = \dot{u} + 4u$$

Comparing this equation with the general equation (7.59a), we observe that

$$n = 3,\ \alpha_1 = 6,\ \alpha_2 = 11,\ \alpha_3 = 6,\ \beta_1 = 0,\ \beta_2 = 1 \text{ and } \beta_3 = 4$$

The state model of the form (7.73) is given below:

$$\begin{bmatrix} \dot{x}_1 \\ \dot{x}_2 \\ \dot{x}_3 \end{bmatrix} = \begin{bmatrix} 0 & 1 & 0 \\ 0 & 0 & 1 \\ -6 & -11 & -6 \end{bmatrix}\begin{bmatrix} x_1 \\ x_2 \\ x_3 \end{bmatrix} + \begin{bmatrix} 0 \\ 0 \\ 1 \end{bmatrix} u$$

$$y = [4 \quad 1 \quad 0]\mathbf{x}$$

Using eqn. (7.74), we get the initial condition vector as

$$\mathbf{x}(0) = \begin{bmatrix} 4 & 1 & 0 \\ 0 & 4 & 1 \\ -6 & -11 & -2 \end{bmatrix}^{-1}\begin{bmatrix} y(0) \\ \dot{y}(0) \\ \ddot{y}(0) - u(0) \end{bmatrix}$$

$$= \begin{bmatrix} \frac{1}{6}(\ddot{y}(0) + 2\dot{y}(0) + 3y(0) - u(0)) \\ \frac{1}{3}(-2\ddot{y}(0) - 4\dot{y}(0) - 3y(0) + 2u(0)) \\ \frac{1}{3}(8\ddot{y}(0) + 19\dot{y}(0) + 12y(0) - 8u(0)) \end{bmatrix}$$

Example 7.6: Let us now consider the differential equation

$$\dddot{y} + 6\ddot{y} + 11\dot{y} + 6y = \dot{u} + u$$

The state model of the form (7.73) is

$$\dot{\mathbf{x}} = \begin{bmatrix} 0 & 1 & 0 \\ 0 & 0 & 1 \\ -6 & -11 & -6 \end{bmatrix} \mathbf{x} + \begin{bmatrix} 0 \\ 0 \\ 1 \end{bmatrix} u$$

$$y = [1 \quad 1 \quad 0]\,\mathbf{x}$$

The matrix

$$\begin{bmatrix} \mathbf{c} \\ \mathbf{cA} \\ \mathbf{cA}^2 \end{bmatrix} = \begin{bmatrix} 1 & 1 & 0 \\ 0 & 1 & 1 \\ -6 & -11 & -5 \end{bmatrix}$$

is singular. Therefore the vector $\mathbf{x}(0)$ cannot be obtained in terms of y, u and their derivatives at $t = 0$.

The transfer function $\hat{h}(s)$ corresponding to the given differential equation is

$$\hat{h}(s) = \frac{\hat{y}(s)}{\hat{u}(s)} = \frac{s+1}{s^3 + 6s^2 + 11s + 6}$$

$$= \frac{s+1}{(s+1)(s+2)(s+3)}$$

$$= \frac{1}{(s+2)(s+3)}$$

As was expected, the transfer function is reducible.

Jordan Canonical Form: We shall first obtain the state model in the Jordan canonical form for differential equation (7.59); the realization for transfer function (7.60) directly follows from this result.

The differential equation (7.59) can be written as

$$y = \frac{\beta_1 p^{n-1} + \ldots + \beta_{n-1} p + \beta_n}{p^n + \alpha_1 p^{n-1} + \ldots + \alpha_{n-1} p + \alpha_n} u \tag{7.79}$$

Consider first the case wherein the denominator polynomial factors into distinct roots λ_i; $i = 1, 2, \ldots, n$. A partial fraction expansion can now be made having the form

$$y = \frac{k_1}{p-\lambda_1} u + \frac{k_2}{p-\lambda_2} u + \ldots + \frac{k_n}{p-\lambda_n} u \tag{7.80}$$

This partial fraction expansion gives a very simple state diagram shown in Fig. 7.8. In Fig. 7.8a, the coefficients k_i; $i = 1, 2, \ldots, n$ are associated with the output. In Fig. 7.8b, they are associated with the input. With the state variables chosen as shown, from Fig. 7.8a we obtain the following state model.

$$\dot{\mathbf{x}}(t) = \mathbf{J}\mathbf{x}(t) + \mathbf{b}u(t) \tag{7.81a}$$

$$y(t) = \mathbf{c}\mathbf{x}(t) \tag{7.81b}$$

where

$$\mathbf{J} = \begin{bmatrix} \lambda_1 & 0 & 0 \ldots 0 \\ 0 & \lambda_2 & 0 \ldots 0 \\ \vdots & \vdots & \vdots \\ 0 & 0 & 0 \ldots \lambda_n \end{bmatrix}; \quad \mathbf{b} = \begin{bmatrix} 1 \\ 1 \\ \vdots \\ 1 \end{bmatrix}; \quad \mathbf{c} = [k_1 \quad k_2 \ldots k_n$$

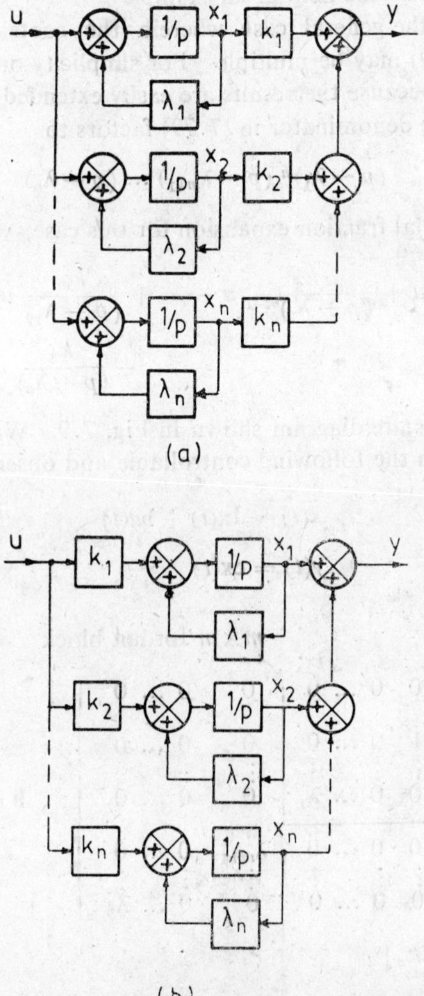

Fig 7.8 Two different state diagrams for equation (7.80)

Similarly the state model from Fig. 7.8b is

$$\dot{\mathbf{x}}(t) = \mathbf{J}\mathbf{x}(t) + \mathbf{b}u(t) \tag{7.82a}$$

$$y(t) = \mathbf{c}\mathbf{x}(t) \tag{7.82b}$$

where

$$\mathbf{J} = \begin{bmatrix} \lambda_1 & 0 & 0 \;\dots\; 0 \\ 0 & \lambda_2 & 0 \;\dots\; 0 \\ \vdots & \vdots & \vdots \qquad \vdots \\ 0 & 0 & 0 \;\dots\; \lambda_n \end{bmatrix}; \quad \mathbf{b} = \begin{bmatrix} k_1 \\ k_2 \\ \vdots \\ k_n \end{bmatrix}; \quad \mathbf{c} = [1 \;\; 1 \;\dots\; 1]$$

The models (7.81) and (7.82) are clearly controllable and observable.

The procedure of obtaining the relations between x_i, y and u will be illustrated shortly with the help of an example.

Consider now the general case wherein the roots of the denominator polynomial of (7.79) may be multiple For simplicity only one multiple root will be considered because the results are easily extended to the general case. We assume that the denominator in (7.79) factors to

$$(p - \lambda_1)^m \, (p - \lambda_{m+1}) \dots (p - \lambda_n)$$

Performing the partial fraction expansion for this case, we get

$$y = \frac{k_1}{(p-\lambda_1)^m}\, u + \frac{k_2}{(p-\lambda_1)^{m-1}}\, u + \dots + \frac{k_m}{(p-\lambda_1)}\, u + \frac{k_{m+1}}{(p-\lambda_{m+1})}\, u + \dots$$
$$+ \frac{k_n}{(p-\lambda_n)}\, u \tag{7.83}$$

This gives a simple state diagram shown in Fig. 7.9. With the state variables as shown, we obtain the following controllable and observable state model.

$$\dot{\mathbf{x}}(t) = \mathbf{J}\mathbf{x}(t) + \mathbf{b}u(t) \tag{7.84a}$$

$$y(t) = \mathbf{c}\mathbf{x}(t) \tag{7.84b}$$

where

$m \times m$ Jordan block

$$\mathbf{J} = \left[\begin{array}{cccccc|ccc} \lambda_1 & 1 & 0 & 0 & \dots & 0 & 0 & 0 & \dots\; 0 \\ 0 & \lambda_1 & 1 & 0 & \dots & 0 & 0 & 0 & \dots\; 0 \\ \vdots & \vdots & \vdots & \vdots & & \vdots & \vdots & \vdots & \quad \vdots \\ 0 & 0 & 0 & 0 & \dots & \lambda_1 & 0 & 0 & \dots\; 0 \\ \hline 0 & 0 & 0 & 0 & \dots & 0 & \lambda_{m+1} & 0 & \dots\; 0 \\ \vdots & \vdots & \vdots & \vdots & & \vdots & \vdots & \vdots & \quad \vdots \\ 0 & 0 & 0 & 0 & \dots & 0 & 0 & 0 & \dots\; \lambda_n \end{array}\right]; \quad \mathbf{b} = \begin{bmatrix} 0 \\ 0 \\ \vdots \\ 1 \\ 1 \\ \vdots \\ 1 \end{bmatrix}$$

$$\mathbf{c} = [k_1 \;\; k_2 \dots k_n]$$

It may be noted here that if the roots of the denominator polynomial in

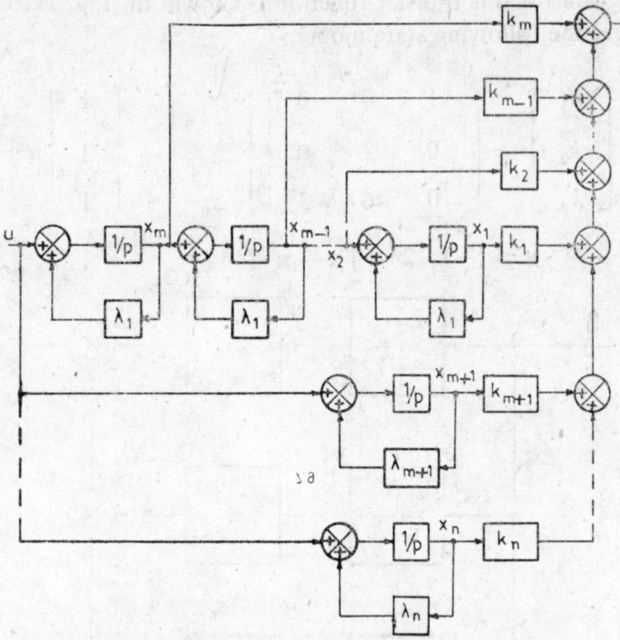

Fig. 7.9 State diagram for equation (7.83)

eqn. (7.79) are complex, the matrices **J**, **b** and **c** in the Jordan canonical forms (7.81), (7.82) and (7.84) will consist of complex numbers. In this case, the Jordan canonical model connot be simulated on an analog computer, for complex numbers cannot be generated in the real world. However, this can be taken care of by introducing some equivalence transformation (refer Problem 2.22).

Example 7.7: Consider the differential equation

$$\dddot{y} + 6\ddot{y} + 11\dot{y} + 6y = \dddot{u} + 8\ddot{u} + 17\dot{u} + 8u \tag{7.85}$$

The corresponding transfer function is

$$\hat{h}(s) = \frac{\hat{y}(s)}{\hat{u}(s)} = \frac{s^3 + 8s^2 + 17s + 8}{s^3 + 6s^2 + 11s + 6}$$

To obtain a state model in Jordan canonical form, we rearrange this transfer function as

$$\hat{h}(s) = \frac{\hat{y}(s)}{\hat{u}(s)} = \frac{s^3 + 8s^2 + 17s + 8}{(s+1)(s+2)(s+3)} \tag{7.86a}$$

$$= 1 + \frac{-1}{s+1} + \frac{2}{s+2} + \frac{1}{s+3} \tag{7.86b}$$

A state diagram for this transfer function is shown in Fig. 7.10. From this figure, we get the following state model.

$$\begin{bmatrix} \dot{x}_1 \\ \dot{x}_2 \\ \dot{x}_3 \end{bmatrix} = \begin{bmatrix} -1 & 0 & 0 \\ 0 & -2 & 0 \\ 0 & 0 & -3 \end{bmatrix} \begin{bmatrix} x_1 \\ x_2 \\ x_3 \end{bmatrix} + \begin{bmatrix} 1 \\ 1 \\ 1 \end{bmatrix} u \tag{7.87a}$$

$$y = [-1 \quad 2 \quad 1]\,\mathbf{x} + u \tag{7.87b}$$

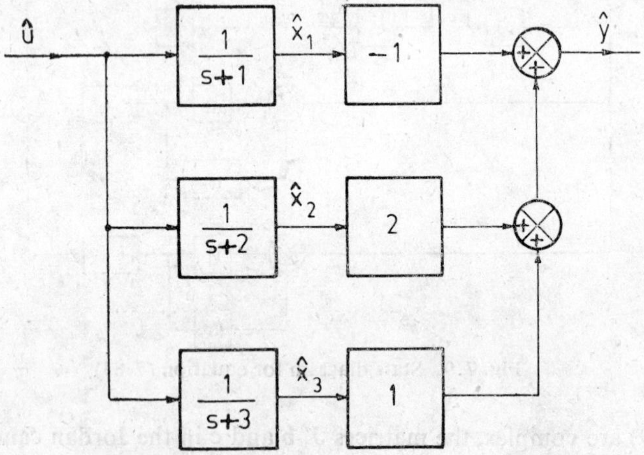

Fig. 7.10 State diagram for equation (7.86b)

In the following, we obtain a relationship between the state vector **x**, output y and input u.

From eqn. (7.86a), we can write

$$\frac{s\hat{y}(s)}{\hat{u}(s)} = s + 2 + \frac{1}{s+1} + \frac{-4}{s+2} + \frac{-3}{s+3} \tag{7.88a}$$

$$\frac{s^2\hat{y}(s)}{\hat{u}(s)} = s^2 + 2s - 6 + \frac{-1}{s+1} + \frac{8}{s+2} + \frac{9}{s+3} \tag{7.88b}$$

Equations (7.86b), (7.88a) and (7.88b) can be rearranged as

$$\begin{bmatrix} \dfrac{\hat{y}(s) - \hat{u}(s)}{\hat{u}(s)} \\[2ex] \dfrac{s\hat{y}(s) - s\hat{u}(s) - 2\hat{u}(s)}{\hat{u}(s)} \\[2ex] \dfrac{s^2\hat{y}(s) - s^2\hat{u}(s) - 2s\hat{u}(s) + 6\hat{u}(s)}{\hat{u}(s)} \end{bmatrix} = \begin{bmatrix} -1 & 2 & 1 \\ 1 & -4 & -3 \\ -1 & 8 & 9 \end{bmatrix} \begin{bmatrix} \dfrac{1}{s+1} \\[2ex] \dfrac{1}{s+2} \\[2ex] \dfrac{1}{s+3} \end{bmatrix}$$

$$= \begin{bmatrix} -1 & 2 & 1 \\ 1 & -4 & -3 \\ -1 & 8 & 9 \end{bmatrix} \begin{bmatrix} \hat{x}_1(s)/\hat{u}(s) \\ \hat{x}_2(s)/\hat{u}(s) \\ \hat{x}_3(s)/\hat{u}(s) \end{bmatrix}$$

This gives

$$\begin{bmatrix} y-u \\ \dot{y}-\dot{u}-2u \\ \ddot{y}-\ddot{u}-2\dot{u}+6u \end{bmatrix} = \begin{bmatrix} -1 & 2 & 1 \\ 1 & -4 & -3 \\ -1 & 8 & 9 \end{bmatrix} \begin{bmatrix} x_1 \\ x_2 \\ x_3 \end{bmatrix} \tag{7.89}$$

Note that the initial conditions $x_1(0)$, $x_2(0)$, $x_3(0)$ may be obtained in terms of specified initial conditions $y(0)$, $\dot{y}(0)$, $\ddot{y}(0)$, $u(0)$, $\dot{u}(0)$, $\ddot{u}(0)$ from relation (7.89).

Realization of Transfer Function Matrices

Consider a $q \times p$ proper rational transfer function matrix $\hat{\mathbf{H}}_1(s)$. It may be expanded into the series

$$\hat{\mathbf{H}}_1(s) = \hat{\mathbf{H}}_1(\infty) + \mathbf{K}_0 s^{-1} + \mathbf{K}_1 s^{-2} + \ldots \tag{7.90}$$

where $\mathbf{K}_i$; $i = 0, 1, 2, \ldots$ are $q \times p$ constant matrices.

Suppose a state variable realization of $\hat{\mathbf{H}}_1(s)$ is found to be

$$\dot{\mathbf{x}} = \mathbf{A}\mathbf{x} + \mathbf{B}\mathbf{u} \tag{7.91a}$$

$$\mathbf{y} = \mathbf{C}\mathbf{x} + \mathbf{D}\mathbf{u} \tag{7.91b}$$

From eqn. (7.17), we have

$$\mathbf{D} = \hat{\mathbf{H}}_1(\infty)$$

Since $\hat{\mathbf{H}}_1(\infty)$ immediately gives the matrix $\mathbf{D}$ of (7.91b), in the following we study only the strictly proper rational function matrix $\hat{\mathbf{H}}(s)$ with series expansion given by

$$\hat{\mathbf{H}}(s) = \mathbf{K}_0 s^{-1} + \mathbf{K}_1 s^{-2} + \ldots \tag{7.92}$$

We seek a realization of $\hat{\mathbf{H}}(s)$ of the form

$$\dot{\mathbf{x}} = \mathbf{A}\mathbf{x} + \mathbf{B}\mathbf{u} \tag{7.93a}$$

$$\mathbf{y} = \mathbf{C}\mathbf{x} \tag{7.93b}$$

If (7.93) is the realization of $\hat{\mathbf{H}}(s)$, then by definition

$$\hat{\mathbf{H}}(s) = \mathbf{C}(s\mathbf{I} - \mathbf{A})^{-1}\mathbf{B}$$

which, with the aid of (5.60), can be expanded into

$$\hat{\mathbf{H}}(s) = \mathbf{CB}s^{-1} + \mathbf{CAB}s^{-2} + \ldots \tag{7.94}$$

Thus (7.93) is the realization of $\hat{\mathbf{H}}(s)$ in (7.92) if and only if

$$\mathbf{K}_i = \mathbf{CA}^i\mathbf{B};\ i = 0, 1, 2, \ldots \tag{7.95}$$

For the transfer function

$$\hat{h}(s) = \frac{\beta_1 s^{n-1} + \beta_2 s^{n-2} + \ldots + \beta_n}{s^n + \alpha_1 s^{n-1} + \ldots + \alpha_n},$$

the companion-form realizations are given in (7.68) and (7.73). We may rearrange the given $\hat{\mathbf{H}}(s)$ in the form

$$\hat{\mathbf{H}}(s) = \frac{\boldsymbol{\beta}_1 s^{n-1} + \boldsymbol{\beta}_2 s^{n-2} + \ldots + \boldsymbol{\beta}_n}{s^n + \alpha_1 s^{n-1} + \ldots + \alpha_n} \tag{7.96}$$

and expand upon (7.68) to realize $\hat{\mathbf{H}}(s)$ in the observable companion form (Fortmann and Hitz 1977):

$$\dot{\mathbf{x}} = \begin{bmatrix} \mathbf{0} & \mathbf{0} \ldots \mathbf{0} & -\alpha_n\mathbf{I} \\ \mathbf{I} & \mathbf{0} \ldots \mathbf{0} & -\alpha_{n-1}\mathbf{I} \\ \mathbf{0} & \mathbf{I} \ldots \mathbf{0} & -\alpha_{n-2}\mathbf{I} \\ \vdots & \vdots \quad \vdots & \vdots \\ \mathbf{0} \ldots \mathbf{0} & \ldots \mathbf{I} & -\alpha_1\mathbf{I} \end{bmatrix} \mathbf{x} + \begin{bmatrix} \boldsymbol{\beta}_n \\ \boldsymbol{\beta}_{n-1} \\ \boldsymbol{\beta}_{n-2} \\ \vdots \\ \boldsymbol{\beta}_1 \end{bmatrix} \mathbf{u}$$

$$\mathbf{y} = [\mathbf{0} \quad \mathbf{0} \ldots \mathbf{0} \quad \mathbf{I}]\mathbf{x} \tag{7.97b}$$

where $\mathbf{0}$ and $\mathbf{I}$ are the $q \times q$ zero and identity matrices respectively. Note that the dimension of the system is nq where n is the order of the denominator polynomial of $\hat{\mathbf{H}}(s)$ in (7.96) and q is the number of outputs. This, of course, reduces to the n-dimensional form (7.68) when $p = q = 1$.

We may expand upon (7.73) to realize $\hat{\mathbf{H}}(s)$ in the controllable companion form:

$$\dot{\mathbf{x}} = \begin{bmatrix} \mathbf{0} & \mathbf{I} & \mathbf{0} & \ldots \mathbf{0} \\ \mathbf{0} & \mathbf{0} & \mathbf{I} & \ldots \mathbf{0} \\ & \vdots & \vdots & \vdots \\ & \mathbf{0} & \mathbf{0} & \ldots \mathbf{I} \\ -\alpha_n\mathbf{I} & -\alpha_{n-1}\mathbf{I} & -\alpha_{n-2}\mathbf{I} & \ldots -\alpha_1\mathbf{I} \end{bmatrix} \mathbf{x} + \begin{bmatrix} \mathbf{0} \\ \mathbf{0} \\ \vdots \\ \mathbf{0} \\ \mathbf{I} \end{bmatrix} \mathbf{u} \tag{7.98a}$$

$$\mathbf{y} = [\boldsymbol{\beta}_n \quad \boldsymbol{\beta}_{n-1} \ldots \boldsymbol{\beta}_1]\mathbf{x} \tag{7.98b}$$

where $\mathbf{0}$ and $\mathbf{I}$ denote the $p \times p$ zero and identity matrices respectively. The dimension of the system is np where n is the order of denominator polynomial of $\hat{\mathbf{H}}(s)$ and p is the number of inputs. (7.98) reduces to the n-dimensional from (7.73) when $p = q = 1$. Further, note that (7.98) is not simply the transpose of (7.97).

It is left to the reader to verify that (7.97) and (7.98) both realize $\hat{\mathbf{H}}(s)$ (Problem 7.18) and that the pair {**A**, **C**} in (7.97) is observable and the pair {**A**, **B**} in (7.98) controllable.

The realizations (7.97)-(7.98) are not irreducible. First we construct (7.97) and then apply Theorem 6.13 to remove the uncontrollable states. This results in controllable and observable (and hence irreducible) realization. Alternatively, and preferably if $p < q$, we may construct (7.98) and then eliminate unobservable states using Theorem 6.14.

Although, this procedure is straightforward, the computational requirements are considerable. For efficient realization algorithms, refer Kalman et al. (1969), Wolovich and Falb (1969), Chen (1970) and Fossard and Gueguen (1977).

Realization of Pulse Transfer Functions/Difference Equations

Consider the general time-invariant linear discrete-time system given by the nth-order linear difference equation

$$\begin{aligned} &y(k+n)+\alpha_1 y(k+n-1)+\ldots+\alpha_{n-1} y(k+1)+\alpha_n y(k)\\ &=\beta_0 u(k+n)+\beta_1 u(k+n-1)+\ldots+\beta_{n-1} u(k+1)+\beta_n u(k) \end{aligned} \tag{7.99}$$

The initial conditions are expressed in terms of $y(0), y(1), \ldots, y(n-1)$. The pulse transfer function obtained from (7.99) is

$$\hat{h}(z)=\frac{\hat{y}(z)}{\hat{u}(z)}=\frac{\beta_0 z^n+\beta_1 z^{n-1}+\ldots+\beta_{n-1} z+\beta_n}{z^n+\alpha_1 z^{n-1}+\ldots+\alpha_{n-1} z+\alpha_n} \tag{7.100}$$

Methods of realization of (7.99)/(7.100) (and the pulse transfer function matrices) are analogous to the ones discussed earlier in this section for linear differential equations/transfer functions (and transfer function matrices). Let us highlight these similarities through some examples.

Example 7.8: Consider a third-order system with pulse transfer function

$$\begin{aligned} \hat{h}_1(z) &= \frac{\beta_0' z^3+\beta_1' z^2+\beta_2' z+\beta_3'}{z^3+\alpha_1 z^2+\alpha_2 z+\alpha_3}\\ &=\beta_0'+\frac{(\beta_1'-\beta_0'\alpha_1) z^2+(\beta_2'-\beta_0'\alpha_2) z+(\beta_3'-\beta_0'\alpha_3)}{z^3+\alpha_1 z^2+\alpha_2 z+\alpha_3}\\ &=\beta_0'+\frac{\beta_1 z^2+\beta_2 z+\beta_3}{z^3+\alpha_1 z^2+\alpha_2 z+\alpha_3}\\ &=\beta_0'+\hat{h}(z) \end{aligned} \tag{7.101}$$

The realization of $\hat{h}_1(z)$ is of the form

$$\mathbf{x}(k+1)=\mathbf{F}\mathbf{x}(k)+\mathbf{g}\,u(k) \tag{7.102a}$$

$$y(k)=\mathbf{c}\mathbf{x}(k)+d\,u(k) \tag{7.102b}$$

where from eqn. (7.29a),

$$\hat{h}_1(z) \quad = \mathbf{c}(z\mathbf{I} - \mathbf{F})^{-1}\,\mathbf{g} + d \tag{7.103}$$

Comparing (7.103) with (7.101), we find that

$$d = \beta_0'$$

and

$$\hat{h}(z) = \mathbf{c}(z\mathbf{I} - \mathbf{F})^{-1}\mathbf{g}$$

Thus, with the value of d in our realization (7.102) known, we have now to realize $\hat{h}(z)$ in the form

$$\mathbf{x}(k+1) = \mathbf{F}\mathbf{x}(k) + \mathbf{g}u(k)$$

$$y'(k) = \mathbf{c}\mathbf{x}(k)$$

From (7.101), we may write

$$\hat{h}(z) = \frac{\beta_1 z^2 + \beta_2 z + \beta_3}{z^3 + \alpha_1 z^2 + \alpha_2 z + \alpha_3}$$

$$= \frac{\beta_1 z^{-1} + \beta_2 z^{-2} + \beta_3 z^{-3}}{1 - (-\alpha_1 z^{-1} - \alpha_2 z^{-2} - \alpha_3 z^{-3})} \tag{7.104}$$

State diagram for this transfer function is shown in Fig. 7.11. Note that this state diagram is analogous to the one shown in Fig. 7.6 for continuous-time case.

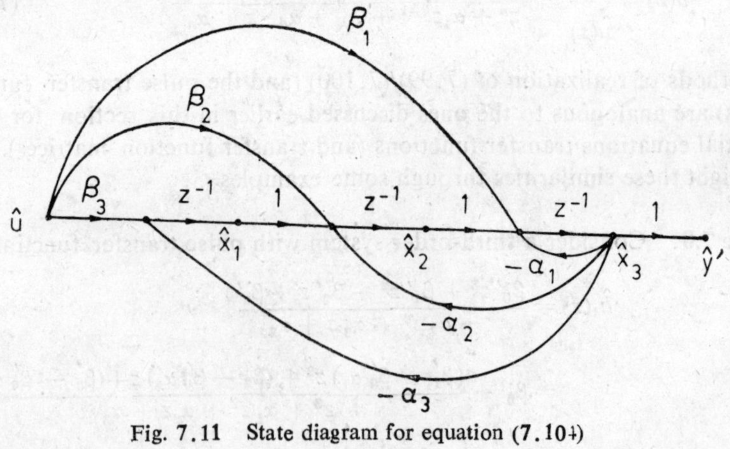

Fig. 7.11 State diagram for equation (7.104)

Taking outputs of delay units as state variables, we can write from Fig. 7.11,

$$x_1(k+1) = -\alpha_3 x_3(k) + \beta_3 u(k)$$

$$x_2(k+1) = x_1(k) - \alpha_2 x_3(k) + \beta_2 u(k)$$

$$x_3(k+1) = x_2(k) - \alpha_3 x_3(k) + \beta_1 u(k)$$

$$y'(k) = x_3(k)$$

In the vector-matrix notation

$$\mathbf{x}(k+1) = \begin{bmatrix} 0 & 0 & -\alpha_3 \\ 1 & 0 & -\alpha_2 \\ 0 & 1 & -\alpha_1 \end{bmatrix} \mathbf{x}(k) + \begin{bmatrix} \beta_3 \\ \beta_2 \\ \beta_1 \end{bmatrix} u(k) \tag{7.105a}$$

$$y'(k) = [0 \quad 0 \quad 1]\,\mathbf{x}(k) \tag{7.105b}$$

This, as we know, is the observable canonical form.

Figure 7.12 gives an alternate state diagram for the transfer function (7.104). This is analogous to Fig. 7.7. Again, taking outputs of delay units as state variables, we obtain the following state model from Fig. 7.12.

$$\mathbf{x}(k+1) = \begin{bmatrix} 0 & 1 & 0 \\ 0 & 0 & 1 \\ -\alpha_3 & -\alpha_2 & -\alpha_1 \end{bmatrix} \mathbf{x}(k) + \begin{bmatrix} 0 \\ 0 \\ 1 \end{bmatrix} u(k) \tag{7.106n}$$

$$y'(k) = [\beta_3 \quad \beta_2 \quad \beta_1]\,\mathbf{x}(k) \tag{7.106b}$$

This is the controllable canonical form.

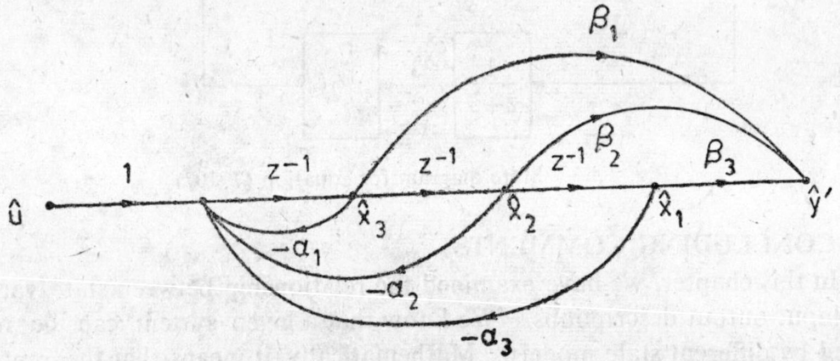

Fig. 7.12 Alternate state diagram for equation (7.104)

The results in (7.105) and (7.106) can easily be generalized for an nth-order pulse tranfser function. Furthermore, because of symmetry, these results can be written down directly by inspection of the pulse transfer function (7.104) without drawing the state diagrams.

Example 7.9: Let us obtain the Jordan canonical form realization for the pulse transfer function

$$\hat{h}(z) = \frac{\hat{y}(z)}{\hat{u}(z)} = \frac{z+6}{z^3+5z^2+7z+3}$$

To obtain the state model in the Jordan canonical form, we rearrange this transfer function as

$$\hat{y}(z) = \left[\frac{5/2}{(z+1)^2} + \frac{-3/4}{z+1} + \frac{3/4}{z+3}\right]\hat{u}(z) \tag{7.107}$$

Figure 7.13 shows a state diagram. Taking outputs of delay units as state variables, we obtain the following state model.

$$\mathbf{x}(k+1) = \begin{bmatrix} -1 & 1 & 0 \\ 0 & -1 & 0 \\ 0 & 0 & -3 \end{bmatrix}\mathbf{x}(k) + \begin{bmatrix} 0 \\ 1 \\ 1 \end{bmatrix} u(k)$$

$$y(k) = [\tfrac{5}{2} \quad -\tfrac{3}{4} \quad \tfrac{3}{4}]\,\mathbf{x}(k)$$

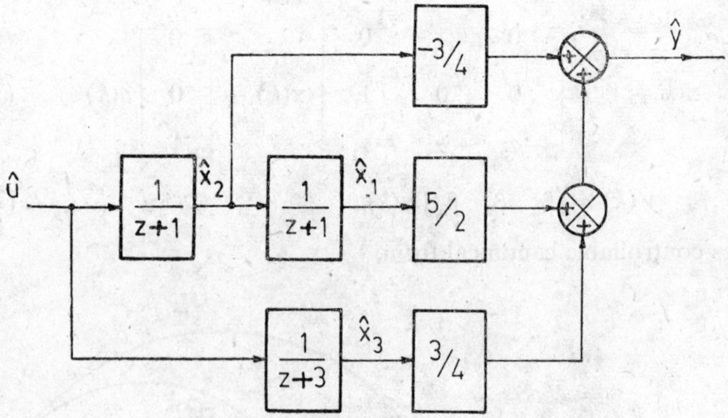

Fig. 7.13 State diagram for equation (7.107)

7.6 CONCLUDING COMMENTS

In this chapter, we have examined the relationship between state variable and input-output descriptions. We know that a given system can be represented by different state models. Mathematically, it means that the representation depends upon the basis chosen for the state space. However, we noted in this chapter that input-output description has nothing to do with the basis; all state models of a given system lead to the same input-output description. Section 7.2 gave the methods of obtaining input-output description from state models. This section also showed that every proper rational function is realizable by a finite-dimensional linear time-invariant state model. For input-output descriptions, we defined output controllability in Section 7.3. In Section 7.4, the underlying structure of a linear system with regard to controllability, observability and transfer function matrices was explored. It was shown that a transfer function matrix for a given state model depends only on the controllable and observable part of the model.

In Section 7.5, we turned to the problem of finding a state-space realization for a given input-output description. Two new canonical forms—controllable companion form and observable companion form—were introduced for this purpose. It was found that these forms can be read off directly from the coefficients of given transfer function/differential equation/difference equation. The Jordan canonical form realization is comparatively more difficult to obtain. The realization of transfer function matrices has not been treated in detail. For further study on this topic, references have been given.

Additional useful references on the subject of realization are as follows: Ackermann and Bucy (1971), Anderson et al. (1966), Mayne (1968, 1973), Montes (1976), Rissanen (1971), Silvermann (1971) and Wolovich (1974).

PROBLEMS

7.1 Determine the transfer function and impulse response of the scalar system described by the state model of eqns. (4.27).

7.2 Determine the transfer function matrix and impulse response matrix of the multivariable system described by the state model of eqns. (4.38).

7.3 Using relations (7.20), prove the Cayley-Hamilton theorem: Every square matrix **A**, satisfies its own characteristic equation.
(Hint: Let

$$\lambda^n + \alpha_1\lambda^{n-1} + \dots + \alpha_{n-1}\lambda + \alpha_n = 0$$

be the characteristic equation of **A**. Using relations (7.20) prove that

$$\mathbf{A}^n + \alpha_1\mathbf{A}^{n-1} + \dots + \alpha_{n-1}\mathbf{A} + \alpha_n\mathbf{I} = \mathbf{0}.)$$

7.4 A discretized version of eqns. (4.27) is given in Problem 5.15. Find the pulse transfer function and pulse response matrix. Compare the pulse response with the impulse response of Problem 7.1.

7.5 Repeat Problem 7.4 for the discretized version of eqns. (4.38) given in Problem 5.16.

7.6 Show that the system

$$\dot{\mathbf{x}} = \mathbf{Ax} + \mathbf{Bu}$$

$$\mathbf{y} = \mathbf{Cx} + \mathbf{Du}$$

where **A, B, C** and **D** are, respectively, $n\times n$, $n\times p$, $q\times n$ and $q\times p$ matrices, is completely output controllable if and only if the composite $q\times(n+1)p$ matrix

$$[\mathbf{CB} \mid \mathbf{CAB} \mid \dots \mid \mathbf{CA}^{n-1}\mathbf{B} \mid \mathbf{D}]$$

is of rank q.

7.7 Consider a linear system described by the differential equation

$$\ddot{y} + 2\dot{y} + y = \dot{u} + u$$

Choose $x_1 = y$ and $x_2 = \dot{y} - u$
Show that the system is completely output controllable but not state controllable.

7.8 Show that the system

$$\dot{\mathbf{x}} = \begin{bmatrix} 0 & 1 \\ -1 & -2 \end{bmatrix} \mathbf{x} + \begin{bmatrix} 1 \\ -1 \end{bmatrix} u$$

$$\mathbf{y} = \begin{bmatrix} 1 & 0 \\ 1 & 1 \end{bmatrix} \mathbf{x}$$

is neither state controllable, nor output controllable.

7.9 The state model of Example 6.17 is reproduced below:

$$\dot{\mathbf{x}} = \begin{bmatrix} -2 & 1 \\ 1 & -2 \end{bmatrix} \mathbf{x} + \begin{bmatrix} 1 \\ 0 \end{bmatrix} u$$

$$y = [1 \quad -1]\,\mathbf{x}$$

Find the transfer function and therefrom show that the state model is reducible.

7.10 Given the transfer function

$$\hat{h}(s) = \frac{1}{s+2}$$

Find

(i) an irreducible realization

(ii) a controllable but unobservable realization

(iii) an observable but uncontrollable realization

(iv) an unobservable and uncontrollable realization

7.11 Show that the state model

$$\dot{\mathbf{x}} = \begin{bmatrix} -1 & 0 & 0 \\ 0 & -2 & 0 \\ 0 & 0 & -3 \end{bmatrix} \mathbf{x} + \begin{bmatrix} 1 & 0 \\ 1 & 2 \\ 2 & 1 \end{bmatrix} \mathbf{u}$$

$$\mathbf{y} = \begin{bmatrix} 1 & 1 & 2 \\ 3 & 1 & 5 \end{bmatrix} \mathbf{x}$$

realizes the transfer function matrix

$$\hat{\mathbf{H}}(s) = \frac{1}{(s+1)(s+2)(s+3)} \begin{bmatrix} 6s^2+21s+17 & 4s^2+14s+10 \\ 14s^2+49s+41 & 7s^2+23s+16 \end{bmatrix}$$

Is this realization reducible? If yes, find an irreducible realization.

7.12 Show that the state model

$$\dot{\mathbf{x}} = \begin{bmatrix} 0 & 1 & 0 & 0 \\ 0 & 0 & 1 & 0 \\ -1 & -2 & -2 & 0 \\ 1 & 1 & 1 & -3 \end{bmatrix} \mathbf{x} + \begin{bmatrix} 0 \\ 0 \\ 1 \\ 0 \end{bmatrix} u$$

$$y = [0 \quad 1 \quad 1 \quad 0]\mathbf{x}$$

realizes the transfer function

$$\hat{h}(s) = \frac{s}{s^2 + s + 1}$$

Is this realization reducible? If yes, find an irreducible realization,

7.13 Find state models for the following differential equations. Obtain different canonical form for each system

(i) $\dddot{y} + 3\ddot{y} + 2\dot{y} = \dot{u} + u$

(ii) $\dddot{y} + 6\ddot{y} + 11\dot{y} + 6y = u$

(iii) $\dddot{y} + 6\ddot{y} + 11\dot{y} + 6y = \dddot{u} + 8\ddot{u} + 17\dot{u} + 8u$

7.14 Find the realizations of the transfer functions

(i) $\dfrac{5}{(s+1)^2\,(s+2)}$

(ii) $\dfrac{s^3 + 8s^2 + 17s + 8}{(s+1)(s+2)(s+3)}$

Are these realizations irreducible? Give block diagrams for analog computer simulation of these transfer functions.

7.15 Give three irreducible realizations for the transfer function

$$\hat{h}(s) = \frac{10(s+4)}{s(s+1)(s+3)}$$

7.16 Find state models for the following differnece equations. Obtain different canonical form for each system.

(i) $y(k+3) + 5y(k+2) + 7y(k+1) + 3y(k) = 0$

(ii) $y(k+2) + 3y(k+1) + 2y(k) = 5u(k+1) + 3u(k)$

(iii) $y(k+3) + 5y(k+2) + 7y(k+1) + 3y(k) = u(k+1) + 2u(k)$

7.17 Give three irreducible realizations for the pulse transfer function

$$\hat{h}(z) = \frac{4z^3 - 12z^2 + 13z - 7}{(z-1)^2(z-2)}$$

Give state diagram for each realization.

7.18 Prove that state models (7.97) and (7.98) realize the transfer function matrix (7.96).

(Hint: From (7.94) and (7.96), we may write

$$(s^n + \alpha_1 s^{n-1} + \dots + \alpha_n)[\mathbf{CB}s^{-1} + \mathbf{CAB}s^{-2} + \dots]$$
$$= \boldsymbol{\beta}_1 s^{n-1} + \boldsymbol{\beta}_2 s^{n-2} + \dots + \boldsymbol{\beta}_n$$

which implies, by identifying the coefficients of same power of s, that

$$\mathbf{CB} = \boldsymbol{\beta}_1$$
$$\mathbf{CAB} + \alpha_1 \mathbf{CB} = \boldsymbol{\beta}_2$$
$$\vdots$$

Identify these relations and verify using eqns. (7.96)–(7.98).)

7.19 Find a realization of the transfer function matrix

$$\hat{\mathbf{H}}(s) = \begin{bmatrix} \dfrac{1}{s+1} & \dfrac{1}{s^2+3s+2} \end{bmatrix}$$

Is your realization irreducible? If not, find an irreducible realization.

7.20 Find an irreducible realization for the pulse transfer function matrix

$$\hat{\mathbf{H}}(z) = \begin{bmatrix} \dfrac{2+z}{z+1} & \dfrac{1}{z+4} \\ \dfrac{z}{z+1} & \dfrac{z+1}{z+2} \end{bmatrix}$$

7.21 Reconsider the system of Problem 5.19. Work out a suitable compensator transfer function $\hat{D}(z)$ so that the output y responds to the input r in a deadbeat manner (Fig. P-5.19). Also find an irreducible realization of the pulse transfer function of the compensator.

7.22 A necessary and sufficient condition for the *cyclicity* of matrix $\mathbf{A}$ is that the resolvent matrix $\Phi(s) = (s\mathbf{I}-\mathbf{A})^{-1}$ is irreducible, i.e., all n^2 elements of the numerator polynomial matrix $(s\mathbf{I}-\mathbf{A})^+$ do not have factors in common with denominator polynomial $|s\mathbf{I}-\mathbf{A}|$ (for proof, refer Seraji (1975)).

Apply this test to show that matrix $\mathbf{A}$ of Example 2.14 is cyclic while that of Example 2.12 is noncyclic.

REFERENCES

1. Ackermann, J.E., and R.S. Bucy, "Canonical minimal realization of a matrix of impulse response sequences", *Info. and Control*, vol. 9, pp. 224–231, **1971.**
2. Anderson, B.D.O., R.W. Newcomb, R.E. Kalman, and D.C. Youla, "Equivalence of

linear time-invariant dynamical systems", *J. Franklin Inst.*, vol. 281, pp. 371–378, **1966.**

3. Cadzow, J.A., and H.R. Martens, *Discrete-Time and Computer Control Systems*, Englewood Cliffs, N.J.: Prentice-Hall, **1970.**
4. Chen, C.T., *Introduction to Linear System Theory*, New York: Holt, Rinehart and Winston, **1970.**
5. Faddeev, V.N., *Computational Methods in Linear Algebra*, New York: Dover, **1959.**
6. Fortmann, T.E., and K.L. Hitz, *An Introduction to Linear Control Systems*, New York: Marcel Dekker, **1977.**
7. Fossard, A., and C. Gueguen, *Multivariable System Control*, Amsterdam: North-Holland Publishing Company, **1977.**
8. Kalman, R.E., P.L. Falb, and M.A. Arbib, *Topics in Mathematical System Theory*, New York: McGraw-Hill, **1969.**
9. Mayne, D.Q., "Computational procedure for the minimal realization of transfer function matrices", *Proc. IEE*, vol. 115, pp. 1363–1368, **1968.**
10. Mayne, D.Q., "An elementary derivation of Rosenbrock's minimum realization algorithm". *IEEE Trans. Automat. Contr.*, vol. AC-18, pp. 306–307, **1973.**
11. Melsa, J.L., and S.K. Jones, *Computer Programs for Computational Assistance in the Study of Linear Control Systems*, 2nd Edition, New York: McGraw-Hill, **1973.**
12. Montes, C.G., "Minimal realization of transfer function matrix in canonical from" *IEEE Trans. Automat. Contr.*, vol. AC-21, pp. 399–401, **1976.**
13. Nagrath, I.J., and M. Gopal, *Systems: Modelling and Analysis*, New Delhi: Tata McGraw-Hill, **1982.**
14. Rissanen, J., "Recursive identification of linear systems", *SIAM J. Control*, vol. 9, pp. 420-430, **1971.**
15. Rosenbrock, H.H., *State Space and Multivariable Theory*, London: Nelson, **1970.**
16. Rugh, W.J., *Mathematical Description of Linear Systems*, New York: Marcel Dekker, **1975.**
17. Seraji, H., "Cyclicity of linear multivariable systems", *Int. J. Control*, vol. 21, pp. 497-504, **1975.**
18. Silverman, L.M., "Realization of linear dynamical systems", *IEEE Trans. Automat. Contr.*, vol. AC-16, pp. 554-567, **1971.**
19. Wolovich, W.A., *Linear Multivariable Systems*, Berlin: Springer Verlag, **1974.**
20. Wolovich, W.A., and P.L. Falb, "On the structure of multivariable systems", *SIAM J. Control*, vol. 7, pp. 437-451, **1969.**
21. Zadeh, L., and C.A. Desoer, *Linear System Theory*, New York: McGraw-Hill, **1963.**

8. STABILITY

8.1 INTRODUCTION

Roughly speaking, stability in a system implies that small changes in the system inputs, in initial conditions or in system parameters do not result in large changes in system behaviour. Stability is a very important characteristic of the transient response of a system. Most of the working systems are designed to be stable. Within the boundaries of parameter variations permitted by stability considerations, we then seek to improve system performance.

In linear time-invariant systems, stability is relatively easy to determine. Powerful tests such as the Routh-Hurwitz criterion yield the necessary and sufficient conditions of stability. As we shall see later in this chapter, in a stable linear time-invariant system,

(i) with bounded input, the output is bounded, and
(ii) with zero input and with arbitrary initial conditions, the output tends towards zero—the equilibrium state of the system.

The second notion of stability concerns a free system relative to its transient behaviour. The first notion concerns a system under the influence of inputs. Clearly, if a system is subjected to an unbounded input and produces an unbounded response, nothing can be said about its stability. But if it is subjected to a bounded input and produces an unbounded response, it is, by definition, unstable. Actually, the output of an unstable system may increase to a certain extent and then the system may break down or become nonlinear after the output exceeds a certain magnitude, so that the linear mathematical model no longer applies.

As we shall see in this chapter, the two notions of stability are essentially equivalent in linear time-invariant systems.

In nonlinear systems, unfortunately, there is no definite correspondence between the two notions. For a free stable nonlinear system, there is no guarantee that output will be bounded whenever input is bounded. Also, if the output is bounded for a particular input, it may not be bounded for other inputs. Most of the important results obtained thus for concern the stability of nonlinear systems in the sense of the second notion above, i.e., when the system has no input. It may be noted that even for this class of nonlinear systems, the concept of stability is not clear cut. The linear autonomous[1] systems (with nonzero eigenvalues) have only one equilibrium state and their behaviour about the equilibrium state completely determines the qualitative behaviour in the entire state-plane. In nonlinear systems, on the other hand, system behaviour for small deviations about the equilibrium point may be different from that for large deviations. Therefore *local stability* does not imply stability in the overall state-plane and the two concepts should be considered separately. Secondly, in nonlinear systems, because of possible existence of multiple equilibrium states, the system trajectories may move away from one equilibrium state to the other as time progresses. Thus it appears that in the case of nonlinear systems, it is simpler to speak of system stability relative to the equilibrium state rather than using the general term 'stability of a system'. Another important point to be kept in mind is that in linear autonomous systems, when oscillations occur, the resulting trajectories will be closed curves. The amplitude of the oscillations is not fixed. It changes with the size of the initial conditions. Slight changes in system parameters (shifting the eigenvalues from the imaginary axis of the complex plane) will destroy the oscillations. In nonlinear systems, on the other hand, there can be oscillations that are independent of the size of initial conditions and these oscillations (*limit cycles*) are usually much less sensitive to parameter variations. Limit cycles of fixed amplitude and period can be sustained over a finite range of system parameters.

The purpose of this chapter is to introduce the reader to the more rigorous definitions of stability and to Lyapunov's methods, and to the role that the state variable approach plays. The word 'introduce' is carefully chosen since there already exist books exclusively devoted to stability theory and Lyapunov's methods (Hahn 1963, LaSalle and Lefschetz 1961, Lefschetz 1965, Aizerman and Gantmacher 1964).

8.2 EQUILIBRIUM POINTS

Consider a linear system with state model

$$\dot{\mathbf{x}}(t) = \mathbf{A}\mathbf{x}(t) + \mathbf{B}\mathbf{u}(t) \tag{8.1a}$$

$$\mathbf{y}(t) = \mathbf{C}\mathbf{x}(t) + \mathbf{D}\mathbf{u}(t) \tag{8.1b}$$

where $\mathbf{x}$ is the $n\times 1$ state vector, $\mathbf{u}$ is the $p\times 1$ input vector, $\mathbf{y}$ is the $q\times 1$ output vector and $\mathbf{A}$, $\mathbf{B}$, $\mathbf{C}$ and $\mathbf{D}$ are, respectively, $n\times n$, $n\times p$, $q\times n$ and $q\times p$ real constant matrices. The time interval of interest is $[0, \infty)$.

If, for any constant input vector $\mathbf{u}(t) = \mathbf{u}^c$, there exists a point $\mathbf{x}(t) = \mathbf{x}^e =$ constant in state space such that at this point $\dot{\mathbf{x}}(t) = \mathbf{0}$ for all t, then this point is called an *equilibrium point* of the system corresponding to the particular input $\mathbf{u}^c$. Applying this definition to system (8.1), we require

[1]An *autonomous system* is one that is both free and time-invariant (refer Section 5.2).

$$\mathbf{A}\mathbf{x}^e + \mathbf{B}\mathbf{u}^c = \mathbf{0} \tag{8.2}$$

Assuming **A** to be nonsingular, we have from eqn. (8.2)

$$\mathbf{x}^e = -\mathbf{A}^{-1}\mathbf{B}\mathbf{u}^c \tag{8.3}$$

If **A** is singular, then $\mathbf{x}^e$ is not a discrete point but a continuum of points (refer Section 2.6).

We shall deal with systems for which the equilibrium points are discrete points.

In stability analysis, quite often it is convenient to shift the origin of the state space of the system (8.1) to the equilibrium point (8.3). Defining

$$\hat{\mathbf{x}}(t) = \mathbf{x}(t) - \mathbf{x}^e \tag{8.4a}$$

the state equation (8.1a), referred to $\mathbf{x} = \mathbf{x}^e$ as origin, becomes

$$\dot{\hat{\mathbf{x}}}(t) = \mathbf{A}\hat{\mathbf{x}}(t) \tag{8.4b}$$

It is obvious that $\hat{\mathbf{x}} = \mathbf{0}$ is the equilibrium point. Hence forth we shall take origin as the equilibrium point.

Consider now a linear time-varying system with state model

$$\dot{\mathbf{x}}(t) = \mathbf{A}(t)\mathbf{x}(t) + \mathbf{B}(t)\mathbf{u}(t) \tag{8.5a}$$

$$\mathbf{y}(t) = \mathbf{C}(t)\mathbf{x}(t) + \mathbf{D}(t)\mathbf{u}(t) \tag{8.5b}$$

where $\mathbf{A}(t)$, $\mathbf{B}(t)$, $\mathbf{C}(t)$ and $\mathbf{D}(t)$ are, respectively, $n \times n$, $n \times p$, $q \times n$ and $q \times p$ real matrices whose entries are continuous functions of t defined over $(-\infty, \infty)$. In a time-varying system, it is probable that $\mathbf{A}^{-1}(t)\,\mathbf{B}(t)$ will be time-varying; in this case therefore the only solution of (8.3) valid for all t will be $\mathbf{x}^e = \mathbf{0} = \mathbf{u}^c$, i.e., the origin of the state space with zero input.

For a nonlinear system represented by

$$\dot{\mathbf{x}}(t) = \mathbf{f}(\mathbf{x}, \mathbf{u}, t) \tag{8.6}$$

any equilibrium point must satisfy

$$\mathbf{f}(\mathbf{x}^e, \mathbf{u}^c, t) = \mathbf{0} \text{ for all } t \tag{8.7}$$

The number of solutions depends entirely upon the nature of $\mathbf{f}(\cdot)$ and no general statement is possible. Stability analysis of a particular equilibrium point of (8.6) is conveniently carried out by shifting the origin of the state space to this point.

8.3 STABILITY CONCEPTS AND DEFINITIONS

In stability study, we are generally concerned with the following questions:

(i) If the system with zero input is perturbed from the equilibrium point $\mathbf{x}^e$ at $t = t_0$, will the state $\mathbf{x}(t)$ return to $\mathbf{x}^e$, remain 'close' to $\mathbf{x}^e$, or diverge from $\mathbf{x}^e$?

(ii) If the system is relaxed, will a bounded input produce a bounded output?

The first question is concerned with the boundedness of free response of a system. This concept of stability was formulated by the Russian mathematician A.M. Lyapunov. The second question is concerned with stability in terms of input-output description of a system. Since the input-output description is applicable only when the system is relaxed, the second question is applicable to relaxed systems.

Stability in the Sense of Lyapunov

Consider a zero-input system described by the state equation

$$\dot{\mathbf{x}}(t) = \mathbf{f}(\mathbf{x}, t) \tag{8.8}$$

where $\mathbf{x}$ is $n \times 1$ state vector.

Assume that the system has only one equilibrium point (This is the case with all properly designed systems). Furthermore, without loss of generality, let the origin of state space be taken as the equilibrium point.

The system described by eqn. (8.8) is *stable in the sense of Lyapunov* at the origin if, for every initial state $\mathbf{x}(t_0)$ which is sufficiently close to the origin, $\mathbf{x}(t)$ remains near the origin for all t. It is *asymptotically stable* if $\mathbf{x}(t)$ in fact approaches the origin as $t \to \infty$. It is *asymptotically stable in-the-large* if it is asymptotically stable for every initial state regardless of how near or far it is from the origin.

Following are the mathematically precise definitions of different types of stability (Hsu and Meyer 1968).

The system described by eqn. (8.8) is *stable in the sense of Lyapunov* at the origin if, for every real number $\varepsilon > 0$, there exists a real number $\delta(\varepsilon) > 0$ such that $\|\mathbf{x}(t_0)\| \leqslant \delta$ results in $\|\mathbf{x}(t)\| \leqslant \varepsilon$ for all $t \geqslant t_0$.

This definition uses the concept of vector norm. The euclidean norm for a vector $\mathbf{x}$ with n components $x_1, x_1, \ldots, x_n$ is (refer Section 2.5)

$$\|\mathbf{x}\| = (x_1^2 + x_2^2 + \ldots + x_n^2)^{1/2}$$

$\|\mathbf{x}\| \leqslant R$ defines a hyper-spherical region $S(R)$ of radius R surrounding the equilibrium point $\mathbf{x} = \mathbf{0}$. In terms of euclidean norm, the above definition of stability implies that for any $S(\varepsilon)$ that we may designate, the designer must produce $S(\delta)$ so that system state initially in $S(\delta)$ will never leave $S(\varepsilon)$. This is illustrated in Fig. 8.1a.

Consider, for example, the linear oscillator described by the differential equation

$$\ddot{y}(t) + \omega^2 y(t) = 0$$

where ω is the frequency of oscillation.

Define the state variables as

$$x_1(t) = y(t),\ x_2(t) = \dot{y}(t)$$

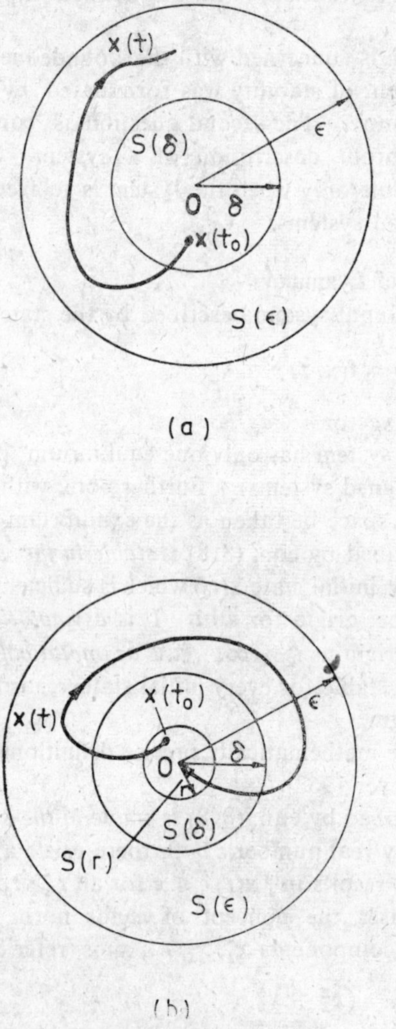

Fig. 8.1 Stability definitions

This gives

$$\dot{x}_1(t) = x_2(t)$$

$$\dot{x}_2(t) = -\omega^2 x_1(t)$$

or

$$\frac{dx_2}{dx_1} = -\omega^2 \frac{x_1}{x_2}$$

or

$$x_2^2 + \omega^2 x_1^2 = c^2 \; ; \quad c = \text{constant}$$

Several solutions for various values of c, corresponding to various initial conditions of x_1 and x_2 are shown in Fig. 8.2. For a specified value of ε,

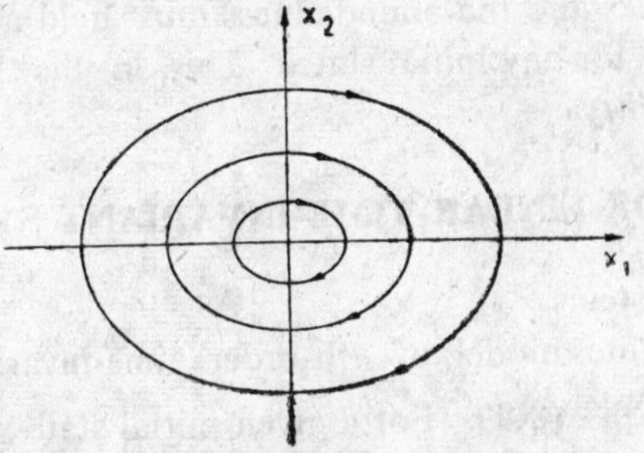

Fig. 8.2

we can find a closed phase trajectory whose maximum distance from the origin is ε. We then select a value of δ which is less than the minimum distance from that curve to the origin. The δ(ε) so chosen, will satisfy the conditions that guarantee stability in the sense of Lyapunov.

The system (8.8) is said to be *locally stable* (or *stable in-the-small*) at the origin if the region $S(\varepsilon)$ is small.

Let us now turn to the other two definitions of stability.

The system (8.8) is *asymptotically stable* at the origin if

(a) it is stable in the sense of Lyapunov,

(b) there exists a real number $r > 0$ such that

$$\| \mathbf{x}(t_0) \| \leqslant r \text{ results in } \mathbf{x}(t) \to \mathbf{0} \text{ as } t \to \infty.$$

Property (b) implies that every motion starting in $S(r)$ con erges to the origin as $t \to \infty$. This is illustrated in Fig. 8.1b.

The system (8.8) is *asymptotically stable in-the-large* (*globally asymptotically stable*) at the origin if

(a) it is stable in the sense of Lyapunov,

(b) every initial state $\mathbf{x}(t_0)$ results in

$$\mathbf{x}(t) \to \mathbf{0} \text{ as } t \to \infty.$$

The asymptotic stability in-the-large guarantees that every motion will approach the origin.

Bounded-Input, Bounded-Output Stability

A relaxed system is said to be *bounded-input, bounded-output (BIBO) stable* if and only if for any bounded input, the output is bounded.

Note that a BIBO stable system under the assumption of relaxedness, might not be BIBO stable if it is not initially relaxed.

So far we have studied the stability of zero-input response and that of zero-state response. We now consider the entire response of a system.

A minimum stability requirement for the entire response is to ask that the system output as well as all the state variables be bounded for all initial states and for any bounded input. This requirement is obviously more stringent than BIBO stability; it requires not only boundedness of output but also of state variables and the boundedness must hold not only for the zero initial state but also for any initial state. This, in the Russian literature, is known as *total stability*.

8.4 STABILITY OF LINEAR TIME-INVARIANT SYSTEMS

Continuous-Time Systems

The general state model of *n*th-order time-invariant continuous-time system is given by eqns. (8.1). For a given initial state $\mathbf{x}(0) \triangleq \mathbf{x}^0$, the solution of the state equation (8.1a) is (refer equation (5.55))

$$\mathbf{x}(t) = e^{\mathbf{A}t}\mathbf{x}^0 + \int_0^t e^{\mathbf{A}(t-\tau)}\mathbf{B}\mathbf{u}(\tau)\, d\tau \tag{8.9a}$$

The output

$$\mathbf{y}(t) = \mathbf{C}\, e^{\mathbf{A}t}\mathbf{x}^0 + \mathbf{C}\int_0^t e^{\mathbf{A}(t-\tau)}\mathbf{B}\mathbf{u}(\tau)\, d\tau \tag{8.9b}$$

Since state model (8.1) possesses an analytical solution, the question of stability can be answered from this solution. However, as we shall see shortly, simple and powerful tools are available to determine the stability without resorting to the complete solution of state equation. In addition, we shall prove in this section that the two notions of stability (asymptotic stability and BIBO stability) are essentially the same for linear time-invariant systems (Rugh 1975).

Theorem 8.1: The state equation (8.1a) is asymptotically stable if and only if all the eigenvalues of matrix **A** have negative real parts.

Proof:

For asymptotic stability, we examine the homogeneous part of the solution (8.9a), i.e.,

$$\mathbf{x}(t) = e^{\mathbf{A}t}\mathbf{x}^0\ ; \quad t \geqslant 0 \tag{8.10}$$

$$e^{\mathbf{A}t} = \mathscr{L}^{-1}\left[(s\mathbf{I}-\mathbf{A})^{-1}\right]$$

$$= \mathscr{L}^{-1}\left[\frac{(s\mathbf{I}-\mathbf{A})^{+}}{|s\mathbf{I}-\mathbf{A}|}\right] = \mathscr{L}^{-1}\left[\frac{\mathbf{Q}(s)}{|s\mathbf{I}-\mathbf{A}|}\right]$$

Assume that $\lambda_1, \lambda_2, \ldots, \lambda_m$ are the distinct eigenvalues of **A** with multiplicities $n_1, n_2, \ldots, n_m$ respectively; $n = \sum_{i=1}^{m} n_i$. Then

$$|s\mathbf{I}-\mathbf{A}| = \prod_{i=1}^{m} (s-\lambda_i)^{n_i}$$

and the resolvent matrix

$$(s\mathbf{I} - \mathbf{A})^{-1} = \frac{\mathbf{Q}(s)}{\prod_{i=1}^{m} (s - \lambda_i)^{n_i}} \tag{8.11}$$

Note that the resolvent matrix given by (8.11) is a matrix of strictly proper rational functions (eqn. (7.18)). Thus we can expand each element

$$\frac{Q_{vw}(s)}{\prod_{i=1}^{m} (s-\lambda_i)^{n_i}}; \; v, w = 1, 2, \ldots, n,$$

in a partial fraction expansion to obtain

$$\frac{Q_{vw}(s)}{\prod_{i=1}^{m} (s-\lambda_i)^{n_i}} = \frac{K_{vw}^{11}}{(s-\lambda_1)} + \frac{{}^{12}_{vw}}{(s-\lambda_1)^2} + \ldots + \frac{K_{vw}^{1n_1}}{(s-\lambda_1)^{n_1}} + \frac{K_{vw}^{21}}{(s-\lambda_2)} + \ldots$$

$$+ \frac{K_{vw}^{2n_2}}{(s-\lambda_2)^{n_2}} + \ldots + \frac{K_{vw}^{m1}}{(s-\lambda_m)} + \ldots + \frac{K_{vw}^{mn_m}}{(s-\lambda_m)^{n_m}}$$

Performing this expansion for each $v, w = 1, \ldots, n$ we define the following matrices of the partial fraction expansion coefficients. Let $\mathbf{K}_{11}$ be the $n \times n$ matrix with v, w entries K_{vw}^{11}; $\mathbf{K}_{12}$ be the $n \times n$ matrix with v, w entries K_{vw}^{12} and so on. Using this matrix notation, we can arrange all the partial fraction expansions into the equation

$$(s\mathbf{I}-\mathbf{A})^{-1} = \sum_{i=1}^{m} \sum_{j=1}^{n_i} \mathbf{K}_{ij} \frac{1}{(s-\lambda_i)^j} \tag{8.12}$$

This gives

$$\mathbf{x} = e^{\mathbf{A}t} \mathbf{x}^0$$

$$= \left\{ \mathcal{L}^{-1} \left[\sum_{i=1}^{m} \sum_{j=1}^{n_i} \mathbf{K}_{ij} \frac{1}{(s-\lambda_i)^j} \right] \right\} \mathbf{x}^0$$

$$= \left[\sum_{i=1}^{m} \sum_{j=1}^{n_i} \mathbf{K}_{ij} \frac{t^{j-1}}{(j-1)!} e^{\lambda_i t} \right] \mathbf{x}^0 \tag{8.13}$$

From eqn. (8.13) it can readily be shown that if each λ_i has negative real part, then $t^{j-1} e^{\lambda_i t}$ is bounded on $[0, \infty)$ for any integer j. Further, using L' Hospitals rule, we can show that

$$\lim_{t \to \infty} \mathbf{x}(t) = \mathbf{0}$$

and hence the state equation (8.1a) is asymptotically stable.

Now suppose that a λ_i has non negative real part. Then

$$\lim_{t \to \infty} (t^{n_i - 1} e^{\lambda_i t}) \neq 0$$

and since $\mathbf{K}_{in_i}$ has at least one nonzero element, we can find an $\mathbf{x}^0$ such that

$$\lim_{t \to \infty} \mathbf{x}(t) \neq \mathbf{0}$$

This proves the theorem.

Theorem 8.2: The state model (8.1) is BIBO stable if and only if $\mathbf{H}(t) = \mathbf{C}\, e^{\mathbf{A}t}\mathbf{B}$ satisfies

$$\int_0^\infty |h_{ij}(\tau)|\, d\tau = N < \infty \text{ for all } \begin{array}{l} i = 1, 2, \ldots, q \\ i = 1, 2, \ldots, p \end{array} \tag{8.14}$$

Proof:

The response of system (8.1) with $\mathbf{x}(0) = \mathbf{0}$ is given by (eqn. (8.9b))

$$\mathbf{y}(t) = \int_0^t \mathbf{C}\, e^{\mathbf{A}(t-\tau)}\, \mathbf{B}\mathbf{u}(\tau) d\tau + \mathbf{D}\mathbf{u}(t)$$

$$= \int_0^t \mathbf{H}(t-\tau)\, \mathbf{u}(\tau) d\tau + \mathbf{D}\mathbf{u}(t)$$

We may assume that there is no direct transmission in (8.1) (i.e., $\mathbf{D} = \mathbf{0}$) because it does not play any role in the stability study. Under this assumption

$$\mathbf{y}(t) = \int_0^t \mathbf{H}(t-\tau)\, \mathbf{u}(\tau) d\tau$$

The ith-element of $\mathbf{y}(t)$ may be expressed as

$$y_i(t) = \sum_{j=1}^{p} \int_0^t h_{ij}(t-\tau) u_j(\tau) d\tau;\; i = 1, 2, \ldots, q \tag{8.15}$$

Assume that (8.14) is true and $u_j(t)$ is any input signal with $|u_j(t)| \leqslant M$, $j = 1, 2, \ldots, p$ $(0 < M < \infty)$ for all $t \geqslant 0$. Taking the absolute value on both sides of (8.15) we get

$$|y_i(t)| = \sum_{j=1}^{p} \left| \int_0^t h_{ij}(\tau) u_j(t-\tau) d\tau \right|$$

Since absolute value of an integral is not greater than the integral of the absolute value of integrand, we have

$$|y_i(t)| \leqslant \sum_{j=1}^{p} \left\{ \int_0^t |h_{ij}(\tau)\, u_j(t-\tau)|\, d\tau \right\}$$

$$\leqslant \sum_{j=1}^{p} \left\{ \int_0^t |h_{ij}(\tau)|\, M d\tau \right\}$$

$$\leqslant pNM$$

and thus (8.1) is BIBO stable.

Now suppose that (8.14) does not hold, i.e.,

$$\lim_{t\to\infty} \int_0^t | h_{ij}(\tau) | \, d\tau = \infty \text{ for some } i, j$$

It is obvious that

$$\lim_{t\to\infty} | y_i(t) | \to \infty$$

Therefore the system is not BIBO stable.

Theorem 8.3: If the state model (8.1) is asymptotically stable, then it is BIBO stable.

Proof:
From eqn. (8.13) we have

$$e^{\mathbf{A}t} = \sum_{i=1}^{m} \sum_{j=1}^{n_i} \mathbf{K}_{ij} \frac{t^{j-1}}{(j-1)!} e^{\lambda_i t}$$

Therefore,

$$\mathbf{H}(t) = \mathbf{C}\, e^{\mathbf{A}t}\, \mathbf{B}$$

$$= \sum_{i=1}^{m} \sum_{j=1}^{n_i} \frac{\mathbf{C}\mathbf{K}_{ij}\mathbf{B}}{(j-1)!} t^{j-1} e^{\lambda_i t}$$

The state model (8.1) is asymptotically stable; therefore by Theorem 8.1 all the eigenvalues of matrix **A** have negative real parts. This ensures that

$$\int_0^\infty | t^{j-1} e^{\lambda_i t} | \, dt < \infty$$

and hence condition (8.14) on $\mathbf{H}(t)$ is satisfied. This proves the theorem.

Theorem 8.4: If the state model (8.1) is completely controllable, completely observable and BIBO stable, then it is asymptotically stable.

Proof:
If (8.1) is BIBO stable, then

$$\lim_{t\to\infty} \mathbf{H}(t) = \lim_{t\to\infty} \mathbf{C}\, e^{\mathbf{A}t}\mathbf{B}$$

$$= \mathbf{0}$$

Since the entries of $\mathbf{H}(t)$ are analytic, we can write

$$\lim_{t\to\infty} \left(\frac{d}{dt} \mathbf{H}(t) \right) = \lim_{t\to\infty} \mathbf{C}\mathbf{A}\, e^{\mathbf{A}t}\mathbf{B} = \lim_{t\to\infty} \mathbf{C}\, e^{\mathbf{A}t}\mathbf{A}\mathbf{B}$$

$$= \mathbf{0}$$

$$\lim_{t\to\infty}\left(\frac{d^2}{dt^2}\mathbf{H}(t)\right) = \lim_{t\to\infty}\mathbf{C}\mathbf{A}^2 e^{\mathbf{A}t}\mathbf{B}$$

$$= \lim_{t\to\infty}\mathbf{C}\mathbf{A}\, e^{\mathbf{A}t}\mathbf{A}\mathbf{B}$$

$$= \lim_{t\to\infty}\mathbf{C}\, e^{\mathbf{A}t}\mathbf{A}^2\mathbf{B}$$

$$= \mathbf{0}$$

$$\vdots$$

By successive differentiation of $\mathbf{H}(t)$, the following result can easily be established.

$$\lim_{t\to\infty}\begin{bmatrix} \mathbf{C}\, e^{\mathbf{A}t}\mathbf{B} & \mathbf{C}\, e^{\mathbf{A}t}\mathbf{A}\mathbf{B} & \dots & \mathbf{C}\, e^{\mathbf{A}t}\mathbf{A}^{n-1}\mathbf{B} \\ \mathbf{C}\mathbf{A}\, e^{\mathbf{A}t}\mathbf{B} & \mathbf{C}\mathbf{A}\, e^{\mathbf{A}t}\mathbf{A}\mathbf{B} & \dots & \mathbf{C}\mathbf{A}\, e^{\mathbf{A}t}\mathbf{A}^{n-1}\mathbf{B} \\ \vdots & \vdots & & \vdots \\ \mathbf{C}\mathbf{A}^{n-1} e^{\mathbf{A}t}\mathbf{B} & \mathbf{C}\mathbf{A}^{n-1} e^{\mathbf{A}t}\mathbf{A}\mathbf{B} & \dots & \mathbf{C}\mathbf{A}^{n-1} e^{\mathbf{A}t}\mathbf{A}^{n-1}\mathbf{B} \end{bmatrix}$$

$$= \lim_{t\to\infty}\begin{bmatrix} \mathbf{C} \\ \mathbf{C}\mathbf{A} \\ \vdots \\ \mathbf{C}\mathbf{A}^{n-1} \end{bmatrix} e^{\mathbf{A}t}\,[\mathbf{B} \quad \mathbf{A}\mathbf{B} \dots \mathbf{A}^{n-1}\mathbf{B}] \tag{8.16}$$

$$= \mathbf{0}$$

Since (8.1) is completely controllable and completely observable, we can premultiply and postmultiply (8.16) by appropriate inverses to obtain

$$\lim_{t\to\infty} e^{\mathbf{A}t} = \mathbf{0}$$

Thus (8.1) is asymptotically stable, This completes the proof of the theorem. □

From the above theorems, we conclude that in general asymptotic stability implies BIBO stability but that a BIBO stable system is not necessarily asymptotically stable. However, under certain conditions, BIBO stability and asymptotic stability are completely equivalent.

From the definition of total stability and Theorems 8.1—8.4, it can be deduced that the system (8.1) is totally stable if all the eigenvalues of the matrix $\mathbf{A}$ have negative real parts.

Thus the problem of ascertaining stability of (8.1) involves checking the root locations of the characteristic polynomial $|\lambda\mathbf{I} - \mathbf{A}|$. The following two cases may arise:

(i) The system is asymptotically stable if $\text{Re}(\lambda_i) < 0$ for all $i = 1, 2, \dots, n$.

(ii) The system is unstable if one or more of the eigenvalues have positive real part.

A dividing case occurs when one or more of the eigenvalues of the system matrix **A** have zero real parts. In this case, the $j\omega$-axis of the complex plane acts as the dividing line between asymptotic stability and instability. The following results can easily be established from eqn. (8.13):

(i) Single eigenvalue at origin of the complex plane and others in the left half plane : $x_i(t)$ neither goes to zero nor to infinity; it simply approaches a constant value.

(ii) A pair of complex conjugate eigenvalues on the $j\omega$-axis of the complex plane ($\lambda_{1,2} = \pm j\omega$) and others in the left half plane : $x_i(t)$ neither goes to zero nor to infinity; it tends to a sinusoid of frequency ω.

(iii) Many distinct eigenvalues at $j\omega$-axis of complex plane in addition to the eigenvalues in left half plane : $x_i(t)$ may have a constant component as well as the sum of sinusoids at the frequencies of imaginary eigenvalues.

(iv) Repeated eigenvalues at the $j\omega$-axis of complex plane in addition to the eigenvalues in left half plane : $x_i(t)$ may grow without bound or may have constant and/or sinusoidal components.

Cases (i)-(iii) correspond to systems which are not asymptotically stable but are stable in the sense of Lyapunov. Stability in the case of case (iv) depends upon the modes corresponding to repeated eigenvalues, present in the system response. If the Jordan form block order for a repeated eigenvalue is one, we have stability in the sense of Lyapunov, otherwise we have instability. Repeated eigenvalues on $j\omega$-axis usually lead to unstable systems.

Linear systems, which are stable in the sense of Lyapunov may be treated as acceptable or nonacceptable depending upon the ultimate response.

For stability analysis of linear time-invariant systems, all we have to know is the eigenvalues of system matrix **A**. The fact that the determination of the eigenvalues of a matrix is a relatively simple task with current digital computers, means that tests of stability for linear time-invariant systems are very easy to make even for high dimensional systems. Computer programs for determination of eigenvalues of a matrix are readily available in the subroutine libraries of most major computer installations. Cadzow and Martens (1970) and Melsa and Jones (1973) give listing of FORTRAN computer program.

The numerical solution of a characteristic equation to determine eigenvalues provides information on absolute stability in 'yes' or 'no' form. Sometimes we need solutions in the form of functions of system parameters so that stable ranges can be determined. Several stability criteria are available for this purpose. In this section, we shall discuss the Routh-Hurwitz criterion.

Routh-Hurwitz Criterion: Consider the nth-order characteristic equation[2]

---

[2]Without loss of generality, we assume that the characteristic polynomial is *monic* (the coefficient of λ^n is equal to 1).

$$\Delta(\lambda) = \lambda^n + \alpha_1\lambda^{n-1} + \ldots + \alpha_{n-1}\lambda + \alpha_n = 0 \tag{8.17}$$

where α_i; $i = 1, 2, \ldots, n$ are constant scalars.

Certain conclusions regarding the stability of a system can be drawn by merely inspecting the coefficients of its characteristic equation in the form (8.17). It can be established that a necessary (but not sufficient) condition for asymptotic stability of linear system (8.1) is that all the coefficients α_i in its characteristic equation (8.17) be real and positive. None of the coefficients can be zero or negative unless one (or more than one) of the following occurs:

(i) one or more roots have positive real parts;

(ii) a root (or roots) at origin;

(iii) roots on $j\omega$-axis.

We therefore conclude that the absence or negativeness of any of the coefficients α_i in (8.17) indicates that the system is either unstable or at most stable in the sense of Lyapunov. If the coefficients are all positive, no conclusion can be drawn (except for $n = 2$) and one should proceed further to examine the sufficient conditions of stability.

A. Hurwitz and E. J. Routh independently published the method of investigating the sufficient conditions of stability of a system. The Hurwitz criterion is in terms of determinants and the Routh criterion is in terms of array formulation which is more convenient to handle. We first present the Hurwitz criterion. The Routh criterion, which is derivable from that of Hurwitz is then presented. We shall prove the criterion by means of the Lyapunov theory later in this chapter.

Theorem 8.5: The system (8.1) with characteristic equation (8.17) is asymptotically stable if and only if the n determinants formed from the coefficients α_i; $i = 1, 2, \ldots, n$, are positive, where these determinants are taken as the principal minors of the following arrangement (called the *Hurwitz arrangement*)

$$\begin{vmatrix} \alpha_1 & 1 & 0 & 0 & 0 & 0 \ldots 0 & 0 \\ \alpha_3 & \alpha_2 & \alpha_1 & 1 & 0 & 0 \ldots 0 & 0 \\ \alpha_5 & \alpha_4 & \alpha_3 & \alpha_2 & \alpha_1 & 1 \ldots 0 & 0 \\ \vdots & \vdots & \vdots & \vdots & \vdots & \vdots & \vdots \\ \alpha_{2n-1} & \alpha_{2n-2} & \alpha_{2n-3} & \cdot & \cdot & \cdot & \alpha_{n+1} & \alpha_n \end{vmatrix} \tag{8.18}$$

The coefficients with indices larger than n or with negative indices are replaced by zeros.

In other words, system (8.1) is asymptotically stable if and only if

$$\Delta_1 = \alpha_1 > 0;$$

$$\Delta_2 = \begin{vmatrix} \alpha_1 & 1 \\ \alpha_3 & \alpha_2 \end{vmatrix} > 0;$$

$$\Delta_3 = \begin{vmatrix} \alpha_1 & 1 & 0 \\ \alpha_3 & \alpha_2 & \alpha_1 \\ \alpha_5 & \alpha_4 & \alpha_3 \end{vmatrix} > 0; \ldots$$

$$\Delta_n = \text{entire arrangement of (8.18)} > 0.$$

Theorem 8.6: The system (8.1) with characteristic equation (8.17) is asymptotically stable if and only if each term of the first column of Routh array given below is positive.

Routh array

$$\begin{array}{l|llllll}
\lambda^n & 1 & \alpha_2 & \alpha_4 & \alpha_6 & \cdot & \cdot \\
\lambda^{n-1} & \alpha_1 & \alpha_3 & \alpha_5 & \cdot & \cdot & \cdot \\
\lambda^{n-2} & b_1 & b_2 & b_3 & \cdot & \cdot & \cdot \\
\lambda^{n-3} & c_1 & c_2 & & & & \\
\lambda^{n-4} & d_1 & d_2 & & & & \\
\vdots & \vdots & \vdots & & & & \\
\lambda^2 & e_1 & \alpha_n & & & & \\
\lambda^1 & f_1 & & & & & \\
\lambda^0 & \alpha_n & & & & &
\end{array}$$

The coefficients b_1, $b_2 \ldots$, are evaluated as follows:

$$b_1 = (\alpha_1\alpha_2 - \alpha_3)/\alpha_1$$

$$b_2 = (\alpha_1\alpha_4 - \alpha_5)/\alpha_1$$

$$\vdots$$

This process is continued till we get a zero as the last coefficient in the third row. In a similar way, the coefficients of 4th, 5th, ..., nth and $(n+1)$th rows are evaluated, e.g.,

$$c_1 = (b_1\alpha_3 - \alpha_1 b_2)/b_1$$

$$c_2 = (b_1\alpha_5 - \alpha_1 b_3)/b_1$$

$$\vdots$$

$$d_1 = (c_1 b_2 - b_1 c_2)/c_1$$

$$\vdots$$

It is to be noted here that in the process of generating the Routh array, the missing terms are regarded as zeros. Also all the elements of any row can be divided by a positive constant during the process to simplify the computational work.

□

The *Routh criterion* stated in Theorem 8.6 and the *Hurwitz criterion* stated in Theorem 8.5 are equivalent, as is shown below.

Elements of the first column of the Routh array can be interpreted in terms of Hurwitz determinants as follows.

$$b_1 = \frac{\alpha_1\alpha_2 - \alpha_3}{\alpha_1} = \frac{\begin{vmatrix} \alpha_1 & 1 \\ \alpha_3 & \alpha_2 \end{vmatrix}}{\alpha_1} = \frac{\Delta_2}{\Delta_1}$$

Similarly

$$c_1 = \Delta_3/\Delta_2;$$

$$d_1 = \Delta_4/\Delta_3; \ldots$$

Therefore the condition of positiveness of the Hurwitz determinants corresponds to the condition of positiveness of the elements of the first column of the Routh array.

Example 8.1: Consider the characteristic equation

$$3\lambda^4 + 10\lambda^3 + 5\lambda^2 + 5\lambda + 2 = 0$$

The Routh array is given below:

λ^4	3	5	2
λ^3	10	5	
	2	1 (after dividing by 5)	
λ^2	$\frac{2\times5-3\times1}{2} = 3.5$	$\frac{2\times2-0\times3}{2} = 2$	
λ^1	$\frac{3.5\times1-2\times2}{3.5} = -0.1429$		
λ^0	2		

It may be noted that in order to simplify computational work, the λ^3-row in formation of the Routh array has been modified by dividing it by 5 throughout. The modified λ^3-row is then used to complete the process of array formation.

Examining the first column of the Routh array, it is found that all the terms are not positive. Therefore by Theorem 8.6, the system under consideration is unstable.

□

If the condition of asymptotic stability, given by Theorem 8.6, is not met, the system is unstable and the number of sign changes of the terms of the first column of the Routh array corresponds to the number of roots of the characteristic equation in the right half of the complex plane. It is to be noted that the Routh criterion gives only the number of roots in the right half of complex plane. It gives no information as regards the values of the roots and also does not distinguish between real and complex roots.

Examining the first column of the Routh array of Example 8.1, it is found that there are two sign changes (from 3.5 to -0.1429 and from -0.1429 to 2). Therefore two roots of the characteristic equation lie in right half of the complex plane.

Whenever two roots of a characteristic equation are negatives of each other, the corresponding Routh array has an all-zero row. At best a system which produces this result will be stable in the sense of Lyapunov, having a pair of imaginary roots which obviously must be negatives of each other. Because of the all-zero row in the array, the Routh test breaks down. It may, however, be desirable to be able to continue the test to determine whether the system is stable in the sense of Lyapunov or unstable. The procedure involves forming a subsidiary equation from the coefficients in the row just above the all-zero row and using this equation as the basis for continuing the test. The following example illustrates the procedure.

Example 8.2: Consider the characteristic equation

$$\lambda^6 + 2\lambda^5 + 8\lambda^4 + 12\lambda^3 + 20\lambda^2 + 16\lambda + 16 = 0$$

The Routh array is

λ^6	1	8	20	16	
λ^5	2	12	16		
	1	6	8		(after dividing by 2)
λ^4	2	12	16		
	1	6	8		(after dividing by 2)
λ^3	0	0			

Since λ^3-row is an all-zero row, the Routh test breaks down. Now the subsidiary polynomial is formed from the coefficients of the λ^4-row, which is given by

$$\lambda^4 + 6\lambda^2 + 8$$

Note that the subsidiary polynomial consists of only even powers of λ.

The Routh array is completed by differentiating the subsidiary polynomial with respect to λ and using the resulting coefficients to form a row to replace the all-zero row. The array is completed from this point onwards by the usual procedure. For the example under consideration, we have the following array completed by the procedure outlined above.

λ^6	1	8	20	16
λ^5	1	6	8	
λ^4	1	6	8	
λ^3	4	12		
	1	3 (after dividing by 4)		
λ^2	3	8		
λ^1	1/3			
λ^0	8			

We see that there is no change of sign in the first column of the new array. We must find the roots of the subsidiary equation to determine whether the system is stable in the sense of Lyapunov or unstable. These roots are easily found to be $\pm j\sqrt{2}$ and $\pm j2$. These two pairs of roots are also the roots of the original characteristic equation. Since there is no sign change in the first column of the new array formed with the help of subsidiary polynomial, we conclude that no root of the characteristic equation has a positive real part. Therefore the system under consideration is stable in the sense of Lyapunov.

□

Another situation which necessitates the use of special procedures occurs when only leading zeros (zeros which occur in the first column of rows containing at least one non-zero element) are present. The leading zeros are easily handled by replacing them with a small positive constant ε, completing the array and taking the limits of the terms involving ε as ε is made to approach zero. This procedure is based on the observation that introducing a parameter ε amounts to a small perturbation in some or all of the coefficients of the characteristic polynomial. The perturbed polynomial will have the same number of roots in the right half plane as the original one provided the perturbation is small and provided also that the original polynomial has no pure imaginary roots which are moved to the right half plane by the perturbation (Chang and Chen 1974, Hostetter 1975). Therefore to apply the ε-limiting method for a characteristic polynomial with the imaginary-axis roots, one must first extract these roots before the test is applied on the remainder polynomial having no imaginary-axis roots.

Example 8.3: Consider the characteristic equation

$$\Delta(\lambda) = \lambda^6 + \lambda^5 + 3\lambda^4 + 3\lambda^3 + 3\lambda^2 + 2\lambda + 1 = 0$$

The Routh array is

λ^6	1	3	3	1
λ^5	1	3	2	
λ^4	0	1	1	
	ε	1	1 (after replacing 0 by ε)	
λ^3	$(3\varepsilon-1)/\varepsilon$	$(2\varepsilon-1)/\varepsilon$		
λ^2	$(-2\varepsilon^2+4\varepsilon-1)/(3\varepsilon-1)$	1		
λ^1	$(4\varepsilon^2-\varepsilon)/(2\varepsilon^2-4\varepsilon+1)$			
λ^0	1			

As $\varepsilon \to 0$, all the elements of λ^1-row tend to zero. This indicates that there are symmetrically located roots in the complex plane. We therefore need to examine the subsidiary equation to find out the possibility of the imaginary-axis roots. If no such roots exist, the usual procedure of replacing the all-zero row by coefficients of the derivative of the subsidiary polynomial is adopted. If the imaginary-axis roots are found to exist, the original polynomial is divided out by the subsidiary polynomial and the test is performed on the remainder polynomial.

For the example under consideration, the subsidiary equation is

$$\lambda^2 + 1 = 0$$

yielding two roots on the imaginary axis. Dividing the original polynomial by $(\lambda^2 + 1)$ we get

$$\Delta'(\lambda) = \lambda^4 + \lambda^3 + 2\lambda^2 + 2\lambda + 1$$

The Routh array for this polynomial is

λ^4	1	2	1
λ^3	1	2	
λ^2	0	1	
	ε	1 (after replacing 0 by ε)	
λ^1	$(2\varepsilon - 1)/\varepsilon$		
λ^0	1		

As $\varepsilon \to 0$, there are two sign changes in the first column elements. This indicates that there are two roots in the right half of complex plane. (Recently, Fahmy and O'Reilly (1982) have demonstrated that ε-method can be applied without factoring out a common divisor).

Discrete-Time Systems

Consider a linear system with state model

$$\mathbf{x}(k+1) = \mathbf{F}\mathbf{x}(k) + \mathbf{G}\mathbf{u}(k) \tag{8.19a}$$

$$\mathbf{y}(k) = \mathbf{C}\mathbf{x}(k) + \mathbf{D}\mathbf{u}(k) \tag{8.19b}$$

where $k = 0, 1, 2, \ldots$ and **F**, **G**, **C** and **D** are, respectively, $n \times n$, $n \times p$, $q \times n$ and $q \times p$ real constant matrices.

The definitions of stability are identical with those given earlier if the k replaces t.

We state stability theorems for the discrete-time system (8.19) in the following. The proofs of these theorems are analogous to those in the continuous-time case and are left as exercises.

Theorem 8.7: The state equation (8.19a) is asymptotically stable if and only if all the eigenvalues of matrix **F** have magnitude less than unity.

Theorem 8.8: The state model (8.19) is BIBO stable if and only if $\mathbf{H}(k) = \mathbf{C}\mathbf{F}^k\mathbf{G};\ k \geqslant 0$ satisfies

$$\sum_{k=0}^{\infty} |\, h_{ij}(k) \,| = N < \infty \quad \text{for all} \quad i = 1, 2, \ldots, q$$
$$j = 1, 2, \ldots, p.$$

Theorem 8.9: If the system (8.19) is asymptotically stable, then it is BIBO stable. If it is completely controllable, completely observable and BIBO stable, then it is asymptotically stable.

□

Like the continuous-time case, there are three distinct regions of complex plane for consideration: the inside ($|\lambda_i| < 1$), the outside ($|\lambda_i| > 1$) and the boundary ($|\lambda_i| = 1$) of the unit circle. If all the eigenvalues of **F** are inside the unit circle, the system is asymptotically stable. Multiple eigenvalues inside the unit circle do not affect the stability. If any eigenvalue of **F** is outside the unit circle, the system is unstable. For single eigenvalues or pairs on the unit circle, we have stability in the sense of Lyapunov. For multiple eigenvalues or pairs, we may have stability in the sense of Lyapunov but we usually obtain unstable systems. (If the Jordan block for multiple eigenvalue is of order one, stability in the sense of Lyapunov results.)

Earlier we remarked that Routh-Hurwitz tests are very useful for checking the stability of a continuous-time system. These tests operate on the coefficients of the characteristic polynomial and determine whether all of its roots lie in the left half of the complex plane. In the discrete-time case, we must determine whether the roots of characteristic polynomial lie within the unit circle in the complex plane. Stability tests to determine this can be broadly categorized into two groups. The first consists of those which are applied directly to the characteristic polynomial $\Delta(\lambda)$. The other group includes those tests which first require a transformation before they can be applied.

Two well known stability tests which work directly with $\Delta(\lambda) = 0$ are the Schur-Cohn test and the Jury test (Kuo 1963, Jury 1974). These tests, in general, reveal only absolute stability versus instability.

The Routh-Hurwitz test can be applied to the stability analysis of discrete-time systems if a transformation of $\Delta(\lambda) = 0$ can be found which fulfils the following requirements:

(i) The outside of the unit circle in the λ-plane is transformed to the right half of new complex plane.

(ii) The boundary of the unit circle in the λ-plane is transformed into imaginary axis of the new complex plane.

(lii) The inside of the unit circle in the λ-plane is transformed into the left half of new complex plane.

The following bilinear transformation can be shown to meet the above stated requirements:

$$\lambda = \frac{1+\omega}{1-\omega}$$

When solved for ω, this gives

$$\omega = \frac{\lambda - 1}{\lambda + 1}$$

Example 8.4: Consider the discrete-time system with the characteristic equation

$$\Delta(\lambda) = 4\lambda^3 - 4\lambda^2 - 7\lambda - 2 = 0$$

Applying the bilinear transformation we obtain

$$\Delta(\omega) = 4\left[\frac{1+\omega}{1-\omega}\right]^3 - 4\left[\frac{1+\omega}{1-\omega}\right]^2 - 7\left[\frac{1+\omega}{1-\omega}\right] - 2$$

$$= \frac{3\omega^3 + 17\omega^2 + 21\omega - 9}{(1-\omega)^3} = 0$$

Since there is a negative sign in the numerator of $\Delta(\omega)$, there is at least one root of $\Delta(\omega) = 0$ in the right half of ω-plane. We need not proceed with the formation of the Routh array.

Effect of Sampling on Stability

It was observed in Section 5.7 that sampling rate selection for sampled-data systems is a complicated problem involving compromise among many factors. *Sampling has a destabilizing effect on a system.* Let us illustrate this with the help of an example.

Example 8.5 Consider the sampled-data system of Fig. 8.3. The plant is described by the transfer function

$$\frac{\hat{y}(s)}{\hat{u}(s)} = \frac{K}{s(s+2)}$$

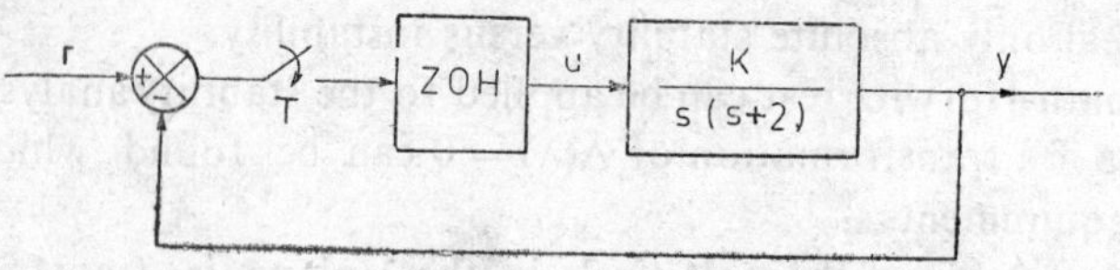

Fig. 8.3 A sampled-data system

A state variable realization of this transfer function is

$$\dot{\mathbf{x}} = \mathbf{A}\mathbf{x} + \mathbf{b}u$$

$$y = \mathbf{c}\mathbf{x}$$

where

$$\mathbf{A} = \begin{bmatrix} 0 & 1 \\ 0 & -2 \end{bmatrix};\ \mathbf{b} = \begin{bmatrix} 0 \\ K \end{bmatrix};\ \mathbf{c} = [\,1 \quad 0\,]$$

The closed-loop continuous-time system (without sampler and zero-order hold) is stable for all values of $K > 0$.

Let us now introduce sampler and zero-order hold as is shown in Fig. 8.3. The discrete-time description of the plant is given by (using eqn. (5.64))

$$\mathbf{x}(k+1) = \mathbf{F}\mathbf{x}(k) + \mathbf{g}u(k)$$

$$y(k) = \mathbf{c}\mathbf{x}(k)$$

where

$$\mathbf{F} = e^{\mathbf{A}T} = \begin{bmatrix} 1 & \frac{1}{2}(1-e^{-2T}) \\ 0 & e^{-2T} \end{bmatrix}$$

$$\mathbf{g} = \int_0^T e^{\mathbf{A}\theta}\,\mathbf{b}\,d\theta = \tfrac{1}{2}K \begin{bmatrix} T-\frac{1}{2}+\frac{1}{2}e^{-2T} \\ 1-e^{-2T} \end{bmatrix}$$

The state variable description of the closed-loop sampled-data system becomes

$$\mathbf{x}(k+1) = \begin{bmatrix} 1-\frac{1}{2}K(T-\frac{1}{2}+\frac{1}{2}e^{-2T}) & \frac{1}{2}(1-e^{-2T}) \\ -\frac{1}{2}K(1-e^{-2T}) & e^{-2T} \end{bmatrix}\mathbf{x}(k)$$

$$+\ \tfrac{1}{2}K \begin{bmatrix} T-\frac{1}{2}+\frac{1}{2}e^{-2T} \\ 1-e^{-2T} \end{bmatrix} r(k)$$

The characteristic equation is given by

$$\lambda^2 + \lambda[\tfrac{1}{2}K(T-\tfrac{1}{2}+\tfrac{1}{2}e^{-2T})-1-e^{-2T}]$$
$$+ e^{-2T} + \tfrac{1}{2}K(\tfrac{1}{2}-\tfrac{1}{2}e^{-2T}-Te^{-2T}) = 0$$

Applying the bilinear transformation we obtain

$$\omega^2(2+e^{-2T}-\tfrac{1}{2}K(T-\tfrac{1}{2}+\tfrac{1}{2}e^{-2T}))+2\omega[1-e^{-2T}-\tfrac{1}{2}K(\tfrac{1}{2}-\tfrac{1}{2}e^{-2T}-Te^{-2T})]$$
$$+\tfrac{1}{2}KT(1-e^{-2T})=0$$

Case 1: $T=0.4$ sec

For this value of T, the system is found to be stable for $0<K<11.52$.

Case 2: $T=3$ sec

For this value of T, the system is found to be stable for $0<K<1.6$.

Thus the system which is stable for $K<11.52$ when $T=0.4$ sec, becomes unstable for $K>1.6$ when $T=3$ sec. It means that increasing the sampling period (or decreasing the sampling rate) reduces the margin of stability. A closed-loop stable linear continuous-time system may become even unstable upon introduction of sampling and zero-order hold in the forward loop (Problem 8.1).

8.5 EQUILIBRIUM STABILITY OF NONLINEAR CONTINUOUS-TIME AUTONOMOUS SYSTEMS

The general state equation for a nonlinear system can be expressed as

$$\dot{\mathbf{x}} = \mathbf{f}(\mathbf{x}(t), \mathbf{u}(t), t); \ \mathbf{x}(t_0) = \mathbf{x}^0 \tag{8.20}$$

The analytical solution to this equation is rarely possible. It is possible, however, to ask for less than a complete solution for $\mathbf{x}$ and still reach conclusions of practical significance regarding the stability of equilibrium points.

For an autonomous system, stability in-the-small (local stability) of an isolated equilibrium point can be obtained through linearization provided that (a) linearization is possible and (b) the linearized system does not have one or more eigenvalues with zero real part. The region of validity of local stability is, however, generally not known. In some cases, the region may be too small to be of any use practically; while in others the region may be much larger than the one assumed by the designer giving rise to systems that are too conservatively designed. We therefore need information about the domain of stability. We give here a method of the investigation of stability based upon the approach first suggested by Russian mathematician A.M. Lyapunov and called the '*second method of Lyapunov*' or '*direct method of Lyapunov*'.

The Lyapunov method is based upon the concept of energy and the relation of stored energy with system stability. We first give an example to motivate the discussion.

Consider the spring-mass-damper system of Fig. 8.4. The governing equation of the system is

$$\ddot{x}_1 + B\dot{x}_1 + Kx_1 = 0$$

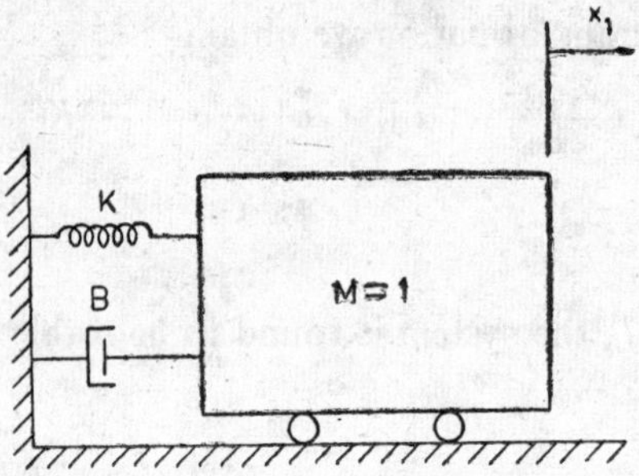

Fig. 8.4 Spring-mass-damper system

A corresponding state model is

$$\begin{aligned} \dot{x}_1 &= x_2 \\ \dot{x}_2 &= -Kx_1 - Bx_2 \end{aligned} \tag{8.21}$$

At any instant, the total energy V in the system consists of the kinetic energy of the moving mass and the potential energy stored in the spring.

$$V(x_1, x_2) = \tfrac{1}{2}x_2^2 + \tfrac{1}{2}Kx_1^2$$

Thus

$$V(\mathbf{x}) > 0 \quad \text{when} \quad \mathbf{x} \neq \mathbf{0}$$

$$V(\mathbf{0}) = 0$$

This means that total energy is positive unless the system is at rest at the equilibrium point $\mathbf{x}^e = \mathbf{0}$, where the energy is zero.

The rate of change of energy is given by

$$\frac{d}{dt} V(x_1, x_2) = \frac{\partial V}{\partial x_1}\frac{dx_1}{dt} + \frac{\partial V}{\partial x_2}\frac{dx_2}{di}$$

$$= -Bx_2^2$$

Thus $\frac{dV}{dt}$ is negative at all points except where x_2 is zero, at which $\frac{dV}{dt}$ is zero. Therefore, under positive damping the system energy cannot increase. From eqns. (8.21), we observe that $\dot{x}_2 = -Kx_1$ at the points where $x_2 = 0$. Thus the system cannot remain in the nonequilibrium state for which $x_2 = 0$ and therefore energy cannot remain constant except at equilibrium point where it will be zero.

A visual analogy may be obtained by considering the surface

$$V = \tfrac{1}{2}x_2^2 + \tfrac{1}{2}Kx_1^2$$

This is a cup-shaped surface as shown in Fig. 8.5a. The constant-V loci are ellipses on the surface of the cup. Let (x_1^0, x_2^0) be the initial condition. If one plots a trajectory on the surface shown, the representative point $\mathbf{x}(t)$ crosses the constant-V curves and moves towards the lowest point of the cup, which is the equilibrium point. Figure 8.5b shows the projection of the tvpical trajectory on the x_1-x_2 plane.

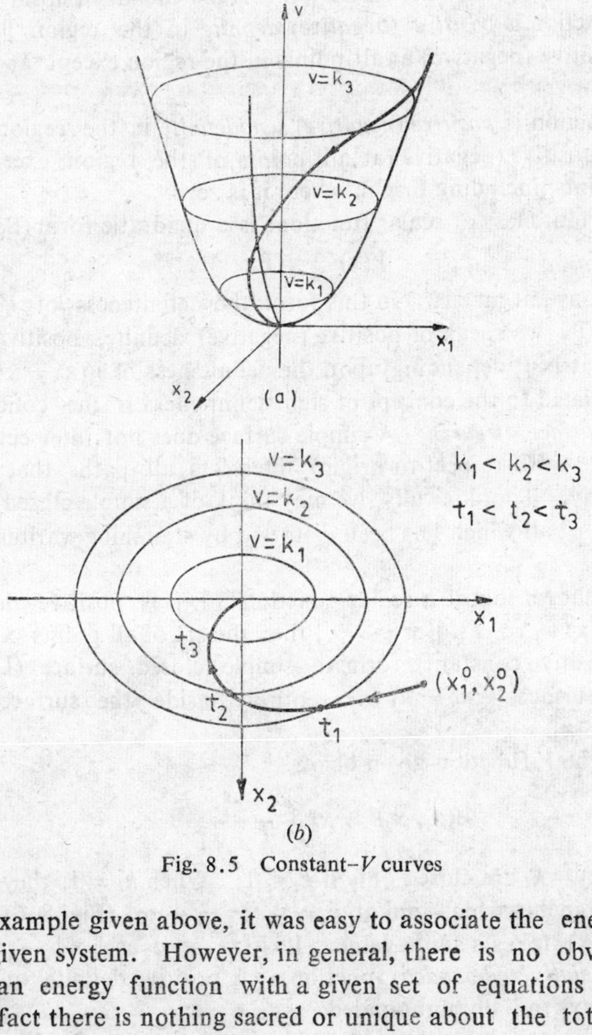

Fig. 8.5 Constant-V curves

In the example given above, it was easy to associate the energy function V with the given system. However, in general, there is no obvious way of associating an energy function with a given set of equations describing a system. In fact there is nothing sacred or unique about the total energy of the system which allows us to determine system stability in the way described above. Other non-negative scalar functions of the system state can also answer the question of stability. This idea was introduced and formalized by Lyapunov; the scalar function is now known as the *Lyapunov function* and the method of investigating stability using Lyapunov function is known as the *Lyapunov's direct method*.

In Section 2.5, we introduced the concept of sign definiteness of scalar functions. Let us examine here the scalar function $V(x_1, x_2, ..., x_n) \triangleq V(\mathbf{x})$ for which $V(\mathbf{0}) = 0$ and the function is continuous in a certain region surrounding the origin in state space. Because of the manner in which these V-functions are used later, we define the sign definiteness with respect to a spherical region around the origin. Such a region can be represented as $\| \mathbf{x} \| \leqslant K$ (a positive constant) where $\| \mathbf{x} \|$ is the euclidean norm of $\mathbf{x}$.

The V-function is *positive (negative) definite* in the region $\| \mathbf{x} \| \leqslant K$ if its value is positive (negative) at all points in the region except at the origin, where it is zero.

The V-function is *positive (negative) semidefinite* in the region $\| \mathbf{x} \| \leqslant K$ if its value is positive (negative) at all points of the region except at finite number of points, including origin, where it is zero.

An important class of scalar functions is a quadratic form (Section 2.5):

$$V(\mathbf{x}) = \mathbf{x}^T\mathbf{P}\mathbf{x}$$

where $\mathbf{P}$ is a constant matrix. In this form, the definiteness of V is usually attributed to $\mathbf{P}$; we speak of positive (negative) definite, positive (negative) semidefinite matrix $\mathbf{P}$ depending upon the definiteness of $V(\mathbf{x}) = \mathbf{x}^T\mathbf{P}\mathbf{x}$.

Closely related to the concept of sign definiteness is the concept of the *simple closed surface* or *curve*. A simple surface does not intersect itself and a closed surface (with respect to origin) intersects all paths that lead from origin to infinity. Topologically, we may think of a simple closed surface as a hollow rubber ball which has been distorted by stretching various segments of its surface.

It can be shown that if a scalar function $V(\mathbf{x})$ is positive definite and, in addition, $V(\mathbf{x}) \to \infty$ as $\| \mathbf{x} \| \to \infty$, then the set of all points $\mathbf{x}$ such that $V(\mathbf{x}) = k$ (a positive constant), forms a simple closed surface (Letov 1961). Further, the surface $V(\mathbf{x}) = k_1$ lies entirely inside the surface $V(\mathbf{x}) = k_2$ whenever $k_1 < k_2$.

Consider the V-function given below.

$$V(x_1, x_2) = x_2^2 + \frac{x_1^2}{1 + x_1^2}$$

The curves $V(\mathbf{x}) = k_i$ are closed only if $k_i \leqslant 1$. When $k_i = 1$, they will have two separate branches which join at $x_2 = 0$, $x_1 = \pm\infty$ (Fig. 8.6). Thus as $\| \mathbf{x} \| \to \infty$, $V(\mathbf{x})$ takes a finite value. If $V(\mathbf{x}) \to \infty$ as $\| \mathbf{x} \| \to \infty$, the V-function is *radially unbounded* function. A positive definite quadratic V-function is always radially unbounded.

In the following, we shall see that the properties of the surfaces $V(\mathbf{x}) = k_i$ determine the stability of a system.

We shall confine our attention to nonlinear autonomous systems described by state equation of the form

$$\dot{\mathbf{x}}(t) = \mathbf{f}(\mathbf{x})\,; \quad \mathbf{f}(\mathbf{0}) = \mathbf{0} \tag{8.22}$$

Note that the origin of the state space has been taken as the equilibrium state of the system. There is no loss in generality in this assumption since any

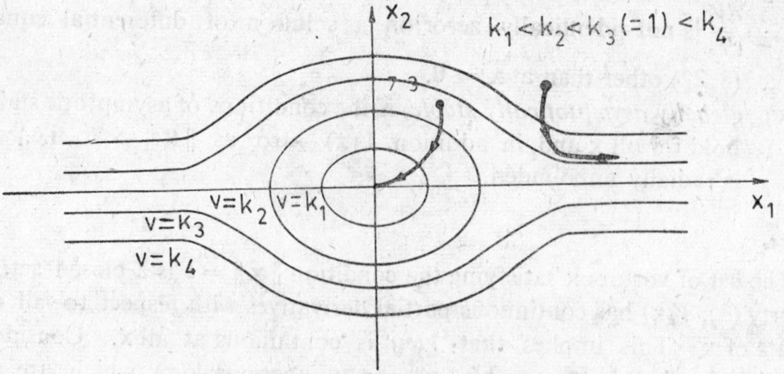

Fig. 8.6 A radially bounded V-function

non-zero equilibrium state can be shifted to the origin by appropriate transformation of the variables. Further, we have taken $t_0 = 0$ in (8.22), which is a convenient choice for time-invariant systems.

In the discussion that follows, we shall assume that $\mathbf{f}(\mathbf{x})$ is continuous and satisfies a Lipschitz condition (refer Section 5.2), so that the differential equation (8.22) has only one solution.[3]

Lyapunov's Stability Theorem

Example 8.10: For the autonomous system (8.22), the sufficient conditions of stability are as follows:

Suppose that there exists a scalar function $V(\mathbf{x})$ which, for some real number $\varepsilon > 0$, satisfies the following properties for all $\mathbf{x}$ in the region $\| \mathbf{x} \| \leqslant \varepsilon$:

(1) $V(\mathbf{x}) > 0; \mathbf{x} \neq \mathbf{0}$
(2) $V(\mathbf{0}) = 0$
} (i.e., $V(\mathbf{x})$ is positive definite function)

(3) $V(\mathbf{x})$ has continuous partial derivatives with respect to all components of $\mathbf{x}$.

Then the equilibrium $\mathbf{x} = \mathbf{0}$ of system (8.22) is

(a) *stable*, if the derivative $\dfrac{dV}{dt} \leqslant 0$, i.e., $\dfrac{dV}{dt}$ is a negative semidefinite scalar function;

(b) *asymptotically stable*, if $\dfrac{dV}{dt} < 0$, $\mathbf{x} \neq \mathbf{0}$, i.e., $\dfrac{dV}{dt}$ is a negative definite scalar function; or if $\dfrac{dV}{dt} \leqslant 0$ (i.e., $\dfrac{dV}{dt}$ is negative semidefinite) and

[3]Lyapunov stability theorems are also available for *nonautonomous* systems of the form (8.8) (Kalman and Bertram 1960). Consideration of these theorems will take us deep into mathematics. Since such systems occur relatively infrequently, we shall concentrate on autonomous systems.

$\frac{dV}{dt}$ is not identically zero on a solution of differential equation (8.22) other than at $\mathbf{x} = \mathbf{0}$.

(c) *globally asymptotically stable,* if the conditions of asymptotic stability hold for all $\mathbf{x}$ and, in addition, $V(\mathbf{x}) \to \infty$ as $\| \mathbf{x} \| \to \infty$, i.e., $V(\mathbf{x})$ is radially unbounded.

Proof[4]:

The set of vectors $\mathbf{x}$ satisfying the condition $\| \mathbf{x} \| = \varepsilon$ is a closed set. By property (3), $V(\mathbf{x})$ has continuous partial derivatives with respect to all components of $\mathbf{x}$. This implies that $V(\mathbf{x})$ is continuous at all $\mathbf{x}$. Consider in particular the closed set given by $\| \mathbf{x} \| = \varepsilon$. Functions $V(\mathbf{x})$ which are continuous everywhere on this closed set, must assume minimum and maximum values on that set, i.e., there exist real constants M and m such that $V(\mathbf{x}) = M$ for some value of $\mathbf{x}$ in the closed set, $V(\mathbf{x}) = m$ for some other value of $\mathbf{x}$ in the closed set and $m \leqslant V(\mathbf{x}) \leqslant M$. The minimum value of $V(\mathbf{x})$ on the closed set is indicated by a contour $C_1(V = m)$ in Fig. 8.7 for a two-dimensional case. By property (1), $m > 0$.

Since $V(\mathbf{x})$ is continuous at all $\mathbf{x}$ given by $\| \mathbf{x} \| \leqslant \varepsilon$, it is continuous at $\mathbf{x} = \mathbf{0}$, where $V = 0$. As per the mathematical properties of continuous functions, for every real number $r > 0$, there exists a real number $\delta > 0$ such that $| V(\mathbf{x}) - V(\mathbf{0}) | \leqslant r$ for all $\mathbf{x}$ satisfying $\| \mathbf{x} - \mathbf{0} \| \leqslant \delta$. In particular, let us take $r = k < m$. This means that there exists a $\delta > 0$ such that $V(\mathbf{x}) \leqslant k$ for all $\| \mathbf{x} \| \leqslant \delta$. This is shown by contour $C_2(V = k)$ and δ-circle in Fig. 8.7.

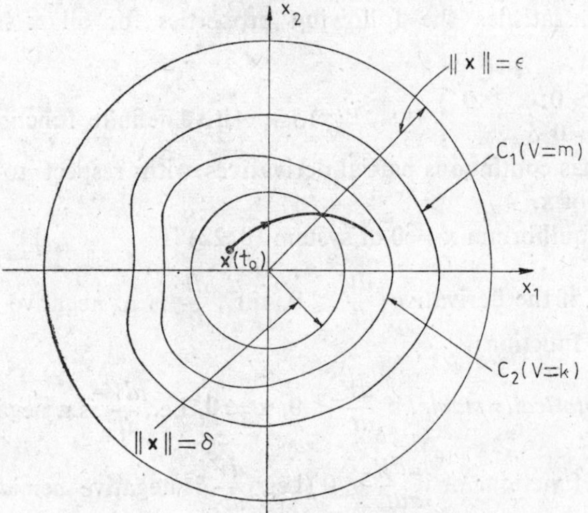

Fig. 8.7

[4]We give here an outline of the proof with the objective of illustrating the stability theorem. For rigorous proof, the reader may refer Kalman and Bertram (1960).

Obviously $\delta < \varepsilon$, otherwise the assumption of $\min_{\|x\|=\varepsilon} V(\mathbf{x}) = m$ will be contradicted.

Consider any $\mathbf{x}(t_0)$ such that $\| \mathbf{x}(t_0) \| \leqslant \delta$. Since for stability, $\frac{dV}{dt} \leqslant 0$, we have

$$V(\mathbf{x}(t)) \leqslant V(\mathbf{x}(t_0)) \leqslant k \text{ for all } t > t_0$$

Thus $V(\mathbf{x}(t))$ can never reach the contour C_1 and hence $\mathbf{x}(t)$ cannot reach the closed set $\| \mathbf{x} \| = \varepsilon$ since $V(\mathbf{x}) \geqslant m$ on this set. Therefore $\| \mathbf{x}(t) \| \leqslant \varepsilon$ for all $\| \mathbf{x}(t_0) \| \leqslant \delta$ and $t \geqslant t_0$. This guarantees stability in the sense of Lyapunov, as per the definition given in Section 8.3.

If a continuous V-function ($V > 0$ except at $\mathbf{x} = \mathbf{0}$) satisfies the condition $\frac{dV}{dt} < 0$, we expect that $\mathbf{x}$ will eventually approach the origin. As time progresses, the closed surfaces $V(\mathbf{x}) = k$ will shrink to the origin and hence $\mathbf{x} \to \mathbf{0}$. This result is somewhat restrictive because of the fact that $\frac{dV}{dt}$ is required to be negative definite. It is possible to replace this requirement by negative semidefiniteness of $\frac{dV}{dt}$ if the system cannot remain in the non-equilibrium state at which $\frac{dV}{dt} = 0$. It must move to some state where $\frac{dV}{dt} < 0$. The mass-spring-damper system discussed earlier in this section satisfies this alternative condition of asymptotic stability.

In order to test this alternative condition, the solution to the equation $\frac{dV}{dt} = 0$ is substituted into eqn. (8.22) to be sure that this equation is not satisfied.

The radial unboundedness condition on $V(\mathbf{x})$ ensures that $V(\mathbf{x}) = k$ are closed surfaces in the entire state space. If $V(\mathbf{x})$ does not form closed surfaces, then it is possible for system trajectories to go towards infinity. Consider, for example, the V-function shown in Fig. 8.6. A solution beginning inside the contour $V(\mathbf{x}) = 1$ will move with V decreasing towards zero, while a solution beginning outside the contour $V(\mathbf{x}) = 1$ may move with V decreasing towards unity. Therefore, the system will be asymptotically stable inside the contour $V(\mathbf{x}) = 1$ but not outside (Koppel 1968, Csaki 1972).

In Theorem 8.10, the condition of radial unboundedness has been imposed for asymptotic stability in-the-large. It may also be imposed for stability in-the small, it merely depends on our understanding of 'smallness' or 'largeness'.

□

The determination of stability via Lyapunov's stability theorem centres around the choice of a positive definite function $V(\mathbf{x})$ called the *Lyapunov function*. Unfortunately, there is no universal method for selecting Lyapunov function which is unique for a specific problem. (Several techniques have been devised for a systematic construction of Lyapunov functions; some of these techniques will be discussed later in this chapter). If a certain V-function is not suitable to show stability, another V-function may still lead to the desired result. Thus if stability conditions are not satisfied, this does not mean that the system is unstable, but that a further attempt is needed. On the other hand, if certain V-function proves stability for a specified parameter region, this does not necessarily mean that leaving that region will result in system instability.

Further, for a given V-function, there is no general method which will easily allow us to ascertain whether it is positive definite. However, if $V(\mathbf{x})$ is in the quadratic form in x_i's, we can use Sylvester's theorem (Section 2.5) to ascertain definiteness of the function.

In spite of all these difficulties, Lyapunov's direct method is the most powerful technique available today for stability analysis of nonlinear systems.

Example 8.6: Let us reconsider the mass-spring-damper system of Fig. 8.4. The spring and the damper are now assumed to be nonlinear. For displacement x_1, the exerted spring force is $g(x_1)$: $g(x)$ is a single-valued centrally symmetrical nonlinearity shown in Fig. 8.8; $g(0) = 0$ and $xg(x) > 0$ if $x \neq 0$. The damper exerts a force proportional to instantaneous velocity $x_2 = \dot{x}_1$; the proportionality factor being an arbitrary function $h(\cdot)$ of displacement x_1.

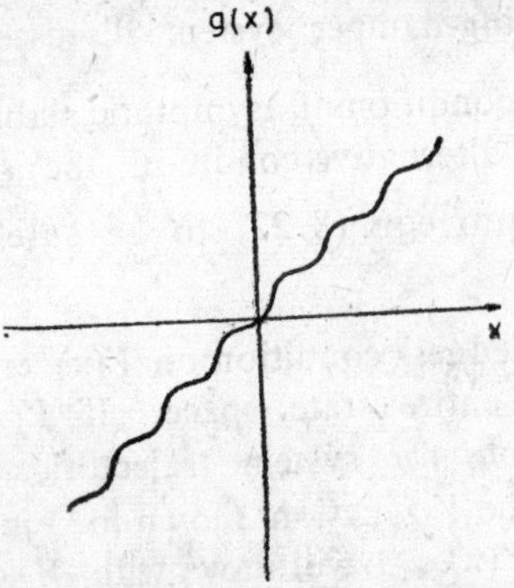

Fig. 8 8 Single-valued centrally symmetrical nonlinearity

For the nonlinear mass-spring-damper system, the governing equations become (refer eqns. (8.21))

$$\begin{aligned} \dot{x}_1 &= x_2 \\ \dot{x}_2 &= -g(x_1) - x_2 h(x_1) \end{aligned} \tag{8.23}$$

Once again, we take total energy as the V-function.

$$V(x_1, x_2) = \tfrac{1}{2}x_2^2 + \int_0^{x_1} g(x)\, dx \tag{8.24}$$

We test this function for Lyapunov properties.

For the nonlinearity shown in Fig. 8.8, the integral in (8.24) will always be nonnegative and therefore

$$V(\mathbf{x}) > 0; \qquad \mathbf{x} \neq \mathbf{0}$$

$$V(\mathbf{0}) = 0$$

Now

$$\frac{dV}{dt} = \frac{\partial V}{\partial x_1}\frac{dx_1}{dt} + \frac{\partial V}{\partial x_2}\frac{dx_2}{dt}$$

$$= g(x_1)\,x_2 + x_2\,[-\,g(x_1) - x_2\,h(x_1)]$$

$$= -\,x_2^2\,h(x_1) \qquad (8.25)$$

If $h(x_1) > 0$, then $\frac{dV}{dt}$ is negative semidefinite.

The solution to the equation $\frac{dV}{dt} = 0$ does not satisfy system equations (8.23) other than at $\mathbf{x} = \mathbf{0}$. Therefore $\frac{dV}{dt} = 0$ is no trajectory of the differential equations (8.23). As per Theorem 8.10, the nonlinear system under consideration is asymptotically stable.

Example 8.7: Consider the nonlinear servomechanism shown in Fig. 8.9. Assume that the linear part is described by the transfer function

$$G(s) = \frac{K}{s(s+\alpha)} \qquad (8.26)$$

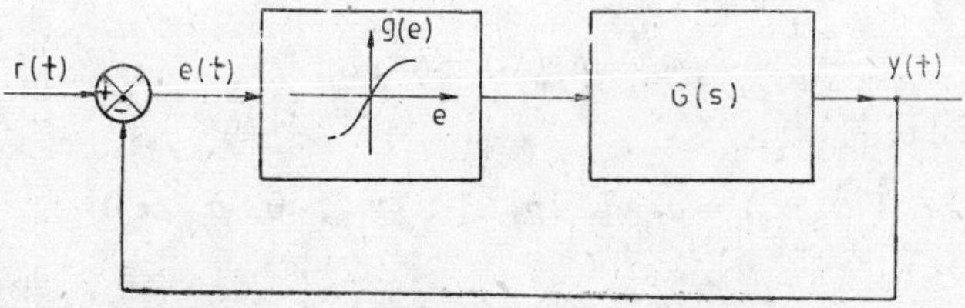

Fig. 8.9 A nonlinear servomechanism

If the reference variable $r(t) = 0$, then the differential equation for the actuating error will be

$$\ddot{e} + \alpha\dot{e} + Kg(e) = 0 \qquad (8.27)$$

Let us first examine the linearized system. With $g(e) = e$, eqn. (8.27) becomes

$$\ddot{e} + \alpha\dot{e} + Ke = 0 \qquad (8.28)$$

Taking e and $\dot{e}$ as state variables x_1 and x_2 respectively, we obtain the following state equations:

$$\dot{x}_1 = x_2 \qquad (8.29)$$

$$\dot{x}_2 = -Kx_1 - \alpha x_2$$

We test

$$V(x_1, x_2) = p_{11}x_1^2 + p_{22}x_2^2; \quad p_{11} > 0, p_{22} > 0 \tag{8.30}$$

for Lyapunov properties. Properties 1, 2 and 3 of Theorem 8.10 are obviously satisfied.

$$\frac{dV}{dt} = 2(p_{11} - p_{22}K)\, x_1x_2 - 2p_{22}\alpha x_2^2$$

$\dfrac{dV}{dt}$ is negative semidefinite if

$$p_{11} - p_{22}K = 0, \quad p_{22}\alpha > 0$$

If we take $p_{11} = Kp_{22}$ with $K > 0$, $\alpha > 0$, then sufficient conditions of stability given in Theorem 8.10 are satisfied. In fact, we can easily show that the linear system with $K > 0$, $\alpha > 0$ satisfies the sufficient conditions for global asymptotic stability.

As per the Routh-Hurwitz criterion, the necessary and sufficient conditions for stability are $K > 0$, $\alpha > 0$. Thus, in this case, the Lyapunov method has given not only the sufficient but also the necessary conditions for stability. As will be seen in the next section, this is always the case with the linear time-invariant systems.

Let the nonlinear element in Fig. 8.9 now be a single-valued centrally symmetrical nonlinearity (Fig. 8.8): $g(0) = 0$ and $eg(e) > 0$ if $e \neq 0$. The state model (8.29) now becomes

$$\dot{x}_1 = x_2$$

$$\dot{x}_2 = -Kg(x_1) - \alpha x_2$$

Choosing again

$$V(x_1, x_2) = p_{11}x_1^2 + p_{22}x_2^2\ ; \quad p_{11} > 0, \ p_{22} > 0$$

gives

$$\frac{dV}{dt} = 2p_{11}x_1x_2 - 2p_{22}Kg(x_1)\, x_2 - 2p_{22}\alpha x_2^2$$

Since $g(x_1)$ is nonlinear, there is no pair $p_{11} > 0$, $p_{22} > 0$ which would make $\dfrac{dV}{dt}$ at least negative semidefinite. Therefore, no decision about system stability can be reached.

Let us now test the following energy expression for Lyapunov properties.

$$V(x_1, x_2) = \tfrac{1}{2}x_2^2 + K\int_0^{x_1} g(x)\, dx$$

V is positive definite if $K > 0$.

$$\frac{dV}{dt} = Kg(x_1)\,x_2 + x_2\,[-Kg(x_1) - \alpha x_2]$$

$$= -\,\alpha x_2^2$$

Thus $\frac{dV}{dt}$ is negative semidefinite if $\alpha > 0$. Further the system trajectory cannot remain in any nonequilibrium state with $x_2 = 0$; the system is therefore asymptotically stable.

Example 8.8: Reconsider the system of Fig. 8.9 with

$$G(s) = \frac{K}{s^2 + \alpha^2}$$

and

$$g(e) = e$$

We know that the closed-loop system will exhibit sustained oscillations.

For the new transfer function, the state model (8.29) now becomes

$$\begin{aligned} \dot{x}_1 &= x_2 \\ \dot{x}_2 &= -\,(K + \alpha^2)\,x_1 \end{aligned} \tag{8.31}$$

For the V-function given by (8.30),

$$\frac{dV}{dt} = 2p_{11}x_1x_2 - 2p_{22}\,(K + \alpha^2)x_1x_2$$

If we set $p_{11} = p_{22}\,(K + \alpha^2)$, $\frac{dV}{dt} = 0$ and therefore as per Theorem 8.10, the system is stable. This result corresponds to the physical situation reflected by the sustained oscillations.

It is of interest to compare this Lyapunov result with that obtained by a direct solution of this simple system and application of the definition of stability.

The system (8.31), for an initial state $\mathbf{x}^0$, has the solution

$$\mathbf{x}(t) = \begin{bmatrix} x_1^0 \cos\sqrt{K + \alpha^2}\,t + \dfrac{x_2^0}{\sqrt{K + \alpha^2}} \sin\sqrt{K + \alpha^2}\,t \\ -\sqrt{K + \alpha^2}\,x_1^0 \sin\sqrt{K + \alpha^2}\,t + x_2^0 \cos\sqrt{K + \alpha^2}\,t \end{bmatrix}$$

Using the norm defined in (2.27a), we may write

$$\|\mathbf{x}(t)\| = \left| x_1^0 \cos\sqrt{K + \alpha^2}\,t + \frac{x_2^0}{\sqrt{K + \alpha^2}} \sin\sqrt{K + \alpha^2}\,t \right| + \left| -\sqrt{K + \alpha^2}\,x_1^0 \sin\sqrt{K + \alpha^2}\,t + x_2^0 \cos\sqrt{K + \alpha^2}\,t \right|$$

$$\leqslant |x_1^0| + \frac{|x_2^0|}{|\sqrt{K+\alpha^2}|} + |\sqrt{K+\alpha^2}|\,|x_1^0| + |x_2^0|$$

$$\leqslant \left(1 + \frac{1}{|\sqrt{K+\alpha^2}|} + |\sqrt{K+\alpha^2}|\right)(|x_1^0| + |x_2^0|)$$

$$\leqslant \left(1 + \frac{1}{|\sqrt{K+\alpha^2}|} + |\sqrt{K+\alpha^2}|\right) \|\mathbf{x}^0\|$$

Therefore, for a given ε, we choose

$$\delta(\varepsilon) = \frac{\varepsilon}{1 + \dfrac{1}{|\sqrt{K+\alpha^2}|} + |\sqrt{K+\alpha^2}|}$$

to satisfy the definition of stability.

Example 8.9: Consider the nonlinear system of equations

$$\dot{x}_1 = -2x_1 + x_1x_2 \tag{8.32}$$

$$\dot{x}_2 = -x_2 + x_1x_2$$

Note that there are two equilibrium points: $\mathbf{x} = \begin{bmatrix} 0 \\ 0 \end{bmatrix}$ and $\mathbf{x} = \begin{bmatrix} 1 \\ 2 \end{bmatrix}$. We shall study the stability of the origin.

A candidate for a Lyapunov function is

$$V = p_{11}x_1^2 + p_{22}x_2^2;\ p_{11} > 0, p_{22} > 0$$

Then

$$\frac{dV}{dt} = 2p_{11}x_1\dot{x}_1 + 2p_{22}x_2\,\dot{x}_2$$

$$= 2p_{11}x_1(-2x_1 + x_1x_2) + 2p_{22}x_2(-x_2 + x_1x_2)$$

$$= 2p_{11}x_1^2\,(x_2-2) + 2p_{22}x_2^2\,(x_1-1)$$

We cannot select any values of $p_{11} > 0$ and $p_{22} > 0$ to make $\frac{dV}{dt}$ globally negative definite. Let us take $p_{11} = p_{22} = 1$ and study the resulting function

$$\frac{dV}{dt} = 2x_1^2\,(x_2-2) + 2x_2^2\,(x_1-1)$$

For asymptotic stability we require that

$$x_1^2\,(x_2-2) + x_2^2\,(x_1-1) < 0 \tag{8.33}$$

The region of state-space where this condition is not satisfied is *possibly* the region of instability. Let us concentrate on the region of state-space where this condition is satisfied. The limiting condition for such a region is

$$x_1^2(x_2-2)+x_2^2(x_1-1)=0$$

The dividing line in the first quadrant is shown in Fig. 8.10. Further, we can easily verify that condition (8.33) is satisfied in the entire third quadrant. Information about second and fourth quadrants can be obtained on similar lines.

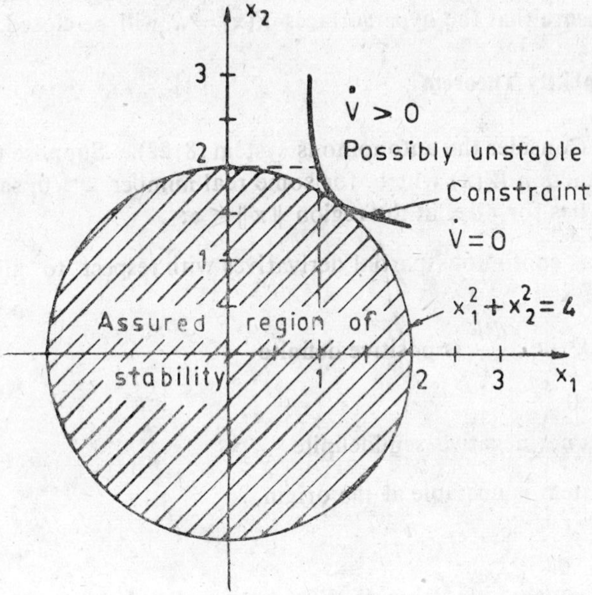

Fig. 8.10

Now, all those initial states **x**(0) for which **x**(t); $t \geqslant 0$ lies in the region where $\dot{V} < 0$, will lead to asymptotic stability. Let δ_k be the region (set of values of **x** including origin) in which $V(\mathbf{x})$ is positive definite, $\frac{dV}{dt}$ is negative definite and $V(\mathbf{x}) < k$. Then every solution with **x** in δ_k, approaches the origin asymptotically as $t \to \infty$ (refer proof of Theorem 8.10). Obviously, we would like to find the largest such region. To do this, we investigate the equation for the contour of V:

$$V = x_1^2 + x_2^2 = k$$

for various values of k, selecting that k which gives the largest region satisfying the condition $\frac{dV}{dt} < 0$. For the example under consideration, we find that $k = 4$ gives the largest region (Fig. 8.10).

□

We now summarize the method of constructing the largest region of asymptotic stability. First establish the constraints on negative definiteness of $\frac{dV}{dt}$. Then search for the hypersurface $V(\mathbf{x}) = k$ which is tangent to at least one constraint hypersurface and lies inside all other constraint hypersurfaces. This gives the largest region of asymptotic stability for a selected V-function. The condition of radial unboundedness may be imposed on V-function to ensure that the hypersurfaces $V(\mathbf{x}) = k$ will be closed.

Lyapunov's Instability Theorem

Theorem 8.11: Consider the autonomous system (8.22). Suppose that there exists a scalar function $W(\mathbf{x})$ which, for some real number $\varepsilon > 0$, satisfies the following properties for all $\mathbf{x}$ in the region $\|\mathbf{x}\| \leqslant \varepsilon$:

(1) $W(\mathbf{x})$ has continuous partial derivatives with respect to all components of $\mathbf{x}$

(2) $\frac{dW}{dt} > 0$, i.e., $\frac{dW}{dt}$ is positive definite

(3) $W(\mathbf{0}) = 0$

(4) $W(\mathbf{x})$ is not negative semidefinite.

Then the system is unstable at the origin.

Proof:

Let M be the maximum value of $W(\mathbf{x})$ on the closed set $\|\mathbf{x}\| = \varepsilon$. This is indicated by contour $C(W = M)$ in Fig. 8.11 for a two-dimensional case. The conditions of the theorem guarantee the existence of a point $\mathbf{x}^0$ in the region $\|\mathbf{x}\| \leqslant \varepsilon$ where $W(\mathbf{x}^0) > 0$. Further, the continuity of $W(\mathbf{x})$ at origin assures that there exists a real number $\delta > 0$ such that $W(\mathbf{x}) < W(\mathbf{x}^0)$ for all $\mathbf{x}$ in the region $\|\mathbf{x}\| \leqslant \delta$; $\delta \leqslant \|\mathbf{x}^0\|$, otherwise there may be contradiction (refer Fig. 8.11).

Consider the solution $\mathbf{x}(t)$ of eqn. (8.22) with $\mathbf{x}(0) = \mathbf{x}^0$. Since $\frac{dW}{dt} > 0$, $W(\mathbf{x}(t)) \geqslant W(\mathbf{x}^0)$ as long as $\|\mathbf{x}(t)\| \leqslant \varepsilon$ and hence $\|\mathbf{x}(t)\| \geqslant \delta$. Let m be the minimum value of $\frac{dW}{dt}$ on the closed set $\delta \leqslant \|\mathbf{x}\| \leqslant \varepsilon$. Then

$$W(\mathbf{x}(t)) = W(\mathbf{x}^0) + \int_0^t \frac{dW(\mathbf{x}(\tau))}{d\tau}\, d\tau$$

$$\geqslant W(\mathbf{x}^0) + mt\,; \quad t \geqslant 0$$

$W(\mathbf{x}(t))$ will eventually exceed M which implies $\|\mathbf{x}\| > \varepsilon$. This proves the theorem (Koppel 1968, Csaki 1972).

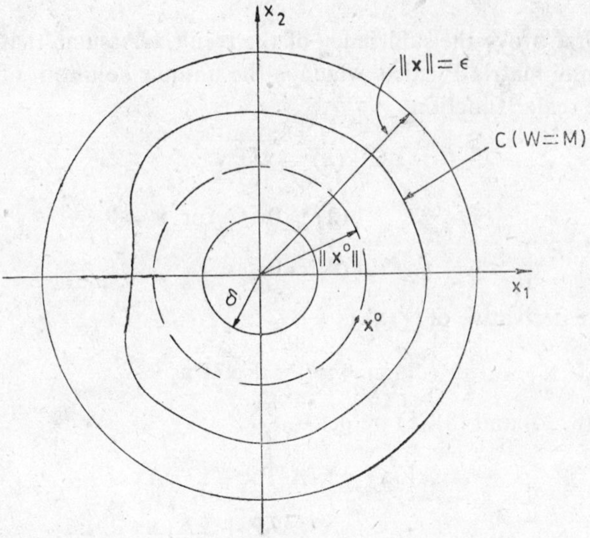

Fig. 8.11

□

Note that it requires as much ingenuity to devise a suitable W function as to devise a Lyapunov function V. In stability analysis of nonlinear systems, it is valuable to establish conditions for which the system is unstable. Then the regions of asymptotic stability need not be sought for such conditions and the analyst is saved of this fruitless effort.

8.6 THE DIRECT METHOD OF LYAPUNOV AND THE LINEAR CONTINUOUS-TIME AUTONOMOUS SYSTEMS

In the case of linear systems, the direct method of Lyapunov provides a simple approach to stability analysis. It must be noted here that compared to the results presented in Section 8.4, no new results are obtained by the use of the direct method of Lyapunov for stability analysis of linear systems. However, the study of linear systems using the direct method is quite useful because it extends our thinking to nonlinear systems (Section 8.7). In addition, Lyapunov functions supply certain performance indices and synthesis data of the system concerned (Chapters 11 and 12).

Consider a linear autonomous system described by the state equation

$$\dot{\mathbf{x}} = \mathbf{A}\mathbf{x} \tag{8.34}$$

where $\mathbf{A}$ is $n \times n$ real constant matrix.

Theorem 8.12: The linear system (8.34) is globally asymptotically stable at the origin if and only if for any given symmetric positive definite matrix $\mathbf{Q}$, there exists a symmetric positive definite matrix $\mathbf{P}$ that satisfies the matrix equation

$$\mathbf{A}^T\mathbf{P} + \mathbf{P}\mathbf{A} = -\mathbf{Q} \tag{8.35}$$

Proof:

Let us first prove the sufficiency of the result. Assume that a symmetric positive definite matrix **P** exists which is the unique solution of eqn. (8.35). Consider the scalar function

$$V(\mathbf{x}) = \mathbf{x}^T\mathbf{P}\mathbf{x}$$

Note that

$$V(\mathbf{x}) > 0 \qquad \text{for } \mathbf{x} \neq \mathbf{0}$$

and

$$V(\mathbf{0}) = 0$$

The time derivative of $V(\mathbf{x})$ is

$$\dot{V}(\mathbf{x}) = \dot{\mathbf{x}}^T\mathbf{P}\mathbf{x} + \mathbf{x}^T\mathbf{P}\dot{\mathbf{x}}$$

Using eqns. (8.34) and (8.35) we get

$$\begin{aligned}\dot{V}(\mathbf{x}) &= \mathbf{x}^T\mathbf{A}^T\mathbf{P}\mathbf{x} + \mathbf{x}^T\mathbf{P}\mathbf{A}\mathbf{x}\\ &= \mathbf{x}^T(\mathbf{A}^T\mathbf{P} + \mathbf{P}\mathbf{A})\,\mathbf{x}\\ &= -\,\mathbf{x}^T\,\mathbf{Q}\mathbf{x}\end{aligned}$$

Since **Q** is positive definite, $\dot{V}(\mathbf{x})$ is negative definite. Norm of **x** may be defined as (eqn. (2.27e))

$$\|\mathbf{x}\| = (\mathbf{x}^T\,\mathbf{P}\mathbf{x})^{1/2}$$

Then

$$V(\mathbf{x}) = \|\mathbf{x}\|^2$$

$$V(\mathbf{x}) \to \infty \;\text{ as }\; \|\mathbf{x}\| \to \infty$$

Therefore as per Theorem 8.10, the system is globally asymptotically stable at the origin.

Let us now prove the necessity of the result. We shall prove it in two parts.

(i) If (8.34) is asymptotically stable, then for any **Q** there exists a matrix **P** satisfying (8.35).

(ii) If **Q** is positive definite, then **P** is also positive definite.

In Chapter 2 we observed that a set of all $n \times n$ matrices with real elements forms a linear space of dimension n^2 (Problem 2.11). Denote such a linear space by $\mathcal{M}(\mathcal{R})$. Let **A** and $\mathbf{P} \in \mathcal{M}(\mathcal{R})$. The operator **L** defined by

$$\mathbf{L}(\mathbf{P}) \triangleq \mathbf{A}\mathbf{P} + \mathbf{P}\mathbf{A}^T$$

maps $\mathcal{M}(\mathcal{R})$ into itself and is a linear operator.

Let λ_i, $i = 1, 2, \ldots, m \leqslant n$ be the distinct eigenvalues of **A** and μ_k, $k = 1, 2, \ldots, p \leqslant n^2$ be the distinct eigenvalues of **L**. Then for each k,

$$\mu_k = \lambda_i + \lambda_j \qquad \text{for some } i \text{ and some } j.$$

This result can be proved as follows (Chen 1970):

Assume that μ_k is an eigenvalue of **L**. Then by definition, there exists a $\mathbf{P} \neq \mathbf{0}$ such that (refer Section 2.7)

$$\mathbf{L}(\mathbf{P}) = \mu_k \mathbf{P}$$

Therefore

$$\mu_k \mathbf{P} = \mathbf{A}\mathbf{P} + \mathbf{P}\mathbf{A}^T$$

or

$$(\mu_k \mathbf{I} - \mathbf{A})\, \mathbf{P} = \mathbf{P}\mathbf{A}^T \tag{8.36}$$

The matrices $(\mu_k \mathbf{I} - \mathbf{A})$ and $\mathbf{A}^T$ have at least one eigenvalue in common. We prove this by contradiction. Suppose $(\mu_k \mathbf{I} - \mathbf{A})$ and $\mathbf{A}^T$ have no eigenvalue in common. Let

$$\Delta(\lambda) = \lambda^n + \alpha_1 \lambda^{n-1} + \dots + \alpha_n = \prod_{i=1}^{l} (\lambda - \eta_i)^{q_i}$$

be the characteristic polynomial of $(\mu_k \mathbf{I} - \mathbf{A})$. By the Cayley-Hamilton theorem,

$$\Delta(\mu_k \mathbf{I} - \mathbf{A}) = \mathbf{0} \tag{8.37}$$

From eqn. (8.36), it follows that

$$(\mu_k \mathbf{I} - \mathbf{A})^2 \mathbf{P} = \mathbf{P}(\mathbf{A}^T)^2$$
$$\vdots$$
$$(\mu_k \mathbf{I} - \mathbf{A})^n \mathbf{P} = \mathbf{P}(\mathbf{A}^T)^n$$

Consequently, we have

$$\Delta(\mu_k \mathbf{I} - \mathbf{A})\, \mathbf{P} = \mathbf{P}\, \Delta(\mathbf{A}^T)$$

where

$$\Delta(\mathbf{A}^T) = \prod_{i=1}^{l} (\mathbf{A}^T - \eta_i \mathbf{I})^{q_i} \tag{8.38}$$

From (8.37) and (8.38), we have

$$\mathbf{P}\, \Delta(\mathbf{A}^T) = \mathbf{0}$$

Since, by assumption, $(\mu_k \mathbf{I} - \mathbf{A})$ and $\mathbf{A}^T$ have no eigenvalue in common, $|\mathbf{A}^T - \eta_i \mathbf{I}| \neq 0$ for all i and hence $\Delta(\mathbf{A}^T)$ is nonsingular. This implies that $\mathbf{P} = \mathbf{0}$, which contradicts the hypothesis that $\mathbf{P} \neq \mathbf{0}$. Therefore, the matrices $(\mu_k \mathbf{I} - \mathbf{A})$ and $\mathbf{A}^T$ have at least one common eigenvalue.

We know that if λ_i is an eigenvalue of **A**, then $f(\lambda_i)$ is an eigenvalue of $f(\mathbf{A})$ (Problem 2.23). Therefore eigenvalue of $(\mu_k \mathbf{I} - \mathbf{A})$ is of the form $(\mu_k - \lambda_i)$. Consequently for some i and for some j,

$$\mu_k - \lambda_i = \lambda_j$$

or

$$\mu_k = \lambda_i + \lambda_j \tag{8.39}$$

Since the linear operator **L** maps n^2 dimensional linear space into itself, it has a matrix representation (Section 2.4). We can obtain the matrix representation by writing the n^2 equations (8.35) in the form

$$\mathbf{M}\mathbf{p} = \mathbf{q} \tag{8.40}$$

where $\mathbf{p}$ is $n^2 \times 1$ column vector consisting of all the n^2 elements of $\mathbf{P}$. Since the determinant of a matrix is equal to the product of its eigenvalues (Problem 2.17b), the matrix representation $\mathbf{M}$ is nonsingular if and only if

$$\lambda_i + \lambda_j \neq 0 \qquad \text{for all } i, j.$$

The condition is assured because of the assumption of asymptotic stability of $\mathbf{A}$ (all the eigenvalues of $\mathbf{A}$ have negative real parts); therefore for a given $\mathbf{Q}$ there exists a $\mathbf{P}$ which satisfies (8.35). This is evident from (8.40).

Let us now go to the second part: if $\mathbf{Q}$ is positive definite, then $\mathbf{P}$ is also positive definite. Suppose this is not the case; $\mathbf{P}$ is negative definite. Consider the scalar function

$$V(\mathbf{x}) = -\,\mathbf{x}^T \mathbf{P}\mathbf{x} \tag{8.41}$$

Then

$$\begin{aligned}\dot{V}(\mathbf{x}) &= -\,[\dot{\mathbf{x}}^T \mathbf{P}\mathbf{x} + \mathbf{x}^T \mathbf{P}\dot{\mathbf{x}}] \\ &= \mathbf{x}^T \mathbf{Q}\mathbf{x} \\ &> 0\end{aligned}$$

There is a contradiction since $V(\mathbf{x})$ given by (8.41) satisfies instability Theorem 8.11.

□

The implication of Theorem 8.12 is that if $\mathbf{A}$ is asymptotically stable and if $\mathbf{Q}$ is positive definite, then the solution $\mathbf{P}$ of (8.35) must be positive definite. Note that it does not say that if $\mathbf{A}$ is asymptotically stable and if $\mathbf{P}$ is positive definite, then $\mathbf{Q}$ computed from (8.35) is positive definite. For an arbitrary $\mathbf{P}$, $\mathbf{Q}$ may be positive definite (semidefinite) or negative definite (semidefinite).

Since Theorem 8.12 holds for any positive definite symmetric matrix $\mathbf{Q}$, the matrix $\mathbf{Q}$ in (8.35) is often chosen to be a unit matrix. Since matrix $\mathbf{P}$ is known to be symmetric, there are only $n(n+1)/2$ independent equations to be solved, rather than n^2. Several numerical methods of solution of the *Lyapunov equation* (8.35) are available. Hagander (1972), Barnett and Storey (1970) and Rothschild and Jameson (1970) review several methods of solution.

From part (b) of Theorem 8.10, it is observed that while applying Theorem 8.12 for stability analysis of linear systems, a positive semidefinite matrix may also be taken as $\mathbf{Q}$ matrix of the Lyapunov equation (8.35).

Example 8.10: Let us determine the stability of the system described by the following equation:

$$\dot{\mathbf{x}} = \mathbf{A}\mathbf{x}$$

where

$$\mathbf{A} = \begin{bmatrix} -1 & -2 \\ 1 & -4 \end{bmatrix}$$

We will first solve eqn. (8.35) for **P** for an arbitrary choice of real symmetric positive definite/semidefinite matrix **Q**. We may choose $\mathbf{Q} = \mathbf{I}$, the identity matrix. Equation (8.35) then becomes

$$\mathbf{A}^T\mathbf{P} + \mathbf{P}\mathbf{A} = -\mathbf{I}$$

or

$$\begin{bmatrix} -1 & 1 \\ -2 & -4 \end{bmatrix}\begin{bmatrix} p_{11} & p_{12} \\ p_{12} & p_{22} \end{bmatrix} + \begin{bmatrix} p_{11} & p_{12} \\ p_{12} & p_{22} \end{bmatrix}\begin{bmatrix} -1 & -2 \\ 1 & -4 \end{bmatrix} = \begin{bmatrix} -1 & 0 \\ 0 & -1 \end{bmatrix} \tag{8.42}$$

Note that we have taken $p_{12} = p_{21}$; this is because the solution matrix **P** is known to be a positive definite real symmetric matrix for a stable system.

From eqn. (8.42), we get

$$-2p_{11} + 2p_{12} = -1$$
$$-2p_{11} - 5p_{12} + p_{22} = 0$$
$$-4p_{12} - 8p_{22} = -1$$

Solving for p_{ij}'s we obtain

$$\mathbf{P} = \begin{bmatrix} p_{11} & p_{12} \\ p_{12} & p_{22} \end{bmatrix} = \begin{bmatrix} \frac{23}{60} & -\frac{7}{60} \\ -\frac{7}{60} & \frac{11}{60} \end{bmatrix}$$

Using Sylvester's theorem (Section 2.5) we find that **P** is positive definite. Therefore, the origin of the system under consideration is asymptotically stable in the large.

In order to illustrate the arbitrariness of **Q**, consider

$$\mathbf{Q} = \begin{bmatrix} 0 & 0 \\ 0 & 1 \end{bmatrix}$$

It can be verified that with this choice of **Q**, we derive the same conclusion about the stability of the system as obtained earlier with $\mathbf{Q} = \mathbf{I}$.

A Proof of the Routh-Hurwitz Criterion

In this subsection, we shall use Lyapunov's direct method (Theorem 8.12) to prove Theorem 8.5 giving Hurwitz stability criterion. The Routh stability criterion can be derived from that of Hurwitz, as was shown in Section 8.4.

Consider a linear autonomous system in the phase-variable canonical form (refer eqn. (7.65))

$$\dot{\mathbf{x}} = \mathbf{A}\mathbf{x}$$

$$= \begin{bmatrix} 0 & 1 & 0 & \dots & 0 \\ 0 & 0 & 1 & \dots & 0 \\ \vdots & \vdots & \vdots & & \vdots \\ 0 & 0 & 0 & \dots & 1 \\ -\alpha_n & -\alpha_{n-1} & -\alpha_{n-2} & \dots & -\alpha_1 \end{bmatrix} \mathbf{x} \tag{8.43}$$

The characteristic equation of this system is

$$\Delta(\lambda) = |\lambda \mathbf{I} - \mathbf{A}| = \lambda^n + \alpha_1 \lambda^{n-1} + \alpha_2 \lambda^{n-2} + \dots + \alpha_n = 0 \tag{8.44}$$

Let us convert the phase equation (8.43) by a transformation $\mathbf{x} = \mathbf{T}\mathbf{y}$, to the *Schwartz form*

$$\begin{aligned} \dot{\mathbf{y}} &= \mathbf{T}^{-1}\mathbf{A}\mathbf{T}\mathbf{y} \\ &= \mathbf{B}\mathbf{y} \end{aligned} \tag{8.45}$$

where

$$\mathbf{B} = \begin{bmatrix} 0 & 1 & 0 \dots & 0 & 0 \\ -b_n & 0 & 1 \dots & 0 & 0 \\ 0 & -b_{n-1} & 0 \dots & 0 & 0 \\ \vdots & \vdots & \vdots & \vdots & \vdots \\ 0 & 0 & 0 \dots & 0 & 1 \\ 0 & 0 & 0 \dots & -b_2 & -b_1 \end{bmatrix}$$

Comparing the coefficients of the characteristic equation

$$|\lambda \mathbf{I} - \mathbf{B}| = 0$$

with equation (8.44), we obtain

$$b_1 = \Delta_1,\ b_2 = \frac{\Delta_2}{\Delta_1},\ b_3 = \frac{\Delta_3}{\Delta_1 \Delta_2}, \dots,$$

$$b_i = \frac{\Delta_{i-3}\,\Delta_i}{\Delta_{i-2}\,\Delta_{i-1}} \quad (i = 4, 5, \dots, n)$$

where Δ_i, $i = 1, 2, \dots, n$ represent the Hurwitz determinants taken as principal minors of the arrangement given by (8.18).

Consider the scalar function

$$V(\mathbf{y}) = \mathbf{y}^T \mathbf{P}\mathbf{y}$$

where

$$\mathbf{P} = \begin{bmatrix} b_1b_2...b_n & 0 & ... & 0 & 0 & 0 \\ 0 & b_1b_2...b_{n-1} & ... & 0 & 0 & 0 \\ \vdots & \vdots & & \vdots & \vdots & \vdots \\ 0 & 0 & ... & b_1b_2b_3 & 0 & 0 \\ 0 & 0 & ... & 0 & b_1b_2 & 0 \\ 0 & 0 & ... & 0 & 0 & b_1 \end{bmatrix}$$

The derivative of V is obtained as

$$\begin{aligned} \dot{V}(\mathbf{y}) &= \mathbf{y}^T(\mathbf{A}^T\mathbf{P} + \mathbf{PA})\,\mathbf{y} \\ &= -2b_1^2y_n^2 \end{aligned}$$

It is observed that $V(\mathbf{y})$ is positive definite and $\dot{V}(\mathbf{y})$ is negative definite, if all the elements in the main diagonal of matrix $\mathbf{P}$ are positive. Therefore, conditions of asymptotic stability of linear system under consideration are

$$b_i > 0 \ (i = 1, 2, ..., n)$$

or equivalently

$$\Delta_i > 0 \ (i = 1, 2, ..., n)$$

These conditions are the same as given by the Hurwitz stability criterion in Theorem 8.5.

□

While it may appear that the Routh-Hurwitz approach is considerably more direct, it must be remembered that Routh-Hurwitz criterion can be applied only to the characteristic polynomial of a system. Therefore, if the system model is known in the state variable form with the state variables chosen as physical variables of the system, we have to first convert the system model into phase-variable canonical form and therefrom obtain the characteristic polynomial. This conversion is not so easy for systems of high-order. Lyapunov's direct method does not require this conversion and stability can therefore be studied directly in terms of physical variables.

8.7 AIDS TO FINDING LYAPUNOV FUNCTIONS FOR NONLINEAR CONTINUOUS-TIME AUTONOMOUS SYSTEMS

As has been said earlier in this chapter, the Lyapunov theorems give only sufficient conditions on system stability and, furthermore, there is no unique way of constructing a Lyapunov function except in the case of linear systems where a Lyapunov function can always be constructed and both necessary and sufficient conditions established. Because of this limitation, a host of methods have become available in literature and many refinements have been suggested to enlarge the region in which the system is found to be stable. Since this treatise is meant as a first exposure of the student to the Lyapunov's direct method, only two of the relatively simpler techniques of constructing a Lyapunov function are advanced here.

The Krasovskii Method

In the following we state and prove a limited form of the theorem due to Krasovskii (Kalman 1960). We consider the nonlinear autonomous system

$$\dot{\mathbf{x}} = \mathbf{f}(\mathbf{x});\ \mathbf{f}(\mathbf{0}) = \mathbf{0} \tag{8.46}$$

We assume that $\mathbf{f}$ has continuous first partial derivatives. Let $\mathbf{J}(\mathbf{x})$ be the Jacobian matrix of $\mathbf{f}$ (refer Appendix III):

$$\mathbf{J}(\mathbf{x}) = \frac{\partial \mathbf{f}(\mathbf{x})}{\partial \mathbf{x}} = \begin{bmatrix} \frac{\partial f_1}{\partial x_1} & \frac{\partial f_1}{\partial x_2} & \cdots & \frac{\partial f_1}{\partial x_n} \\ \frac{\partial f_2}{\partial x_1} & \frac{\partial f_2}{\partial x_2} & \cdots & \frac{\partial f_2}{\partial x_n} \\ \vdots & \vdots & & \vdots \\ \frac{\partial f_n}{\partial x_1} & \frac{\partial f_n}{\partial x_2} & \cdots & \frac{\partial f_n}{\partial x_n} \end{bmatrix} \tag{8.47}$$

Theorem 8.13: The nonlinear system (8.46) is asymptotically stable at the origin if there exists a constant, positive definite, symmetric matrix $\mathbf{P}$ such that the matrix

$$\mathbf{F}(\mathbf{x}) = \mathbf{J}^T(\mathbf{x})\,\mathbf{P} + \mathbf{P}\mathbf{J}(\mathbf{x}) \tag{8.48}$$

is negative definite for all $\mathbf{x}$. Furthermore

$$V(\mathbf{x}) = \mathbf{f}^T\mathbf{P}\mathbf{f} \tag{8.49}$$

Proof:

From eqn. (8.49), we may write

$$\frac{dV}{dt} = \mathbf{f}^T(\mathbf{x})\,\mathbf{P}\mathbf{f}(\mathbf{x}) + \mathbf{f}^T(\mathbf{x})\,\mathbf{P}\dot{\mathbf{f}}(\mathbf{x})$$

$$= \left(\frac{\partial \mathbf{f}(\mathbf{x})}{\partial \mathbf{x}}\dot{\mathbf{x}}\right)^T \mathbf{P}\mathbf{f}(\mathbf{x}) + \mathbf{f}^T(\mathbf{x})\,\mathbf{P}\,\frac{\partial \mathbf{f}(\mathbf{x})}{\partial \mathbf{x}}\dot{\mathbf{x}}$$

$$= \mathbf{f}^T(\mathbf{x})\,\mathbf{J}^T(\mathbf{x})\,\mathbf{P}\mathbf{f}(\mathbf{x}) + \mathbf{f}^T(\mathbf{x})\,\mathbf{P}\mathbf{J}(\mathbf{x})\,\mathbf{f}(\mathbf{x})$$

$$= \mathbf{f}^T(\mathbf{x})\,[\mathbf{J}^T(\mathbf{x})\,\mathbf{P} + \mathbf{P}\mathbf{J}(\mathbf{x})]\,\mathbf{f}(\mathbf{x})$$

$$= \mathbf{f}^T(\mathbf{x})\,\mathbf{F}(\mathbf{x})\,\mathbf{f}(\mathbf{x})$$

Since $\mathbf{F}(\mathbf{x})$ is negative definite by hypothesis, $\frac{dV}{dt}$ is negative definite.

From (8.49), it is clear that $V(\mathbf{0}) = 0$. Further, $V(\mathbf{x}) > 0$ for all $\mathbf{x} \neq \mathbf{0}$, unless $\mathbf{f}(\mathbf{x}) = \mathbf{0}$ for some $\mathbf{x} \neq \mathbf{0}$, i.e., unless the system has another equilibrium point. In the following we show that if $\mathbf{F}(\mathbf{x})$ is negative definite, the system (8.46) cannot have an equilibrium point other than the origin.

Consider the vector

$$\frac{d\mathbf{f}(\alpha\mathbf{x})}{d\alpha} = \frac{\partial \mathbf{f}(\alpha\mathbf{x})}{\partial \mathbf{x}}\,\mathbf{x}$$

Integrating both sides from $\alpha = 0$ to $\alpha = 1$ we obtain

$$\mathbf{f}(\mathbf{x}) = \int_0^1 \frac{\partial \mathbf{f}(\alpha\mathbf{x})}{\partial \mathbf{x}}\,\mathbf{x}\,d\alpha$$

Consider now a vector $\mathbf{x} \neq \mathbf{0}$ for which $\mathbf{f}(\mathbf{x}) = \mathbf{0}$. Then

$$\mathbf{x}^T\mathbf{P}\mathbf{f}(\mathbf{x}) = 0 = \int_0^1 \mathbf{x}^T\mathbf{P}\frac{\partial \mathbf{f}(\alpha\mathbf{x})}{\partial \mathbf{x}}\,\mathbf{x}\,d\alpha$$

$$= \tfrac{1}{2}\int_0^1 \mathbf{x}^T\mathbf{F}(\alpha\mathbf{x})\,\mathbf{x}\,d\alpha$$

Since $\mathbf{F}(\alpha\mathbf{x})$ is negative definite for all values of α, the integral is negative and cannot equal zero. This is a contradiction.

Thus the conditions of asymptotic stability in Lyapunov's stability theorem are satisfied.

Example 8.11: Consider the nonlinear system shown in Fig. 8.12 where the nonlinear element is described as

$$u = g(e)$$

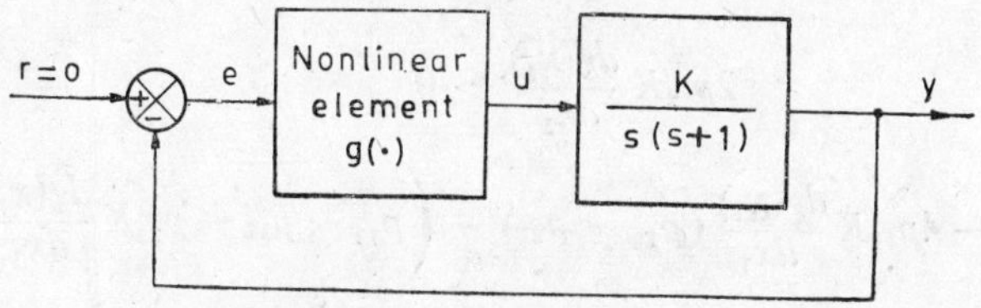

Fig. 8.12 A nonlinear system

With $r = 0$, $y = -e$. Therefore

$$Ku = -\ddot{e} - \dot{e}$$

Defining $x_1 = e$ and $x_2 = \dot{e}$, we get the following state equations:

$$\begin{aligned} \dot{x}_1 &= f_1(\mathbf{x}) = x_2 \\ \dot{x}_2 &= f_2(\mathbf{x}) = -x_2 - Kg(x_1) \end{aligned} \tag{8.50}$$

The equilibrium point lies at the origin if $g(0) = 0$. Now

$$\mathbf{J}(\mathbf{x}) = \begin{bmatrix} \dfrac{\partial f_1}{\partial x_1} & \dfrac{\partial f_1}{\partial x_2} \\ \dfrac{\partial f_2}{\partial x_1} & \dfrac{\partial f_2}{\partial x_2} \end{bmatrix} = \begin{bmatrix} 0 & 1 \\ -K\dfrac{dg(x_1)}{dx_1} & -1 \end{bmatrix} \tag{8.51}$$

Let

$$\mathbf{P} = \begin{bmatrix} p_{11} & p_{12} \\ p_{12} & p_{22} \end{bmatrix} \tag{8.52a}$$

For this matrix to be positive definite,

$$p_{11} > 0 \tag{8.52b}$$

$$p_{11}p_{22} - p_{12}^2 > 0 \tag{8.52c}$$

The matrix

$$\mathbf{F}(\mathbf{x}) = \mathbf{J}^T(\mathbf{x})\,\mathbf{P} + \mathbf{P}\mathbf{J}(\mathbf{x})$$

$$= \begin{bmatrix} 0 & -K\dfrac{dg(x_1)}{dx_1} \\ 1 & -1 \end{bmatrix} \begin{bmatrix} p_{11} & p_{12} \\ p_{12} & p_{22} \end{bmatrix} + \begin{bmatrix} p_{11} & p_{12} \\ p_{12} & p_{22} \end{bmatrix} \begin{bmatrix} 0 & 1 \\ -K\dfrac{dg(x_1)}{dx_1} & -1 \end{bmatrix}$$

$$= \begin{bmatrix} -2p_{12}K\dfrac{dg(x_1)}{dx_1} & p_{11} - p_{12} - p_{22}K\dfrac{dg(x_1)}{dx_1} \\ p_{11} - p_{12} - p_{22}K\dfrac{dg(x_1)}{dx_1} & 2(p_{12} - p_{22}) \end{bmatrix} \tag{8.53}$$

For the system to be asymptotically stable, $-\mathbf{F}(\mathbf{x})$ should be positive definite, i.e.,

$$2p_{12}K\frac{dg(x_1)}{dx_1} > 0 \tag{8.54}$$

$$-4p_{12}K\frac{dg(x_1)}{dx_1}(p_{12} - p_{22}) - \left(p_{11} - p_{12} - p_{22}K\frac{dg(x_1)}{dx_1}\right)^2 > 0$$

or

$$4p_{12}K\frac{dg(x_1)}{dx_1}(p_{22} - p_{12}) > \left(p_{11} - p_{12} - p_{22}K\frac{dg(x_1)}{dx_1}\right)^2 \tag{8.55}$$

Assume that $K > 0$ and choose $p_{12} > 0$. Inequality (8.54) then yields the condition

$$\frac{dg(x_1)}{dx_1} > 0 \tag{8.56}$$

Choose $p_{11} = p_{12}$ and $p_{22} = \beta p_{12}$, where $\beta > 1$. Inequality (8.55) then gives the condition

$$4(\beta - 1) > K\beta^2 \frac{dg(x_1)}{dx_1} \qquad (8.57)$$

The inequalities (8.56) and (8.57) together constitute the conditions under which the system is asymptotically stable. Take, for example,
$g(x_1) = x_1^3$; nonlinearity symmetrically lies in first and third quadrants.
Then

$$\frac{dg(x_1)}{dx_1} = 3x_1^2 > 0$$

The condition (8.57) gives

$$x_1^2 < \frac{4}{3K}\left(\frac{1}{\beta} - \frac{1}{\beta^2}\right)$$

It can be easily shown that the largest value of x_1 occurs when $\beta = 2$. Then

$$x_1^2 < \frac{1}{3K}$$

or

$$-\frac{1}{\sqrt{3K}} < x_1 < \frac{1}{\sqrt{3K}}$$

This region of asymptotic stability is shown in Fig. 8.13.

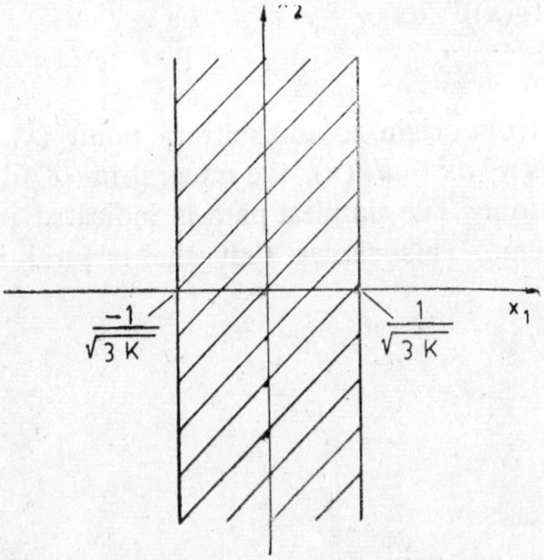

Fig. 8.13 Region of asymptotic stability for system of Example 8.11 with $g(e) = e^3$

The Variable-Gradient Method

The quadratic form approach used so far to Lyapunov function formulation is too restrictive. Schultz and Gibson (1962) suggested the variable gradient method for generating Lyapunov function which provides considerable flexibility in selecting a suitable function.

For the autonomous system (8.46), let $V(\mathbf{x})$ be a candidate for a Lyapunov function. The time derivative of V can be expressed as

$$\dot{V}(\mathbf{x}) = \frac{\partial V}{\partial x_1}\dot{x}_1 + \frac{\partial V}{\partial x_2}\dot{x}_2 + \ldots + \frac{\partial V}{\partial x_n}\dot{x}_n \tag{8.58}$$

Let (refer Appendix III)

$$\mathbf{g}(\mathbf{x}) = \text{grad } V(\mathbf{x}) = \begin{bmatrix} \dfrac{\partial V}{\partial x_1} \\ \dfrac{\partial V}{\partial x_2} \\ \vdots \\ \dfrac{\partial V}{\partial x_n} \end{bmatrix} = \begin{bmatrix} g_1(\mathbf{x}) \\ g_2(\mathbf{x}) \\ \vdots \\ g_n(\mathbf{x}) \end{bmatrix} \tag{8.59}$$

Then eqn. (8.58) may be written as

$$\dot{V}(\mathbf{x}) = (\mathbf{g}(\mathbf{x}))^T \dot{\mathbf{x}} \tag{8.60}$$

The Lyapunov function can be generated by integrating (8.60) on both sides.

$$V(\mathbf{x}) = \int_0^t \frac{dV(\mathbf{x})}{dt}\, dt = \int_0^t (\mathbf{g}(\mathbf{x}))^T \frac{d\mathbf{x}}{dt}\, dt$$

$$= \int_0^{\mathbf{x}} (\mathbf{g}(\mathbf{x}))^T \, d\mathbf{x} \tag{8.61}$$

This is a line integral from origin to an arbitrary point $(x_1, x_2, \ldots, x_n)$ in the state space. Since $(\mathbf{g}(\mathbf{x}))^T d\mathbf{x} = dV(\mathbf{x})$, the integral in (8.61) is independent of the path of integration. The simplest path is indicated in Fig. 8.14 for a three-dimensional system. This shows that the integral in (8.61) can be

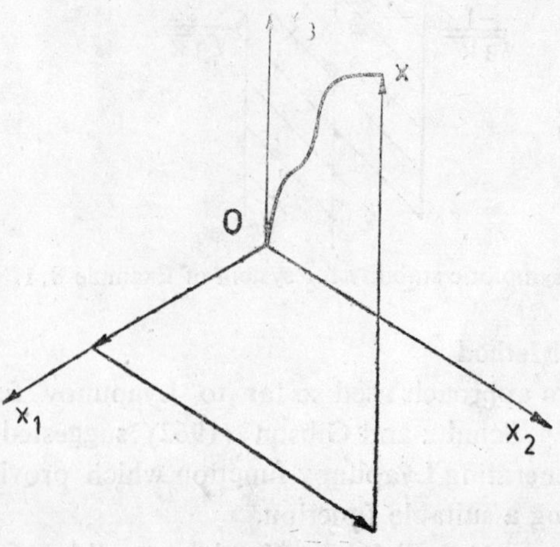

Fig. 8.14

evaluated sequentially along the component directions $\{x_1, x_2, \ldots, x_n\}$ of the state vector x, i.e.,

$$V(\mathbf{x}) = \int_0^{\mathbf{x}} (\mathbf{g}(\mathbf{x}))^T \, d\mathbf{x}$$

$$= \int_0^{x_1} g_1(\theta_1, 0, 0, \ldots, 0, 0)\, d\theta_1 + \int_0^{x_2} g_2(x_1, \theta_2, 0, \ldots, 0, 0)\, d\theta_2 + \ldots$$

$$\ldots + \int_0^{x_n} g_n(x_1, x_2, \ldots, x_{n-1}, \theta_n)\, d\theta_n \qquad (8.62)$$

The variable-gradient method then consists of selecting a vector function g(x) and integrating this function as per (8.62) to obtain the scalar function $V(\mathbf{x})$. However, for a continuous vector function g(x) to be gradient of a scalar $V(\mathbf{x})$, we must have

$$\frac{\partial g_i}{\partial x_j} = \frac{\partial g_j}{\partial x_i}; \quad i, j = 1, 2, \ldots, n \qquad (8.63)$$

This can be easily established. From the results of Appendix III, we can write

$$\frac{\partial^2 V(\mathbf{x})}{d\mathbf{x}^2} = \begin{bmatrix} \frac{\partial^2 V}{\partial x_1^2} & \frac{\partial^2 V}{\partial x_1\,\partial x_2} & \cdots & \frac{\partial^2 V}{\partial x_1\,\partial x_n} \\ \frac{\partial^2 V}{\partial x_2\,\partial x_1} & \frac{\partial^2 V}{\partial x_2^2} & \cdots & \frac{\partial^2 V}{\partial x_2\,\partial x_n} \\ \vdots & \vdots & & \vdots \\ \frac{\partial^2 V}{\partial x_n\,\partial x_1} & \frac{\partial^2 V}{\partial x_n\,\partial x_2} & \cdots & \frac{\partial^2 V}{\partial x_n^2} \end{bmatrix}$$

Since $\frac{\partial^2 V}{\partial x_i\,\partial x_j} = \frac{\partial^2 V}{\partial x_j\,\partial x_i}$, this matrix is symmetric.

$$\frac{\partial \mathbf{g}(\mathbf{x})}{\partial \mathbf{x}} = \begin{bmatrix} \frac{\partial g_1}{\partial x_1} & \frac{\partial g_1}{\partial x_2} & \cdots & \frac{\partial g_1}{\partial x_n} \\ \frac{\partial g_2}{\partial x_1} & \frac{\partial g_2}{\partial x_2} & \cdots & \frac{\partial g_2}{\partial x_n} \\ \vdots & \vdots & & \vdots \\ \frac{\partial g_n}{\partial x_1} & \frac{\partial g_n}{\partial x_2} & \cdots & \frac{\partial g_n}{\partial x_n} \end{bmatrix}$$

For g(x) to be equal to grad $V(\mathbf{x})$, this matrix must be symmetric which implies (8.63). These are total $\frac{n(n-1)}{2}$ equations.

The procedure to formulate a Lyapunov function is given in the following steps.

1. To begin with, assume a completely general form given below, for a gradient vector g(x).

$$\mathbf{g}(\mathbf{x}) = \begin{bmatrix} g_1(\mathbf{x}) \\ g_2(\mathbf{x}) \\ \vdots \\ g_n(\mathbf{x}) \end{bmatrix} = \begin{bmatrix} a_{11}x_1 + a_{12}x_2 + \dots + a_{1n}x_n \\ a_{21}x_1 + a_{22}x_2 + \dots + a_{2n}x_n \\ \vdots \\ a_{n1}x_1 + a_{n2}x_2 + \dots + a_{nn}x_n \end{bmatrix} \tag{8.64}$$

   The a_{ij}'s are completely undetermined quantities and could be constants or functions of both state variables and t. It is convenient, however, to choose a_{nn} as a constant.
2. Form $\dot{V}$ as per equation (8.60) with g(x) as in eqn. (8.64). Choose a_{ij}'s to constrain it to be negative definite or at least negative semi-definite.
3. Determine the remaining a_{ij}'s to satisfy the equations (8.63).
4. Recheck $\dot{V}$, in case step (3) has altered its definiteness.
5. Determine V by integrating as in eqn. (8.62).
6. Determine the region of stability where V is positive definite.

Example 8.12: Consider the nonlinear system shown in Fig. 8.15. The system is described by the state equations

$$\begin{aligned} \dot{x}_1 &= -3x_2 - h(x_1) \\ \dot{x}_2 &= -x_2 + h(x_1) \end{aligned} \tag{8.65}$$

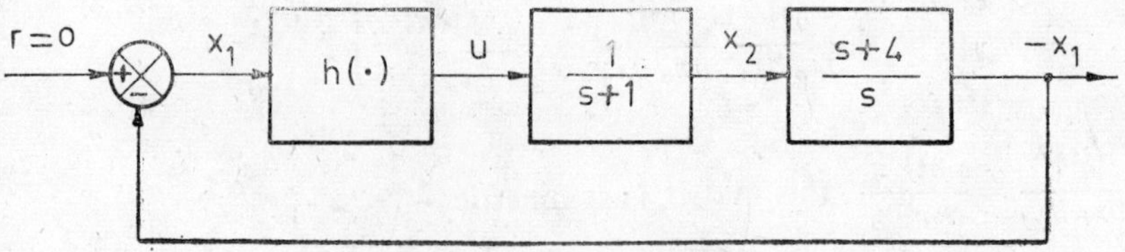

Fig. 8.15 A nonlinear system

Let us consider the special case wherein the nonlinearity can be expressed as

$$h(x_1) = h'(x_1)x_1$$

Therefore, the state description of the system becomes

$$\begin{aligned} \dot{x}_1 &= -3x_2 - h'(x_1)\,x_1 \\ \dot{x}_2 &= -x_2 + h'(x_1)\,x_1 \end{aligned} \tag{8.66}$$

It is immediately obvious that the equilibrium point lies at the origin. Assume

$$\mathbf{g}(\mathbf{x}) = \begin{bmatrix} a_{11}x_1 + a_{12}x_2 \\ a_{21}x_1 + a_{22}x_2 \end{bmatrix} \tag{8.67}$$

Form $\dot{V}$ as per eqn. (8.60).

$$\dot{V} = [a_{11}x_1 + a_{12}x_2 \quad a_{21}x_1 + a_{22}x_2] \begin{bmatrix} -3x_2 - h'(x_1)\,x_1 \\ -x_2 + h'(x_1)\,x_1 \end{bmatrix}$$

$$= -x_1^2(a_{11} - a_{21})\,h'(x_1) + x_1x_2\,[-3a_{11} - a_{12}\,h'(x_1) - a_{21} + a_{22}\,h'(x_1)] - x_2^2(3a_{12} + a_{22})$$

One of the ways of keeping $\dot{V}$ negative definite is to choose a_{ij}'s such that

$$(a_{11} - a_{21})\,h'(x_1) > 0$$

$$-3a_{11} - a_{12}\,h'(x_1) - a_{21} + a_{22}\,h'(x_1) = 0$$

$$3a_{12} + a_{22} > 0$$

These conditions get simplified by making the choice

$$a_{12} = a_{21} = 0$$

so that

$$a_{11}\,h'(x_1) > 0$$

$$a_{11} = \frac{a_{22}}{3}\,h'(x_1)$$

$$a_{22} > 0$$

The first of these conditions is satisfied when the other two are met with because

$$\frac{a_{22}}{3}\,(h'(x_1))^2 > 0$$

We can therefore choose a_{22} to be any positive constant. This gives

$$\mathbf{g}(\mathbf{x}) = \begin{bmatrix} \frac{a_{22}}{3}\,h'(x_1)\,x_1 \\ a_{22}x_2 \end{bmatrix} \tag{8.68}$$

It may also be noted that the gradient vector in (8.68) meets the conditions (8.63).

The Lyapunov function can now be obtained by taking the line integral of the gradient vector along the path defined in (8.62).

$$V = \tfrac{1}{3}a_{22} \int_0^{x_1} h'(\theta_1)\,\theta_1\,d\theta_1 + a_{22} \int_0^{x_2} \theta_2\,d\theta_2$$

$$= \tfrac{1}{2}a_{22}\int_0^{x_1} h'(\theta_1)\,\theta_1\,d\theta_1 + \tfrac{1}{2}a_{22}x_2^2 \tag{8.69}$$

If $h'(x_1) > 0$, i.e., $h(x_1) = h'(x_1)\,x_1$ lies in the first and third quadrants, V is positive definite. Under this condition, the system is asymptotically stable. Also if $\lim_{x_1\to\infty}\int_0^{x_1} h'(\theta_1)\,\theta_1\,d\theta_1 \to \infty$, the system would be globally asymptotically stable.

8.8 USE OF LYAPUNOV FUNCTIONS TO ESTIMATE TRANSIENTS

One of the first uses of the Lyapunov functions, outside the realm of stability, was in the estimation of transient behaviour of dynamic systems. Other uses include the parameter optimization, design of a class of suboptimal control systems etc. Some of these design methods will be discussed in Chapter 11. In this section, we shall discuss the relationship between Lyapunov functions and the transient behaviour of dynamic systems.

For a given Lyapunov function, we define

$$\eta = \min_{\mathbf{x}}\left[\frac{-\dot{V}(\mathbf{x})}{V(\mathbf{x})}\right] \tag{8.70}$$

where the minimization is carried out over all values of $\mathbf{x} \neq \mathbf{0}$ in the region for which the system is asymptotically stable. In the present analysis we are assuming that for asymptotically stable systems, a Lyapunov function with negative definite $\dot{V}$ can be found; we therefore, exclude the case where $\dot{V}(\mathbf{x}) = 0$ for $\mathbf{x} \neq \mathbf{0}$.

From (8.70) it follows that $\eta \geqslant 0$ and

$$\dot{V}(\mathbf{x}) \leqslant -\eta\, V(\mathbf{x})$$

Dividing both sides by $V(\mathbf{x})$ and integrating with respect to time from t_0 to t, we obtain

$$\ln\frac{V[\mathbf{x}(t)]}{V[\mathbf{x}(t_0)]} \leqslant -\eta(t - t_0) \tag{8.71a}$$

or

$$V[\mathbf{x}(t)] \leqslant V[\mathbf{x}(t_0)]\, e^{-\eta(t-t_0)} \tag{8.71b}$$

For any motion starting at some point $\mathbf{x}(t_0)$ in the region of asymptotic stability, inequality (8.71a) gives an upper bound on $V(\mathbf{x})$ for all $t \geqslant t_0$; at $t_1 \geqslant t_0$, the state of the system is known to lie inside the contour $V[\mathbf{x}(t)] = V[\mathbf{x}(t_0)]\, e^{-\eta(t_1-t_0)}$. $1/\eta$ corresponds to the largest time-constant relating to the changes in Lyapunov function $V(\mathbf{x})$. Larger values of η correspond to faster response.

Since there may be a number of Lyapunov functions for a given system, a number of different values of η may be obtained. It is not known how to find a Lyapunov function which yields the smallest value of η. Further, for a given Lyapunov function, the computational effort to obtain η is considerable. In general, η is to be obtained numerically from definition (8.70).

The problem is comparatively simpler for linear systems. Consider the linear system $\dot{\mathbf{x}} = \mathbf{A}\mathbf{x}$ and assume that the $n \times n$ matrix $\mathbf{A}$ has all negative eigenvalues. Then a Lyapunov function $V(\mathbf{x})$ and its derivative $\dot{V}(\mathbf{x})$ are given by

$$V(\mathbf{x}) = \mathbf{x}^T \mathbf{P}\mathbf{x}$$

$$\dot{V}(\mathbf{x}) = -\mathbf{x}^T \mathbf{Q}\mathbf{x}$$

where $\mathbf{P}$ = positive definite real symmetric matrix and

$$\mathbf{Q} = -(\mathbf{A}^T \mathbf{P} + \mathbf{P}\mathbf{A})$$

For this case, η becomes

$$\eta = \min_{\mathbf{x}} \left[\frac{\mathbf{x}^T \mathbf{Q}\mathbf{x}}{\mathbf{x}^T \mathbf{P}\mathbf{x}} \right]$$

At a minimum point η, we must have (refer Appendix III)

$$\frac{\partial \eta}{\partial \mathbf{x}} = \frac{(2\mathbf{Q}\mathbf{x})(\mathbf{x}^T \mathbf{P}\mathbf{x}) - (\mathbf{x}^T \mathbf{Q}\mathbf{x})\, 2\mathbf{P}\mathbf{x}}{(\mathbf{x}^T \mathbf{P}\mathbf{x})^2} = 0 \qquad (8.72)$$

There can be many points $\mathbf{x}$ which satisfy this relation. Let $\gamma = \dfrac{\mathbf{x}^T \mathbf{Q}\mathbf{x}}{\mathbf{x}^T \mathbf{P}\mathbf{x}}$ at a typical point of this kind. Then from (8.72), we obtain

$$(\mathbf{Q} - \gamma\mathbf{P})\,\mathbf{x} = 0$$

or

$$(\mathbf{P}^{-1}\mathbf{Q} - \gamma\mathbf{I})\,\mathbf{x} = 0$$

This shows that γ must be an eigenvalue of the matrix $\mathbf{P}^{-1}\mathbf{Q}$. It therefore follows that

$$\eta = \text{minimum eigenvalue of } [\mathbf{P}^{-1}\mathbf{Q}] \qquad (8.73)$$

Example 8.13: Consider the linear system of Example 8.10:

$$\dot{\mathbf{x}} = \mathbf{A}\mathbf{x}$$

where

$$\mathbf{A} = \begin{bmatrix} -1 & -2 \\ 1 & -4 \end{bmatrix}$$

From the results of Example 8.10 we have

$$\mathbf{Q} = \begin{bmatrix} 1 & 0 \\ 0 & 1 \end{bmatrix}$$

$$\mathbf{P} = \begin{bmatrix} 23/60 & -7/60 \\ -7/60 & 11/60 \end{bmatrix}$$

η is the minimum eigenvalue of $\mathbf{P}^{-1}\mathbf{Q} = \mathbf{P}^{-1}$, since $\mathbf{Q} = \mathbf{I}$. Finding the eigenvalues of $\mathbf{P}^{-1}$ is equivalent to solving

$$|\mathbf{I} - \mathbf{P}\lambda| = 0$$

or

$$\begin{vmatrix} 1-\frac{23}{60}\lambda & \frac{7}{60}\lambda \\ \frac{7}{60}\lambda & 1-\frac{11}{60}\lambda \end{vmatrix} = 0$$

This gives

$$\lambda_1 = 2.288, \ \lambda_2 = 7.71$$

Hence

$$\eta = 2.288$$

From eqn. (8.71), we get

$$V[\mathbf{x}(t)] \leqslant V[\mathbf{x}(t_0)]\, e^{-2.288\,(t-t_0)} \tag{8.74}$$

Suppose that it is desired to find an upper bound on time that it takes the system to get from the initial state $\mathbf{x}(0) = \begin{bmatrix} 1 \\ 0 \end{bmatrix}$ to within the area defined by $x_1^2 + x_2^2 = (0.25)^2$. First, we have to find the largest value of K such that the surface $V(\mathbf{x}) = \mathbf{x}^T\mathbf{P}\mathbf{x} = K$ lies entirely within, or at most tangent to the surface $x_1^2 + x_2^2 = (0.25)^2$. A value of K satisfying this requirement is found to be 0.009 (see Fig. 8.16). Let $t = t_s$ be the upper bound on time that is

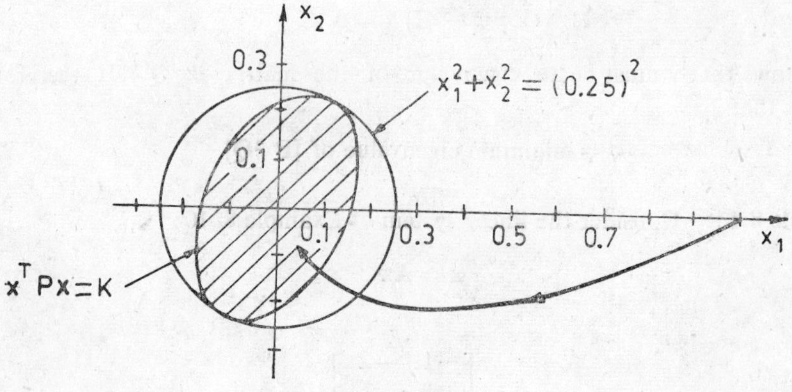

Fig. 8.16

taken by the system to settle down within the area $x_1^2 + x_2^2 = (0.25)^2$. From (8.71a), we obtain

$$t_s \leqslant -\frac{1}{\eta}\left[\ln \frac{K}{V(\mathbf{x}(0))}\right] \tag{8.75}$$

$$V(\mathbf{x}(0)) = [1 \quad 0]^T \begin{bmatrix} 23/60 & -7/60 \\ -7/60 & 11/60 \end{bmatrix} \begin{bmatrix} 1 \\ 0 \end{bmatrix}$$

$$= 23/60$$

Therefore from (8.75),

$$t_s \leqslant -\frac{1}{2.288}\left[\ln \frac{0.009 \times 60}{23}\right]$$

$$\leqslant 1.64$$

8.9 THE DIRECT METHOD OF LYAPUNOV AND THE DISCRETE-TIME AUTONOMOUS SYSTEMS

In this section, we develop the requirements for stability of the autonomous discrete-time system

$$\mathbf{x}(k+1) = \mathbf{f}(\mathbf{x}(k)) \; ; \quad \mathbf{f}(\mathbf{0}) = \mathbf{0} \tag{8.76}$$

using Lyapunov's direct method. Our discussion will be brief because of the strong analogy between the discrete-time and continuous-time cases.

Lyapunov's Stability Theorem

Theorem 8.14: For autonomous system (8.76), the sufficient conditions of stability are as follows:

Suppose that there exists a scalar function $V(\mathbf{x}(k))$ which, for some real number $\varepsilon > 0$, satisfies the following properties for all $\mathbf{x}$ in the region $\|\mathbf{x}\| \leqslant \varepsilon$:

(1) $V(\mathbf{x}) > 0 \; ; \quad \mathbf{x} \neq \mathbf{0}$

(2) $V(\mathbf{0}) = 0$

(3) $V(\mathbf{x})$ is continuous for all $\mathbf{x}$

Then the equilibrium $\mathbf{x} = \mathbf{0}$ of the system (8.76) is

(a) *stable* if the difference $\Delta V(\mathbf{x}(k))$

$$= V(\mathbf{x}(k+1)) - V(\mathbf{x}(k)) \leqslant 0;$$

(b) *asymptotically stable* if $\Delta V(\mathbf{x}(k)) < 0, \; \mathbf{x} \neq \mathbf{0}$
or if $\Delta V(\mathbf{x}(k)) \leqslant 0$ and $\Delta V(\mathbf{x}(k))$ is not identically zero on a solution of difference equation (8.76) other than at $\mathbf{x} = \mathbf{0}$.

(c) *globally asymptotically stable*, if the conditions of asymptotic stability hold for all $\mathbf{x}$ and, in addition, $V(\mathbf{x}) \to \infty$ as $\|\mathbf{x}\| \to \infty$.

Proof:

The proof is essentially identical with that for the continuous-time case.

Stability of Linear Systems

Consider a linear autonomous system described by the state equation

$$\mathbf{x}(k+1) = \mathbf{F}\mathbf{x}(k) \tag{8.77}$$

where $\mathbf{F}$ is $n \times n$ real constant matrix.

Theorem 8.15: The linear system (8.77) is globally asymptotically stable at the origin if and only if for any given symmetric positive definite matrix $\mathbf{Q}$, there exists a symmetric positive definite matrix $\mathbf{P}$ that satisfies the matrix equation

$$\mathbf{F}^T\mathbf{P}\mathbf{F} - \mathbf{P} = -\mathbf{Q} \tag{8.78}$$

Proof:

Let us first prove the sufficiency of the result. Assume that a symmetric positive definite matrix $\mathbf{P}$ exists which is the unique solution of eqn. (8.78). Consider the scalar function

$$V(\mathbf{x}) = \mathbf{x}^T\mathbf{P}\mathbf{x}$$

Note that

$$V(\mathbf{x}) > 0 \quad \text{for} \quad \mathbf{x} \neq \mathbf{0}$$

and

$$V(\mathbf{0}) = 0$$

The difference

$$\begin{aligned}\Delta V(\mathbf{x}) &= V(\mathbf{x}(k+1)) - V(\mathbf{x}(k)) \\ &= \mathbf{x}^T(k+1)\,\mathbf{P}\mathbf{x}(k+1) - \mathbf{x}^T(k)\mathbf{P}\mathbf{x}(k)\end{aligned}$$

Using eqns. (8.77) and (8.78) we get

$$\begin{aligned}\Delta V(\mathbf{x}) &= \mathbf{x}^T(k)\,\mathbf{F}^T\,\mathbf{P}\mathbf{F}\mathbf{x}(k) - \mathbf{x}^T(k)\,\mathbf{P}\mathbf{x}(k) \\ &= \mathbf{x}^T(k)[\mathbf{F}^T\mathbf{P}\mathbf{F} - \mathbf{P}]\,\mathbf{x}(k) \\ &= -\,\mathbf{x}^T(k)\,\mathbf{Q}\mathbf{x}(k)\end{aligned}$$

Since $\mathbf{Q}$ is positive definite, $\Delta V(\mathbf{x})$ is negative definite. Further $V(\mathbf{x}) \to \infty$ as $\|\mathbf{x}\| \to \infty$. Therefore, the system is globally asymptotically stable at the origin.

The proof of necessity is analogous to that of continuous-time case.

□

Berger (1971) gives a numerical method for solving Lyapunov equation (8.78). Power (1969) gives a transformation that brings equations of the type (8.78) into the form given by (8.35).

8.10 CONCLUDING COMMENTS

In this chapter, we studied various stability concepts of linear and non-linear systems. For relaxed systems with inputs or disturbances, we have bounded-input, bounded-output (BIBO) stability; for the zero-input response we have stability in the sense of Lyapunov and asymptotic stability and for unrelaxed systems with inputs or disturbances, we have total stability. We observed that for linear time-invariant systems, BIBO stability and asymptotic stability are completely equivalent under certain conditions. The problem of ascertaining the stability of linear time-invariant systems involves checking the root locations of characteristic polynomial. Solving for the roots of a polynomial is a difficult task. Routh-Hurwitz criterion provides a simple test for stability, without the need to solve for the roots.

Equilibrium stability of nonlinear autonomous systems was studied using direct (second) method of Lyapunov. We observed that Lyapunov function is basically a generalization of total system energy. If a Lyapunov function can be found in a region including the origin (equilibrium point) of state space, then stability or asymptotic stability of the origin is assured. This, however, does not necessarily mean that the trajectories starting from a state outside the region approach infinity, since Lyapunov's method establishes only sufficient conditions for stability. Some Lyapunov functions may yield larger region of assured stability compared to others. Though number of systematic methods for generating Lyapunov functions are available in literature, we have discussed only two of these methods for want of space. Numerical analysis techniques have also been utilized to generate a Lyapunov function and then to test it for regions of stability (Rodden 1964; Hewit and Storey 1969).

Some standard or canonic forms of systems have been proposed by Lur'e and Letov (Letov 1961) for which Lyapunov functions have been developed. For these canonic forms, simplified stability criteria are available which may significantly simplify the analysis.

The application of Lyapunov's direct method to linear time-invariant systems provides necessary and sufficient conditions for stability. The Lyapunov functions for linear systems result in a simple method for studying transient behaviour of systems. Further, as we shall see in Chapter 11, these results can be used for solving optimization problems based on quadratic performance index.

PROBLEMS

8.1 (a) A unity feedback system has the open-loop transfer function

$$\frac{\hat{y}(s)}{\hat{u}(s)} = \frac{5}{s(s+1)(s+2)}$$

Using the Routh-Hurwitz criterion, show that the closed-loop system is asymptotically stable.

(b) A sampler and zero-order hold are now introduced in the forward loop (Fig. P-8.1). Study the stability of the sampled-data system via bilinear transformation and show that the stable linear continuous-time system becomes unstable upon the introduction of a sampler and ZOH.

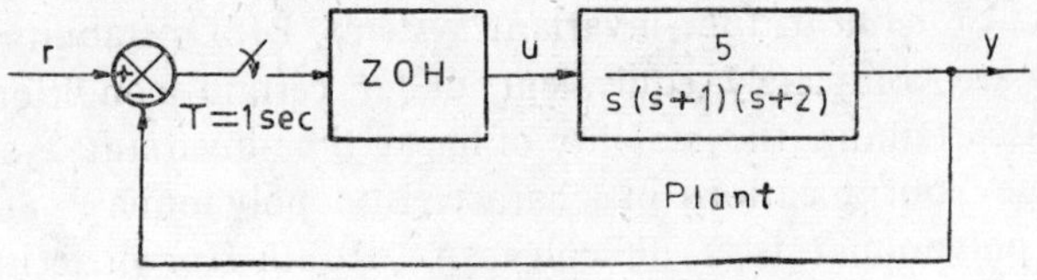

Fig. P-8.1

8.2 A linear autonomous system is described by the state equation

$$\dot{\mathbf{x}} = \mathbf{A}\mathbf{x}$$

where

$$\mathbf{A} = \begin{bmatrix} -4K & 4K \\ 2K & -6K \end{bmatrix}$$

Find restrictions on the parameter K to guarantee the stability of the system.

8.3 Consider the linear autonomous system

$$\mathbf{x}(k+1) = \begin{bmatrix} 0.5 & 1 \\ -1 & -1 \end{bmatrix} \mathbf{x}(k)$$

Using direct method of Lyapunov, determine stability of the equilibrium state.

8.4 For the system

$$\dot{\mathbf{x}} = \begin{bmatrix} 0 & 1 \\ -1 & -1 \end{bmatrix} \mathbf{x}$$

find a suitable Lyapunov function $V(\mathbf{x})$. Obtain an upper bound on the response time such that it takes the system to go from a point on the boundary of the closed curve $V(\mathbf{x}) = 100$ to a point within the closed curve $V(\mathbf{x}) = 0.05$.

8.5 For the system

$$\dot{\mathbf{x}} = \begin{bmatrix} 0 & 1 \\ -2 & -3 \end{bmatrix} \mathbf{x}$$

find a suitable Lyapunov function $V(\mathbf{x})$. Find an upper bound on time that it takes the system to get from the initial condition $\mathbf{x}(0) = \begin{bmatrix} 1 \\ 1 \end{bmatrix}$ to within the area defined by $x_1^2 + x_2^2 = 0.1$.

8.6 Consider the nonlinear system described by the equations

$$\dot{x}_1 = x_2$$

$$\dot{x}_2 = -(1 - |x_1|)\, x_2 - x_1$$

Find the region in the state plane for which the equilibrium state of

the system is asymptotically stable.
(Hint: A Lyapunov function is $V = x_1^2 + x_2^2$).)

8.7 Consider the system shown in Fig. P-8.7. Determine the restrictions which must be placed on the nonlinear function $\phi(\cdot)$ in order for the system state with $r(t) = 0$ to be asymptotically stable in-the-large.
(Hint: Barbashin's result (Ogata 1967) may be used:
The system

$$\dot{x}_1 = x_2$$

$$\dot{x}_2 = x_3$$

$$\dot{x}_3 = -f(x_1) - g(x_2) - ax_3$$

is asymptotically stable in-the-large at the origin provided that

(i) $f(0) = 0 = g(0)$ and f, g are differentiable functions

(ii) $a > 0$

(iii) $\dfrac{f(x_1)}{x_1} \geqslant \varepsilon_1 > 0$ if $x_1 \neq 0$

(iv) $\dfrac{ag(x_2)}{x_2} - \dfrac{df(x_1)}{dx_1} \geqslant \varepsilon_2 > 0$ if $x_2 \neq 0$

A Lyapunov function is

$$V(\mathbf{x}) = aF(x_1) + f(x_1)\, x_2 + G(x_2) + \frac{(ax_2 + x_3)^2}{2}$$

where

$$F(x_1) = \int_0^{x_1} f(\tau)\, d\tau\ ; \quad G(x_2) = \int_0^{x_2} g(\tau)\, d\tau.)$$

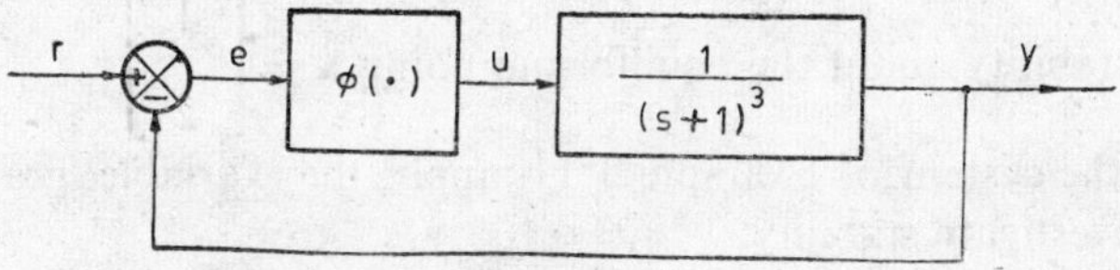

Fig. P-8.7

8.8 Consider the system of Fig. P-8.7. The nonlinear function is replaced by the linear function $\phi(e) = Ke$. Find the restrictions on the parameter K to guarantee system stability. Use the Lyapunov's direct method and the Routh-Hurwitz criterion. Compare the results.

8.9 Check the stability of the system described by

$$\dot{x}_1 = x_2$$

$$\dot{x}_2 = -x_1 - x_1^2 x_2$$

8.10 Consider a nonlinear system described by the equations

$$\dot{x}_1 = -3x_1 + x_2$$

$$\dot{x}_2 = x_1 - x_2 - x_2^3$$

Investigate the stability of equilibrium state.
(Hint: Use Krasovskii method with **P** as identity matrix.)

8.11 Consider the nonlinear system

$$\dot{x}_1 = -x_1 - x_2^2$$

$$\dot{x}_2 = -x_2$$

Find a region of asymptotic stability using Krasovskii method.

8.12 Check the stability of the system described by

$$\dot{x}_1 = x_2$$

$$\dot{x}_2 = -x_1 - b_1x_2 - b_2x_2^3\,; \quad b_1, b_2 > 0$$

(Hint: Use the variable gradient method with a_{ij}'s as constants. Also try by choosing $a_{12} = x_1/x_2$; $a_{21} = x_2/x_1$.)

8.13 In Example 8.9, we considered the Lyapunov function $V = x_1^2 + x_2^2$ and obtained a region of stability. In order to further explore the regions of stability, use the gradient method to generate a Lyapunov function and obtain stability regions.

8.14 For the system

$$\dot{x}_1 = -2x_1 + x_1x_2$$

$$\dot{x}_2 = -x_2 + x_1x_2$$

there are two equilibrium points; $\mathbf{x} = \begin{bmatrix} 0 \\ 0 \end{bmatrix}$ and $\mathbf{x} = \begin{bmatrix} 1 \\ 2 \end{bmatrix}$. The stability about the origin was studied in Example 8.9. Investigate the stability about the equilibrium point $\mathbf{x} = \begin{bmatrix} 1 \\ 2 \end{bmatrix}$.

8.15 For the system of Example 8.11, apply the variable gradient method to investigate stability.

REFERENCES

1. Aizerman, M.A , and F.R. Gantmacher, *Absolute Stability of Regulator Systems*, San Francisco: Holden-Day, **1964**.
2. Barnett, S., and S. Storey, *Matrix Methods in Stability Theory*, London: Nelson, **1970**.
3. Berger, C.S., "A numerical solution of the matrix equation $\mathbf{P} = \Phi\mathbf{P}\Phi^T + \mathbf{S}$" *IEEE Trans. Automat. Contr.*, vol. 14, pp. 381–383, **1971**.
4. Cadzow, J.A., and H.R. Martens, *Discrete-Time and Computer Control Systems*, Englewood Cliffs, N.J.: Prentice-Hall, **1970**.

5. Chang, T.S., and C.T. Chen, "On the Routh-Hurwitz criterion," *IEEE Trans. Automat. Contr.*, vol. 19, pp. 250–251, **1974.**

6. Chen, C.T., *Introduction to Linear System Theory*, New York: Holt, Rinehart and Winston, **1970.**

7. Csaki, F., *Modern Control Theories*, Budapest: Akademiai Kiado, **1972.**

8. Fahmy, M.M., and J. O'Reilly, "A note on the Routh-Hurwitz test," *IEEE Trans. Automat. Contr.*, vol. 27, pp. 483–485, **1982.**

9. Hagander, P., "Numerical solution of $\mathbf{A}^T\mathbf{S}+\mathbf{S}\mathbf{A}+\mathbf{Q}=\mathbf{0}$," *Information Sci.*, vol. 4, pp. 35–50, **1972.**

10. Hahn, W., *Theory and application of Lyapunov's Direct Method*, Englewood Cliffs, N.J.: Prentice-Hall, **1963.**

11. Hewit, J.R., and C. Storey, "Comparison of numerical methods for stability analysis," *Int. J. Control*, vol. 10, pp. 687–701, **1969.**

12. Hostetter, G.H., "Additional comments on 'On the Routh-Hurwitz criterion'," *IEEE Trans. Automat. Contr.*, vol. 20, pp. 296-297, **1975.**

13. Hsu, J.C., and A.U. Meyer, *Modern Control Principles and Applications*, New York: McGraw-Hill, **1968.**

14. Jury, E.I., *Inners and Stability of Dynamic Systems*, New York: Wiley, **1974.**

15. Kalman, R.E., and J.E. Bertram, "Control system analysis and design via the 'second method' of Lyapunov," *J. Basic Engineering*, vol. 82, pp. 371–399, **1960.**

16. Koppel, L.B., *Introduction to Control Theory*, Englewood Cliffs, N.J.: Prentice-Hall, **1968.**

17. Kuo, B.C., *Analysis and Synthesis of Sampled-Data Control Systems*, Englewood Cliffs, N.J.: Prentice Hall, **1963.**

18. LaSalle, J.P., and S. Lefschetz, *Stability by Liapunov's Direct Method with Applications*, New York: Academic Press, **1961.**

19. Lefschetz, S., *Stability of Nonlinear Control Systems*, New York: Academic Press, **1965.**

20. Letov, A.M., *Stability in Nonlinear Control Systems*, Princeton, N.J.: Princeton University Press, **1961.**

21. Melsa, J.L., and S K. Jones, *Computer Programs for Computational Assistance in the Study of Linear Control Systems*, 2nd Edition, New York: McGraw-Hill, **1973.**

22. Ogata, K., *State Space Analysis of Control Systems*, Englewood Cliffs, N.J.: Prentice-Hall, **1967.**

23. Power, H.M., "A note on the matrix equation $\mathbf{A}^T\mathbf{L}\mathbf{A}-\mathbf{L}=-\mathbf{K}$", *IEEE Trans. Automat. Contr.*, vol. 14, pp 411-412, **1969.**

24. Rodden, J.J., "Numerical applications of Liapunov stability theory", *Proc. of J.A.C.C.*, pp. 261–268, June **1964.**

25. Rothschild, D., and A. Jameson, "Comparison of four numerical algorithms for solving the Liapunov matrix equation," *Int. J. Control*, vol. 11, pp. 181–198, **1970.**

26. Rugh, W.J., *Methematical Description of Linear Systems*, New York: Marcel Dekker, **1975,**

27. Schultz, D.G., and J.E. Gibson, "The variable gradient method for generating Liapunov functions," *AIEE Trans.*, Part II, pp. 203-210, **1962.**

9. MODAL CONTROL

9.1 INTRODUCTION

Chapters 3 and 4 of this book dealt with the problem of modelling of systems. Chapters 5 to 8 primarily dealt with the problem of analysis, though some of the design techniques which evolved from analysis were also discussed. In this chapter and the next three, we are mainly concerned with the design of multivariable controllers for given plants.

Important requirements of a control system may be outlined as follows:

(i) Most of the working systems are designed to be stable.
(ii) The output $\mathbf{y}(t)$ of the plant should approximate to its desired value within a prescribed tolerance.
(iii) The output $\mathbf{y}(t)$ should approach its desired value reasonably rapidly without excessive overshoots.
(iv) The degree of interaction in the system should be minimized. In general, depending upon the structure of the system, one input will affect all outputs. It is desirable that one input affects one output only. Undue interaction may reduce the stability margins of system operation. Further, the well established design techniques for scalar systems can be applied to multivariable systems if the system is made non-interacting.
(v) The system should be reasonably insensitive to parameter variations.
(vi) The system should be reasonably immune to the effects of disturbances or noise.
(vii) The structure of the controller should be as simple as possible.

It is clear that the designer has to opt for a compromise solution to get a satisfactory design since many of the above-said requirements are conflicting, e.g., high accuracy demands high gains which tend to produce instability. In Chapters 9–12, some of the methods leading to a satisfactory design are presented.

In this chapter, we assume a simple structure of the controller and study its use to 'improve the performance' of the plant. Consider the linear time-invariant system

$$\dot{\mathbf{x}}(t) = \mathbf{A}\mathbf{x}(t) + \mathbf{B}\mathbf{u}(t) \quad (9.1a)$$

$$\mathbf{y}(t) = \mathbf{C}\mathbf{x}(t) + \mathbf{D}\mathbf{u}(t) \quad (9.1b)$$

where $\mathbf{x}$ is the $n\times 1$ state vector, $\mathbf{u}$ is the $p\times 1$ input vector, $\mathbf{y}$ is the $q\times 1$ output vector and $\mathbf{A}$, $\mathbf{B}$, $\mathbf{C}$ and $\mathbf{D}$ are, respectively, $n\times n$, $n\times p$, $q\times n$ and $q\times p$ real constant matrices. The time interval of interest is $[0, \infty)$.

For this system, assume that the control law depends linearly on $\mathbf{r}(t)$ and $\mathbf{x}(t)$ and is of the form

$$\mathbf{u}(t) = \mathbf{r}(t) + \mathbf{K}\mathbf{x}(t) \tag{9.2}$$

where $\mathbf{r}(t)$ stands for reference input and $\mathbf{K}$ is some real constant matrix called the feedback matrix. Finding the best $\mathbf{K}$ is the scope of optimal control theory (Chapter 11). Here we shall discuss the effect of introducing linear feedback of the form (9.2) and study what can be achieved by introducing this state feedback.

The state model of the closed-loop system obtained from eqns. (9.1) and (9.2) is

$$\dot{\mathbf{x}}(t) = (\mathbf{A} + \mathbf{B}\mathbf{K})\,\mathbf{x}(t) + \mathbf{B}\mathbf{r}(t) \tag{9.3a}$$

$$\mathbf{y}(t) = (\mathbf{C} + \mathbf{D}\mathbf{K})\,\mathbf{x}(t) + \mathbf{D}\mathbf{r}(t) \tag{9.3b}$$

This closed-loop system is represented schematically in Fig. 9.1, with dotted lines enclosing the uncompensated portion of the system, i.e., the plant.

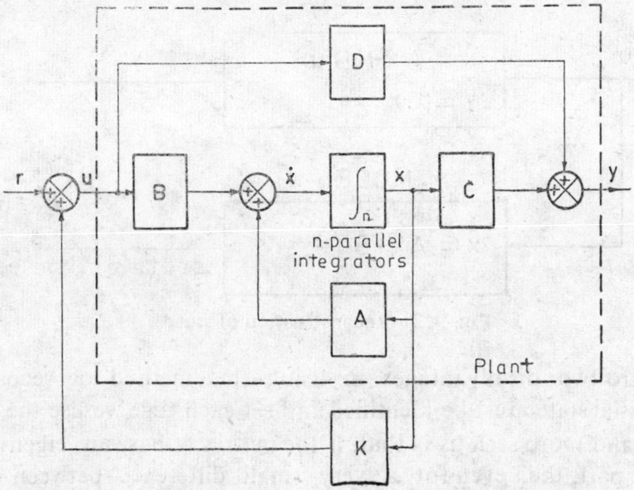

Fig. 9.1 Structure of state feedback control system

general, it is not at all clear from (9.3), what effect the control law (9.2) has on the plant (9.1). However, if (9.1) is in the controllable companion form, the effect of linear state variable feedback on the closed-loop system can be clarified considerably. This we shall discuss in Section 9.3, after introducing companion forms of multivariable state models in Section 9.2. We shall prove that if all the states of a controllable system are available for feedback, then a feedback gain matrix **K** may be chosen so as to place the eigenvalues of the closed-loop system anywhere in the complex plane (subject to conjugate pairing). This allows the designer a great deal of freedom to specify the transient behaviour of the closed-loop system.

The other, but related, problem taken up in this chapter is that of the implementation of the linear state feedback control law (9.2), which requires the ability to directly measure the entire state vector $\mathbf{x}(t)$. In general, however, only the input vector $\mathbf{u}(t)$ and the output vector $\mathbf{y}(t)$ are directly measurable and so the state feedback control law cannot be implemented. Thus, either a new approach that directly accounts for the nonavailability of entire state vector (Chapter 11) is to be devised or a suitable approximation (reconstruction) of the state vector must be determined. The latter approach is much simpler in many situations.

There are two simple solutions to the problem of the reconstruction of a state vector. One is to construct a model of the system which has all of its state variables directly measurable. Any input signal is applied to the system as well as to the model. Thus, even if the state vector of the original system cannot be measured, the model's state variables which are equivalent to those of the system are available (Fig. 9.2). It is also necessary to set the current initial conditions on the model. The initial state of the system may be determined using the relation (6.35).

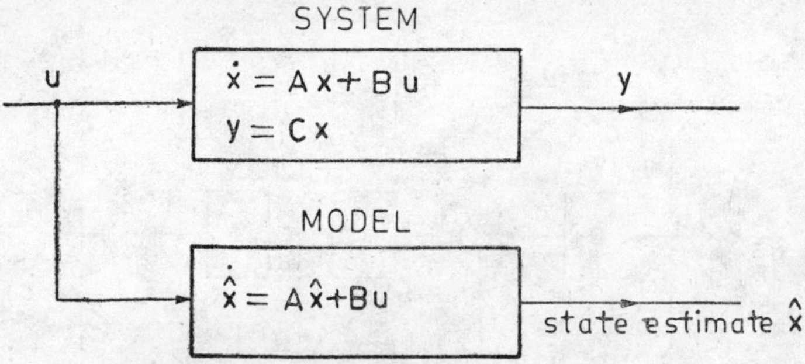

Fig. 9.2 Reconstruction of state

There are two disadvantages in using this method of reconstruction. First, the initial state must be identified and set each time we use the estimator. The second and more serious is that, if the matrix **A** has an eigenvalue with positive real part, then even for a very small difference between $\mathbf{x}(t_0)$ and $\hat{\mathbf{x}}(t_0)$ at some $t = t_0$, which may be caused by a disturbance entering the system but not the model or by incorrect estimation of the initial state, the difference between the real state $\mathbf{x}(t)$, and the reconstructed state $\hat{\mathbf{x}}(t)$ will increase with time.

Another method consists of differentiating the output (and input) variables a certain number of times and combining these terms to form the unmeasured state variables (eqn. 7.74). The poor noise characteristics and the difficulty of building good differentiators generally make this method impractical.

In this chapter, we shall discuss an alternative method due to Luenberger for state vector reconstruction. Not surprisingly, his scheme which is commonly known as 'Luenberger state observer' relies on the complete state observability of a system.

9.2 CONTROLLABLE AND OBSERVABLE COMPANION FORMS

If a system in the state form (9.1) is controllable/observable, it can be reduced via a nonsingular transformation **P** to an equivalent controllable/observable system in a certain structured form which we call a controllable companion form/observable companion form. As we shall see later in this chapter, it is advantageous to work with companion forms of the system, rather than the original system, to study the effect of state variable feedback on system performance. The purpose of this section is to introduce the dual notions of controllable and observable companion forms and to present a constructive procedure for reducing systems to these forms. We begin by considering the single-input/single-output systems.

Single-Input/Single-Output Systems

Given a single-input controllable system

$$\dot{\mathbf{x}} = \mathbf{A}\mathbf{x} + \mathbf{b}u; \tag{9.4}$$

it is always possible to find a transformation

$$\bar{\mathbf{x}} = \mathbf{P}\mathbf{x}; \; \mathbf{P} \text{ a nonsingular constant matrix} \tag{9.5}$$

which puts the matrix A in companion form given in eqn. (7.65a).

Under the transformation (9.5), system (9.4) is transformed to

$$\dot{\bar{\mathbf{x}}} = \bar{\mathbf{A}}\,\bar{\mathbf{x}} + \bar{\mathbf{b}}u$$

where

$$\bar{\mathbf{A}} = \mathbf{P}\mathbf{A}\mathbf{P}^{-1}; \; \bar{\mathbf{b}} = \mathbf{P}\mathbf{b} \tag{9.6}$$

Theorem 9.1: If the n-dimensional linear time-invariant single-input system (9.4) is controllable, then it can be transformed, by an equivalence transformation, into the controllable companion form

$$\dot{\bar{\mathbf{x}}} = \begin{bmatrix} 0 & 1 & 0 & \cdots & 0 & 0 \\ 0 & 0 & 1 & \cdots & 0 & 0 \\ \vdots & \vdots & \vdots & & \vdots & \vdots \\ 0 & 0 & 0 & \cdots & 0 & 1 \\ -\alpha_n & -\alpha_{n-1} & -\alpha_{n-2} & \cdots & -\alpha_2 & -\alpha_1 \end{bmatrix} \bar{\mathbf{x}} + \begin{bmatrix} 0 \\ 0 \\ \vdots \\ 0 \\ 1 \end{bmatrix} u \tag{9.7}$$

Proof:

Let us assume that such a transformation exists and is given by

$$\bar{\mathbf{x}} = \mathbf{P}\mathbf{x} \tag{9.8}$$

where

$$\mathbf{P} = \begin{bmatrix} p_{11} & p_{12} & \cdots & p_{1n} \\ p_{21} & p_{22} & \cdots & p_{2n} \\ \vdots & \vdots & & \vdots \\ p_{n1} & p_{n2} & \cdots & p_{nn} \end{bmatrix} = \begin{bmatrix} \mathbf{p}_1 \\ \mathbf{p}_2 \\ \vdots \\ \mathbf{p}_n \end{bmatrix}$$

$$\mathbf{p}_i = [\, p_{i1} \quad p_{i2} \ \cdots \ p_{in} \,]; \quad i = 1, 2, \ldots, n$$

The first equation in the set (9.8) is given by

$$\bar{x}_1(t) = p_{11}x_1 + p_{12}x_2 + \ldots + p_{1n}x_n$$
$$= \mathbf{p}_1 \mathbf{x}(t)$$

Taking derivative on both sides of this equation, we get

$$\dot{\bar{x}}_1(t) = \mathbf{p}_1\dot{\mathbf{x}}(t) = \mathbf{p}_1\mathbf{A}\mathbf{x} + \mathbf{p}_1\mathbf{b}\,u$$

But $\dot{\bar{x}}_1(t)(=\bar{x}_2(t))$ is a function of $\mathbf{x}$ only as per the transformation (9.8). Therefore

$$\mathbf{p}_1\,\mathbf{b} = 0$$

and

$$\bar{x}_2(t) = \mathbf{p}_1\mathbf{A}\mathbf{x}$$

Taking derivative on both sides once again, we get

$$\mathbf{p}_1\mathbf{A}\mathbf{b} = 0 \ \text{ and } \ \bar{x}_3(t) = \mathbf{p}_1\mathbf{A}^2\mathbf{x}$$

Continuing the process, we obtain

$$\mathbf{p}_1\mathbf{A}^{n-2}\mathbf{b} = 0$$

and

$$\bar{x}_n = \mathbf{p}_1\mathbf{A}^{n-1}\mathbf{x}$$

Taking derivative once again, we obtain $\mathbf{p}_1\mathbf{A}^{n-1}\mathbf{b} = 1$

Thus

$$\bar{\mathbf{x}}(t) = \mathbf{P}\mathbf{x} = \begin{bmatrix} \mathbf{p}_1 \\ \mathbf{p}_1\mathbf{A} \\ \vdots \\ \mathbf{p}_1\mathbf{A}^{n-1} \end{bmatrix} \mathbf{x}$$

where $\mathbf{p}_1$ must satisfy the conditions

$$\mathbf{p}_1\mathbf{b} = \mathbf{p}_1\mathbf{A}\mathbf{b} = \ldots = \mathbf{p}_1\mathbf{A}^{n-2}\mathbf{b} = 0;\ \mathbf{p}_1\mathbf{A}^{n-1}\mathbf{b} = 1$$

From (9.6) and (9.7) we have

$$\mathbf{P}\mathbf{b} = \begin{bmatrix} 0 \\ 0 \\ \vdots \\ 1 \end{bmatrix} = \begin{bmatrix} \mathbf{p}_1\mathbf{b} \\ \mathbf{p}_1\mathbf{A}\mathbf{b} \\ \vdots \\ \mathbf{p}_1\mathbf{A}^{n-1}\mathbf{b} \end{bmatrix}$$

or

$$\mathbf{p}_1\,[\mathbf{b} \,\vert\, \mathbf{A}\mathbf{b} \,\vert\, \ldots \,\vert\, \mathbf{A}^{n-1}\mathbf{b}] = [0 \quad 0 \ldots 1]$$

This gives

$$\mathbf{p}_1 = [0 \quad 0 \ldots 1]\,\mathbf{U}^{-1}$$

where

$$\mathbf{U} = [\mathbf{b} \mid \mathbf{Ab} \mid \dots \mid \mathbf{A}^{n-1}\mathbf{b}]$$

is the controllability matrix, which is nonsingular because of the assumption of controllability of the system. Therefore, the controllable state model (9.4) can be transformed to the companion form (9.7) by the transformation

$$\bar{\mathbf{x}} = \mathbf{P}\mathbf{x} \tag{9.9}$$

where

$$\mathbf{P} = \begin{bmatrix} \mathbf{p}_1 \\ \mathbf{p}_1\mathbf{A} \\ \vdots \\ \mathbf{p}_1\mathbf{A}^{n-1} \end{bmatrix}; \ \mathbf{p}_1 = [0 \quad 0 \dots 1]\mathbf{U}^{-1}$$

□

Now that an algorithm for reducing single-input controllable systems to the controllable companion form has been presented, we can employ duality to directly achieve an analogous result for single-output observable systems. In particular, we now consider the observable system

$$\begin{aligned} \dot{\mathbf{x}} &= \mathbf{A}\mathbf{x} \\ y &= \mathbf{c}\mathbf{x} \end{aligned} \tag{9.10}$$

The input $\mathbf{u}(t)$ has been omitted for convenience, since it plays no role in the discussion which follows. It is always possible to find a transformation

$$\bar{\mathbf{x}} = \mathbf{Q}\mathbf{x}\ ; \quad \mathbf{Q} \text{ a nonsingular constant matrix} \tag{9.11}$$

which puts the matrix **A** in the companion form given in eqns. (7.69). Under the transformation (9.11), system (9 10) is transformed to

$$\begin{aligned} \dot{\bar{\mathbf{x}}} &= \bar{\mathbf{A}}\bar{\mathbf{x}} \\ y &= \bar{\mathbf{c}}\bar{\mathbf{x}} \end{aligned} \tag{9.12}$$

where $\bar{\mathbf{A}} = \mathbf{Q}\mathbf{A}\mathbf{Q}^{-1}\ ; \quad \bar{\mathbf{c}} = \mathbf{c}\mathbf{Q}^{-1}$

Theorem 9.2: If the n-dimensional linear time-invariant single-output system (9.10) is observable, then it can be transformed, by an equivalence transformation, into the observable companion form

$$\begin{aligned} \dot{\bar{\mathbf{x}}} &= \begin{bmatrix} 0 & 0 & 0 \dots 0 & -\alpha_n \\ 1 & 0 & 0 \dots 0 & -\alpha_{n-1} \\ 0 & 1 & 0 \dots 0 & -\alpha_{n-2} \\ \vdots & \vdots & \vdots \quad\ \ \vdots & \vdots \\ 0 & 0 & 0 \dots 0 & -\alpha_2 \\ 0 & 0 & 0 \dots 1 & -\alpha_1 \end{bmatrix} \bar{\mathbf{x}} \\ y &= [\,0 \quad 0 \quad 0 \dots 0 \quad 1\,]\,\bar{\mathbf{x}} \end{aligned} \tag{9.13}$$

Proof:

System (9.10) is assumed to be observable; therefore, the observability matrix has rank n, i.e.,

$$\rho(\mathbf{V}) = \rho \begin{bmatrix} \mathbf{c} \\ \mathbf{cA} \\ \vdots \\ \mathbf{cA}^{n-1} \end{bmatrix} = n$$

Equivalently,

$$\rho\, [\mathbf{c}^T \mid \mathbf{A}^T\mathbf{c}^T \mid \dots \mid (\mathbf{A}^T)^{n-1}\mathbf{c}^T] = n$$

Therefore, the pair $\{\mathbf{A}^T, \mathbf{c}^T\}$ is controllable and as per Theorem 9.1, there exists a nonsingular constant matrix $\mathbf{P}$ such that

$$\mathbf{PA}^T\mathbf{P}^{-1} = \begin{bmatrix} 0 & 1 & 0 & \cdots & 0 & 0 \\ 0 & 0 & 1 & \cdots & 0 & 0 \\ \vdots & \vdots & \vdots & & \vdots & \vdots \\ 0 & 0 & 0 & \cdots & 0 & 1 \\ -\alpha_n & -\alpha_{n-1} & -\alpha_{n-2} & \cdots & -\alpha_2 & -\alpha_1 \end{bmatrix} \tag{9.14a}$$

$$\mathbf{Pc}^T = \begin{bmatrix} 0 \\ 0 \\ \vdots \\ 0 \\ 1 \end{bmatrix} \tag{9.14b}$$

Taking the transpose on both sides of eqns. (9.14), we obtain

$$(\mathbf{P}^{-1})^T\mathbf{AP}^T = \begin{bmatrix} 0 & 0 & \cdots & 0 & -\alpha_n \\ 1 & 0 & \cdots & 0 & -\alpha_{n-1} \\ 0 & 1 & \cdots & 0 & -\alpha_{n-2} \\ \vdots & \vdots & & \vdots & \vdots \\ 0 & 0 & \cdots & 0 & -\alpha_2 \\ 0 & 0 & \cdots & 1 & -\alpha_1 \end{bmatrix} = \mathbf{QAQ}^{-1}$$

$$\mathbf{cP}^T = [\,0 \quad 0 \cdots 0 \quad 1\,] = \mathbf{cQ}^{-1}$$

where

$$\mathbf{Q} = (\mathbf{P}^T)^{-1} \tag{9.15}$$

Thus, the observable state model (9.10) can be transformed to the observable companion from (9.13) by the transformation

$$\bar{\mathbf{x}} = \mathbf{Qx}$$

where $\mathbf{Q}$ is given by (9.15).

Multi-Input/Multi-Output Systems

Consider the n-dimensional multi-input linear time-invariant system

$$\dot{\mathbf{x}} = \mathbf{A}\mathbf{x} + \mathbf{B}\mathbf{u} \tag{9.16}$$

Let $\mathbf{b}_i$ be the ith column of $\mathbf{B}$, i.e.,

$$\mathbf{B} = [\mathbf{b}_1 \,\vdots\, \mathbf{b}_2 \,\vdots \cdots \vdots\, \mathbf{b}_p]$$

If the dynamical equation (9.16) is controllable, then the controllability matrix

$$\mathbf{U} = [\mathbf{b}_1 \,\vdots\, \mathbf{b}_2 \,\vdots \cdots \vdots\, \mathbf{b}_p \,\vdots\, \mathbf{A}\mathbf{b}_1 \,\vdots \cdots \vdots\, \mathbf{A}\mathbf{b}_p \,\vdots \cdots \vdots\, \mathbf{A}^{n-1}\mathbf{b}_1 \,\vdots \cdots \vdots\, \mathbf{A}^{n-1}\mathbf{b}_p] \tag{9.1}$$

has rank n. Consequently, there are n linearly independent vectors in $\mathbf{U}$. There are many ways to choose n linearly independent vectors from $n \times np$ composite matrix $\mathbf{U}$; thus many controllable companion forms can be obtained (Luenberger 1967, Bucy 1968, Jordan and Sridhar 1973, Daly 1976). One controllable companion form that has been extensively used for multi-input systems is

$$\dot{\bar{\mathbf{x}}} = \bar{\mathbf{A}}\bar{\mathbf{x}} + \bar{\mathbf{B}}\mathbf{u} \; ; \tag{9.18a}$$

$\bar{\mathbf{A}}$ and $\bar{\mathbf{B}}$ have respectively the forms ($\times$ represents a possibly nonzero element)

$$\bar{\mathbf{A}} = \begin{bmatrix} \bar{\mathbf{A}}_{11} & \bar{\mathbf{A}}_{12} & \cdots & \bar{\mathbf{A}}_{1p} \\ \bar{\mathbf{A}}_{21} & \bar{\mathbf{A}}_{22} & \cdots & \bar{\mathbf{A}}_{2p} \\ \vdots & \vdots & & \vdots \\ \bar{\mathbf{A}}_{p1} & \bar{\mathbf{A}}_{p2} & \cdots & \bar{\mathbf{A}}_{pp} \end{bmatrix}$$

$$= \left[\begin{array}{ccc|ccc|c|ccc}
0 & 1 \cdots & 0 & 0 & 0 \cdots & 0 & & 0 & 0 \cdots & 0 \\
0 & 0 \cdots & 0 & 0 & 0 \cdots & 0 & & 0 & 0 \cdots & 0 \\
\vdots & \vdots & \vdots & \vdots & \vdots & \vdots & & \vdots & \vdots & \vdots \\
0 & 0 \cdots & 1 & 0 & 0 \cdots & 0 & & 0 & 0 \cdots & 0 \\
\times & \times \cdots & \times & \times & \times \cdots & \times & & \times & \times \cdots & \times \\
\hline
0 & 0 \cdots & 0 & 0 & 1 \cdots & 0 & \cdots & 0 & 0 \cdots & 0 \\
0 & 0 \cdots & 0 & 0 & 0 \cdots & 0 & & 0 & 0 \cdots & 0 \\
\vdots & \vdots & \vdots & \vdots & \vdots & \vdots & & \vdots & \vdots & \vdots \\
0 & 0 \cdots & 0 & 0 & 0 \cdots & 1 & & 0 & 0 \cdots & 0 \\
\times & \times \cdots & \times & \times & \times \cdots & \times & & \times & \times \cdots & \times \\
\hline
\cdot & \cdot & \cdot & \cdot & \cdot & \cdot & \cdots & \cdot & \cdot & \cdot \\
\hline
0 & 0 \cdots & 0 & 0 & 0 \cdots & 0 & & 0 & 1 \cdots & 0 \\
0 & 0 \cdots & 0 & 0 & 0 \cdots & 0 & & 0 & 0 \cdots & 0 \\
\vdots & \vdots & \vdots & \vdots & \vdots & \vdots & & \vdots & \vdots & \vdots \\
0 & 0 \cdots & 0 & 0 & 0 \cdots & 0 & & 0 & 0 \cdots & 1 \\
\times & \times \cdots & \times & \times & \times \cdots & \times & & \times & \times \cdots & \times
\end{array}\right] \tag{9.18b}$$

$$
\bar{\mathbf{B}} = \begin{bmatrix} \bar{\mathbf{B}}_1 \\ \bar{\mathbf{B}}_2 \\ \vdots \\ \bar{\mathbf{B}}_p \end{bmatrix} = \left[\begin{array}{ccc}
0 & 0 & \dots\ 0 \\
0 & 0 & \dots\ 0 \\
\vdots & \vdots & \vdots \\
0 & 0 & \dots\ 0 \\
1 & \times & \dots\ \times \\
\hline
0 & 0 & \dots\ 0 \\
0 & 0 & \dots\ 0 \\
\vdots & \vdots & \vdots \\
0 & 0 & \dots\ 0 \\
0 & 1 & \times \dots \times \\
\hline
\cdot & \cdot & \cdot\ \ \cdot \\
\hline
0 & 0 & \dots\ 0 \\
0 & 0 & \dots\ 0 \\
\vdots & \vdots & \vdots \\
0 & 0 & \dots\ 0 \\
0 & 0 & \dots\ 1
\end{array}\right] \tag{9.18c}
$$

The system (9.16) has p inputs and the p diagonal blocks in (9.18b) have dimensions $\gamma_1, \gamma_2, \dots, \gamma_p$ respectively, which sum to n—the dimension of the system. Each diagonal block $\bar{\mathbf{A}}_{ii}$ is a companion form of eqn. (9.7); off-diagonal blocks $\bar{\mathbf{A}}_{ij}$, $i \neq j$, are each identically zero except for their respective last rows.

The method of transforming (9.16) to the controllable companion form (9.18) is as follows. From (9.17), we first select n linearly independent vectors in the order shown by (9.17), i.e., we start from $\mathbf{b}_1, \mathbf{b}_2, \dots, \mathbf{b}_p$ and then go to $\mathbf{A}\mathbf{b}_1, \mathbf{A}\mathbf{b}_2, \dots, \mathbf{A}\mathbf{b}_p$ and then to $\mathbf{A}^2\mathbf{b}_1, \mathbf{A}^2\mathbf{b}_2, \dots$ and so forth until we obtain n linearly independent vectors. Note that if the vector $\mathbf{A}\mathbf{b}_2$ is skipped because of linear dependence on the vectors $\{\mathbf{b}_1, \mathbf{b}_2, \dots, \mathbf{b}_p, \mathbf{A}\mathbf{b}_1\}$, then all vectors of the form $\mathbf{A}^k\mathbf{b}_2$ for $k \geqslant 1$ can also be skipped because they must also be dependent on the previous columns.

After choosing n linearly independent vectors in this order, we arrange them as follows. First take γ_1 columns involving $\mathbf{b}_1$—the first column of $\mathbf{B}$, and then take γ_2 columns which involve $\mathbf{b}_2$ and so forth. The rearrangement gives

$$
\bar{\mathbf{U}} = [\mathbf{b}_1 \,\vdots\, \mathbf{A}\mathbf{b}_1 \,\vdots \dots \vdots\, \mathbf{A}^{\gamma_1 - 1}\mathbf{b}_1 \,\vdots\, \mathbf{b}_2 \,\vdots \dots \vdots\, \mathbf{A}^{\gamma_2 - 1}\mathbf{b}_2 \,\vdots \dots \vdots\, \mathbf{b}_p \,\vdots \dots \vdots\, \mathbf{A}^{\gamma_p - 1}\mathbf{b}_p] \tag{9.19}
$$

If all the columns of $\mathbf{B}$ are linearly independent (i.e., all available inputs are mutually independent, which is usually the case in practice), then the matrix $\bar{\mathbf{U}}$ contains all the $\mathbf{b}_i$'s. Otherwise, the linearly dependent vector will not appear in $\bar{\mathbf{U}}$. Note that

$$\gamma_1 + \gamma_2 + \dots + \gamma_p = n$$

We now set

$$\sigma_k = \sum_{i=1}^{k} \gamma_i \quad \text{for } k = 1, 2, \dots, p$$

which implies that

$$\sigma_1 = \gamma_1,\ \sigma_2 = \gamma_1 + \gamma_2,\ \dots,\ \sigma_p = \gamma_1 + \gamma_2 + \dots \gamma_p = n$$

We can now enlarge the algorithm employed in single-input case to determine an appropriate equivalence transformation matrix $\mathbf{P}$ in the case of multi-input system, i.e., we set $\mathbf{p}_k$ equal to the σ_k-th row of $\bar{\mathbf{U}}^{-1}$ for $k = 1, 2, \dots, p$ and consider the following $n \times n$ matrix.

$$\mathbf{P} = \begin{bmatrix} \mathbf{p}_1 \\ \mathbf{p}_1\mathbf{A} \\ \vdots \\ \mathbf{p}_2\mathbf{A}^{\gamma_1-1} \\ \mathbf{p}_2 \\ \vdots \\ \mathbf{p}_2\mathbf{A}^{\gamma_2-1} \\ \vdots \\ \mathbf{p}_p\mathbf{A}^{\gamma_p-1} \end{bmatrix} \tag{9.20}$$

If $\mathbf{P}$ thus defined is post multiplied by $\bar{\mathbf{U}}$, it can be shown that $|\mathbf{P}\bar{\mathbf{U}}| = 1$:

$$\mathbf{P}\bar{\mathbf{U}} = \begin{bmatrix} \mathbf{p}_1 \\ \mathbf{p}_1\mathbf{A} \\ \vdots \\ \mathbf{p}_1\mathbf{A}^{\gamma_1-1} \\ \mathbf{p}_2 \\ \vdots \\ \mathbf{p}_2\mathbf{A}^{\gamma_2-1} \\ \vdots \\ \mathbf{p}_p\mathbf{A}^{\gamma_p-1} \end{bmatrix} [\mathbf{b}_1 \mid \mathbf{A}\mathbf{b}_1 \mid \dots \mid \mathbf{A}^{\gamma_1-1}\mathbf{b}_1 \mid \mathbf{b}_2 \mid \dots \mid \mathbf{A}^{\gamma_2-1}\mathbf{b}_2 \mid \dots \mid \mathbf{A}^{\gamma_p-1}\mathbf{b}_p]$$

$$= \begin{bmatrix} \mathbf{p}_1\mathbf{b}_1 & \mathbf{p}_1\mathbf{A}\mathbf{b}_1 \dots \mathbf{p}_1\mathbf{A}^{\gamma_1-1}\mathbf{b}_1 & \mathbf{p}_1\mathbf{b}_2 \dots \mathbf{p}_1\mathbf{A}^{\gamma_2-1}\mathbf{b}_2 \dots \mathbf{p}_1\mathbf{A}^{\gamma_p-1}\mathbf{b}_p \\ \mathbf{p}_1\mathbf{A}\mathbf{b}_1 & \mathbf{p}_1\mathbf{A}^2\mathbf{b}_1 \dots \mathbf{p}_1\mathbf{A}^{\gamma_1}\mathbf{b}_1 & \mathbf{p}_1\mathbf{A}\mathbf{b}_2 \dots \mathbf{p}_1\mathbf{A}^{\gamma_2}\mathbf{b}_2 \dots \mathbf{p}_1\mathbf{A}^{\gamma_p}\mathbf{b}_p \\ & \vdots & \\ \mathbf{p}_p\mathbf{A}^{\gamma_p-1}\mathbf{b}_1 & \mathbf{p}_p\mathbf{A}^{\gamma_p}\mathbf{b}_1 \dots \mathbf{p}_p\mathbf{A}^{\gamma_p+\gamma_1-2}\mathbf{b}_1 & \mathbf{p}_p\mathbf{A}^{\gamma_p-1}\mathbf{b}_2 \dots \mathbf{p}_p\mathbf{A}^{\gamma_p+\gamma_2-2}\mathbf{b}_2 \dots \mathbf{p}_p\mathbf{A}^{2\gamma_p-2}\mathbf{b}_p \end{bmatrix} \tag{9.21}$$

Define

$$\overline{\mathbf{U}}^{-1} = \begin{bmatrix} \mathbf{e}_{11} \\ \mathbf{e}_{12} \\ \vdots \\ \mathbf{e} \\ \mathbf{e}_{21} \\ \mathbf{e}_{22} \\ \vdots \\ \mathbf{e}_{2\gamma_2} \\ \vdots \\ \mathbf{e}_{p1} \\ \vdots \\ \mathbf{e}_{p\gamma_p} \end{bmatrix}$$

From the fact that $\overline{\mathbf{U}}^{-1}\overline{\mathbf{U}} = \mathbf{I}$, where $\mathbf{I}$ is a unit matrix, it can easily be verified that

$$\begin{aligned} \mathbf{e}_{ij}\mathbf{A}^k\mathbf{b}_l &= 1 \text{ if } i = l \text{ and } j = k+1 \\ &= 0 \text{ otherwise} \end{aligned}$$

Therefore

$$\begin{aligned} \mathbf{e}_{1\gamma_1}\mathbf{A}^k\mathbf{b}_l &= 1 \text{ for } l = 1 \text{ and } k = \gamma_1 - 1 \\ &= 0 \text{ otherwise} \end{aligned}$$

But $\quad \mathbf{e}_{1\gamma_1} = \mathbf{p}_1$; therefore

$$\mathbf{p}_1\mathbf{A}^{\gamma_1 - 1}\mathbf{b}_1 = 1$$

and

$$\mathbf{p}_1\mathbf{A}^k\mathbf{b}_l = 0 \text{ for } k \neq \gamma_1 - 1 \text{ and } l \neq 1$$

Similarly

$$\mathbf{p}_2\mathbf{A}^{\gamma_2 - 1}\mathbf{b}_2 = 1$$

and

$$\mathbf{p}_2\mathbf{A}^k\mathbf{b}_l = 0 \text{ for } k \neq \gamma_2 - 1 \text{ and } l \neq 2$$

$$\vdots$$

The matrix $\mathbf{P}\overline{\mathbf{U}}$ is therefore

$$\mathbf{P}\overline{\mathbf{U}} = \begin{bmatrix} \mathbf{M}_{11} & & \mathbf{0} \\ & \mathbf{M}_{22} & \\ & & \ddots \\ \mathbf{0} & & \mathbf{M}_{pp} \end{bmatrix}$$

where

$$\underset{(\gamma_i \times \gamma_i)}{\mathbf{M}_{ii}} = \begin{bmatrix} 0 & 0 \ldots 0 & 1 \\ 0 & 0 \ldots 1 & 0 \\ \vdots & \vdots & \vdots \\ 1 & 0 \ldots 0 & 0 \end{bmatrix} \text{ for all } i = 1, 2, \ldots, p$$

$$|\mathbf{P}\overline{\mathbf{U}}| = 1$$

This proves that matrix **P** is nonsingular.

If matrix **P** is used as transformation matrix, we obtain $\bar{\mathbf{A}} = \mathbf{PAP}^{-1}$ in the form given in (9.18b).

$$\bar{\mathbf{B}} = \mathbf{PB} = \begin{bmatrix} \mathbf{p}_1\mathbf{b}_1 & \mathbf{p}_1\mathbf{b}_2 & \cdots & \mathbf{p}_1\mathbf{b}_p \\ \mathbf{p}_1\mathbf{A}\mathbf{b}_1 & \mathbf{p}_1\mathbf{A}\mathbf{b}_2 & \cdots & \mathbf{p}_1\mathbf{A}\mathbf{b}_p \\ \vdots & \vdots & & \vdots \\ \mathbf{p}_1\mathbf{A}^{\gamma_1-1}\mathbf{b}_1 & \mathbf{p}_1\mathbf{A}^{\gamma_1-1}\mathbf{b}_2 & \cdots & \mathbf{p}_1\mathbf{A}^{\gamma_1-1}\mathbf{b}_p \\ \mathbf{p}_2\mathbf{b}_1 & \mathbf{p}_2\mathbf{b}_2 & \cdots & \mathbf{p}_2\mathbf{b}_p \\ \vdots & \vdots & & \vdots \\ \mathbf{p}_p\mathbf{A}^{\gamma_p-1}\mathbf{b}_1 & \mathbf{p}_p\mathbf{A}^{\gamma_p-1}\mathbf{b}_2 & \cdots & \mathbf{p}_p\mathbf{A}^{\gamma_p-1}\mathbf{b}_p \end{bmatrix}$$

From the result $\bar{\mathbf{U}}^{-1}\bar{\mathbf{U}} = \mathbf{I}$, we can easily obtain the following form for $\bar{\mathbf{B}}$ ($\times$'s are possibly nonzero entries):

$$\bar{\mathbf{B}} = \left[\begin{array}{ccc} 0 & 0 & \cdots\ 0 \\ 0 & 0 & \cdots\ 0 \\ \vdots & \vdots & \vdots \\ 0 & 0 & \cdots\ 0 \\ 1 & \times & \cdots\ \times \\ \hline 0 & 0 & \cdots\ 0 \\ 0 & 0 & \cdots\ 0 \\ \vdots & \vdots & \vdots \\ 0 & 0 & \cdots\ 0 \\ 0 & 1 & \cdots\ \times \\ \hline \cdot & \cdot & \cdot\ \ \cdot\ \ \cdot \\ \hline 0 & 0 & \cdots\ 0 \\ 0 & 0 & \cdots\ 0 \\ \vdots & \vdots & \vdots \\ 0 & 0 & \cdots\ 0 \\ 0 & 0 & \cdots\ 1 \end{array}\right] \tag{9.22}$$

Further, if $\gamma_1 \leqslant \gamma_2 \leqslant \gamma_3 \cdots \gamma_{p-1} \leqslant \gamma_p$, then the '$\times$' entries in (9.22) become '0' entries. The condition $\gamma_1 \leqslant \gamma_2 \leqslant \gamma_3 \cdots \gamma_{p-1} \leqslant \gamma_p$ may be satisfied by suitably rearranging the columns of **B** in (9.16), which is equivalent to rearranging the input vector.

The p integers γ_i are known as *controllability indices* of the system and $\gamma : \max \gamma_i$ for $i = 1, 2, \ldots, p$ is known as *controllability index* of the system, i.e.,

$$\gamma = \max \{\gamma_i : i = 1, 2, \ldots, p\} \tag{9.23}$$

We will now employ the duality principle to achieve an analogous result for observable systems. Consider the observable multi-output system

$$\begin{aligned} \dot{\mathbf{x}} &= \mathbf{A}\mathbf{x} \\ \mathbf{y} &= \mathbf{C}\mathbf{x} \end{aligned} \tag{9.24}$$

The transformation

$$\bar{\mathbf{x}} = \mathbf{Q}\mathbf{x}\,; \quad \mathbf{Q} = (\mathbf{P}^T)^{-1} \tag{9.25}$$

where $\mathbf{P}$ is a matrix analogous to (9.20) obtained by applying the algorithm to the pair $\{\mathbf{A}^T, \mathbf{C}^T\}$; gives the following equivalent observable companion form of (9.24):

$$\begin{aligned} \dot{\bar{\mathbf{x}}} &= \bar{\mathbf{A}}\bar{\mathbf{x}} \\ \mathbf{y} &= \bar{\mathbf{C}}\bar{\mathbf{x}} \end{aligned} \tag{9.26a}$$

where

$$\bar{\mathbf{A}} = \mathbf{Q}\mathbf{A}\mathbf{Q}^{-1} = \begin{bmatrix} \bar{\mathbf{A}}_{11} & \bar{\mathbf{A}}_{12} & \cdots & \bar{\mathbf{A}}_{1q} \\ \bar{\mathbf{A}}_{21} & \bar{\mathbf{A}}_{22} & \cdots & \bar{\mathbf{A}}_{2q} \\ \vdots & \vdots & & \vdots \\ \bar{\mathbf{A}}_{q1} & \bar{\mathbf{A}}_{q2} & & \bar{\mathbf{A}}_{qq} \end{bmatrix}$$

$$= \left[\begin{array}{cccc|cccc|c|cccc}
0 & 0 \ldots & 0 & \times & 0 & 0 \ldots & 0 & \times & & 0 & 0 \ldots & 0 & \times \\
1 & 0 \ldots & 0 & \times & 0 & 0 \ldots & 0 & \times & & 0 & 0 \ldots & 0 & \\
\vdots & \vdots & \vdots & \vdots & \vdots & \vdots & \vdots & \vdots & & \vdots & \vdots & \vdots & \vdots \\
0 & 0 \ldots & 1 & \times & 0 & 0 \ldots & 0 & \times & & 0 & 0 \ldots & 0 & \times \\
\hline
0 & 0 \ldots & 0 & \times & 0 & 0 \ldots & 0 & \times & \cdots & 0 & 0 \ldots & 0 & \times \\
0 & 0 \ldots & 0 & \times & 1 & 0 \ldots & 0 & \times & & 0 & 0 \ldots & 0 & \times \\
\vdots & \vdots & \vdots & \vdots & \vdots & \vdots & \vdots & \vdots & & \vdots & \vdots & \vdots & \vdots \\
0 & 0 \ldots & 0 & \times & 0 & 0 \ldots & 1 & \times & & 0 & 0 \ldots & 0 & \times \\
\hline
\cdot & \cdot & \cdot & \cdot & \cdot & \cdot & \cdot & \cdot & \cdot & \cdot & \cdot & \cdot & \cdot \\
\hline
0 & 0 \ldots & 0 & \times & 0 & 0 \ldots & 0 & \times & & 0 & 0 \ldots & 0 & \times \\
0 & 0 \ldots & 0 & \times & 0 & 0 \ldots & 0 & \times & & 1 & 0 \ldots & 0 & \times \\
\vdots & \vdots & \vdots & \vdots & \vdots & \vdots & \vdots & \vdots & & \vdots & \vdots & \vdots & \vdots \\
0 & 0 \ldots & 0 & \times & 0 & 0 \ldots & 0 & \times & & 0 & 0 \ldots & 1 & \times
\end{array}\right] \tag{9.26b}$$

$$\bar{\mathbf{C}} = \mathbf{C}\mathbf{Q}^{-1} = \left[\begin{array}{cccc|cccc|c|cccc} 0 & 0 & \dots 0 & 1 & 0 & 0 & \dots 0 & 0 & & 0 & 0 & \dots 0 & 0 \\ 0 & 0 & \dots 0 & \times & 0 & 0 & \dots 0 & 1 & \dots & 0 & 0 & \dots 0 & 0 \\ \vdots & \vdots & \vdots & \vdots & \vdots & \vdots & \vdots & \vdots & & \vdots & \vdots & \vdots & \vdots \\ 0 & 0 & \dots 0 & \times & 0 & 0 & \dots 0 & \times & & 0 & 0 & \dots 0 & 1 \end{array}\right] \tag{9.26c}$$

This form closely resembles the controllable companion form and in particular a transposed version of that form. Each diagonal submatrix $\bar{\mathbf{A}}_{ii}$ is a μ_i dimensional companion matrix of the form given in eqn. (9.13) while the off-diagonal blocks $\bar{\mathbf{A}}_{ij}$ for $i \neq j$ are each identically zero except for their respective last columns.

As in the controllable case, we define the q integers μ_i as the *observability indices* of the system and

$$\mu = \max\{\mu_i : i = 1, 2, \dots, q\} \tag{9.27}$$

as the *observability index* of the system.

Example 9.1: Consider the system

$$\dot{\mathbf{x}} = \mathbf{A}\mathbf{x} + \mathbf{B}\mathbf{u} \tag{9.28}$$

where

$$\mathbf{A} = \begin{bmatrix} 1 & 0 & 0 \\ 0 & 2 & 0 \\ 0 & 0 & 3 \end{bmatrix}; \mathbf{B} = \begin{bmatrix} 1 & 0 \\ 0 & 1 \\ 1 & 1 \end{bmatrix} = [\mathbf{b}_1 \quad \mathbf{b}_2]$$

From **A** and **B** matrices, just by inspection, we find that system (9.28) is controllable (Section 6.7). The controllability matrix

$$\mathbf{U} = [\mathbf{B} \quad \mathbf{AB} \quad \mathbf{A}^2\mathbf{B}]$$

$$= \begin{bmatrix} 1 & 0 & 1 & 0 & 1 & 0 \\ 0 & 1 & 0 & 2 & 0 & 4 \\ 1 & 1 & 3 & 3 & 9 & 9 \end{bmatrix}$$

We obtain three linearly independent columns of **U** matrix and place them in the order suggested in (9.19):

$$\bar{\mathbf{U}} = \begin{bmatrix} 1 & 1 & 0 \\ 0 & 0 & 1 \\ 1 & 3 & 1 \end{bmatrix} = [\mathbf{b}_1 \quad \mathbf{A}\mathbf{b}_1 \quad \mathbf{b}_2]$$

The controllability indices γ_1 and γ_2 are, therefore,

$$\gamma_1 = 2 \quad \text{and} \quad \gamma_2 = 1$$

This gives

$$\sigma_1 = 2 \quad \text{and} \quad \sigma_2 = 3$$

Let $\mathbf{p}_1$ be the σ_1-th row of $\overline{\mathbf{U}}^{-1}$ and $\mathbf{p}_2$ be the σ_2-th row of $\overline{\mathbf{U}}^{-1}$, where $\overline{\mathbf{U}}^{-1}$ is obtained as

$$\overline{\mathbf{U}}^{-1} = \begin{bmatrix} 1.5 & 0.5 & -0.5 \\ -0.5 & -0.5 & 0.5 \\ 0 & 1 & 0 \end{bmatrix}$$

Therefore

$$\mathbf{p}_1 = [-0.5 \quad -0.5 \quad 0.5]$$

and

$$\mathbf{p}_2 = [\,0 \quad 1 \quad 0\,]$$

The transformation matrix (refer eqn. (9.20))

$$\mathbf{P} = \begin{bmatrix} \mathbf{p}_1 \\ \mathbf{p}_1\mathbf{A} \\ \mathbf{p}_2 \end{bmatrix} = \begin{bmatrix} -0.5 & -0.5 & 0.5 \\ -0.5 & -1 & 1.5 \\ 0 & 1 & 0 \end{bmatrix}$$

$$\overline{\mathbf{A}} = \mathbf{PAP}^{-1} = \left[\begin{array}{cc|c} 0 & 1 & 0 \\ -3 & 4 & 0.5 \\ \hline 0 & 0 & 2 \end{array}\right]$$

$$\overline{\mathbf{B}} = \mathbf{PB} = \left[\begin{array}{cc} 0 & 0 \\ 1 & 0.5 \\ \hline 0 & 1 \end{array}\right]$$

The controllable companion form of (9.28) is

$$\dot{\overline{\mathbf{x}}} = \overline{\mathbf{A}}\,\overline{\mathbf{x}} + \overline{\mathbf{B}}\mathbf{u}$$

9.3 THE EFFECT OF STATE FEEDBACK ON CONTROLLABILITY AND OBSERVABILITY

While designing a feedback system, it is necessary to determine for the contemplated design whether it is controllable, observable and stable. In this section, we shall study the effect of feedback on the controllability and observability properties of a system. The stability property is discussed in the next section.

Controllability

Theorem 9.3: The state feedback system given by (9.3) is controllable for any feedback gain matrix $\mathbf{K}$ if and only if the system (9.1) is controllable.

Proof:

Let us first show that the controllability of (9.1) implies controllability of (9.3). Let $\mathbf{x}(0) \triangleq \mathbf{x}^0$ be an arbitrary state in state space Σ. By the controllability assumption of (9.1), there exists an input $\mathbf{u}_{(0,t_1]}$, that will transfer $\mathbf{x}^0$ to any arbitrary state $\mathbf{x}^1 \in \Sigma$ in finite time t_1.

Now for the feedback system (9.3), consider an input

$$\mathbf{r}(t) = \mathbf{u}(t) - \mathbf{K}\mathbf{x}(t);\ 0 \leqslant t \leqslant t_1$$

This input causes the system to follow precisely the same trajectory as in the case of (9.1), because the term $-\mathbf{Kx}$ just cancels out the feedback. Thus the input $\mathbf{r}(t)$ will transfer $\mathbf{x}^0$ to $\mathbf{x}^1$ in finite time. Therefore (9.3) is controllable.

If (9.1) is not controllable, then (9.3) is also not controllable. This is obvious from Fig. 9.1; if $\mathbf{u}$ cannot control $\mathbf{x}$, neither can $\mathbf{r}$.

□

Theorem 9.3 is concerned with the overall controllability of a system. We shall now see that feedback can be used to alter controllability with respect to particular components of the input vector.

Theorem 9.4: If system (9.1) is controllable and $\mathbf{b}_i\ (\neq \mathbf{0})$ is the ith column of $\mathbf{B}$, then there exists a feedback matrix $\mathbf{K}_i$ such the single-input system

$$\dot{\mathbf{x}} = (\mathbf{A} + \mathbf{B}\mathbf{K}_i)\,\mathbf{x} + \mathbf{b}_i r_i$$

is controllable.

Proof (Chen 1970):

Without loss of generality, we prove the theorem for $i = 1$. The controllability matrix of system (9.1) is

$$\mathbf{U} = [\mathbf{B}\,|\,\mathbf{AB}\,|\ldots|\,\mathbf{A}^{n-1}\mathbf{B}]$$

Since (9.1) is controllable, $\rho(\mathbf{U}) = n$, and hence there are n linearly independent column vectors in $\mathbf{U}$. Let us select and arrange these linearly independent columns of $\mathbf{U}$ as per the scheme suggested in (9.19). Assuming γ_1, γ_2 and γ_3 to be the controllability indices with $\gamma_1 + \gamma_2 + \gamma_3 = n$, we obtain

$$\bar{\mathbf{U}} = [\mathbf{b}_1 \,\vdots\, \mathbf{A}\mathbf{b}_1 \,|\ldots|\, \mathbf{A}^{\gamma_1 - 1}\mathbf{b}_1 \,\vdots\, \mathbf{b}_2 \,|\ldots|\, \mathbf{A}^{\gamma_2 - 1}\mathbf{b}_2 \,\vdots\, \mathbf{b}_3 \,|\ldots\vdots\, \mathbf{A}^{\gamma_3 - 1}\mathbf{b}_3]$$

$$\sigma_1 = \gamma_1,\ \sigma_2 = \gamma_1 + \gamma_2,\ \sigma_3 = \gamma_1 + \gamma_2 + \gamma_3$$

Define the $p \times n$ matrix

$$\mathbf{Q} \triangleq [\,\mathbf{0}\ \ \mathbf{0}\ \ldots\ \mathbf{0}\ \ \underset{\substack{\uparrow \\ \text{The } \sigma_1\text{-th} \\ \text{column}}}{\mathbf{e}_2}\ \ \mathbf{0}\ \ldots\ \mathbf{0}\ \ \underset{\substack{\uparrow \\ \text{The } \sigma_2\text{-th} \\ \text{column}}}{\mathbf{e}_3}\ \ \mathbf{0}\ \ldots\ \mathbf{0}\ \ \underset{\substack{\uparrow \\ \text{The } \sigma_3\text{-th} \\ \text{column}}}{\mathbf{0}}\,]$$

where $\mathbf{e}_i$ is the i-th column of a $p \times p$ unit matrix. Now, we claim that the matrix $\mathbf{K}_1$ given by

$$\mathbf{K}_1 = \mathbf{Q}\bar{\mathbf{U}}^{-1}$$

satisfies the theorem. This result can be established as follows

$$\mathbf{K}_1\bar{\mathbf{U}} = \mathbf{Q}$$

or

$$\mathbf{K}_1 \, [\mathbf{b}_1 \,\vdots\, \mathbf{A}\mathbf{b}_1 \,\vdots \ldots \vdots\, \mathbf{A}^{\gamma_1-1}\mathbf{b}_1 \,\vdots\, \mathbf{b}_2 \,\vdots \ldots \vdots\, \mathbf{A}^{\gamma_2-1}\mathbf{b}_2 \,\vdots\, \mathbf{b}_3 \,\vdots \ldots \vdots\, \mathbf{A}^{\gamma_3-1}\mathbf{b}_3]$$

$$= [\mathbf{0} \;\; \mathbf{0} \; \ldots \; \mathbf{0} \;\; \mathbf{e}_2 \;\; \mathbf{0} \; \ldots \; \mathbf{0} \;\; \mathbf{e}_3 \;\; \mathbf{0} \; \ldots \; \mathbf{0} \;\; \mathbf{0}]$$

This gives

$$\mathbf{K}_1\mathbf{b}_1 = \mathbf{K}_1\mathbf{A}\mathbf{b}_1 = \ldots = \mathbf{K}_1\mathbf{A}^{\gamma_1-2}\mathbf{b}_1 = \mathbf{0},\; \mathbf{K}_1\mathbf{A}^{\gamma_1-1}\mathbf{b}_1 = \mathbf{e}_2$$

$$\mathbf{K}_1\mathbf{b}_2 = \mathbf{K}_1\mathbf{A}\mathbf{b}_2 = \ldots = \mathbf{K}_1\mathbf{A}^{\gamma_2-2}\mathbf{b}_2 = \mathbf{0},\; \mathbf{K}_1\mathbf{A}^{\gamma_2-1}\mathbf{b}_2 = \mathbf{e}_3$$

$$\mathbf{K}_1\mathbf{b}_3 = \mathbf{K}_1\mathbf{A}\mathbf{b}_3 = \ldots = \mathbf{K}_1\mathbf{A}^{\gamma_3-1}\mathbf{b}_3 = \mathbf{0}$$

Using these results, we get

$$\begin{aligned}
(\mathbf{A} + \mathbf{B}\mathbf{K}_1)\,\mathbf{b}_1 &= \mathbf{A}\mathbf{b}_1 \\
(\mathbf{A} + \mathbf{B}\mathbf{K}_1)^2\mathbf{b}_1 &= (\mathbf{A} + \mathbf{B}\mathbf{K}_1)\,\mathbf{A}\mathbf{b}_1 = \mathbf{A}^2\mathbf{b}_1 \\
&\;\;\vdots \\
(\mathbf{A} + \mathbf{B}\mathbf{K}_1)^{\gamma_1-1}\mathbf{b}_1 &= (\mathbf{A} + \mathbf{B}\mathbf{K}_1)\,\mathbf{A}^{\gamma_1-2}\mathbf{b}_1 = \mathbf{A}^{\gamma_1-1}\mathbf{b}_1 \\
(\mathbf{A} + \mathbf{B}\mathbf{K}_1)^{\gamma_1}\mathbf{b}_1 &= (\mathbf{A} + \mathbf{B}\mathbf{K}_1)\,\mathbf{A}^{\gamma_1-1}\mathbf{b}_1 = \mathbf{A}^{\gamma_1}\mathbf{b}_1 + \mathbf{B}\mathbf{e}_2 = \mathbf{b}_2 + \mathbf{A}^{\gamma_1}\mathbf{b}_1 \\
(\mathbf{A} + \mathbf{B}\mathbf{K}_1)^{\gamma_1+1}\mathbf{b}_1 &= (\mathbf{A} + \mathbf{B}\mathbf{K}_1)\,(\mathbf{b}_2 + \mathbf{A}^{\gamma_1}\mathbf{b}_1) = \mathbf{A}\mathbf{b}_2 + \ldots \\
&\;\;\vdots \\
(\mathbf{A} + \mathbf{B}\mathbf{K}_1)^{n-1}\mathbf{b}_1 &= (\mathbf{A} + \mathbf{B}\mathbf{K}_1)\,(\mathbf{A}^{\gamma_3-2}\mathbf{b}_3 + \ldots) = \mathbf{A}^{\gamma_3-1}\mathbf{b}_3 + \ldots
\end{aligned}$$

The (...) terms in the above expressions denote linear combinations of the preceding vectors. From these expressions, it is easily established that the pair $\{\mathbf{A} + \mathbf{B}\mathbf{K}_1, \mathbf{b}_1\}$ is controllable.

Observability

The observability of a system is not preserved under state feedback. This is obvious from the following example.

Example 9.2: Consider a single-input/single-output system

$$\dot{\mathbf{x}} = \begin{bmatrix} 0 & 1 \\ -1 & -2 \end{bmatrix} \mathbf{x} + \begin{bmatrix} 0 \\ 1 \end{bmatrix} u$$

$$y = [1 \;\; 2]\,\mathbf{x}$$

The system without feedback is both controllable and observable.

Introducing a feedback signal

$$u = r + [2 \;\; -1]\,\mathbf{x}$$

we get the following feedback system

$$\dot{\mathbf{x}} = \begin{bmatrix} 0 & 1 \\ 1 & -3 \end{bmatrix} \mathbf{x} + \begin{bmatrix} 0 \\ 1 \end{bmatrix} r$$

$$y = [1 \quad 2]\, \mathbf{x}$$

This system is both controllable and observable.

For the feedback signal

$$u = r + [-1 \quad -5/2]\, \mathbf{x}$$

the feedback system becomes

$$\dot{\mathbf{x}} = \begin{bmatrix} 0 & 1 \\ -2 & -9/2 \end{bmatrix} \mathbf{x} + \begin{bmatrix} 0 \\ 1 \end{bmatrix} r$$

$$y = [1 \quad 2]\, \mathbf{x}$$

This system is controllable but not observable.

Thus it is possible to destroy the observability property of a system by some choice of $\mathbf{K}$.

9.4 POLE PLACEMENT BY STATE FEEDBACK

In Chapter 8 we observed that an important aspect of control system is stability. Whatever we want to achieve with the control system, its stability must be assured. Sometimes the main goal of a feedback design is to stabilize an unstable plant or to improve the transient behaviour of a plant. We shall see in this section, how state feedback can fix arbitrarily the modes of a system and hence control the transient behaviour.

The fundamental result on pole placement by state feedback was presented by Wonham (1967). He showed that a controllable system is always pole-assignable by appropriate state feedback. He made use of certain linear algebra concepts like cyclic subspaces, decomposition, module etc. Anderson and Luenberger (1967) developed time-domain methods based on Wonham's result making use of canonical form transformation. Power (1971) extended their method exploiting extra degrees of freedom. Sundareswaren and Bayoumi (1971) treated partially controllable systems through Luenberger's result. Fallside and Seraji (1971, 1973) have given a simple method of pole placement assuming the $p \times n$ feedback matrix $\mathbf{K}$ to be of unity rank. Munro (1979) has reviewed various methods of pole assignment.

During the last decade, many more papers have been published dealing with the problem of pole placement. The purpose of this (small) section is to introduce the problem to the reader. The pole placement methods we deal with here are based on the utilisation of particular structural decompositions of the system. The problem will be separated into three stages.

(i) Casting the system into the appropriate canonical form.

(ii) Finding a feedback controller associated with this form.
(iii) Transforming the controller to the original coordinates.

Single-Input Systems

Suppose that the nth-order system with the state equation

$$\dot{\mathbf{x}} = \mathbf{A}\mathbf{x} + \mathbf{b}u \tag{9.29}$$

where u is a scalar input, is completely controllable. The problem consists of finding a control law

$$u = \mathbf{k}\mathbf{x} + r \tag{9.30}$$

such that the feedback system

$$\dot{\mathbf{x}} = (\mathbf{A} + \mathbf{b}\mathbf{k})\mathbf{x} + \mathbf{b}r \tag{9.31}$$

has an arbitrarily given characteristic equation of degree n.

The first step in the algorithm that follows, is to cast system (9.29) into a controllable companion form. Since $\{\mathbf{A}, \mathbf{b}\}$ is controllable, there exists a transformation matrix $\mathbf{P}(\bar{\mathbf{x}} = \mathbf{P}\mathbf{x})$ which converts (9.29) into the form (Theorem 9.1)

$$\dot{\bar{\mathbf{x}}} = \bar{\mathbf{A}}\bar{\mathbf{x}} + \bar{\mathbf{b}}u \tag{9.32}$$

where

$$\bar{\mathbf{A}} = \mathbf{P}\mathbf{A}\mathbf{P}^{-1} = \begin{bmatrix} 0 & 1 & 0 & \cdots & 0 & 0 \\ 0 & 0 & 1 & \cdots & 0 & 0 \\ \vdots & \vdots & \vdots & & \vdots & \vdots \\ 0 & 0 & 0 & \cdots & 0 & 1 \\ -\alpha_n & -\alpha_{n-1} & -\alpha_{n-2} & \cdots & -\alpha_2 & -\alpha_1 \end{bmatrix}; \bar{\mathbf{b}} = \mathbf{P}\mathbf{b} = \begin{bmatrix} 0 \\ 0 \\ \vdots \\ 0 \\ 1 \end{bmatrix}$$

Here the numbers α_i; $i = 1, 2, \ldots, n$ are the coefficients of the characteristic polynomial of (9.29), i.e.,

$$|s\mathbf{I} - \mathbf{A}| = s^n + \alpha_1 s^{n-1} + \ldots + \alpha_{n-1}s + \alpha_n$$
$$= |s\mathbf{I} - \bar{\mathbf{A}}|$$

Because of the equivalence transformation, the feedback control law (9.30) becomes

$$\begin{aligned} u &= \mathbf{k}\mathbf{x} + r \\ &= \mathbf{k}\mathbf{P}^{-1}\bar{\mathbf{x}} + r \\ &\triangleq \bar{\mathbf{k}}\,\bar{\mathbf{x}} + r \end{aligned} \tag{9.33}$$

where

$$\bar{\mathbf{k}} = \mathbf{k}\mathbf{P}^{-1} = [\bar{k}_1 \quad \bar{k}_2 \cdots \bar{k}_n]$$

With this control law, system (9.32) becomes

$$\dot{\bar{\mathbf{x}}} = (\bar{\mathbf{A}} + \bar{\mathbf{b}}\,\bar{\mathbf{k}})\,\bar{\mathbf{x}} + \bar{\mathbf{b}}\,r$$

$$= \begin{bmatrix} 0 & 1 & 0 & \cdots & 0 & 0 \\ 0 & 0 & 1 & \cdots & 0 & 0 \\ \vdots & \vdots & \vdots & & \vdots & \vdots \\ 0 & 0 & 0 & \cdots & 0 & 1 \\ -\alpha_n + \bar{k}_1 & -\alpha_{n-1} + \bar{k}_2 & -\alpha_{n-2} + \bar{k}_3 & \cdots & -\alpha_2 + \bar{k}_{n-1} & -\alpha_1 + \bar{k}_n \end{bmatrix}$$

$$+ \begin{bmatrix} 0 \\ 0 \\ \vdots \\ 0 \\ 1 \end{bmatrix} r \qquad (9.34)$$

Since the coefficients $\bar{k}_i$ are arbitrarily chosen real numbers, the coefficients of the characteristic polynomial of $(\bar{\mathbf{A}} + \bar{\mathbf{b}}\bar{\mathbf{k}})$ can be given any desired values i.e., the closed-loop poles can be located to arbitrary locations (subject to conjugate pairing: coefficients of a characteristic equation will be real only if the complex poles are present in conjugate pairs).

Assume that the desired characteristic polynomial of $(\mathbf{A} + \mathbf{bk})$ and hence $(\bar{\mathbf{A}} + \bar{\mathbf{b}}\bar{\mathbf{k}})$ is

$$s^n + \bar{\alpha}_1 s^{n-1} + \ldots + \bar{\alpha}_n \qquad (9.35)$$

From (9.34), it is obvious that this requirement is met if $\bar{\mathbf{k}}$ is chosen as

$$\bar{\mathbf{k}} = [\alpha_n - \bar{\alpha}_n \quad \alpha_{n-1} - \bar{\alpha}_{n-1} \ldots \alpha_1 - \bar{\alpha}_1] \qquad (9.36)$$

The last step in the design algorithm is to transform the feedback controller (9.33) to the original coordinates, i.e., to obtain $\mathbf{k}$ from $\bar{\mathbf{k}}$. This is given as

$$\bar{\mathbf{k}} = \mathbf{kP} \qquad (9.37)$$

This proves that if (9.29) is controllable, the closed-loop poles may be arbitrarily assigned. For a given set of poles, a unique gain vector $\mathbf{k}$ is found. Melsa and Jones (1973) list a FORTRAN computer program to determine this vector.

Stabilizability: The result derived above implies that it is always possible to stabilize a completely controllable linear system. Suppose, however, we are confronted with a time-invariant system that is not completely controllable. It was shown in Theorem 6.13 that an uncontrollable system can be transformed into the controllability canonical form

$$\dot{\hat{\mathbf{x}}} = \begin{bmatrix} \hat{\mathbf{A}}_{11} & \hat{\mathbf{A}}_{12} \\ \mathbf{0} & \hat{\mathbf{A}}_{22} \end{bmatrix} \hat{\mathbf{x}} + \begin{bmatrix} \hat{\mathbf{b}}_1 \\ \mathbf{0} \end{bmatrix} u = \hat{\mathbf{A}}\hat{\mathbf{x}} + \hat{\mathbf{b}}u$$

where the pair $\{\hat{\mathbf{A}}_{11}, \hat{\mathbf{b}}_1\}$ is completely controllable.

The set of eigenvalues of $\hat{\mathbf{A}}$ is the union of the sets of eigenvalues of $\hat{\mathbf{A}}_{11}$ and $\hat{\mathbf{A}}_{22}$. In view of the form of $\hat{\mathbf{b}}$, it is obvious that the matrix $\hat{\mathbf{A}}_{22}$ is not affected by the introduction of any state feedback of the form $u = \mathbf{k}\hat{\mathbf{x}} + r$. Therefore, the eigenvalues of $\hat{\mathbf{A}}_{22}$ cannot be controlled. On the other hand, since the pair $\{\hat{\mathbf{A}}_{11}, \hat{\mathbf{b}}_1\}$ is controllable, all the eigenvalues of $\hat{\mathbf{A}}_{11}$ can be arbitrarily assigned.

Thus, a linear system which is not completely controllable, can be stabilized by linear state feedback if and only if the uncontrollable system poles are all stable. In this case, the system is said to be *stabilizable.*

Example 9.3: Consider the Inverted Pendulum system discussed in Section 4.6. The state model of this system is (eqn. (4.45a))

$$\dot{\mathbf{x}} = \begin{bmatrix} 0 & 1 & 0 & 0 \\ 0 & 0 & -0.5809 & 0 \\ 0 & 0 & 0 & 1 \\ 0 & 0 & 4.4537 & 0 \end{bmatrix} \mathbf{x} + \begin{bmatrix} 0 \\ 0.9211 \\ 0 \\ -0.3947 \end{bmatrix} u \tag{9.38}$$

This system is unstable. In Example 6.6, we found this system to be completely controllable. Therefore there exists a feedback control law of the form

$$u = \mathbf{kx} + r = [k_1 \quad k_2 \quad k_3 \quad k_4]\mathbf{x} + r \tag{9.39}$$

which will stabilize this system. In the following, we determine such a control law.

The characteristic polynomial of (9.38) is

$$s^4 + \alpha_1 s^3 + \alpha_2 s^2 + \alpha_3 s + \alpha_4$$
$$= s^4 + 0s^3 - 4.4537s^2 + 0s + 0$$

Suppose that we wish to assign all closed-loop poles to the location -1. Then the closed-loop characteristic polynomial should be given by

$$s^4 + \bar{\alpha}_1 s^3 + \bar{\alpha}_2 s^2 + \bar{\alpha}_3 s + \bar{\alpha}_4$$
$$= s^4 + 4s^3 + 6s^2 + 4s + 1$$

Using (9.36), we obtain

$$\bar{\mathbf{k}} = [-1 \quad -4 \quad -10.4537 \quad -4]$$

The controllability matrix for (9.38) is (Example 6.6)

$$\mathbf{U}=\begin{bmatrix} 0 & 0.9211 & 0 & 0.2293 \\ 0.9211 & 0 & 0.2293 & 0 \\ 0 & -0.3947 & 0 & -1.7579 \\ -0.3947 & 0 & -1.7579 & 0 \end{bmatrix}$$

$$\mathbf{U}^{-1}=\begin{bmatrix} 0 & 1.1499 & 0 & -0.15 \\ 1.1499 & 0 & 0.15 & 0 \\ 0 & -0.2582 & 0 & -0.6025 \\ -0.2582 & 0 & -0.6025 & 0 \end{bmatrix}$$

Therefore

$$\mathbf{p}_1=[-0.2582 \quad 0 \quad -0.6025 \quad 0]$$

and the transformation matrix

$$\mathbf{P}=\begin{bmatrix} \mathbf{p}_1 \\ \mathbf{p}_1\mathbf{A} \\ \mathbf{p}_1\mathbf{A}^2 \\ \mathbf{p}_1\mathbf{A}^3 \end{bmatrix}=\begin{bmatrix} -0.2582 & 0 & -0.6025 & 0 \\ 0 & -0.2582 & 0 & -0.6025 \\ 0 & 0 & -2.533 & 0 \\ 0 & 0 & 0 & -2.533 \end{bmatrix}$$

$$\mathbf{k}=\bar{\mathbf{k}}\mathbf{P}$$
$$=[0.2582 \quad 1.0328 \quad 27.0817 \quad 12.542]$$

With these values of k_1, k_2, k_3 and k_4 the feedback control law (9.39) yields a stable closed-loop system.

Since matrix **A** in system (9.38) is a sparse matrix (i.e., with number of zero entries), it is possible to work out the state-feedback control law without transformation to the canonical form. This is left as an exercise for the reader,

Multi-Input Systems

Consider now the general case of an n-dimensional system having several inputs:

$$\dot{\mathbf{x}}=\mathbf{A}\mathbf{x}+\mathbf{B}\mathbf{u} \tag{9.40}$$

The control matrix **B** is of dimension $(n \times p)$, p being the number of inputs. The system is assumed to be controllable. It is desired to obtain a feedback control law

$$\mathbf{u}=\mathbf{K}\mathbf{x}+\mathbf{r} \tag{9.41}$$

so that the matrix $(\mathbf{A}+\mathbf{B}\mathbf{K})$ has desired characteristic polynomial. Two cases can arise:

1. *The system is completely controllable from a single input u_i*

If $\rho[\mathbf{b}_i \quad \mathbf{A}\mathbf{b}_i \quad \dots \quad \mathbf{A}^{n-1}\mathbf{b}_i] = n,$

where $\mathbf{b}_i$ is the ith column of $\mathbf{B}$, then the system is completely controllable from the single input u_i. The method of previous subsection is then directly applicable. $\bar{\mathbf{A}}$ and $\bar{\mathbf{B}}$ will appear in the form

$$\bar{\mathbf{A}} = \begin{bmatrix} 0 & 1 & 0 & \dots & 0 \\ 0 & 0 & 1 & \dots & 0 \\ \vdots & \vdots & \vdots & & \vdots \\ 0 & 0 & 0 & \dots & 1 \\ -\alpha_n & -\alpha_{n-1} & -\alpha_{n-2} & \dots & -\alpha_1 \end{bmatrix}; \quad \bar{\mathbf{B}} = \begin{bmatrix} & & 0 & & \\ & & 0 & & \\ \bar{\mathbf{b}}_1 \; \bar{\mathbf{b}}_2 \dots \bar{\mathbf{b}}_{i-1} & \vdots & & \vdots & \bar{\mathbf{b}}_{i+1} \dots \bar{\mathbf{b}}_p \\ & & 0 & & \\ & & 1 & & \end{bmatrix} \tag{9.42}$$

It will then be sufficient to take $\bar{\mathbf{K}}$ in the form

$$\bar{\mathbf{K}} = \begin{bmatrix} 0 & 0 & \dots & 0 \\ \vdots & \vdots & & \vdots \\ \alpha_n - \bar{\alpha}_n & \alpha_{n-1} - \bar{\alpha}_{n-1} & \dots & \alpha_1 - \bar{\alpha}_1 \\ \vdots & \vdots & & \vdots \\ 0 & 0 & \dots & 0 \end{bmatrix} \longleftarrow i\text{th row} \tag{9.43}$$

in order for $\bar{\mathbf{A}} + \bar{\mathbf{B}}\bar{\mathbf{K}}$ to have the characteristic polynomial

$$s^n + \bar{\alpha}_1 s^{n-1} + \dots + \bar{\alpha}_n$$

2. *The system is controllable from whole set of inputs*

In general, the system will be controllable through the action of several inputs. The algorithm for pole assignment essentially remains the same as in the case of single-input. We first transform (9.40) into the controllable canonical form given by (9.18). From (9.18b) we observe that the only elements of $\bar{\mathbf{A}}$ to be adjusted by feedback matrix $\bar{\mathbf{K}}$ are the (np) significant elements of the p significant rows of $\bar{\mathbf{A}}$. The $p\times n$ feedback matrix $\bar{\mathbf{K}}$ has np elements, the decision about the choice of these elements is unique corresponding to a particular choice of $(\bar{\mathbf{A}}+\bar{\mathbf{B}}\bar{\mathbf{K}})$ which can give the desired characteristic polynomial. Note that because of the particular form of $\bar{\mathbf{B}}$ in (9.18c), the nonsignificant rows of $\bar{\mathbf{A}}$ are not affected by adding $\bar{\mathbf{B}}\bar{\mathbf{K}}$ to it.

Since, for a given characteristic polynomial, many choices of $(\bar{\mathbf{A}}+\bar{\mathbf{B}}\bar{\mathbf{K}})$ in the form (9.18b) are possible, many solutions for the feedback matrix $\bar{\mathbf{K}}$ can be obtained. We will shortly illustrate this point with the help of an example.

An alternative method of pole placement in the case of multi-input system is to introduce a state feedback

$$\mathbf{u} = \mathbf{K}_1\mathbf{x} + \mathbf{r}_1$$

so that the resulting system

$$\dot{\mathbf{x}} = (\mathbf{A} + \mathbf{B}\mathbf{K}_1)\,\mathbf{x} + \mathbf{B}\mathbf{r}_1$$

is controllable by a single component of $\mathbf{r}_1$ (Theorem 9.4), and then apply the result established for a single-input state equation.

Example 9.4: Given the system

$$\dot{\mathbf{x}} = \mathbf{A}\mathbf{x} + \mathbf{B}\mathbf{u} \tag{9.44}$$

where

$$\mathbf{A} = \begin{bmatrix} 1 & 0 & 0 \\ 0 & 2 & 0 \\ 0 & 0 & 3 \end{bmatrix}, \quad \mathbf{B} = \begin{bmatrix} 1 & 0 \\ 0 & 1 \\ 1 & 1 \end{bmatrix};$$

design a linear state variable feedback such that the closed-loop poles are located at -1, -2, and -3.

Solution:

The system of this example is the same as that of Example 9.1, for which the following controllable canonical form was obtained:

$$\dot{\bar{\mathbf{x}}} = \bar{\mathbf{A}}\bar{\mathbf{x}} + \bar{\mathbf{B}}\mathbf{u} \tag{9.45}$$

where

$$\bar{\mathbf{A}} = \left[\begin{array}{cc|c} 0 & 1 & 0 \\ -3 & 4 & 0.5 \\ \hline 0 & 0 & 2 \end{array}\right], \quad \bar{\mathbf{B}} = \left[\begin{array}{cc} 0 & 0 \\ 1 & 0.5 \\ \hline 0 & 1 \end{array}\right]$$

This was done with the transformation

$$\bar{\mathbf{x}} = \mathbf{P}\mathbf{x}$$

where

$$\mathbf{P} = \begin{bmatrix} -0.5 & -0.5 & 0.5 \\ -0.5 & -1 & 1.5 \\ 0 & 1 & 0 \end{bmatrix}$$

The system under consideration is obviously unstable.

The required characteristic polynomial of the compensated system is

$$(s+1)(s+2)(s+3) = s^3 + 6s^2 + 11s + 6 \tag{9.46}$$

Let the compensated system have its evolution matrix $\bar{\mathbf{A}}_{\text{com}}$ in the form

$$\bar{\mathbf{A}}_{\text{com}} = \begin{bmatrix} 0 & 1 & 0 \\ 0 & 0 & 1 \\ -6 & -11 & -6 \end{bmatrix}$$

It can easily be checked that

$$| s\mathbf{I} - \bar{\mathbf{A}}_{\text{com}} | = s^3 + 6s^2 + 11s + 6$$

The feedback matrix $\bar{\mathbf{K}}$ should be such that

$$\bar{\mathbf{A}}_{\text{com}} = \bar{\mathbf{A}} + \bar{\mathbf{B}}\bar{\mathbf{K}}$$

This gives

$$\begin{bmatrix} 0 & 0 \\ 1 & 0.5 \\ 0 & 1 \end{bmatrix} \begin{bmatrix} \bar{k}_{11} & \bar{k}_{12} & \bar{k}_{13} \\ \bar{k}_{21} & \bar{k}_{22} & \bar{k}_{23} \end{bmatrix}$$

$$= \begin{bmatrix} 0 & 1 & 0 \\ 0 & 0 & 1 \\ -6 & -11 & -6 \end{bmatrix} - \begin{bmatrix} 0 & 1 & 0 \\ -3 & 4 & 0.5 \\ 0 & 0 & 2 \end{bmatrix}$$

The unique solution of the above equation is

$$\bar{\mathbf{K}} = \begin{bmatrix} 6 & 1.5 & 4.5 \\ -6 & -11 & -8 \end{bmatrix}$$

The state-feedback control law for the system (9.44) becomes

$$\mathbf{u} = \mathbf{K}\mathbf{x} + \mathbf{r} \tag{9.47}$$

where

$$\mathbf{K} = \bar{\mathbf{K}}\mathbf{P}$$

$$= \begin{bmatrix} 6 & 1.5 & 4.5 \\ -6 & -11 & -8 \end{bmatrix} \begin{bmatrix} -0.5 & -0.5 & 0.5 \\ -0.5 & -1 & 1.5 \\ 0 & 1 & 0 \end{bmatrix}$$

$$= \begin{bmatrix} -3.75 & 0 & 5.25 \\ 8.5 & 6 & -19.5 \end{bmatrix}$$

We can consider different forms for $\bar{\mathbf{A}}_{\text{com}}$, having the same characteristic Polynomial given by (9.46). Take, for example,

$$\bar{\mathbf{A}}_{\text{com}} = \begin{bmatrix} 0 & 1 & 0 \\ -18 & -7 & 1 \\ -24 & 0 & 1 \end{bmatrix}$$

Using the procedure given above, we obtain the following value of **K** matrix for the feedback control law (9.47).

$$\mathbf{K} = \begin{bmatrix} 7 & 13.5 & -18 \\ 12 & 11 & -12 \end{bmatrix}$$

□

A careful study of the following points regarding the pole placement problem is very necessary to get a feel of the powers of the linear state variable feedback control law derived in this section. Detailed discussion of these topics could not be included in this book for want of space.

Zeros of Multivariable Systems

In Problem 9.5, the reader is asked to show that in single-input/single-output systems, while the poles are shifted by state feedback, the zeros remain unchanged after the introduction of state feedback. This is, however, not so in the multivariable case; the numerators of some elements of the transfer function matrix will be changed after introducing state feedback. The zeros can have a profound effect upon the shape of the transient response. Thus the placement of closed-loop poles does not guarantee the desired transient behaviour. Zeros of multivariable systems are the subject of a good deal of current research, though there is not even complete agreement on the basic definition, particularly when the input and output vectors have different dimensions (Kwakernaak and Sivan 1972, Wolovich 1973b, Davison and Wang 1974, Desoer and Schulman 1974, Patel 1975).

Extra Degrees of Freedom

We observed that the state feedback control law for pole placement is unique for single-input systems. For a multi-input system, however, there will be many control laws which achieve the same pole configuration, We have illustrated this point in Example 9.4 by choosing different $\bar{A}_{com}$ matrices for the same characteristic polynomial.

In Section 9.2, we pointed out the fact that many controllable companion forms are possible for multi-input models. Some companion forms give the designer free parameters after achieving the desired pole positions. This indicates that apart from closed-loop pole assignment, a state feedback could satisfy additional performance requirements. Some possible forms of additional requirements are sensitivity minimization, minimization of quadratic performance index (Chapter 11) etc. Shankar and Ramar (1976) and Gomathi et al. (1980) deal with the problem of minimization of eigenvalue sensitivity. Various optimization algorithms for minimizing a quadratic performance index in addition to satisfying arbitrary pole assignment have been reported (Anderson and Moore 1969, Vegte and Maki 1973, Maki and Vegte 1974, Solheim 1972, Pugh and Shelton 1975, Sebakhy and Sorial 1979, Abdel-

Moneim 1980). We shall discuss the result by Anderson and Moore in Chapter 11.

Noninteracting Control

Consider the p-input, p-output system

$$\dot{\mathbf{x}} = \mathbf{A}\mathbf{x} + \mathbf{B}\mathbf{u};\ \mathbf{x}(0) = \mathbf{0}$$

$$\mathbf{y} = \mathbf{C}\mathbf{x}$$

The transfer function matrix of this system is

$$\hat{\mathbf{H}}(s) = \mathbf{C}(s\mathbf{I}-\mathbf{A})^{-1}\mathbf{B}$$

$$= \begin{bmatrix} \hat{h}_{11}(s) & \hat{h}_{12}(s) & \dots & \hat{h}_{1p}(s) \\ \vdots & \vdots & & \vdots \\ \hat{h}_{p1}(s) & \hat{h}_{p2}(s) & \dots & \hat{h}_{pp}(s) \end{bmatrix}$$

The input-output relations are

$$\hat{y}_1(s) = \hat{h}_{11}(s)\ \hat{u}_1(s) + \dots + \hat{h}_{1p}(s)\ \hat{u}_p(s)$$
$$\hat{y}_2(s) = \hat{h}_{21}(s)\ \hat{u}_1(s) + \dots + \hat{h}_{2p}(s)\ \hat{u}_p(s)$$
$$\vdots$$
$$\hat{y}_p(s) = \hat{h}_{p1}(s)\ \hat{u}_1(s) + \dots + \hat{h}_{pp}(s)\ \hat{u}_p(s)$$

These equations are obviously coupled, since each input influences all of the outputs. Because of this *coupling* (*interaction*), it is generally difficult to control a multivariable system. For example, if we want to control $\hat{y}_1(s)$ without affecting the other outputs, one has to simultaneously manipulate all the inputs and the required manipulation cannot be readily found. We would like to introduce a compensator so that the multivariable system becomes *decoupled* (*noninteracting*) in the sense that every input controls only one output and every output is controlled by only one input.

It is obvious that the transfer function matrix of the decoupled system will be diagonal. Use of state feedback in decoupling a system and stabilizing the decoupled system has been extensively studied (Falb and Wolovich 1967, Gilbert 1969, Morse and Wonham 1970, Wonham and Morse 1970, Morse and Wonham 1971, Wolovich 1973a).

Incomplete State Feedback

In our discussion, we have assumed that feedback would be applied from the state to the input. Often, however, all the states will not be available. In such cases we may reconstruct the state using an observer (discussed in the next section). An alternative approach is to design a feedback control law of the form

$$\mathbf{u} = \mathbf{K}\mathbf{y} + \mathbf{r}$$

for pole placement. The pioneering work in this direction is by Davison (1970) and Jameson (1970). They proved that a constant output feedback controller can always be found so that $q(q =$ number of independent outputs) out of the $n(n =$ number of state variables) poles of closed-loop system can be assigned at specified locations. However, nothing is said about the remaining $(n-q)$ poles. A number of extensions of this result are now available in the literature. Davison and Wang (1975) gave a further result which places min $\{n, p+q-1\}$ poles ($p =$ number of inputs) arbitrarily close to any preassigned symmetric configuration. Seraji (1975) gives a method for calculating output feedback matrix for pole assignment using dyadic structure.

9.5 FULL-ORDER OBSERVERS

Consider the plant model of an nth order linear time-invariant system, given by

$$\dot{\mathbf{x}} = \mathbf{A}\mathbf{x} + \mathbf{B}\mathbf{u}; \quad \mathbf{x}(0) = \mathbf{x}^0 \tag{9.48a}$$

where $\mathbf{A}$ and $\mathbf{B}$ are, respectively, $n \times n$ and $n \times p$ real constant matrices. The measurement model may be written as

$$\mathbf{y} = \mathbf{C}\mathbf{x} \tag{9.48b}$$

where $\mathbf{C}$ is $q \times n$ real constant matrix. Without loss of generality, the direct transmission part has been assumed to be zero. Further, we assume that (9.48) is observable.

Figure 9.3 shows the schematic of a Luenberger observer (Luenberger 1966, 1971). The observer is driven by the input as well as the output of the original system. The output $\mathbf{y} = \mathbf{C}\mathbf{x}$ (eqn. (9.48b)) is compared with $\hat{y} = \mathbf{C}\hat{\mathbf{x}}$ and this difference is used to serve as a correcting term. The difference $\tilde{\mathbf{y}}(t) = \mathbf{y}(t) - \hat{\mathbf{y}}(t)$ is multiplied by an $n \times q$ real constant matrix $\mathbf{M}$ and fed into the input of the integrators of the observer. This observer will be called the *asymptotic estimator* for reasons to be seen later.

The dynamical equation of the asymptotic estimator shown in Fig. 9.3, is given by

$$\dot{\hat{\mathbf{x}}}(t) = \mathbf{A}\hat{\mathbf{x}}(t) + \mathbf{B}\mathbf{u}(t) - \mathbf{M}\tilde{\mathbf{y}}(t) \tag{9.49}$$

The state error vector

$$\tilde{\mathbf{x}}(t) = \mathbf{x}(t) - \hat{\mathbf{x}}(t)$$

Differentiating both sides, we get

$$\dot{\tilde{\mathbf{x}}}(t) = \dot{\mathbf{x}}(t) - \dot{\hat{\mathbf{x}}}(t)$$

Substituting for $\dot{\mathbf{x}}(t)$ and $\dot{\hat{\mathbf{x}}}(t)$ from eqns. (9.48a) and (9.49) respectively, we obtain

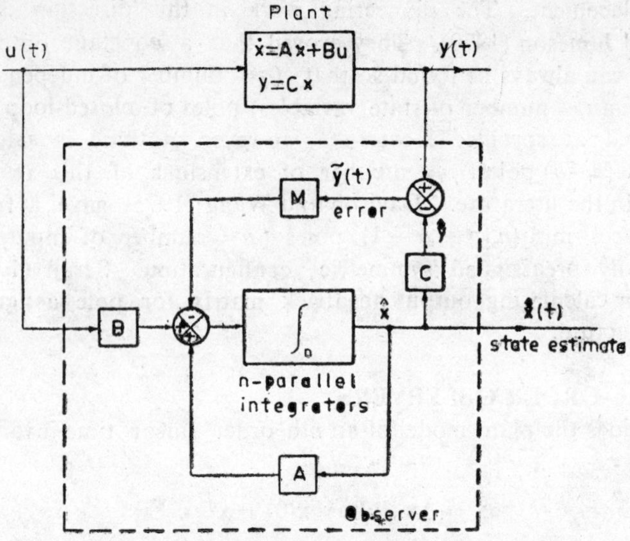

Fig 9.3 Full-order state observer

$$\dot{\tilde{\mathbf{x}}}(t) = \mathbf{Ax}(t) + \mathbf{Bu}(t) - \mathbf{A}\hat{\mathbf{x}}(t) - \mathbf{Bu}(t) + \mathbf{MC}(\mathbf{x}-\hat{\mathbf{x}})$$

$$= \mathbf{A}(\mathbf{x}-\hat{\mathbf{x}}) + \mathbf{MC}(\mathbf{x}-\hat{\mathbf{x}})$$

$$= (\mathbf{A} + \mathbf{MC})\,\tilde{\mathbf{x}}(t) \qquad (9.50)$$

Thus, the error in state vector is given by

$$\tilde{\mathbf{x}}(t) = \exp\,[(\mathbf{A} + \mathbf{MC})\,t]\,\tilde{\mathbf{x}}(0) \qquad (9.51)$$

and is independent of the applied control.

The error $\tilde{\mathbf{x}}$ will decay to $\mathbf{0}$ if $\mathbf{M}$ is chosen such that (9.50) is asymptotically stable, i.e., all the eigenvalues of the matrix $(\mathbf{A} + \mathbf{MC})$ lie in the left-half plane. It can be easily proved that if (9.48) is completely observable, the matrix $\mathbf{M}$ may be chosen so as to place the eigenvalues of $(\mathbf{A} + \mathbf{MC})$ in any desired configuration (subject to conjugate pairing).

By the assumption of observability of (9.48), the pair $\{\mathbf{A}, \mathbf{C}\}$ is observable. Then by duality, the pair $\{\mathbf{A}^T, \mathbf{C}^T\}$ is controllable. Using the result of Section 9.4, we say that eigenvalues of $(\mathbf{A}^T + \mathbf{C}^T\,\mathbf{K})$ or $(\mathbf{A}^T + \mathbf{C}^T\,(\mathbf{M}^T\,))$ can be arbitrarily assigned. Since the eigenvalues do not change under transpose, we may say that the eigenvalues of $[\mathbf{A}^T + \mathbf{C}^T\,\mathbf{M}^T\,]^T$ or $(\mathbf{A} + \mathbf{MC})$ can be arbitrarily assigned.

Thus, if the system (9.48) is observable, its state can be estimated with an n-dimensional observer of the form

$$\dot{\hat{\mathbf{x}}} = (\mathbf{A} + \mathbf{MC})\,\hat{\mathbf{x}} + \mathbf{Bu} - \mathbf{My}\,; \quad \hat{\mathbf{x}}(0) = \hat{\mathbf{x}}^0 \qquad (9.52)$$

as shown in Fig. 9.3. The matrix $\mathbf{M}$ may be chosen so as to place the eigenvalues of (9.52) in any desired configuration. (In the next section, we shall study the possibility of reducing the dimension of the observer.) It may be noted that if the eigenvalues of $(\mathbf{A}+\mathbf{MC})$ have negative real parts, then no matter what $\hat{\mathbf{x}}^0$ is, $\hat{\mathbf{x}}$ will approach $\mathbf{x}$ asymptotically. For an asymptotic estimator (9.52), there is no need of setting an initial state, because no matter what the initial state is, the estimator output will tend to the real state.

Example 9.5: In Example 9.3, we obtained a linear state feedback control law for stabilizing the Inverted Pendulum system. To implement this control law, all the four states: $x_1 = z(t)$, $x_2 = \dot{z}(t)$, $x_3 = \phi(t)$, $x_4 = \dot{\phi}(t)$ (refer Section 4.6) must be available for feedback.

Suppose that the only variable available for measurement is $z(t)$, the position of the cart. The linearized equations governing this system are

$$\dot{\mathbf{x}} = \begin{bmatrix} 0 & 1 & 0 & 0 \\ 0 & 0 & -0.5809 & 0 \\ 0 & 0 & 0 & 1 \\ 0 & 0 & 4.4537 & 0 \end{bmatrix} \mathbf{x} + \begin{bmatrix} 0 \\ 0.9211 \\ 0 \\ -0.3947 \end{bmatrix} u$$

$$= \mathbf{Ax} + \mathbf{b}u \tag{9.53a}$$

$$y = [1 \quad 0 \quad 0 \quad 0]\,\mathbf{x}$$

$$= \mathbf{cx} \tag{9.53b}$$

In Example 6.7, we had established that this system is completely observable. In the following, we design a full-order observer for this system.

Following the algorithm given in Section 9.4, we shall first obtain the vector $\mathbf{k}$ such that the eigenvalues of $(\mathbf{A}^T + \mathbf{c}^T\mathbf{k})$ are placed at the desired locations.

The characteristic polynomial of (9.53) is

$$s^4 - 4.4537s^2 = s^4 + \alpha_1 s^3 + \alpha_2 s^2 + \alpha_3 s + \alpha_4 \tag{9.54a}$$

Therefore the eigenvalues of matrix $\mathbf{A}$ are located at 0, 0, -2.11 and 2.11.

Suppose the observer is required to have the poles located at

$$-2, -2 \pm j1, -3$$

The corresponding characteristic polynomial is

$$s^4 + 9s^3 + 31s^2 + 49s + 30 = s^4 + \bar{\alpha}_1 s^3 + \bar{\alpha}_2 s^2 + \bar{\alpha}_3 s + \bar{\alpha}_4 \tag{9.54b}$$

The vector $\mathbf{k}$ that makes $(\mathbf{A}^T + \mathbf{c}^T\mathbf{k})$ to have this characteristic polynomial is derived in the following:

$$\mathbf{A}^T = \begin{bmatrix} 0 & 0 & 0 & 0 \\ 1 & 0 & 0 & 0 \\ 0 & -0.5809 & 0 & 4.4537 \\ 0 & 0 & 1 & 0 \end{bmatrix};$$

$$\mathbf{c}^T = \begin{bmatrix} 1 \\ 0 \\ 0 \\ 0 \end{bmatrix}$$

$$\mathbf{U} = [\mathbf{c}^T \quad \mathbf{A}^T\mathbf{c}^T \quad (\mathbf{A}^T)^2\mathbf{c}^T \quad (\mathbf{A}^T)^3\mathbf{c}^T]$$

$$= \begin{bmatrix} 1 & 0 & 0 & 0 \\ 0 & 1 & 0 & 0 \\ 0 & 0 & -0.5809 & 0 \\ 0 & 0 & 0 & -0.5809 \end{bmatrix}$$

The last row of $\mathbf{U}^{-1}$ is given by

$$\mathbf{p}_1 = [0 \quad 0 \quad 0 \quad -1.7215]$$

$$\mathbf{P} = \begin{bmatrix} \mathbf{p}_1 \\ \mathbf{p}_1\mathbf{A}^T \\ \mathbf{p}_1(\mathbf{A}^T)^2 \\ \mathbf{p}_1(\mathbf{A}^T)^3 \end{bmatrix} = \begin{bmatrix} 0 & 0 & 0 & -1.7215 \\ 0 & 0 & -1.7215 & 0 \\ 0 & 1 & 0 & -7.667 \\ 1 & 0 & -7.667 & 0 \end{bmatrix} \tag{9.54c}$$

From (9.54a) and (9.54b), we obtain

$$\bar{\mathbf{k}} = [\alpha_4 - \bar{\alpha}_4 \quad \alpha_3 - \bar{\alpha}_3 \quad \alpha_2 - \bar{\alpha}_2 \quad \alpha_1 - \bar{\alpha}_1]$$

$$= [-30 \quad -49 \quad -35.4537 \quad -9]$$

$$\mathbf{k} = \bar{\mathbf{k}}\mathbf{P} = [-9 \quad -35.4537 \quad 153.3565 \quad 323.4685]$$

The observer is therefore described by the equation (refer eqn. (9.52))

$$\dot{\hat{\mathbf{x}}} = (\mathbf{A} + \mathbf{mc})\,\hat{\mathbf{x}} + \mathbf{b}u - \mathbf{m}y$$

where

$$\mathbf{m} = \mathbf{k}^T$$

Substituting the values, we get

$$\dot{\hat{\mathbf{x}}} = \begin{bmatrix} -9 & 1 & 0 & 0 \\ -35.4537 & 0 & -0.5809 & 0 \\ 153.3565 & 0 & 0 & 1 \\ 323.4685 & 0 & 4.4537 & 0 \end{bmatrix} \hat{\mathbf{x}}$$

$$+ \begin{bmatrix} 0 & 9 \\ 0.9211 & 35.4537 \\ 0 & -153.3565 \\ -0.3947 & -323.4685 \end{bmatrix} \begin{bmatrix} u \\ y \end{bmatrix} \tag{9.55}$$

The observer (9.55) will process the cart position $z(t) = y(t)$ and the input $u(t)$ to continuously provide an estimate $\hat{\mathbf{x}}(t)$ of the entire state vector. Errors in the estimate will decay at least as fast as e^{-2t}.

The reader is advised to work out **m** vector without transformation to the canonical form.

□

The choice of observer poles is completely arbitrary in principle. However, some caution must be used in drawing conclusions about the resulting transient behaviour. One reason for caution is that we have not taken into account the zeros of the system and these can have a profound effect on the shape of the transient response.

To obtain a fast convergence of the estimation error $\dot{\tilde{\mathbf{x}}}$ to zero, we may be tempted to choose **M** so that the observer poles are quite deep in the left-half plane. This, however, must generally be achieved by making the gain matrix **M** quite large, which in turn makes the observer very sensitive to any observation noise. In short, what the best poles of an observer are is not known at present.

The Separation Principle

In section 9.4, we observed that state feedback control law can be used on a linear system to place its closed loop poles in any desired configuration. If the entire state is not available for feedback, it looks reasonable to estimate the state employing an observer and using the estimate in the control law. In this subsection, we consider control systems requiring both the feedback control and the state estimation.

Consider a controllable and observable time-invariant n-dimensional system

$$\dot{\mathbf{x}} = \mathbf{A}\mathbf{x} + \mathbf{B}\mathbf{u} \tag{9.56a}$$

$$\mathbf{y} = \mathbf{C}\mathbf{x} \tag{9.56b}$$

It is assumed that a control law of the form

$$\mathbf{u} = \mathbf{K}\mathbf{x} + \mathbf{r} \tag{9.57}$$

has been found to place the poles of the closed-loop system in any desired configuration. If the state is not directly available for measurement, we propose to construct an observer of the form

$$\dot{\hat{\mathbf{x}}} = (\mathbf{A} + \mathbf{MC})\,\hat{\mathbf{x}} + \mathbf{Bu} - \mathbf{My} \tag{9.58}$$

and interconnect the control law with the reconstructed state $\hat{\mathbf{x}}$:

$$\mathbf{u} = \mathbf{K}\hat{\mathbf{x}} + \mathbf{r} \tag{9.59}$$

Figure 9.4 depicts the interconnection of the plant, the observer and the control law. For the purpose of analysis, we may look upon the plant (9.56) and observer (9.58) as a composite system of dimension $2n$,

$$\begin{bmatrix} \dot{\mathbf{x}} \\ \dot{\hat{\mathbf{x}}} \end{bmatrix} = \begin{bmatrix} \mathbf{A} & \mathbf{BK} \\ -\mathbf{MC} & \mathbf{A}+\mathbf{MC}+\mathbf{BK} \end{bmatrix} \begin{bmatrix} \mathbf{x} \\ \hat{\mathbf{x}} \end{bmatrix} + \begin{bmatrix} \mathbf{B} \\ \mathbf{B} \end{bmatrix} \mathbf{r} \tag{9.60}$$

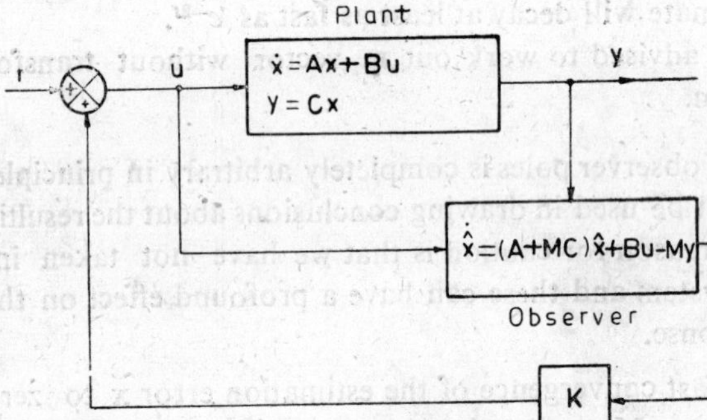

Fig. 9.4 Use of observer to implement state feedback control law

By the equivalence transformation

$$\begin{bmatrix} \mathbf{x} \\ \tilde{\mathbf{x}} \end{bmatrix} = \begin{bmatrix} \mathbf{I} & \mathbf{0} \\ \mathbf{I} & -\mathbf{I} \end{bmatrix} \begin{bmatrix} \mathbf{x} \\ \hat{\mathbf{x}} \end{bmatrix} = \begin{bmatrix} \mathbf{x} \\ \mathbf{x} - \hat{\mathbf{x}} \end{bmatrix},$$

eqn. (9.60) becomes

$$\begin{bmatrix} \dot{\mathbf{x}} \\ \dot{\tilde{\mathbf{x}}} \end{bmatrix} = \begin{bmatrix} \mathbf{A}+\mathbf{BK} & -\mathbf{BK} \\ \mathbf{0} & \mathbf{A}+\mathbf{MC} \end{bmatrix} \begin{bmatrix} \mathbf{x} \\ \tilde{\mathbf{x}} \end{bmatrix} + \begin{bmatrix} \mathbf{B} \\ \mathbf{0} \end{bmatrix} \mathbf{r} \tag{9.61}$$

The eigenvalues of (9.61), which are the eigenvalues of (9.60), are the zeros of

$$\det \begin{bmatrix} s\mathbf{I} - (\mathbf{A}+\mathbf{BK}) & \mathbf{BK} \\ \mathbf{0} & s\mathbf{I} - (\mathbf{A}+\mathbf{MC}) \end{bmatrix}$$

$$= \det\,(s\mathbf{I} - (\mathbf{A}+\mathbf{BK}))\,\det\,(s\mathbf{I} - (\mathbf{A}+\mathbf{MC})) \tag{9.62}$$

Consequently, the set of closed-loop eigenvalues comprises the eigenvalues of

$\mathbf{A}+\mathbf{BK}$ and the eigenvalues of $\mathbf{A}+\mathbf{MC}$. Thus, we can consider the problem of determining a stable observer and the state feedback control law separately, since their interconnection results in a stable control system. This property is often called the separation principle.

9.6 REDUCED-ORDER OBSERVERS

The state observer discussed in the earlier section, was derived by setting up a model of the plant and feeding back a 'correction term' proportional to the difference between the actual and estimated outputs. Such an observer contains redundancy because q state variables can be directly obtained from q outputs which are available for measurement and need not be estimated. The remaining $(n-q)$ state variables can be estimated using an observer of order $(n-q)$, as will be seen below.

Let the given observable plant be

$$\dot{\mathbf{x}} = \mathbf{A}\mathbf{x} + \mathbf{B}\mathbf{u} \tag{9.63a}$$

$$\mathbf{y} = \mathbf{C}\mathbf{x} \tag{9.63b}$$

where A and C are in the multivariable observable companion form given by (9.26). Assuming C to be of full rank, there will be q observability indices, $\mu_1, \mu_2, \ldots, \mu_q$. Thus A and C each have q significant columns $\sigma_k = \sum_{i=1}^{k} \mu_i$ for $k = 1, 2, \ldots, q$.

Let

$$\mathbf{x}_2 \triangleq \begin{bmatrix} x_{\sigma_1} \\ x_{\sigma_2} \\ \vdots \\ x_{\sigma_q} \end{bmatrix} \text{ and } \mathbf{x}_1 \triangleq \begin{bmatrix} x_1 \\ \vdots \\ x_{\sigma_1-1} \\ x_{\sigma_1+1} \\ \vdots \\ x_{\sigma_q-1} \end{bmatrix} \tag{9.64}$$

i.e., the state vector is partitioned into two groups.

The system (9.63), in the observable companion form (9.26), may be rearranged as follows:

$$\begin{bmatrix} \dot{\mathbf{x}}_1 \\ \hline \dot{\mathbf{x}}_2 \end{bmatrix} = \left[\begin{array}{c|c} \mathbf{A}_{11} & \mathbf{A}_{12} \\ \hline \mathbf{A}_{21} & \mathbf{A}_{22} \end{array}\right] \begin{bmatrix} \mathbf{x}_1 \\ \hline \mathbf{x}_2 \end{bmatrix} + \begin{bmatrix} \mathbf{B}_1 \\ \hline \mathbf{B}_2 \end{bmatrix} \mathbf{u} \tag{9.65a}$$

$$\mathbf{y} = [\, \mathbf{0} \mid \mathbf{C}_q \,] \begin{bmatrix} \mathbf{x}_1 \\ \hline \mathbf{x}_2 \end{bmatrix} \tag{9.65b}$$

The q state variables $\mathbf{x}_2$ can be directly obtained from (9.65b) as

$$\mathbf{x}_2 = [\, \mathbf{C}_q \,]^{-1} \mathbf{y} \tag{9.66}$$

The remaining $(n-q)$ state variables require an observer for estimation.

By manipulating the equations in (9.65), $\mathbf{x}_1$ may be viewed as the state of an $(n-q)$ dimensional subsystem

$$\dot{\mathbf{x}}_1 = \mathbf{A}_{11}\mathbf{x}_1 + \mathbf{A}_{12}\mathbf{x}_2 + \mathbf{B}_1\mathbf{u}$$

$$= \mathbf{A}_{11}\mathbf{x}_1 + \mathbf{v} \tag{9.67}$$

$$\mathbf{z} = \mathbf{A}_{21}\mathbf{x}_1 \tag{9.68}$$

where $\mathbf{v} = \mathbf{A}_{12}\mathbf{x}_2 + \mathbf{B}_1\mathbf{u}$

$$= \mathbf{A}_{12}\,[\mathbf{C}_q]^{-1}\,\mathbf{y} + \mathbf{B}_1\mathbf{u} \tag{9.69}$$

$\mathbf{v}$ can be treated as a known input since $\mathbf{u}$ is known and $\mathbf{y}$ is directly measurable.

The 'output vector' $\mathbf{z}$ may be expressed as

$$\mathbf{z} = \mathbf{A}_{21}\mathbf{x}_1$$

$$= \dot{\mathbf{x}}_2 - \mathbf{A}_{22}\mathbf{x}_2 - \mathbf{B}_2\mathbf{u}$$

$$= [\mathbf{C}_q]^{-1}\,\dot{\mathbf{y}} - \mathbf{A}_{22}\,[\mathbf{C}_q]^{-1}\,\mathbf{y} - \mathbf{B}_2\mathbf{u} \tag{9.70}$$

We can estimate $\mathbf{x}_1$ with an observer (refer eqn. (9.52))

$$\dot{\hat{\mathbf{x}}}_1 = (\mathbf{A}_{11} + \mathbf{M}_1\mathbf{A}_{21})\,\hat{\mathbf{x}}_1 + \mathbf{v} - \mathbf{M}_1\mathbf{z} \tag{9.71}$$

where the $(n-q)\times q$ matrix $\mathbf{M}_1$ may be chosen so as to place the poles of (9.71) in any desired configuration. Substituting for $\mathbf{v}$ and $\mathbf{z}$ from (9.69) and (9.70) respectively, we have

$$\dot{\hat{\mathbf{x}}}_1 = (\mathbf{A}_{11} + \mathbf{M}_1\mathbf{A}_{21})\,\hat{\mathbf{x}}_1 + \mathbf{A}_{12}\,[\mathbf{C}_q]^{-1}\,\mathbf{y} + \mathbf{B}_1\mathbf{u}$$

$$-\mathbf{M}_1\,[\mathbf{C}_q]^{-1}\,\dot{\mathbf{y}} + \mathbf{M}_1\mathbf{A}_{22}\,[\mathbf{C}_q]^{-1}\,\mathbf{y} + \mathbf{M}_1\mathbf{B}_2\mathbf{u} \tag{9.72}$$

which is an $(n\times q)$ dimensional observer for the system (9.63).

The only apparent difficulty in implementing the observer (9.72) is that differentiation of the output $\mathbf{y}$ is required. This can be avoided by redefining the state of the observer to be

$$\bar{\mathbf{x}}(t) = \hat{\mathbf{x}}_1(t) + \mathbf{M}_1\,[\mathbf{C}_q]^{-1}\,\mathbf{y}(t) \tag{9.73}$$

Substituting (9.73) in (9.72), we get

$$\dot{\bar{\mathbf{x}}} = (\mathbf{A}_{11} + \mathbf{M}_1\mathbf{A}_{21})\,\bar{\mathbf{x}} + (\mathbf{B}_1 + \mathbf{M}_1\mathbf{B}_2)\,\mathbf{u}$$

$$+ [\mathbf{A}_{12} + \mathbf{M}_1\mathbf{A}_{22} - (\mathbf{A}_{11} + \mathbf{M}_1\mathbf{A}_{21})\,\mathbf{M}_1]\,[\mathbf{C}_q]^{-1}\,\mathbf{y} \tag{9.74}$$

The estimate of the full state $\mathbf{x}$ is given by

$$\hat{\mathbf{x}} = \left[\begin{array}{c} \hat{\mathbf{x}}_1 \\ \hline [\mathbf{C}_q]^{-1}\mathbf{y} \end{array}\right] = \left[\begin{array}{c} \bar{\mathbf{x}} - \mathbf{M}_1[\mathbf{C}_q]^{-1}\mathbf{y} \\ \hline [\mathbf{C}_q]^{-1}\mathbf{y} \end{array}\right]$$

$$= \left[\begin{array}{c} \mathbf{I} \\ \hline \mathbf{0} \end{array}\right]\bar{\mathbf{x}} + \left[\begin{array}{c} -\mathbf{M}_1 \\ \hline \mathbf{I} \end{array}\right][\mathbf{C}_q]^{-1}\mathbf{y} \tag{9.75}$$

Since the reduced order observer has a direct link from the observed variable $\mathbf{y}(t)$ to the estimated state $\hat{\mathbf{x}}(t)$, the estimate $\hat{\mathbf{x}}(t)$ will be more sensitive to measurement errors in $\mathbf{y}(t)$ than the estimate generated by a full-order observer. This is because the noise bypasses the natural filtering action of the observer dynamics. If output noise appears to be a significant problem, then a full-order observer is probably needed with its gain matrix chosen via the Kalman filter theory (Chapter 12).

It may be verified that the separation principle holds for observer of any dimension.

Example 9.6: In Example 9.5, an observer of dimension 4 was constructed for Inverted Pendulum system with four states and one output. However, this would just as well be done with an observer of dimension $n - q = 4 - 1 = 3$. In the following we construct such an observer. We shall follow the procedure outlined earlier in this section, though for this particular case of Inverted Pendulum, it is possible to work out the observer model without transformation to canonical form (Problem 9.15).

The state model of Inverted Pendulum, reproduced from (9.53), is given by

$$\dot{\mathbf{x}} = \begin{bmatrix} 0 & 1 & 0 & 0 \\ 0 & 0 & -0.5809 & 0 \\ 0 & 0 & 0 & 1 \\ 0 & 0 & 4.4537 & 0 \end{bmatrix}\mathbf{x} + \begin{bmatrix} 0 \\ 0.9211 \\ 0 \\ -0.3947 \end{bmatrix}u$$

$$= \mathbf{A}\mathbf{x} + \mathbf{b}u$$

$$\mathbf{y} = [1 \quad 0 \quad 0 \quad 0]\,\mathbf{x}$$

$$= \mathbf{c}\mathbf{x}$$

Using the transformation

$$\bar{\mathbf{x}} = \mathbf{Q}\mathbf{x}, \tag{9.76}$$

this model is transformed to the following observable companion form:

$$\dot{\bar{\mathbf{x}}} = \mathbf{Q}\mathbf{A}\mathbf{Q}^{-1}\bar{\mathbf{x}} + \mathbf{Q}\mathbf{b}u$$

$$= \bar{\mathbf{A}}\bar{\mathbf{x}} + \bar{\mathbf{b}}u \tag{9.77a}$$

$$y = \mathbf{c}\mathbf{Q}^{-1}\overline{\mathbf{x}}$$
$$= \overline{\mathbf{c}}\,\overline{\mathbf{x}} \tag{9.77b}$$

where

$$\overline{\mathbf{A}} = \left[\begin{array}{ccc|c} 0 & 0 & 0 & 0 \\ 1 & 0 & 0 & 0 \\ 0 & 1 & 0 & 4.4537 \\ \hline 0 & 0 & 1 & 0 \end{array}\right]; \quad \overline{\mathbf{b}} = \begin{bmatrix} -3.873 \\ 0 \\ 0.9211 \\ \hline 0 \end{bmatrix}$$

$$\overline{\mathbf{c}} = [\begin{array}{ccc|c} 0 & 0 & 0 & 1 \end{array}]$$

$$\mathbf{Q} = (\mathbf{P}^T)^{-1}$$

$$= \begin{bmatrix} 0 & -4.4537 & 0 & -0.5809 \\ -4.4537 & 0 & -0.5809 & 0 \\ 0 & 1 & 0 & 0 \\ 1 & 0 & 0 & 0 \end{bmatrix}$$

This is as per eqn. (9.15); the matrix **P** is given by eqn. (9.54c).

Matrices $\overline{\mathbf{A}}$ and $\overline{\mathbf{c}}$ in (9.76) have been partitioned, giving

$$\mathbf{A}_{11} = \begin{bmatrix} 0 & 0 & 0 \\ 1 & 0 & 0 \\ 0 & 1 & 0 \end{bmatrix}; \quad \mathbf{A}_{21} = [\,0 \quad 0 \quad 1\,]$$

$$\mathbf{A}_{12} = \begin{bmatrix} 0 \\ 0 \\ 4.4537 \end{bmatrix}; \quad \mathbf{A}_{22} = 0\,; \quad c_q = 1$$

$$\mathbf{b}_1 = \begin{bmatrix} -3.873 \\ 0 \\ 0.9211 \end{bmatrix}; \quad b_2 = 0$$

Let us select a vector **m** such that the eigenvalues of the matrix $(\mathbf{A}_{11} + \mathbf{m}\mathbf{A}_{21})$ are at the locations, say $-2 \pm j1$, -3, i.e., the characteristic polynomial of $(\mathbf{A}_{11} + \mathbf{m}\mathbf{A}_{21})$ is

$$(s + 2 + j1)\,(s + 2 - j1)\,(s + 3)$$
$$= s^3 + 7s^2 + 17s + 15$$

The vector **m** is easily found to be

$$\mathbf{m} = \begin{bmatrix} -15 \\ -17 \\ -7 \end{bmatrix}$$

From eqns. (9.74)-(9.75), we obtain the following observer model:

$$\hat{\mathbf{x}} = \begin{bmatrix} \mathbf{I} \\ \hline \mathbf{0} \end{bmatrix} \mathbf{w} + \begin{bmatrix} -\mathbf{m} \\ \hline 1 \end{bmatrix} c_q^{-1} y$$

$$\begin{aligned} \dot{\mathbf{w}} = & (\mathbf{A}_{11} + \mathbf{m}\mathbf{A}_{21})\, \mathbf{w} + (\mathbf{b}_1 + \mathbf{m}b_2)\, u \\ & + [\mathbf{A}_{12} + \mathbf{m}\mathbf{A}_{22} - (\mathbf{A}_{11} + \mathbf{m}\mathbf{A}_{21})\, \mathbf{m}]\, c_q^{-1} y \end{aligned}$$

Substituting the values, we obtain

$$\hat{\mathbf{x}} = \begin{bmatrix} 1 & 0 & 0 \\ 0 & 1 & 0 \\ 0 & 0 & 1 \\ 0 & 0 & 0 \end{bmatrix} \mathbf{w} + \begin{bmatrix} 15 \\ 17 \\ 7 \\ 1 \end{bmatrix} y \qquad (9.78\text{a})$$

$$\dot{\mathbf{w}} = \begin{bmatrix} 0 & 0 & -15 \\ 1 & 0 & -17 \\ 0 & 1 & -7 \end{bmatrix} \mathbf{w} + \begin{bmatrix} -3.873 \\ 0 \\ 0.9211 \end{bmatrix} u + \begin{bmatrix} -105 \\ -104 \\ -27.5463 \end{bmatrix} y \qquad (9.78\text{b})$$

Using transformation (9.76), we obtain $\hat{\mathbf{x}}$ as

$$\hat{\mathbf{x}} = \mathbf{Q}^{-1} \hat{\mathbf{x}}$$

The observer given by eqns. (9.78) will provide a continuous estimate $\hat{\mathbf{x}}$ of the state vector with estimation errors decaying at least as fast as e^{-2t}. □

In this section, we have given a procedure of obtaining an observer of dimension $(n - q)$. In some cases, the dimension of the observer can be reduced still further. Luenberger (1966, 1971) has showed that in the case of a single-input system, the estimate of **kx** could always be provided by an observer of dimension $(\mu - 1)$ where μ is the observability index of the system.

9.7 DEADBEAT CONTROL BY STATE FEEDBACK

In this and the next section, we consider control and estimation problems of linear time-invariant discrete-time systems

$$\mathbf{x}(k+1) = \mathbf{F}\mathbf{x}(k) + \mathbf{G}\mathbf{u}(k)\,;\;\; \mathbf{x}(0) = \mathbf{x}^0 \tag{9.79a}$$

$$\mathbf{y}(k) = \mathbf{C}\mathbf{x}(k) \tag{9.79b}$$

where **F**, **G** and **C** are $n \times n$, $n \times p$ and $q \times n$ real constant matrices respectively.

Feedback control laws may be derived for discrete-time systems using the techniques that have been developed for the continuous-time case. Linear feedback from state to input has the form

$$\mathbf{u}(k) = \mathbf{K}\mathbf{x}(k) + \mathbf{r}(k) \tag{9.80}$$

where **K** is a constant $p \times n$ feedback gain matrix and $\mathbf{r}(k)$ is an external input. Substituting (9.80) into (9.79), we get the following closed-loop system.

$$\mathbf{x}(k+1) = (\mathbf{F} + \mathbf{G}\mathbf{K})\,\mathbf{x}(k) + \mathbf{G}\mathbf{r}(k) \tag{9.81a}$$

$$\mathbf{y}(k) = \mathbf{C}\mathbf{x}(k) \tag{9.81b}$$

The closed-loop poles, i.e., the eigenvalues of $(\mathbf{F} + \mathbf{G}\mathbf{K})$ can be arbitrarily located in the complex plane (subject to conjugate pairing) by choosing **K** suitably if and only if (9.79) is completely controllable. It is possible to choose **K** such that the closed-loop system (9.81) is stable if and only if (9.79) is stabilizable.

The computational methods of assigning closed-loop poles are identical to those developed for continuous-time case.

A case of special interest occurs when a state feedback control law is chosen which places all the closed-loop poles at the origin, i.e.,

$$\det(\lambda\mathbf{I} - (\mathbf{F} + \mathbf{G}\mathbf{K})) = \lambda^n = 0$$

According to the Cayley-Hamilton theorem (Section 2.8), any matrix satisfies its own characteristic equation. Therfore

$$(\mathbf{F} + \mathbf{G}\mathbf{K})^n = \mathbf{0}$$

i.e., $(\mathbf{F} + \mathbf{G}\mathbf{K})$ is a nilpotent matrix of index n (Problem 2.24). This result implies that the force-free response of closed-loop system (9.81),

$$\mathbf{x}(k) = (\mathbf{F} + \mathbf{G}\mathbf{K})^k\,\mathbf{x}^0 = \mathbf{0} \quad \text{for } k \geqslant n$$

In other words, any initial state $\mathbf{x}^0$ is driven to zero in (at most) n steps. The feedback control law that assigns all the closed-loop poles to origin, is therefore a deadbeat control law.

Example 9.7: In the following, we determine deadbeat control law for Position Servo system discussed in Section 4.6. The discrete-time model of the system is (eqn. (6.60))

$$\mathbf{x}(k+1) = \mathbf{F}\mathbf{x}(k) + \mathbf{g}u(k) \tag{9.82}$$

where

$$\mathbf{F} = \begin{bmatrix} 1 & 0.0787 \\ 0 & 0.6065 \end{bmatrix};\; \mathbf{g} = \begin{bmatrix} 0.0043 \\ 0.0787 \end{bmatrix}$$

The system characteristic polynomial is

$$(z-1)(z-0.6065) = z^2 - 1.6065z + 0.6065$$
$$= z^2 + \alpha_1 z + \alpha_2$$

The controllability matrix is (refer Example 6.10)

$$\mathbf{U} = \begin{bmatrix} 0.0043 & 0.0105 \\ 0.0787 & 0.0477 \end{bmatrix}$$

The vector $\mathbf{p}_1$ which is the last row of $\mathbf{U}^{-1}$ is obtained as

$$\mathbf{p}_1 = [\ 126.9355 \quad -6.9355\ ]$$

The matrix

$$\mathbf{P} = \begin{bmatrix} \mathbf{p}_1 \\ \mathbf{p}_1\mathbf{F} \end{bmatrix} = \begin{bmatrix} 126.9355 & -6.9355 \\ 126.9355 & 5.7834 \end{bmatrix}$$

The transformation

$$\bar{\mathbf{x}} = \mathbf{P}\mathbf{x}$$

converts the system (9.82) to the following controllable companion form.

$$\dot{\bar{\mathbf{x}}} = \begin{bmatrix} 0 & 1 \\ -0.6065 & 1.6065 \end{bmatrix} \bar{\mathbf{x}} + \begin{bmatrix} 0 \\ 1 \end{bmatrix} u \qquad (9.83)$$

It is immediately seen that in terms of the transformed state, the deadbeat control law is given by

$$u = \bar{\mathbf{k}}\,\bar{\mathbf{x}} + r$$

where

$$\bar{\mathbf{k}} = [\ 0.6065 \quad -1.6065\ ]$$

In terms of the original state, we have

$$u = \mathbf{k}\mathbf{x} + r \qquad (9.84)$$

where

$$\mathbf{k} = \bar{\mathbf{k}}\mathbf{P}$$
$$= [\ -126.9395 \quad 13.4974\ ]$$

9.8 DEADBEAT OBSERVERS

In this section, we consider discrete-time observers that are able to reconstruct the state of the system (9.79). The results are identical to those of the continuous-time case. The system

$$\hat{\mathbf{x}}(k+1) = (\mathbf{F} + \mathbf{MC})\,\hat{\mathbf{x}}(k) + \mathbf{G}u(k) - \mathbf{M}y(k);\ \ \hat{\mathbf{x}}(0) = \hat{\mathbf{x}}^0 \qquad (9.85)$$

is a full-order observer for the observable system (9.79). The observer poles, i.e., eigenvalues of $(\mathbf{F}+\mathbf{MC})$ can be arbitrarily located in the complex plane (subject to conjugate pairing) by suitably choosing the gain matrix $\mathbf{M}$.

As obtained from (9.79) and (9.85), the estimation error is governed by the equation

$$\begin{aligned}\tilde{\mathbf{x}}(k+1) &= \mathbf{x}(k+1) - \hat{\mathbf{x}}(k+1)\\ &= (\mathbf{F}+\mathbf{MC})\,\tilde{\mathbf{x}}(k)\end{aligned} \qquad (9.86)$$

The techniques of obtaining the $\mathbf{M}$ matrix are identical to the continuous-time case. The block diagram realization is also identical except that the integrators are replaced by one-period delay.

A case of special interest occurs when all the observer poles are located at the origin, i.e., all the eigenvalues of $(\mathbf{F}+\mathbf{MC})$ are zero. Then

$$\det(\lambda\mathbf{I} - (\mathbf{F}+\mathbf{MC})) = \lambda^n = 0$$

This implies that, as per Cayley-Hamilton theorem

$$(\mathbf{F}+\mathbf{MC})^n = \mathbf{0}$$

Therefore, from (9.86) we have

$$\tilde{\mathbf{x}}(n) = (\mathbf{F}+\mathbf{MC})^n\,\tilde{\mathbf{x}}^0 = \mathbf{0}$$

Thus, every initial value of the estimation error is reduced to zero in (at most) n steps. In analogy with the deadbeat control law, we refer to observers with this property as deadbeat observers.

The reduced-order discrete-time observers can be obtained analogously to the continuous-time case. The separation principle also carries over to the discrete-time systems.

Example 9.8: Reconsider the Position Servo system of Example 9.7. To implement the deadbeat control law (9.84), both the state variates; $x_1 = \theta$, $x_2 = \dot{\theta}$, are required for feedback. We assume that the observed variable is $x_1 = \theta$ only and design a deadbeat observer to reconstruct the state $\mathbf{x}$ from measurements of θ. The observer model is of the form (eqn. (9.85))

$$\hat{\mathbf{x}}(k+1) = (\mathbf{F}+\mathbf{mc})\,\hat{\mathbf{x}}(k) + \mathbf{g}u(k) - \mathbf{m}y(k) \qquad (9.87)$$

where

$$\mathbf{F} = \begin{bmatrix} 1 & 0.0787 \\ 0 & 0.6065 \end{bmatrix},\ \mathbf{g} = \begin{bmatrix} 0.0043 \\ 0.0787 \end{bmatrix},\ \mathbf{c} = [1 \quad 0]$$

and the vector $\mathbf{m}$ is to be selected such that the poles of $(\mathbf{F}+\mathbf{mc})$ are located at the origin. Let us write

$$\mathbf{m} = \begin{bmatrix} m_1 \\ m_2 \end{bmatrix}$$

Then we find

$$\mathbf{F}+\mathbf{mc}=\begin{bmatrix} 1+m_1 & 0.0787 \\ m_2 & 0.6065 \end{bmatrix}$$

This matrix has the characteristic polynomial

$$z^2-(1+m_1+0.6065)\,z+0.6065m_1-0.0787m_2+0.6065$$

We obtain a deadbeat observer by setting

$$m_1+1.6065=0$$

$$0.6065m_1-0.0787m_2+0.6065=0$$

This results in the gain vector

$$\mathbf{m}=\begin{bmatrix} -1.6065 \\ -4.674 \end{bmatrix} \tag{9.88}$$

The observer (9.87) with the gain given by (9.88) reduces any initial estimation error to zero in at most two steps.

9.9 CONCLUDING COMMENTS

In this chapter, we studied the practical implications of controllability and observability. We proved that in a linear time-invariant controllable system we can, by introducing state feedback, arbitrarily assign the poles of the resulting closed-loop system. We also examined the problem of state estimation and proved that for a linear time-invariant observable system, an observer with a set of arbitrary eigenvalues can be constructed.

This chapter provided an introduction to Linear Multivariable Control Theory. A large number of references have been suggested for further study.

PROBLEMS

9.1 Consider the state model

$$\dot{\mathbf{x}}=\begin{bmatrix} 0 & 0 & 0 \\ 0 & -1 & 0 \\ 0 & 0 & -3 \end{bmatrix}\mathbf{x}+\begin{bmatrix} 1 \\ 1 \\ 1 \end{bmatrix}u$$

$$y=[40/3 \quad -15 \quad 5/3]$$

Determine the transfer function $\hat{y}(s)/\hat{u}(s)$. Using the results of Section 7.5, realize this transfer function in controllable and observable companion forms.

9.2 Reconsider the state model of Problem 9.1. Using the results of Section 9.2, transform this model into controllable and observable companion forms.

9.3 Consider the systems with

(a) $\mathbf{A}=\begin{bmatrix} 1 & 0 & 0 \\ 0 & 2 & 0 \\ 0 & 0 & 3 \end{bmatrix}$; $\mathbf{B}=\begin{bmatrix} 0 & 1 \\ 1 & 0 \\ 1 & 1 \end{bmatrix}$

(b) $\mathbf{A}=\begin{bmatrix} 0 & 0 & -3 \\ 2 & 0 & -7 \\ 0 & -1 & 0 \end{bmatrix}$; $\mathbf{B}=\begin{bmatrix} 1 & 1 \\ -1 & 0 \\ 1 & 0 \end{bmatrix}$

Obtain equivalent systems in controllable companion form (9.18).

9.4 Transform the system with

$$\mathbf{A}=\begin{bmatrix} 0 & 1 & 0 \\ 3 & 2 & 0 \\ 1 & 1 & 1 \end{bmatrix}; \mathbf{B}=\begin{bmatrix} 0 & 0 \\ 1 & 0 \\ 0 & 1 \end{bmatrix}; \mathbf{C}=\begin{bmatrix} 1 & 2 & 0 \\ 0 & 0 & 1 \end{bmatrix}$$

into equivalent observable companion form (9.26).

9.5 Show that the zeros of a scalar system are invariant under linear state feedback to the input.

9.6 Consider a linear system described by the transfer function

$$\frac{\hat{y}(s)}{\hat{u}(s)}=\frac{10}{s(s+1)(s+2)}$$

Design a feedback controller with a state feedback so that the eigenvalues of the closed-loop system are at -2, $-1 \pm j1$.

9.7 A single-input system is described by the following state equation.

$$\dot{\mathbf{x}}=\begin{bmatrix} -1 & 0 & 0 \\ 1 & -2 & 0 \\ & 1 & -3 \end{bmatrix}\mathbf{x}+\begin{bmatrix} 10 \\ 1 \\ 0 \end{bmatrix}u$$

Design a state feedback controller which will give closed-loop poles at $-1 \pm j2$, -6.

Draw a block diagram of the resulting closed-loop system.

9.8 Reconsider the system of Problem 9.4. Design a state feedback control law for this system so that the closed-loop system has poles at -1, -2, -3.

9.9 Convert the system of Problem 9.8 into a single-input controllable system and repeat the problem.

9.10 A linear time-invariant system is described by the state equation

$$\dot{\mathbf{x}}(t) = \mathbf{A}\mathbf{x}(t) + \mathbf{b}u(t)$$

where

$$\mathbf{A} = \begin{bmatrix} 1 & 0 \\ 0 & 0 \end{bmatrix}; \quad \mathbf{b} = \begin{bmatrix} 1 \\ 1 \end{bmatrix}$$

$$y(t) = [\,2 \quad -1\,]\,\mathbf{x}(t)$$

Design a state observer that makes the estimation error to decay at least as fast as e^{-10t}.

9.11 Consider the system described by

$$\dot{\mathbf{x}} = \begin{bmatrix} 0 & 1 \\ -1 & -2 \end{bmatrix} \mathbf{x} + \begin{bmatrix} 1 \\ 1 \end{bmatrix} u$$

$$y = [\,1 \quad 0\,]\,\mathbf{x}$$

Design a state observer so that the estimation error will decay in less than 4 seconds.

9.12 Find a three-dimensional observer with eigenvalues $-2, -2, -3$, for the system

$$\dot{\mathbf{x}} = \begin{bmatrix} -1 & -2 & -2 \\ 0 & -1 & 1 \\ 1 & 0 & -1 \end{bmatrix} \mathbf{x} + \begin{bmatrix} 2 \\ 0 \\ 1 \end{bmatrix} u$$

$$y = [\,1 \quad 1 \quad 0\,]\,\mathbf{x}$$

9.13 Find a two-dimensional observer with eigenvalues $-2, -3$, for the system in Problem 9.12.

9.14 Consider the system of Problem 9.4. For this system, design an observer of order one. The observer pole is required to be located at -4.

9.15 Solve the problem of Example 9.6 without transformation of the given system to canonical form.

9.16 Design a deadbeat state feedback controller for the Mixing Tank system discussed in Section 4.6. A discrete-time model of the Mixing Tank system is given in eqn. (6.62).

9.17 For the Mixing Tank system of Problem 9.16, design a deadbeat observer.

REFERENCES

1. Abdel-Moneim, T.M., "Optimal compensators with pole constraints", *IEEE Trans. Automat. Contr.*, vol. AC-25, pp. 596–598, 1980.

2. Anderson, B.D.O., and D. G. Luenberger, "Design of multivariable feedback systems", *Proc. IEE*, vol. 114, pp. 395–399, **1967**.

3. Anderson, B.D.O., and J.B. Moore, "Linear system optimization with prescribed degree of stability", *Proc. IEE*, vol. 116, pp. 2083–87, **1969**.

4. Bucy, R.S., "Canonical forms for multivariable systems", *IEEE Trans. Automat. Contr.*, vol. AC–13, pp. 567–569, **1968**.

5. Chen, C.T., *Introduction to Linear System Theory*, New York: Holt, Rinehart and Winston, **1970**.

6. Daly, K.C., "The computation of Luenberger canonical forms using elementary similarity transformations", *Int. J. Syst. Sci.*, vol. 7, pp. 1–15, **1976**.

7. Davison, E.J., "On pole assignment in linear systems with incomplete state feedback", *IEEE Trans. Automat. Contr.*, vol. AC–15, pp. 348–351, **1970**.

8. Davison, E.J., and S.H. Wang, "Properties and calculation of transmission zeros of linear multivariable systems", *Automatica*, vol. 10, pp. 643–658, **1974**.

9. Davison, E.J., and S.H. Wang, "On pole assignment in linear multivariable systems using output feedback", *IEEE Trans. Automat. Contr.*, vol. AC–20, pp. 516-518, **1975**.

10. Desoer, C.A., and J.D. Schulman, "Zeros and poles of matrix transfer functions and their dynamical interpretation", *IEEE Trans. Circuits and Systems*, vol. CAS–21, pp. 3–8, **1974**.

11. Falb, P.L., and W.A. Wolovich, "Decoupling in the design and synthesis of multivariable systems", *IEEE Trans. Automat. Contr.*, vol. AC–12, pp. 651–659, **1967**.

12. Fallside, F., and H. Seraji, "Direct design procedure for multivariable systems", *Proc. IEE*, vol. 118, pp. 797–801, **1971**.

13. Fallside, F., and H. Seraji, "Design of multivariable systems using unity rank feedback", *Int. J. Control*, vol. 17, pp. 351–364, **1973**.

14. Gilbert, E., "The decoupling of multivariable systems by state feedback", *SIAM J. Control*, vol. 7, pp. 50–63, **1969**.

15. Gomathi, K., S.S. Prabhu, and M.A. Pai, "A suboptimal controller for minimum sensitivity of closed-loop eigenvalues to parameter variations", *IEEE Trans. Automat. Contr.*, vol. AC–25, pp. 587–588, **1980**.

16. Jameson, A., "Design of a single-input system for specified roots using output feedback", *IEEE Trans. Automat. Control*, vol. AC–15, pp, 345–348, **1970**.

17. Jordan, D., and B. Sridhar, "An efficient algorithm for calculation of Luenberger canonical form", *IEEE Trans. Automat. Contr.*, vol. AC–18, pp. 292–295, **1973**.

18. Kwakernaak, H., and R. Sivan, *Linear Optimal Control Systems*, New York: Wiley, **1972**.

19. Luenberger, D.G., "Observers for multivariable systems", *IEEE Trans. Automat. Contr.*, vol. AC–11, pp. 190–197, **1966**.

20. Luenberger, D.G., "Canonical forms for linear multivariable systems", *IEEE Trans. Automat. Contr.*, vol. AC–12, pp. 290–293, **1967**.

21. Luenberger, D.G., "An introduction to observers", *IEEE Trans. Automat. Contr.*, vol. AC–16, pp. 596–602, **1971**.

22. Maki, M.C., and J.V.D. Vegte, "Optimization of multi-input systems with assigned poles", *IEEE Trans. Automat. Contr.*, vol. AC–19, pp. 130-133, **1974**.

23. Melsa, J.L., and S.K. Jones, *Computer Programs for Computational Assistance in the Study of Linear Control Theory*, 2nd Edition, New York: McGraw-Hill, **1973**.

24. Morse, A.S., and W.M. Wonham, "Decoupling and pole assignment by dynamic compensation", *SIAM J. Control*, vol. 8, pp. 317-337, **1970**.

25. Morse, A.S., and W.M. Wonham, "Status of non-interacting control", *IEEE Trans. Automat. Contr.*, vol. AC–16, pp. 568–581, **1971.**

26. Munro, N., "Pole assignment", *Proc. IEE*, vol. 126, pp, 549-554, **1979.**

27. Patel, R.V., "On zeros of multivariable systems", *Int. J. Control*, vol. 21, pp. 599–608, **1975.**

28. Power, H.M., "Extension to the method of Anderson and Luenberger for eigenvalue assignment", *Electron. Lett.*, vol. 7, pp. 158–160, **1971.**

29. Pugh, A.C., and A.K. Shelton, "Improved algorithm for optimal stabilization of linear systems", *Electron. Lett.*, vol. 11, pp. 528–529, **1975.**

30. Sebakhy, O.A., and N.N. Sorial, "Optimization of linear multivariable systems with prespecified closed-loop eigenvalues", *IEEE Trans. Automat. Contr.*, vol. AC–24, pp. 355–357, **1979.**

31. Seraji, H., "Pole assignment technique for multivariable systems using unity rank output feedback", *Int. J. Control*, vol. 21, pp. 945-954, **1975.**

32. Shankar, V.G., and K. Ramar, "Pole assignment with minimum eigenvalue sensitivity to parameter variations", *Int. J. Control*, vol 23, pp. 493–504, **1976.**

33. Solheim, O.A., "Design of optimal control systems with prescribed eigenvalues", *Int. J. Control*, vol. 15, pp. 143–160, **1972.**

34. Sundareswaren, K.K., and M.M. Bayoumi, "Eigenvalue assignment in linear multivariable systems", *Electron. Lett.*, vol. 7, pp. 573–574, **1971.**

35. Vegte, J.V.D., and M.C. Maki, "Optimisation of systems with assigned poles", *Int. J. Control*, vol. 18, pp. 1105-1112, **1973.**

36. Wolovich, W.A., "Static decoupling", *IEEE Trans. Automat. Contr.*, vol. AC–18, pp. 536-537, **1973a.**

37. Wolovich, W.A., "On the numerators and zeros of rational transfer matrices", *IEEE Trans. Automat. Contr.*, vol. AC–18, pp. 544–546, **1973b.**

38. Wonham, W.M., "On pole assignment in multi-input controllable linear systems", *IEEE Trans. Automat., Contr.*, vol. AC–12, pp. 660-665, **1967.**

39. Wonham, W.M., and A.S. Morse, "Decoupling and pole assignment in linear multivariable systems: a geometric approach", *SIAM J. Control*, vol. 8, pp. 1–18, **1970.**

10. OPTIMAL CONTROL: GENERAL MATHEMATICAL PROCEDURES

10.1 INTRODUCTION

In Chapter 9, we discussed some techniques of designing feedback controllers for multivariable systems to get *satisfactory* properties in certain areas of performance, e.g., stability, steady-state accuracy, absence of interaction etc.

In the majority of industrial control systems, the cost of a control system is weighed against the expected benefits of its installation. The economic factors lead in practice to a compromise solution for the controller, which must be reasonably cheap to install, yet satisfactory within certain margins of tolerance in its performance. Clearly, in the absence of economic factors there will be a trend towards designing a controller which is not merely satisfactory but which is the best that can be designed whatever criterion of 'bestness' the designer has in mind. It is not surprising, therefore, to find that such fields of development as space travel, guided weaponry etc., wherein the cost was insignificant compared with the benefits obtained, should have stimulated the development of optimal control theory. In contrast, industrial applications of optimal control systems are still comparatively rare, but with the rapid development of powerful small digital computers, they are undoubtedly becoming increasingly important.

Many excellent reviews of Optimal Control Theory have been written and the reader may care to begin with that by Athans (1966). Much of the fundamental theory has crystallised out from the research and is available in comprehensive texts including Merriam (1964), Athans and Falb (1966), Lapidus and Luus (1967), Bryson and Ho (1969), Kirk (1970), Anderson and Moore (1971), Kwakernaak and Sivan (1972) and Sage and White (1977).

The purpose of this chapter and the next two is to develop the concepts behind optimal control theory with the object of providing a basis for the design of optimal control systems.

10.2 FORMULATION OF THE OPTIMAL CONTROL PROBLEM

The starting point of optimal control theory is the set of state equations which describes the behaviour of dynamic system (plant) to be controlled.

For a continuous-time system, the state equations are a set of first-order differential equations

$$\dot{\mathbf{x}}(t) = \mathbf{f}(\mathbf{x}, \mathbf{u}, t) \; ; \quad t \in [t_0, t_1] \tag{10.1}$$

where $\mathbf{x}(t)$ is $n \times 1$ state vector, $\mathbf{u}(t)$ is $p \times 1$ input vector, $\mathbf{f}$ is a vector-valued function and $[t_0, t_1]$ is the control interval.

For discrete-time systems, the state equations are a set of first-order difference equations

$$\mathbf{x}(k+1) = \mathbf{f}(\mathbf{x}, \mathbf{u}, k) \; ; \quad k \in [k_0, k_1] \tag{10.2}$$

The problem in hand is to control the plant so as to achieve certain specified objectives while satisfying the operational constraints. Following steps are involved in solution of an optimal control problem:

(i) For the given plant, find a control function $\mathbf{u}^*$ which will act upon the given plant in what is, in some known sense, the best possible way.

(ii) Realize the control function obtained from step (i) with the help of a controller.

The design of an optimum controller is based on the following factors relating to the plant and to the nature of its connection with the controller:

(i) The characteristics of the plant.

(ii) The requirements made upon the plant.

(iii) The nature of information about the plant supplied to the controller.

The Characteristics of the Plant

Because of physical limitations of system components, constraints on state variables and control variables are frequently necessary. The plant inputs $u_1(t), u_2(t), \ldots, u_p(t)$ cannot have unrestricted values. They must not exceed certain limits. A control which satisfies the *control constraints* during the entire control interval $[t_0, t_1]$ of interest will be called *admissible control.* We shall denote the set of admissible controls by $\mathbf{U}$. Then $\mathbf{u}(t)$ is admissible if

$$\mathbf{u}(t) \in \mathbf{U} \quad \text{for all} \quad t \in [t_0, t_1] \tag{10.3a}$$

Constraints can be laid on $\mathbf{x}(t)$ also. A state trajectory which satisfies the state variable constraints during the entire time interval $[t_0, t_1]$ will be called the *admissible trajectory.* The set of admissible state trajectories will be denoted by $\mathbf{X}$. Then $\mathbf{x}(t)$ is admissible if

$$\mathbf{x}(t) \in \mathbf{X} \quad \text{for all} \quad t \in [t_0, t_1] \tag{10.3b}$$

The Requirements Made upon the Plant

The designer translates the requirements made upon the plant into a mathematical *performance criterion* or *index* and then tries to find solutions that optimize this performance measure. In certain cases, the problem statement may clearly indicate what to select for a performance measure; whereas in other problems the selection is a subjective matter. For example 'Transfer the system from state $\mathbf{x}^0$ to state $\mathbf{x}^1$ as quickly as possible' clearly indicates that elapsed time is the performance measure to be minimized. On the other hand 'Maintain state $\mathbf{x}$ of the system near zero with a small expenditure of control energy' does not instantly suggest a unique performance measure. In such problems, the designer may be required to try several performance measures before selecting one which yields what he considers to be the optimal performance.

Let us now discuss some typical control problems to provide some physical motivation for the selection of a performance measure.

Minimum-Time Problem: The reader is familiar with minimum-time problems (Chapter 6). In the following, we give the problem formulation in a more general way.

We are given the time t_0 and the initial state $\mathbf{x}(t_0) \overset{\Delta}{=} \mathbf{x}^0$. The final state is required to lie in a specified region S of the $(n \times 1)$ dimensional state-time space. We shall call S the *target set* (If the final state is fixed, then S is a straight line). The objective is to transfer a system from the initial state $\mathbf{x}^0$ to the specified target set S in the minimum time. The performance index to be minimized is

$$J = t_1 - t_0$$
$$= \int_{t_0}^{t_1} dt \tag{10.4}$$

where t_1 is the first instant of time when $\mathbf{x}(t)$ and S intersect. The interception of attacking aircraft and missiles is an example of such a control problem.

Note that the problem is almost always associated with inequality constraints on the magnitude of the input-vector elements, for in the absence of such constraints the change can usually be accomplished in zero time with control signals of infinite magnitude (refer Section 6.4).

Minimum-Energy Problem: The objective is to transfer a system from a given initial state $\mathbf{x}(t_0) \overset{\Delta}{=} \mathbf{x}^0$ to a specified target set S with a minimum expenditure of energy.

As was pointed out in Section 6.4, in a number of problems $u^2(t)$ is a measure of instantaneous rate of expenditure of energy. To minimize energy expenditure, we minimize

$$J = \int_{t_0}^{t_1} u^2(t)\, dt \tag{10.5a}$$

For several control inputs, performance index (10.5a) takes the form

$$J = \int_{t_0}^{t_1} (\mathbf{u}^T(t)\, \mathbf{u}(t))\, dt \tag{10.5b}$$

which has already been used in Chapter 6.

To allow greater generality, we can insert a real symmetric positive definite constant[1] matrix **R** to obtain

$$J = \int_{t_0}^{t_1} (\mathbf{u}^T(t)\, \mathbf{R}\mathbf{u}(t))\, dt \tag{10.5c}$$

Suppose that **R** is a diagonal matrix. The assumption that **R** is positive definite implies that all of its diagonal elements are positive. By adjusting the element values, we can weigh the relative importance of each of the control variables $u_i(t)$ in the expenditure of energy.

Minimum-Fuel Problem: This problem is important in connection with the analysis of rocket-propelled space craft where clearly any reduction which can be made in the mass of the fuel to be carried results in a possible increase in pay-load (passengers, instrumentation, etc.). The rate of fuel consumption of a jet engine is proportional to the thrust (the input to space vehicle) developed. Thus in order to minimize fuel consumption of a jet engine, the performance index

$$J = \int_{t_0}^{t_1} |u(t)|\, dt \tag{10.6a}$$

would be selected.

Allowing the various jets present to have different thrust-to-consumption ratio, the criterion (10.6a) will take the form

$$J = \int_{t_0}^{t_1} (K_1\, |u_1(t)| + K_2\, |u_2(t)| + \ldots)\, dt \tag{10.6b}$$

where K_i are non-negative weighing factors.

State Regulator Problem: The objective is to transfer a system from initial state $\mathbf{x}(t) \triangleq \mathbf{x}^0$ to the desired state $\mathbf{x}^1$ ($\mathbf{x}^1$ may in many cases be the equilibrium point of the system) with the minimum integral-square error.

Relative to the desired state $\mathbf{x}^1$, quantity $(\mathbf{x}(t)-\mathbf{x}^1)$ can be viewed as the instantaneous system error. If we transform the system coordinates such that $\mathbf{x}^1$ becomes the origin (Footnote 1, Chapter 6), then the new state $\mathbf{x}(t)$ is itself the error.

---

[1]The elements of R may be functions of time if it is desired to vary the weightage on energy expenditure during the interval $[t_0, t_1]$.

Investigators have advanced the argument that the integral-square error

$$J = \int_{t_0}^{t_1} \left[\sum_{i=1}^{n} (x_i(t))^2 \right] dt$$

$$= \int_{t_0}^{t_1} (\mathbf{x}^T(t)\, \mathbf{x}(t))\, dt$$

is a reasonable measure of the system transient response from time t_0 to t_1. To be more general,

$$J = \int_{t_0}^{t_1} (\mathbf{x}^T(t)\, \mathbf{Q}\mathbf{x}(t))\, dt$$

with **Q** a real, symmetric, positive semidefinite, constant matrix can be used as performance measure. The simplest form of **Q** one can use is a diagonal matrix:

$$\mathbf{Q} = \begin{bmatrix} q_1 & & \mathbf{0} \\ & q_2 & \\ \mathbf{0} & & \ddots \; q_n \end{bmatrix}$$

The i-th entry of **Q** represents the amount of weight the designer places on the constraint on the state variable $x_i(t)$. The larger the value of q_i relative to the other values of q, the more control effort is spent to regulate $x_i(t)$.

To minimize the deviation of final state $\mathbf{x}(t_1)$ of the system from the desired state $\mathbf{x}^1 = \mathbf{0}$, a possible performance measure is

$$J = \mathbf{x}^T(t_1)\, \mathbf{H}\mathbf{x}(t_1)$$

where **H** is a positive semidefinite, real, symmetric, constant matrix.

The design obtained by minimizing

$$J = \mathbf{x}^T(t_1)\mathbf{H}\mathbf{x}(t_1) + \int_{t_0}^{t_1} (\mathbf{x}^T(t)\, \mathbf{Q}\mathbf{x}(t))\, dt$$

may be unsatisfactory in practice. A more realistic solution to the problem is obtained if the performance index is modified by adding a penalty term for physical constraints on **u**. One of the ways of accomplishing this is to introduce the following quadratic control term in the performance index:

$$J = \int_{t_0}^{t_1} (\mathbf{u}^T(t)\, \mathbf{R}\mathbf{u}(t))\, dt$$

where **R** is a positive definite, real, symmetric, constant matrix. By giving sufficient weight to control terms, the amplitudes of control signals which minimize the overall performance index may be kept within practical bounds, although at the expense of increased error in $\mathbf{x}(t)$.

For the state regulator problem, a useful performance measure is therefore[2]

$$J = \tfrac{1}{2}\mathbf{x}^T(t_1)\,\mathbf{H}\mathbf{x}(t_1) + \tfrac{1}{2}\int_{t_0}^{t_1} (\mathbf{x}^T(t)\,\mathbf{Q}\mathbf{x}(t) + \mathbf{u}^T(t)\,\mathbf{R}\mathbf{u}(t))\,dt \qquad (10.7a)$$

Infinite-time state regulator problem

If the terminal time is not constrained ($t_1 \to \infty$), then the final state should approach the equilibrium state $\mathbf{x}^1 = \mathbf{0}$ (assuming a stable system). So the terminal constraint in J is not necessary. For an infinite-time state regulator problem, the performance index is[3]

$$J = \tfrac{1}{2}\int_{t_0}^{\infty} (\mathbf{x}^T(t)\,\mathbf{Q}\mathbf{x}(t) + \mathbf{u}^T(t)\mathbf{R}\mathbf{u}(t))\,dt \qquad (10.7b)$$

Output Regulator Problem: In the state regulator problem, we are concerned with making all the components of the state vector $\mathbf{x}(t)$ small. In the output regulator problem, on the other hand, we are concerned with making the components of the output vector $\mathbf{y}(t)$ small. A useful performance measure for the output regulator problem is

$$J = \tfrac{1}{2}\mathbf{y}^T(t_1)\,\mathbf{H}\mathbf{y}(t_1) + \tfrac{1}{2}\int_{t_0}^{t_1} (\mathbf{y}^T(t)\,\mathbf{Q}\mathbf{y}(t) + \mathbf{u}^T(t)\,\mathbf{R}\mathbf{u}(t))\,dt \qquad (10.8)$$

Tracking Problem: The objective is to maintain the system state $\mathbf{x}(t)$ as close as possible to the desired state $\mathbf{r}(t)$ in the interval $[t_0, t_1]$.

As a performance measure, we select

$$J = \tfrac{1}{2}\mathbf{e}^T(t_1)\,\mathbf{H}\mathbf{e}(t_1) + \tfrac{1}{2}\int_{t_0}^{t_1} (\mathbf{e}^T(t)\,\mathbf{Q}\mathbf{e}(t) + \mathbf{u}^T(t)\,\mathbf{R}\mathbf{u}(t))\,dt \qquad (10.9a)$$

where

$$\mathbf{e}(t) = (\mathbf{x}(t) - \mathbf{r}(t)) \qquad (10.9b)$$

□

For discrete-time systems, the performance indices may be defined on similar lines.

Whilst it is not claimed that the performance indices given above are the only possible ones, they cover the majority of practical cases encountered.

---

[2]Note that multiplication by $\frac{1}{2}$ does not affect the minimization problem. The constant helps us in mathematical manipulations as we shall see later.

[3]t_0 may be taken as zero in time-invariant systems.

In all that follows, it will be assumed that the performance of a system is evaluated by a measure of the form

$$J = h(\mathbf{x}(t_1), t_1) + \int_{t_0}^{t_1} g(\mathbf{x}(t), \mathbf{u}(t), t)\, dt \tag{10.10}$$

where t_0 and t_1 are the initial and terminal times; h and g are scalar functions. t_1 may be specified or 'free' depending upon the problem statement.[4]

Choosing a performance measure is a translation of system's physical requirements into mathematical terms. Indeed, it is rather inconceivable that for a complex system a single performance index can encompass all the qualities that are desired. Fortunately, there are many engineering situations which fit into the format of performance indices defined above. We shall discuss a number of system control examples in this and the next chapter to illustrate the method of setting up and solving optimal control problems.

The Nature of Information about the Plant Supplied to the Controller

If the optimal control is determined as a function of time for a specified initial state value, i.e.,

$$\mathbf{u}^*(t) = \mathbf{f}(\mathbf{x}(t_0), t) \tag{10.11}$$

then the optimal control is said to be in the *open-loop* form. Systems with open-loop controllers once started, operate without requiring further knowledge of the process development. In the absence of meaningful disturbances and errors, open-loop control can be very successful. The function $\mathbf{f}$ given by (10.11) is called *optimal control function.*

If a functional relationship of the form

$$\mathbf{u}^*(t) = \mathbf{f}(\mathbf{x}(t), t) \tag{10.12}$$

can be found for the optimal control at time t, then the optimal control is said to be in the *closed-loop* form and the function $\mathbf{f}$ is called the *optimal control law*. In closed-loop control, the controller obtains information by way of feedback lines on the actual state $\mathbf{x}(t)$ of the process. If this does not answer the requirements, the controller acts upon the plant in such a way as to bring the state closer to these requirements.

At this stage we may note that closed-loop controllers are more powerful than open-loop controllers. Closed-loop controllers can accumulate information about the plant during operation and thus are able to reduce the effects of disturbances and compensate for plant parameter variations. Open-

[4]The performance measure J gives a correspondence (or mapping) between the functions of a given class and real numbers Such mathematical entities are called *functionals*. In a gross sense, one might consider a functional as a kind of function whose domain is a class of functions rather than a set of numbers.

loop controllers obviously have no access to any information about the plant except for what is available before the control starts. However, all the optimization problems described earlier, require finding of the plant input vector $\mathbf{u}^*(t)$ minimizing some criterion functional. It may be that only in a few cases will it be possible to generate this input vector by feedback techniques. Apart from these cases, we are therefore concerned basically with open-loop systems.

□

With the background material we have accumulated, it is now possible to present an explicit statement of the optimal control problem:

Find an admissible control $\mathbf{u}^*$ *which causes the system*

$$\dot{\mathbf{x}}(t) = \mathbf{f}(\mathbf{x}(t), \mathbf{u}(t), t)$$

to follow an admissible trajectory $\mathbf{x}^*$ *and minimizes*[5] *the performance measure*

$$J = h(\mathbf{x}(t_1), t_1) + \int_{t_0}^{t_1} g(\mathbf{x}(t), \mathbf{u}(t), t)\, dt$$

The main theoretical approaches to the optimal control system design utilize

(i) Calculus of Variations
(ii) Maximum (or Minimum) Principle
(iii) Dynamic Programming.

We shall study these mathematical procedures in this chapter.

10.3 CALCULUS OF VARIATIONS

Calculus of variations is the branch of mathematics which is concerned with the finding of trajectories that maximize or minimize a given functional. The optimal control theory techniques such as the Minimum Principle of Pontryagin and Dynamic Programming of Bellman, though derived from significantly different points of view, are inspired to a great extent by the classical calculus of variations (Dreyfus 1965).

There is vast literature devoted to the subject of calculus of variations. Some elementary books on the subject are Elsgolc (1962), Gelfand and Fomin (1963), Tou (1964) and Lietman (1966). For a comprehensive treatment, the reader may refer Bliss (1963). Examples of many optimal control problems that can be solved by calculus of variations can be found in Leitman (1962), Kalman (1963), Dreyfus (1965), Hestenas (1966) and Pierre (1969). In this

---

[5]The optimal control problem is standardized by minimization of the performance index. Maximization, if required, is given by

$$(J)_{\max} = (-J)_{\min}$$

book, no attempt at a thorough discussion on variational calculus is made; only those features which have a direct bearing on the optimal control problems are presented. These classical results are then transformed into modern optimal control framework through the introduction of plant equation constraints.

Since the concepts and methods of variational calculus are very much similar to those employed in ordinary maxima and minima theory of differential calculus, it is of value to review briefly the theory of ordinary maxima and minima so that the parallels may be easily drawn. The brief review of maxima and minima theory presented below will also form a basis for developing most of the numerical techniques of optimization. It may be noted that we shall present only that material which is used in the sequel (for detailed study, refer Hancock (1960)).

Minimization of Functions

The optimization of a function is understood to mean its minimization or maximization by the variation of variables upon which the given function depends. Since algebraically, the maximization of a function implies the minimization if its negative, it will usually be convenient to identify optimization with minimization.

Let D be a subset of real numbers x, given by $x^0 < x < x^1$ and let f be a real-valued function defined on D, i.e., $f: D \to \mathcal{R}$. The function f has a *relative* (*local*) *minimum* at point $x^* \in D$ if there exists a neighbourhood N of x^* such that for any $x \in N$, the increment of f,

$$\Delta f = f(x) - f(x^*) \geqslant 0 \tag{10.13}$$

We may define the neighbourhood N of x^* as the set of all points $x \in D$ for which $|x - x^*| < \varepsilon$ where ε is a positive real number.

If the inequality (10.13) holds for all $x \in D$ (i.e., the neighbourhood N contains the whole domain), then f has an *absolute minimum.*

Clearly the definition (10.13) is of little practical use in determining x^* for it implies that the function f must be evaluated for all admissible x before choosing the parameter x^*. However, provided f has continuous derivatives with respect to x, the problem of computing the minimum of f can be facilitated by deriving certain properties that must be satisfied by x^*. These properties may be satisfied by other points as well, but in general the set of points that share these properties with x^* is much smaller than all possible ones.

Consider a point $x_1 \in N$ and define $\Delta x = (x_1 - x^*)$ as the total increment in x. The corresponding total increment in f is given by (Fig. 10.1)

$$\begin{aligned}\Delta f(x^*, \Delta x) &= f(x_1) - f(x^*)\\ &= f(x^* + \Delta x) - f(x^*)\end{aligned}$$

If we expand $f(x^* + \Delta x)$ in a Taylor series about x^*, then Δf becomes

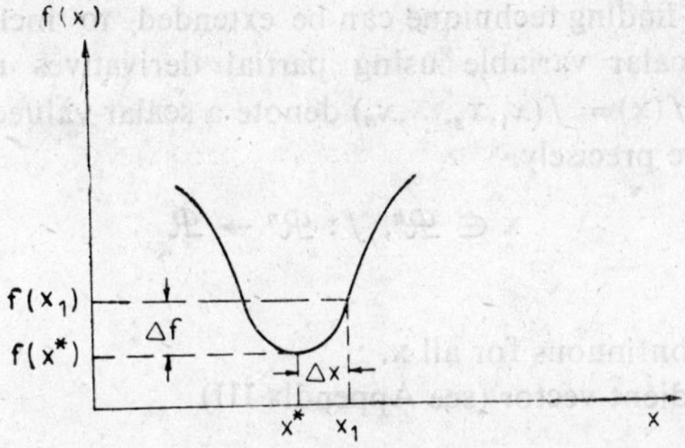

Fig. 10.1 Illustration of minimal of a function

$$\Delta f(x^*, \Delta x) = \left.\frac{df}{dx}\right|_{x^*} \Delta x + \tfrac{1}{2} \left.\frac{d^2 f}{dx^2}\right|_{x^*} \Delta x^2 + \dots \quad (10.14a)$$

$$= df(x^*, \Delta x) + \tfrac{1}{2} d^2 f(x^*, \Delta x) + \dots \quad (10.14b)$$

We know from elementary calculus that a necessary condition for a differentiable function f to take on a minimum value at point x^* is that the *differential* of f at the point x^* be zero, i.e.,

$$df(x^*, \Delta x) = 0 \quad (10.15a)$$

This condition is necessary but not sufficient. df may be zero at point x^* but f may not have a minimum there. If d^2f exists, we know from elementary calculus that another necessary condition for a minimum at x^* is

$$d^2 f(x^*, \Delta x) > 0 \quad (10.15b)$$

Conditions (10.15a) and (10.15b) jointly still do not guarantee that an absolute minimum point has been found, as there may be many points satisfying these conditions. In order to establish which point is the absolute minimum, the values of f at each of these points must be compared and only the smallest one selected.

It is of course clear that if $f(x)$ is not twice differentiable, some or all of the above conditions cannot be applied. Also, if the minimum point of $f(x)$ is to be sought over a closed interval $[x^0, x^1]$, then a minimum point may occur at the boundary. Conditions (10.15a) and (10.15b) give only the interior relative minimum. In such cases, a more involved search procedure must be applied to determine minima.

Thus (10.15a) and (10.15b) give necessary and sufficient conditions for a relative minimum of a twice differentiable function $f(x)$ at an interior point x^*. When there are no constraints, the infinitesimal Δx can be arbitrarily taken and conditions (10.15) reduce themselves to the following:

$$\left.\frac{df}{dx}\right|_{x^*} = 0 \quad (10.16a)$$

$$\left.\frac{d^2 f}{dx^2}\right|_{x^*} > 0 \quad (10.16b)$$

The minima–finding technique can be extended to include functions of more than one scalar variable using partial derivatives rather than total derivatives, Let $f(\mathbf{x}) = f(x_1, x_2, \ldots, x_n)$ denote a scalar-valued function of the n-vector $\mathbf{x}$. More precisely,

$$\mathbf{x} \in \mathcal{R}^n, \; f : \mathcal{R}^n \to \mathcal{R}$$

Suppose that

(i) $f(\mathbf{x})$ is continuous for all $\mathbf{x}$.

(ii) The gradient vector (see Appendix III)

$$\frac{\partial f}{\partial \mathbf{x}} \triangleq \begin{bmatrix} \dfrac{\partial f}{\partial x_1} \\ \dfrac{\partial f}{\partial x_2} \\ \vdots \\ \dfrac{\partial f}{\partial x_n} \end{bmatrix}$$

is continuous for all $\mathbf{x}$.

(iii) The second-derivative (or Hessian) matrix (see Appendix III)

$$\frac{\partial^2 f}{\partial \mathbf{x}^2} \triangleq \begin{bmatrix} \dfrac{\partial^2 f}{\partial x_1^2} & \dfrac{\partial^2 f}{\partial x_1 \partial x_2} & \cdots & \dfrac{\partial^2 f}{\partial x_1 \partial x_n} \\ \dfrac{\partial^2 f}{\partial x_2 \partial x_1} & \dfrac{\partial^2 f}{\partial x_2^2} & \cdots & \dfrac{\partial^2 f}{\partial x_2 \partial x_n} \\ \vdots & \vdots & & \vdots \\ \dfrac{\partial^2 f}{\partial x_n \partial x_1} & \dfrac{\partial^2 f}{\partial x_n \partial x_2} & \cdots & \dfrac{\partial^2 f}{\partial x_n^2} \end{bmatrix}$$

is continuous for all $\mathbf{x}$.

The Taylor series expansion of $f(\mathbf{x})$ about a vector $\mathbf{x}^*$ has the form

$$f(\mathbf{x}) = f(\mathbf{x}^*) + \left.\frac{\partial f}{\partial \mathbf{x}}\right|_{\mathbf{x}^*} (\mathbf{x} - \mathbf{x}^*) + \tfrac{1}{2}(\mathbf{x} - \mathbf{x}^*)^T \left.\frac{\partial^2 f}{\partial \mathbf{x}^2}\right|_{\mathbf{x}^*} (\mathbf{x} - \mathbf{x}^*) + \cdots$$

A necessary condition for a vector $\mathbf{x}^*$ to be an interior minimum is that the gradient vector must be zero at the minimum, i.e.,

$$\left.\frac{\partial f}{\partial \mathbf{x}}\right|_{\mathbf{x}^*} = \mathbf{0} \tag{10.17a}$$

A sufficient condition that $\mathbf{x}^*$ be a relative minimum is that (10.17a) holds and the Hessian matrix

$$\left.\frac{\partial^2 f}{\partial \mathbf{x}^2}\right|_{\mathbf{x}^*} \text{ is positive definite} \tag{10.17b}$$

Note that the neighbourhood N of vector $\mathbf{x}^*$ may be defined as the set of vectors $\mathbf{x} \in \mathcal{R}^n$ for which $\| \mathbf{x} - \mathbf{x}^* \| < \varepsilon$ where ε is a positive real number.

For the case where the Hessian matrix of a function is positive semi-definite, the problem of determining the sufficient conditions for minima can be resolved by investigating the higher-order derivatives in the Taylor series expansion. However, the algebra becomes quite involved.

An important consideration arises when we wish to find a minimum point of a twice continuously differentiable function $f: \mathcal{R}^n \to \mathcal{R}$ under the constraint that this point be located on a subspace defined by

$$\mathbf{g}(\mathbf{x}) = \mathbf{0} \tag{10.18}$$

where

$$\mathbf{g}: \mathcal{R}^n \to \mathcal{R}^m$$

In (10.18), $m < n$ and each of g_i is assumed to be twice continuously differentiable.

A straightforward way to do this, but one that is often technically complex, is to try to eliminate m of the variables $x_1, x_2, \ldots, x_n$ in the function f through the use of m constraint equations and then proceed to minimize the resulting function of $(n - m)$ variables by the usual approach.

A highly useful alternative way is to use the method of *Lagrange multipliers,* which we illustrate below. For simplicity of illustration, we take $n = 2$ and $m = 1$.

The necessary condition for an extremum (minimum or maximum) may be written as

$$df(\mathbf{x}, d\mathbf{x}) = \frac{\partial f}{\partial x_1} dx_1 + \frac{\partial f}{\partial x_2} dx_2 = 0 \tag{10.19a}$$

and the constraint implies the equation

$$\frac{\partial g}{\partial x_1} dx_1 + \frac{\partial g}{\partial x_2} dx_2 = 0 \tag{10.19b}$$

When there are no constraints, the infinitesimals dx_i can be arbitrarily taken and (10.19a) will still hold. This means that $\frac{\partial f}{\partial x_i} = 0$ for all i, as has already been mentioned. With the constraints, however, the dx_i's cannot be chosen arbitrarily but must be chosen as prescribed by (10.19b). A candidate for a constrained extremum must satisfy both (10.19a) and (10.19b).

Assuming that all partial derivatives in (10.19) are nonzero, it follows that

$$\frac{dx_2}{dx_1} = -\frac{\partial f(\mathbf{x})}{\partial x_1} \Big/ \frac{\partial f(\mathbf{x})}{\partial x_2} = -\frac{\partial g(\mathbf{x})}{\partial x_1} \Big/ \frac{\partial g(\mathbf{x})}{\partial x_2}$$

that is, the ratios

$$\frac{\partial f(\mathbf{x})}{\partial x_1} : \frac{\partial f(\mathbf{x})}{\partial x_2} \quad \text{and} \quad \frac{\partial g(\mathbf{x})}{\partial x_1} : \frac{\partial g(\mathbf{x})}{\partial x_2}$$

must be proportional for a vector to be a candidate for an interior extremum. Let λ be the proportionality constant, which we call the Lagrange multiplier. Thus

$$\lambda = -\frac{\partial f(\mathbf{x})}{\partial x_1}\Big/\frac{\partial f(\mathbf{x})}{\partial x_2} = -\frac{\partial g(\mathbf{x})}{\partial x_1}\Big/\frac{\partial g(\mathbf{x})}{\partial x_2} \tag{10.20}$$

We define the Lagrangian

$$L(\mathbf{x}, \lambda) = f(\mathbf{x}) + \lambda g(\mathbf{x})$$

Then, (10.20) is equivalent to $\dfrac{\partial L}{\partial \mathbf{x}} = \mathbf{0}$ and (10.18) is equivalent to $\dfrac{\partial L}{\partial \lambda} = 0$.

The Lagrange multiplier approach is easily generalized to consider the case where m and n are arbitrary, finite positive integers. The Lagrangian is defined as

$$L(\mathbf{x}, \boldsymbol{\lambda}) = f(\mathbf{x}) + \boldsymbol{\lambda}^T \mathbf{g}(\mathbf{x}) \tag{10.21a}$$

where $\boldsymbol{\lambda} \in \mathcal{R}^m$.

If $\mathbf{x}^*$ is a constrained extremum, then there exists a $\boldsymbol{\lambda}^* \in \mathcal{R}^m$ such that

$$\left.\frac{\partial L}{\partial \mathbf{x}}\right|_{\mathbf{x}^*, \boldsymbol{\lambda}^*} = \mathbf{0} \tag{10.21b}$$

and

$$\left.\frac{\partial L}{\partial \boldsymbol{\lambda}}\right|_{\mathbf{x}^*, \boldsymbol{\lambda}^*} = \mathbf{0} \tag{10.21c}$$

Example 10.1: Find the points in the three-dimensional euclidean space that extremize the function

$$f(x_1, x_2, x_3) = x_1^2 + x_2^2 + x_3^2$$

and lie on the intersection of the surfaces

$$x_3 = x_1 x_2 + 5$$

$$x_1 + x_2 + x_3 = 1$$

Solution: Define the Lagrangian

$$L(\mathbf{x}, \boldsymbol{\lambda}) = x_1^2 + x_2^2 + x_3^2 + \lambda_1(x_1 x_2 + 5 - x_3) + \lambda_2(x_1 + x_2 + x_3 - 1)$$

$$\left.\frac{\partial L}{\partial \mathbf{x}}\right|_{\mathbf{x}^*, \boldsymbol{\lambda}^*} = \mathbf{0} \quad \text{gives}$$

$$2x_1^* + \lambda_1^* x_2^* + \lambda_2^* = 0$$

$$2x_2^* + \lambda_1^* x_1^* + \lambda_2^* = 0$$

$$2x_3^* - \lambda_1^* + \lambda_2^* = 0$$

$$\left.\frac{\partial L}{\partial \boldsymbol{\lambda}}\right|_{\mathbf{x}^*, \boldsymbol{\lambda}^*} = \mathbf{0} \quad \text{gives}$$

$$x_1^* x_2^* + 5 - x_3^* = 0$$

$$x_1^* + x_2^* + x_3^* - 1 = 0$$

Solving these five equations, we get

$$\mathbf{x}^* = \begin{cases} (2, -2, 1) \\ (-2, 2, 1) \end{cases}$$

Minimization of Functionals

With the brief review of ordinary maxima and minima theory as background, let us proceed to the problem at hand, the minimization of functionals. The following basic concepts of variational calculus are presented in such a manner as to parallel as closely as possible the concepts of differential calculus.

A functional J is a transformation (or mapping) that assigns to each function $\mathbf{x}$ in a certain class $\mathbf{\Omega}$, a unique real number. $\mathbf{\Omega}$ is called the *domain* of the functional and the set of all real numbers associated with the functions in $\mathbf{\Omega}$ is called the *range* of the functional.

Note that the domain of a functional is a class of functions. We shall be concerned with the functionals with domain $\mathbf{\Omega}$ consisting of continuous functions of time t. The performance indices defined in Section 10.2 are examples of such functionals.

A functional J with domain $\mathbf{\Omega}$ has a relative minimum for $\mathbf{x}^*(t)$ if there is a neighbourhood $\bar{\mathbf{N}}$ of $\mathbf{x}^*(t)$ such that for every $\mathbf{x}(t) \in \bar{\mathbf{N}}$, the increment

$$\Delta J = J(\mathbf{x}) - J(\mathbf{x}^*) \geqslant 0 \tag{10.22}$$

Since we are now dealing with function spaces rather than point spaces, it is necessary to generalize the concept of neighbourhood. Consider a class Ω of continuous scalar functions $x(t)$ defined for $t \in [t_0, t_1]$. Figure 10.2 shows a few of the admissible trajectories. Suppose that $x^*(t)$ is the trajectory which minimizes a functional $J(x)$ in a local sense, We say that another curve $x(t) \in \Omega$ lies in the neighbourhood of $x^*(t)$ if $\| x - x^* \| < \varepsilon$ where ε is a positive real number. For the norm (verify that it satisfies the properties of norm)

$$\| x \| = \max_{t_0 \leqslant t \leqslant t_1} \{ | x(t) | \},$$

some trajectories in the neighbourhood of $x^*(t)$ are shown in Fig. 10.2.

The concept of neighbourhood for scalar functions can easily be extended to vector functions.

Functionals of a Single Function: To illustrate the relevant principles of calculus of variations, we shall consider first, in some details, a relatively simple problem. We shall subdivide this problem into four cases.

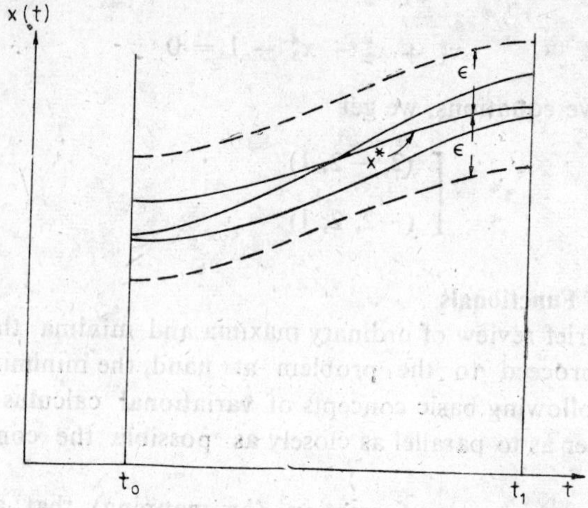

Fig. 10.2 Illustration of neighbourhood of a function

Case I: *Fixed End-Points Problem*

Let $x(t)$ be a scalar function in the class of functions with continuous first derivatives. In the (t, x) plane, given two points (t_0, x^0) and (t_1, x^1): it is required to find a trajectory joining (t_0, x^0) to (t_1, x^1) such that the integral along this trajectory $x = x^*(t)$, given by

$$J(x) = \int_{t_0}^{t_1} g(x, \dot{x}, t)\, dt \tag{10.23}$$

has a relative extremum; g being a function with continuous first and second partial derivatives with respect to all of its arguments.

All continuous curves joining the points (t_0, x^0) and (t_1, x^1) are admissible curves. We are required to find the curve (if any exists) that extremizes $J(x)$. Let x be any curve in the admissible class Ω and $x + \delta x$ be the curve in its neighbourhood (Fig. 10.3). δx represents the *variation* in x which is defined as an infinitesimal, arbitrary change in x for a fixed value of the variable t, i.e., for $\Delta t = 0$ (δ is called the variational operator similar to differential operator d). The operation of variation is commutative with both integration and differentiation, i.e.,

$$\delta\left(\int g\, dt\right) = \int (\delta g)\, dt$$

and

$$\delta\left(\frac{dx}{dt}\right) = \frac{d}{dt}(\delta x)$$

The total increment in J due to variation δx in x is given by

$$\Delta J(x, \delta x) = J(x + \delta x) - J(x)$$

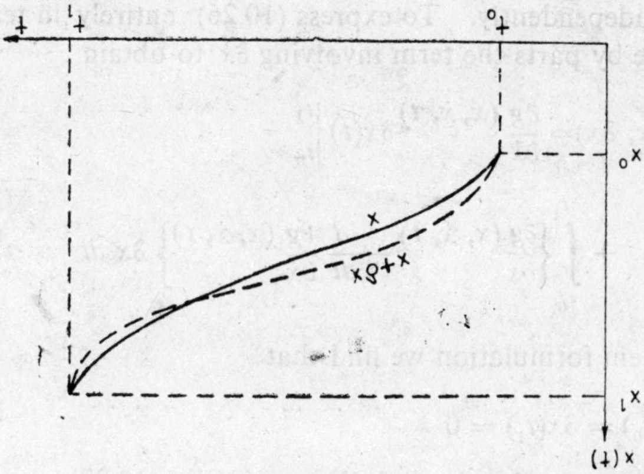

Fig. 10.3 Illustration of fixed end-points problem

$$= \int_{t_0}^{t_1} g(x + \delta x, \dot{x} + \delta \dot{x}, t)\, dt - \int_{t_0}^{t_1} g(x, \dot{x}, t)\, dt$$

$$= \int_{t_0}^{t_1} \Big[g(x + \delta x, \dot{x} + \delta \dot{x}, t) - g(x, \dot{x}, t) \Big]\, dt \tag{10.24}$$

Note that the dependence of J on $\dot{x}$ and $\delta\dot{x}$ is not indicated in the argument of ΔJ because x and $\dot{x}$, δx and $\delta \dot{x}$ are not independent.

Expanding the integrand of (10.24) in a Taylor series about the point $(x(t), \dot{x}(t))$ gives

$$\Delta J(x, \delta x) = \int_{t_0}^{t_1} \Big\{ g(x, \dot{x}, t) + \left(\frac{\partial}{\partial x} g(x, \dot{x}, t)\right) \delta x + \left(\frac{\partial}{\partial \dot{x}} g(x, \dot{x}, t)\right) \delta \dot{x}$$

$$+ \tfrac{1}{2}\left(\frac{\partial^2}{\partial \dot{x}^2} g(x, \dot{x}, t)\right) \delta \dot{x}^2 + \left(\frac{\partial^2}{\partial x \partial \dot{x}} g(x, \dot{x}, t)\right) \delta x\, \delta \dot{x}$$

$$+ \tfrac{1}{2}\left(\frac{\partial^2 g}{\partial \dot{x}^2}(x, \dot{x}, t)\right) \delta \dot{x}^2 + \dots - g(x, \dot{x}, t) \Big\}\, dt \tag{10.25}$$

The (first) variation in J,

$$\delta J(x, \delta x) = \int_{t_0}^{t_1} \left\{ \left(\frac{\partial g}{\partial x}(x, \dot{x}, t)\right) \delta x + \left(\frac{\partial g}{\partial \dot{x}}(x, \dot{x}, t)\right) \delta \dot{x} \right\} dt \tag{10.26}$$

Since δx and $\delta \dot{x}$ are related by

$$\delta x(t) = \int_{t_0}^{t_1} \delta \dot{x}(t)\, dt + \delta x(t_0);$$

selecting δx uniquely determines $\delta \dot{x}$. We shall regard δx as being the function that is varied independently. To express (10.26) entirely in terms containing δx, we integrate by parts the term involving $\delta \dot{x}$ to obtain

$$\delta J(x, \delta x) = \frac{\partial g(x, \dot{x}, t)}{\partial \dot{x}} \delta x(t) \Big|_{t_0}^{t_1} + \int_{t_0}^{t_1} \left\{ \frac{\partial g(x, \dot{x}, t)}{\partial x} - \frac{d}{dt} \frac{\partial g(x, \dot{x}, t)}{\partial \dot{x}} \right\} \delta x \, dt \tag{10.27}$$

From the problem formulation we find that

$$\delta x(t_0) = \delta x(t_1) = 0 \tag{10.28}$$

Therefore, the first term on the right hand side in (10.27) vanishes.

If the variation δJ of a functional J exists and if J takes on a relative extremum for x^*, then

$$\delta J(x^*, \delta x) = 0 \tag{10.29}$$

This result is similar to the necessary condition for extremization of functions and is called *fundamental necessary condition of variational calculus.*

From (10.27)–(10.29) we get

$$\delta J(x^*, \delta x) = \int_{t_0}^{t_1} \left\{ \frac{\partial g(x^*, \dot{x}^*, t)}{\partial x} - \frac{d}{dt} \frac{\partial g(x^*, \dot{x}^*, t)}{\partial \dot{x}} \right\} \delta x \, dt = 0 \tag{10.30}$$

as the necessary condition for $x^*(t)$ to extremize $J(x)$. Equation (10.30) may be expressed as

$$\int_{t_0}^{t_1} \phi(t) \, \delta x \, dt = 0 \tag{10.31}$$

where $\phi(t)$ is a continuous function (as per the assumptions made on g).

Obviously, we can satisfy this equation by making $\phi(t)$ vanish at every point of the trajectory. In the following, we show that this is not merely sufficient but also a necessary condition for satisfying eqn. (10.31). We shall prove this result by contradiction. Assume that (10.31) is satisfied for a $\phi(t)$ not zero for every $t_0 \leqslant t \leqslant t_1$; let $\phi(t)$ be positive (or negative) in some subinterval $t_0 \leqslant a \leqslant t \leqslant b \leqslant t_1$. Since the variation in δx is arbitrary in the interval (t_0, t_1), we may take

$$\delta x = (t-a)^2 (t-b)^2; \; a \leqslant t \leqslant b$$

$$= 0 \quad \text{in the rest of the range.}$$

Since $\delta x(t)$ is positive or zero in $t_0 \leqslant t \leqslant t_1$, the integrand of eqn. (10.31) will have the same sign as $\phi(t)$ (positive or negative) in $a \leqslant t \leqslant b$ and will be zero outside this range. This contradicts the assumption that (10.31) is satisfied.

Therefore, the extremal $x^*(t)$ must satisfy at all points, the equation

$$\frac{\partial g(x^*, \dot{x}^*, t)}{\partial x} - \frac{d}{dt}\frac{\partial g(x^*, \dot{x}^*, t)}{\partial \dot{x}} = 0 \tag{10.32}$$

This equation is called the *Euler-Lagrange equation.*

In general, eqn. (10.32) is a nonlinear, ordinary, time-varying, second-order differential equation which is hard to solve analytically. We may therefore think of numerical integration. There is a difficulty in numerical integration also—the boundary conditions are split; instead of having $x(t_0)$ and $\dot{x}(t_0)$ (or $x(t_1)$ and $\dot{x}(t_1)$) specified, we are given $x(t_0)$ and $x(t_1)$. To integrate numerically, we need values for all the boundary conditions at one end. Therefore, to obtain the extremal $x^*(t)$, a nonlinear *two-point boundary value problem* (TPBVP) is to be solved. We shall discuss the numerical solution of the two-point boundary value problem in Section 10.6.

The trajectory $x^*(t)$ that satisfies the Euler-Lagrange equation is an extremizing trajectory. A way of differentiating between maxima and minima lies in the investigation of second-order terms in eqn. (10.25); but this method is usually laborious. Normally, we content ourselves with investigating a few curves in the neighbourhood of extremal $x^*(t)$ to ascertain whether the solution $x^*(t)$ of the Euler-Lagrange equation yields a minimum (Problem 10.2). In many simple problems, the nature of the extremal is fairly easily determined from the nature of the problem.

Another point to be noted is that the Euler-Lagrange equation may yield multiple solutions. We may, in such a case, evaluate $J(x)$ along each of the extremal trajectories found and differentiate between maxima and minima (Berkovitz (1961) provides a good discussion on necessary and sufficient conditions for a minimum).

Example 10.2: Given

$$\dot{x} = -x + u\,; \quad x(0) = x^0,\ x(2) = x^1$$

Find u^* that minimizes

$$J = \int_0^2 (x^2 + u^2)\, dt$$

Solution:

From the given state equation, we have

$$u = \dot{x} + x$$

Substituting in the functional J, we get

$$J = \int_0^2 (x^2 + (x + \dot{x})^2)\, dt$$

$$= \int_0^2 (2x^2 + 2x\dot{x} + \dot{x}^2)\, dt$$

With $g(x, \dot{x}) = 2x^2 + 2x\dot{x} + \dot{x}^2$, we get from the Euler-Lagrange eqn. (10.32),

$$4x^* + 2\dot{x}^* - \frac{d}{dt}(2x^* + 2\dot{x}^*) = 0$$

or

$$\ddot{x}^* - 2x^* = 0$$

This gives

$x^*(t) = k_1 e^{-\sqrt{2}t} + k_2 e^{\sqrt{2}t}$; k's are constants of integration which can be determined from the given boundary conditions.

$$u^*(t) = x^*(t) + \dot{x}^*(t)$$

$$= k_1(1 - \sqrt{2})\, e^{-\sqrt{2}t} + k_2(1 + \sqrt{2})\, e^{\sqrt{2}t}$$

□

It may be pointed out that the simple procedure for solving optimal control problem described in Example 10.2 may not always be practical as it may often be very difficult to solve for **u** in terms of **x**, **x** and t. An alternate and more useful approach using Lagrange-multiplier method is described later in this section.

Case II: *Terminal Time t_1 Specified, $x(t_1)$ Free*

In the (t, x) plane, given a point (t_0, x^0); it is required to find a trajectory passing through (t_0, x^0) such that the integral along this trajectory, given by

$$J(x) = \int_{t_0}^{t_1} g(x, \dot{x}, t)\, dt \tag{10.33}$$

has a relative extremum (the terminal time t_1 is specified). $x(t)$ and $g(\cdot)$ satisfy the conditions of Case I.

All continuous curves joining the point (t_0, x^0) and the vertical line at $t = t_1$ are admissible curves (Fig. 10.4). Let x be any curve in the admissible class and $x + \delta x$ be the curve in its neighbourhood. Corresponding to these curves, the total increment in J is given by eqn. (10.25). The (first) variation in J as given by eqn. (10.27) is

$$\delta J(x, \delta x) = \frac{\partial g(x, \dot{x}, t)}{\partial \dot{x}}\, \delta x(t)\Big|_{t_0}^{t_1} + \int_{t_0}^{t_1} \left\{ \frac{\partial g}{\partial x}(x, \dot{x}, t) - \frac{d}{dt}\frac{\partial g}{\partial \dot{x}}(x, \dot{x}, t) \right\} \delta x\, dt \tag{10.34}$$

For an extremal $x^*(t)$, we know that $\delta J(x, \delta x)$ must be zero. In the followin

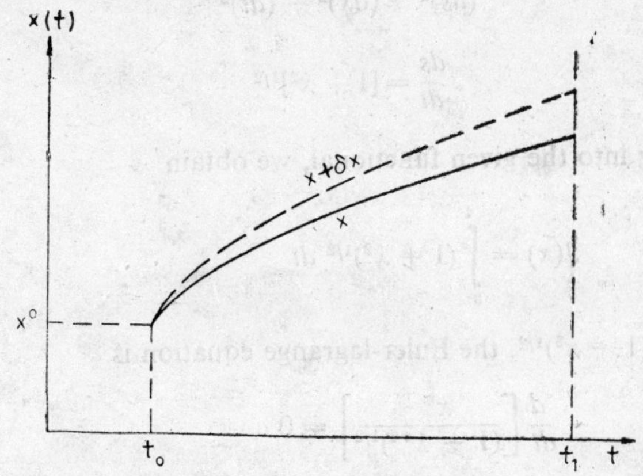

Fig. 10.4 Illustration of the problem with terminal time t_1 fixed and $x(t_1)$ free

we show that the integral in (10.34) must be zero on an extremal.

Suppose that a curve $x^*(t)$ is an extremal for the problem under consideration; t_1 specified and $x(t_1)$ free. The value of $x^*(t)$ at t_1 is say $x^*(t_1) = x^1$. Now consider the fixed end-points problem with the functional (10.33) and the end points (t_0, x^0) and $(t_1, x^*(t_1))$. The curve $x^*(t)$ must be an extremal for this fixed end-points problem and therefore must be a solution of the Euler-Lagrange eqn. (10.32). The integral term in eqn. (10.34) must be zero on an extremal and

$$\left.\frac{\partial g\,(x^*, \dot{x}^*, t)}{\partial \dot{x}}\right|_{t_1} \delta x(t_1) = 0$$

Since $x(t_1)$ is free, $\delta x(t_1)$ is arbitrary; therefore it is necessary that

$$\left.\frac{\partial g\,(x^*, \dot{x}^*, t)}{\partial \dot{x}}\right|_{t_1} = 0 \tag{10.35}$$

Equation (10.35) provides the second required boundary condition (the first being $x(t_0) = x^0$) for the solution of the second-order Euler-Lagrange equation.

Example 10.3: Find the curve with minimum arc length between the point $x(0) = 1$ and the line $t_1 = 4$.

Solution :

If we define the differential arc length as ds, the functional to be minimized is

$$J(x) = \int_0^4 ds\,; \quad x(0) = 1,\ x(2) \text{ unspecified}$$

For a differential arc length,

$$(ds)^2 = (dx)^2 + (dt)^2$$

or

$$\frac{ds}{dt} = [1 + \dot{x}^2]^{1/2}$$

By substituting into the given functional, we obtain

$$J(x) = \int_0^4 (1 + \dot{x}^2)^{1/2}\, dt \tag{10.36}$$

For $g(x, \dot{x}) = (1 + \dot{x}^2)^{1/2}$, the Euler-lagrange equation is

$$-\frac{d}{dt}\left[\frac{\dot{x}^*}{(1 + \dot{x}^{*2})^{1/2}}\right] = 0$$

which reduces to

$$\ddot{x}^*(t) = 0$$

The solution of this equation is

$$x^*(t) = k_1 t + k_2$$

where k_1 and k_2 are constants of integration.

For the problem under consideration,

$$x^*(0) = 1 \quad \text{and}$$

$$\left.\frac{\partial g\,(x^*, \dot{x}^*)}{\partial \dot{x}}\right|_{t_1=4} = \frac{\dot{x}^*(4)}{(1 + \dot{x}^{*2}(4))^{1/2}} = 0$$

which implies that

$$\dot{x}^*(4) = 0$$

Corresponding to these boundary conditions, the constants of integration are obtained as

$$k_1 = 0, \; k_2 = 1$$

Thus

$$x^*(t) = 1$$

Physically, this was of course evident from the statement of the problem.

Case III: *Terminal Time t_1 Free, $x(t_1)$ Specified*

In the (t, x) plane, given a point (t_0, x^0); it is required to find a trajectory passing through (t_0, x^0) such that the integral along the trajectory, given by

$$J(x) = \int_{t_0}^{t_1} g(x, \dot{x}, t)\, dt \tag{10.37}$$

has a relative extremum (the terminal state $x(t_1) \triangleq x^1$ is specified). $x(t)$ and $g(\cdot)$ satisfy the conditions of Case I.

All continuous curves joining the point (t_0, x^0) and the horizontal line at $x = x^1$ are admissible curves. Let x be any curve in admissible class and $x + \delta x$ be the curve in its neighbourhood (Fig. 10.5). Corresponding to these curves, the total increment in J is given by

$$\Delta J(x, \delta x) = \int_{t_0}^{t_1+\Delta t_1} g(x + \delta x, \dot{x} + \delta \dot{x}, t)\, dt - \int_{t_0}^{t_1} g(x, \dot{x}, t)\, dt$$

$$= \int_{t_1}^{t_1+\Delta t_1} g(x + \delta x, \dot{x} + \delta \dot{x}, t)\, dt$$

$$+ \int_{t_0}^{t_1} \Big[g(x + \delta x, \dot{x} + \delta \dot{x}, t) - g(x, \dot{x}, t) \Big]\, dt$$

Applying the mean-value theorem for integrals to the first integral, we obtain

$$\int_{t_1}^{t_1+\Delta t_1} g(x + \delta x, \dot{x} + \delta \dot{x}, t)\, dt \sim g(x, \dot{x}, t)\Big|_{t_1} \Delta t_1$$

where the symbol $\sim$ indicates equality except for terms of order higher than one in δx, $\delta \dot{x}$ and Δt_1. The integrand of the second integral can be expanded about $(x, \dot{x})$ in a Taylor series to give (refer eqns. (10.24)–(10.27))

$$\Delta J(x, \delta x) \sim g(x, \dot{x}, t)\Big|_{t_1} \Delta t_1 + \frac{\partial g}{\partial \dot{x}}(x, \dot{x}, t)\Big|_{t_1} \delta x(t_1)$$

$$+ \int_{t_0}^{t_1} \left[\frac{\partial g(x, \dot{x}, t)}{\partial x} - \frac{d}{dt} \frac{\partial g}{\partial \dot{x}}(x, \dot{x}, t) \right] \delta x\, dt \qquad (10.38)$$

The increment $\Delta x(t_1)$ shown in Fig. 10.5 depends on Δt_1. This dependence can be linearly approximated as follows:

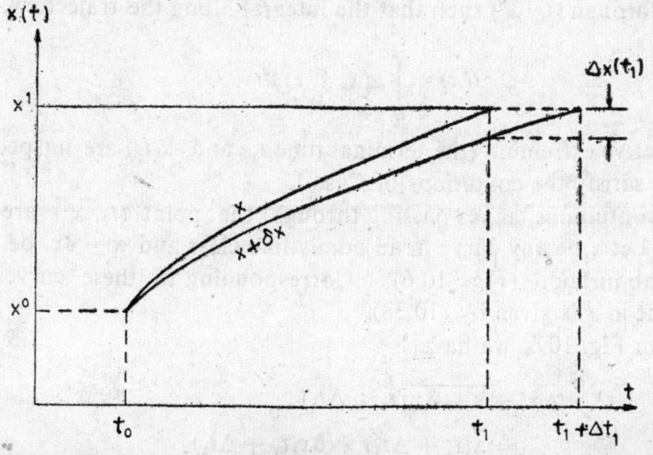

Fig. 10.5 Illustration of the problem with terminal time t_1 free and $x(t_1)$ fixed

$$x(t_1) = x^1$$

$$\overline{x + \delta x}\,(t_1 + \Delta t_1) = x^1$$

$$= x(t_1 + \Delta t_1) + \delta x(t_1 + \Delta t_1)$$

$$= x(t_1) + \dot{x}(t_1)\,\Delta t_1 + \delta x(t_1); \text{ to a first-order}$$

This gives

$$\delta x(t_1) = -\,\dot{x}(t_1)\,\Delta t_1$$

Substituting in eqn. (10.38) and applying fundamental necessary condition of variational calculus, we have for an extremal $x^*(t)$,

$$\delta J(x^*, \delta x) = 0 = \left[g(x^*, \dot{x}^*, t) - \dot{x}^* \frac{\partial}{\partial \dot{x}} g(x^*, \dot{x}^*, t) \right]\Bigg|_{t_1} \Delta t_1$$

$$+ \int_{t_0}^{t_1} \left[\frac{\partial g}{\partial x}(x^*, \dot{x}^*, t) - \frac{d}{dt}\frac{\partial}{\partial \dot{x}} g(x^*, \dot{x}^*, t) \right] \delta x\, dt \tag{10.39}$$

As in Case II, we argue that the extremal for this free end point problem is also an extremal for a particular fixed end point problem and therefore must be a solution of the Euler-Lagrange equation (10.32). The integral term in eqn. (10.39) must be zero on an extremal and

$$\left[g(x^*, \dot{x}^*, t) - \dot{x}^* \frac{\partial}{\partial \dot{x}} g(x^*, \dot{x}^*, t) \right]\Bigg|_{t_1} \Delta t_1 = 0$$

Since t_1 is free, Δt_1 is arbitrary and the required boundary condition at t_1 is

$$\left[g(x^*, \dot{x}^*, t) - \dot{x}^* \frac{\partial g}{\partial \dot{x}}(x^*, \dot{x}^*, t) \right]\Bigg|_{t_1} = 0 \tag{10.40}$$

Case IV: *Both the Terminal Time t_1 and $x(t_1)$ Free*

In the (t, x) plane, given a point (t_0, x^0); it is required to find a trajectory passing through (t_0, x^0) such that the integral along the trajectory, given by

$$J(x) = \int_{t_0}^{t_1} g(x, \dot{x}, t)\, dt \tag{10.41}$$

has a relative extremum (the terminal time t_1 and $x(t_1)$ are unspecified). $x(t)$ and $g(\cdot)$ satisfy the conditions of Case I.

All continuous curves passing through the point (t_0, x^0) are admissible curves. Let x be any curve in an admissible class aud $x + \delta x$ be the curve in its neighbourhood (Fig. 10.6). Corresponding to these curves, the total increment in J is given by (10.38).

From Fig. 10.6, we have

$$x^1 + \Delta x^1 = \overline{x + \delta x}\,(t_1 + \Delta t_1)$$

$$= x(t_1 + \Delta t_1) + \delta x(t_1 + \Delta t_1)$$

$$= x(t_1) + \dot{x}(t_1)\Delta t_1 + \delta x(t_1)\,; \text{ to a first-order}$$

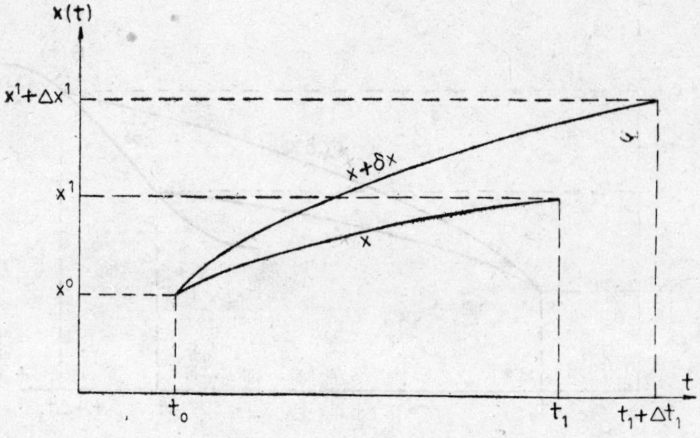

Fig. 10.6 Illustration of the problem with both the terminal time t_1 and $\mathbf{x}(t_1)$ free

Therefore

$$\delta x(t_1) = \Delta x^1 - \dot{x}(t_1)\,\Delta t_1$$

Substituting this in eqn. (10.38), we obtain for an extremal

$$\delta J(x^*, \delta x) = 0 = \frac{\partial g\,(x^*, \dot{x}^*, t)}{\partial \dot{x}}\bigg|_{t_1} \Delta x^1$$

$$+ \left[g(x^*, \dot{x}^*, t) - \dot{x}^* \frac{\partial g\,(x^*, \dot{x}^*, t)}{\partial \dot{x}} \right]_{t_1} \Delta t_1$$

$$+ \int_{t_0}^{t_1} \left[\frac{\partial g\,(x^*, \dot{x}^*, t)}{\partial x} - \frac{d}{dt}\frac{\partial g\,(x^*, \dot{x}^*, t)}{\partial \dot{x}} \right] \delta x\, dt \qquad (10.42)$$

As before, we argue that the Euler-Lagrange equation must be satisfied; therefore the integral is zero. There may be a variety of end conditions in practice. In the following we consider a case wherein the final value of x is constrained to lie on a specified moving point $\theta(t)$, i.e.,

$$x(t_1) = \theta(t_1)$$

An extremal x^* and a neighbouring curve $x^* + \delta x$ for this case are shown in Fig. 10.7. From this figure we have

$$\Delta x^1 = \dot{\theta}(t_1)\,\Delta t_1\,; \quad \text{to a first-order.}$$

Substituting this into eqn. (10.42), we get the following end condition.

$$\left\{ \left[\frac{\partial g\,(x^*, \dot{x}^*, t)}{\partial \dot{x}} \right] \left[\dot{\theta} - \dot{x}^* \right] \right\}\bigg|_{t_1} + g(x^*, \dot{x}^*, t)\bigg|_{t_1} = 0 \qquad (10.43)$$

This equation is called the *transversality condition*.

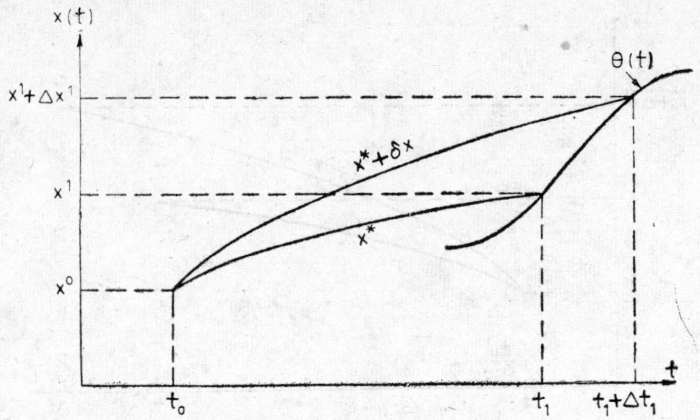

Fig. 10.7 Illustration of the problem with both the terminal time t_1 and $x(t_1)$ free but related

Example 10.4: Find the curve with the minimum arc length joining the point (0, 0) and the line $\theta(t) = 2 - t$.

Solution :

From Example 10.3, we know that extremals for the minimum arc length problem are straight lines

$$x^*(t) = k_1 t + k_2$$

Applying the boundary condition $x(0) = 0$, we get $k_2 = 0$. To evaluate the other constant of integration, we use the transversality condition. From eqn. (10.43), we have

$$\frac{\dot{x}^*}{(1 + \dot{x}^{*2})^{1/2}} [\dot{\theta} - \dot{x}^*] \Big|_{t_1} + (1 + \dot{x}^{*2})^{1/2} \Big|_{t_1} = 0$$

Since $\dot{x}^*(t_1) = k_1$ and $\dot{\theta}(t_1) = -1$, we have

$$\frac{k_1}{\sqrt{1 + k_1^2}} (-1 - k_1) + \sqrt{1 + k_1^2} = 0$$

Solving for k_1, we get

$$k_1 = 1$$

and the extremal trajectory is given by

$$x^*(t) = t$$

The value of t_1 is obtained from the equation

$$x^*(t_1) = \theta(t_1)$$

which gives

$$t_1 = 1$$

Figure 10.8 shows what we knew already; the shortest path is along the perpendicular to the line and passes through the origin.

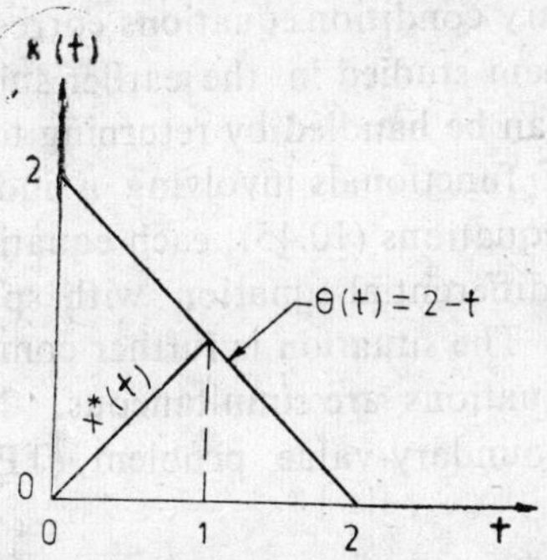

Fig. 10.8 Extremal curve for Example 10.4

Functionals Involving n Independent Functions: The 2-dimensional problem discussed above, can now be generalized to a $(n+1)$-dimensional problem. Consider the functional

$$J(\mathbf{x}) = \int_{t_0}^{t_1} g(\mathbf{x}, \dot{\mathbf{x}}, t)\, dt \tag{10.44}$$

where $\mathbf{x}(t) \overset{\Delta}{=} [x_1(t) \quad x_2(t) \; \dots \; x_n(t)]^T$;

$x_i(t)$ are independent functions with continuous first derivatives and g has continuous first and second derivatives with respect to all of its arguments.

In the $(n+1)$-dimensional space $(t, \mathbf{x})$, given a point $(t_0, \mathbf{x}^0)$, it is required to find a trajectory $\mathbf{x} = \mathbf{x}^*(t)$, such that the integral along the trajectory given by (10.44) has a relative extremum (the terminal time t_1 and $\mathbf{x}(t_1)$ are unspecified).

The reader might have noted that this problem is the extension of the problem in Case IV of earlier subsection. The results obtained in the earlier subsection can easily be generalized.

The extremal $\mathbf{x}^*$ for the free end point problem under consideration must be a solution of the Euler-Lagrange equations (refer eqn. (10.32))

$$\frac{\partial g(\mathbf{x}^*, \dot{\mathbf{x}}^*, t)}{\partial \mathbf{x}} - \frac{d}{dt}\left[\frac{\partial g(\mathbf{x}^*, \dot{\mathbf{x}}^*, t)}{\partial \dot{\mathbf{x}}}\right] = \mathbf{0} \tag{10.45}$$

The boundary conditions at the terminal time t_1 are specified by the relationship (refer eqn. (10.42))

$$\left[\left.\frac{\partial g(\mathbf{x}^*, \dot{\mathbf{x}}^*, t)}{\partial \dot{\mathbf{x}}}\right|_{t_1}\right]^T \Delta \mathbf{x}^1$$
$$+ \left[g(\mathbf{x}^*, \dot{\mathbf{x}}^*, t) - \left(\frac{\partial g(\mathbf{x}^*, \dot{\mathbf{x}}^*, t)}{\partial \dot{\mathbf{x}}}\right)^T \dot{\mathbf{x}}^* \right]\Bigg|_{t_1} \Delta t_1 = 0 \tag{10.46}$$

Equations (10.45) and (10.46) summarize necessary conditions that must be satisfied by an extremal curve. The boundary condition equations are obtained by making appropriate substitutions for $\Delta \mathbf{x}^1$ and Δt_1 in equation (10.46). Table 10.1 gives boundary condition equations corresponding to the four cases of the variational problem studied in the earlier subsection. Situations not included in Table 10.1 can be handled by returning to eqn. (10.46).

Thus, in the case of functionals involving n independent functions, we have n Euler-Lagrange equations (10.45); each equation in general, a nonlinear ordinary, second-order differential equation with split boundary conditions, which is hard to solve. The situation is further complicated by the fact that the n Euler-Lagrange equations are simultaneous. Numerical techniques for solving a two-point boundary-value problem (TPBVP) are discussed in Section 10.6.

Table 10.1: Summary of Boundary-Value Relationships

Problem Description	Boundary conditions	Remarks
1. $t_1, \mathbf{x}(t_1)$ both specified	$\mathbf{x}^*(t_0)=\mathbf{x}^\circ$ $\mathbf{x}^*(t_1)=\mathbf{x}^1$	$2n$ equations for $2n$ constants of integration
2. t_1 specified, $\mathbf{x}(t_1)$ free	$\mathbf{x}^*(t_0)=\mathbf{x}^\circ$ $\left.\dfrac{\partial g(\mathbf{x}^*, \dot{\mathbf{x}}^*, t)}{\partial \dot{\mathbf{x}}}\right\rvert_{t_1}=\mathbf{0}$	$2n$ equations for $2n$ constants of integration
3. t_1 free, $\mathbf{x}(t_1)$ specified	$\mathbf{x}^*(t_0)=\mathbf{x}^\circ$ $\mathbf{x}^*(t_1)=\mathbf{x}^1$ $\left.\left[g(\mathbf{x}^*,\dot{\mathbf{x}}^*,t)-\left[\dfrac{\partial g(\mathbf{x}^*, \dot{\mathbf{x}}^*, t)}{\partial \dot{\mathbf{x}}}\right]^T \dot{\mathbf{x}}^*\right]\right\rvert_{t_1}=0$	$(2n+1)$ equations for $2n$ constants of integration and t_1
4. $t_1, \mathbf{x}(t_1)$ free but related by $\mathbf{x}(t_1)=\boldsymbol{\theta}(t_1)$	$\mathbf{x}^*(t_0)=\mathbf{x}^\circ$ $\mathbf{x}^*(t_1)=\boldsymbol{\theta}(t_1)$ $\left.\left[g(\mathbf{x}^*, \dot{\mathbf{x}}^*, t_1) + \left[\dfrac{\partial g(\mathbf{x}^*, \dot{\mathbf{x}}^*, t)}{\partial \dot{\mathbf{x}}}\right]^T [\dot{\boldsymbol{\theta}}-\dot{\mathbf{x}}^*]\right]\right\rvert_{t_1}=0$	$(2n+1)$ equations to determine $2n$ constants of intergration and t_1

Example 10.5: Find the extremal of the functional

$$J(x) = \int_0^{\pi/4} (x_1^2 + \dot{x}_2^2 + \dot{x}_1\dot{x}_2)\, dt$$

The boundary conditions are

$$x_1(0) = 0,\ x_1\left(\frac{\pi}{4}\right) = 1,\ x_2(0) = 0,\ x_2\left(\frac{\pi}{4}\right) = -1$$

Solution:

For $$g(\mathbf{x}, \dot{\mathbf{x}}) = x_1^2 + \dot{x}_2^2 + \dot{x}_1\dot{x}_2$$

the Euler-Lagrange equations are

$$2x_1^* - \ddot{x}_2^* = 0$$

$$-2\ddot{x}_2^* - \ddot{x}_1^* = 0$$

From these equations, we obtain

$$\ddot{x}_1^* + 4x_1^* = 0$$

which has the solution

$$x_1^*(t) = k_1 \cos 2t + k_2 \sin 2t$$

Now

$$\ddot{x}_2^* = 2x_1^* = 2k_1 \cos 2t + 2k_2 \sin 2t$$

which has the solution

$$x_2^*(t) = -\frac{k_1}{2} \cos 2t - \frac{k_2}{2} \sin 2t + k_3 t + k_4$$

From the boundary conditions, we find that

$$k_1 = 0,\ k_2 = 1,\ k_3 = -\frac{2}{\pi},\ k_4 = 0$$

The extremal curve is then

$$x_1^*(t) = \sin 2t$$

$$x_2^*(t) = -\tfrac{1}{2} \sin 2t - \frac{2t}{\pi}$$

Constrained Minimization: We now examine the following constrained problem:

Determine a set of necessary conditions for an $(n+p)\times 1$ vector function $\mathbf{z}^*$ to be an extremal for a functional of the form

$$J(\mathbf{z}) = \int_{t_0}^{t_1} g(\mathbf{z}, \dot{\mathbf{z}}, t)\, dt \tag{10.47}$$

subject to n-vector equality constraint

$$\boldsymbol{\eta}(\mathbf{z}, \dot{\mathbf{z}}, t) = \mathbf{0} \quad \text{for all} \quad t \in [t_0, t_1] \tag{10.48}$$

Note that because of n differential equation constraints, only p of the $(n+p)$ components of $\mathbf{z}$ are independent.

Let us form the *augmented functional* by adjoining the constraining relationships to J.

$$J_a(\mathbf{z}, \boldsymbol{\lambda}) = \int_{t_0}^{t_1} \left[g(\mathbf{z}, \dot{\mathbf{z}}, t) + \boldsymbol{\lambda}^T(t)\, [\boldsymbol{\eta}(\mathbf{z}, \dot{\mathbf{z}}, t)] \right] dt \tag{10.49}$$

where $\lambda_i(t)$, $i = 1, 2, \ldots, n$, are Lagrange multiplier functions (Since the constraints are to be satisfied for all $t \in [t_0, t_1]$, the Lagrange multipliers are assumed to be functions of time. This allows the flexibility of multiplying the constraining relationship by a different real number for each value of t).

By introducing the variation in the functions $\mathbf{z}$, $\dot{\mathbf{z}}$ and $\boldsymbol{\lambda}$, we get the variation in the functional J_a,

$$\begin{aligned}\delta J_a(\mathbf{z}, \delta\mathbf{z}, \boldsymbol{\lambda}, \delta\boldsymbol{\lambda}) = \int_{t_0}^{t_1} \Bigg\{ &\left[\left(\frac{\partial g(\mathbf{z}, \dot{\mathbf{z}}, t)}{\partial \mathbf{z}} \right)^T + \boldsymbol{\lambda}^T \frac{\partial \boldsymbol{\eta}(\mathbf{z}, \dot{\mathbf{z}}, t)}{\partial \mathbf{z}} \right] \delta\mathbf{z} \\ &+ \left[\left(\frac{\partial g(\mathbf{z}, \dot{\mathbf{z}}, t)}{\partial \dot{\mathbf{z}}} \right)^T \right. \\ &\left. + \boldsymbol{\lambda}^T \frac{\partial \boldsymbol{\eta}(\mathbf{z}, \dot{\mathbf{z}}, t)}{\partial \dot{\mathbf{z}}} \right] \delta\dot{\mathbf{z}} \\ &+ \left[\boldsymbol{\eta}(\mathbf{z}, \dot{\mathbf{z}}, t)^T \right] \delta\boldsymbol{\lambda} \Bigg\}\, dt \end{aligned} \tag{10.50}$$

where (refer Appendix III)

$$\underset{((n+p)\times 1)}{\frac{\partial g}{\partial \mathbf{z}}} \triangleq \begin{bmatrix} \dfrac{\partial g}{\partial z_1} \\ \dfrac{\partial g}{\partial z_2} \\ \vdots \\ \dfrac{\partial g}{\partial z_{n+p}} \end{bmatrix}; \quad \underset{(n\times(n+p))}{\frac{\partial \boldsymbol{\eta}}{\partial \mathbf{z}}} \triangleq \begin{bmatrix} \dfrac{\partial \eta_1}{\partial z_1} & \dfrac{\partial \eta_1}{\partial z_2} & \cdots & \dfrac{\partial \eta_1}{\partial z_{n+p}} \\ \vdots & \vdots & & \vdots \\ \dfrac{\partial \eta_n}{\partial z_1} & \dfrac{\partial \eta_n}{\partial z_2} & \cdots & \dfrac{\partial \eta_n}{\partial z_{n+p}} \end{bmatrix}$$

In the previous subsections, we observed that the Euler-Lagrange equation must be satisfied by an extremal regardless of the boundary conditions. For the time being, we shall ignore terms that enter only into the determination of boundary conditions.

Integrating by parts the terms containing $\delta\dot{\mathbf{z}}$ in (10.50) and retaining only the terms inside the integral (the terms outside the integral enter only into the determination of boundary conditions) we obtain

$$\delta J_a(\mathbf{z}, \delta\mathbf{z}, \boldsymbol{\lambda}. \delta\boldsymbol{\lambda}) = \int_{t_0}^{t_1} \Bigg[\Bigg\{\left(\frac{\partial}{\partial \mathbf{z}} g(\mathbf{z}, \dot{\mathbf{z}}, t)\right)^T + \boldsymbol{\lambda}^T \frac{\partial}{\partial \mathbf{z}} \boldsymbol{\eta}(\mathbf{z}, \dot{\mathbf{z}}, t)$$

$$- \frac{d}{dt}\left[\left(\frac{\partial}{\partial \dot{\mathbf{z}}} g(\mathbf{z}, \dot{\mathbf{z}}, t)\right)^T + \boldsymbol{\lambda}^T \frac{\partial}{\partial \dot{\mathbf{z}}} \boldsymbol{\eta}(\mathbf{z}, \dot{\mathbf{z}}, t)\right]\Bigg\} \delta\mathbf{z}$$

$$+ [\boldsymbol{\eta}(\mathbf{z}, \dot{\mathbf{z}}, t)]^T \delta\boldsymbol{\lambda}\Bigg] dt$$

An extremal $\mathbf{z}^*(t)$ must satisfy the following necessary conditions:

$$\delta J_a(\mathbf{z}^*, \delta\mathbf{z}, \boldsymbol{\lambda}, \delta\boldsymbol{\lambda}) = 0$$

$$\boldsymbol{\eta}(\mathbf{z}^*, \dot{\mathbf{z}}^*, t) \qquad = \mathbf{0}$$

Therefore

$$\int_{t_0}^{t_1} \Bigg\{\left(\frac{\partial g(\mathbf{z}^*, \dot{\mathbf{z}}^*, t)}{\partial \mathbf{z}}\right)^T + \boldsymbol{\lambda}^T \frac{\partial \boldsymbol{\eta}(\mathbf{z}^*, \dot{\mathbf{z}}^*, t)}{\partial \mathbf{z}} - \frac{d}{dt}\left[\left(\frac{\partial g(\mathbf{z}^*, \dot{\mathbf{z}}^*, t)}{\partial \dot{\mathbf{z}}}\right)^T\right.$$

$$\left. + \boldsymbol{\lambda}^T \frac{\partial \boldsymbol{\eta}(\mathbf{z}^*, \dot{\mathbf{z}}^*, t)}{\partial \dot{\mathbf{z}}}\right]\Bigg\} \delta\mathbf{z}\, dt = 0$$

Once the constraints are satisfied, we can choose the n Lagrange multipliers $\lambda_i(t)$ arbitrarily. (From eqn. (10.49) we see that $J_a = J$ for any function $\boldsymbol{\lambda}$ if the constraints are satisfied.) We choose λ_i's so that the coefficients of n of the components of $\delta\mathbf{z}$ are zero for all $t \in [t_0, t_1]$. The remaining p components of $\delta\mathbf{z}$ are then independent and as per the result of eqn. (10.31), the coefficients of these components of $\delta\mathbf{z}$ must be zero.

Therefore the necessary conditions for $\mathbf{z}^*(t)$ to extremize $J(\mathbf{z})$ of (10.47) under the constraint (10.48) are

$$\boldsymbol{\eta}(\mathbf{z}^*, \dot{\mathbf{z}}^*, t) = \mathbf{0} \quad \text{for all } t \in [t_0, t_1] \tag{10.51}$$

$$\frac{\partial g(\mathbf{z}^*, \dot{\mathbf{z}}^*, t)}{\partial \mathbf{z}} + \left(\frac{\partial \boldsymbol{\eta}(\mathbf{z}^*, \dot{\mathbf{z}}^*, t)}{\partial \mathbf{z}}\right)^T \boldsymbol{\lambda}^*(t)$$

$$- \frac{d}{dt}\left[\frac{\partial g(\mathbf{z}^*, \dot{\mathbf{z}}^*, t)}{\partial \dot{\mathbf{z}}} + \left(\frac{\partial \boldsymbol{\eta}(\mathbf{z}^*, \dot{\mathbf{z}}^*, t)}{\partial \dot{\mathbf{z}}}\right)^T \boldsymbol{\lambda}^*(t)\right] = \mathbf{0} \tag{10.52}$$

Defining *augmented integrand function*

$$g_a(\mathbf{z}, \dot{\mathbf{z}}, \boldsymbol{\lambda}, t) = g(\mathbf{z}, \dot{\mathbf{z}}, t) + \boldsymbol{\lambda}^T \boldsymbol{\eta}(\mathbf{z}, \dot{\mathbf{z}}, t); \tag{10.53}$$

eqn. (10.52) may be written as

$$\frac{\partial g_a(\mathbf{z}^*, \dot{\mathbf{z}}^*, \boldsymbol{\lambda}^*, t)}{\partial \mathbf{z}} - \frac{d}{dt}\left(\frac{\partial g_a(\mathbf{z}^*, \dot{\mathbf{z}}^*, \boldsymbol{\lambda}^*, t)}{\partial \dot{\mathbf{z}}}\right) = \mathbf{0} \tag{10.54}$$

Example 10.6: Suppose that the system

$$\dot{x}_1(t) = x_2(t)$$

$$\dot{x}_2(t) = u(t)$$

is to be controlled to minimize the performance measure

$$J(\mathbf{x}, u) = \tfrac{1}{2} \int_0^2 u^2 \, dt$$

Find a set of necessary conditions for optimal control.

Solution:

If we define $x_1 = z_1$, $x_2 = z_2$ and $u = z_3$; our problem becomes:

Find the equations that must be satisfied for a function $\mathbf{z}^*$ to be the extremal for the functional

$$J(\mathbf{z}) = \tfrac{1}{2} \int_0^2 z_3^2 \, dt$$

where the function $\mathbf{z}$ must satisfy the differential equation constraints

$$\dot{z}_1 = z_2$$
$$\dot{z}_2 = z_3$$

The augmented integrand of the functional is

$$g_a(\mathbf{z}, \dot{\mathbf{z}}, \boldsymbol{\lambda}) = \tfrac{1}{2} z_3^2 + \lambda_1(z_2 - \dot{z}_1) + \lambda_2(z_3 - \dot{z}_2)$$

From eqn. (10.54), we obtain

$$\dot{\lambda}_1^*(t) = 0$$
$$\lambda_1^*(t) + \dot{\lambda}_2^*(t) = 0 \qquad (10.55a)$$
$$z_3^*(t) + \lambda_2^*(t) = 0$$

The two additional equations that must be satisfied by the extremal are the constraints (eqn. (10.51))

$$\dot{z}_1^*(t) = z_2^*(t)$$
$$\dot{z}_2^*(t) = z_3^*(t) \qquad (10.55b)$$

We thus have 5 equations in (10.55) and we have to solve these equations for 5 unknowns—z_1^*, z_2^*, z_3^*, λ_1^*, λ_2^*. If the boundary conditions are

$$x_1(0) = 1 \ ; \quad x_1(2) = 0$$
$$x_2(0) = 1 \ ; \quad x_2(2) = 0$$

we obtain

$$z_1^*(t) = x_1^*(t) = \tfrac{1}{2}t^3 - \tfrac{7}{4}t^2 + t + 1$$
$$z_2^*(t) = x_2^*(t) = \tfrac{3}{2}t^2 - \tfrac{7}{2}t + 1$$
$$z_3^*(t) = u^*(t) = 3t - \tfrac{7}{2}$$

The optimal control and the corresponding trajectories are plotted in Fig. 10.9.

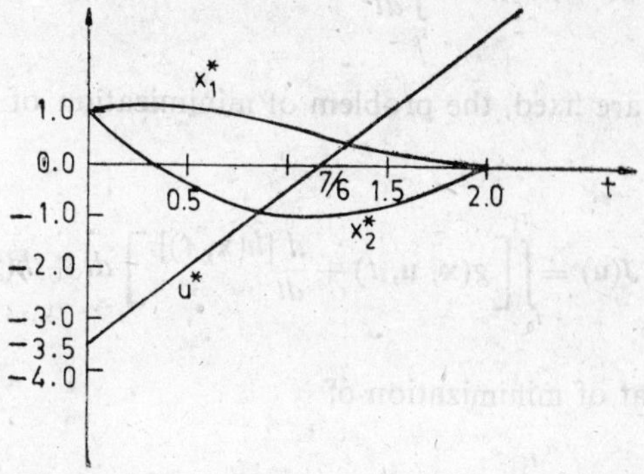

Fig. 10.9 Optimal control and corresponding trajectories for system of Example 10.6

Formulation of Variational Calculus Using Hamiltonian Method

In the previous two subsections, we formulated many problems in classical calculus of variations. Some simple optimal control examples were stated and solved. In this subsection, we wish to re-examine these problems to provide more convenient methods to handle them. To achieve this objective, we will present a formulation of variational calculus using the Hamiltonian method. From this formulation, we shall describe Pontryagin's Minimum Principle in the next section.

In Section 10.2, we presented the following formulation of a general optimal control problem.

Find an admissible control $\mathbf{u}^*(t)$ which causes the system

$$\dot{\mathbf{x}}(t) = \mathbf{f}(\mathbf{x}(t), \mathbf{u}(t), t)\ ; \quad \mathbf{x} \in \mathcal{R}^n,\ \mathbf{u} \in \mathcal{R}^p \tag{10.56}$$

to follow an admissible trajectory $\mathbf{x}^*(t)$ and minimizes the performance measure

$$J = h(\mathbf{x}(t_1), t_1) + \int_{t_0}^{t_1} g(\mathbf{x}(t), \mathbf{u}(t), t)\, dt \tag{10.57}$$

We shall assume that the admissible control and admissible state trajectories are not bounded and that the initial state $\mathbf{x}(t_0) \triangleq \mathbf{x}^0$ and initial time t_0 are specified.

In the terminology of the previous subsection (Constrained Minimization), we have a problem involving $(n + p)$ functions which must satisfy the n differential equation constraints given by (10.56). The p control inputs are the independent functions. The only difference is in the performance index (functional); the term $h(\mathbf{x}(t_1), t_1)$ appears in (10.57) while such a term did not appear in (10.47). However, it is possible to transform the performance index (10.57) into a purely integral type of performance index. Assuming h to be a differentiable function, we can write

$$h(\mathbf{x}(t_1), t_1) = \int_{t_0}^{t_1} \frac{d}{dt} [h(\mathbf{x}(t), t)]\, dt + h(\mathbf{x}(t_0), t_0)$$

Since $\mathbf{x}(t_0)$ and t_0 are fixed, the problem of minimization of the performance index

$$J(\mathbf{u}) = \int_{t_0}^{t_1} \left[g(\mathbf{x}, \mathbf{u}, t) + \frac{d}{dt} [h(\mathbf{x}, t)] \right] dt + h(\mathbf{x}(t_0), t_0)$$

is equivalent to that of minimization of

$$J(\mathbf{u}) = \int_{t_0}^{t_1} \left[g(\mathbf{x}, \mathbf{u}, t) + \frac{d}{dt} [h(\mathbf{x}, t)] \right] dt \tag{10.58}$$

Thus the performance index (10.57) is transformed into integral performance index (10.58). It is, however, convenient to rewrite (10.58) by making use of the chain rule for differentiation to express $\dfrac{dh(\mathbf{x}, t)}{dt}$ as

$$\frac{dh(\mathbf{x}, t)}{dt} = \left(\frac{\partial h(\mathbf{x}, t)}{\partial \mathbf{x}} \right)^T \dot{\mathbf{x}} + \frac{\partial h(\mathbf{x}, t)}{\partial t}$$

The performance index then becomes

$$J(\mathbf{u}) = \int_{t_0}^{t_1} \left[g(\mathbf{x}, \mathbf{u}, t) + \left(\frac{\partial h(\mathbf{x}, t)}{\partial \mathbf{x}} \right)^T \dot{\mathbf{x}} + \frac{\partial h(\mathbf{x}, t)}{\partial t} \right] dt \tag{10.59}$$

To include the differential equation constraints (10.56), we form the augmented integrand of the functional,

$$g_a(\mathbf{x}, \dot{\mathbf{x}}, \mathbf{u}, \boldsymbol{\lambda}, t) = g(\mathbf{x}, \mathbf{u}, t) + \boldsymbol{\lambda}^T[\mathbf{f}(\mathbf{x}, \mathbf{u}, t) - \dot{\mathbf{x}}] + \left(\frac{\partial h(\mathbf{x}, t)}{\partial \mathbf{x}} \right)^T \dot{\mathbf{x}} + \frac{\partial h(\mathbf{x}, t)}{\partial t} \tag{10.60}$$

where $\boldsymbol{\lambda} = [\lambda_1, \lambda_2, \ldots, \lambda_n]^T$ is a Lagrange multiplier.

The problem now is to minimize

$$J_a(\mathbf{u}) = \int_{t_0}^{t_1} g_a(\mathbf{x}, \dot{\mathbf{x}}, \mathbf{u}, \boldsymbol{\lambda}, t)\, dt$$

The general optimal control problem is now in such a form that calculus of variations may be directly applied by defining an $(n + p)$-dimensional vector

$$\mathbf{z}(t) = \begin{bmatrix} x_1(t) \\ x_2(t) \\ \vdots \\ x_n(t) \\ u_1(t) \\ \vdots \\ u_p(t) \end{bmatrix} = \begin{bmatrix} \mathbf{x} \\ \mathbf{u} \end{bmatrix}$$

However, for notational convenience, we continue to treat **x** and **u** separately. This simply yields two sets of Euler-Lagrange equations in terms of $\mathbf{x}^*$ and $\mathbf{u}^*$.

Let us first look at the Euler-Lagrange equations in terms of $\mathbf{x}^*$.

$$\frac{\partial}{\partial \mathbf{x}} g_a(\mathbf{x}^*, \dot{\mathbf{x}}^*, \mathbf{u}^*, \boldsymbol{\lambda}^*, t) - \frac{d}{dt}\left[\frac{\partial}{\partial \dot{\mathbf{x}}} g_a(\mathbf{x}^*, \dot{\mathbf{x}}^*, \mathbf{u}^*, \boldsymbol{\lambda}^*, t)\right] = 0$$

If eqn. (10.60) is substituted for g_a and the terms which are zero are removed, we get

$$\frac{\partial}{\partial \mathbf{x}}\left[g(\mathbf{x}^*, \mathbf{u}^*, t) + \left(\frac{\partial h}{\partial \mathbf{x}}(\mathbf{x}^*, t)\right)^T \dot{\mathbf{x}}^*, + \frac{\partial h}{\partial t}(\mathbf{x}^*, t) + \boldsymbol{\lambda}^{*T}\mathbf{f}(\mathbf{x}^*, \mathbf{u}^*, t)\right]$$
$$- \frac{d}{dt}\left[\frac{\partial h}{\partial \mathbf{x}}(\mathbf{x}^*, t) - \boldsymbol{\lambda}^*\right] = 0 \qquad (10.61)$$

Next, let us carry out the indicated differentiations only on the three terms involving $h(\mathbf{x}^*, t)$.

$$\frac{\partial}{\partial \mathbf{x}}[g(\mathbf{x}^*, \mathbf{u}^*, t) + \boldsymbol{\lambda}^{*T}\mathbf{f}(\mathbf{x}^*, \mathbf{u}^*, t)] + \left(\frac{\partial^2 h}{\partial \mathbf{x}^2}(\mathbf{x}^*, t)\right)\dot{\mathbf{x}}^* + \frac{\partial^2 h}{\partial \mathbf{x}\,\partial t}(\mathbf{x}^*, t)$$
$$- \frac{d}{dt}\left[\frac{\partial h}{\partial \mathbf{x}}(\mathbf{x}^*, t)\right] + \dot{\boldsymbol{\lambda}}^* = 0 \qquad (10.62)$$

Now

$$\frac{d}{dt}\left[\frac{\partial h}{\partial \mathbf{x}}(\mathbf{x}^*, t)\right] = \left(\frac{\partial^2 h}{\partial \mathbf{x}^2}(\mathbf{x}^*, t)\right)\dot{\mathbf{x}}^* + \frac{\partial^2 h}{\partial t\,\partial \mathbf{x}}(\mathbf{x}^*, t)$$

If we assume that the second partial derivatives are continuous, the order of differentiation can be interchanged so that

$$\left(\frac{\partial^2 h}{\partial \mathbf{x}^2}(\mathbf{x}^*, t)\right)\dot{\mathbf{x}}^* + \frac{\partial^2 h}{\partial \mathbf{x}\,\partial t}(\mathbf{x}^*, t) = \frac{d}{dt}\left[\frac{\partial h}{\partial \mathbf{x}}(\mathbf{x}^*, t)\right]$$

This gives from eqn. (10.62),

$$\dot{\boldsymbol{\lambda}}^*(t) = -\frac{\partial}{\partial \mathbf{x}}[g(\mathbf{x}^*, \mathbf{u}^*, t) + \boldsymbol{\lambda}^{*T}\mathbf{f}(\mathbf{x}^*, \mathbf{u}^*, t)] \qquad (10.63)$$

Writing Euler-Lagrange equations in terms of $\mathbf{u}^*$, we get

$$\frac{\partial}{\partial \mathbf{u}}[g_a(\mathbf{x}^*, \dot{\mathbf{x}}^*, \mathbf{u}^*, \boldsymbol{\lambda}^*, t)] = 0$$

or

$$\frac{\partial}{\partial \mathbf{u}} [g(\mathbf{x}^*, \mathbf{u}^*, t) + \boldsymbol{\lambda}^{*T} \mathbf{f}(\mathbf{x}^*, \mathbf{u}^*, t)] = \mathbf{0} \tag{10.64}$$

Equations (10.56), (10.63) and (10.64) provide the necessary conditions for the determination of optimal control. These necessary conditions consist of $(2n + p)$ equations of which $2n$ are first-order differential equations— n equations (10.56) and n equations (10.63); and p are algebraic relations (10.64). The solution of $2n$ equations will contain $2n$ constants of integration. To evaluate these constants, we have the n boundary conditions at $t = t_0$ given by the equation

$$\mathbf{x}(t_0) = \mathbf{x}^0$$

and additional n boundary conditions ($(n + 1)$ conditions if terminal time t_1 is free) at the terminal point. Note that though the data, whatever its nature, is theoretically sufficient to solve the problem (not necessarily uniquely), the actual process of solution may be very tedious, due largely once again to the two-point boundary value problem.

We recognize that the quantities involved in the partial differentiation in (10.63) and (10.64) are identical. This leads to the definition of a new state function, the *Hamiltonian* as

$$\mathcal{H}(\mathbf{x}, \mathbf{u}, \boldsymbol{\lambda}, t) \stackrel{\Delta}{=} g(\mathbf{x}, \mathbf{u}, t) + \boldsymbol{\lambda}^T \mathbf{f}(\mathbf{x}, \mathbf{u}, t) \tag{10.65}$$

In terms of Hamiltonian $\mathcal{H}$, equations (10.63) and (10.64) become

$$\dot{\boldsymbol{\lambda}} = -\frac{\partial \mathcal{H}(\mathbf{x}, \mathbf{u}, \boldsymbol{\lambda}, t)}{\partial \mathbf{x}} \tag{10.66a}$$

$$\mathbf{0} = \frac{\partial \mathcal{H}(\mathbf{x}, \mathbf{u}, \boldsymbol{\lambda}, t)}{\partial \mathbf{u}} \tag{10.66b}$$

In addition, the constraint equations (10.56) may be written as

$$\dot{\mathbf{x}} = \frac{\partial \mathcal{H}(\mathbf{x}, \mathbf{u}, \boldsymbol{\lambda}, t)}{\partial \boldsymbol{\lambda}} \tag{10.66c}$$

The similarity of form of the two equations (10.66a) and (10.66c) for $\mathbf{x}$ and $\boldsymbol{\lambda}$ has led to the Lagrange multiplier $\boldsymbol{\lambda}(t)$ being termed the *co-state vector*; eqn. (10.66a) is usually referred to as the *co-state equation.* The equation (10.66b) is called the *control equation.*

In order to solve the optimal control problem we must solve equations (10.66) simultaneously subject to the boundary conditions. This is generally and most conveniently done by first solving the algebraic equations (10.66b) for the optimal control $\mathbf{u}^*$ in terms of $\mathbf{x}$, $\mathbf{u}$ and t,

$$\mathbf{u}^* = \mathbf{u}^*(\mathbf{x}, \boldsymbol{\lambda}, t) \tag{10.67}$$

This result is then substituted into the Hamiltonian $\mathcal{H}(\mathbf{x}, \boldsymbol{\lambda}, \mathbf{u}, t)$ to get

$$\mathcal{H}^*(\mathbf{x}, \boldsymbol{\lambda}, t) = \mathcal{H}(\mathbf{x}, \mathbf{u}^*(\mathbf{x}, \boldsymbol{\lambda}, t), \boldsymbol{\lambda}, t) \tag{10.68}$$

The differential equations (10.66a)–(10.66c) now become a set of $2n$ first-order simultaneous equations

$$\dot{\mathbf{x}} = \frac{\partial \mathcal{H}^* (\mathbf{x}, \boldsymbol{\lambda}, t)}{\partial \boldsymbol{\lambda}} \tag{10.69a}$$

$$\dot{\boldsymbol{\lambda}} = -\frac{\partial \mathcal{H}^* (\mathbf{x}, \boldsymbol{\lambda}, t)}{\partial \mathbf{x}} \tag{10.69b}$$

The trajectories $\mathbf{x}^*(t)$, and $\boldsymbol{\lambda}^*(t)$ are found by solving these equations with $2n$ boundary conditions. The optimal control law is then obtained using eqn. (10.67).

Boundary Conditions: The generalized boundary condition is (refer eqn. (10.46))

$$\left[\frac{\partial g_a (\mathbf{x}, \dot{\mathbf{x}}, \mathbf{u}, \boldsymbol{\lambda}, t)}{\partial \dot{\mathbf{x}}}\bigg|_{t_1}\right]^T \Delta \mathbf{x}^1 + \left[g_a(\mathbf{x}, \dot{\mathbf{x}}, \mathbf{u}, \boldsymbol{\lambda}, t) - \left(\frac{\partial g_a (\mathbf{x}, \dot{\mathbf{x}}, \mathbf{u}, \boldsymbol{\lambda}, t)}{\partial \dot{\mathbf{x}}}\right)^T \dot{\mathbf{x}}\right]\bigg|_{t_1} \Delta t_1 = 0 \tag{10.70}$$

For the optimal control problem, we have from eqn. (10.61),

$$\frac{\partial g_a (\mathbf{x}, \dot{\mathbf{x}}, \mathbf{u}, \boldsymbol{\lambda}, t)}{\partial \dot{\mathbf{x}}} = \frac{\partial h (\mathbf{x}, t)}{\partial \mathbf{x}} - \boldsymbol{\lambda}$$

Substituting in eqn. (10.70), we get

$$\left[\frac{\partial h (\mathbf{x}, t)}{\partial \mathbf{x}} - \boldsymbol{\lambda}\right]^T\bigg|_{t_1} \Delta \mathbf{x}^1 + \left[g(\mathbf{x}, \mathbf{u}, t) + \left(\frac{\partial h (\mathbf{x}, t)}{\partial \mathbf{x}}\right)^T \dot{\mathbf{x}} + \frac{\partial h (\mathbf{x}, t)}{\partial t} + \boldsymbol{\lambda}^T \mathbf{f}(\mathbf{x}, \mathbf{u}, t) - \boldsymbol{\lambda}^T \dot{\mathbf{x}} - \left(\frac{\partial h (\mathbf{x}, t)}{\partial \mathbf{x}} - \boldsymbol{\lambda}\right)^T \dot{\mathbf{x}}\right]\bigg|_{t_1} \Delta t_1 = 0$$

which may be reduced to

$$\left[\frac{\partial h (\mathbf{x}, t)}{\partial \mathbf{x}} - \boldsymbol{\lambda}\right]^T\bigg|_{t_1} \Delta \mathbf{x}^1 + \left[g(\mathbf{x}, \mathbf{u}, t) + \boldsymbol{\lambda}^T \mathbf{f}(\mathbf{x}, \mathbf{u}, t) + \frac{\partial h (\mathbf{x}, t)}{\partial t}\right]\bigg|_{t_1} \Delta t_1 = 0$$

or

$$\left[\frac{\partial h (\mathbf{x}, t)}{\partial \mathbf{x}} - \boldsymbol{\lambda}\right]^T\bigg|_{t_1} \Delta \mathbf{x}^1 + \left[\mathcal{H}^*(\mathbf{x}, \boldsymbol{\lambda}, t) + \frac{\partial h (\mathbf{x}, t)}{\partial t}\right]\bigg|_{t_1} \Delta t_1 = 0 \tag{10.71}$$

This is the generalized boundary condition. Boundary conditions for four specific cases studied earlier are summarized in Table 10.3. These conditions have been obtained from (10.71) by appropriate substitutions. Kirk (1970) has derived boundary conditions for the cases wherein $\mathbf{x}(t_1)$ is required to lie on a fixed surface or on a moving surface in free/specified time. These are derived from the generalized boundary condition (10.71).

Table 10.2 summarizes the procedure of solving optimal control problem.

Table 10.2: Summary of the Procedure for Solving Optimal Control Problems Using Hamiltonian Formulation of Variational Calculus

Given plant equation	$\dot{\mathbf{x}}(t)=\mathbf{f}(\mathbf{x},\mathbf{u},t)$
Given performance index	$J=h(\mathbf{x}(t_1),\, t_1)+\int_{t_0}^{t_1} g(\mathbf{x},\mathbf{u},t)\, dt$
Step 1	Form the Hamiltonian $\mathcal{H}(\mathbf{x},\mathbf{u},\boldsymbol{\lambda},t)=g(\mathbf{x},\mathbf{u},t)+\boldsymbol{\lambda}^T\mathbf{f}(\mathbf{x},\mathbf{u},t)$
Step 2	Solve the equation $\frac{\partial\mathcal{H}(\mathbf{x},\mathbf{u},\boldsymbol{\lambda},t)}{\partial\mathbf{u}}=0$ (control equation) to obtain $\mathbf{u}^*=\mathbf{u}^*(\mathbf{x},\boldsymbol{\lambda},t)$
Step 3	Find Hamiltonian $\mathcal{H}^*(\mathbf{x},\boldsymbol{\lambda},t)=\mathcal{H}(\mathbf{x},\mathbf{u}^*,\boldsymbol{\lambda},t)$
Step 4	Solve the set of $2n$ equations $\dot{\mathbf{x}}(t)=\frac{\partial\mathcal{H}^*(\mathbf{x},\boldsymbol{\lambda},t)}{\partial\boldsymbol{\lambda}}$ (state equations) $\dot{\boldsymbol{\lambda}}(t)=-\frac{\partial\mathcal{H}^*(\mathbf{x},\boldsymbol{\lambda},t)}{\partial\mathbf{x}}$ (co-state equations) with the given boundary conditions (see Table 10.3)
Step 5	Substitute the results of Step 4 into the expression for $\mathbf{u}^*$ to obtain the optimal control

Example 10.7: Reconsider the system of Example 10.6. We shall solve this example using the Hamiltonian method summarized in Tables 10.2 and 10.3.

Table 10.3: Summary of Boundary Conditions in Optimal Control Problems

Problem	Boundary conditions	Remarks
t_1 and $\mathbf{x}(t_1)$ both specified	$\mathbf{x}(t_0)=\mathbf{x}^0$ $\mathbf{x}(t_1)=\mathbf{x}^1$	$2n$ equations to determine $2n$ constants of integration
t_1 is fixed and $\mathbf{x}(t_1)$ is free	$\mathbf{x}(t_0)=\mathbf{x}^0$ $\left[\frac{\partial h(\mathbf{x},t)}{\partial \mathbf{x}}-\boldsymbol{\lambda}(t)\right]\Big\rvert_{t_1}=0$	$2n$ equations to determine $2n$ constants of integration
t_1 is free and $\mathbf{x}(t_1)$ is fixed	$\mathbf{x}(t_0)=\mathbf{x}^0$, $\mathbf{x}(t_1)=\mathbf{x}^1$ $\left[\mathscr{H}^*(\mathbf{x},\mathbf{u},t)+\frac{\partial h(\mathbf{x},t)}{\partial t}\right]\Big\rvert_{t_1}=0$	$(2n+1)$ equations to determine $2n$ constants of integration and t_1
t_1 and $\mathbf{x}(t_1)$ are free but related by $\mathbf{x}(t_1)=\boldsymbol{\theta}(t_1)$	$\mathbf{x}(t_0)=\mathbf{x}^0$ $\mathbf{x}(t_1)=\boldsymbol{\theta}(t_1)$ $\left[\mathscr{H}^*(\mathbf{x},\boldsymbol{\lambda},t)+\frac{\partial h(\mathbf{x},t)}{\partial t}\right]\Big\rvert_{t_1}$ $+\left[\frac{\partial h(\mathbf{x},t)}{\partial \mathbf{x}}-\boldsymbol{\lambda}(t)\right]^T\dot{\boldsymbol{\theta}}(t)\Big\rvert_{t_1}=0$	$(2n+1)$ equations to determine $2n$ constants of integration and t_1

Plant equations : $\dot{x}_1(t)=x_2(t)$

$$\dot{x}_2(t)=u(t)$$

Performance index: $J=\int_0^2 \tfrac{1}{2}u^2\,dt$

Step 1 : The Hamiltonian

$$\mathscr{H}(\mathbf{x},u,\boldsymbol{\lambda},t)=\tfrac{1}{2}u^2+\lambda_1x_2+\lambda_2u$$

Step 2 : $\frac{\partial \mathscr{H}(\mathbf{x},u,\boldsymbol{\lambda},t)}{\partial u}=0$ gives

$$u^*=-\lambda_2$$

Since $\frac{\partial^2\mathscr{H}}{\partial u^2}=1$, u^* does minimize the Hamiltonian.

Step 3 : The Hamiltonian

$$\mathcal{H}^*(\mathbf{x}, \boldsymbol{\lambda}, t) = \tfrac{1}{2}\lambda_2^2 + \lambda_1 x_2 - \lambda_2^2$$
$$= \lambda_1 x_2 - \tfrac{1}{2}\lambda_2^2$$

Step 4 : The state equations are

$$\dot{x}_1 = \frac{\partial \mathcal{H}^*(\mathbf{x}, \boldsymbol{\lambda}, t)}{\partial \lambda_1} = x_2$$

$$\dot{x}_2 = \frac{\partial \mathcal{H}^*(\mathbf{x}, \boldsymbol{\lambda}, t)}{\partial \lambda_2} = -\lambda_2$$

The co-state equations are

$$\dot{\lambda}_1 = -\frac{\partial \mathcal{H}^*(\mathbf{x}, \boldsymbol{\lambda}, t)}{\partial x_1} = 0$$

$$\dot{\lambda}_2 = -\frac{\partial \mathcal{H}^*(\mathbf{x}, \boldsymbol{\lambda}, t)}{\partial x_2} = -\lambda_1$$

The boundary conditions are

$$x_1(0) = x_2(0) = 1;\ x_1(2) = x_2(2) = 0$$

The solution of state and co-state equations subject to the given boundary conditions gives

$$x_1^* = \tfrac{1}{2}t^3 - \tfrac{7}{4}t^2 + t + 1$$

$$x_2^* = \tfrac{3}{2}t^2 - \tfrac{7}{2}t + 1$$

$$\lambda_1^* = 3;\quad \lambda_2^* = -3t + \tfrac{7}{2}$$

Step 5 : $u^* = -\lambda_2^*$

$$= 3t - \tfrac{7}{2}$$

10.4 MINIMUM PRINCIPLE

Perhaps the most useful single technique in control theory is the calculus of variations. The Hamiltonian formulation of the variational calculus has existed since the early nineteenth century. The most significant contribution in recent times was made by L.S. Pontryagin and his students (Pontryagin et al. 1962). Their work made it possible to handle control problems wherein the control and state vectors are bounded.

Control Variable Inequality Constraints

So far in our discussion in earlier sections, we have assumed that the control **u** was unbounded. Because of this assumption, the variational calculus could be directly applied to obtain the solution of optimal control problems. In order to write the Euler-Lagrange equations

$$\dot{\boldsymbol{\lambda}} = -\frac{\partial \mathcal{H}(\mathbf{x}, \mathbf{u}, \boldsymbol{\lambda}, t)}{\partial \mathbf{x}} \tag{10.72a}$$

$$\mathbf{0} = \frac{\partial \mathcal{H}(\mathbf{x}, \mathbf{u}, \boldsymbol{\lambda}, t)}{\partial \mathbf{u}} \tag{10.72b}$$

it was necessary to take the partial derivative of $\mathcal{H}$ with respect to **u**. Equation (10.72b) may be viewed as minimization of $\mathcal{H}(\mathbf{x}, \mathbf{u}, \boldsymbol{\lambda}, t)$ with respect to **u**. This operation of partial differentiation is only possible for arbitrary **x**, **λ**, and t if **u** is unconstrained.

In the situation presently under discussion, this is not the case. We now have the control variable constraint

$$\mathbf{u}(t) \in \mathbf{U} \text{ for all } t \in [t_0, t_1] \tag{10.73}$$

where **U** is a given subset of $\mathcal{R}^p$.

Equation (10.73) distinguishes the problem now under consideration from the problems we considered in earlier sections. Such a restricting assumption has important significance since the controls that can be applied to many physical systems must be constrained in amplitude and/or in the number of feasible control settings.

It is obvious that we can no longer proceed with the methods presented in the previous section. Pontryagin has shown that regardless of any constraints that exist on **u**, **u*** must still be chosen to minimize $\mathcal{H}$. A rigorous proof of the fact that **u*** must be chosen to minimize $\mathcal{H}$ is perhaps Pontryagin's most significant contribution to optimal control theory. For this reason, the approach presented here is often referred to as *Pontryagin's Minimum Principle.*[6]

In Pontryagin's original work, the result is referred to as the *Maximum Principle* because of the sign difference in the definition of Hamiltonian.

In the case of closed and bounded control region, the optimal control $\mathbf{u}^*(\mathbf{x}, \boldsymbol{\lambda}, t)$ is found by minimizing $\mathcal{H}(\mathbf{x}, \mathbf{u}, \boldsymbol{\lambda}, t)$ with respect to controls **u** in the given control region U, while treating the other variables as constants. In other words, $\mathbf{u}^*(\mathbf{x}, \boldsymbol{\lambda}, t)$ is the admissible control vector for which $\mathcal{H}(\mathbf{x}, \mathbf{u}, \boldsymbol{\lambda}, t)$ has its minimum value. The Pontryagin's state function then becomes

$$\begin{aligned} \mathcal{H}^*(\mathbf{x}, \boldsymbol{\lambda}, t) &= \mathcal{H}(\mathbf{x}, \mathbf{u}^*(\mathbf{x}, \boldsymbol{\lambda}, t), \boldsymbol{\lambda}, t) \\ &= \min_{\mathbf{u} \in \mathbf{U}} \mathcal{H}(\mathbf{x}, \mathbf{u}, \boldsymbol{\lambda}, t) \end{aligned} \tag{10.74}$$

If **U** is unbounded, obviously the control equation

$$\frac{\partial \mathcal{H}(\mathbf{x}, \mathbf{u}, \boldsymbol{\lambda}, t)}{\partial \mathbf{u}} = \mathbf{0}$$

in Table 10.2 is a special case of above procedure.

Table 10.4 summarizes the procedure of solving optimal control problems using Pontryagin's Minimum Principle.

[6]For proof of the Pontryagin's Minimum Principle refer Baum and Cesari (1972).

Table 10.4: Summary of the Procedure for Solving Optimal Control Problems Using Pontryagin's Minimum Principle

Given plant equation	$\dot{\mathbf{x}}(t)=\mathbf{f}(\mathbf{x}, \mathbf{u}, t)$
Given performance index	$J=h(\mathbf{x}(t_1), t_1)+\int_{t_0}^{t_1} g(\mathbf{x},\mathbf{u},t)\, dt$
Given control variable constraints	$\mathbf{u} \in \mathbf{U}$ for all $t \in [t_0, t_1]$
Step 1	Form the Pontryagin function $\mathcal{H}(\mathbf{x},\mathbf{u},\boldsymbol{\lambda},t)=g(\mathbf{x},\mathbf{u},t)+\boldsymbol{\lambda}^T\mathbf{f}(\mathbf{x},\mathbf{u},t)$
Step 2	Minimize $\mathcal{H}(\mathbf{x},\mathbf{u},\boldsymbol{\lambda},t)$ with respect to all admissible control vectors to find $\mathbf{u}^*=\mathbf{u}^*(\mathbf{x}\ \boldsymbol{\lambda},t)$
Step 3	Find Pontryagin function $\mathcal{H}^*(\mathbf{x},\boldsymbol{\lambda},t)=\min_{\mathbf{u}\in\mathbf{U}} \mathcal{H}(\mathbf{x},\mathbf{u},\boldsymbol{\lambda},t)$
Step 4	Solve the set of $2n$ equations $\dot{\mathbf{x}}(t)=\frac{\partial\mathcal{H}^*(\mathbf{x},\boldsymbol{\lambda},t)}{\partial\boldsymbol{\lambda}}$ (state equations) $\dot{\boldsymbol{\lambda}}(t)=-\frac{\partial\mathcal{H}^*(\mathbf{x},\boldsymbol{\lambda},t)}{\partial\mathbf{x}}$ (co-state equations) with the given boundary conditions (see Table 10.3)
Step 5	Substitute the results of Step 4 into the expression for $\mathbf{u}^*$ to obtain the optimal control

□

Once again, although the formulation of the solution is quite easy, the actual computation problem of obtaining it is far more difficult. The control variable inequality constraints further add to the computation difficulties. The solution may be obtained by a digital computer using complex computational procedures. This probably is the price one has to pay for demanding the best performance from the system.

Example 10.8: Consider a system with state equations

$$\dot{x}_1(t) = x_2(t)$$

$$\dot{x}_2(t) = -x_2(t) + u(t)$$

The performance index to be minimized is

$$J = \frac{1}{2} \int_{t_0}^{t_1} (x_1^2 + u^2)\, dt$$

The control inequality constraints are given by

$$|u(t)| \leqslant 1 \quad \text{for} \quad t \in [t_0, t_1]$$

The Pontryagin function

$$\mathcal{H}(\mathbf{x}, u, \boldsymbol{\lambda}, t) = \tfrac{1}{2}x_1^2 + \tfrac{1}{2}u^2 + \lambda_1 x_2 - \lambda_2 x_2 + \lambda_2 u \tag{10.75}$$

To determine the control that minimizes $\mathcal{H}$ subject to the inequality constraints, we first separate all of the terms containing $u(t)$,

$$\tfrac{1}{2}u^2 + \lambda_2 u \tag{10.76}$$

from the $\mathcal{H}$ function. For times when the control is unsaturated, we have

$$\frac{\partial \mathcal{H}(\mathbf{x}, u, \boldsymbol{\lambda}, t)}{\partial u} = 0$$

which gives

$$u^*(t) = -\lambda_2 \tag{10.77a}$$

Thus for $|\lambda_2(t)| \leqslant 1$, the control is given by $u^*(t) = -\lambda_2(t)$. When $|\lambda_2(t)| > 1$, then from (10.76) we find that the control that minimizes $\mathcal{H}$ is

$$u^*(t) = \begin{bmatrix} -1 & \text{for } \lambda_2(t) > 1 \\ +1 & \text{for } \lambda_2(t) < -1 \end{bmatrix} \tag{10.77b}$$

The optimal control strategy is shown in Fig. 10.10. To determine $u(t)$ explicitly, state and co-state equations must be solved subject to the given boundary conditions.

□

Recently, the Minimum Principle has also been applied to problems in discrete-time systems. However, it may be noted that the Minimum Principle is not universally valid for the case of discrete-time systems. Due to restrictions on possible variations of the control signal, the Minimum Principle must be

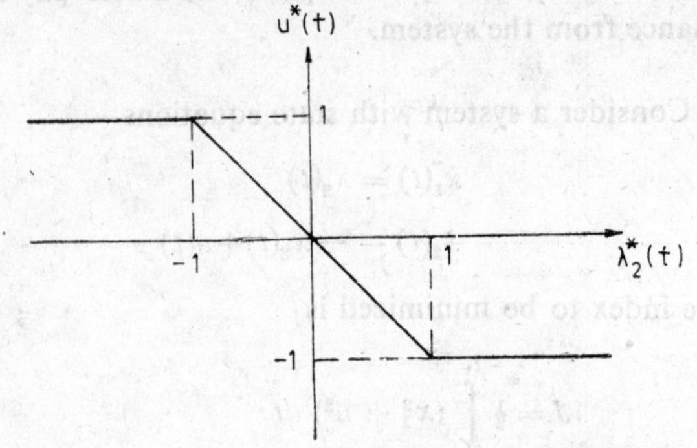

Fig. 10.10 Optimal control strategy for Example 10.8

modified for the general discrete-time problem. A derivation of the modified form of the Minimum Principle is given by Jordan and Polak (1964). Other useful references for Discrete Minimum Principle are Tou (1964), Fan and Wang (1964), Pearson and Sridhar (1966), Holtzman (1966) and Hautus (1973).

Control and State Variable Inepuality Constraints

We now extend the results of the previous subsection to include inequality constraints on the state variables. It will be assumed that the state constraints are of the form

$$\boldsymbol{\eta}(\mathbf{x}(t), t) \geqslant \mathbf{0} \tag{10.78}$$

where $\boldsymbol{\eta}$ is an m-vector function; each component of $\boldsymbol{\eta}$ is assumed to be continuously differentiable in state space.

There are several methods for converting the inequality constraint (10.78) to an equality constraint (Leitman 1962, McGill 1965, Berkovitz 1962). The approach given below is of McGill.

We define a new state variable x_{n+1} by

$$\dot{x}_{n+1} = f_{n+1} \overset{\Delta}{=} [\eta_1(\mathbf{x}, t)]^2\, \mathbf{h}(\eta_1) + [\eta_2(\mathbf{x}, t)]^2\, \mathbf{h}(\eta_2) + \dots$$
$$\dots + [\eta_m(\mathbf{x}, t)]^2\, \mathbf{h}(\eta_m) \tag{10.79a}$$

where $\mathbf{h}(\eta_i)$ is a unit Heaviside step function defined by

$$\mathbf{h}[\eta_i(\mathbf{x}, t)] = \begin{bmatrix} 0\,, & \text{if } \eta_i(\mathbf{x}, t) \geqslant 0 \\ 1\,, & \text{if } \eta_i(\mathbf{x}, t) < 0; \end{bmatrix} \tag{10.79b}$$

$$i = 1, 2, \dots, m$$

Note that $\dot{x}_{n+1}(t) \geqslant 0$ for all t and $\dot{x}_{n+1}(t) = 0$ only for times when all of the constraints (10.78) are satisfied. We specify the boundary conditions as

$$x_{n+1}(t_0) = 0 = x_{n+1}(t_1)$$

This implies that

$$\int_{t_0}^{t_1} \dot{x}_{n+1}(t)\, dt = 0$$

Since $\dot{x}_{n+1}(t) \geqslant 0$ for all t, the above equation implies that $x_{n+1}(t)$ must be zero for all $t \in [t_0, t_1]$ and this occurs only if the constraints are satisfied for all $t \in [t_0, t_1]$. Thus, satisfaction of boundary conditions on $x_{n+1}(t)$ implies satisfaction of the constraints (10.78).

The procedure summarized in Table 10.4 can now be modified to account for the state variable inequality constraints. The modified procedure is given in Table 10.5.

Table 10.5: Summary of the Procedure for Solving Optimal Control Problems with Control Variable and State Variable Inequality Constraints

Given plant equations	$\dot{x}_i(t) = f_i(\mathbf{x}, \mathbf{u}, t); \quad i = 1, 2 \ldots, n$
Given performance index	$J = h(\mathbf{x}(t_1), t_1) + \int_{t_0}^{t_1} g(\mathbf{x}, \mathbf{u}, t)\, dt$
Given control variable constraints	$\mathbf{u} \in \mathbf{U}$ for all $t \in [t_0, t_1]$
Given state variable constraints	$\eta_i(\mathbf{x}, t) \geqslant 0; \quad i = 1, 2, \ldots, m$
Step 1	Form the Pontryagin function $\mathscr{H}(\mathbf{x}, \mathbf{u}, \boldsymbol{\lambda}, t) = g(\mathbf{x}, \mathbf{u}, t) + \lambda_1(t) f_1(\mathbf{x}, \mathbf{u}, t) + \lambda_2(t) f_2(\mathbf{x}, \mathbf{u}, t) + \ldots \ldots + \lambda_n(t) f_n(\mathbf{x}, \mathbf{u}, t) + \lambda_{n+1}(t) f_{n+1}(\mathbf{x}, t)$ where $f_{n+1}(\mathbf{x}, t) \triangleq [\eta_1(\mathbf{x}, t)]^2 \mathbf{h}(\eta_1) + (\eta_2(\mathbf{x}, t)]^2 \mathbf{h}(\eta_2) + \ldots + [\eta_m(\mathbf{x}, t)]^2 \mathbf{h}(\eta_m)$; $\mathbf{h}(\eta_i)$ is defined in (10.79b)
Step 2	Minimize $\mathscr{H}(\mathbf{x}, \mathbf{u}, \boldsymbol{\lambda}, t)$ with respect to all admissible control vectors to find $\mathbf{u}^* = \mathbf{u}^*(\mathbf{x}, \boldsymbol{\lambda}, t)$

Step 3	Find the Pontryagin function $\mathcal{H}^*(\mathbf{x}, \boldsymbol{\lambda}, t) = \min_{\mathbf{u} \in U} \mathcal{H}(\mathbf{x}, \mathbf{u}, \boldsymbol{\lambda}, t)$
Step 4	Solve the set of $(2n+2)$ equations $\dot{x}_1(t) = f_1(\mathbf{x}, \mathbf{u}, t)$ $\vdots$ $\dot{x}_n(t) = f_n(\mathbf{x}, \mathbf{u}, t)$ $\dot{x}_{n+1}(t) = f_{n+1}(\mathbf{x}, t)$ $\dot{\lambda}_1(t) = -\dfrac{\partial \mathcal{H}^*(\mathbf{x}, \boldsymbol{\lambda}, t)}{\partial x_1}$ $\vdots$ $\dot{\lambda}_n(t) = -\dfrac{\partial \mathcal{H}^*(\mathbf{x}, \boldsymbol{\lambda}, t)}{\partial x_n}$ $\dot{\lambda}_{n+1}(t) = -\dfrac{\partial \mathcal{H}^*(\mathbf{x}, \boldsymbol{\lambda}, t)}{\partial x_{n+1}} = 0$ (because function $\mathcal{H}$ does not contain x_{n+1} in its argument) with the given boundary conditions on $x_1, x_2, \ldots, x_n$ (refer Table 10.3). The boundary conditions on $x_{n+1}(t)$ are $x_{n+1}(t_0) = 0 = x_{n+1}(t_1)$
Step 5	Substitute the results of Step 4 into the expression for $\mathbf{u}^*$ to obtain the optimal control

Example 10.9: Reconsider the problem of Example 10.8. Now we assume the presence of state variable inequality constraint

$$-1 \leqslant x_1(t) \leqslant 2 \quad \text{for all } t \in [t_0, t_1]$$

The constraint may be interpreted as

$$\eta_1(\mathbf{x}, t) = [x_1(t) + 1] \geqslant 0$$
$$\eta_2(\mathbf{x}, t) = [2 - x_1(t)] \geqslant 0$$

The Pontryagin function

$$\mathcal{H}(\mathbf{x}, \mathbf{u}, \boldsymbol{\lambda}, t) = \tfrac{1}{2}x_1^2 + \tfrac{1}{2}u^2 + \lambda_1 x_2 - \lambda_2 x_2 + \lambda_2 u + \lambda_3 [(x_1 + 1)^2 \mathbf{h}(\eta_1) + (2 - x_1)^2 \mathbf{h}(\eta_2)]$$

where

$$\mathbf{h}(\eta_1) = \begin{cases} 0, & \text{if } [x_1(t) + 1] \geqslant 0 \\ 1, & \text{if } [x_1(t) + 1] < 0 \end{cases}$$

and

$$\mathbf{h}(\eta_2) = \begin{bmatrix} 0, & \text{if} \quad [2-x_1(t)] \geqslant 0 \\ 1, & \text{if} \quad [2-x_1(t)] < 0 \end{bmatrix}$$

The minimizing control is given by (eqns. (10.77a) and (10.77b))

$$u^*(t) = \begin{bmatrix} -1 & \text{for} \quad \lambda_2(t) > 1 \\ -\lambda_2(t) & \text{for} \quad |\lambda_2(t)| \leqslant 1 \\ 1 & \text{for} \quad \lambda_2(t) < -1 \end{bmatrix}$$

The Pontryagin function for the case of unconstrained control is

$$\mathcal{H}^*(\mathbf{x}, \boldsymbol{\lambda}, t) = \tfrac{1}{2}x_1^2 + \lambda_1 x_2 - \lambda_2 x_2 - \tfrac{1}{2}\lambda_2^2 + \lambda_3[(x_1+1)^2\,\mathbf{h}(\eta_1) + (2-x_1)^2\,\mathbf{h}(\eta_2)]$$

The state and co-state equations are

$$\dot{x}_1 = x_2$$

$$\dot{x}_2 = -x_2 - \lambda_2$$

$$\dot{x}_3 = (x_1+1)^2\,\mathbf{h}(\eta_1) + (2-x_1)^2\,\mathbf{h}(\eta_2)$$

$$\dot{\lambda}_1 = -\frac{\partial \mathcal{H}^*(\mathbf{x}, \boldsymbol{\lambda}, t)}{\partial x_1} = -x_1 - 2\lambda_3[(x_1+1)\,\mathbf{h}(\eta_1)] + 2\lambda_3[(2-x_1)\,\mathbf{h}(\eta_2)]$$

$$\dot{\lambda}_2 = -\frac{\partial \mathcal{H}^*(\mathbf{x}, \boldsymbol{\lambda}, t)}{\partial x_2} = -\lambda_1 + \lambda_2$$

$$\dot{\lambda}_3 = 0$$

The boundary conditions on $x_{n+1}(t)$ are

$$x_{n+1}(t_0) = 0 = x_{n+1}(t_1)$$

10.5 DYNAMIC PROGRAMMING

The theory of dynamic programming was introduced by Bellman in the 1950s (Bellman 1957, Bellman and Kalaba 1965). Although the theory was primarily developed for the solution of certain problems by a digital computer (which implies discrete-time data), it has been extended to continuous-time analysis. We begin with the optimal control problem in discrete-time systems.

Multistage Decision Process in Discrete-Time

The optimization method based on dynamic programming views the control problem as a multistage decision problem and the control input as a time sequence of decisions. A discrete-time system gives rise to a sequence of transformations of the original state vector; the decision process therefore consists of selecting the transformations at each discrete interval of time.

Consider a process described by the state equations

$$\mathbf{x}(k+1)=\mathbf{f}(\mathbf{x}(k), \mathbf{u}(k), k);\ k \in [0, N-1] \tag{10.80}$$

where $\mathbf{x}$ is $n\times 1$ state vector, $\mathbf{u}$ is $p\times 1$ input vector, $\mathbf{f}$ is a vector-valued function and $[0, N-1]$ is the control interval. This process is called multistage decision process of N stages where the choice of $\mathbf{u}(k)$ at each sampling instant is considered the decision of interest.

We shall be interested in selecting the control $\mathbf{u}(k)$; $k=0, 1, \ldots, N-1$ which minimizes a performance criterion[7] of the form

$$J=h(\mathbf{x}(N), N)+\sum_{k=0}^{N-1} g(\mathbf{x}(k), \mathbf{u}(k), k) \tag{10.81}$$

This criterion is the discrete analog of that given by eqn. (10.10); a summation replaces integration. The initial state $\mathbf{x}(0)$ is generally specified. If we select from the myriad of possible choices, the choice of decisions $\{\mathbf{u}(0), \mathbf{u}(1), \ldots, \mathbf{u}(N-1)\} \triangleq \mathbf{u}[0, N-1]$ which minimizes the given performance index J, we have selected the optimal policy or sequence.

The dynamic programming solution is based on the following principles that are a direct consequence of the structure of the problem.

Principle of Causality: The processes which we are analyzing are such that the 'future' state may always be determined by the 'present' state; more specifically, the state $\mathbf{x}(j)$ at the jth stage together with the sequence of controls $\{\mathbf{u}(j), \mathbf{u}(j+1), \ldots, \mathbf{u}(r-1)\}=\mathbf{u}[j, r-1]$ uniquely determines the state $\mathbf{x}(r)$ at the rth stage. As a consequence of this principle, the initial state $\mathbf{x}(0)$ and the control sequence $\mathbf{u}[0, N-1]$ uniquely determine the trajectory $\mathbf{x}[1, N]=\{\mathbf{x}(1), \mathbf{x}(2), \ldots, \mathbf{x}(N)\}$. Figure 10.11 shows a schematic of state transitions.

The performance index J defined by (10.81) may be written as a function of $\mathbf{x}(0)$ and $\mathbf{u}[0, N-1]$:

$$J=J_{0,N}(\mathbf{x}(0), 0, \mathbf{u}[0, N-1])$$

If $\mathbf{x}(0)$ is specified, this relation implies that it is necessary only to determine $\mathbf{u}[0, N-1]$ to minimize J. This assumption is implicit in the statement of the problem.

Principle of Invariant Imbedding: For a given discrete-time system (10.80), the optimal control sequence is a function of the initial state $\mathbf{x}(0)$ and the number of stages N. We take the decision at the stage corresponding to $k=0$ to determine optimal path not merely from the given state $\mathbf{x}(0)$ but from all admissible states which the system can possess. In general, at the

[7]A performance criterion is also called the *cost function* (if it is to be minimized) or the *return function* (if it is to be maximized).

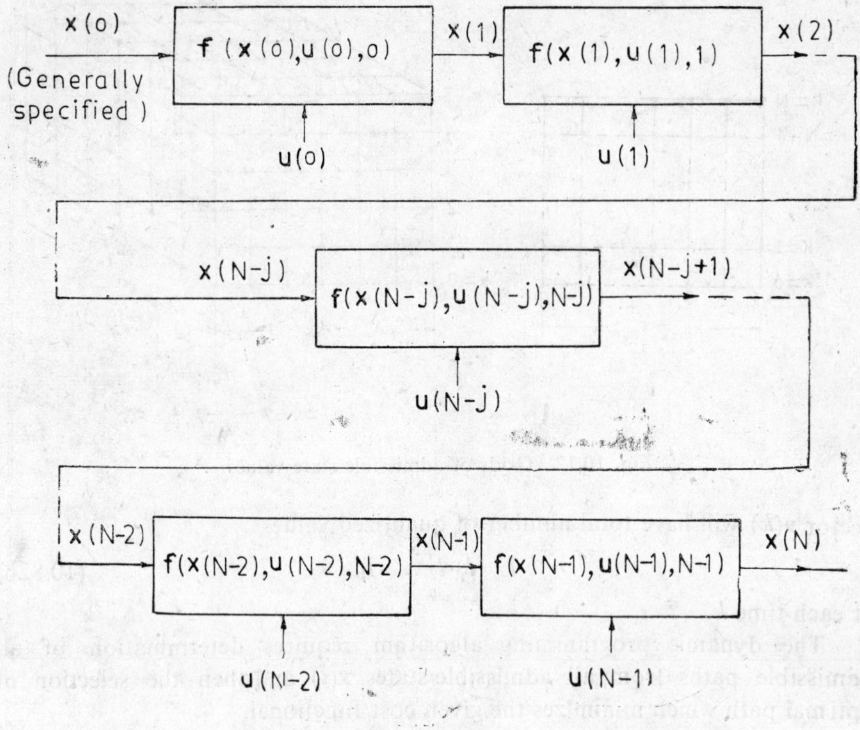

Fig. 10.11 Schematic of state transitions for discrete-time system

ith stage, we determine the optimal path not merely from state $\mathbf{x}(i)$ which is a function of the initial state $\mathbf{x}(0)$ and the controls $\mathbf{u}[0, i-1]$, but from all admissible states of the ith stage. Thus, we do not regard the control problem as an isolated problem with fixed values of $\mathbf{x}(0)$ and N but rather imbed it within a family of problems.

As we shall see later, to make the dynamic programming computational procedure feasible, it will be necessary to quantize the admissible state (and control) values into a finite number of levels. For example, if the system is of first-order, the grid of state values would appear as in Fig. 10.12a. The heavily dotted points, total number say m_1, correspond to admissible quantized state values for each time k. For a second-order system, the total number of state grid points for each time is $m_1 m_2$ (Fig. 10.12b) where m_1 is the number of points in the x_1 coordinate direction and m_2 is the number of points in the x_2 coordinate direction. For an nth-order system, the number of state grid points for each time k is

$$m = m_1 \cdot m_2 \ldots \cdot m_n \tag{10.82a}$$

On similar lines, the dynamic programming computational procedure will require quantization of admissible control values. The p-dimensional

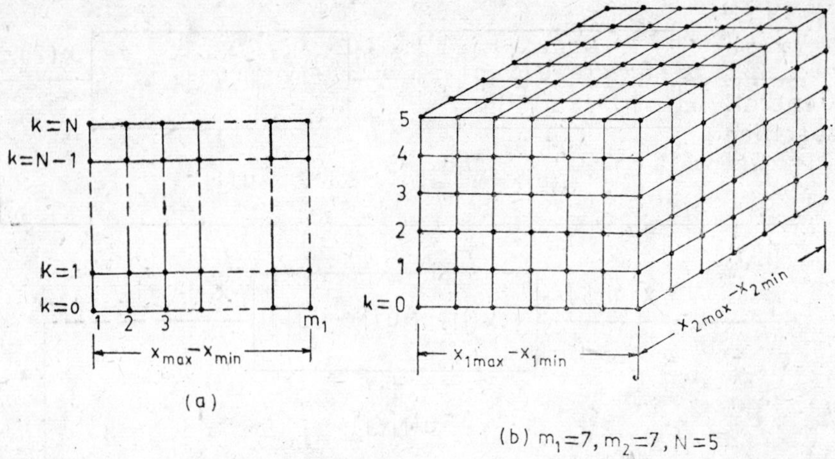

Fig. 10.12 Grids of admissible state values

vector $\mathbf{u}(k)$ will have total number of quantized values

$$l = l_1 \cdot l_2 \ldots \cdot l_p \tag{10.82b}$$

at each time k.

The dynamic programming algorithm requires determination of all admissible paths from all admissible states $\mathbf{x}(i)$ and then the selection of optimal path which minimizes the given cost functional.

Principle of Optimality: Another important feature of the dynamic programming formulation is that we do not attempt to find all the values of optimal control sequence simultaneously; we find one control value at a time until the entire optimal policy is determined. This sequential decision process is based on the principle of optimality which is stated below in a form suitable for our present work.

An optimal sequence of controls in a multistage optimization problem has the property that whatever the initial stage, state and controls are, the remaining controls must constitute an optimal sequence of decisions for the remaining problem with stage and state resulting from previous controls as initial conditions.

Suppose that somehow the control sequence $\mathbf{u}[0, j-1]$ has been chosen. Then from the principle of causality, the trajectory $\mathbf{x}[0, j]$ is determined.

Invoking the principle of optimality, we may define

$$J_{j,N}(\mathbf{x}(j), j, \mathbf{u}[j, N-1]) = h(\mathbf{x}(N), N) + \sum_{k=j}^{N-1} g(\mathbf{x}(k), \mathbf{u}(k), k)$$

Then, for optimization of $J_{0,N}(\mathbf{x}(0), 0, \mathbf{u}[0, N-1])$, it is clear that it is both necessary and sufficient to determine $\mathbf{u}[j, N-1]$ to minimize

$$J_{j,N}(\mathbf{x}(j), j, \mathbf{u}[j, N-1]).$$

The optimal value $\mathbf{u}^*[j, N-1]$ will obviously depend on $\mathbf{x}(j)$ and j.

Thus, in general, optimal control at the kth stage, $\mathbf{u}(k)$, provided it exists and is unique, may be expressed as a function of the state at the kth stage;

$$\mathbf{u}^*(k) = \mathbf{u}^*(\mathbf{x}(k), k) \tag{10.83}$$

The function (10.83), as defined earlier in Section 10.2, is the optimal control law. The optimal control law (10.83) in conjunction with system equations (10.80) may be used to generate the optimal control sequence.

□

We now derive a recursive relation to determine optimal control law (10.83) for minimization of the performance index (10.81).

The principle of optimality reduces the N-stage decision process to N single-stage decision processes. It is possible to find the last decision as the initial calculation step in multistage decision process, i.e., the calculations of optimal decisions may proceed from the last decision back to the first decision.

We begin by defining

$$J_{N,N}(\mathbf{x}(N), N) \overset{\Delta}{=} h(\mathbf{x}(N), N) \tag{10.84a}$$

$$= J^*_{N,N}(\mathbf{x}(N), N), \text{ since no decision is involved} \tag{10.84b}$$

Next define

$$\begin{aligned} J_{N-1,N}(\mathbf{x}(N-1), N-1, \mathbf{u}(N-1)) & \\ &\overset{\Delta}{=} h(\mathbf{x}(N), N) + g(\mathbf{x}(N-1), \mathbf{u}(N-1), N-1) \\ &= g(\mathbf{x}(N-1), \mathbf{u}(N-1), N-1) + J_{N,N}(\mathbf{x}(N), N) \end{aligned}$$

Note that the value of $J_{N-1,N}$ is dependent only on $\mathbf{x}(N-1)$ and $\mathbf{u}(N-1)$ since $\mathbf{x}(N)$ is related to $\mathbf{x}(N-1)$ and $\mathbf{u}(N-1)$ through the state equation (10.80). $J_{N-1,N}$ is the cost of operation during the last stage of the N-stage process with state value $\mathbf{x}(N-1)$ at the beginning of $(N-1)$th stage. However, $J_{N-1,N}$ is also the cost of *one-stage process* with initial state numerically equal to the value $\mathbf{x}(N-1)$. As per the principle of optimality, regardless of how the state $\mathbf{x}(N-1)$ is obtained; once $\mathbf{x}(N-1)$ is known then using $\mathbf{x}(N-1)$ as the initial stage for the one-stage process, the control $\mathbf{u}(N-1)$ must be chosen so that the cost $J_{N-1,N}$ of the one-stage process is minimized.

The minimum cost for the one-stage process is

$$\begin{aligned} J^*_{N-1,N}(\mathbf{x}(N-1), N-1) & \\ &\overset{\Delta}{=} \min_{\mathbf{u}(N-1)} \{g(\mathbf{x}(N-1), \mathbf{u}(N-1), N-1) + J_{N,N}(\mathbf{x}(N), N)\} \end{aligned}$$

The optimal control may be denoted as $\mathbf{u}^*(\mathbf{x}(N-1), N-1)$ since we know that the optimal choice of $\mathbf{u}(N-1)$ will depend on $\mathbf{x}(N-1)$.

As we shall see, the optimal policy and the minimum cost of the one-stage process are imbedded (contained) in the results for the N-stage process.

The cost function for the last two stages is given by

$$J_{N-2,N}(\mathbf{x}(N-2), N-2, \mathbf{u}[N-2, N-1])$$

$$= h(\mathbf{x}(N), N) + \sum_{k=N-2}^{N-1} g(\mathbf{x}(k), \mathbf{u}(k), k)$$

$$= g(\mathbf{x}(N-2), \mathbf{u}(N-2), N-2) + J_{N-1,N}(\mathbf{x}(N-1), N-1, \mathbf{u}(N-1))$$

As before, observe that $J_{N-2,N}$ is the cost of a *two-stage process* with initial stage $\mathbf{x}(N-2)$. The minimum cost for the two-stage process is

$$J^*_{N-2,N}(\mathbf{x}(N-2), N-2) = \min_{\mathbf{u}(N-2),\, \mathbf{u}(N-1)} \{g(\mathbf{x}(N-2), \mathbf{u}(N-2), N-2) + J_{N-1,N}(\mathbf{x}(N-1), N-1, \mathbf{u}(N-1))\}$$

Imbedding the one-stage process into the two-stage process and using the principle of optimality, we have the following result:

Whatever the initial state $\mathbf{x}(N-2)$ and initial control $\mathbf{u}(N-2)$, the remaining control $\mathbf{u}(N-1)$ must be optimal with respect to the value of $\mathbf{x}(N-1)$ that results from the application of $\mathbf{u}(N-2)$. Therefore

$$J^*_{N-2,N}(\mathbf{x}(N-2), N-2) = \min_{\mathbf{u}(N-2)} \{g(\mathbf{x}(N-2), \mathbf{u}(N-2), N-2) + J^*_{N-1,N}(\mathbf{x}(N-1), N-1)\}$$

From this relation, $\mathbf{u}^*(\mathbf{x}(N-2), N-2)$ is determined.

By induction, we obtain for a j-stage process, the result

$$J^*_{N-j,N}(\mathbf{x}(N-j), N-j) = \min_{\mathbf{u}(N-j)} \{g(\mathbf{x}(N-j), \mathbf{u}(N-j), N-j) + J^*_{N-j+1,N}(\mathbf{x}(N-j+1), N-j+1)\} \tag{10.85}$$

This is the recurrence relation that we set out to obtain.

Unfortunately, in general, elegant analytical solutions to the dynamic programming equations (except for linear regulator problem discussed in the next chapter) are not possible. Instead, numerical techniques must be employed. The traditional dynamic programming approach involves a systematic search procedure described below.

We shall use eqns. (10.80)–(10.85). To make the computational procedure feasible, the admissible state and control values are first quantized into a finite number of levels. For nth-order system (10.80), the number of state grid points for each time k may be taken as equal to m (eqn. (10.82a)) and the number of control grid points for each time k equal to l (eqn. (10.82b)).

The first step in computational procedure is to calculate ($j = 0$ in eqn. (10.85)) and store

$$J_{N,N}(\mathbf{x}^{(i)}(N), N) = h(\mathbf{x}^{(i)}(N), N); \quad i = 1, 2, \ldots, m.$$

This calculation provides initial conditions for the dynamic programming recurrence relations.

The next step is to ($j = 1$ in (10.85)) select the first trial state point $\mathbf{x}^{(1)}(N-1)$. Each admissible control value $\mathbf{u}^{(r)}(N-1)$; $r = 1, 2, \ldots, l$ is then used to determine the next state value $\mathbf{x}^{(1, r)}(N)$ from the relation

$$\mathbf{x}^{(1, r)}(N) = \mathbf{f}(\mathbf{x}^{(1)}(N-1), \mathbf{u}^{(r)}(N-1), N-1); \quad r = 1, 2, \ldots, l.$$

The state values $\mathbf{x}^{(1, r)}(N)$ are used to look up appropriate value of $J^*_{N, N}(\mathbf{x}^{(1, r)}(N))$ in the initial conditions already stored (note that interpolation will be required if $\mathbf{x}^{(1, r)}(N)$ does not fall exactly on the grid value). Using the value of $J^*_{N, N}(\mathbf{x}^{(1, r)}(N))$, we calculate

$$\begin{aligned} J_{N-1, N}(\mathbf{x}^{(1)}(N-1), \mathbf{u}^{(r)}(N-1), N-1) \\ = g(\mathbf{x}^{(1)}(N-1), \mathbf{u}^{(r)}(N-1), N-1) + J^*_{N, N}(\mathbf{x}^{(1, r)}(N)); \\ r = 1, 2. \ldots, l. \end{aligned}$$

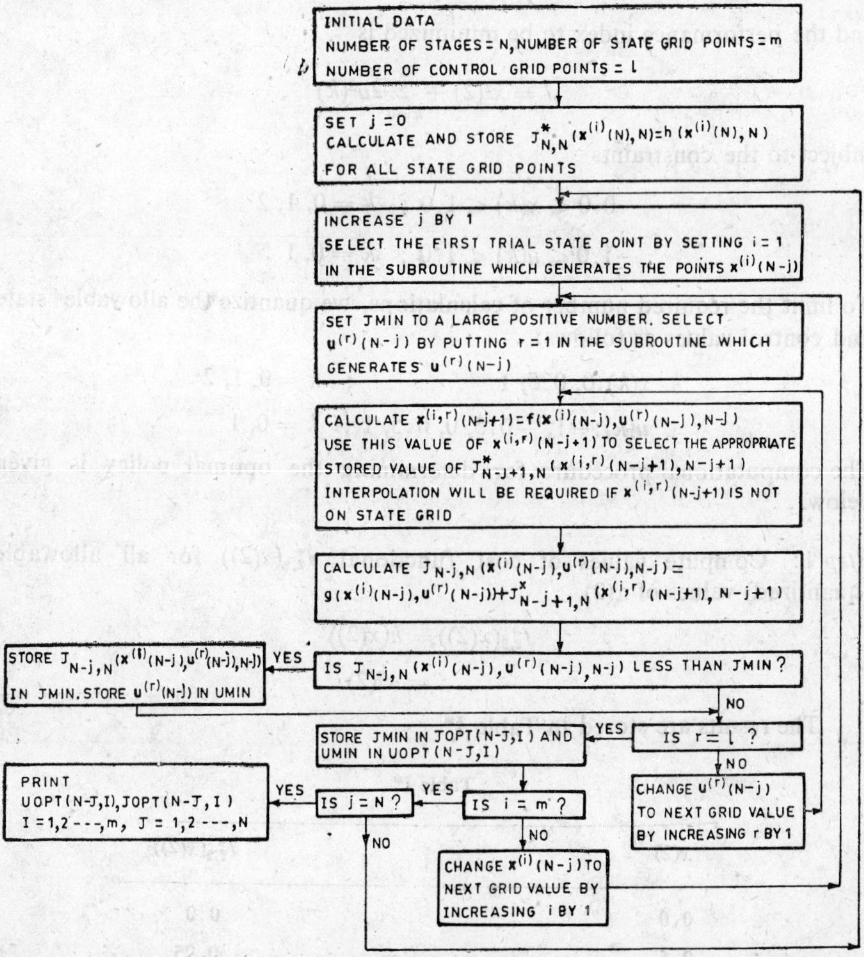

Fig. 10.13 Flow chart of dynamic programming algorithm

The smallest of these r values of $J_{N-1,N}$ and the associated control are then stored, i.e., we store

$$J^*_{N-1,N}(\mathbf{x}^{(1)}(N-1), N-1) = \min_{\mathbf{u}(N-1)} \{J_{N-1,N}(\mathbf{x}^{(1)}(N-1), \mathbf{u}^{(r)}(N-1), N-1)\}$$

and the associated control $\mathbf{u}^*(\mathbf{x}^{(1)}(N-1))$.

This procedure is repeated for other state grid values $\mathbf{x}^{(i)}(N-1)$; $i = 2, 3, \ldots, m$ and then for $j = 2, 3, \ldots, N$. A flow chart of the computational procedure is given in Fig. 10.13.

Example 10.10: Consider a very simple one-dimensional regulator problem. The system equation is

$$x(k+1) = x(k) + u(k)$$

and the performance index to be minimized is

$$J = x^2(2) + \sum_{k=0}^{1} 2u^2(k)$$

subject to the constraints

$$0.0 \leqslant x(k) \leqslant 1.0 \;; \quad k = 0, 1, 2$$

$$-1.0 \leqslant u(k) \leqslant 1.0 \;; \quad k = 0, 1$$

To limit the required number of calculations, we quantize the allowable state and control values as follows:

$$x(k): 0, 0.5, 1 \qquad ; \quad k = 0, 1, 2$$

$$u(k): -1, -0.5, 0, 0.5, 1 \;; \quad k = 0, 1$$

The computational procedure for determining the optimal policy is given below.

Step 1: Compute values of cost functional $J^*_{2,2}(x(2))$ for all allowable (quantized) values of $x(2)$.

$$J^*_{2,2}(x(2)) = h(x(2))$$

$$= x^2(2)$$

The results are stored in Table I*.

Table I*

$x(2)$	$J^*_{2,2}(x(2))$
0.0	0.0
0.5	0.25
1.0	1.0

Step 2: Choose an allowable (quantized) value of $x(1)$. For each allowable (quantized) control value $u(1)$, evaluate $x(2)$ from the relation

$$x(k+1) = x(k) + u(k)$$

and then evaluate

$$J_{1,2}(x(1), u(1)) = g(x(1), u(1), 1) + J^*_{2,2}(x(2))$$
$$= 2u^2(1) + J^*_{2,2}(x(2))$$

where $J^*_{2,2}(x(2))$ is the appropriate stored value of Table I*. Prepare a table such as Table II. Choose the $u(1)$ which produces the minimum value of $J_{1,2}(x(1), u(1))$. Enter this minimum value as $J^*_{1,2}(x(1))$ in Table II* opposite $x(1)$. The minimizing $u(1)$ is also recorded in this table. Choose a new value of $x(1)$ and repeat the computation of Table II to generate a new entry for Table II*.

Table II

$x(1)$	$u(1)$	$x(2)$	$J^*_{2,2}(x(2))$	$J_{1,2}(x(1), u(1))$
0.0	−1	−1	—	—
	−0.5	−0.5	—	—
	0.0	0.0	0.0	0.0
	0.5	0.5	0.25	0.75
	1.0	1.0	1.0	3.0
0.5	−1.0	−0.5	—	—
	−0.5	0.0	0.0	0.5
	0.0	0.5	0.25	0.25
	0.5	1.0	1.0	1.25
	1.0	1.5	—	—
1.0	−1.0	0.0	0.0	2.0
	−0.5	0.5	0.25	0.75
	0.0	1.0	1.0	1.0
	0.5	1.5	—	—
	1.0	2.0	—	—

Table II*

$x(1)$	$u^*(1)$	$J^*_{1,2}(x(1))$
0.0	0.0	0.0
0.5	0.0	0.25
1.0	−0.5	0.75

Note that values of $J^*_{2,2}(x(1)+u(1))$ have to be interpolated if $x(1)+u(1)$ does not equal one of the stored values for $x(2)$. For example, suppose that the trial values of $u(k)$ had been

$$-1,\ -0.75,\ -0.5,\ -0.25,\ 0,\ 0.25,\ 0.5,\ 0.75,\ 1$$

For $x(1)=0$, $u(1)=0.25$, we have $x(1)+u(1)=x(2)=0.25$. However, this $x(2)$ is not available in Table 1*. Then by interpolation we may obtain

$$\begin{aligned} J^*_{2,2}(0.25) &= J^*_{2,2}(0)+\tfrac{1}{2}[J^*_{2,2}(0.5)-J^*_{2,2}(0)] \\ &= 0+\tfrac{1}{2}[0.25-0] \\ &= 0.125 \end{aligned}$$

Step 3: Repeat step 2 at $k=0$ and record the results in Tables III and III*.

Table III

$x(0)$	$u(0)$	$x(1)$	$J^*_{1,2}(x(1))$	$J_{0,2}(x(0), u(0))$
0.0	−1.0	−1.0	—	—
	−0.5	−0.5	—	—
	0.0	0.0	0.0	0.0
	0.5	0.5	0.25	0.75
	1.0	1.0	0.75	2.75
0.5	−1.0	−0.5	—	—
	−0.5	0.0	0.0	0.5
	0.0	0.5	0.25	0.25
	0.5	1.0	0.75	1.75
	1.0	1.5	—	—
1.0	−1	0.0	0.0	2.0
	−0.5	0.5	0.25	0.75
	0.0	1.0	0.75	0.75
	0.5	1.5	—	—
	1.0	2.0	—	—

Table III*

$x(0)$	$u^*(0)$	$J^*_{0,2}(x(0))$
0.0	0.0	0.0
0.5	0.0	0.25
1.0	0.0 or -0.5	0.75

The result of the computational procedure is a number for the optimal control and minimum cost at every point on the 2-dimensional state-time grid (Tables II*, III*). For a specified initial condition $x(0)$, we obtain $u^*(0)$ and the minimum cost from Table III*. Next we determine $x(1)$ from the state equation which is then used to extract $u^*(1)$ from Table II*.

For example if $x(0) = 1.0$, the minimum cost is $J^*_{0,2}(1) = 0.75$ and the optimal policy is

$$\{u^*(0), u^*(1)\} = \{0.0, -0.5\} \text{ or } \{-0.5, 0.0\}$$

□

Some advantages of the dynamic programming algorithm are apparent. This algorithm decomposes the problem of selection of the entire N control vectors $\mathbf{u}(0), \mathbf{u}(1)\ldots, \mathbf{u}(N-1)$ into a sequence of N selections, each of the selections involving only one of the $\mathbf{u}(k)$ vectors. This dramatically reduces the number of calculations required to determine the optimal control law. The solution obtained gives the absolute minimum in contrast to relative minimum obtained by calculus of variations. Further, if there are constraints on state and control variables, the search problem is in fact made easier. The optimization procedure presented earlier becomes quite complicated when constraints on state and control variables are present.

Offsetting these advantages are some disadvantages. The computational requirements become excessive for all but very simple systems. Both storage requirements and computation time can be troublesome. For example, for a third-order system with 100 quantization levels in each state coordinate direction, $10^2 \times 10^2 \times 10^2 = 10^6$ storage locations are required. To mitigate this curse of dimensionality, several methods have been developed. Refer, for example, Larson (1965) and Larson (1967).

Multistage Decision Process in Continuous-Time

Consider the continuous-time process described by the state equations

$$\dot{\mathbf{x}}(t) = \mathbf{f}(\mathbf{x}(t), \mathbf{u}(t), t); \; t \in [t_0, t_1] \tag{10.86}$$

where $\mathbf{x}$ is $n \times 1$ state vector, $\mathbf{u}$ is $p \times 1$ input vector, $\mathbf{f}$ is a vector-valued function and $[t_0, t_1]$ is the control interval.

We shall be interested in selecting the control $\mathbf{u}(t)$; $t \in [t_0, t_1]$ which minimizes a performance criterion of the form

$$J = h(\mathbf{x}(t_1), t_1) + \int_{t_0}^{t_1} g(\mathbf{x}, \mathbf{u}, t)\, dt \tag{10.87}$$

A natural approach is to replace the continuous-time problem by its finite difference approximation. The results of previous subsection are then directly applicable. The interval $[t_0, t_1]$ is divided into N equal intervals of time T. It is assumed that the control $\mathbf{u}(i)$ is constant over the ith interval and the state $\mathbf{x}(i)$ refers to the value of the state at the beginning of the ith interval or stage. The finite difference approximation to (10.86) is

$$\frac{\mathbf{x}(t+T) - \mathbf{x}(t)}{T} \simeq \mathbf{f}(\mathbf{x}(t), \mathbf{u}(t), t)$$

or

$$\mathbf{x}(t+T) = \mathbf{x}(t) + T\mathbf{f}(\mathbf{x}(t), \mathbf{u}(t), t)$$

For $t = kT$,

$$\mathbf{x}(kT+T) = \mathbf{x}(kT) + T\mathbf{f}(\mathbf{x}(kT), \mathbf{u}(kT), kT)$$

or

$$\begin{aligned} \mathbf{x}(k+1) &= \mathbf{x}(k) + T\mathbf{f}(\mathbf{x}(k), \mathbf{u}(k), k) \\ &= \mathbf{f}_D(\mathbf{x}(k), \mathbf{u}(k), k) \end{aligned} \tag{10.88a}$$

In a similar way, the performance measure becomes

$$\begin{aligned} J &= h(\mathbf{x}(N), N) + \int_0^T g(\mathbf{x}, \mathbf{u}, t)\, dt + \int_T^{2T} g(\mathbf{x}, \mathbf{u}, t)\, dt + \dots \\ &\qquad \dots + \int_{(N-1)T}^{NT} g(\mathbf{x}, \mathbf{u}, t)\, dt \\ &\simeq h(\mathbf{x}(N), N) + T \sum_{k=0}^{N-1} g(\mathbf{x}(k), \mathbf{u}(k), k) \\ &= h(\mathbf{x}(N), N) + \sum_{k=0}^{N-1} g_D(\mathbf{x}(k), \mathbf{u}(k), k) \end{aligned} \tag{10.88b}$$

Next we quantize the admissible state and control values into a finite number of levels.

By making the problem discrete, as we have done, it is now required that the optimal control law $\mathbf{u}^*(\mathbf{x}(k), k)$; $k = 0, 1, \dots, N-1$, be determined for the system given by (10.88a) with the performance index given by (10.88b). This approach, as established in the previous subsection, leads to a recurrence relation that is ideally suited for digital computer solution.

In the following we consider an alternative approach for continuous-time systems which leads to a nonlinear partial differential equation—the Hamilton-Jacobi equation.

Hamilton-Jacobi Equation: The problem is to find optimal control for the process described by (10.86) that minimizes the performance index given by (10.87). By the use of imbedding principle, let us include our problem in a larger class of problems by considering the cost functional

$$J(\mathbf{x}(t), t, \mathbf{u}_t) = h(\mathbf{x}(t_1), t_1) + \int_t^{t_1} g(\mathbf{x}(\tau), \mathbf{u}(\tau), \tau)\, d\tau$$

where t can be any value less than or equal to t_1, $\mathbf{u}_t \triangleq \{\mathbf{u}(\tau);\ t \leqslant \tau \leqslant t_1\}$ and $\mathbf{x}(\tau)$, $t \leqslant \tau \leqslant t_1$ is the trajectory associated with a control function, given $\mathbf{x}(t)$. The functional J is therefore the cost of operation over the interval $[t, t_1]$ given initial condition $\mathbf{x}(t)$. The minimum cost

$$J^*(\mathbf{x}(t), t) = \min_{\mathbf{u}_t} \left\{ \int_t^{t_1} g(\mathbf{x}(\tau), \mathbf{u}(\tau), \tau)\, d\tau + h(\mathbf{x}(t_1), t_1) \right\} \tag{10.89}$$

By subdividing the control interval, we obtain

$$J^*(\mathbf{x}(t), t) = \min_{\mathbf{u}_t} \left\{ \int_t^{t+\Delta t} g(\mathbf{x}(\tau), \mathbf{u}(\tau), \tau)\, d\tau \right.$$

$$\left. + \int_{t+\Delta t}^{t_1} g(\mathbf{x}(\tau), \mathbf{u}(\tau), \tau)\, d\tau + h(\mathbf{x}(t_1), t_1) \right\}$$

Using the principle of optimality, we may write

$$J^*(\mathbf{x}(t), t) = \min_{\substack{\mathbf{u}(\tau) \\ t \leqslant \tau \leqslant t+\Delta t}} \left\{ \int_t^{t+\Delta t} g(\mathbf{x}(\tau), \mathbf{u}(\tau), \tau)\, d\tau + J^*(\mathbf{x}(t + \Delta t), t + \Delta t) \right\}$$

where $J^*(\mathbf{x}(t + \Delta t), t + \Delta t)$ is the minimum cost of the process for the time interval $t + \Delta t \leqslant \tau \leqslant t_1$ with the initial state

$$\mathbf{x}(t + \Delta t) = \mathbf{x}(t) + \int_t^{t+\Delta t} \mathbf{f}(\mathbf{x}(\tau), \mathbf{u}^*(\tau), \tau)\, d\tau$$

$$= \mathbf{x}(t) + \Delta \mathbf{x}$$

Assuming that J^* has continuous first and second partial derivatives for all points of interest in $\mathcal{R}^{n+1}$, we may expand $J^*(\mathbf{x} + \Delta\mathbf{x}, t + \Delta t)$ in a Taylor series about the point $(\mathbf{x}(t), t)$ to obtain

$$J^*(\mathbf{x}(t), t) = \min_{\substack{\mathbf{u}(\tau) \\ t \leqslant \tau \leqslant t+\Delta t}} \left\{ \int_t^{t+\Delta t} g(\mathbf{x}(\tau), \mathbf{u}(\tau), \tau)\, d\tau + J^*(\mathbf{x}(t), t) \right.$$

$$+ \left[\frac{\partial J^*}{\partial t}(\mathbf{x}(t), t)\right] \Delta t + \left[\frac{\partial J^*}{\partial \mathbf{x}}(\mathbf{x}(t), t)\right]^T \Delta \mathbf{x}$$

$$\left. + \text{Higher Order Terms } (\Delta t) + \text{Higher Order Terms } (\Delta \mathbf{x}) \right\}$$

$$= \min_{\substack{\mathbf{u}(\tau) \\ t \leqslant \tau \leqslant t+\Delta t}} \left\{ \int_t^{t+\Delta t} g(\mathbf{x}(\tau), \mathbf{u}(\tau), \tau)\, d\tau \right.$$

$$+ J^*(\mathbf{x}(t), t) + \left[\frac{\partial J^*}{\partial t}(\mathbf{x}(t), t)\right] \Delta t$$

$$\left. + \left[\frac{\partial J^*}{\partial \mathbf{x}}(\mathbf{x}(t), t)\right]^T \Delta \mathbf{x} + \text{H.O.T. } (\Delta t) \right\}$$

Removing the terms involving $J^*(\mathbf{x}(t), t)$ and $\dfrac{\partial J^*}{\partial t}(\mathbf{x}(t), t)$ from the minimization (since they do not depend on $\mathbf{u}(\tau)$) we obtain

$$0 = \frac{\partial J^*}{\partial t}(\mathbf{x}(t), t)\, \Delta t + \min_{\substack{\mathbf{u}(\tau) \\ t \leqslant \tau \leqslant t+\Delta t}} \left\{ \int_t^{t+\Delta t} g(\mathbf{x}(\tau), \mathbf{u}(\tau), \tau)\, d\tau \right.$$

$$\left. + \left[\frac{\partial J^*}{\partial \mathbf{x}}(\mathbf{x}(t), t)\right]^T \Delta \mathbf{x} + \text{H.O.T. } (\Delta t) \right\}$$

Dividing both sides of this equation by Δt, taking the limit $\Delta t \to 0$ and noting that

$$\lim_{\Delta t \to 0} \frac{\Delta \mathbf{x}}{\Delta t} = \mathbf{f}(\mathbf{x}(t), \mathbf{u}(t), t)$$

and

$$\lim_{\Delta t \to 0} \frac{1}{\Delta t} \int_t^{t+\Delta t} g(\mathbf{x}(\tau), \mathbf{u}(\tau), \tau)\, d\tau = g(\mathbf{x}(t), \mathbf{u}(t), t)$$

we obtain

$$0 = \frac{\partial J^*}{\partial t}(\mathbf{x}(t), t) + \min_{\mathbf{u}(t)} \left\{ g(\mathbf{x}(t), \mathbf{u}(t), t) + \left[\frac{\partial J^*}{\partial \mathbf{x}}(\mathbf{x}(t), t)\right]^T \mathbf{f}(\mathbf{x}, \mathbf{u}, t) \right\} \tag{10.90}$$

Note that in taking the limit, the functions g, $\mathbf{u}$, J^* and $\mathbf{f}$ are required to be continuous.

From eqn. (10.89) we find that J^* must satisfy the boundary condition

$$J^*(\mathbf{x}(t_1), t_1) = h(\mathbf{x}(t_1), t_1) \tag{10.91}$$

Defining the Hamiltonian $\mathcal{H}$ as

$$\mathcal{H}\left(\mathbf{x}(t), \mathbf{u}(t), \frac{\partial J^* (\mathbf{x}(t), t)}{\partial \mathbf{x}}, t\right)$$

$$\triangleq g(\mathbf{x}(t), \mathbf{u}(t), t) + \left[\frac{\partial J^* (\mathbf{x}(t), t)}{\partial \mathbf{x}}\right]^T \mathbf{f}(\mathbf{x}, \mathbf{u}, t)$$

we rewrite equation (10.90).

$$0 = \frac{\partial J^* (\mathbf{x}(t), t)}{\partial t} + \min_{\mathbf{u}(t)} \left\{\mathcal{H}\left(\mathbf{x}(t), \mathbf{u}(t), \frac{\partial J^* (\mathbf{x}(t), t)}{\partial \mathbf{x}}, t\right)\right\} \tag{10.92}$$

The optimal control is obtained by minimizing the Hamiltonian. If $\mathbf{u}$ is constrained $\{\mathbf{u} \in \mathbf{U}$ for all $t \in [t_0, t_1])$, then an admissible minimizing control is obtained. If $\mathbf{u}$ is not constrained, a necessary condition that the optimal control must satisfy is

$$\frac{\partial \mathcal{H}}{\partial \mathbf{u}}\left(\mathbf{x}(t), \mathbf{u}(t), \frac{\partial J^* (\mathbf{x}(t), t)}{\partial \mathbf{x}}, t\right) = \mathbf{0} \tag{10.93}$$

(For sufficiency, the Hessian matrix $\frac{\partial^2 \mathcal{H}}{\partial \mathbf{u}^2}$ must be positive definite).

The minimizing control is

$$\mathbf{u}^* = \mathbf{u}^* \left(\mathbf{x}(t), \frac{\partial J^* (\mathbf{x}(t), t)}{\partial \mathbf{x}}, t\right) \tag{10.94}$$

Substituting $\mathbf{u}^*$ for $\mathbf{u}$ in (10.92) we get

$$0 = \frac{\partial J^* (\mathbf{x}(t), t)}{\partial t} + \mathcal{H}^* \left(\mathbf{x}(t), \frac{\partial J^* (\mathbf{x}(t), t)}{\partial \mathbf{x}}, t\right) \tag{10.95}$$

This equation, known as *Hamilton-Jacobi equation*, is the continuous-time analog of Bellman's recurrence relation (10.85). The Hamilton-Jacobi procedure for solving optimal control problems is summarized in Table 10.6.

Example 10.11: A first-order system is described by the differential equation

$$\dot{x}(t) = u(t)\ ; \quad x(0) = x^0$$

It is desired to find the control law that minimizes the performance measure

$$J = \int_0^{t_1} (x^2 + u^2)\, dt\ ; \quad t_1 \text{ is specified.}$$

Admissible state and control variables are not constrained by any boundaries.

For this problem, we have

$$J^*(x(t_1), t_1) = h(x(t_1), t_1) = 0$$

$$\mathcal{H}\left(x(t), u(t), \frac{\partial J^*}{\partial x}, t\right) = x^2 + u^2 + \frac{\partial J^*}{\partial x} \cdot u$$

Table 10.6: Summary of the Hamilton-Jacobi Procedure for Solving Optimal Control Problems

Given plant equation	$\dot{\mathbf{x}}(t)=\mathbf{f}(\mathbf{x}, \mathbf{u}, t)$
Given performance index	$J=h(\mathbf{x}(t_1), t_1)+\int_{t_0}^{t_1} g(\mathbf{x}, \mathbf{u}, t)dt$
Given control variable constraints	$\mathbf{u}\in U$ for all $t\in[t_0, t_1]$
Step 1	Form the Hamiltonian $\mathcal{H}(\mathbf{x}, \mathbf{u}, \frac{\partial J^*(\mathbf{x}, t)}{\partial \mathbf{x}}, t)=g(\mathbf{x}, \mathbf{u}, t)+\left[\frac{\partial J^*(\mathbf{x}, t)}{\partial \mathbf{x}}\right]^T \mathbf{f}(\mathbf{x}, \mathbf{u}, t)$
Step 2	Minimize $\mathcal{H}(\mathbf{x}, \mathbf{u}, \frac{\partial J^*}{\partial \mathbf{x}}, t)$ with respect to all admissible control vectors to find $\mathbf{u}^*=\mathbf{u}^*(\mathbf{x}, \frac{\partial J^*}{\partial \mathbf{x}}, t)$
Step 3	Find Hamiltonian $\mathcal{H}^*(\mathbf{x}, \frac{\partial J^*}{\partial \mathbf{x}}, t)=\mathcal{H}(\mathbf{x}, \mathbf{u}^*(\mathbf{x}, \frac{\partial J^*}{\partial \mathbf{x}}, t), t)$
Step 4	Solve the Hamilton-Jacobi equation $\mathcal{H}^*(\mathbf{x}, \frac{\partial J^*}{\partial \mathbf{x}}, t)+\frac{\partial J^*}{\partial t}=0$ with the appropriate boundary conditions to obtain $J^*(\mathbf{x}, t)$
Step 5	Substitute the results of Step 4 into the expression for $\mathbf{u}^*$ to obtain the optimal control

Since the control is unconstrained, a necessary condition that the optimal control must satisfy is

$$\frac{\partial \mathcal{H}}{\partial u} = 2u + \frac{\partial J^*}{\partial x} = 0$$

$$\text{(Observe that } \frac{\partial^2 \mathcal{H}}{\partial u^2} = 2 > 0)$$

This gives

$$u^* = -\frac{1}{2}\frac{\partial J^*}{\partial x}$$

Substituting u^* in Hamilton-Jacobi equation (10.95) we get

$$0 = \frac{\partial J^*}{\partial t} + x^2 + \frac{1}{4}\left(\frac{\partial J^*}{\partial x}\right)^2 + \frac{\partial J^*}{\partial x}\left(-\frac{1}{2}\frac{\partial J^*}{\partial x}\right)$$

or

$$\frac{\partial J^*}{\partial t} + x^2 - \frac{1}{4}\left(\frac{\partial J^*}{\partial x}\right)^2 = 0$$

The boundary value is

$$J^*(x(t_1), t_1) = 0$$

One way to solve the Hamilton-Jacobi equation is to guess a form for the solution and see if it can be made to satisfy the differential equation and the boundary condition. Let us assume a solution for our problem to be of the form

$$J^*(x(t), t) = K(t)\, x^2(t)$$

where $K(t)$ represents an unknown scalar function of t that is to be determined. For the assumed form of the solution

$$\frac{\partial J^*}{\partial x} = 2K(t)\, x(t)$$

$$\frac{\partial J^*}{\partial t} = \dot{K}(t)\, x^2$$

Therefore the Hamilton-Jacobi equation becomes

$$\dot{K}(t)\, x^2 + x^2 - \frac{1}{4}(4K^2(t)\, x^2) = 0$$

or

$$\dot{K}(t) + (1 - K^2(t)) = 0$$

Since $J^*(\mathbf{x}(t_1), t_1) = 0$ we get $K(t_1) = 0$

The solution is

$$K(t) = \tanh\,(t_1 - t)$$

This gives

$$u^* = -\tanh(t_1 - t)\, x(t)$$

$$J^* = \tanh(t_1 - t)\, x^2(t)$$

□

In the simple example discussed above, it was possible to obtain a solution by first guessing its form. Unfortunately, we are normally unable to find solutions for higher-order systems. Therefore, we have to resort to numerical techniques which involve some sort of discrete approximation to the Hamilton-Jacobi equation. An alternative and probably better way would be to use the recurrence relation (10.85) for continuous-time systems also using the procedure described earlier in this subsection.

The solution of the Hamilton-Jacobi equation gives a candidate for optimal control. We have not derived the sufficient conditions. Athans and Falb (1966) give a detailed account of necessary and sufficient conditions embodied in the Hamilton-Jacobi equation. The Hamilton-Jacobi equation is most often used as a check on the optimality of a control derived from the minimum principle. Athans and Falb give several examples.

We may formally obtain the minimum principle by taking appropriate partial derivatives of the Hamilton-Jacobi equation (Problem 10.20). Thus, the Hamilton-Jacobi equation provides us with a bridge from the dynamic programming approach to variational methods. It may however be noted that the minimum principle derived from the Hamilton-Jacobi equation is not applicable to as broad a class of problems as is possible. The reason for this is that the Hamilton-Jacobi equation requires $J^*(\mathbf{x}(t), t)$ to be twice continuously differentiable with respect to $\mathbf{x}$ while there are examples wherein this condition is not met.

10.6 NUMERICAL SOLUTION OF TWO-POINT BOUNDARY VALUE PROBLEM

A rather straightforward direct numerical method which has been used for solving two-point boundary value problems is based on gradient technique. In the following, we first consider the problem of minimization of functions by the gradient technique and then extend the procedure so that it is capable of solving two-point boundary value problems (Refer Kirk (1970) for alternative numerical methods).

Minimization of Functions

Consider the scalar-valued function

$$f(\mathbf{x}) = f(x_1, x_2, \ldots, x_n) \tag{10.96}$$

A necessary condition for a vector $\mathbf{x}^*$ to be an interior minimum is that the gradient vector must be zero at the minimum (eqn. (10.17a)), i.e.,

$$\left.\frac{\partial f}{\partial \mathbf{x}}\right|_{\mathbf{x}^*} = \mathbf{0} \tag{10.97}$$

The direction of gradient at a point gives the direction of greatest change in f (steepest ascent), i.e., the rate of change of a function is maximum along the direction of gradient. As we are interested in minimization, we look for a direction of steepest descent. Obviously, the direction opposite to that of $\partial f/\partial \mathbf{x}$ will give us the direction of steepest descent. Gradient is a local property; it varies from point to point and therefore minimization requires a number of steps. At the initial guess point, the direction of steepest descent can be obtained and we can move along that. An immediate question comes to our mind as to how far one should move. This is decided by the choice of step size. If step size is too small, clearly it will take more computational time, which is not desirable. If step size is too large, we may keep touring back and forth missing the required minimum point.

The steepest descent algorithm is described below.

The Steepest Descent Method: Knowing the function $f(\mathbf{x})$, we obtain $\partial f/\partial \mathbf{x}$ analytically. We then make an initial guess $\mathbf{x}^{(i)}$ $(i = 0)$ and evaluate search direction $\mathbf{s}_i$ given by

$$\mathbf{s}_i = -\left.\frac{\partial f}{\partial \mathbf{x}}\right|_{\mathbf{x}^{(i)}} \tag{10.98}$$

This search direction takes us iteratively towards the minimum point according to the rule

$$\mathbf{x}^{(i+1)} = \mathbf{x}^{(i)} + \alpha_i^* \mathbf{s}_i \tag{10.99}$$

where α_i^* (> 0) is the optimum step size which satisfies the following inequality:

$$f(\mathbf{x}^{(i)} + \alpha_i^* \mathbf{s}_i) \leqslant f(\mathbf{x}^{(i)} + \alpha_i \mathbf{s}_i) \tag{10.100}$$

The flow chart of the steepest descent algorithm is shown in Fig. 10.14.

The reader might have observed that we have replaced the minimization problem given by (10.13) by another given by (10.100). However, (10.100) represents basically a minimization problem with respect to a scalar. Scalar minimization problems are easy to solve numerically. In the following, we give *quadratic interpolation algorithm* to solve this problem (Athans et al. 1974).

(i) Select a reasonable value of α_i so that it provides only a small change in the function, of the order of 1%.

(ii) Change the value of α_i in a doubling fashion as long as the function is decreasing up to and including the first time the function increases.

(iii) Fit a quadratic to the last three values of the function.

The flow chart of the quadratic interpolation algorithm is shown in Fig. 10.15.

The Fletcher-Powell Method: The steepest descent method described above is simple and applicable to all well behaved functions. However, it is obvious

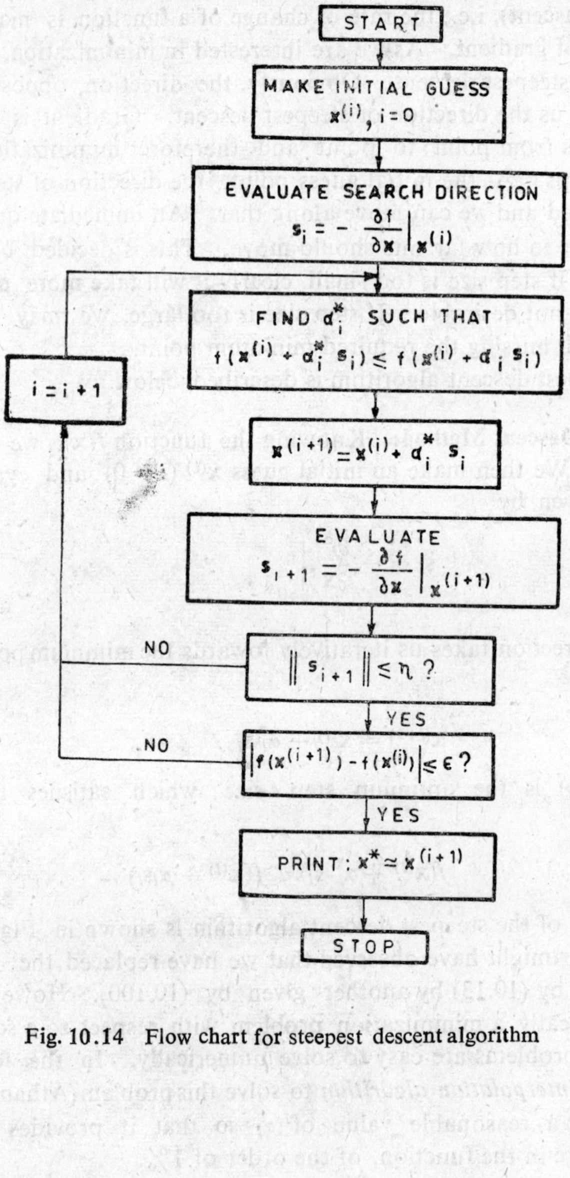

Fig. 10.14 Flow chart for steepest descent algorithm

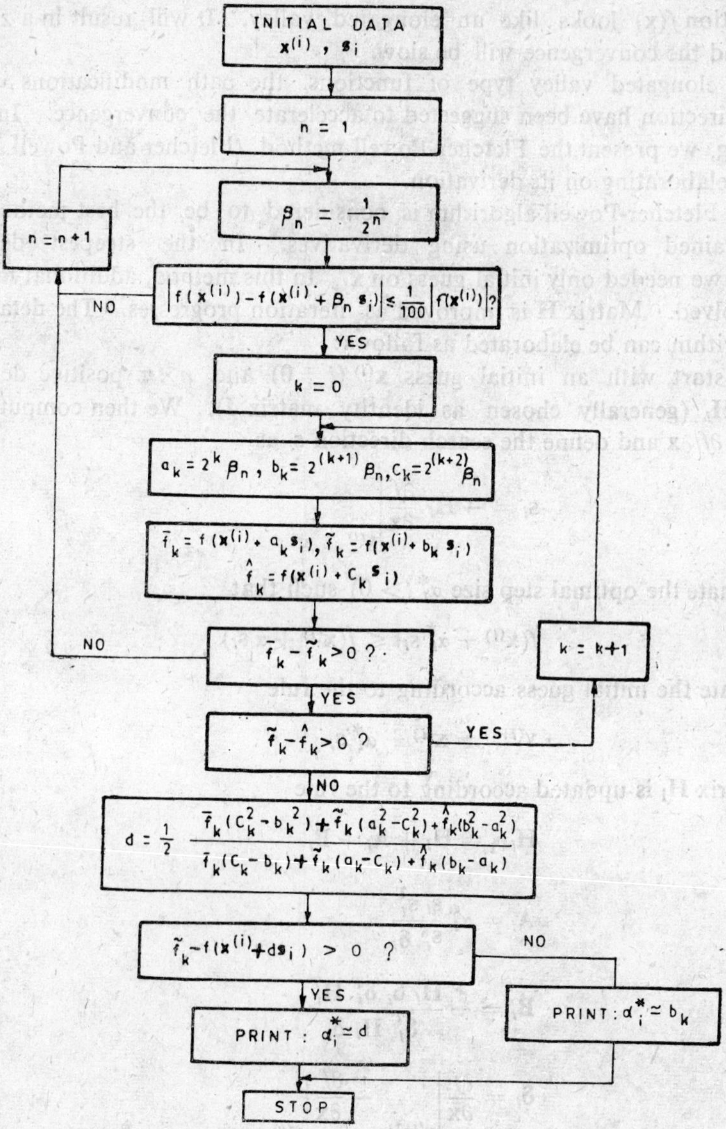

Fig. 10.15 Flow chart for quadratic interpolation algorithm

that the policy of steepest descent will not work well in situations in which the function $f(\mathbf{x})$ looks like an elongated valley. It will result in a zigzag effect and the convergence will be slow.

For elongated valley type of functions, the path modifications in the search direction have been suggested to accelerate the convergence. In the following, we present the Fletcher-Powell method (Fletcher and Powell 1963) without elaborating on its derivation.

The Fletcher-Powell algorithm is considered to be the best method for unconstrained optimization using derivatives. In the steepest descent method, we needed only initial guess on **x**. In this method, additional matrix **H** is involved. Matrix **H** is improved as iteration progresses. The details of the algorithm can be elaborated as follows:

We start with an initial guess $\mathbf{x}^{(i)}$ $(i=0)$ and $n\times n$ positive definite matrix $\mathbf{H}_i$ (generally chosen as identity matrix **I**). We then compute the gradient $\partial f/\partial \mathbf{x}$ and define the search direction $\mathbf{s}_i$ as

$$\mathbf{s}_i = -\mathbf{H}_i \left.\frac{\partial f}{\partial \mathbf{x}}\right|_{\mathbf{x}^{(i)}}$$

We estimate the optimal step size α_i^* (> 0) such that

$$f(\mathbf{x}^{(i)} + \alpha_i^* \mathbf{s}_i) \leqslant f(\mathbf{x}^{(i)} + \alpha_i \mathbf{s}_i)$$

and update the initial guess according to the rule

$$\mathbf{x}^{(i+1)} = \mathbf{x}^{(i)} + \alpha_i^* \mathbf{s}_i$$

The matrix $\mathbf{H}_i$ is updated according to the rule

$$\mathbf{H}_{i+1} = \mathbf{H}_i + \mathbf{A}_i + \mathbf{B}_i$$

where

$$\mathbf{A}_i = \alpha_i^* \frac{\mathbf{s}_i\, \mathbf{s}_i^T}{\mathbf{s}_i^T\, \boldsymbol{\delta}_i}$$

$$\mathbf{B}_i = -\frac{\mathbf{H}_i\, \boldsymbol{\delta}_i\, \boldsymbol{\delta}_i^T\, \mathbf{H}_i^T}{\boldsymbol{\delta}_i^T\, \mathbf{H}_i\, \boldsymbol{\delta}_i}$$

$$\boldsymbol{\delta}_i = \left.\frac{\partial f}{\partial \mathbf{x}}\right|_{\mathbf{x}^{(i+1)}} - \left.\frac{\partial f}{\partial \mathbf{x}}\right|_{\mathbf{x}^{(i)}}$$

It can be shown that $\mathbf{H}_{i+1}$ remains positive definite if $\mathbf{H}_i$ is positive definite. The search direction is based on a gradient rotated by a positive definite matrix **H**. This method converges fast to the solution. The flow chart given in Fig. 10.16 depicts the steps involved.

Solution of Two-Point Boundary-Value Problem

The procedures of function minimization based on gradient techniques can easily be extended for a solution of two-point boundary-value problems.

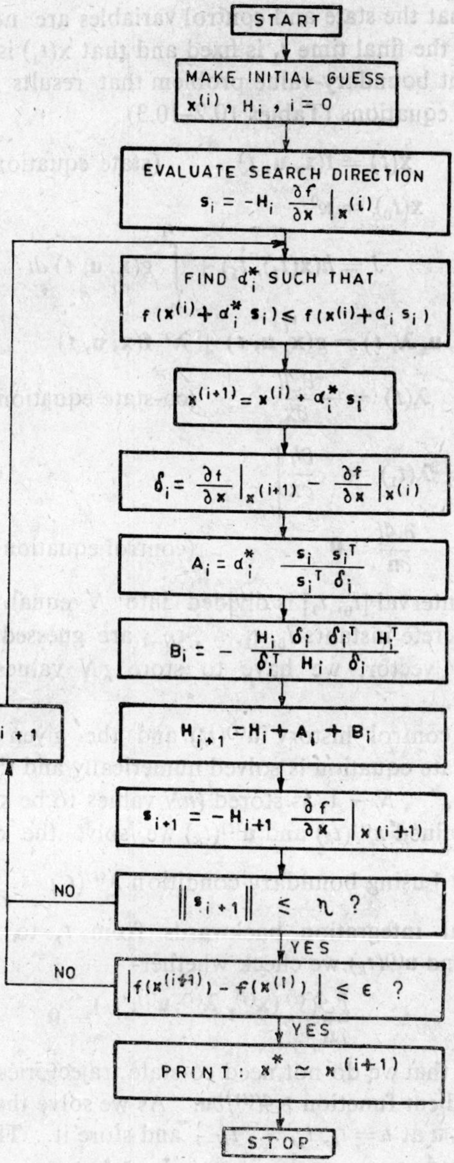

Fig. 10.16 Flow chart for Fletcher-Powell algorithm

Assuming that the state and control variables are not constrained by any boundaries, that the final time t_1 is fixed and that $\mathbf{x}(t_1)$ is free, we can summarize the two-point boundary-value problem that results from the variational approach by the equations (Tables 10.2–10.3)

$$\dot{\mathbf{x}}(t) = \mathbf{f}(\mathbf{x}, \mathbf{u}, t) \qquad \text{(state equation)}$$

$$\mathbf{x}(t_0) = \mathbf{x}^0$$

$$J = h(\mathbf{x}(t_1), t_1) + \int_{t_0}^{t_1} g(\mathbf{x}, \mathbf{u}, t)\, dt$$

$$\mathcal{H}(\mathbf{x}, \mathbf{u}, \boldsymbol{\lambda}, t) = g(\mathbf{x}, \mathbf{u}, t) + \boldsymbol{\lambda}^T \mathbf{f}(\mathbf{x}, \mathbf{u}, t)$$

$$\dot{\boldsymbol{\lambda}}(t) = -\frac{\partial \mathcal{H}}{\partial \mathbf{x}} \qquad \text{(co-state equation)}$$

$$\boldsymbol{\lambda}(t_1) = \left.\frac{\partial h}{\partial \mathbf{x}}\right|_{t_1}$$

$$\frac{\partial \mathcal{H}}{\partial \mathbf{u}} = \mathbf{0} \qquad \text{(control equation)}$$

The given time interval $[t_0, t_1]$ is divided into N equal parts. Initial values of $\mathbf{u}(t)$ at the discrete instants $t_0, t_1, \ldots, t_{N-1}$ are guessed to get $\mathbf{u}^{(i)}(t_k)$, $i = 0$. Since $\mathbf{u}$ is a $p \times 1$ vector, we have to store pN values in the memory of a digital computer.

Using the control history $\mathbf{u}^{(i)}(t_k)$ and the given boundary condition $\mathbf{x}(t_0) = \mathbf{x}^0$, the state equation is solved numerically and the resulting trajectory $\mathbf{x}^{(i)}(t_k)$; $k = 0, 1, \ldots, N-1$, is stored (nN values to be stored).

Having obtained $\mathbf{x}^{(i)}(t_k)$ and $\mathbf{u}^{(i)}(t_k)$ we solve the co-state equations in the interval $[t_0, t_1]$ using boundary condition $\boldsymbol{\lambda}^{(i)}(t_1) = \left.\frac{\partial h(\mathbf{x}^{(i)})}{\partial \mathbf{x}}\right|_{t_1}$. This requires numerical integration backwards from t_1 to t_0. For the values of $\mathbf{x}^{(i)}(t_k)$, $\boldsymbol{\lambda}^{(i)}(t_k)$ and $\mathbf{u}^{(i)}(t_k)$ we check whether

$$\frac{\partial \mathcal{H}^{(i)}(\mathbf{x}^{(i)}, \boldsymbol{\lambda}^{(i)}, \mathbf{u}^{(i)}, t_k)}{\partial \mathbf{u}} = \mathbf{0}$$

It may be noted that we do not need co-state trajectories explicitly. However, we need the gradient function $\partial \mathcal{H}^{(i)}/\partial \mathbf{u}$. As we solve the co-state equation, we compute $\partial \mathcal{H}^{(i)}/\partial \mathbf{u}$ at $t = t_0, t_1, \ldots, t_{N-1}$ and store it. The co-state trajectory need not be stored.

If some norm of gradient $\partial \mathcal{H}^{(i)}/\partial \mathbf{u}$ is less than a certain preselected constant, and if $| \mathcal{H}^{(i+1)} - \mathcal{H}^{(i)} |$ is also less than some preselected constant, we discontinue the iterative procedure and output the extremal state and control.

If the conditions above are not satisfied, we modify our guess on $\mathbf{u}^{(i)}$ according to the following rule:

$$\mathbf{u}^{(i+1)}(t_k) = \mathbf{u}^{(i)}(t_k) - \alpha_i \frac{\partial \mathcal{H}^{(i)}}{\partial \mathbf{u}}(t_k)\,; \quad \alpha_i > 0.$$

On similar lines, the Fletcher-Powell method, discussed earlier for function minimization can easily be extended for the solution of a two-point boundary value problem.

Example 10.12: In the following we shall discuss the problem of the start-up of a Nuclear Reactor. A nonlinear reactor model is given by the following equations:

$$\dot{n}(t) = \frac{\rho_t - \beta}{l} n(t) + \lambda c(t) \tag{10.101a}$$

$$\dot{c}(t) = \frac{\beta}{l} n(t) - \lambda c(t) \tag{10.101b}$$

$$\rho_t = \rho - \alpha n(t) \tag{10.101c}$$

Compared to the model given in eqns. (4.57)–(4.58), the model (10.101) accounts for fuel temperature reactivity feedback; ρ is the external reactivity from control rods, $\rho_{\text{fuel}} = \alpha n$, is fuel temperature (proportional to neutron flux or power) reactivity feedback where α is reactivity coefficient of power, per KW, ρ_t is the total reactivity.

Substituting (10.101c) into (10.101a) we get

$$\dot{n} = \frac{\rho - \beta}{l} n + \lambda c - \frac{\alpha}{l} n^2 \tag{10.102}$$

The start up problem is to increase the power level from n^0 to n^1 in the given time while minimizing certain performance index. The performance index can be chosen so as to minimize the reactivity input (i.e., control rod motion) or to minimize control energy, (i.e., rate of change of reactivity). For minimizing the rate of change of reactivity, we selected the following performance index:

$$J = \frac{1}{2} \int_{t_0}^{t_1} [\dot{\rho}^2 + \sigma(n - n^1)^2] \, dt \tag{10.103}$$

where σ is a weighing factor.

To treat $\dot{\rho}$ as control variable u, we include the following equation in the set of eqns. (10.101).

$$\dot{\rho}(t) = u(t) \tag{10.104}$$

Thus $\rho(t)$ in effect becomes the third state variable; the first two being $n(t)$ and $c(t)$.

For the above system, the Hamiltonian $\mathcal{H}$ is formed as follows (Table 10.2).

$$\mathcal{H} = \frac{1}{2}u^2 + \frac{1}{2}\sigma(n - n^1)^2 + \lambda_1 \left[\frac{\rho - \beta}{l} n + \lambda c - \frac{\alpha}{l} n^2\right] + \lambda_2 \left[\frac{\beta}{l} n - \lambda c\right] + \lambda_3 u$$

where λ_1, λ_2 and λ_3 are the co-state variables.

The co-state equations are given by

$$\dot{\lambda}_1 = -\frac{\partial \mathcal{H}}{\partial n} = -\sigma(n - n^1) - \frac{\rho - \beta}{l}\lambda_1 - \frac{\beta}{l}\lambda_2 + \frac{2\alpha n}{l}\lambda_1$$

$$\dot{\lambda}_2 = -\frac{\partial \mathcal{H}}{\partial c} = -\lambda\lambda_1 + \lambda\lambda_2$$

$$\dot{\lambda}_3 = -\frac{\partial \mathcal{H}}{\partial \rho} = -\frac{n}{l}\lambda_1$$

The derivative of the Hamiltonian with respect to the input is given by

$$\frac{\partial \mathcal{H}}{\partial u} = u + \lambda_3$$

Thus the optimal control which minimizes the Hamiltonian is given by

$$u^* = -\lambda_3$$

The assumed values of various parameters and boundary conditions are as follows :

$\lambda = 0.1 \text{ sec}^{-1}$ $\qquad$ $n^0 = 10$ KW

$\beta = 0.0064$ $\qquad$ $n^1 = 40$ KW

$l = 0.001$ sec $\qquad$ $c^0 = 640$ KW

$t_0 = 0$ sec $\qquad$ $\rho = -\alpha n^0$

$t_1 = 0.5$ sec $\qquad$ $= 0.0001$

$\alpha = 0.00001/\text{KW}$

$\sigma = 10^{-6}$

This problem was solved on a digital computer using the Fletcher-Powell method. The following stopping criterion was used:

$$\left|\frac{\partial \mathcal{H}}{\partial u}\right| \leqslant 0.001$$

$$|\mathcal{H}^{(i+1)} - \mathcal{H}^{(i)}| \leqslant 0.05$$

The forward and backward integration of the state and co-state equations, respectively, was carried out using the fourth-order Runge-Kutta method. In our problem, initially the reactivity has to be increased to increase the power level; thus the control input (rate of change of reactivity) is initially positive. While giving the initial guess, this fact was made use of.

The computer output is shown in Fig. 10.17.

10.7 CONCLUDING COMMENTS

In this chapter, we have studied the general mathematical procedures for the solution of optimal control problems.

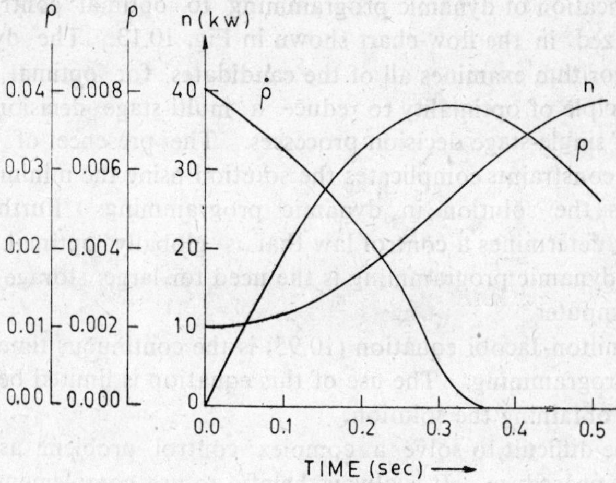

Fig. 10.17

In Section 10.2, we discussed the formulation of some optimal control problems encountered in practice. We observed that, in general, it may not be possible to translate all the performance requirements into a single mathematical criterion. Quite frequently the optimal control format is not directly applicable due to sensitivity and other considerations. In such cases the knowledge gained from the study of optimal control systems often permits us to evolve near-optimal designs.

Sections 10.3–10.5 dealt with the solution problem. Before applying a mathematical procedure for the solution of an optimal control problem, it is natural to ask whether the solution of the problem exists. In case the existence is guaranteed, one would like to determine necessary and sufficient conditions for the optimal solution. Though the existence theorems for some classes of problems are available, these are very difficult to apply and one normally attempts to find an optimal control rather than trying to prove that one exists. The mathematical procedures for the solution of optimal control problems provide necessary but usually not sufficient conditions that the optimum solution must satisfy. With the number of necessary conditions at our disposal, we may try numerical methods to find optimal solutions. Further the optimal solution to a control problem may not be unique. Out of a set of solutions available, the designer may choose one depending upon considerations other than the minimization of the performance index.

The application of calculus of variations to optimal control problems was summarized in Tables 10.2 and 10.3. A more general procedure based on Pontryagin's Minimum Principle which accounts for control and state variable constraints was summarized in Table 10.5. Application of these methods generally leads to a nonlinear two-point boundary-value problem. An iterative numerical technique for the solution of two-point boundary-value problems was discussed in Section 10.6.

The application of dynamic programming to optimal control problems was summarized in the flow-chart shown in Fig. 10.13. The dynamic programming algorithm examines all of the candidates for optimal control law and uses principle of optimality to reduce a multi-stage decision process to a sequence of single-stage decision processes. The presence of control and state variable constraints complicates the solution using the minimum principle but simplifies the solution in dynamic programming. Further, dynamic programming determines a control law that is globally optimal. The main limitation of dynamic programming is the need for large storage capacity in the digital computer.

The Hamilton-Jacobi equation (10.95) is the continuous-time formulation of dynamic programming. The use of this equation is limited because of the difficulties of obtaining the solution.

It may be difficult to solve a complex control problem using a single mathematical procedure. It is always helpful to use complementary features of several solution procedures. For example, suppose that the minimum principle indicates that the only values which the optimal control can take are +1, 0, and −1. This knowledge can be used effectively in obtaining the dynamic programming solution.

Let us admit that the use of the optimal control theory has been very limited in industrial control. The implementation of the optimal control solution requires either an on-line digital computer that calculates optimal control signals, as the process evolves and additional hardware to synthesize the signals, or a special-purpose digital computer to synthesize an optimal control law that has been precomputed off-line with a general purpose digital computer. With the availability of small and powerful digital computers, the use of the optimal control theory in industrial control is bound to increase.

Some of the industrial control problems which have been successfully solved using the optimal control theory are discussed in the next chapter. For other problems also the optimal control theory has a definite role to play—it provides an index of comparison to design a suboptimal controller which is easily implementable and takes care of system requirements other than the minimization of the performance index. The structure of optimal control frequently provides insight as to how suboptimal but practical systems can be designed.

PROBLEMS

10.1 Let $f(\mathbf{x}) = -x_1x_2$ and let $g(\mathbf{x}) = x_1^2 + x_2^2 - 1$. What are the potential candidates for minima of f subject to the constraint $g = 0$?

Show that the points $\left(\frac{1}{\sqrt{2}}, \frac{1}{\sqrt{2}}\right)$ and $\left(-\frac{1}{\sqrt{2}}, -\frac{1}{\sqrt{2}}\right)$ actually provide the minima.

10.2 Show that the extremal for the functional

$$J(x) = \int_0^{\pi/2} (\dot{x}^2 - x^2)\, dt$$

which satisfies the boundary conditions

$$x(0) = 0; \quad x(\pi/2) = 1$$

is $$x^*(t) = \sin t$$

By investigating the increments in J for curves in the neighbourhood of $x^*(t)$, demonstrate that $x^*(t)$ minimizes $J(x)$.
(Hint: You may take $\delta x = k \sin 2t$ with k as a real constant).

10.3 Prove that the Euler-Lagrange equation (10.32) reduces to the first-order differential equation

$$g(x^*, \dot{x}^*) - \dot{x}^* \frac{\partial g(x^*, \dot{x}^*)}{\partial \dot{x}} = \text{constant}$$

whenever g is not explicitly a function of time, i.e.,

$$g = g(x, \dot{x})$$

10.4 Find the extremals for the following functionals.

(i) $J(x) = \int_0^{\pi/4} (x^2 - \dot{x}^2)\, dt\,; \quad x(0) = 0,\ x\left(\frac{\pi}{4}\right)$ is free

(ii) $J(x) = \int_1^{t_1} (2x + \frac{1}{2}\dot{x}^2)\, dt\,; \quad x(1) = 2,\ x(t_1) = 2,\ t_1 > 1$ is free

(iii) $J(x) = \int_0^{t_1} \frac{\sqrt{1 + \dot{x}^2}}{x}\, dt\,; \quad x(0) = 0,\ x(t_1) = t_1 - 5$

10.5 Find the curve with minimum arc length between the point $x(0) = 0$ and the curve

$$\theta(t) = t^2 - 10t + 24$$

10.6 Find the trajectories in the (t, x) plane which will extremize

$$J(x) = \int_0^{t_1} (t\dot{x} + \dot{x}^2)\, dt$$

in each of the three cases:

(i) $t_1 = 1,\ x(0) = 1,\ x(1) = 5$
(ii) $t_1 = 1,\ x(0) = 1,\ x(1)$ is free
(iii) t_1 is free, $x(0) = 1,\ x(t_1) = 5$

10.7 Find the Euler-Lagrange equations and boundary conditions for the extremal of the functional

$$J(x) = \int_0^{\pi/2} (\dot{x}_1^2 + 2x_1\, x_2 + \dot{x}_2^2)\, dt$$

$$x_1(0) = 0, \quad x_1\left(\frac{\pi}{2}\right) \text{ is free}$$

$$x_2(0) = 0, \quad x_2\left(\frac{\pi}{2}\right) = -1$$

10.8 The system

$$\dot{x} = -x + u$$

is to be transferred from $x(0) = 5$ to $x(1) = 0$ such that

$$J = \tfrac{1}{2} \int_0^1 (\dot{u})^2\, dt$$

is minimized. Find the optimal control.

10.9 Find the optimal control $u^*(t)$ for the system

$$\dot{x} = u; \quad x(0) = 1$$

which minimizes

$$J = \frac{1}{2} x^2(4) + \frac{1}{2} \int_0^4 u^2\, dt$$

10.10 Find the optimal control $u^*(t)$ for the system

$$\dot{\mathbf{x}} = \begin{bmatrix} 0 & 1 \\ -10 & 0 \end{bmatrix} \mathbf{x} + \begin{bmatrix} 0 \\ 10 \end{bmatrix} u$$

which minimizes the performance index

$$J = \frac{1}{2} \int_0^2 u^2\, dt$$

Given: $\mathbf{x}(0) = \begin{bmatrix} 1 \\ 1 \end{bmatrix}, \mathbf{x}(2) = \begin{bmatrix} 0 \\ 0 \end{bmatrix}$

10.11 Consider the system

$$\dot{x}_1 = x_2 + u_1$$

$$\dot{x}_2 = u_2$$

Find the optimal control $u^*(t)$ for the functional

$$J = \frac{1}{2}\int_0^4 (u_1^2 + u_2^2)\,dt$$

Given: $x_1(0) = x_2(0) = 1$; $x_1(4) = 0$

10.12 Formulate the two-point boundary-value problem which when solved, yields the optimal control $u^*(t)$ for the system

$$\dot{x}_1 = x_2$$

$$\dot{x}_2 = x_1 + (1 - x_1^2)\,x_2 + u$$

$$\mathbf{x}(0) = [1 \quad 0]^T$$

$$J = \frac{1}{2}\int_0^2 (2x_1^2 + x_2^2 + u^2)\,dt$$

when (i) $u(t)$ is not bounded (ii) $|u(t)| \leqslant 1.0$

10.13 Consider a system described by the equations

$$\mathbf{x}(t) = \mathbf{f}(\mathbf{x}) + \mathbf{F}(\mathbf{x})\,\mathbf{u}(t)$$

where $\mathbf{f}(\mathbf{x})$ and the $n \times p$ matrix $\mathbf{F}(\mathbf{x})$ may be nonlinear. Use the Minimum Principle to show that it is necessary to use a Bang-Bang controller to obtain the minimum-time response for this system if the input variables $u_i(t)$; $i = 1, 2, \ldots, p$ are constrained to a magnitude of less than a constant M.

10.14 Assume now that the system defined in Problem 10.13 is to be driven so that

$$J = \int_{t_0}^{t_1} \left[\sum_{i=1}^{p} |u_i(t)|\right] dt$$

is minimized and the admissible controls are to satisfy the constraints

$$-1 \leqslant u_i(t) \leqslant 1;\ i = 1, 2, \ldots, p$$

Use the Minimum Principle to show that it is necessary to use a Bang-off-Bang controller.

10.15 For the system

$$\frac{d^2y}{dt^2} = u$$

with $|u| \leqslant 1$, find the control which drives the system from an arbitrary initial state to the origin in a condition satisfying $|y| \leqslant 0.5$ in the minimum time.

10.16 For the system

$$\dot{x} = u$$

with $|u| \leqslant 1$, find the control which drives the system from an arbitrary initial state to the origin and minimizes

$$J = \int_0^{t_1} |u(t)|\, dt; \quad t_1 \text{ is free.}$$

10.17 Consider the discrete-time system

$$x(k+1) = -0.5x(k) + u(k)$$

The performance index

$$J = \sum_{k=0}^{2} |x(k)|$$

is to be minimized, subject to state and control constraints:

$$-0.2 \leqslant x(k) \leqslant 0.2; \; -0.1 \leqslant u(k) \leqslant 0.1$$

Quantize both $x(k)$ and $u(k)$ in steps of 0.1 and carry out computational steps required to determine the optimal control law using dynamic programming. Also determine the optimal control sequence for an initial state $x(0) = 0$.

10.18 Consider a perfect integrator being governed by the first-order differential equation

$$\dot{x} = u$$

The problem is the minimization of cost function

$$J = \int_0^{t_1} (x^2 + u^2)\, dt$$

subject to the initial condition $x(0) = x^0$. Recast the system formulation in the discrete form with $t_1 = 4$ and sampling interval $T = 1$. Using this formulation, obtain the optimal control sequence to minimize the cost function. Investigate the solution over the range $x = 5$ to $x = -5$ and assume that the magnitude of u is normalized to be less than or equal to unity.

10.19 Find the control vector which minimizes

$$J = \tfrac{1}{2} x^2(t_1) + \tfrac{1}{4} \int_0^{t_1} u^2(t)\, dt; \quad t_1 \text{ specified}$$

for the system described by

$$\dot{x} = x + u$$

Use Hamilton-Jacobi equations to find the optimum control vector.

Comment on your result when $t_1 \to \infty$.

10.20 Derive the Pontryagin's Minimum Principle from the Hamilton-Jacobi equation (refer Kirk (1970)).

REFERENCES

1. Anderson, B.D.O., and J.B. Moore, *Linear Optimal Control,* Englewood Cliffs, N.J.: Prentice Hall, **1971.**
2. Athans, M., "The status of optimal control theory and applications to deterministic systems", *IEEE Trans. Automat. Contr.*, vol. AC-11, pp. 580-596, **1966.**
3. Athans, M., M.L. Dertouzos, R.F. Spann, and S.J. Mason, *Systems, Networks and Computaion: Multivariable Methods,* New York: McGraw-Hill, **1974.**
4. Athans, M., and P.L. Falb, *Optimal Control,* New York: McGraw-Hill, **1966.**
5. Baum, R.F., and L. Cesari, "On a recent proof of Pontryagin's necessary conditions", *SIAM J. Control,* vol. 10, pp. 56-75, **1972.**
6. Bellman, R.E., *Dynamic Programming,* Princeton, N.J.: Princeton University Press, **1957.**
7. Bellman, R.E., and R.E. Kalaba, *Dynamic Programming and Modern Control Theory,* New York: Academic Press, **1965.**
8. Berkovitz, L.D., "Variational methods in problems of control and programming", *J. Math. Anal. Appl.*, vol. 3, pp. 145-69, **1961.**
9. Berkovitz, L.D., "On control problems with bounded state variables," *J. Math. Anal. Appl.*, vol. 5, pp. 488-98, **1962.**
10. Bliss, G.A., *Lectures on the Calculus of Variations,* Chicago: University of Chicago Press, **1963.**
11. Bryson, A.E., and Y.C.Ho, *Applied Optimal Control,* Waltham, Mass.: Blaisdell, **1969.**
12. Dreyfus, S.E., *Dynamic Programming and the Calculus of Variations,* New York: Academic Press, **1965.**
13. Elsgolc, L.E., *Calculus of Variations,* Reading, Mass.: Addison-Wesley, **1962.**
14. Fan, L.T., and C.S. Wang, *The Discrete Maximum Principle,* New York: Wiley, **1964.**
15. Fletcher, R., and M.J.D. Powell, "A rapidly convergent descent method for minimization," *Comput., J.*, vol. 6, pp. 163-168, **1963.**
16. Gelfand, I.M., and S.V. Fomin, *Calculus of Variations,* Englewood Cliffs, N.J.: Prentice-Hall, **1963.**
17. Hancock, H., *Theory of Maxima and Minima,* New York: Dover, **1960.**
18. Hautus, M.L.J., "Necessary conditions for multiple constraint optimization problems", *SIAM J. Control,* vol. 4, pp. 653-69, **1973.**
19. Hestenas, M.R., *Calculus of Variations and Optimal Control Theory,* New York: Wiley, **1966.**
20. Holtzman, J.M., "Convexity and the maximum principle for discrete systems", *IEEE Trans. Automat. Contr.*, vol. AC-11, pp. 30-35, **1966.**
21. Jordan, B.W., and E. Polak, "Theory of a class of discrete optimal control systems" *J. Electron. Control,* vol. 17, **1964.**
22. Kalman, R.E., "The theory of optimal control and the calculus of variations", *Mathematical Optimization Techniques,* R. Bellman (ed.), Berkeley: Univ. of California Press, **1963.**
23. Kirk, D.E., *Optimal Control Theory: An Introduction,* Englewood Cliffs, N.J.: Prentice-Hall, **1970.**
24. Kwakernaak, H., and R. Sivan, *Linear Optimal Control Systems,* New York: Wiley, **1972.**
25. Lapidus, L., and R. Luus, *Optimal Control of Engineering Processes,* Waltham, Mass.: Blaisdell, 1967.
26. Larson, R.E., "Dynamic programming and reduced computational requirements", *IEEE Trans. Automat. Contr.*, vol. AC-10, pp. 135-43, **1965.**
27. Larson, R.E., "A survey of dynamic programming computational procedures," *IEEE*

Trans. Automat. Contr., vol. AC-12, pp. 767-774, 1967.
28. Lietman, G. (ed.), *Optimization Techniques with Applications to Aerospace Systems*, New York: Academic Press, 1962.
29. Lietman, G., *Optimal Control*, New York: McGraw-Hill, 1966.
30. McGill, R., "Optimal Control, inequality state constraints and the generalized Newton-Raphson algorithm", *SIAM J. Control*, vol.3, pp. 291-98, 1965.
31. Merrium, C.W., III, *Optimization Theory and the Design of Feedback Control Systems*, New York: McGraw-Hill, 1964.
32. Pearson, J.B., and R. Sridhar, "A discrete optimal control problem", *IEEE Trans. Automat. Contr.*, vol. AC-11, pp. 171-174, 1966.
33. Pierre, D.A., *Optimization Theory with Applications*, New York: Wiley, 1969.
34. Pontryagin, L.S., V.G. Boltyanskii, R.V. Gamkrelidze, and E.F. Mischenko, *The Mathematical Theory of Optimal Processes*, New York: Wiley, 1962.
35. Sage, A.P., and C.C. White, III, *Optimum Systems Control*, 2nd edition. Englewood Cliffs, N.J.: Prentice-Hall, 1977.
36. Tou, J., *Modern Control Theory*, New York: McGraw-Hill. 1964.

11. OPTIMAL FEEDBACK CONTROL

11.1 INTRODUCTION

There exist two important classes of optimal control problems for which quite general results have been obtained. Both these involve control of linear systems. In the first class, the objective is to minimize the integral of a quadratic performance index. There are no constraints on the control variable. The linear state regulator problem, the linear output regulator problem and the linear tracking problem formulated in Section 10.2 belong to this class. In the second class, we have the time-optimal problems wherein the objective is to minimize the time required to drive the system from an initial state to the origin. The control is constrained.

An important feature of these classes of problems is that optimal control is possible by feedback controllers. The feedback control system requires measurement of the state vector. The state vector may need to be estimated from plant input and output if it is not directly measurable (Chapters 9 and 12). There is obviously a great practical advantage to use a feedback system rather than a *programmed* system in which the optimal control is simply computed and applied to the control process without regard to the behaviour of the state vector during the transient. Slight deviations of the process dynamics will always cause deviations of the physical state $\mathbf{x}(t)$ from $\mathbf{x}^*(t)$. In feedback systems, on the other hand, slight deviations of the state behaviour tend to be self-correcting.

Many problems of industrial control can be formulated as the standard linear regulator and time-optimal problems. This justifies the seemingly excessive attention given to these problems in this book.

11.2 DISCRETE-TIME LINEAR STATE REGULATOR

Consider the plant represented by linear discrete-time state equations of the form

$$\mathbf{x}(k+1) = \mathbf{F}(k)\,\mathbf{x}(k) + \mathbf{G}(k)\,\mathbf{u}(k)\,;\;\; \mathbf{x}(0) = \mathbf{x}^0 \tag{11.1}$$

where $\mathbf{x}$ is the $n\times 1$ state vector, $\mathbf{u}$ is the $p\times 1$ input vecor, F and G a) are respectively, $n\times n$ and $n\times p$ time-varying real matrices and $k = 0, 1, 2, \ldots, N-1$ ($N =$ integer).

We shall be interested in selecting the controls $\mathbf{u}(k)$; $k = 0, 1, \ldots, N-1$ which minimize a performance criterion of the form

$$J = \tfrac{1}{2}\mathbf{x}^T(N)\,\mathbf{H}\,\mathbf{x}(N) + \tfrac{1}{2}\sum_{k=0}^{N-1}[\mathbf{x}^T(k)\,\mathbf{Q}(k)\,\mathbf{x}(k) + \mathbf{u}^T(k)\,\mathbf{R}(k)\,\mathbf{u}(k)] \qquad (11.2)$$

where $\mathbf{H}$ and $\mathbf{Q}$ are real symmetric positive semi definite $n \times n$ matrices and $\mathbf{R}$ is a real symmetric positive definite $p \times p$ matrix. Both $\mathbf{H}$ and $\mathbf{Q}$ should not be zero matrices simultaneously. This criterion is the discrete analog of that given by eqn. (10.7a); a summation replaces integration.

For notational convenience, we shall assume in our derivation that the matrices $\mathbf{F}$, $\mathbf{G}$, $\mathbf{Q}$ and $\mathbf{R}$ are constant matrices; later we shall relax this assumption.

Following the approach of Section 10.5, we begin by defining

$$J_{N,N}(\mathbf{x}(N)) \overset{\Delta}{=} \tfrac{1}{2}\mathbf{x}^T(N)\,\mathbf{H}\,\mathbf{x}(N) \qquad (11.3a)$$

$$= J^*_{N,N}(\mathbf{x}(N))$$

$$\overset{\Delta}{=} \tfrac{1}{2}\mathbf{x}^T(N)\,\mathbf{P}(0)\,\mathbf{x}(N) \qquad (11.3b)$$

where

$$\mathbf{P}(0) \overset{\Delta}{=} \mathbf{H} \qquad (11.3c)$$

Next define

$$J_{N-1,N}(\mathbf{x}(N-1), \mathbf{u}(N-1)) \overset{\Delta}{=} \tfrac{1}{2}\mathbf{x}^T(N)\,\mathbf{P}(0)\,\mathbf{x}(N) + \tfrac{1}{2}\mathbf{x}^T(N-1)\,\mathbf{Q}\,\mathbf{x}(N-1) + \tfrac{1}{2}\mathbf{u}^T(N-1)\,\mathbf{R}\,\mathbf{u}(N-1)$$

Since $\mathbf{x}(N)$ is related to $\mathbf{u}(N-1)$ by the state equation (11.1), we may write

$$\begin{aligned}
J_{N-1,N}(\mathbf{x}(N-1), \mathbf{u}(N-1)) &= \tfrac{1}{2}\mathbf{x}^T(N-1)\,\mathbf{Q}\,\mathbf{x}(N-1) + \tfrac{1}{2}\mathbf{u}^T(N-1)\,\mathbf{R}\,\mathbf{u}(N-1) \\
&\quad + \tfrac{1}{2}[\mathbf{F}\,\mathbf{x}(N-1) + \mathbf{G}\,\mathbf{u}(N-1)]^T\,\mathbf{P}(0)\,[\mathbf{F}\,\mathbf{x}(N-1) + \mathbf{G}\,\mathbf{u}(N-1)] \\
&= \tfrac{1}{2}\mathbf{x}^T(N-1)\,\mathbf{Q}\,\mathbf{x}(N-1) + \tfrac{1}{2}\mathbf{u}^T(N-1)\,\mathbf{R}\mathbf{u}(N-1) \\
&\quad + \tfrac{1}{2}\mathbf{x}^T(N-1)\,\mathbf{F}^T\mathbf{P}(0)\,\mathbf{F}\,\mathbf{x}(N-1) + \tfrac{1}{2}\mathbf{x}^T(N-1)\,\mathbf{F}^T\mathbf{P}(0)\,\mathbf{G}\,\mathbf{u}(N-1) \\
&\quad + \tfrac{1}{2}\mathbf{u}^T(N-1)\,\mathbf{G}^T\mathbf{P}(0)\,\mathbf{F}\mathbf{x}(N-1) + \tfrac{1}{2}\mathbf{u}^T(N-1)\,\mathbf{G}^T\mathbf{P}(0)\,\mathbf{G}\,\mathbf{u}(N-1)
\end{aligned} \qquad (11.4)$$

The minimum cost for the *one-stage process* is

$$J^*_{N-1,N}(\mathbf{x}(N-1)) \overset{\Delta}{=} \min_{\mathbf{u}(N-1)} \{J_{N-1,N}(\mathbf{x}(N-1), \mathbf{u}(N-1))\}$$

Since the admissible controls are not bounded, to minimize $J_{N-1,N}(\mathbf{x}(N-1), \mathbf{u}(N-1))$ with respect to $\mathbf{u}(N-1)$, we need to consider only those control values for which

$$\frac{\partial J_{N-1,N}(\mathbf{x}(N-1), \mathbf{u}(N-1))}{\partial \mathbf{u}} = \mathbf{0}$$

Evaluating the indicated partial derivative we get (refer Appendix III)

$$\mathbf{R}\,\mathbf{u}(N-1) + \tfrac{1}{2}\mathbf{G}^T\mathbf{P}(0)\,\mathbf{F}\,\mathbf{x}(N-1) + \tfrac{1}{2}\mathbf{G}^T\mathbf{P}(0)\,\mathbf{F}\,\mathbf{x}(N-1) + \mathbf{G}^T\mathbf{P}(0)\,\mathbf{G}\,\mathbf{u}(N-1) = \mathbf{0}$$

or

$$\mathbf{R}\,\mathbf{u}(N-1) + \mathbf{G}^T\mathbf{P}(0)\,[\mathbf{F}\,\mathbf{x}(N-1) + \mathbf{G}\,\mathbf{u}(N-1)] = \mathbf{0} \qquad (11.5)$$

Since the Hessian matrix

$$\frac{\partial^2 J_{N-1,\,N}(\mathbf{x}(N-1),\,\mathbf{u}(N-1))}{\partial \mathbf{u}^2} = \mathbf{R} + \mathbf{G}^T\mathbf{P}(0)\,\mathbf{G}$$

is a positive definite matrix under the assumptions of positive semidefinite $\mathbf{H}$ (and hence $\mathbf{P}(0)$) and positive definite $\mathbf{R}$, the control values given by eqn. (11.5) yield the absolute minimum of $J_{N-1,\,N}(\mathbf{x}(N-1),\,\mathbf{u}(N-1))$. Solving eqn. (11.5) for the optimal control, we get

$$\mathbf{u}^*(N-1) = -\,[\mathbf{R} + \mathbf{G}^T\mathbf{P}(0)\,\mathbf{G}]^{-1}\,\mathbf{G}^T\mathbf{P}(0)\,\mathbf{F}\,\mathbf{x}(N-1) \qquad (11.6a)$$

$$\triangleq \mathbf{K}(N-1)\,\mathbf{x}(N-1) \qquad (11.6b)$$

Since $[\mathbf{R} + \mathbf{G}^T\mathbf{P}(0)\,\mathbf{G}]$ is positive definite, the inverse in eqn. (11.6a) is guaranteed to exist.

The minimum value of the performance index obtained from eqns. (11.4) and (11.6) is

$$J^*_{N-1,\,N}(\mathbf{x}(N-1)) = \tfrac{1}{2}\mathbf{x}^T(N-1)\mathbf{Q}\mathbf{x}(N-1) + \tfrac{1}{2}\mathbf{x}^T(N-1)\mathbf{K}^T(N-1)\mathbf{R}\mathbf{K}(N-1)\mathbf{x}(N-1)$$
$$+ \tfrac{1}{2}[\mathbf{F}\mathbf{x}(N-1) + \mathbf{G}\mathbf{K}(N-1)\mathbf{x}(N-1)]^T\,\mathbf{P}(0)\,[\mathbf{F}\mathbf{x}(N-1) + \mathbf{G}\mathbf{K}(N-1)\mathbf{x}(N-1)]$$
$$= \tfrac{1}{2}\mathbf{x}^T(N-1)\{[\mathbf{F} + \mathbf{G}\mathbf{K}(N-1)]^T\,\mathbf{P}(0)\,[\mathbf{F} + \mathbf{G}\mathbf{K}(N-1)] + \mathbf{K}^T(N-1)\,\mathbf{R}\mathbf{K}(N-1) + \mathbf{Q}\}\,\mathbf{x}(N-1) \qquad (11.7a)$$

$$\triangleq \tfrac{1}{2}\mathbf{x}^T(N-1)\,\mathbf{P}(1)\,\mathbf{x}(N-1) \qquad (11.7b)$$

where

$$\mathbf{P}(1) = [\mathbf{F} + \mathbf{G}\mathbf{K}(N-1)]^T\,\mathbf{P}(0)\,[\mathbf{F} + \mathbf{G}\mathbf{K}(N-1)] + \mathbf{K}^T(N-1)\,\mathbf{R}\mathbf{K}(N-1) + \mathbf{Q} \qquad (11.7c)$$

From eqns. (11.3) and (11.7) we observe that $J^*_{N-1,\,N}(\mathbf{x}(N-1))$ is of exactly the same form as $J^*_{N,\,N}(\mathbf{x}(N))$. We can write the solution for the *two-stage process*:

$$\mathbf{u}^*(N-2) = -\,[\mathbf{R} + \mathbf{G}^T\mathbf{P}(1)\,\mathbf{G}]^{-1}\,\mathbf{G}^T\mathbf{P}(1)\,\mathbf{F}\mathbf{x}(N-2) \qquad (11.8a)$$

$$\triangleq \mathbf{K}(N-2)\,\mathbf{x}(N-2) \qquad (11.8b)$$

$$J^*_{N-2,\,N}(\mathbf{x}(N-2)) = \tfrac{1}{2}\mathbf{x}^T(N-2)\,\mathbf{P}(2)\,\mathbf{x}(N-2) \qquad (11.8c)$$

where

$$\mathbf{P}(2) = [\mathbf{F} + \mathbf{G}\mathbf{K}(N-2)]^T\,\mathbf{P}(1)\,[\mathbf{F} + \mathbf{G}\mathbf{K}(N-2)] + \mathbf{K}^T(N-2)\mathbf{R}\mathbf{K}(N-2) + \mathbf{Q} \qquad (11.8d)$$

The correctness of this solution can easily be verified. By induction, we obtain for a j-stage process, the result

$$\mathbf{u}^*(N-j) = \mathbf{K}(N-j)\,\mathbf{x}(N-j);\; j = 1, 2, \ldots, N \tag{11.9a}$$

$$\mathbf{K}(N-j) = -[\mathbf{R} + \mathbf{G}^T\mathbf{P}(j-1)\mathbf{G}]^{-1}\,\mathbf{G}^T\mathbf{P}(j-1)\mathbf{F} \tag{11.9b}$$

$$\mathbf{P}(0) = \mathbf{H} \tag{11.9c}$$

$$\mathbf{P}(j) = [\mathbf{F} + \mathbf{G}\mathbf{K}(N-j)]^T\,\mathbf{P}(j-1)\,[\mathbf{F} + \mathbf{G}\mathbf{K}(N-j)] + \mathbf{K}^T(N-j)\,\mathbf{R}\mathbf{K}(N-j) + \mathbf{Q} \tag{11.9d}$$

$$J^*_{N-j,\,N}(\mathbf{x}(N-j)) = \tfrac{1}{2}\mathbf{x}^T(N-j)\,\mathbf{P}(j)\,\mathbf{x}(N-j) \tag{11.9e}$$

Equations (11.9) form the recursive relations for determination of **K** and **P** matrices. Starting with $j = 1$, $\mathbf{K}(N-1)$ is evaluated from eqn. (11.9b) with $\mathbf{P}(0) = \mathbf{H}$, a given matrix. Equation (11.9d) is then solved for $\mathbf{P}(1)$. This constitutes one cycle of the procedure, which we then continue by calculating $\mathbf{K}(N-2)$, $\mathbf{P}(2)$ and so on. The matrices $\mathbf{K}(N-1)$, $\mathbf{K}(N-2)$, ..., $\mathbf{K}(0)$ are stored. The optimal controller is then realized by determining the appropriate gain settings (eqn. (11.9a)) as the system transits from stage to stage. Table 11.1 summarizes the algorithm.[1]

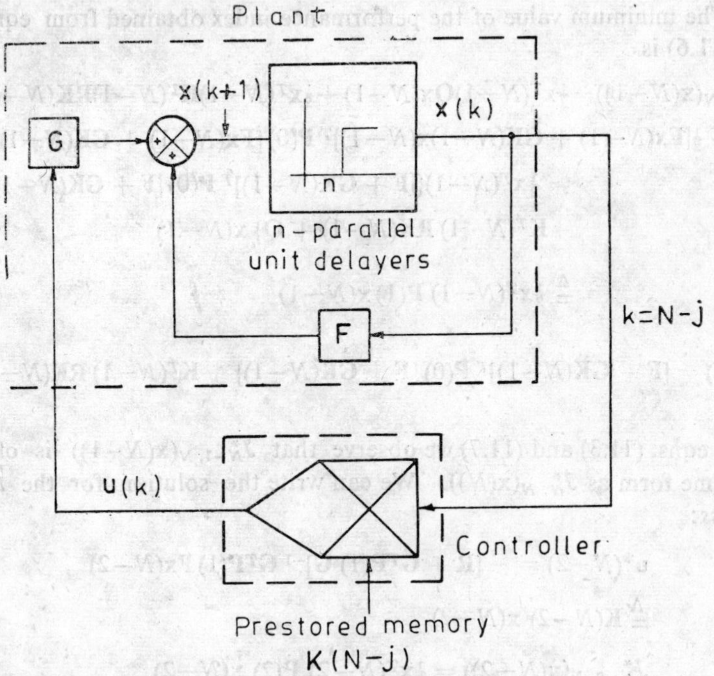

Fig. 11.1 The feedback structure of the optimal linear regulator

[1]The symmetry of $\mathbf{P}(j)$ tends to be destroyed because of numerical errors. This difficulty can be taken care of by replacing $\mathbf{P}(j)$ with $\tfrac{1}{2}[\mathbf{P}(j)+\mathbf{P}^T(j)]$ after each step.

Table 11.1: Summary of Algorithm for the Solution of Discrete-time Linear State Regulator Problem

Plant model	$\mathbf{x}(k+1)=\mathbf{F}\mathbf{x}(k)+\mathbf{G}\mathbf{u}(k)$
Performance index	$J=\frac{1}{2}\,\mathbf{x}^T(N)\mathbf{H}\,\mathbf{x}(N)+\frac{1}{2}\sum_{k=0}^{N-1}[\mathbf{x}^T(k)\,\mathbf{Q}\,\mathbf{x}(k)+\mathbf{u}^T(k)\,\mathbf{R}\,\mathbf{u}(k)]$
Feedback gain algorithm	$\mathbf{K}(N-j)=-[\mathbf{R}+\mathbf{G}^T\,\mathbf{P}(j-1)\,\mathbf{G}]^{-1}\,\mathbf{G}^T\mathbf{P}(j-1)\,\mathbf{F}$ $\mathbf{P}(0)=\mathbf{H}$ $\mathbf{P}(j)=[\mathbf{F}+\mathbf{G}\mathbf{K}(N-j)]^T\,\mathbf{P}(j-1)\,[\mathbf{F}+\mathbf{G}\,\mathbf{K}\,(N-j)]+\mathbf{K}^T(N-j)\,\mathbf{R}\,\mathbf{K}(N-j)+\mathbf{Q}$ $=1, 2, \ldots, N$
Optimal oontrol law	$\mathbf{u}^*(N-j)=\mathbf{K}(N-j)\,\mathbf{x}(N-j)$ $j=1, 2, \ldots, N$
Optimal performance index	$J^*_{0,N}(\mathbf{x}(0))=\frac{1}{2}\mathbf{x}^T(0)\mathbf{P}(N)\,\mathbf{x}(0)$

For the N-stage process with specified initial state $\mathbf{x}(0)=\mathbf{x}^0$, the minimum value of performance index is (eqn. (11.9e))

$$J^*_{0,N}\,(\mathbf{x}(0))=\tfrac{1}{2}\mathbf{x}^T(0)\;\mathbf{P}(N)\;\mathbf{x}(0) \tag{11.10}$$

The following important points are observed in the result given by eqn. (11.9):

(i) The optimal control at each stage is a linear combination of the states; thus giving a linear state variable feedback control policy. Figure 11.1 shows the structure of the optimal feedback control system. The realization of the optimal control law given by eqn. (11.9a) seeks the feedback of state variables. If the plant states are available for measurement, $\mathbf{u}^*$ is implemented by feedback of all state variables through time-varying gain elements; the engineering construction of time-varying functions is easily done by means of a digital computer.

If the plant states are not available for measurement, then it is possible to construct a physical device—*state observer*, which produces at its output the plant states, when driven by both plant input and output. When all inputs can be specified exactly and all outputs can be measured with unlimited precision, we can go for the Luenberger observer (discussed in Chapter 9). Plant equations must satisfy the conditions of observability for the construction of the Luenberger observer. Figure 11.2 shows the structure of feedback control system wherein the states are estimated using the Luenberger observer.

In case the input and output transducers are subject to unpredictable fluctuations and disturbances, we may go for the Kalman filter (discussed in Chapter 12) for the estimation of state $\mathbf{x}(t)$.

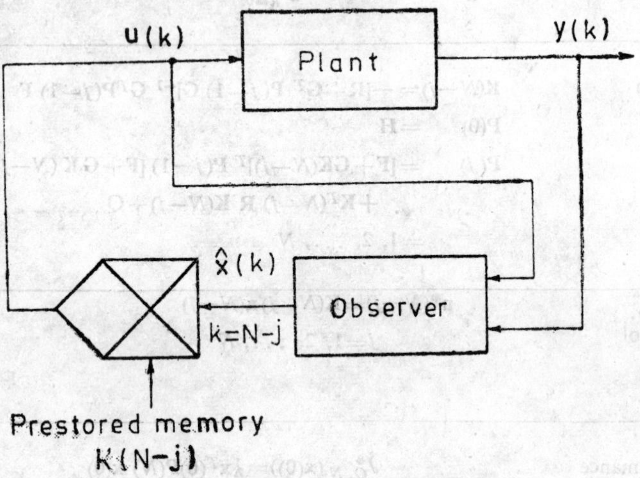

Fig. 11.2 Use of observer in implementing a control law

An alternative to the use of a state observer is to design a *suboptimal feedback controller* which requires only the measurable outputs of the plant to be fedback. This design approach will be discussed in Section 11.7.

(ii) From eqns. (11.9), we observe that feedback is time-varying although **F**, **G**, **R** and **Q** are all constant matrices, i.e., optimal control policy converts a linear time-invariant plant with time-invariant quadratic performance index into a linear time-varying feedback system. The engineer may not quarrel with the necessity of time-varying controllers for time-varying systems. However, he may like time-invariant controllers for time-invariant systems. In Section 11.4, we shall study the conditions which lead to time-invariant feedback.

(iii) In our derivation of the result (11.9), the controllability of the plant is not necessary. Assume that some states of the plant are uncontrollable. We expect that this will impose problems in the regulator system if the uncontrollable states are unstable because these unstable states will be reflected in the performance index. However, the contribution of the unstable uncontrollable (unstable controllable states are stabilizable) states to the performance index is always finite provided that the control interval $[0, N-1]$ is finite. We shall see in Section 11.4 that we require controllability as $N \rightarrow \infty$ to ensure that the value of the performance index is finite.

(iv) Following the procedure given above, results for optimal control of linear time-varying plant (11.1) with time-varying matrices **Q** and **R** in the performance index (11.2) can easily be obtained.

$$\mathbf{u}^*(N-j) = \mathbf{K}(N-j)\,\mathbf{x}(N-j)\,;\; j=1, 2, \ldots, N \tag{11.11a}$$

$$\mathbf{K}(N-j) = -[\mathbf{R}(N-j) + \mathbf{G}^T(N-j)\,\mathbf{P}(j-1)\,\mathbf{G}(N-j)]^{-1}[\mathbf{G}^T(N-j)\,\mathbf{P}(j-1)\,\mathbf{F}(N-j)] \tag{11.11b}$$

$$\mathbf{P}(0) = \mathbf{H} \tag{11.11c}$$

$$\mathbf{P}(j)=[\mathbf{F}(N-j)+\mathbf{G}(N-j)\mathbf{K}(N-j)]^T\mathbf{P}(j-1)[\mathbf{F}(N-j)+\mathbf{G}(N-j)\mathbf{K}(N-j)] + \mathbf{K}^T(N-j)\,\mathbf{R}(N-j)\,\mathbf{K}(N-j)+\mathbf{Q}(N-j) \tag{11.11d}$$

$$J^*_{0,N}(\mathbf{x}(0))=\tfrac{1}{2}\mathbf{x}^T(0)\,\mathbf{P}(N)\,\mathbf{x}(0) \tag{11.11e}$$

(v) Examination of the right hand side of eqn. (11.11d) reveals that $\mathbf{P}(j)$ is a symmetric matrix. Further, it is obvious from our derivation of the recursive relations, that the matrix $\mathbf{P}(j)$ must be nonnegative definite.

Example 11.1: Consider the Mixing Tank System discussed in Section 4.6. The continuous-time system is described by the state model (eqns. (4.38))

$$\dot{\mathbf{x}}(t) = \begin{bmatrix} -0.01 & 0 \\ 0 & -0.02 \end{bmatrix} \mathbf{x}(t) + \begin{bmatrix} 1 & 1 \\ -0.004 & 0.002 \end{bmatrix} \mathbf{u}(t) \tag{11.12a}$$

$$\mathbf{y}(t) = \begin{bmatrix} 0.01 & 0 \\ 0 & 1 \end{bmatrix} \mathbf{x}(t) \tag{11.12b}$$

where x_1 and x_2 are incremental changes in volume V and concentration C respectively about the steady-state values

$$V_0 = 1500 \text{ litres}$$

$$C_0 = 15 \text{ g-moles/litre}$$

and u_1 and u_2 are incremental changes in inflow rates Q_1 and Q_2 respectively about the steady-state values

$$Q_{10} = 10 \text{ litres/sec}$$

$$Q_{20} = 20 \text{ litres/sec}$$

The variables y_1 and y_2 are incremental changes in outgoing flow and outgoing concentration respectively.

For a sampling interval of $T = 5$ sec, the discretized equations for mixing tank are given by (eqn. (6.62))

$$\mathbf{x}(k+1) = \mathbf{F}\,\mathbf{x}(k) + \mathbf{G}\,\mathbf{u}(k) \tag{11.13a}$$

$$\mathbf{y}(k) = \mathbf{C}\,\mathbf{x}(k) \tag{11.13b}$$

where

$$\mathbf{F}=\begin{bmatrix} 0.9512 & 0 \\ 0 & 0.9048 \end{bmatrix}; \quad \mathbf{G}=\begin{bmatrix} 4.88 & 4.88 \\ -0.019 & 0.0095 \end{bmatrix}$$

$$\mathbf{C}=\begin{bmatrix} 0.01 & 0 \\ 0 & 1 \end{bmatrix}$$

Suppose that it is desired to improve the transients in volume and concentration of fluid in the tank. Let us assume a performance measure of the form

$$J=\sum_{k=0}^{N-1}[\mathbf{x}^T(k)\,\mathbf{Q}\,\mathbf{x}(k)+\mathbf{u}^T(k)\,\mathbf{R}\,\mathbf{u}(k)]$$

We assume the diagonal form of **Q** and **R** matrices:

$$\mathbf{Q}=\begin{bmatrix} q_1 & 0 \\ 0 & q_2 \end{bmatrix}; \quad \mathbf{R}=\begin{bmatrix} r_1 & 0 \\ 0 & r_2 \end{bmatrix}$$

From the steady-state values we observe that a 1% change in volume corresponds to 15 litres and a 1% change in concentration corresponds to 0.15 g-moles/litre. Let us assume that a 1% change in concentration is to make about the same contribution to the performance criterion as 1% change in the volume. Then we must have

$$q_2(0.15)^2 \simeq q_1(15)^2 \quad \text{or} \quad \frac{q_2}{q_1} \simeq \frac{100}{0.01}$$

or

$$\mathbf{Q}=\begin{bmatrix} 0.01 & 0 \\ 0 & 100 \end{bmatrix}$$

We use a similar approach for the selection of **R**. A 1% change in Q_1 corresponds to 0.1 litres/sec and a 1% change in Q_2 corresponds to 0.2 litres/sec. If both the terms contribute equally,

$$r_1(0.1)^2 \simeq r_2(0.2)^2$$

or

$$\frac{r_1}{r_2} \simeq \frac{2}{0.5}$$

or

$$\mathbf{R}=\rho\begin{bmatrix} 2 & 0 \\ 4 & 0.5 \end{bmatrix}$$

where ρ is a scalar constant to be determined.

We know that the smaller the value of ρ, the faster will be the response. Thus faster response is obtained at the cost of larger input amplitudes. We shall consider three values of ρ in our study:

$$\rho = 10,\ 1,\ 0.1.$$

For $N = 10$ ($t_1 = 50$ sec), the behaviour of gain $\mathbf{K}(k)$, the trajectory $\mathbf{x}^*(k)$ and the optimal control $\mathbf{u}^*(k)$ are shown in Table 11.2 and Figs. 11.3–11.5 for various numerical values. The calculations were done using the algorithm given in Table 11.1.

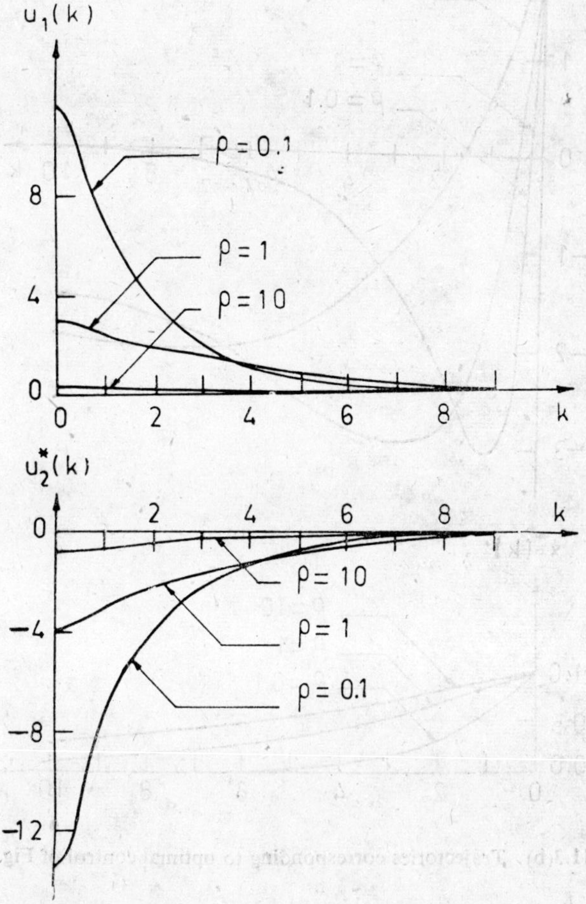

Fig. 11.3(a) Optimal controls:

$$\mathbf{Q} = \begin{bmatrix} 0.01 & 0 \\ 0 & 100 \end{bmatrix}; \mathbf{R} = \rho \begin{bmatrix} 2 & 0 \\ 0 & 0.5 \end{bmatrix}$$

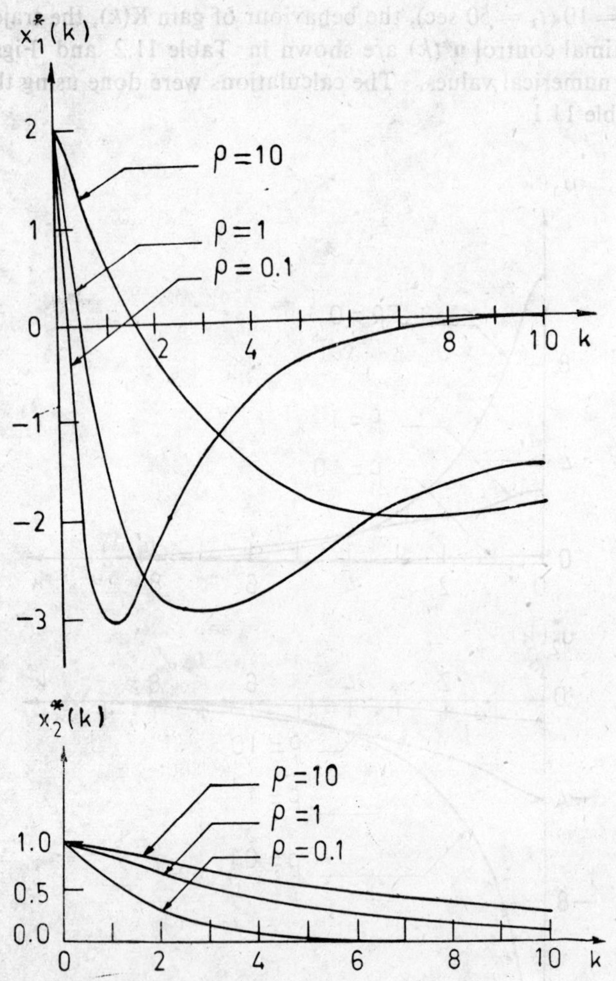

Fig. 11.3(b) Trajectories corresponding to optimal control of Fig. 11.3a.

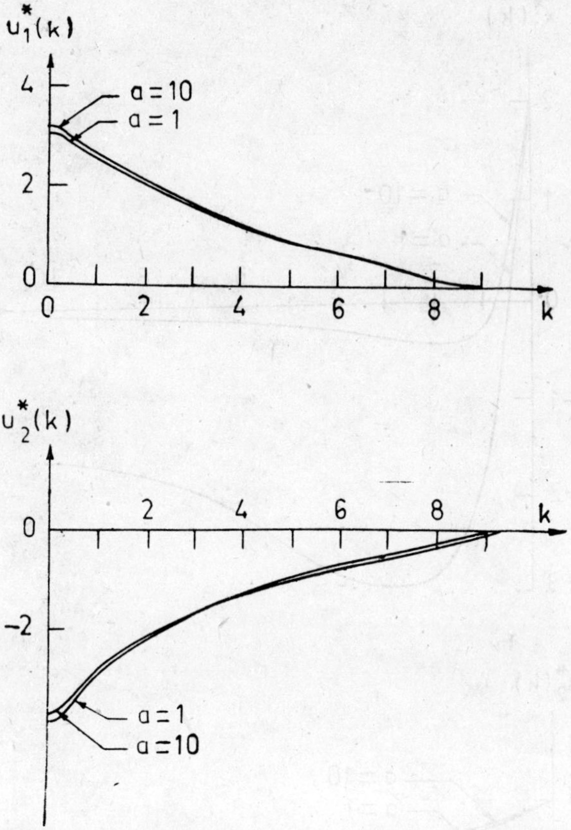

Fig. 11.4(a) Optimal controls:

$$\mathbf{Q} = \begin{bmatrix} 0.01a & 0 \\ 0 & 100 \end{bmatrix}; \ \mathbf{R} = \begin{bmatrix} 2 & 0 \\ 0 & 0.5 \end{bmatrix}$$

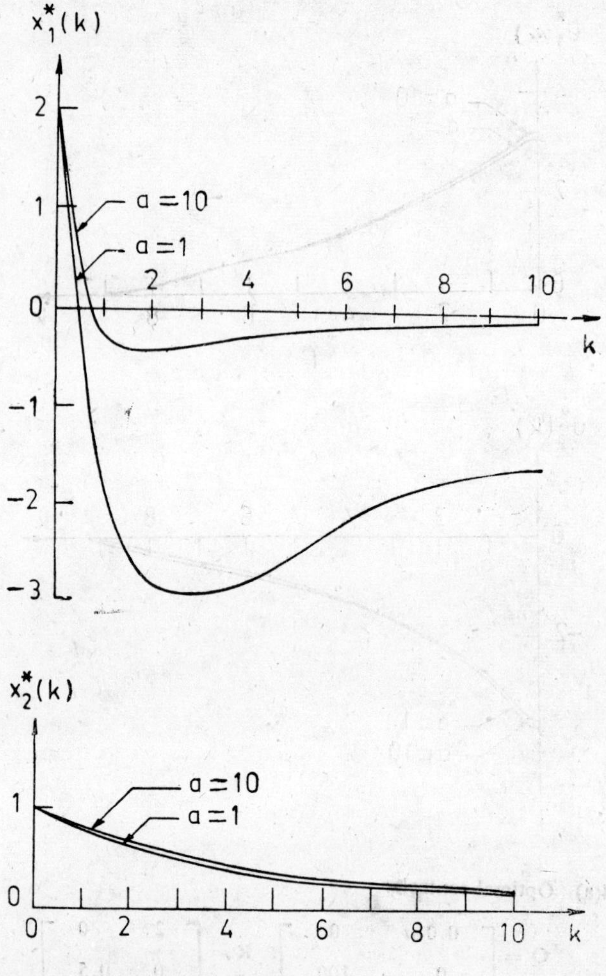

Fig. 11.4(b) Trajectories corresponding to optimal controls of Fig. 11.4a

Table 11.2: Feedback gain coefficients for optimal control (Prestored memory of digital computer);

$$\mathbf{Q} = \begin{bmatrix} 0.01 & 0 \\ 0 & 100 \end{bmatrix}; \ \mathbf{R} = \begin{bmatrix} 2 & 0 \\ 0 & 0.5 \end{bmatrix}$$

$$\mathbf{u}^*(k) = \begin{bmatrix} K_{11} & K_{12} \\ K_{21} & K_{22} \end{bmatrix} \mathbf{x}^*(k)$$

k	K_{11}	K_{12}	K_{21}	K_{22}
0	−0.02223	3.211	−0.07789	−3.729
1	−0.02218	3.160	−0.07723	−3.683
2	−0.02211	3.084	−0.07799	−3.614
3	−0.02200	2.968	−0.07808	−3.509
4	−0.02183	2.793	−0.07819	−3.350
5	−0.02156	2.534	−0.07822	−3.114
6	−0.02101	2.156	−0.07769	−2.763
7	−0.01958	1.618	−0.07423	−2.239
8	−0.01482	0.8927	−0.05776	−1.413
9	0	0	0	0

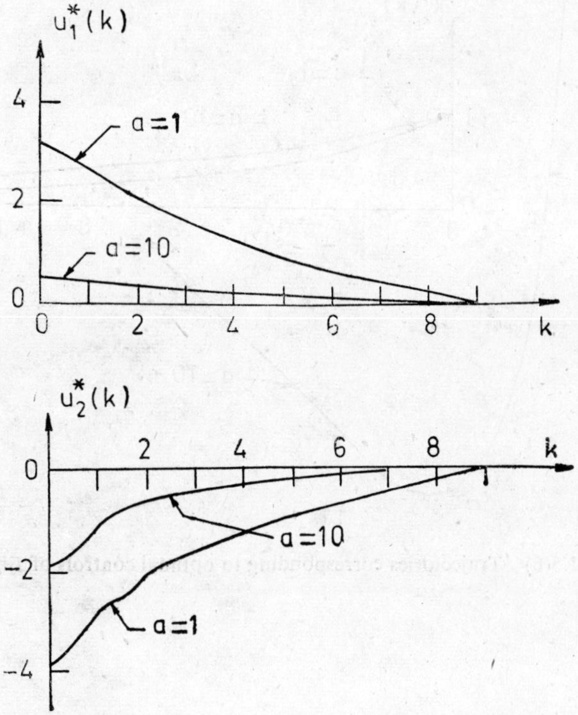

Fig. 11.5(a) Optimal controls;

$$\mathbf{Q} = \begin{bmatrix} 0.01 & 0 \\ 0 & 100 \end{bmatrix}; \ \mathbf{R} = \begin{bmatrix} 2a & 0 \\ 0 & 0.5 \end{bmatrix}$$

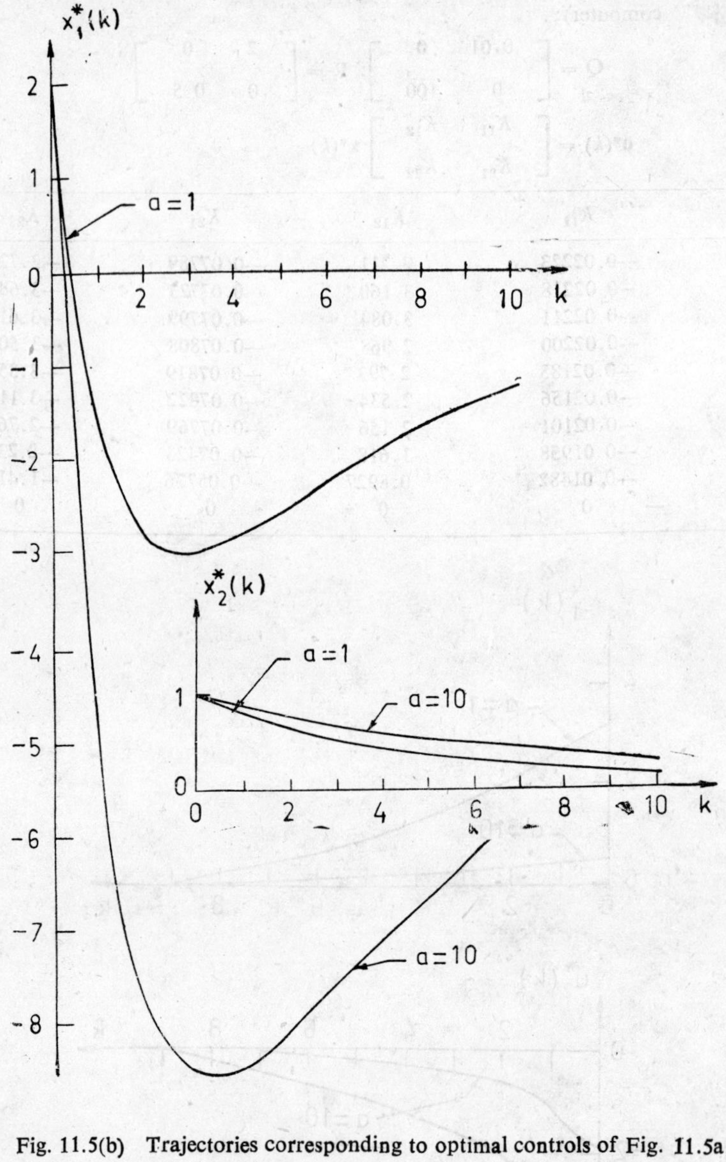

Fig. 11.5(b) Trajectories corresponding to optimal controls of Fig. 11.5a

The block diagram of the optimal system is shown in Fig. 11.6.

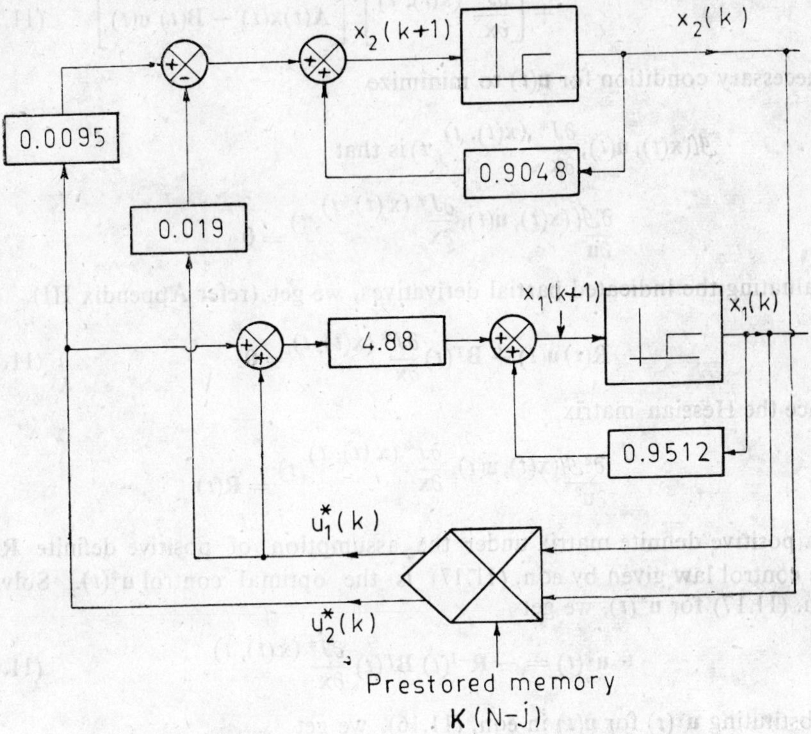

Fig. 11.6 Structure of controller for system of Example 11.1

11.3 CONTINUOUS-TIME LINEAR STATE REGULATOR

Consider the plant represented by linear continuous-time state equations of the form

$$\dot{\mathbf{x}}(t) = \mathbf{A}(t)\,\mathbf{x}(t) + \mathbf{B}(t)\,\mathbf{u}(t);\quad \mathbf{x}(t_0) = \mathbf{x}^0 \tag{11.14}$$

where $\mathbf{x}$ is the $n\times 1$ state vector, $\mathbf{u}$ is the $p\times 1$ input vector, $\mathbf{A}$ and $\mathbf{B}$ are, respectively, $n\times n$ and $n\times p$ time-varying real matrices and $t\in[t_0, t_1]$, the control interval.

We shall be interested in selecting the controls $\mathbf{u}(t)$; $t\in[t_0, t_1]$ which minimize a performance criterion of the form

$$J = \tfrac{1}{2}\mathbf{x}^T(t_1)\,\mathbf{H}\,\mathbf{x}(t_1) + \tfrac{1}{2}\int_{t_0}^{t_1} (\mathbf{x}^T(t)\,\mathbf{Q}(t)\,\mathbf{x}(t) + \mathbf{u}^T(t)\,\mathbf{R}(t)\,\mathbf{u}(t))\,dt \tag{11.15}$$

where $\mathbf{H}$ and $\mathbf{Q}(t)$ are real symmetric positive semidefinite $n\times n$ matrices and $\mathbf{R}(t)$ is a real symmetric positive definite $p\times p$ matrix.

We use the Hamilton-Jacobi equation (Section 10.5) to determine optimal control. The Hamiltonian

$$\mathcal{H}\left(\mathbf{x}(t), \mathbf{u}(t), \frac{\partial J^*(\mathbf{x}(t), t)}{\partial \mathbf{x}}, t\right) = \tfrac{1}{2}\mathbf{x}^T(t)\,\mathbf{Q}(t)\,\mathbf{x}(t) + \tfrac{1}{2}\mathbf{u}^T(t)\,\mathbf{R}(t)\,\mathbf{u}(t) + \left[\frac{\partial J^*(\mathbf{x}(t), t)}{\partial \mathbf{x}}\right]^T \left[\mathbf{A}(t)\mathbf{x}(t) + \mathbf{B}(t)\,\mathbf{u}(t)\right] \quad (11.16)$$

A necessary condition for $\mathbf{u}(t)$ to minimize

$\mathcal{H}\left(\mathbf{x}(t), \mathbf{u}(t), \frac{\partial J^*(\mathbf{x}(t), t)}{\partial \mathbf{x}}, t\right)$ is that

$$\frac{\partial \mathcal{H}\left(\mathbf{x}(t), \mathbf{u}(t), \frac{\partial J^*(\mathbf{x}(t), t)}{\partial \mathbf{x}}, t\right)}{\partial \mathbf{u}} = 0$$

Evaluating the indicated partial derivatives, we get (refer Appendix III)

$$\mathbf{R}(t)\,\mathbf{u}(t) + \mathbf{B}^T(t)\frac{\partial J^*(\mathbf{x}(t), t)}{\partial \mathbf{x}} = 0 \quad (11.17)$$

Since the Hessian matrix

$$\frac{\partial^2 \mathcal{H}\left(\mathbf{x}(t), \mathbf{u}(t), \frac{\partial J^*(\mathbf{x}(t), t)}{\partial \mathbf{x}}, t\right)}{\partial \mathbf{u}^2} = \mathbf{R}(t)$$

is a positive definite matrix under the assumption of positive definite $\mathbf{R}(t)$, the control law given by eqn. (11.17) is the optimal control $\mathbf{u}^*(t)$. Solving eqn. (11.17) for $\mathbf{u}^*(t)$, we get

$$\mathbf{u}^*(t) = -\,\mathbf{R}^{-1}(t)\,\mathbf{B}^T(t)\,\frac{\partial J^*(\mathbf{x}(t), t)}{\partial \mathbf{x}} \quad (11.18)$$

Substituting $\mathbf{u}^*(t)$ for $\mathbf{u}(t)$ in eqn. (11.16), we get

$$\begin{aligned}\mathcal{H}^*\left(\mathbf{x}(t), \frac{\partial J^*(\mathbf{x}(t), t)}{\partial \mathbf{x}}, t\right) &= \tfrac{1}{2}\mathbf{x}^T(t)\,\mathbf{Q}(t)\,\mathbf{x}(t) \\ &\quad + \tfrac{1}{2}\left[\frac{\partial J^*(\mathbf{x}(t), t)}{\partial \mathbf{x}}\right]^T \mathbf{B}(t)\mathbf{R}^{-1}(t)\mathbf{R}(t)\mathbf{R}^{-1}(t)\mathbf{B}^T(t)\,\frac{\partial J^*(\mathbf{x}(t), t)}{\partial \mathbf{x}} \\ &\quad + \left(\frac{\partial J^*(\mathbf{x}(t), t)}{\partial \mathbf{x}}\right)^T \mathbf{A}(t)\,\mathbf{x}(t) \\ &\quad - \left(\frac{\partial J^*(\mathbf{x}(t), t)}{\partial \mathbf{x}}\right)^T \left[\mathbf{B}(t)\,\mathbf{R}^{-1}(t)\,\mathbf{B}^T(t)\,\frac{\partial J^*(\mathbf{x}(t), t)}{\partial \mathbf{x}}\right] \\ &= \tfrac{1}{2}\mathbf{x}^T(t)\mathbf{Q}(t)\,\mathbf{x}(t) - \tfrac{1}{2}\left(\frac{\partial J^*(\mathbf{x}(t), t)}{\partial \mathbf{x}}\right)^T \mathbf{B}(t)\mathbf{R}^{-1}(t)\mathbf{B}^T(t)\,\frac{\partial J^*(\mathbf{x}(t), t)}{\partial \mathbf{x}} \\ &\quad + \left(\frac{\partial J^*(\mathbf{x}(t), t)}{\partial \mathbf{x}}\right)^T \mathbf{A}(t)\,\mathbf{x}(t)\end{aligned}$$

The Hamilton-Jacobi equation is

$$\begin{aligned}0 &= \frac{\partial J^*(\mathbf{x}(t), t)}{\partial t} + \tfrac{1}{2}\mathbf{x}^T(t)\,\mathbf{Q}(t)\,\mathbf{x}(t) \\ &\quad - \tfrac{1}{2}\left(\frac{\partial J^*(\mathbf{x}(t), t)}{\partial \mathbf{x}}\right)^T \left[\mathbf{B}(t)\,\mathbf{R}^{-1}(t)\,\mathbf{B}^T(t)\,\frac{\partial J^*(\mathbf{x}(t), t)}{\partial \mathbf{x}}\right] + \left(\frac{\partial J^*(\mathbf{x}(t), t)}{\partial \mathbf{x}}\right)^T \mathbf{A}(t)\,\mathbf{x}(t)\end{aligned} \quad (11.19a)$$

The boundary condition is

$$J^*(\mathbf{x}(t_1), t_1) = \tfrac{1}{2}\mathbf{x}^T(t_1)\,\mathbf{H}\mathbf{x}(t_1) \tag{11.19b}$$

Earlier in Section 11.2 we found that in the linear regulator problem for discrete-time systems, the minimum value of the performance index is a time-varying quadratic function of state (eqn. (11.11e)). Therefore, it is reasonable to assume

$$J^*(\mathbf{x}(t), t) = \tfrac{1}{2}\mathbf{x}^T(t)\,\mathbf{P}(t)\,\mathbf{x}(t) \tag{11.20a}$$

where $\mathbf{P}(t)$ is a real symmetric matrix and

$$\mathbf{P}(t_1) = \mathbf{H} \tag{11.20b}$$

Substituting the assumed solution in eqn. (11.19a), we get

$$0 = \tfrac{1}{2}\mathbf{x}^T(t)\dot{\mathbf{P}}(t)\mathbf{x}(t) + \tfrac{1}{2}\mathbf{x}^T(t)\,\mathbf{Q}(t)\,\mathbf{x}(t)$$
$$-\tfrac{1}{2}\mathbf{x}^T(t)\,\mathbf{P}(t)\,\mathbf{B}(t)\,\mathbf{R}^{-1}(t)\,\mathbf{B}^T(t)\,\mathbf{P}(t)\,\mathbf{x}(t) + \mathbf{x}^T(t)\,\mathbf{P}(t)\,\mathbf{A}(t)\,\mathbf{x}(t)$$

or

$$0 = \mathbf{x}^T(t)\,[2\mathbf{P}(t)\,\mathbf{A}(t) - \mathbf{P}(t)\,\mathbf{B}(t)\,\mathbf{R}^{-1}(t)\,\mathbf{B}^T(t)\,\mathbf{P}(t) + \mathbf{Q}(t) + \dot{\mathbf{P}}(t)]\,\mathbf{x}(t) \tag{11.21}$$

The only way this equation can be satisfied for an arbitrary $\mathbf{x}(t)$ is that the quantity inside the brackets be equal to zero. However, we know that in the scalar function $\mathbf{z}^T\mathbf{W}\mathbf{z}$, only the symmetric part of the matrix $\mathbf{W}_s$,

$$\mathbf{W}_s = \frac{\mathbf{W} + \mathbf{W}^T}{2}$$

is of importance (refer Appendix III).

Examination of eqn. (11.21) reveals that all the terms within brackets are already symmetric except the first term.

$$\text{Symmetric part of } 2\mathbf{P}(t)\,\mathbf{A}(t) = 2\,\frac{\mathbf{P}(t)\,\mathbf{A}(t) + \mathbf{A}^T(t)\,\mathbf{P}(t)}{2}$$
$$= \mathbf{P}(t)\,\mathbf{A}(t) + \mathbf{A}^T(t)\,\mathbf{P}(t)$$

Therefore, in order for eqn. (11.21) to be satisfied, it is necessary that the differential equation

$$\dot{\mathbf{P}}(t) + \mathbf{Q}(t) - \mathbf{P}(t)\,\mathbf{B}(t)\,\mathbf{R}^{-1}(t)\,\mathbf{B}^T(t)\,\mathbf{P}(t) + \mathbf{P}(t)\,\mathbf{A}(t) + \mathbf{A}^T(t)\,\mathbf{P}(t) = \mathbf{0} \tag{11.22}$$

is satisfied subject to the boundary condition (11.20b). The matrix differential equation (11.22) for $\mathbf{P}(t)$ is symmetric and the matrix $\mathbf{H}$ that occurs in the boundary condition for $\mathbf{P}(t)$ is also symmetric, therefore the solution $\mathbf{P}(t)$ for all $t_0 \leqslant t \leqslant t_1$ must be symmetric. This symmetry will often be used, especially when computing $\mathbf{P}(t)$. At first glance, it may appear that since $\mathbf{P}(t)$ is $n \times n$ matrix, eqn. (11.22) represents a system of n^2 first-order non-linear time-varying ordinary differential equations. However, because of symmetry of $\mathbf{P}(t)$, eqn. (11.22) represents a system of $\frac{n(n+1)}{2}$ first-order differential equations.

The matrix differential equation (11.22) resembles the well-known scalar differential equation

$$\frac{dy}{dx} + \alpha(x)y + \beta(x)y^2 = \gamma(x)$$

where x is independent variable and y is dependent variable; $\alpha(x)$, $\beta(x)$ and $\gamma(x)$ are known functions of x. This equation is known as the *Riccati equation* (Davis 1962); consequently eqn. (11.22) is referred to as the *matrix Riccati equation* (Kalman 1960).

The reader might have noted from the derivation of the matrix Riccati equation that $\mathbf{P}(t)$ must be non-negative definite. Rather, if $\mathbf{u}(t) \neq \mathbf{0}$ for all states, then $\mathbf{P}(t)$ is positive definite for all $t \in [t_0, t_1]$ and $\mathbf{P}(t_1)$ is positive semi definite. This is because for nonzero $\mathbf{u}(t)$, the cost J must be positive and hence from eqn. (11.20a), $\mathbf{P}(t)$ must be positive definite for $t < t_1$.

Once $\mathbf{P}(t)$ has been determined, the optimal control law is given by

$$\begin{aligned} \mathbf{u}^*(t) &= -\mathbf{R}^{-1}(t)\,\mathbf{B}^T(t)\,\mathbf{P}(t)\,\mathbf{x}(t) \\ &= \mathbf{K}(t)\,\mathbf{x}(t) \end{aligned} \tag{11.23}$$

where

$$\mathbf{K}(t) = -\mathbf{R}^{-1}(t)\,\mathbf{B}^T(t)\,\mathbf{P}(t)$$

Thus by assuming a solution of the form (11.20a), the optimal control law is linear, time-varying state feedback. However, it may be noted that other forms are also possible as solutions to the Hamilton-Jacobi equation. Johnson and Gibson (1964) give an alternative approach which under certain conditions leads to a nonlinear but time-invariant form for the optimal control law.

We conclude our derivation with the remark that under the conditions imposed on matrices of the performance index (11.15), the deterministic linear regulator problem always has a unique solution. The existence of solution of the regulator problem guarantees the existence of a unique solution of the matrix Riccati equation (11.22) with boundary condition (11.20b). The reader may refer Kalman (1960), Athans and Falb (1966) or Anderson and Moore (1971) for the existence of a solution of the regulator problem.

Table 11.3 summarizes the results for continuous-time linear regulator problem.

Example 11.2: A first-order system is described by the differential equation

$$\dot{x}(t) = 2x(t) + u(t)$$

It is desired to find the control law that minimizes the performance index

Table 11.3: Summary of Results for Continuous-time Linear Regulator Problem

Plant model	$\dot{\mathbf{x}}(t)=\mathbf{A}(t)\,\mathbf{x}(t)+\mathbf{B}(t)\,\mathbf{u}(t);\ \mathbf{x}(t_0)=\mathbf{x}^e$
Performance index	$J=\frac{1}{2}\mathbf{x}^T(t_1)\,\mathbf{H}\,\mathbf{x}(t_1)+\frac{1}{2}\int_{t_0}^{t_1}(\mathbf{x}^T(t)\,\mathbf{Q}(t)\,\mathbf{x}(t)+\mathbf{u}^T(t)\,\mathbf{R}(t)\,\mathbf{u}(t))\,dt$
Optimal control law	$\mathbf{u}^*(t)=\mathbf{K}(t)\,\mathbf{x}(t)$
Feedback gain algorithm	$\mathbf{K}(t)=-\mathbf{R}^{-1}(t)\,\mathbf{B}^T(t)\,\mathbf{P}(t)$ $-\dot{\mathbf{P}}(t)=\mathbf{Q}(t)-\mathbf{P}(t)\,\mathbf{B}(t)\,\mathbf{R}^{-1}(t)\,\mathbf{B}^T(t)\,\mathbf{P}(t)+\mathbf{P}(t)\,\mathbf{A}(t)+\mathbf{A}^T(t)\,\mathbf{P}(t)$ $\mathbf{P}(t_1)=\mathbf{H}$
Optimal performance index	$J^*=\frac{1}{2}\mathbf{x}^T(t)\,\mathbf{P}(t)\,\mathbf{x}(t)$

$$J=\frac{1}{2}\int_0^{t_1}(3x^2+\tfrac{1}{4}u^2)\,dt\ ;\quad t_1 \text{ is specified}$$

For this problem, the matrices **A**, **B**, **H**, **Q** and **R** reduce to scalars and are given by

$$\mathbf{A}=2,\ \mathbf{B}=1,\ \mathbf{H}=0,\ \mathbf{Q}=3,\ \mathbf{R}=\tfrac{1}{4}$$

In addition, the matrix $\mathbf{P}(t)$ also reduces itself to a scalar function of time $p(t)$. The matrix Riccati equation then becomes the scalar differential equation

$$\dot{p}(t)+3-4p^2(t)+4p(t)=0$$

with the boundary condition $p(t_1)=0$.

The solution $p(t)$ can be obtained by the separation of variables.

$$\int_t^{t_1}\frac{dp}{4(p-\frac{3}{2})(p+\frac{1}{2})}=\int_t^{t_1}dt$$

or

$$\frac{1}{8}\left\{\ln\left[\frac{p(t)-\frac{3}{2}}{p(t_1)-\frac{3}{2}}\right]-\ln\left[\frac{p(t)+\frac{1}{2}}{p(t_1)+\frac{1}{2}}\right]\right\}=t-t_1$$

This gives

$$p(t) = \frac{\frac{3}{2}(1 - e^{8(t-t_1)})}{1 + 3e^{8(t-t_1)}} \tag{11.24}$$

The optimal control law is

$$u^*(t) = -4p(t)\, x(t) = K(t)\, x(t) \tag{11.25}$$

The block diagram of the optimal system is shown in Fig. 11.7.

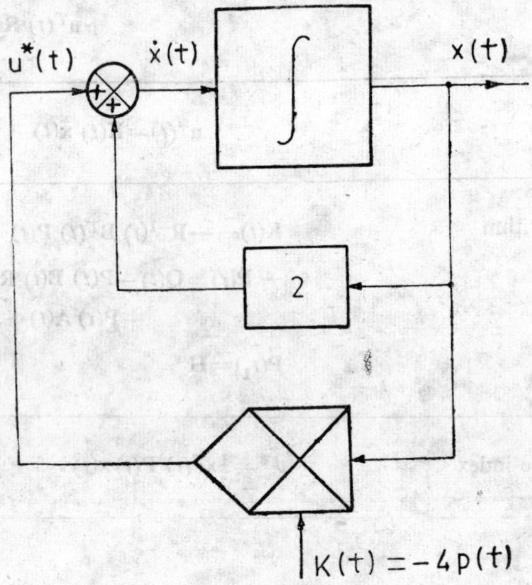

Fig. 11.7 Structure of controller for system of Example 11.2

□

Let us now derive the matrix Riccati equation using Pontryagin's minimum principle. This will strengthen our earlier statement made in Chapter 10 that there exists a bridge between the dynamic programming approach and the calculus of variations/minimum principle. In addition, this will provide us with some information useful for the solution of the matrix Riccati equation.

Table 10.4 provides the summary of the procedure for solving optimal control problems using Pontryagin's minimum principle. For the system (11.14) and the performance index (11.15), the Pontryagin function

$$\begin{aligned}\mathcal{H}(\mathbf{x}, \mathbf{u}, \boldsymbol{\lambda}, t) = \tfrac{1}{2}\mathbf{x}^T(t)\,\mathbf{Q}(t)\,\mathbf{x}(t) + \tfrac{1}{2}\mathbf{u}^T(t)\,\mathbf{R}(t)\,\mathbf{u}(t) \\ + \boldsymbol{\lambda}^T(t)\,\mathbf{A}(t)\,\mathbf{x}(t) + \boldsymbol{\lambda}^T(t)\,\mathbf{B}(t)\,\mathbf{u}(t)\end{aligned} \tag{11.26}$$

Along the optimal trajectory, we must have

$$\frac{\partial \mathcal{H}(\mathbf{x}, \mathbf{u}, \boldsymbol{\lambda}, t)}{\partial \mathbf{u}} = \mathbf{0}$$

which implies that

$$\mathbf{R}(t)\,\mathbf{u}(t) + \mathbf{B}^T(t)\,\boldsymbol{\lambda}(t) = \mathbf{0}$$

and

$$\mathbf{u}^*(t) = -\,\mathbf{R}^{-1}(t)\,\mathbf{B}^T(t)\,\boldsymbol{\lambda}(t) \tag{11.27}$$

The assumption of positive definiteness of $\mathbf{R}(t)$ for all $t \in [t_0, t_1]$ guarantees the existence of $\mathbf{R}^{-1}(t)$ for all $t \in [t_0, t_1]$.

Note that since the Hessian matrix

$$\frac{\partial^2 \mathscr{H}(\mathbf{x}, \mathbf{u}, \boldsymbol{\lambda}, t)}{\partial \mathbf{u}^2} = \mathbf{R}(t)$$

is positive definite, $\mathbf{u}^*(t)$ given by (11.27) minimizes the Pontryagin function $\mathscr{H}(\mathbf{x}, \mathbf{u}, \boldsymbol{\lambda}, t)$.

Substituting $\mathbf{u}^*(t)$ in eqn. (11.26), we get

$$\begin{aligned}\mathscr{H}^*(\mathbf{x}, \boldsymbol{\lambda}, t) &= \tfrac{1}{2}\mathbf{x}^T(t)\mathbf{Q}(t)\mathbf{x}(t) + \tfrac{1}{2}\boldsymbol{\lambda}^T(t)\mathbf{B}(t)\,\mathbf{R}^{-1}(t)\,\mathbf{R}(t)\,\mathbf{R}^{-1}(t)\mathbf{B}^T(t)\boldsymbol{\lambda}(t)\\ &\quad + \boldsymbol{\lambda}^T(t)\mathbf{A}(t)\,\mathbf{x}(t) - \boldsymbol{\lambda}^T(t)\,\mathbf{B}(t)\,\mathbf{R}^{-1}(t)\,\mathbf{B}^T(t)\,\boldsymbol{\lambda}(t)\\ &= \tfrac{1}{2}\mathbf{x}^T(t)\,\mathbf{Q}(t)\,\mathbf{x}(t) + \boldsymbol{\lambda}^T(t)\,\mathbf{A}(t)\,\mathbf{x}(t)\\ &\quad - \tfrac{1}{2}\boldsymbol{\lambda}^T(t)\,\mathbf{B}(t)\,\mathbf{R}^{-1}(t)\,\mathbf{B}^T(t)\,\boldsymbol{\lambda}(t)\end{aligned}$$

The $2n$ state and co-state equations are

$$\dot{\mathbf{x}}(t) = \mathbf{A}(t)\,\mathbf{x}(t) - \mathbf{B}(t)\,\mathbf{R}^{-1}(t)\,\mathbf{B}^T(t)\,\boldsymbol{\lambda}(t)$$

$$\dot{\boldsymbol{\lambda}}(t) = -\,\mathbf{Q}(t)\,\mathbf{x}(t) - \mathbf{A}^T(t)\,\boldsymbol{\lambda}(t)$$

or

$$\begin{bmatrix}\dot{\mathbf{x}}(t)\\ \dot{\boldsymbol{\lambda}}(t)\end{bmatrix} = \begin{bmatrix}\mathbf{A}(t) & -\mathbf{B}(t)\,\mathbf{R}^{-1}(t)\,\mathbf{B}^T(t)\\ -\mathbf{Q}(t) & -\mathbf{A}^T(t)\end{bmatrix}\begin{bmatrix}\mathbf{x}(t)\\ \boldsymbol{\lambda}(t)\end{bmatrix} \tag{11.28}$$

The $2n$ boundary conditions for these differential equations are

$$\mathbf{x}(t_0) = \mathbf{x}^0 \tag{11.29a}$$

and (refer Table 10.3; t_1 is fixed and $\mathbf{x}(t_1)$ is free)

$$\boldsymbol{\lambda}(t_1) = \mathbf{H}\mathbf{x}(t_1) \tag{11.29b}$$

We are, therefore, faced with a two-point boundary value problem.

Let $\Phi(t, t_0)$ be the state transition matrix of the $2n$-dimensional linear system (11.28). We partition this transition matrix corresponding to (11.28) as

$$\Phi(t, t_0) = \begin{bmatrix}\Phi_{11}(t, t_0) & \Phi_{12}(t, t_0)\\ \Phi_{21}(t, t_0) & \Phi_{22}(t, t_0)\end{bmatrix} \tag{11.30}$$

If we let $\boldsymbol{\lambda}(t_0)$ be the (unknown) initial co-state, then the solution of eqn. (11.28) is of the form

$$\begin{bmatrix} \mathbf{x}(t) \\ \boldsymbol{\lambda}(t) \end{bmatrix} = \Phi(t, t_0) \begin{bmatrix} \mathbf{x}(t_0) \\ \boldsymbol{\lambda}(t_0) \end{bmatrix}$$

Which implies that

$$\begin{bmatrix} \mathbf{x}(t_0) \\ \boldsymbol{\lambda}(t_0) \end{bmatrix} = \Phi(t_0, t) \begin{bmatrix} \mathbf{x}(t) \\ \boldsymbol{\lambda}(t) \end{bmatrix}$$

with the partitioning given by (11.30), we can write

$$\mathbf{x}(t) = \Phi_{11}(t, t_1)\,\mathbf{x}(t_1) + \Phi_{12}(t, t_1)\,\boldsymbol{\lambda}(t_1)$$

$$\boldsymbol{\lambda}(t) = \Phi_{21}(t, t_1)\,\mathbf{x}(t_1) + \Phi_{22}(t, t_1)\,\boldsymbol{\lambda}(t_1)$$

With the terminal condition (11.29b), it follows that

$$\mathbf{x}(t) = [\Phi_{11}(t, t_1) + \Phi_{12}(t, t_1)\,\mathbf{H}]\,\mathbf{x}(t_1) \tag{11.31a}$$

$$\boldsymbol{\lambda}(t) = [\Phi_{21}(t, t_1) + \Phi_{22}(t, t_1)\,\mathbf{H}]\,[\Phi_{11}(t, t_1) + \Phi_{12}(t, t_1)\,\mathbf{H}]^{-1}\,\mathbf{x}(t) \tag{11.31b}$$

provided the indicated inverse exists.

Equation (11.31b) suggests that the co-state $\boldsymbol{\lambda}(t)$ and the state $\mathbf{x}(t)$ are related by an equation of the form

$$\boldsymbol{\lambda}(t) = \mathbf{P}(t)\,\mathbf{x}(t) \tag{11.32a}$$

where

$$\mathbf{P}(t) = [\Phi_{21}(t, t_1) + \Phi_{22}(t, t_1)\,\mathbf{H}]\,[\Phi_{11}(t, t_1) + \Phi_{12}(t, t_1)\,\mathbf{H}]^{-1} \tag{11.32b}$$

From eqns. (11.27) and (11.32) we obtain

$$\mathbf{u}^*(t) = -\mathbf{R}^{-1}(t)\,\mathbf{B}^T(t)\,\mathbf{P}(t)\,\mathbf{x}(t)$$

$$= \mathbf{K}(t)\,\mathbf{x}(t) \tag{11.33}$$

where

$$\mathbf{K}(t) = -\mathbf{R}^{-1}(t)\,\mathbf{B}^T(t)\,\mathbf{P}(t)$$

Earlier in this section, we stated the result that under the assumptions made on the various matrices of the performance index given by (11.15), the solution of the regulator problem exists. The existence of the solution of regulator problem guarantees the existence of an inverse matrix in (11.32b). Therefore, eqns. (11.32)–(11.33) provide the optimal solution to the regulator problem.

Ths matrix $\mathbf{P}(t)$ can be computed using eqn. (11.32b). This $\mathbf{P}(t)$ is, in fact, the solution of the matrix Riccati equation (11.22). (In problem 11.18, the reader is advised to verify this). Therefore, alternatively, $\mathbf{P}(t)$ may be obtained by direct integration of the matrix Riccati equation.

Example 11.3: Reconsider the problem of Example 11.2. The plant equation is

$$\dot{x}(t) = 2x(t) + u(t)$$

and the performance index is

$$J = \tfrac{1}{2}\int_0^{t_1} (3x^2 + \tfrac{1}{4}u^2)\,dt; \quad t_1 \text{ is specified}$$

The combined state and co-state equations (11.28) are now given by

$$\begin{bmatrix} \dot{x}(t) \\ \dot{\lambda}(t) \end{bmatrix} = \begin{bmatrix} 2 & -4 \\ -3 & -2 \end{bmatrix} \begin{bmatrix} x(t) \\ \lambda(t) \end{bmatrix}$$

The transition matrix corresponding to this system of differential equations can be found to be

$$\Phi(t, t_0) = \left[\begin{array}{c|c} \tfrac{1}{4}(3e^{4(t-t_0)} + e^{-4(t-t_0)}) & \tfrac{1}{2}(-e^{4(t-t_0)} + e^{-4(t-t_0)}) \\ \hline \tfrac{3}{8}(-e^{4(t-t_0)} + e^{-4(t-t_0)}) & \tfrac{1}{4}(e^{4(t-t_0)} + 3e^{-4(t-t_0)}) \end{array}\right]$$

$$= \left[\begin{array}{c|c} \Phi_{11}(t, t_0) & \Phi_{12}(t, t_0) \\ \hline \Phi_{21}(t, t_0) & \Phi_{22}(t, t_0) \end{array}\right]$$

From eqn. (11.32b), we get for our system

$$\begin{aligned} p(t) &= [\Phi_{21}(t, t_1)][\Phi_{11}(t, t_1)]^{-1} \\ &= \tfrac{3}{8}[-e^{4(t-t_1)} + e^{-4(t-t_1)}][\tfrac{1}{4}(3e^{4(t-t_1)} + e^{-4(t-t_1)}]^{-1} \\ &= \frac{\tfrac{3}{2}(1-e^{8(t-t_1)})}{1+3e^{8(t-t_1)}} \end{aligned}$$

The same result is given by eqn. (11.24) obtained by direct integration of the matrix Riccati equation.

11.4 TIME-INVARIANT LINEAR STATE REGULATORS

In the previous two sections, we observed that the optimal feedback regulator system turns out to be a linear but time-varying system. This is the case even if the plant and the cost functional are time-invariant.

It is worthwhile to ask the following question: Under what circumstances is the optimal state regulator system time-invariant? In this section, we shall provide the answer to this question.

Continuous-Time Systems

Consider the linear time-invariant system

$$\mathbf{x}(t) = \mathbf{A}\,\mathbf{x}(t) + \mathbf{B}\,\mathbf{u}(t) \qquad (11.34)$$

where **A** and **B** are respectively $n \times n$ and $n \times p$ constant matrices; and the performance index

$$J = \frac{1}{2} \int_{t_0}^{\infty} (\mathbf{x}^T(t)\, \mathbf{Q}\, \mathbf{x}(t) + \mathbf{u}^T(t)\, \mathbf{R}\, \mathbf{u}(t))\, dt \tag{11.35}$$

where **Q** and **R** are $n \times n$ and $p \times p$ positive semidefinite and positive definite constant matrices respectively.

This is infinite-time state regulator problem, formulated earlier in Section 10.2. In many industrial processes, we wish to guarantee that the state stays near zero after an initial transient interval; specification of terminal time t_1 is of no importance. This situation is realized by setting $t_1 = \infty$ in the performance index.

In the finite-time problem, the optimal J is always finite; this may not be so in the infinite-time case. J will become infinite if the unstable part of the system trajectory is reflected in the system performance index. We expect that the problem of the infinite value of the performance index will not arise for stabilizable plants. If the plant given by (11.34) is completely controllable, then it is stabilizable (Section 9.4). If the plant is not completely controllable, transformation of (11.34) into the controllability canonical form gives (eqn. (6.74a))

$$\begin{bmatrix} \dot{\hat{\mathbf{x}}}_1(t) \\ \dot{\hat{\mathbf{x}}}_2(t) \end{bmatrix} = \begin{bmatrix} \hat{\mathbf{A}}_{11} & \hat{\mathbf{A}}_{12} \\ \mathbf{0} & \hat{\mathbf{A}}_{22} \end{bmatrix} \begin{bmatrix} \hat{\mathbf{x}}_1(t) \\ \hat{\mathbf{x}}_2(t) \end{bmatrix} + \begin{bmatrix} \hat{\mathbf{B}}_1 \\ \mathbf{0} \end{bmatrix} \mathbf{u}(t)$$

where the pair $(\hat{\mathbf{A}}_{11}, \hat{\mathbf{B}}_1)$ is completely controllable. Suppose now that the system is not stabilizable so that $\hat{\mathbf{A}}_{22}$ is not asymptotically stable. Then obviously there exist initial state vectors of the form $[\mathbf{0} \;\; \hat{\mathbf{x}}_2^0]^T$ such that $\mathbf{x}(t) \to \infty$ no matter how $\mathbf{u}(t)$ is chosen. For such initial states, the performance index given by (11.35) will never converge to a finite value. However, if $\hat{\mathbf{A}}_{22}$ is asymptotically stable (i.e., system (11.34) is stabilizable), we can always find a feedback law

$$\mathbf{u}^*(t) = \mathbf{K}\, \mathbf{x}(t) \tag{11.36a}$$

that makes the closed-loop system

$$\dot{\mathbf{x}}(t) = (\mathbf{A} + \mathbf{BK})\, \mathbf{x}(t) \tag{11.36b}$$

asymptotically stable and thus results in a finite value of the performance index

$$J^* = \frac{1}{2} \int_{t_0}^{\infty} (\mathbf{x}^T\mathbf{Q}\mathbf{x} + \mathbf{x}^T\mathbf{K}^T\mathbf{R}\mathbf{K}\mathbf{x})\, dt$$

$$= \frac{1}{2} \int_{t_0}^{\infty} \mathbf{x}^T(\mathbf{Q} + \mathbf{K}^T\mathbf{R}\mathbf{K})\,\mathbf{x}\,dt \tag{11.36c}$$

We have thus established the result that the solution (11.36a) to the infinite-time regulator problem exists when the plant is stabilizable. This proves that the solution $\mathbf{P}(t)$ to the matrix Riccati equation (11.22) exists when $t_1 = \infty$ under the assumption of stabilizability of the plant (11.34). In fact $\mathbf{P}(t)$ is a constant matrix $\mathbf{P}^0$ when $t_1 = \infty$. This can easily be established.

From eqn. (11.20a), we have

$$J^*(\mathbf{x}(t), t) = \tfrac{1}{2}\mathbf{x}^T(t)\,\mathbf{P}(t)\,\mathbf{x}(t) \tag{11.37a}$$

Assuming for the time being that $\mathbf{P}(t)$ is a constant matrix $\mathbf{P}^0$ in the infinite-time case, we may write

$$J^* = \int_{t_0}^{\infty} -\tfrac{1}{2}\frac{d}{dt}[\mathbf{x}^T(t)\,\mathbf{P}^0\mathbf{x}(t)]\,dt$$

$$= -\tfrac{1}{2}\mathbf{x}^T(t)\,\mathbf{P}^0\mathbf{x}(t)\Big|_{t_0}^{\infty}$$

$$= \tfrac{1}{2}\mathbf{x}^T(t_0)\,\mathbf{P}^0\mathbf{x}(t_0) \tag{11.37b}$$

where it has been assumed that $\lim_{t\to\infty} \mathbf{x}(t) = \mathbf{0}$, i.e., the controlled linear system is asymptotically stable.

The plant (11.34) is time-invariant and the function under the integral sign of the performance index (11.35) is not specifically time dependent. This means that the choice of the initial time is arbitrary, i.e., all initial times must give the same performance index. The optimal performance index (11.37a) becomes constant (eqn. (11.37b)) for all initial times if $\mathbf{P}(t)$ is a constant matrix $\mathbf{P}^0$. Thus the feedback matrix

$$\mathbf{K} = -\mathbf{R}^{-1}\mathbf{B}^T\mathbf{P}^0$$

is constant in the infinite-time regulator problem and hence the implementation of the optimal control law (11.36a) is considerably easier.

The constant matrix $\mathbf{P}^0$ may be found in several ways. One way is to perform a limit operation on the solution for the time-varying problem.

Example 11.4: Reconsider the system of Example 11.3. For the time varying system of this example with time-invariant performance index, the solution of matrix Riccati equation is given by (eqn. (11.24))

$$p(t) = \frac{\frac{3}{2}(1 - e^{8(t-t_1)})}{1 + 3e^{8(t-t_1)}}$$

Think of time-varying solution $p(t)$ as a function of time t, initial time $t_0 = 0$

and terminal time t_1. We can write $p(t)$ as $p(t, 0, t_1)$ to indicate this dependence.

The solution for the infinite-time case is then given by

$$p^0 = \lim_{t_1 \to \infty} p(0, 0, t_1) = 3/2$$

In other words, p^0 is the initial value of the function $p(t)$ for the control interval 0 to t_1 as $t_1 \to \infty$. For this case, the optimal feedback control law becomes

$$u(t) = -6x(t) \tag{11.38}$$

There is a major difference in the nature of the optimal control represented by eqn. (11.38) as compared with eqn. (11.25). In the present case, the feedback coefficients are constant and hence the implementation of the optimal control law is considerably easier.

□

Note that in practice the technique of limit operation on the solution for the time-varying problem is not utilized because the analytic expression for $\mathbf{P}(t)$ is usually not known. Normally, we integrate the matrix Riccati equation (11.22) backward in time from the known terminal condition $\lim_{t_1 \to \infty} \mathbf{P}(t_1) = \mathbf{0}$. We take t_1 sufficiently large for the matrix $\mathbf{P}(t)$ to converge to a constant matrix $\mathbf{P}^0$.

The $\mathbf{P}^0$ matrix may also be found by utilizing the fact that $\mathbf{P}^0$ is constant and therefore its time derivative is zero. Substituting this result into the Riccati equation (11.22) we obtain

$$\mathbf{A}^T\mathbf{P}^0 + \mathbf{P}^0\mathbf{A} - \mathbf{P}^0\mathbf{B}\mathbf{R}^{-1}\mathbf{B}^T\mathbf{P}^0 + \mathbf{Q} = \mathbf{0} \tag{11.39}$$

which is often referred to as the *reduced matrix Riccati equation.* In (11.39) we have $\dfrac{n(n+1)}{2}$ nonlinear algebraic equations to be solved for the elements of $\mathbf{P}^0$.

Unfortunately, the solution of eqn. (11.39) is not unique. Of the several possible solutions, the desired answer is obtained by enforcing the requirement that $\mathbf{P}^0$ be positive definite. The positive definite solution of (11.39) is unique and is identical with the solution obtained by the limiting procedure discussed earlier.

Throughout this development it has been assumed that the controlled linear system (11.36b) is stable so that $\lim_{t \to \infty} \mathbf{x}(t) = \mathbf{0}$. However, the system (11.36b) is not always stable. Consider for example the open-loop system (Example 11.2)

$$\dot{x} = 2x + u$$

with performance index

$$J = \tfrac{1}{2} \int_0^\infty u^2 \, dt$$

The optimal control

$$u^* \equiv 0$$

and the closed-loop system is

$$\dot{x} = 2x$$

which is unstable. The factor which contributes to the instability of the closed-loop system is that the unstable trajectories do not contribute in any way to the performance index; in a sense the unstable states are not *observed* by the performance index. If all the trajectories (or at least unstable ones) show up in the $\mathbf{x}^T\mathbf{Q}\mathbf{x}$ part of the integrand of the performance index, it ensures the asymptotic stability of the optimal feedback system because if some of the state variables would not go to zero, the cost J^* would be infinite. All the trajectories of the system show up in the $\mathbf{x}^T\mathbf{Q}\mathbf{x}$ part of the integrand of the performance index if $\mathbf{Q}$ is positive definite. Therefore positive definiteness of $\mathbf{Q}$ is a sufficient condition for the asymptotic stability of the optimal regulator. Let us establish this result using the Lyapunov stability theorem.

Assume that

$$\mathbf{u}^*(t) = \mathbf{K}\mathbf{x}(t) = -\mathbf{R}^{-1}\mathbf{B}^T\mathbf{P}^0\mathbf{x}(t)$$

is the optimal control law (existence is assumed) for the system (11.34) that minimizes the performance index (11.35). Substituting $\mathbf{u}^*$ for $\mathbf{u}$ in (11.34), we obtain the state equation for the closed-loop system

$$\begin{aligned}\dot{\mathbf{x}}(t) &= (\mathbf{A} + \mathbf{B}\mathbf{K})\,\mathbf{x}(t)\\ &= (\mathbf{A} - \mathbf{B}\mathbf{R}^{-1}\,\mathbf{B}^T\mathbf{P}^0)\,\mathbf{x}(t) \qquad (11.40)\end{aligned}$$

Take $V(\mathbf{x}) = \mathbf{x}^T\mathbf{P}^0\mathbf{x}$ as the prospective Lyapunov function for the closed-loop system (11.40). Since $\mathbf{P}^0$ is positive definite, $V(\mathbf{x}) > 0$ for all non-zero $\mathbf{x}$.

$$\begin{aligned}\frac{dV(\mathbf{x})}{dt} &= \mathbf{x}^T(t)\,\mathbf{P}^0\dot{\mathbf{x}}(t) + \dot{\mathbf{x}}^T(t)\,\mathbf{P}^0\mathbf{x}(t)\\ &= \mathbf{x}^T(t)\mathbf{P}^0(\mathbf{A}-\mathbf{B}\mathbf{R}^{-1}\mathbf{B}^T\mathbf{P}^0)\mathbf{x}(t)+\mathbf{x}^T(t)(\mathbf{A}-\mathbf{B}\mathbf{R}^{-1}\mathbf{B}^T\mathbf{P}^0)^T\mathbf{P}^0\mathbf{x}(t)\\ &= \mathbf{x}^T(t)\,[\mathbf{A}^T\mathbf{P}^0 + \mathbf{P}^0\mathbf{A} - \mathbf{P}^0\mathbf{B}\mathbf{R}^{-1}\mathbf{B}^T\mathbf{P}^0]\,\mathbf{x}(t)\\ &\qquad - \mathbf{x}^T(t)\,\mathbf{P}^0\mathbf{B}\mathbf{R}^{-1}\mathbf{B}^T\mathbf{P}^0\mathbf{x}(t)\\ &= -\mathbf{x}^T(t)\,\mathbf{Q}\mathbf{x}(t) - \mathbf{x}^T(t)\,\mathbf{P}^0\mathbf{B}\mathbf{R}^{-1}\mathbf{B}^T\mathbf{P}^0\mathbf{x}(t) \qquad (11.41)\end{aligned}$$

If $\mathbf{Q}$ is positive definite, then $\dfrac{dV}{dt} < 0$ for all $\mathbf{x} \neq \mathbf{0}$ (note that $\mathbf{P}^0\mathbf{B}\mathbf{R}^{-1}\mathbf{B}^T\mathbf{P}^0$ is non-negative definite); therefore asymptotic stability is guaranteed.

Suppose now that $\mathbf{Q}$ is non-negative definite. The property of asymptotic stability of closed-loop system will be retained if all trajectories show up in the $\mathbf{x}^T\mathbf{Q}\mathbf{x}$ part of the integrand of the performance index. This requirement is met if the pair $(\mathbf{A}, \mathbf{F}^T)$ is completely observable where $\mathbf{F}^T$ is any matrix such that $\mathbf{F}\mathbf{F}^T = \mathbf{Q}$. Let us investigate the stability of the closed-loop system (11.40) under this assumption.

Inspection of eqn. (11.41) reveals that $\dot{V} \leqslant 0$. Suppose that $\dot{V}$ is identically zero along a trajectory starting from a nonzero initial state $\mathbf{x}^0$. Then $\mathbf{x}^T\mathbf{Q}\mathbf{x}$ and $\mathbf{x}^T\,\mathbf{P}^0\,\mathbf{B}\mathbf{R}^{-1}\mathbf{B}^T\mathbf{P}^0\mathbf{x}$ are identically zero and $-\mathbf{R}^{-1}\mathbf{B}^T\mathbf{P}^0\mathbf{x}$, the optimal control for the open-loop system, is also identically zero. Therefore, the trajectories of the closed-loop system are same as those of the open-loop system and are given by

$$\mathbf{x}(t) = e^{\mathbf{A}t}\mathbf{x}^0$$

Now

$$\mathbf{x}^T\mathbf{Q}\mathbf{x} = \mathbf{x}^{0T}\,e^{\mathbf{A}^Tt}\mathbf{Q}\,e^{\mathbf{A}t}\mathbf{x}^0$$
$$= \mathbf{x}^{0T}\,e^{\mathbf{A}^Tt}\,\mathbf{\Gamma}\mathbf{\Gamma}^T\,e^{\mathbf{A}t}\mathbf{x}^0$$

must be identically zero. This contradicts the assumption that the pair $(\mathbf{A}, \mathbf{\Gamma}^T)$ is completely observable because observability of $(\mathbf{A}, \mathbf{\Gamma}^T)$ implies that $\mathbf{\Gamma}^T\,e^{\mathbf{A}t}\,\mathbf{x}^0 = \mathbf{0}$ for any $t \in [0, \infty)$ if and only if $\mathbf{x}^0 = \mathbf{0}$ (Theorem 6.4). Consequently, it is impossible to have $\dot{V}$ identically zero along a trajectory starting at the nonzero state. The asymptotic stability of (11.40) is thus established.

If we define a *synthetic output* as (Kalman 1964)

$$\mathbf{y} = \mathbf{\Gamma}^T\mathbf{x} \tag{11.42}$$

then the observability of the pair $(\mathbf{A}, \mathbf{\Gamma}^T)$ implies that the synthetic system given by eqns. (11.34) and (11.42) be completely observable.

The requirement of observability of the pair $(\mathbf{A}, \mathbf{\Gamma}^T)$ is a sufficient condition to ensure asymptotic stability of the closed-loop system. Assume now that the pair $(\mathbf{A}, \mathbf{\Gamma}^T)$ is not completely observable. Then we can transform eqns. (11.34) and (11.42) into the following observability canonical form (eqn. (6.80)):

$$\begin{bmatrix} \dot{\hat{\mathbf{x}}}_1(t) \\ \dot{\hat{\mathbf{x}}}_2(t) \end{bmatrix} = \begin{bmatrix} \hat{\mathbf{A}}_{11} & 0 \\ \hat{\mathbf{A}}_{21} & \hat{\mathbf{A}}_{22} \end{bmatrix} \begin{bmatrix} \hat{\mathbf{x}}_1(t) \\ \hat{\mathbf{x}}_2(t) \end{bmatrix} + \begin{bmatrix} \hat{\mathbf{B}}_1 \\ \hat{\mathbf{B}}_2 \end{bmatrix} \mathbf{u}(t)$$

$$\mathbf{y}(t) = [\hat{\mathbf{\Gamma}} \quad \mathbf{0}] \begin{bmatrix} \hat{\mathbf{x}}_1 \\ \hat{\mathbf{x}}_2 \end{bmatrix}$$

The pair $(\hat{\mathbf{A}}_{11}, \hat{\mathbf{\Gamma}})$ is completely observable. We have seen that the control law does not affect the unobservable part; therefore if the observable part is stable (i.e., the synthetic system given by eqns. (11.34) and (11.42) is detectable), the asymptotic stability of the closed-loop system is guaranteed (For further details, refer Wonham (1968), and Martenssan (1971).

The results of this subsection are summarized below:

Given the stabilizable linear time-invariant plant (eqn. (11.34))

$$\dot{\mathbf{x}}(t) = \mathbf{A}\mathbf{x}(t) + \mathbf{B}\mathbf{u}(t)$$

with the performance index (eqn. (11.35))

$$J = \tfrac{1}{2}\int_0^\infty (\mathbf{x}^T(t)\,\mathbf{Q}\mathbf{x}(t) + \mathbf{u}^T(t)\,\mathbf{R}\mathbf{u}(t))\,dt$$

A unique optimal control that minimizes J exists and is given by (eqn. (11.36a))

$$\mathbf{u}^*(t) = \mathbf{K}\mathbf{x}(t)\ ;\ \ \mathbf{K} = -\mathbf{R}^{-1}\mathbf{B}^T\mathbf{P}^0$$

where $\mathbf{P}^0$ is a constant positive definite matrix which is the solution of the reduced matrix Riccati equation (eqn. (11.39))

$$\mathbf{A}^T\mathbf{P}^0 + \mathbf{P}^0\mathbf{A} - \mathbf{P}^0\mathbf{B}\mathbf{R}^{-1}\mathbf{B}^T\mathbf{P}^0 + \mathbf{Q} = \mathbf{0}$$

The closed-loop system (eqn. (11.36b))

$$\dot{\mathbf{x}}(t) = (\mathbf{A} - \mathbf{B}\mathbf{R}^{-1}\mathbf{B}^T\mathbf{P}^0)\ \mathbf{x}(t)$$

is asymptotically stable if the synthetic system (eqn. (11.42))

$$\dot{\mathbf{x}}(t) = \mathbf{A}\mathbf{x} + \mathbf{B}\mathbf{u}$$

$$\mathbf{y} = \mathbf{\Gamma}^T\mathbf{x}\ ;\ \ \mathbf{Q} = \mathbf{\Gamma}\mathbf{\Gamma}^T$$

is detectable.

The minimum value of the performance index is (eqn. (11.37b))

$$J^* = \tfrac{1}{2}\mathbf{x}^T(0)\ \mathbf{P}^0\mathbf{x}(0)$$

Example 11.5: Let us obtain the control law which minimizes the performance index

$$J = \int_0^\infty (x_1^2 + u^2)\ dt$$

for the system

$$\dot{\mathbf{x}} = \begin{bmatrix} 0 & 1 \\ 0 & 0 \end{bmatrix}\mathbf{x} + \begin{bmatrix} 0 \\ 1 \end{bmatrix} u$$

We have for this problem

$$\mathbf{A} = \begin{bmatrix} 0 & 1 \\ 0 & 0 \end{bmatrix},\ \mathbf{b} = \begin{bmatrix} 0 \\ 1 \end{bmatrix},$$

$$\mathbf{Q} = \begin{bmatrix} 2 & 0 \\ 0 & 0 \end{bmatrix},\ \mathbf{R} = [2]$$

A matrix $\mathbf{\Gamma}^T$ such that $\mathbf{\Gamma}\mathbf{\Gamma}^T = \mathbf{Q}$ is given by

$$\mathbf{\Gamma}^T = [\sqrt{2}\quad 0]$$

It can easily be verified that $(\mathbf{A}, \mathbf{\Gamma}^T)$ pair is observable. Consequently, the closed-loop system (optimal) will be asymptotically stable.

The reduced matrix Riccati equation is given as

$$\begin{bmatrix} 0 & 0 \\ 1 & 0 \end{bmatrix} \begin{bmatrix} p_{11} & p_{12} \\ p_{12} & p_{22} \end{bmatrix} + \begin{bmatrix} p_{11} & p_{12} \\ p_{12} & p_{22} \end{bmatrix} \begin{bmatrix} 0 & 1 \\ 0 & 0 \end{bmatrix}$$

$$- \begin{bmatrix} p_{11} & p_{12} \\ p_{12} & p_{22} \end{bmatrix} \begin{bmatrix} 0 \\ 1 \end{bmatrix} [\tfrac{1}{2}] [0 \quad 1] \begin{bmatrix} p_{11} & p_{12} \\ p_{12} & p_{22} \end{bmatrix} + \begin{bmatrix} 2 & 0 \\ 0 & 0 \end{bmatrix}$$

$$= \begin{bmatrix} 0 & 0 \\ 0 & 0 \end{bmatrix}$$

(Note that we have utilized the fact that $\mathbf{P}^0$ is symmetric). Upon simplification we get,

$$-\frac{p_{12}^2}{2} + 2 = 0$$

$$p_{11} - \frac{p_{12}\, p_{22}}{2} = 0$$

$$-\frac{p_{22}^2}{2} + 2p_{12} = 0$$

The solution of these equations yields the positive definite matrix

$$\mathbf{P}^0 = \begin{bmatrix} 2\sqrt{2} & 2 \\ 2 & 2\sqrt{2} \end{bmatrix}$$

The optimal control law is given by

$$u^*(t) = -\mathbf{R}^{-1}\mathbf{b}^T\mathbf{P}^0\mathbf{x}(t) = -[\tfrac{1}{2}]\,[0 \quad 1] \begin{bmatrix} 2\sqrt{2} & 2 \\ 2 & 2\sqrt{2} \end{bmatrix} \begin{bmatrix} x_1(t) \\ x_2(t) \end{bmatrix}$$

$$= -x_1(t) - \sqrt{2}\, x_2(t)$$

Discrete-Time Systems

All the results of the preceding subsection have obvious extensions for linear discrete-time systems (Caines and Mayne 1970, 1971). The summary is given below.

Given the discrete-time system

$$\mathbf{x}(k+1) = \mathbf{F}\mathbf{x}(k) + \mathbf{G}\mathbf{u}(k) \tag{11.43}$$

and the performance index

$$J = \tfrac{1}{2} \sum_{k=0}^{\infty} [\mathbf{x}^T(k)\,\mathbf{Q}\mathbf{x}(k) + \mathbf{u}^T(k)\,\mathbf{R}\mathbf{u}(k)] \tag{11.44}$$

where $\mathbf{Q}$ is a positive semidefinite constant matrix and $\mathbf{R}$ is a positive definite constant matrix.

If the system (11.43) is stabilizable, a unique optimal control that minimizes J exists. The results for the optimal control of infinite-stage regulator problem directly follow from those of finite-stage regulator problem. The matrices **P** and **K** in the recursive relations (11.9) are time-invariant when $N \to \infty$. Therefore,

$$\mathbf{u}^*(k) = \mathbf{K}\mathbf{x}(k) \tag{11.45a}$$

$$\mathbf{K} = -[\mathbf{R} + \mathbf{G}^T\mathbf{P}\mathbf{G}]^{-1}\mathbf{G}^T\mathbf{P}\mathbf{F} \tag{11.45b}$$

$$\mathbf{P} = [\mathbf{F} + \mathbf{G}\mathbf{K}]^T\mathbf{P}[\mathbf{F} + \mathbf{G}\mathbf{K}] + \mathbf{K}^T\mathbf{R}\mathbf{K} + \mathbf{Q} \tag{11.45c}$$

The optimal control sequence for the infinite-stage process may be obtained from the algebraic equations given above. An alternative way is to solve the recursive relations (11.9) using a digital computer for as many stages as required for $\mathbf{P}(N-k)$ to converge to a constant matrix.

Substituting $\mathbf{u}^*$ from (11.45a) for $\mathbf{u}$ in (11.43) we obtain the closed-loop system

$$\mathbf{x}(k+1) = (\mathbf{F} + \mathbf{G}\mathbf{K})\,\mathbf{x}(k) \tag{11.46}$$

This closed-loop system is asymptotically stable if the synthetic system, i.e.,

$$\mathbf{x}(k+1) = \mathbf{F}\mathbf{x}(k) + \mathbf{G}\mathbf{u}(k) \tag{11.47a}$$

$$\mathbf{y}(k) = \mathbf{F}^T\mathbf{x}(k);\ \mathbf{Q} = \mathbf{F}\mathbf{F}^T \tag{11.47b}$$

is detectable.

The minimum value of the performance index is

$$J^* = \tfrac{1}{2}\mathbf{x}^T(0)\,\mathbf{P}\mathbf{x}(0) \tag{11.48}$$

Example 11.6: Reconsider the system of Example 11.1. Let

$$\mathbf{Q} = \begin{bmatrix} 0.01 & 0 \\ 0 & 100 \end{bmatrix}; \quad \mathbf{R} = \begin{bmatrix} 2 & 0 \\ 0 & 0.5 \end{bmatrix}$$

The optimal feedback gain matrix $\mathbf{K}(k)$ for $N = 200$ is shown in Fig. 11.8a. Looking backward from $k = 199$, we observe that at $k = 170$, the matrix has reached its steady-state value

$$\mathbf{K} = \begin{bmatrix} -0.22 & 3.3 \\ -0.78 & -3.82 \end{bmatrix}$$

The optimal control and the corresponding trajectory for $\mathbf{x}(0) = [2 \quad 1]^T$ are shown in Fig. 11.8b. Notice that the trajectory has essentially reached 0 at $k = 25$.

Discretization of Performance Index: For continuous-time plant (11.34) with performance index (11.35), the optimal control law is obtained by solving

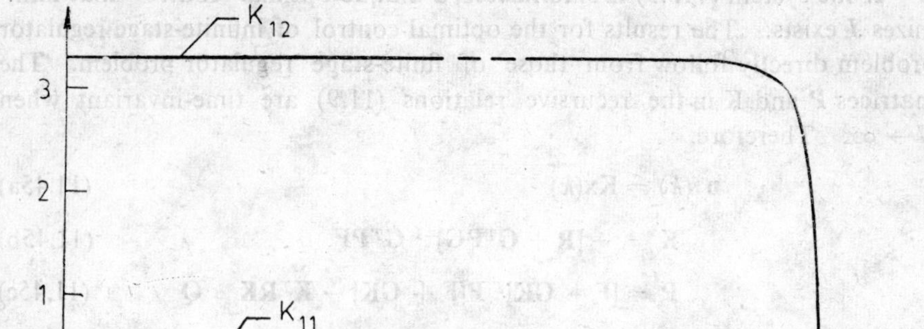

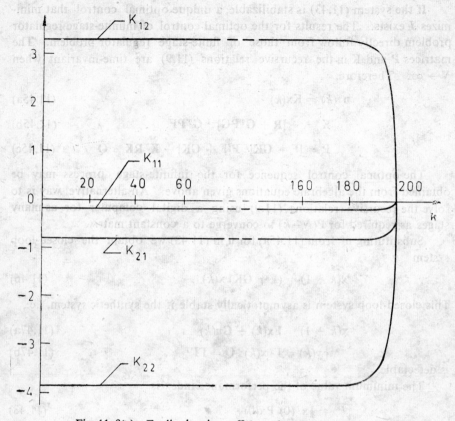

Fig. 11.8(a) Feedback gain coefficients for optimal control

the reduced matrix Riccati equation. The continuous-time regulator problem can be transformed into a discrete-time regulator problem by discretizing the plant model (11.34) and the performance index (11.35). Discretization of the plant model has already been discussed in Section 5.7. In this subsection, we discuss the procedure of discretizing the performance index (Dorato and Levis Alexander 1971; Levis Alexander et al. 1971).

For the system

$$\dot{\mathbf{x}} = \mathbf{A}\mathbf{x} + \mathbf{B}\mathbf{u}, \ \ \mathbf{x}(t_0) = \mathbf{x}^0;$$

the solution is

$$\mathbf{x}(t) = e^{\mathbf{A}(t-t_0)}\,\mathbf{x}(t_0) + \int_{t_0}^{t} e^{\mathbf{A}(t-\tau)}\,\mathbf{B}\mathbf{u}(\tau)\,d\tau$$

The continuous-time performance index

$$J = \tfrac{1}{2}\int_{t_0}^{\infty} (\mathbf{x}^T(t)\,\mathbf{Q}\mathbf{x}(t) + \mathbf{u}^T(t)\,\mathbf{R}\mathbf{u}(t))\,dt$$

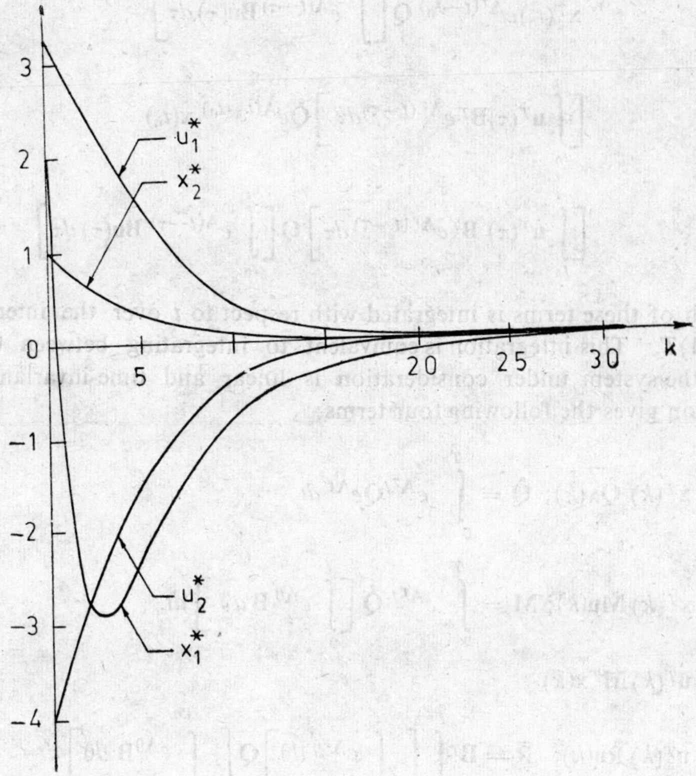

Fig. 11.8(b) Optimal control and corresponding trajectories

may be written as

$$J \simeq \frac{1}{2} \sum_{k=0}^{\infty} \int_{kT}^{(k+1)T} (\mathbf{x}^T(t)\, \mathbf{Q}\mathbf{x}(t) + \mathbf{u}^T(t)\, \mathbf{R}\mathbf{u}(t))\, dt$$

by splitting the integral into constant intervals of length T.

First consider the '**u** part' of the integral which can be written as

$$\int_{kT}^{(k+1)T} \mathbf{u}^T(t)\, \mathbf{R}\mathbf{u}(t)\, dt = [\mathbf{u}^T(k)\, \mathbf{R}\mathbf{x}(k)]T$$

as per the assumption that

$$\mathbf{u}(t) = \mathbf{u}(kT); \;\; kT \leqslant t < (k+1)T$$

Now substituting the solution $\mathbf{x}(t)$ in the term $\mathbf{x}^T\mathbf{Q}\mathbf{x}$, we get $\mathbf{x}^T\mathbf{Q}\mathbf{x}$ as the sum of the following four terms:

$$\mathbf{x}^T(t_0)\, e^{\mathbf{A}^T(t-t_0)}\mathbf{Q}e^{\mathbf{A}(t-t_0)}\, \mathbf{x}(t_0)$$

$$\mathbf{x}^T(t_0)e^{\mathbf{A}^T(t-t_0)}\,\mathbf{Q}\left[\int_{t_0}^{t} e^{\mathbf{A}(t-\tau)}\,\mathbf{B}\mathbf{u}(\tau)\,d\tau\right]$$

$$\left[\int_{t_0}^{t} \mathbf{u}^T(\tau)\mathbf{B}^T e^{\mathbf{A}^T(t-\tau)}d\tau\right]\mathbf{Q}e^{\mathbf{A}(t-t_0)}\,\mathbf{x}(t_0)$$

$$\left[\int_{t_0}^{t} \mathbf{u}^T(\tau)\,\mathbf{B}^T e^{\mathbf{A}^T(t-\tau)}\,d\tau\right]\mathbf{Q}\left[\int_{t_0}^{t} e^{\mathbf{A}(t-\tau)}\,\mathbf{B}\mathbf{u}(\tau)\,d\tau\right]$$

Each of these terms is integrated with respect to t over the interval kT to $(k+1)T$. This integration is equivalent to integrating between 0 and T because the system under consideration is linear and time-invariant. The integration gives the following four terms:

(i) $\mathbf{x}^T(k)\,\hat{\mathbf{Q}}\mathbf{x}(k)$; $\hat{\mathbf{Q}} = \int_0^T e^{\mathbf{A}^T t}\mathbf{Q}e^{\mathbf{A}t}\,dt$

(ii) $\mathbf{x}^T(k)\,\mathbf{M}\mathbf{u}(k)$; $\mathbf{M} = \int_0^T e^{\mathbf{A}^T t}\,\mathbf{Q}\left[\int_0^t e^{\mathbf{A}\theta}\,\mathbf{B}\,d\theta\right]dt$

(iii) $\mathbf{u}^T(k)\,\mathbf{M}^T\mathbf{x}(k)$

(iv) $\mathbf{u}^T(k)\,\overline{\mathbf{R}}\mathbf{u}(k)$; $\overline{\mathbf{R}} = \mathbf{B}^T\int_0^T\left[\int_0^t e^{\mathbf{A}^T\theta}\,d\theta\right]\mathbf{Q}\left[\int_0^t e^{\mathbf{A}\theta}\mathbf{B}\,d\theta\right]dt$

The terms (ii) and (iii) are identical, since each is a scalar. We can now write the performance index as

$$J = \tfrac{1}{2}\sum_{k=0}^{\infty}\left[\mathbf{x}^T(k)\,\hat{\mathbf{Q}}\mathbf{x}(k) + 2\mathbf{x}^T(k)\,\mathbf{M}\mathbf{u}(k) + \mathbf{u}^T(k)\,\hat{\mathbf{R}}\mathbf{u}(k)\right] \tag{11.49}$$

where

$$\hat{\mathbf{R}} = \overline{\mathbf{R}} + (\mathbf{R})\,T$$

Matrices $\overline{\mathbf{R}}$, $\hat{\mathbf{Q}}$ and $\mathbf{M}$ have been defined earlier.

The performance index (11.49) can be reduced to the standard form by the substitution

$$\hat{\mathbf{u}}(k) = \mathbf{u}(k) + \hat{\mathbf{R}}^{-1}\,\mathbf{M}^T\mathbf{x}(k)$$

which changes the plant equation from

$$\mathbf{x}(k+1) = \mathbf{F}\mathbf{x}(k) + \mathbf{G}\mathbf{u}(k)$$

to

$$\mathbf{x}(k+1) = \hat{\mathbf{F}}\mathbf{x}(k) + \mathbf{G}\hat{\mathbf{u}}(k) \tag{11.50a}$$

where

$$\hat{\mathbf{F}} = \mathbf{F} - \mathbf{G}\hat{\mathbf{R}}^{-1}\mathbf{M}^T$$

and the performance index (11.49) to

$$J = \tfrac{1}{2}\sum_{k=0}^{\infty} [\mathbf{x}^T(k)\,\tilde{\mathbf{Q}}\mathbf{x}(k) + \hat{\mathbf{u}}^T(k)\,\hat{\mathbf{R}}\hat{\mathbf{u}}(k)] \tag{11.50b}$$

where

$$\tilde{\mathbf{Q}} = \hat{\mathbf{Q}} - \mathbf{M}\hat{\mathbf{R}}^{-1}\mathbf{M}^T$$

For the system (11.50a) with the performance index (11.50b), we can obtain optimal control law in the form

$$\hat{\mathbf{u}}(k) = \mathbf{K}\mathbf{x}(k) \tag{11.50c}$$

which can be expressed as

$$\mathbf{u}(k) = (\mathbf{K} - \hat{\mathbf{R}}^{-1}\mathbf{M}^T)\,\mathbf{x}(k) \tag{11.51}$$

Note that given the positive definite $\mathbf{R}$, it can easily be verified that $\hat{\mathbf{R}}$ will be positive definite. Similarly, the definiteness of $\mathbf{Q}$ is also retained in $\hat{\mathbf{Q}}$ and $\tilde{\mathbf{Q}}$.

Two simple methods for evaluating the various matrices of this subsection are given by VanLoan (1978) and Armstrong (1978).

11.5 NUMERICAL SOLUTION OF THE RICCATI EQUATION

The solution of the Riccati equation is of fundamental importance for the regulator problems and, as we shall see in Chapter 12, also for state reconstruction problems. In this section we discuss some methods for the numerical solution of the Riccati equation.

Direct Integration

It is possible to solve the Riccati equation

$$-\dot{\mathbf{P}}(t) = \mathbf{Q}(t) - \mathbf{P}(t)\,\mathbf{B}(t)\,\mathbf{R}^{-1}(t)\,\mathbf{B}^T(t)\,\mathbf{P}(t) + \mathbf{P}(t)\,\mathbf{A}(t) + \mathbf{A}^T(t)\,\mathbf{P}(t); \tag{11.52a}$$

$$\mathbf{P}(t_1) = \mathbf{H} \tag{11.52b}$$

directly by appropriately programming a digital computer. This is a commonly used procedure, despite the availability of many other procedures. Equations (11.52) are a set of n^2 simultaneous nonlinear first-order differential equations and any standard numerical integration algorithm can be used to integrate these equations backward from t_1.

A disadvantage of direct integration is that for sufficient accuracy, discretization of time into quite small intervals is required, which results in a large number of steps. Also the symmetry of $\mathbf{P}(t)$ tends to be destroyed because of numerical errors. This difficulty can be overcome by replacing $\mathbf{P}(t)$ with $\tfrac{1}{2}[\mathbf{P}(t) + \mathbf{P}^T(t)]$ after each step or by reducing (11.52) to a set of

$\frac{n(n+1)}{2}$ simultaneous first-order differential equations. Kalman (1960) discusses how the numerical errors introduced at one point in the calculation propagate. It is shown that if the closed-loop system resulting from the application of the optimal control law is asymptotically stable, then the Riccati differential equation is numerically stable; as an error propagates, it is attenuated. For the time-invariant case, we have discussed the conditions for asymptotic stability of the closed-loop system. More complex conditions apply in the time-varying case. Refer Bucy and Joseph (1968) for further discussion.

The direct method of integrating the Riccati equation is applicable in the time-invariant case also. We choose a large value for t_1 and integrate the Riccati equation with $\mathbf{P}(t_1) = \mathbf{0}$ and obtain the constant matrix $\mathbf{P}^0$. For time-invariant problems, we have more effective methods of obtaining $\mathbf{P}^0$; two of them will be discussed in this section.

A Negative Exponential Method

This method is quite convenient for the computation of $\mathbf{P}^0$.

From eqn. (11.32b) we have

$$\mathbf{P}(t) = [\Phi_{21}(t, t_1) + \Phi_{22}(t, t_1)\mathbf{H}]\,[\Phi_{11}(t, t_1) + \Phi_{12}(t, t_1)\mathbf{H}]^{-1}$$

Discretizing time t, we may write

$$\mathbf{P}(t_{i+1}) = [\Phi_{21}(t_{i+1}, t_i) + \Phi_{22}(t_{i+1}, t_i)\,\mathbf{P}(t_i)]\,[\Phi_{11}(t_{i+1}, t_i) + \Phi_{12}(t_{i+1}, t_i)\,\mathbf{P}(t_i)]^{-1} \tag{11.53}$$

where

$$t_{i+1} = t_i - \Delta t$$

The matrices $\Phi_{ij}(t, t_0)$ are obtained by partitioning the transition matrix $\Phi(t, t_0)$ of the system (eqn. (11.28))

$$\begin{bmatrix} \dot{\mathbf{x}}(t) \\ \dot{\boldsymbol{\lambda}}(t) \end{bmatrix} = \begin{bmatrix} \mathbf{A} & -\mathbf{B}\mathbf{R}^{-1}\mathbf{B}^T \\ -\mathbf{Q} & -\mathbf{A}^T \end{bmatrix} \begin{bmatrix} \mathbf{x}(t) \\ \boldsymbol{\lambda}(t) \end{bmatrix}$$

$$= \mathbf{M} \begin{bmatrix} \mathbf{x}(t) \\ \boldsymbol{\lambda}(t) \end{bmatrix}$$

We can compute $\Phi(t_{i+1}, t_i)$ as

$$\Phi(t_{i+1}, t_i) = e^{-\mathbf{M}\Delta t}$$

The exponential can be evaluated according to the power series evaluation algorithm of Section 5.5. Repeated application of (11.53) then gives the constant matrix $\mathbf{P}^0$. Symmetrizing $\mathbf{P}$ after each step is advantageous.

Vaughan (1969) discusses some of the numerical dlfficulties associated with this method.

An Iterative Method

In this subsection, we present a method which is based upon the repeated solution of a Lyapunov type equation (refer Section 8.6).

The reduced matrix Riccati equation is (eqn. (11.39))

$$\mathbf{0} = \mathbf{A}^T\mathbf{P}^0 + \mathbf{P}^0\mathbf{A} - \mathbf{P}^0\mathbf{B}\mathbf{R}^{-1}\mathbf{B}^T\mathbf{P}^0 + \mathbf{Q}$$

or

$$\mathbf{0} = \mathbf{Q} - \mathbf{P}^0\mathbf{S}\mathbf{P}^0 + \mathbf{A}^T\mathbf{P}^0 + \mathbf{P}^0\mathbf{A}$$

where

$$\mathbf{S} = \mathbf{B}\mathbf{R}^{-1}\mathbf{B}^T$$

Consider the matrix function

$$f(\mathbf{P}^0) = \mathbf{Q} - \mathbf{P}^0\mathbf{S}\mathbf{P}^0 + \mathbf{A}^T\mathbf{P}^0 + \mathbf{P}^0\mathbf{A} \tag{11.54}$$

We want to find a positive definite matrix $\mathbf{P}^0$ that satisfies

$$f(\mathbf{P}^0) = \mathbf{0}$$

We go for an iterative procedure. Suppose in ith iteration, a solution $\mathbf{P}_i$, close to $\mathbf{P}^0$, has been obtained and let

$$\mathbf{P}^0 = \mathbf{P}_i + \breve{\mathbf{P}}$$

Substituting in (11.54), we obtain

$$f(\mathbf{P}^0) \simeq \mathbf{Q} - \mathbf{P}_i\mathbf{S}\mathbf{P}_i - \mathbf{P}_i\mathbf{S}\breve{\mathbf{P}} - \breve{\mathbf{P}}\mathbf{S}\mathbf{P}_i + \mathbf{A}^T(\mathbf{P}_i + \breve{\mathbf{P}}) + (\mathbf{P}_i + \breve{\mathbf{P}})\mathbf{A} \tag{11.55}$$

Note that we have omitted quadratic terms in $\breve{\mathbf{P}}$

Suppose $\breve{\mathbf{P}}_i$ yields

$$f(\mathbf{P}^0) = \mathbf{0} \quad \text{in (11.55);}$$

then (11.55) may be written as

$$\mathbf{0} = \mathbf{Q} + \mathbf{P}_i\mathbf{S}\mathbf{P}_i + \mathbf{P}_{i+1}\mathbf{A}_i + \mathbf{A}_i^T\mathbf{P}_{i+1} \tag{11.56}$$

where

$$\mathbf{A}_i = \mathbf{A} - \mathbf{S}\mathbf{P}_i$$

$$\mathbf{P}_{i+1} = \mathbf{P}_i + \breve{\mathbf{P}}_i$$

Equation (11.56) is of the form of a Lyapunov equation for which efficient methods of solution exist (Section 8.6).

The iterative algorithm discussed above is summarized in the flow chart of Fig. 11.9.

Kleinman (1968) and McClamroch (1969) have shown that if the reduced matrix Riccati equation has a unique positive definite solution, $\mathbf{P}_i \to \mathbf{P}^0$ provided $\mathbf{P}_0$ is so chosen that

$$\mathbf{A}_0 = \mathbf{A} - \mathbf{S}\mathbf{P}_0$$

is asymptotically stable. Thus, convergence is assured if the initial value $\mathbf{P}_0$

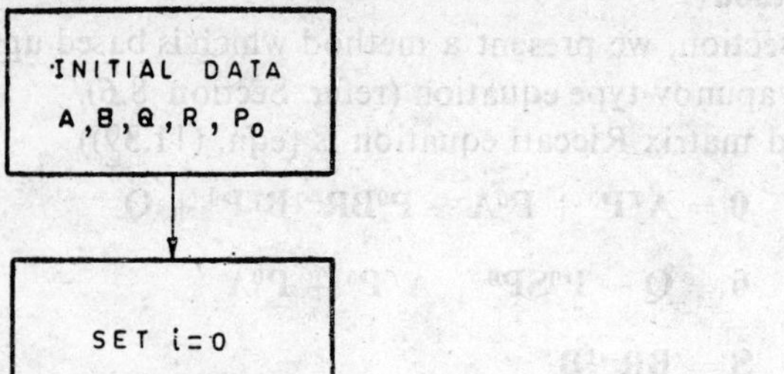

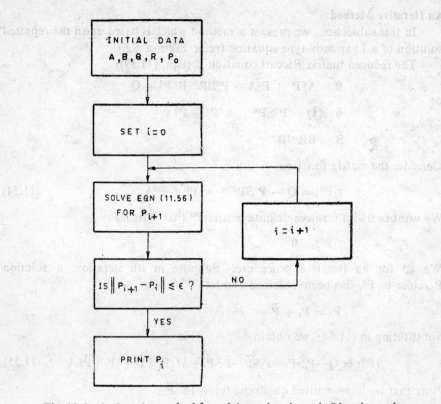

Fig. 11.9 An iterative method for solving reduced matrix Riccati equation

is suitably chosen. If **A** is asymptotically stable, a safe choice is $\mathbf{P}_0 = \mathbf{0}$. Wonham and Cashman (1968) and Man and Smith (1969) present methods for selecting $\mathbf{P}^0$ when **A** is not asymptotically stable. Results of Chapter 9 could also be used for selecting $\mathbf{P}^0$.

□

Some other useful results for the numerical solution of a Riccati equation have been reported by O'Donnell (1966), Potter (1966), McFarlane (1963), VanNess (1969) and Fath (1969).

Example 11.7: In example 10.12, we discussed the start up problem of a Nuclear Reactor. The optimal open-loop control and the resulting state trajectories are shown in Fig. 10.17.

We shall now develop a method of feedback control about the nominal trajectories which minimizes a quadratic performance index—quadratic in deviation from the nominal trajectory and control. Let us express the deviation of the state and control variables about the nominal trajectories by

$$n(t) = n_n(t) + \tilde{n}(t)$$
$$c(t) = c_n(t) + \tilde{c}(t)$$

$$\rho(t) = \rho_n(t) + \tilde{\rho}(t)$$

$$u(t) = u_n(t) + \tilde{u}(t)$$

The linearized equations obtained from (10.101) are (refer Section 4.6)

$$\begin{bmatrix} \dot{\tilde{n}} \\ \dot{\tilde{c}} \\ \dot{\tilde{\rho}} \end{bmatrix} = \begin{bmatrix} \dfrac{\rho_n(t)-\beta}{l} & \lambda & \dfrac{n_n(t)}{l} \\ \beta/l & -\lambda & 0 \\ 0 & 0 & 0 \end{bmatrix} \begin{bmatrix} \tilde{n} \\ \tilde{c} \\ \tilde{\rho} \end{bmatrix} + \begin{bmatrix} 0 \\ 0 \\ 1 \end{bmatrix} \tilde{u}$$

or

$$\dot{\tilde{\mathbf{x}}} = \mathbf{A}(t)\,\tilde{\mathbf{x}} + \mathbf{b}\,\tilde{u}$$

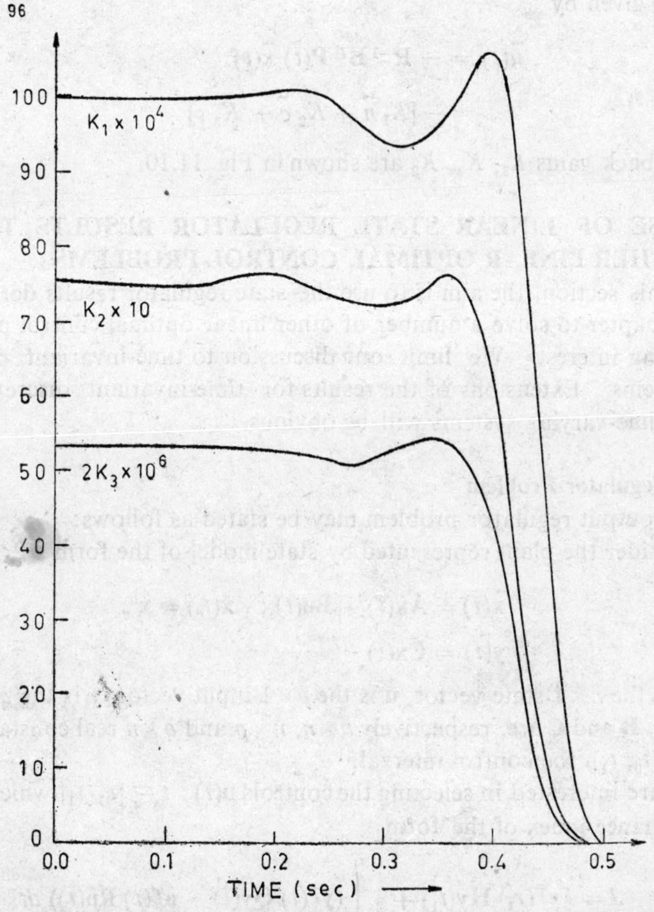

Fig. 11.10

Let us take the performance index to be

$$J=\frac{1}{2}\int_0^{0.5}(\tilde{\mathbf{x}}^T\mathbf{Q}\,\tilde{\mathbf{x}}+r\,\tilde{u}^2)\,dt$$

where

$$r=1,\ \mathbf{Q}=\begin{bmatrix}1 & 0 & 0\\ 0 & 0 & 0\\ 0 & 0 & 10^4\end{bmatrix}$$

To complete the design of closed-loop controller, the matrix $\mathbf{A}(t)$ is evaluated about the nominal trajectories and the associated Riccati equation is solved. The problem is thus a time-varying linear regulator problem.

The third-order matrix Riccati equation was solved using the Runge-Kutta fourth-order integration method to compute matrix **P**. The feedback gains are given by

$$\tilde{u}(t)=-\mathbf{R}^{-1}\mathbf{B}^T\mathbf{P}(t)\,\tilde{\mathbf{x}}(t)$$

$$=-[K_1\,\tilde{n}+K_2\,\tilde{c}+K_3\,\tilde{\rho}]$$

The feedback gains K_1, K_2, K_3 are shown in Fig. 11.10.

11.6 USE OF LINEAR STATE REGULATOR RESULTS TO SOLVE OTHER LINEAR OPTIMAL CONTROL PROBLEMS

In this section, the aim is to use the state regulator results derived earlier in this chapter to solve a number of other linear optimal control problems of engineering interest. We limit our discussion to time-invariant continuous-time systems. Extensions of the results for time-invariant discrete-time systems or time-varying systems will be obvious.

Output Regulator Problem

The output regulator problem may be stated as follows:

Consider the plant represented by state model of the form

$$\dot{\mathbf{x}}(t)=\mathbf{A}\mathbf{x}(t)+\mathbf{B}\mathbf{u}(t)\ ;\quad \mathbf{x}(t_0)=\mathbf{x}^0 \tag{11.57a}$$

$$\mathbf{y}(t)=\mathbf{C}\mathbf{x}(t) \tag{11.57b}$$

where **x** is the $n\times 1$ state vector, **u** is the $p\times 1$ input vector, **y** is the $q\times 1$ output vector, **A**, **B** and **C** are, respectively $n\times n$, $n\times p$ and $q\times n$ real constant matrices and $t\in[t_0, t_1]$, the control interval.

We are interested in selecting the controls $\mathbf{u}(t)$; $t\in[t_0, t_1]$ which minimize a performance index of the form

$$J=\frac{1}{2}\mathbf{y}^T(t_1)\,\mathbf{H}\mathbf{y}(t_1)+\frac{1}{2}\int_{t_0}^{t_1}(\mathbf{y}^T(t)\,\mathbf{Q}\mathbf{y}(t)+\mathbf{u}^T(t)\,\mathbf{R}\mathbf{u}(t))\,dt \tag{11.58}$$

where **H** and **Q** are positive semidefinite real constant symmetric matrices and **R** is a positive definite real constant symmetric matrix.

From the performance index given by (11.58) it is obvious that in the output regulator problem, we desire to bring and keep output $\mathbf{y}(t)$ near zero without using an excessive amount of control energy.

Substituting $\mathbf{y}(t) = \mathbf{Cx}(t)$ in eqn. (11.58) we get

$$J = \tfrac{1}{2}\mathbf{x}^T(t_1)\,\mathbf{C}^T\mathbf{HCx}(t_1) + \tfrac{1}{2}\int_{t_0}^{t_1} (\mathbf{x}^T(t)\,\mathbf{C}^T\mathbf{QCx}(t) + \mathbf{u}^T(t)\,\mathbf{Ru}(t))\,dt \qquad (11.59)$$

Comparing eqn. (11.59) with (11.15) we observe that the two indices are identical in form; **H** and **Q** in (11.15) are replaced, respectively, by $\mathbf{C}^T\mathbf{HC}$ and $\mathbf{C}^T\mathbf{QC}$ in (11.59). If we assume that the system (11.57) is observable, then **C** cannot be zero; $\mathbf{C}^T\mathbf{HC}$ and $\mathbf{C}^T\mathbf{QC}$ will be positive definite (or semidefinite) matrices whenever **H** and **Q** are positive definite (or semidefinite).

We have the following solution for the output regulator problem which directly follows from the results of Section 11.3.

For the observable system (11.57) with the performance index (11.58), a unique optimal control exists and is given by

$$\mathbf{u}^*(t) = \mathbf{K}(t)\,\mathbf{x}(t) \qquad (11.60a)$$

where

$$\mathbf{K}(t) = -\mathbf{R}^{-1}\mathbf{B}^T\mathbf{P}(t) \qquad (11.60b)$$

Here the symmetric non-negative definite matrix $\mathbf{P}(t)$ is the solution of the matrix Riccati equation

$$\dot{\mathbf{P}}(t) + \mathbf{C}^T\mathbf{QC} - \mathbf{P}(t)\,\mathbf{BR}^{-1}\mathbf{B}^T\mathbf{P}(t) + \mathbf{P}(t)\,\mathbf{A} + \mathbf{A}^T\mathbf{P}(t) = \mathbf{0} \qquad (11.61a)$$

$$\mathbf{P}(t_1) = \mathbf{C}^T\mathbf{HC} \qquad (11.61b)$$

Results for the output regulator problem for the case when final time is not constrained, can easily be obtained from the results of Section 11.4.

Accommodation of Non-zero Regulation Set Points: The preceding methods for designing optimum controllers implicitly assumed that the null state $\mathbf{x} = \mathbf{0}$ was the desired operating set point for the process. In some applications, that special assumption is not realistic. For instance, one may require that the output $\mathbf{y}(t)$ should quickly move to some specified non-null value $\mathbf{y}_{SP}$ for any initial perturbation in the system state. To accommodate non-null set points, we proceed as follows (Refer Foot Note 1 in Chapter 6):

Let us assume that for the system given by (11.57), set point of controlled variable is given by $\mathbf{y}_{SP}$. Then in order to maintain the system at this point, a constant input $\mathbf{u}_0$ must be found that holds the state at $\mathbf{x}_d$ such that

$$\mathbf{y}_{SP} = \mathbf{Cx}_d \qquad (11.62)$$

It follows from eqn. (11.57a), that $\mathbf{x}_d$ and $\mathbf{u}_0$ must be related by

$$\mathbf{0} = \mathbf{A}\mathbf{x}_d + \mathbf{B}\mathbf{u}_0 \tag{11.63}$$

Assuming for the present that a $\mathbf{u}_0$ exists that satisfies eqns. (11.62)–(11.63), we define shifted input, shifted state and shifted controlled variable as

$$\begin{aligned} \mathbf{u}'(t) &= \mathbf{u}(t) - \mathbf{u}_0 \\ \mathbf{x}'(t) &= \mathbf{x}(t) - \mathbf{x}_d \\ \mathbf{y}'(t) &= \mathbf{y}(t) - \mathbf{y}_{SP} \end{aligned} \tag{11.64}$$

The shifted variables satisfy the equations

$$\dot{\mathbf{x}}'(t) = \mathbf{A}\mathbf{x}'(t) + \mathbf{B}\mathbf{u}'(t) \tag{11.65a}$$

$$\mathbf{y}'(t) = \mathbf{C}\mathbf{x}'(t) \tag{11.65b}$$

Suppose that at a given time, the set point is suddenly shifted from one value to another. Then in terms of the shifted equations (11.65), the system acquires a non-zero initial state. Assume that this shifted regulator problem possesses a time-invariant asymptotically stable control law

$$\mathbf{u}'(t) = \mathbf{K}\mathbf{x}'(t) \tag{11.66}$$

The application of this control law ensures that $\mathbf{x}'(t) \to \mathbf{0}$ ($\mathbf{x}(t) \to \mathbf{x}_d$; $\mathbf{y}(t) \to \mathbf{y}_{SP}$) as quickly as possible without excessively large input amplitudes.

In terms of the original system variables, total control effort

$$\mathbf{u}(t) = \mathbf{K}\mathbf{x}(t) + \mathbf{u}_0 - \mathbf{K}\mathbf{x}_d \tag{11.67a}$$

$$= \mathbf{u}_R + \mathbf{u}_S \tag{11.67b}$$

where

$$\mathbf{u}_R = \mathbf{K}\mathbf{x}(t) \tag{11.67c}$$

$$\mathbf{u}_S = \mathbf{u}_0 - \mathbf{K}\mathbf{x}_d \tag{11.67d}$$

Substituting this control law into (11.57a), we get

$$\dot{\mathbf{x}}(t) = (\mathbf{A} + \mathbf{B}\mathbf{K})\,\mathbf{x}(t) + \mathbf{B}\mathbf{u}_S$$

Since the closed-loop system is asymptotically stable, as $t \to \infty$ the state reaches a steady-state value $\mathbf{x}_d$ that satisfies

$$\mathbf{0} = (\mathbf{A} + \mathbf{B}\mathbf{K})\,\mathbf{x}_d + \mathbf{B}\mathbf{u}_S$$

or

$$\mathbf{x}_d = -(\mathbf{A} + \mathbf{B}\mathbf{K})^{-1}\,\mathbf{B}\mathbf{u}_S$$

To achieve the set point $\mathbf{y}_{SP}$, we must have

$$\mathbf{C}(\mathbf{A} + \mathbf{B}\mathbf{K})^{-1}\mathbf{B}\mathbf{u}_S = -\,\mathbf{y}_{SP} \tag{11.68}$$

Equation (11.68) has a unique solution for $\mathbf{u}_S$ if the dimensions of $\mathbf{u}$ and $\mathbf{y}$ are same (provided $\mathbf{C}(\mathbf{A}+\mathbf{BK})^{-1}\mathbf{B}$ is nonsingular). If the dimension of $\mathbf{y}$ is less than that of $\mathbf{u}$, then $\mathbf{u}_S$ is nonunique. When the dimension of $\mathbf{y}$ is greater than that of $\mathbf{u}$, then (11.63) has a solution for special values of $\mathbf{y}_{SP}$ only; in general no solution exists and we may obtain the least-squares solution for $\mathbf{u}_S$ (refer Section 2.6).

Accommodation of External Disturbances Acting on the Process: In practical applications of control theory, one of the few things engineers can assume for certain is that the signals to be encountered are not known for certain. When dealing with uncertain external influences acting on a controlled process, it is important to distinguish between *noise* and *disturbances*. Noise is an erratic, uncertain signal having no distinguishing waveform properties. The accommodation of noise in a control problem is best handled by treating the noise as a random process (Chapter 12). Disturbances, on the other hand, are well behaved uncertain signals having distinguishing waveform properties. For accommodating such influences one should employ a method which makes use of the known waveform properties which characterize the disturbances. In the following, we shall consider the case of constant disturbances over long time-periods compared with time-constants of the system.

Consider first, the plant

$$\dot{\mathbf{x}}(t)=\mathbf{A}\mathbf{x}(t)+\mathbf{B}\mathbf{u}(t);\ \mathbf{x}(0)=\mathbf{x}^0 \tag{11.69a}$$

$$\mathbf{y}(t)=\mathbf{C}\mathbf{x}(t) \tag{11.69b}$$

and the optimal control law

$$\mathbf{u}(t)=\mathbf{K}\mathbf{x}(t) \tag{11.70}$$

which minimizes the performance index

$$J=\tfrac{1}{2}\int_0^\infty (\mathbf{x}^T(t)\,\mathbf{Q}\mathbf{x}(t)+\mathbf{u}^T(t)\,\mathbf{R}\mathbf{u}(t))\,dt \tag{11.71}$$

The closed-loop system is described by the equation

$$\dot{\mathbf{x}}(t)=(\mathbf{A}+\mathbf{BK})\,\mathbf{x}(t) \tag{11.72}$$

At steady-state, $\dot{\mathbf{x}}(t)=\mathbf{0}$ and therefore $\lim\limits_{t\to\infty}\mathbf{x}(t)\to\mathbf{0}$.

Assume now that a constant disturbance occurs in the system, so that we must replace the state equation (11.69a) with

$$\dot{\mathbf{x}}(t)=\mathbf{A}\mathbf{x}(t)+\mathbf{B}\mathbf{u}(t)+\mathbf{\Gamma w} \tag{11.73}$$

where $\mathbf{w}$ is constant vector.

The closed-loop system with the control law (11.70) is described by the equation

$$\dot{\mathbf{x}}(t)=(\mathbf{A}+\mathbf{BK})\,\mathbf{x}(t)+\mathbf{\Gamma w} \tag{11.74}$$

At steady-state, $\dot{\mathbf{x}}(t) = \mathbf{0}$ and the system state reaches a nonzero state $\mathbf{x}_{ss}$ governed by the relation

$$\mathbf{0} = (\mathbf{A} + \mathbf{BK})\,\mathbf{x}_{ss} + \mathbf{\Gamma w}$$

This gives

$$\mathbf{x}_{ss} = -(\mathbf{A} + \mathbf{BK})^{-1}\,\mathbf{\Gamma w} \tag{11.75}$$

Thus, the disturbance will result in an off-set in the system's steady-state; the magnitude of off-set depends upon the magnitude of the disturbance. The off-set will shift from one value to another with change in the step value of the disturbance.

One simple and obvious way of suppressing this off-set is to add a feed-forward control term to the closed-loop system. Consider the control law

$$\mathbf{u}(t) = \mathbf{Kx}(t) + \mathbf{u}_0 \tag{11.76}$$

where $\mathbf{u}_0$ is a contribution to the input that counteracts the effect of the constant disturbance. With this control law, the closed-loop system is governed by the equation

$$\dot{\mathbf{x}}(t) = (\mathbf{A} + \mathbf{BK})\,\mathbf{x}(t) + \mathbf{\Gamma w} + \mathbf{Bu}_0 \tag{11.77}$$

If $\mathbf{u}_0$ is generated such that

$$\mathbf{\Gamma w} + \mathbf{Bu}_0 = \mathbf{0}, \tag{11.78}$$

then the system state will approach the origin as $t \to \infty$. The least squares solution of (11.78) is (refer Foot Note 1 in Chapter 6).

$$\mathbf{u}_0 = -(\mathbf{B}^T\mathbf{B})^{-1}\,\mathbf{B}^T\,\mathbf{\Gamma w} \tag{11.79}$$

From (11.79), it is obvious that there is a problem of estimating the disturbance, which is a complicated proposition.

The problem of suppression of disturbance effects in linear regulators has been studied by Johnson (1968, 1971), Yore (1968), West and McGurie (1969), Smith and Murrill (1969), Markland (1970), Young and Willems (1972), Zinober (1970), Shih (1970), Newell and Fisher (1972), Newell et al. (1972) in different perspectives. A number of techniques have been developed and experimented with.

In this subsection, we discuss the use of integral feedback for improving steady-state performance of the system with a disturbance.

Consider the system (11.69) where $\mathbf{x}$ is $n \times 1$ state vector, $\mathbf{u}$ is $p \times 1$ control vector and $\mathbf{y}$ is $q \times 1$ output vector. The output $\mathbf{y}(t)$ is required to remain as close as possible to the reference output $\mathbf{y}_r = \mathbf{0}$. We assume the system (11.69) to be controllable.

We add to the system variables, the 'integral state' $\mathbf{z}(t)$ defined by

$$\mathbf{z}(t) = \int_0^t \mathbf{y}(\tau)\,d\tau$$

Since $\mathbf{z}(t)$ satisfies the differential equation

$$\dot{\mathbf{z}} = \mathbf{y}(t)$$
$$= \mathbf{Cx}(t); \quad \mathbf{z}(0) = \mathbf{0} \tag{11.80}$$

it is easily included by augmenting the original system (11.69) as follows:

$$\begin{bmatrix} \dot{\mathbf{x}} \\ \dot{\mathbf{z}} \end{bmatrix} = \begin{bmatrix} \mathbf{A} & \mathbf{0} \\ \mathbf{C} & \mathbf{0} \end{bmatrix} \begin{bmatrix} \mathbf{x} \\ \mathbf{z} \end{bmatrix} + \begin{bmatrix} \mathbf{B} \\ \mathbf{0} \end{bmatrix} \mathbf{u} \tag{11.81a}$$

$$\mathbf{y} = [\mathbf{C} \quad \mathbf{0}] \begin{bmatrix} \mathbf{x} \\ \mathbf{z} \end{bmatrix} \tag{11.81b}$$

The new state vector

$$\hat{\mathbf{x}} = \begin{bmatrix} \mathbf{x} \\ \mathbf{z} \end{bmatrix} \tag{11.82}$$

has dimension $n + q$.

The first question which arises is whether the new system (11.81) is controllable from the input $\mathbf{u}$. In the following, we prove that (11.81) is controllable if and only if (11.69) is controllable and

$$\rho \begin{bmatrix} \mathbf{A} & \mathbf{B} \\ \mathbf{C} & \mathbf{0} \end{bmatrix} = n + q \tag{11.83}$$

The system (11.81) is controllable if and only if the $(n + q)\times(n + q)p$ controllability matrix

$$\mathbf{U} = \begin{bmatrix} \mathbf{B} & \mathbf{AB} & \mathbf{A}^2\mathbf{B} & \dots & \mathbf{A}^{n+q-1}\mathbf{B} \\ \mathbf{0} & \mathbf{CB} & \mathbf{CAB} & \dots & \mathbf{CA}^{n+q-2}\mathbf{B} \end{bmatrix} \text{ has rank } n + q.$$

Now

$$\mathbf{U} = \begin{bmatrix} \mathbf{B} & \mathbf{AU}' \\ \mathbf{0} & \mathbf{CU}' \end{bmatrix}$$

$$= \begin{bmatrix} \mathbf{A} & \mathbf{B} \\ \mathbf{C} & \mathbf{0} \end{bmatrix} \begin{bmatrix} \mathbf{0} & \mathbf{U}' \\ \mathbf{I} & \mathbf{0} \end{bmatrix} \tag{11.84}$$

where

$$\mathbf{U}' = [\mathbf{B} \quad \mathbf{AB} \dots \mathbf{A}^{n+q-2}\mathbf{B}]$$

The augmented system (11.81) cannot be controllable unless the original system (11.69) is controllable. Therefore we assume

$$\rho\,[\mathbf{B} \quad \mathbf{AB} \dots \mathbf{A}^{n-1}\mathbf{B}] = n$$

This implies that

$$\rho\,[\mathbf{U}'] = n$$

since additional columns cannot reduce the rank of a matrix. This result further proves that

$$\rho \begin{bmatrix} \mathbf{0} & \mathbf{U}' \\ \mathbf{I} & \mathbf{0} \end{bmatrix} = n + q$$

Using the result (Appendix III)

$$\rho\,(\mathbf{AB}) = \rho\,(\mathbf{A}) \quad \text{if } \mathbf{B} \text{ is nonsingular}$$

we have from (11.84),

$$\rho\,(\mathbf{U}) = \rho \begin{bmatrix} \mathbf{A} & \mathbf{B} \\ \mathbf{C} & \mathbf{0} \end{bmatrix}$$

Therefore, for the augmented system (11.81) to be controllable, the condition (11.83) must be satisfied. Moreover, since the matrix in (11.83) is $(n+q)\times(n+p)$, its rank can be $n+q$ only if $p \geqslant q$ and $\rho\,(\mathbf{C}) = q$.

Having determined a test for the controllability of the composite system (11.81), we can find an optimal control law

$$\mathbf{u} = \mathbf{K}\hat{\mathbf{x}} \tag{11.85}$$

which minimizes the performance index

$$J = \tfrac{1}{2} \int_0^\infty (\hat{\mathbf{x}}^T \mathbf{Q} \hat{\mathbf{x}} + \mathbf{u}^T \mathbf{R} \mathbf{u})\, dt \tag{11.86}$$

where $\hat{\mathbf{x}}$ is defined in (11.82).

The control law (11.85) may be split as

$$\begin{aligned} \mathbf{u} &= \mathbf{K}_1\mathbf{x} + \mathbf{K}_2\mathbf{z} \\ &= \mathbf{K}_1\mathbf{x} + \mathbf{K}_2 \int_0^t \mathbf{y}(\tau)\, d\tau \end{aligned} \tag{11.87}$$

This is obviously a proportional plus integral control law.

Suppose now that a constant disturbance occurs in the system; we must therefore replace eqn. (11.69a) with eqn. (11.73) in our plant model. The closed-loop system with a disturbance is described by the equations

$$\dot{\hat{\mathbf{x}}} = \begin{bmatrix} \dot{\mathbf{x}} \\ \dot{\mathbf{z}} \end{bmatrix} = \begin{bmatrix} \mathbf{A} + \mathbf{BK}_1 & \mathbf{BK}_2 \\ \mathbf{C} & \mathbf{0} \end{bmatrix} \begin{bmatrix} \mathbf{x} \\ \mathbf{z} \end{bmatrix} + \begin{bmatrix} \mathbf{\Gamma} \\ \mathbf{0} \end{bmatrix} \mathbf{w} \tag{11.88a}$$

$$\mathbf{y} = [\mathbf{C} \quad \mathbf{0}] \begin{bmatrix} \mathbf{x} \\ \mathbf{z} \end{bmatrix} \tag{11.88b}$$

At steady-state $\dot{\hat{\mathbf{x}}} = \mathbf{0}$; therefore

$$\lim_{t \to \infty} \dot{\mathbf{z}}(t) \to \mathbf{0}$$

or

$$\lim_{t \to \infty} \mathbf{y}(t) \to \mathbf{0} \tag{11.89}$$

Thus by integrating action, the output **y** is driven to the no-offset condition.

□

Quite often, the use of only $\mathbf{u}^T\mathbf{R}\mathbf{u}$ term in the performance index results in extremely high gains and demands the transition of **u** from a low level to a high level in a very short duration. This difficulty can be overcome by including a term of the form

$$\dot{\mathbf{u}}^T \mathbf{R}' \dot{\mathbf{u}}$$

in the performance index. A solution of such a problem is given by Smith and Murrill (1969) and is outlined below.

Consider the system (assumed controllable)

$$\dot{\mathbf{x}} = \mathbf{A}\mathbf{x} + \mathbf{B}\mathbf{u} + \mathbf{\Gamma}\mathbf{w} \tag{11.90a}$$

$$\mathbf{y} = \mathbf{C}\mathbf{x} \tag{11.90b}$$

where **x** is $n \times 1$ state vector, **u** is $p \times 1$ input vector, **y** is $p \times 1$ output vector and **w** is $p \times 1$ constant disturbance vector.

Define the vectors $\hat{\mathbf{x}}$ and **v** as follows:

$$\hat{\mathbf{x}} = \begin{bmatrix} \dot{\mathbf{x}} \\ \mathbf{y} \end{bmatrix} \tag{11.91a}$$

$$\mathbf{v} = \dot{\mathbf{u}} \tag{11.91b}$$

From (11.90)–(11.91) we obtain

$$\dot{\hat{\mathbf{x}}} = \begin{bmatrix} \ddot{\mathbf{x}} \\ \dot{\mathbf{y}} \end{bmatrix} = \begin{bmatrix} \mathbf{A}\dot{\mathbf{x}} + \mathbf{B}\dot{\mathbf{u}} \\ \dot{\mathbf{y}} \end{bmatrix} = \begin{bmatrix} \mathbf{A}\dot{\mathbf{x}} \\ \dot{\mathbf{y}} \end{bmatrix} + \begin{bmatrix} \mathbf{B} \\ \mathbf{0} \end{bmatrix} \dot{\mathbf{u}}$$

$$= \hat{\mathbf{A}}\hat{\mathbf{x}} + \hat{\mathbf{B}}\mathbf{v} \tag{11.92}$$

where

$$\hat{\mathbf{A}} = \begin{bmatrix} \mathbf{A} & \mathbf{0} \\ \mathbf{C} & \mathbf{0} \end{bmatrix}, \hat{\mathbf{B}} = \begin{bmatrix} \mathbf{B} \\ \mathbf{0} \end{bmatrix}; \hat{\mathbf{x}}(0) = \begin{bmatrix} \dot{\mathbf{x}}(0) \\ \mathbf{y}(0) \end{bmatrix} = \begin{bmatrix} \mathbf{A} & \mathbf{B} & \mathbf{\Gamma} \\ \mathbf{C} & \mathbf{0} & \mathbf{0} \end{bmatrix} \begin{bmatrix} \mathbf{x}(0) \\ \mathbf{u}(0) \\ \mathbf{w} \end{bmatrix}$$

If the system (11.92) is controllable, we can obtain a control law

$$\mathbf{v} = \mathbf{K}\,\hat{\mathbf{x}} \tag{11.93}$$

that minimizes the performance index of the form

$$J = \tfrac{1}{2}\int_0^\infty (\hat{\mathbf{x}}^T\mathbf{Q}\,\hat{\mathbf{x}} + \mathbf{v}^T\mathbf{R}\mathbf{v})\,dt \tag{11.94}$$

In terms of the original variables, the index (11.94) is

$$J = \tfrac{1}{2}\int_0^\infty (\dot{\mathbf{x}}^T\,\mathbf{Q}_1\,\dot{\mathbf{x}} + \mathbf{y}^T\,\mathbf{Q}_2\mathbf{y} + \dot{\mathbf{u}}^T\,\mathbf{R}\,\dot{\mathbf{u}})\,dt \tag{11.95}$$

This index is explicitly a function of state derivatives, output and control effort derivatives. The control effort is penalized indirectly by penalizing its derivatives. The new form of the performance index allows the possibility of penalizing the rates of change of state variables.

The control law (11.93) may be expressed as

$$\mathbf{u} = \mathbf{K}_1\mathbf{x} + \mathbf{K}_2\int_0^t \mathbf{y}(\tau)\,d\tau \tag{11.96}$$

where $\mathbf{K}_1$ and $\mathbf{K}_2$ are appropriate matrices obtained by partitioning the feedback gain matrix $\mathbf{K}$.

Example 11.8: Consider the Power System discussed in Section 4.6. The following plant model was obtained.

$$\dot{\mathbf{x}} = \mathbf{A}\mathbf{x} + \mathbf{b}u + \gamma w \tag{11.97}$$

where

$$\mathbf{A} = \begin{bmatrix} -0.05 & 0.1 & 0 \\ 0 & -0.3610 & 0.3610 \\ -200 & 0 & -10 \end{bmatrix};\ \mathbf{b} = \begin{bmatrix} 0 \\ 0 \\ 10 \end{bmatrix};\ \gamma = \begin{bmatrix} -0.1 \\ 0 \\ 0 \end{bmatrix}$$

$$\mathbf{x} = [\,x_1 \quad x_2 \quad x_3\,]^T$$
$$= [\,\Delta f \quad \Delta P_G \quad \Delta x_D\,]^T$$
$$u = \Delta P_C$$
$$w = \Delta P_D$$

It can easily be checked that the pair $\{\mathbf{A}, \mathbf{b}\}$ is controllable.

We shall design a controller for controlling the load frequency; $x_1 = \Delta f =$ deviation in the load frequency is required to be maintained as close as possible to zero. The best method of load frequency control is proportional plus integral control. We add to the system variables, the integral state $z(t)$ defined by

$$z(t) = x_4(t) = \int_0^t x_1(\tau)\, d\tau \tag{11.98}$$

The augmented system is described by the equations

$$\begin{bmatrix} \dot{x}_1 \\ \dot{x}_2 \\ \dot{x}_3 \\ \dot{x}_4 \end{bmatrix} = \begin{bmatrix} -0.05 & 0.1 & 0 & 0 \\ 0 & -0.3610 & 0.3610 & 0 \\ -200 & 0 & -10 & 0 \\ 1 & 0 & 0 & 0 \end{bmatrix} \begin{bmatrix} x_1 \\ x_2 \\ x_3 \\ x_4 \end{bmatrix} + \begin{bmatrix} 0 \\ 0 \\ 10 \\ 0 \end{bmatrix} u + \begin{bmatrix} -0.1 \\ 0 \\ 0 \\ 0 \end{bmatrix} w \tag{11.99a}$$

or

$$\dot{\hat{\mathbf{x}}} = \hat{\mathbf{A}}\hat{\mathbf{x}} + \hat{\mathbf{b}}u + \hat{\gamma}w \tag{11.99b}$$

It can easily be checked that the condition (11.83) is satisfied for the system under consideration. Therefore, the augmented system is controllable.

We select the following performance index, whose minimization is expected to yield the desired result.

$$J = \frac{1}{2}\int_0^\infty (\hat{\mathbf{x}}^T \mathbf{Q}\, \hat{\mathbf{x}} + ru^2)\, dt \tag{11.100}$$

where

$$\mathbf{Q} = \begin{bmatrix} 1 & 0 & 0 & 0 \\ 0 & 0 & 0 & 0 \\ 0 & 0 & 0 & 0 \\ 0 & 0 & 0 & 1 \end{bmatrix}; \quad r = 1$$

A matrix $\mathbf{\Gamma}^T$ such that $\mathbf{\Gamma}\mathbf{\Gamma}^T = \mathbf{Q}$ is given by

$$\mathbf{\Gamma}^T = \begin{bmatrix} 1 & 0 & 0 & 0 \\ 0 & 0 & 0 & 1 \end{bmatrix}$$

It can easily be verified that $\{\hat{\mathbf{A}}, \mathbf{\Gamma}^T\}$ pair is observable. Consequently, the optimal closed-loop system will be asymptotically stable.

Using the matrices $\hat{\mathbf{A}}$, $\hat{\mathbf{b}}$, $\mathbf{Q}$ and r defined above, the following equations were solved on digital computer:

$$\hat{\mathbf{A}}^T \mathbf{P}^0 + \mathbf{P}^0\hat{\mathbf{A}} - \mathbf{P}^0\, \hat{\mathbf{b}}r^{-1}\, \hat{\mathbf{b}}^T \mathbf{P}^0 + \mathbf{Q} = \mathbf{0}$$

$$\mathbf{K} = -r^{-1}\hat{\mathbf{b}}^T \mathbf{P}^0$$

The result obtained is given below.

$$\mathbf{K} = [-0.5703 \quad -0.1501 \quad -0.0054 \quad -0.9998]$$

The optimal control law is therefore

$$u = -0.5703x_1 - 0.1501x_2 - 0.0054x_3 - 0.9998 \int_0^t x_1(\tau)\, d\tau$$

Now, we convert the continuous-time linear regulator problem to an equivalent discrete-time linear regulator problem. This will be done by discretizing the plant model (11.99) and the performance index (11.100).

The plant model was discretized using the results of Section 5.6. The series evaluation method was used to compute the exponential function. The result obtained is given below for $T = 1$ sec.

$$\hat{\mathbf{x}}(k+1) = \begin{bmatrix} 0.7024 & 0.0744 & 0.0025 & 0 \\ -5.008 & 0.4711 & 0.0194 & 0 \\ -15.04 & -1.387 & -0.046 & 0 \\ 0.8953 & 0.0419 & 0.0013 & 1 \end{bmatrix} \hat{\mathbf{x}}(k) + \begin{bmatrix} 0.0126 \\ 0.2567 \\ 0.7931 \\ 0.0041 \end{bmatrix} u \tag{11.101a}$$

or

$$\hat{\mathbf{x}}(k+1) = \mathbf{F}\hat{\mathbf{x}}(k) + \mathbf{g}u(k) \tag{11.101b}$$

The performance index (11.100) was discretized using the method by VanLoan (1978) and the following result was obtained (refer eqn. (11.49)):

$$J = \tfrac{1}{2} \sum_{k=0}^{\infty} [\hat{\mathbf{x}}^T(k)\hat{\mathbf{Q}}\hat{\mathbf{x}}(k) + 2\hat{\mathbf{x}}^T(k)\mathbf{m}u(k) + \hat{r}u^2(k)]$$

$$\hat{\mathbf{Q}} = \begin{bmatrix} 1.101 & 0.0459 & 0.0014 & 0.4727 \\ 0.0459 & 0.0026 & 0.00008 & 0.0147 \\ 0.0014 & 0.00008 & 0 & 0.0004 \\ 0.4727 & 0.0147 & 0.0004 & 1 \end{bmatrix}$$

$$\hat{r} = 1$$

$$\mathbf{m} = \begin{bmatrix} 0.004 \\ 0.0003 \\ 0 \\ 0.001 \end{bmatrix}$$

Transformation of variables resulted in the following performance index (analogous to (11.50b)) and the plant (analogous to (11.50a)):

$$J = \frac{1}{2} \sum_{k=0}^{\infty} [\hat{\mathbf{x}}^T(k)\, \tilde{\mathbf{Q}} \hat{\mathbf{x}}(k) + \hat{r}\, \hat{u}^2(k)]$$

$$\tilde{\mathbf{Q}} \simeq \hat{\mathbf{Q}}$$

$$\hat{\mathbf{x}}(k+1) = \hat{\mathbf{F}}\hat{\mathbf{x}}(k) + \mathbf{g}\hat{u}(k)$$

$$\hat{\mathbf{F}} = \begin{bmatrix} 0.7024 & 0.0744 & 0.0025 & -0.1217\times 10^{-4} \\ -5.009 & 0.4710 & 0.01936 & -0.2471\times 10^{-3} \\ -15.05 & -1.388 & -0.046 & -0.7635\times 10^{-3} \\ 0.8953 & 0.0419 & 0.013 & 1 \end{bmatrix}$$

This problem was solved using the discrete-time linear regulator algorithm and the following result was obtained.

$$\begin{aligned} \mathbf{u}^*(k) &= [-0.5011 \quad -0.1457 \quad -0.0053 \quad -0.9735]\, \hat{\mathbf{x}}(k) \\ &= -0.5011 x_1(k) - 0.1457 x_2(k) - 0.0053 x_3(k) \\ &\qquad - 0.9735 \sum_{i=0}^{k} x_1(i)\, T \end{aligned} \tag{11.102}$$

With the sampled-data load frequency control scheme, it is possible to use a digital computer in the feedback loop so that the controller need not be updated when control errors lie within acceptable magnitude bounds. This avoids the unnecessary hunting of the controller and therefore, lessens the wear and tear of the control equipment. One such scheme is shown in the flow chart of Fig. 11.11. In this scheme, the control error (CE) is computed after every small interval of t_i seconds, the accumulation of this control error (ACE) during the sampling interval of T seconds is calculated. If ACE is less than a pre-defined threshold value K_t, no new control signal is given to the controller; otherwise the control law (11.102) is used to generate a new control signal for the controller.

Before concluding this example, let us compare the frequency errors for continuous-time LFC (load frequency control) and sampled-data LFC schemes. The Δf versus t curves, shown in Fig. 11.12, give us confidence in our selection of the sampling interval $T = 1$ sec. However, the comparison of Fig. 11.12 with Fig. 5.9 gives a little disappointing result. Reduction in the error of a controlled system compared to that of an uncontrolled system is not very significant. This simply indicates that we have not made a proper choice of **Q** and **R** matrices in the performance index (11.100). Placing more weight on the x_1-variable will obviously reduce the error in frequency. This will be an interesting and educating exercise for the reader to try suitable values of **Q** and **R** matrices so that frequency error is significantly reduced compared to the one shown in Fig. 11.12.

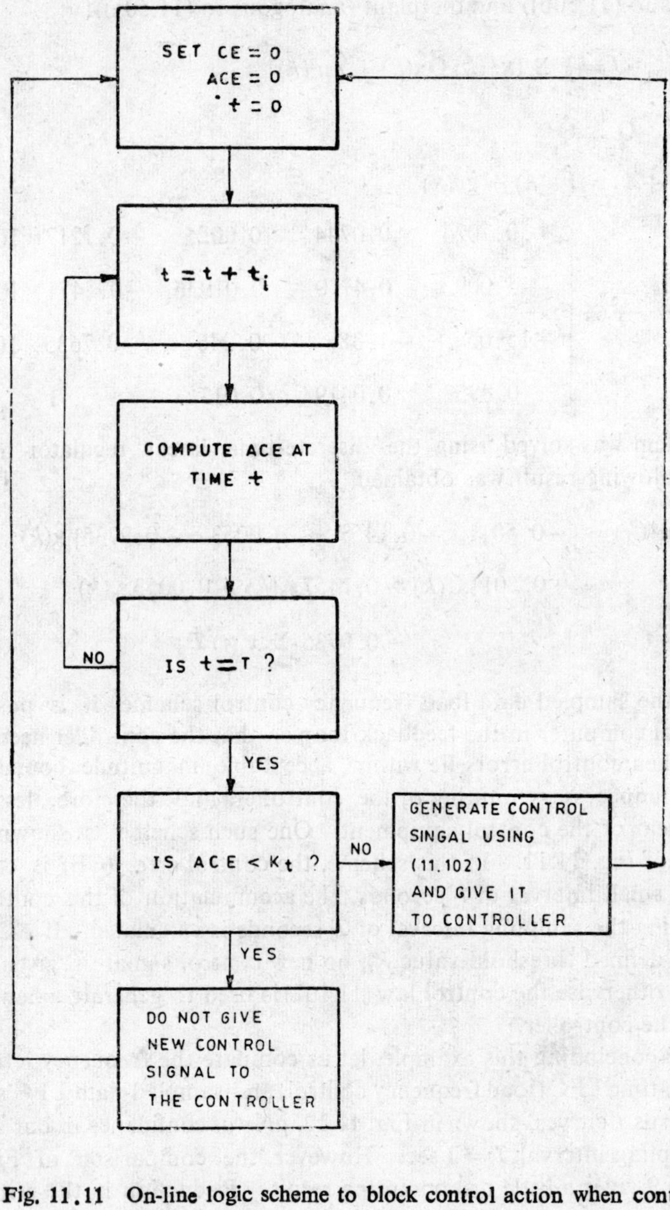

Fig. 11.11 On-line logic scheme to block control action when control errors lie in acceptable magnitude bound

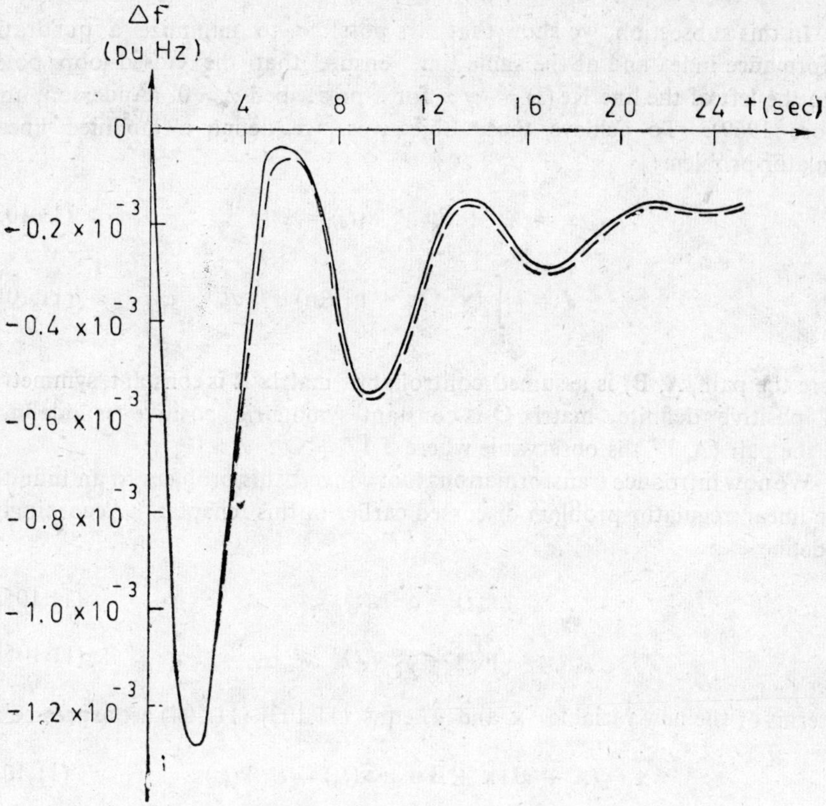

Fig. 11.12

Linear Regulator with a Prescribed Degree of Stability

Consider the linear time-invariant controllable system

$$\dot{\mathbf{x}} = \mathbf{A}\mathbf{x} + \mathbf{B}\mathbf{u}\ ; \quad \mathbf{x}(t_0) \triangleq \mathbf{x}^0$$

with linear control law of the form

$$\mathbf{u} = \mathbf{K}\mathbf{x}$$

The feedback gain matrix **K** may be selected to

(i) minimize the quadratic performance index of the form

$$J = \frac{1}{2}\int_{t_0}^{\infty} (\mathbf{x}^T\mathbf{Q}\mathbf{x} + \mathbf{u}^T\mathbf{R}\mathbf{u})\, dt$$

where **R** is positive definite constant symmetric matrix and **Q** is positive semidefinite constant symmetric matrix; the pair $(\mathbf{A}, \mathbf{\Gamma}^T)$ is observable where $\mathbf{\Gamma}\mathbf{\Gamma}^T = \mathbf{Q}$ (Section 11.4);

(ii) place the poles of the closed-loop system

$$\dot{\mathbf{x}} = (\mathbf{A} + \mathbf{B}\mathbf{K})\,\mathbf{x}$$

at certain desired locations (Section 9.4).

In this subsection, we show that it is possible to minimize a quadratic performance index and at the same time ensure that the closed-loop poles lie to the left of the line $\text{Re}(s) = -\alpha$ for a prescribed $\alpha > 0$ (Anderson and Moore 1969). To achieve these objectives, we define a modified linear regulator problem:

$$\dot{\mathbf{x}} = \mathbf{A}\mathbf{x} + \mathbf{B}\mathbf{u} \; ; \quad \mathbf{x}(t_0) = \mathbf{x}^0 \tag{11.103}$$

$$J = \frac{1}{2}\int_{t_0}^{\infty} (\mathbf{x}^T \mathbf{Q}\mathbf{x} + \mathbf{u}^T \mathbf{R}\mathbf{u})\, e^{2\alpha t}\, dt \tag{11.104}$$

where the pair $\{\mathbf{A}, \mathbf{B}\}$ is assumed controllable, matrix $\mathbf{R}$ is constant, symmetric and positive definite, matrix $\mathbf{Q}$ is constant symmetric positive semidefinite and the pair $\{\mathbf{A}, \mathbf{\Gamma}^T\}$ is observable where $\mathbf{\Gamma}\mathbf{\Gamma}^T = \mathbf{Q}$; $\alpha > 0$.

We now introduce transformations that convert this problem to an infinite-time linear regulator problem discussed earlier in this chapter. Accordingly, we define

$$\hat{\mathbf{x}}(t) = e^{\alpha t}\, \mathbf{x}(t) \tag{11.105a}$$

$$\hat{\mathbf{u}}(t) = e^{\alpha t}\, \mathbf{u}(t) \tag{11.105b}$$

In terms of the new variables $\hat{\mathbf{x}}$ and $\hat{\mathbf{u}}$, eqns. (11.103)–(11.104) are expressed as

$$\dot{\hat{\mathbf{x}}} = (\mathbf{A} + \alpha\mathbf{I})\,\hat{\mathbf{x}} + \mathbf{B}\,\hat{\mathbf{u}} \; ; \quad \hat{\mathbf{x}}(t_0) = e^{\alpha t_0}\mathbf{x}(t_0) \tag{11.106}$$

$$\hat{J} = \frac{1}{2}\int_{t_0}^{\infty} (\hat{\mathbf{x}}^T \mathbf{Q}\,\hat{\mathbf{x}} + \hat{\mathbf{u}}^T \mathbf{R}\hat{\mathbf{u}})\, dt \tag{11.107}$$

Equations (11.106)–(11.107) are in the form of infinite-time linear regulator problem. It can easily be established that given the controllable pair $\{\mathbf{A}, \mathbf{B}\}$, the pair $\{\mathbf{A} + \alpha\mathbf{I}, \mathbf{B}\}$ will also be controllable. Similarly if the pair $\{\mathbf{A}, \mathbf{\Gamma}^T\}$ is observable, the pair $\{\mathbf{A} + \alpha\mathbf{I}, \mathbf{\Gamma}^T\}$ will also be observable. Therefore, for the system (11.106) with performance index (11.107), there exists a feedback control law

$$\hat{\mathbf{u}} = \mathbf{K}_\alpha \hat{\mathbf{x}} \tag{11.108}$$

such that the resulting closed-loop system

$$\dot{\hat{\mathbf{x}}} = (\mathbf{A} + \mathbf{B}\mathbf{K}_\alpha + \alpha\mathbf{I})\,\hat{\mathbf{x}} \tag{11.109}$$

is asymptotically stable.

The feedback matrix

$$\mathbf{K}_\alpha = -\mathbf{R}^{-1}\mathbf{B}^T \mathbf{P}_\alpha \tag{11.110}$$

where $\mathbf{P}_\alpha$ is given by the solution of the algebraic Riccati equation

$$\mathbf{P}_\alpha(\mathbf{A}+\alpha\mathbf{I})+(\mathbf{A}^T+\alpha\mathbf{I})\,\mathbf{P}_\alpha-\mathbf{P}_\alpha\mathbf{B}\mathbf{R}^{-1}\mathbf{B}^T\mathbf{P}_\alpha+\mathbf{Q}=\mathbf{0} \qquad (11.111)$$

Using the transformations (11.105), we obtain from (11.108) and (11.110)

$$\begin{aligned}\mathbf{u}(t)&=\mathbf{K}_\alpha\mathbf{x}\\ &=-\mathbf{R}^{-1}\mathbf{B}^T\mathbf{P}_\alpha\mathbf{x}\end{aligned} \qquad (11.112)$$

Thus the control law has constant feedback gains; the resulting closed-loop system

$$\dot{\mathbf{x}}(t)=(\mathbf{A}+\mathbf{B}\mathbf{K}_\alpha)\,\mathbf{x}(t) \qquad (11.113)$$

will therefore be time-invariant.

Since the poles of the system (11.109), given by the eigenvalues of $(\mathbf{A}+\mathbf{B}\mathbf{K}_\alpha+\alpha\mathbf{I})$, have negative real parts, it follows that poles of the system (11.113), being given by the eigenvalues of $(\mathbf{A}+\mathbf{B}\mathbf{K}_\alpha)$ (which are less by α than the eigenvalues of $(\mathbf{A}+\mathbf{B}\mathbf{K}_\alpha+\alpha\mathbf{I})$), all possess real parts less than $-\alpha$.

11.7 SUBOPTIMAL LINEAR REGULATORS

So far we have studied the optimal regulator problem wherein (i) for a given plant, we find a control function $\mathbf{u}^*$ which is optimal with respect to the given performance criterion and then (ii) realize the control function $\mathbf{u}^*(t)=\mathbf{K}\mathbf{x}(t)$.

An optimal solution obtained through steps (i) and (ii) may not be the best solution in all circumstances. For example, all the elements of matrix $\mathbf{K}$ may not be free; some gains are fixed by physical constraints of the system and are therefore relatively inflexible. Similarly if all the states $\mathbf{x}(t)$ are not accessible for feedback, one has to go for a state observer whose complexity is comparable to that of the system itself. It is natural to seek a procedure that relies on the use of feedback from only the accessible state variables, constraining the gain elements of matrix $\mathbf{K}$ corresponding to the inaccessible state variables to have zero value. Thus, whether one chooses an optimal or suboptimal solution depends on many factors in addition to the performance required out of the system.

In this section, we present a method of obtaining the solution of a control problem when some elements of the feedback matrix $\mathbf{K}$ are constrained.

Continuous-Time Systems

The system considered here may be represented by

$$\dot{\mathbf{x}}=\mathbf{A}\mathbf{x}+\mathbf{B}\mathbf{u}\;;\quad \mathbf{x}(0)\overset{\Delta}{=}\mathbf{x}^0 \qquad (11.114)$$

where $\mathbf{x}=n\times 1$ state vector, $\mathbf{u}=p\times 1$ control vector, $\mathbf{A}=n\times n$ constant matrix and $\mathbf{B}=n\times p$ constant matrix. The system is assumed to be completely controllable.

The performance index is

$$J=\tfrac{1}{2}\int_0^\infty(\mathbf{x}^T\mathbf{Q}\mathbf{x}+\mathbf{u}^T\mathbf{R}\mathbf{u})\,dt \qquad (11.115)$$

where **R** is positive definite real symmetric constant matrix and **Q** is positive semidefinite real symmetric constant matrix. The pair $\{\mathbf{A}, \mathbf{\Gamma}^T\}$ is assumed to be observable, where $\mathbf{\Gamma}\mathbf{\Gamma}^T = \mathbf{Q}$.

We know that the optimal control law is a linear combination of the state variables, i.e.,

$$\mathbf{u}(t) = \mathbf{K}\mathbf{x}(t) \tag{11.116}$$

where **K** is $p \times n$ constant matrix.

With the linear feedback law of (11.116), the closed-loop system is described by

$$\dot{\mathbf{x}}(t) = (\mathbf{A} + \mathbf{B}\mathbf{K})\,\mathbf{x}(t) \tag{11.117}$$

Substituting for control vector **u** from eqn. (11.116) in the performance index J of eqn. (11.115), we have

$$J = \tfrac{1}{2}\int_0^\infty (\mathbf{x}^T\mathbf{Q}\mathbf{x} + \mathbf{x}^T\mathbf{K}^T\mathbf{R}\mathbf{K}\mathbf{x})\,dt$$

$$= \tfrac{1}{2}\int_0^\infty \mathbf{x}^T(\mathbf{Q} + \mathbf{K}^T\mathbf{R}\mathbf{K})\,\mathbf{x}\,dt \tag{11.118}$$

Let us assume a Lyapunov function

$$V(\mathbf{x}(t)) = \tfrac{1}{2}\int_t^\infty \mathbf{x}^T(\mathbf{Q} + \mathbf{K}^T\mathbf{R}\mathbf{K})\,\mathbf{x}\,dt$$

Thus the value of performance index for system trajectory starting at $\mathbf{x}(0)$ is $V(\mathbf{x}(0))$. The time derivative of the Lyapunov function is

$$\dot{V}(\mathbf{x}) = -\tfrac{1}{2}\mathbf{x}^T(\mathbf{Q} + \mathbf{K}^T\mathbf{R}\mathbf{K})\,\mathbf{x} \tag{11.119a}$$

(Note that $\mathbf{x}(\infty)$ has been taken as zero because the optimal control law (11.116) will result in an asymptotically stable closed-loop system.)

Since $\dot{V}(\mathbf{x})$ is a quadratic in **x** and the plant equation is linear, let us assume that $V(\mathbf{x})$ is also given by the quadratic form

$$V(\mathbf{x}) = \tfrac{1}{2}\mathbf{x}^T\mathbf{P}\mathbf{x} \tag{11.119b}$$

where **P** is a positive definite real symmetric constant matrix. Therefore,

$$\dot{V}(\mathbf{x}) = \tfrac{1}{2}(\dot{\mathbf{x}}^T\mathbf{P}\mathbf{x} + \mathbf{x}^T\mathbf{P}\dot{\mathbf{x}})$$

Substituting for $\dot{\mathbf{x}}$ from eqn. (11.114), we have

$$\dot{V}(\mathbf{x}) = \tfrac{1}{2}\mathbf{x}^T\,[(\mathbf{A} + \mathbf{B}\mathbf{K})^T\mathbf{P} + \mathbf{P}(\mathbf{A} + \mathbf{B}\mathbf{K})]\,\mathbf{x}$$

Comparing this result with eqn. (11.119a), we get

$$-\tfrac{1}{2}\mathbf{x}^T(\mathbf{Q} + \mathbf{K}^T\mathbf{R}\mathbf{K})\,\mathbf{x} = \tfrac{1}{2}\mathbf{x}^T\,[(\mathbf{A} + \mathbf{B}\mathbf{K})^T\mathbf{P} + \mathbf{P}(\mathbf{A} + \mathbf{B}\mathbf{K})]\,\mathbf{x}$$

Since the above equality holds for arbitrary $\mathbf{x}(t)$, we have

$$(\mathbf{A}+\mathbf{BK})^T\,\mathbf{P}+\mathbf{P}(\mathbf{A}+\mathbf{BK})+\mathbf{K}^T\,\mathbf{RK}+\mathbf{Q}=\mathbf{0} \tag{11.120}$$

This equation is of the form of the Lyapunov equation defined in Section 8.6. From this equation, we can determine elements of $\mathbf{P}$ as functions of the feedback matrix $\mathbf{K}$. As pointed out earlier, $V(\mathbf{x}(0))$ is the value of performance index starting at $\mathbf{x}(0)$. From eqn. (11.119b) we get

$$J=\tfrac{1}{2}\mathbf{x}^T(0)\,\mathbf{Px}(0) \tag{11.121}$$

A suboptimal control law may be obtained by minimizing J with respect to all the available elements k_{ij} of $\mathbf{K}$, i.e., by setting

$$\frac{\partial[\mathbf{x}^T(0)\,\mathbf{Px}(0)]}{\partial k_{ij}}=0 \tag{11.122}$$

Matrix $\mathbf{K}$ corresponding to suboptimal solution has to satisfy the further constraint that the closed-loop system be asymptotically stable.

It is thus observed that when the configuration of the controller is constrained, a solution which is independent of initial conditions can no longer be found. If a system is to operate satisfactorily for a range of initial disturbances, it may not be clear which is the most suitable for optimization. The dependence on initial conditions can be avoided by the method given by Levine and Athans (1970) (Also see Levine et al. 1971). Before we discuss this method, we have some comments to offer.

(i) If the feedback matrix $\mathbf{K}$ is unconstrained, then the control law is optimal and will be independent of the initial conditions. The optimal k_{ij}'s can be obtained from

$$\frac{\partial\mathbf{P}}{\partial k_{ij}}=\mathbf{0}\quad\text{for all } i,j \tag{11.123}$$

However, this procedure for obtaining the optimal unconstrained feedback matrix $\mathbf{K}$ gets very much involved for higher-order systems. An indirect route to the solution is through the reduced matrix Riccati equation discussed earlier.

(ii) In the special case where the performance index is independent of control $\mathbf{u}$, we have

$$J=\tfrac{1}{2}\int_0^\infty (\mathbf{x}^T\,\mathbf{Q}\,\mathbf{x})\,dt \tag{11.124}$$

In this case, the matrix $\mathbf{P}$ is obtained from eqn. (11.120) by putting $\mathbf{R}=\mathbf{0}$, resulting in the modified Lyapunov equation

$$(\mathbf{A}+\mathbf{BK})^T\mathbf{P}+\mathbf{P}(\mathbf{A}+\mathbf{BK})+\mathbf{Q}=\mathbf{0} \tag{11.125}$$

Even though $\mathbf{R}$ is originally assumed to be positive definite, substituting $\mathbf{R}=\mathbf{0}$ is a valid operation here as the positive definiteness of $\mathbf{R}$ has not been used in the derivation so far.

Let us now come back to the problem of avoiding dependence on initial conditions. To eliminate this dependence, a simple way is to average the performance obtained for a linearly independent set of initial conditions. This is equivalent to assuming the initial state $\mathbf{x}(0)$ to be a random variable (refer Section 12.2) uniformly distributed on the surface of the n-dimensional unit sphere. That is, we define a new performance index

$$\bar{J} = E\{J\} = E\{\tfrac{1}{2}\mathbf{x}^T(0)\,\mathbf{P}\mathbf{x}(0)\} \tag{11.126a}$$

with the random variable $\mathbf{x}(0)$ satisfying

$$E\{\mathbf{x}(0)\,\mathbf{x}^T(0)\} = \frac{1}{n}\mathbf{I} \tag{11.126b}$$

where $E\{\cdot\}$ denotes the expected value.

An alternative simpler expression for $\bar{J}$ can be determined directly from (11.126a)–(11.126b) as follows:

$$\begin{aligned}\bar{J} &= E\{\tfrac{1}{2}\mathbf{x}^T(0)\,\mathbf{P}\mathbf{x}(0)\}\\ &= \tfrac{1}{2}\mathbf{E}\{\text{trace}\,(\mathbf{P}\mathbf{x}(0)\,\mathbf{x}^T(0))\}\\ &= \tfrac{1}{2}\,\text{trace}\,(\mathbf{P}\mathbf{E}\,\{\mathbf{x}(0)\,\mathbf{x}^T(0)\})\\ &= \frac{1}{2n}\,\text{trace}\,\mathbf{P}\end{aligned}$$

Let us assume that the performance index to be minimized is

$$\hat{J} = \tfrac{1}{2}\,\text{trace}\,\mathbf{P} \tag{11.127}$$

This is equivalent to assuming that

$$E\{\mathbf{x}(0)\,\mathbf{x}^T(0)\} = \mathbf{I}, \tag{11.128}$$

i.e., random variable $\mathbf{x}(0)$ is uniformly distributed on the surface of an n-dimensional sphere of non-unity radius.

The reader is reminded here that the matrix $\mathbf{K}$ corresponding to sub-optimal solution has to satisfy the further constraint that the closed-loop system be asymptotically stable. The question of existence of $\mathbf{K}$ satisfying this constraint has, as yet, no straight-forward answer. We now turn to the problem of finding such a $\mathbf{K}$, assuming that one exists. Let us first take a simple example

Example 11.9: Consider the second-order system

$$\begin{bmatrix}\dot{x}_1\\ \dot{x}_2\end{bmatrix} = \begin{bmatrix}0 & 1\\ 0 & 0\end{bmatrix}\begin{bmatrix}x_1\\ x_2\end{bmatrix} + \begin{bmatrix}0\\ 1\end{bmatrix}u$$

It is desired to find optimal control

$$u = -[k_1 \;\; k_2]\begin{bmatrix}x_1\\ x_2\end{bmatrix}$$

which minimizes the performance index

$$J = \int_0^{\infty} x_1^2 \, dt$$

under the constraint that $k_1 = 1$.

Substituting various values in eqn. (11.125), we get

$$\begin{bmatrix} 0 & -1 \\ 1 & -k_2 \end{bmatrix}\begin{bmatrix} p_{11} & p_{12} \\ p_{12} & p_{22} \end{bmatrix} + \begin{bmatrix} p_{11} & p_{12} \\ p_{12} & p_{22} \end{bmatrix}\begin{bmatrix} 0 & 1 \\ -1 & -k_2 \end{bmatrix}$$

$$+ \begin{bmatrix} 2 & 0 \\ 0 & 0 \end{bmatrix} = \begin{bmatrix} 0 & 0 \\ 0 & 0 \end{bmatrix}$$

Solving we get

$$\mathbf{P} = \begin{bmatrix} \dfrac{1+k_2^2}{k_2} & 1 \\ 1 & \dfrac{1}{k_2} \end{bmatrix}$$

$$\text{Trace } \mathbf{P} = \frac{1}{k_2} + \frac{1+k_2^2}{k_2} = \frac{2+k_2^2}{k_2}$$

The optimum value of k_2 is obtained by setting

$$\frac{\partial \hat{J}}{\partial k_2} = \tfrac{1}{2}\,\frac{\partial\,(\text{trace } \mathbf{P})}{\partial k_2} = 0$$

This gives

$$k_2 = \sqrt{2}$$

Suboptimal control law is therefore

$$u = -[1 \quad \sqrt{2}]\begin{bmatrix} x_1 \\ x_2 \end{bmatrix}$$

(The resulting closed-loop system is asymptotically stable). The corresponding minimum value of performance index is (eqn. (11.127))

$$\hat{J} = \tfrac{1}{2} \text{ trace } \mathbf{P}$$
$$= 2\sqrt{2}$$

□

From this simple example it is obvious that for higher-order systems, mathematical programming techniques have to be employed for minimization of $\hat{J}$

given by (11.127) (The Fletcher-Powell method (refer Section 10.6) will be described in this section). The efficiency of these techniques can be improved if the partial derivatives $\partial \hat{J}/\partial k_{ij}$ are provided in the analytical form. For the case of incomplete state feedback (some of the entries of **K** forced to zero), the analytical expression for derivatives is derived below using perturbation analysis (Jameson 1970).

Let **H** be the gradient matrix with elements

$$h_{ij} = \frac{\partial \hat{J}}{\partial k_{ij}} \tag{11.129}$$

For the constrained feedback configuration to be optimal, it is necessary that h_{ij} vanish for each allowed feedback element k_{ij}. To determine **H**, it is convenient to introduce the outer product

$$\mathbf{X} = \mathbf{x}\,\mathbf{x}^T \tag{11.130}$$

The closed-loop system differential equation (11.117) in terms of **X** then becomes

$$\dot{\mathbf{X}} = \mathbf{D}\mathbf{X} + \mathbf{X}\mathbf{D}^T\ ; \quad \mathbf{X}(0) = \mathbf{x}^0\mathbf{x}^{0T} \tag{11.131}$$

where $\mathbf{D} = \mathbf{A} + \mathbf{B}\mathbf{K}$

Using (11.130), the performance index J in (11.118) becomes

$$J = \tfrac{1}{2}\int_0^\infty \mathbf{x}^T(\mathbf{Q} + \mathbf{K}^T\mathbf{R}\mathbf{K})\,\mathbf{x}\,dt$$

$$= \tfrac{1}{2}\int_0^\infty \mathbf{x}^T\mathbf{S}\mathbf{x}\,dt$$

$$= \tfrac{1}{2}\int_0^\infty \text{trace}\,(\mathbf{S}\mathbf{X})\,dt \tag{11.132}$$

where $\mathbf{S} = \mathbf{Q} + \mathbf{K}^T\mathbf{R}\mathbf{K}$

The perturbation of J gives

$$\delta J = \tfrac{1}{2}\int_0^\infty [\text{tr}\,(\mathbf{S}\,\delta\mathbf{X}) + \text{tr}\,(\delta\mathbf{S}\,\mathbf{X})]\,dt$$

$$= \tfrac{1}{2}\int_0^\infty [\text{tr}\,(\mathbf{S}\,\delta\mathbf{X}) + 2\,\text{tr}\,(\delta\mathbf{K}^T\mathbf{R}\mathbf{K}\mathbf{X})]\,dt \tag{11.133}$$

The perturbation of (11.131) gives

$$\delta\dot{\mathbf{X}} = \mathbf{D}\,\delta\mathbf{X} + \delta\mathbf{X}\,\mathbf{D}^T + \mathbf{B}\,\delta\mathbf{K}\,\mathbf{X} + \mathbf{X}\,\delta\mathbf{K}^T\,\mathbf{B}^T\,; \quad \delta\mathbf{X}(0) = \mathbf{0} \qquad (11.134)$$

Now multiplying both sides of (11.134) by $\mathbf{P}$ we get

$$\mathbf{P}\delta\dot{\mathbf{X}} = \mathbf{P}\mathbf{D}\delta\mathbf{X} + \mathbf{P}\delta\mathbf{X}\mathbf{D}^T + \mathbf{P}\mathbf{B}\delta\mathbf{K}\mathbf{X} + \mathbf{P}\mathbf{X}\delta\mathbf{K}^T\mathbf{B}^T$$

or

$$\mathrm{tr}\,(\mathbf{P}\delta\dot{\mathbf{X}}) = \mathrm{tr}\,(\mathbf{P}\mathbf{D}\delta\mathbf{X}) + \mathrm{tr}\,(\mathbf{P}\delta\mathbf{X}\mathbf{D}^T) + \mathrm{tr}\,(\mathbf{P}\mathbf{B}\delta\mathbf{K}\mathbf{X}) + \mathrm{tr}\,(\mathbf{P}\mathbf{X}\delta\mathbf{K}^T\mathbf{B}^T) \quad (11.135)$$

From eqn. (11.120), we have

$$\mathbf{D}^T\mathbf{P} + \mathbf{P}\mathbf{D} + \mathbf{S} = \mathbf{0}$$

or

$$\mathbf{D}^T\mathbf{P}\delta\mathbf{X} + \mathbf{P}\mathbf{D}\delta\mathbf{X} + \mathbf{S}\delta\mathbf{X} = \mathbf{0}$$

or

$$\mathrm{tr}\,(\mathbf{D}^T\mathbf{P}\delta\mathbf{X}) + \mathrm{tr}\,(\mathbf{P}\mathbf{D}\delta\mathbf{X}) = -\,\mathrm{tr}\,(\mathbf{S}\,\delta\mathbf{X})$$

Substituting this result in (11.135), we get

$$\mathrm{tr}\,(\mathbf{P}\delta\dot{\mathbf{X}}) = -\,\mathrm{tr}\,(\mathbf{D}^T\mathbf{P}\delta\mathbf{X}) + \mathrm{tr}\,(\mathbf{P}\delta\mathbf{X}\mathbf{D}^T) + \mathrm{tr}\,(\mathbf{P}\mathbf{B}\delta\mathbf{K}\mathbf{X})$$
$$+\,\mathrm{tr}\,(\mathbf{P}\mathbf{X}\delta\mathbf{K}^T\mathbf{B}^T) - \mathrm{tr}\,(\mathbf{S}\delta\mathbf{X})$$

Using the trace identities given in Appendix III, we obtain the following result.

$$\mathrm{tr}\,(\mathbf{P}\delta\dot{\mathbf{X}}) = 2\,\mathrm{tr}\,(\delta\mathbf{K}^T\mathbf{B}^T\mathbf{P}\mathbf{X}) - \mathrm{tr}\,(\mathbf{S}\delta\mathbf{X})$$
$$= \frac{d}{dt}\,\mathrm{tr}\,(\mathbf{P}\delta\mathbf{X})$$

Since $\mathbf{P}\delta\mathbf{X}$ vanishes at both the boundaries, it follows that

$$\int_0^\infty \mathrm{tr}\,(\mathbf{S}\delta\mathbf{X})\,dt = 2\int_0^\infty \mathrm{tr}\,(\delta\mathbf{K}^T\mathbf{B}^T\mathbf{P}\mathbf{X})\,dt$$

Substituting this result is (11.133), we get

$$\delta J = \int_0^\infty \mathrm{tr}\,[\delta\mathbf{K}^T(\mathbf{B}^T\mathbf{P} + \mathbf{R}\mathbf{K})\,\mathbf{X}]\,dt$$

Hence

$$\mathbf{H} = \frac{\partial J}{\partial \mathbf{K}} = \int_0^\infty (\mathbf{B}^T\mathbf{P} + \mathbf{R}\mathbf{K})\,\mathbf{X}\,dt$$

(Refer Athans (1968) for trace identities)

or

$$\mathbf{H} = (\mathbf{B}^T\mathbf{P} + \mathbf{R}\mathbf{K})\int_0^\infty \mathbf{X}\,dt$$

$$= (\mathbf{B}^T\mathbf{P} + \mathbf{R}\mathbf{K})\,\mathbf{W}$$

where

$$\mathbf{W} = \int_0^\infty \mathbf{X}\, dt$$

Integrating (11.131), it is found that

$$\int_0^\infty \dot{\mathbf{X}}\, dt = \mathbf{D}\int_0^\infty \mathbf{X}\, dt + \left(\int_0^\infty \mathbf{X}\, dt\right)\mathbf{D}^T$$

or

$$-\mathbf{X}(0) = \mathbf{D}\mathbf{W} + \mathbf{W}\mathbf{D}^T$$

or

$$(\mathbf{A} + \mathbf{B}\mathbf{K})\mathbf{W} + \mathbf{W}(\mathbf{A} + \mathbf{B}\mathbf{K})^T + \mathbf{x}^0\mathbf{x}^{0T} = 0$$

If the performance index to be minimized is $\hat{J} = \frac{1}{2}\,\text{tr}\,[\mathbf{P}]$, then we have

$$\mathbf{H} = (\mathbf{B}^T\mathbf{P} + \mathbf{R}\mathbf{K})\mathbf{W} \qquad (11.136a)$$

where

$$(\mathbf{A} + \mathbf{B}\mathbf{K})\mathbf{W} + \mathbf{W}(\mathbf{A} + \mathbf{B}\mathbf{K})^T + \mathbf{I} = \mathbf{0} \qquad (11.136b)$$

The gradient of $\hat{J}$ with respect to unknown elements in **K**, given by eqn. (11.136a), is set equal to zero to get the necessary conditions for a minimum of $\hat{J}$. These necessary conditions of optimality cannot directly give the unknown elements in **K** as it involves expressing **W** and **P** in terms of unknown elements of **K** from eqns. (11.136b) and (11.125) respectively. Moreover, the resulting equations are nonlinear.

However, since explicit expressions are available for $\hat{J}$ and $\frac{\partial \hat{J}}{\partial \mathbf{K}}$, a gradient technique such as the Fletcher-Powell algorithm may be employed to determine **K** by direct minimization of $\hat{J}$. Note that for a fixed **K**, the computation of $\hat{J}$ and $\frac{\partial \hat{J}}{\partial \mathbf{K}}$ requires the solution of only linear equations (11.125) and (11.136). Before we give the solution algorithm, some remarks are in order:

(i) If there are no constraints on **K**, then the **K** which minimizes the index (11.127) associated with the suboptimal control problem is the same resulting from the standard regulator problem with index (11.121). Thus, by using the average performance criterion (11.127), one

- (a) retains many of the properties of linear systems which are optimal with respect to standard quadratic criterion;
- (b) obtains a design which is optimal in an average sense.

(ii) The conditions $\frac{\partial \hat{J}}{\partial \mathbf{K}} = \mathbf{0}$ provide only necessary conditions.

(iii) There exists a **K** which minimizes $\hat{J}$ if and only if there exists a $\mathbf{K}_0$ such that $(\mathbf{A} + \mathbf{B}\mathbf{K}_0)$ is asymptotically stable (Horisberger and Belanger 1974).

From this remark, it is obvious that a primary value of $\mathbf{K}$ that makes $(\mathbf{A}+\mathbf{BK})$ stable is required. If the open-loop matrix $\mathbf{A}$ is unstable, then the procedure of Fortmann (1973) may be employed to find such a value if one exists. All values of $\mathbf{K}$ tried during the subsequent unidirectional searches must also stabilize the system if the initial value is a stabilizing one.

We now give the solution algorithm (refer Section 10.6 and Horisberger and Belanger (1974)).

Let $k_1, k_2, \ldots, k_m$ be the unknown elements of matrix $\mathbf{K}$. Put them in the form of a column vector

$$\mathbf{z} = \text{col}\,\mathbf{K};$$

$$\mathbf{z}^T = [k_1 \quad k_2 \ldots k_m]^T$$

Then

$$\frac{\partial \hat{J}}{\partial \mathbf{z}} = \text{col}\,\frac{\partial \hat{J}}{\partial \mathbf{K}}$$

Algorithm

Step 0: Find $\mathbf{K}_0$ such that $(\mathbf{A}+\mathbf{BK}_0)$ is stable

Set $i = k = 0$

Step 1: Find $\hat{J}(\mathbf{K}_i)$ from (11.125) and (11.127).

Step 2: Compute

$$\mathbf{h}_i = \text{col}\,\frac{\partial \hat{\mathbf{J}}}{\partial \mathbf{K}}\bigg|_{\mathbf{K}_i} \quad \text{from (11.136a) and (11.136b)}$$

If $\|\mathbf{h}_i\|$ is sufficiently small, stop; otherwise go on.

Step 3: If $k \neq 0$, define

$$\mathbf{M}_i = \mathbf{M}_{i-1} + \frac{(\mathbf{z}_i - \mathbf{z}_{i-1})(\mathbf{z}_i - \mathbf{z}_{i-1})^T}{(\mathbf{z}_i - \mathbf{z}_{i-1})^T(\mathbf{h}_i - \mathbf{h}_{i-1})} - \frac{\mathbf{M}_{i-1}(\mathbf{h}_i - \mathbf{h}_{i-1})(\mathbf{h}_i - \mathbf{h}_{i-1})^T\mathbf{M}_{i-1}^T}{(\mathbf{h}_i - \mathbf{h}_{i-1})^T\mathbf{M}_{i-1}(\mathbf{h}_i - \mathbf{h}_{i-1})}$$

or else set $\mathbf{M}_i = \mathbf{I}$.

Determine

$$\mathbf{s}_i = -\mathbf{M}_i\mathbf{h}_i$$

Step 4: Perform a one-dimensional minimization,

$$\hat{J}(\mathbf{z}_i + \alpha_i\mathbf{s}_i) = \min_{\alpha \geqslant 0} \hat{J}(\mathbf{z}_i + \alpha\mathbf{s}_i)$$

Let $\quad \mathbf{z}_{i+1} \to \mathbf{z}_i + \alpha_i\mathbf{s}_i\,; \quad i \to i+1$

If $k = 2\dim(\mathbf{z})$, set $k = 0$ or else $k \to k+1$

Return to Step 2.

Discrete-Time Systems

The system considered here may be represented by

$$\mathbf{x}(k+1) = \mathbf{F}\mathbf{x}(k) + \mathbf{G}\mathbf{u}(k)\ ; \quad \mathbf{x}(0) = \mathbf{x}^0 \tag{11.137}$$

where $\mathbf{x} = n \times 1$ vector, $\mathbf{u} = p \times 1$ vector, $\mathbf{F} = n \times n$ constant matrix and $\mathbf{G} = n \times p$ constant matrix. The system is assumed to be completely controllable.

The performance index is

$$J = \tfrac{1}{2} \sum_{k=0}^{\infty} [\mathbf{x}^T(k)\, \mathbf{Q}\mathbf{x}(k) + \mathbf{u}^T(k)\, \mathbf{R}\mathbf{u}(k)] \tag{11.138}$$

where $\mathbf{R}$ is a positive definite real symmetric constant matrix and $\mathbf{Q}$ is a positive semidefinite constant real symmetric matrix. The pair $\{\mathbf{F}, \boldsymbol{\Gamma}^T\}$ is assumed to be observable, where $\mathbf{F}\,\boldsymbol{\Gamma}^T = \mathbf{Q}$.

We know that the optimal control law is a linear combination of the state variables, i.e.,

$$\mathbf{u}(k) = \mathbf{K}\mathbf{x}(\mathrm{k}) \tag{11.139}$$

where $\mathbf{K}$ is a $p \times n$ constant matrix.

With the linear feedback law of (11.139), the closed-loop system is described by

$$\mathbf{x}(k+1) = (\mathbf{F} + \mathbf{G}\mathbf{K})\, \mathbf{x}(k) \tag{11.140}$$

Substituting for the control vector $\mathbf{u}(k)$ from eqn. (11.139) in the performance index J of eqn. (11.138), we have

$$J = \tfrac{1}{2} \sum_{k=0}^{\infty} \mathbf{x}^T(k)\, (\mathbf{Q} + \mathbf{K}^T\mathbf{R}\mathbf{K})\, \mathbf{x}(k) \tag{11.141}$$

Let us assume a Lyapunov function

$$V(\mathbf{x}(k)) = \tfrac{1}{2} \sum_{i=k}^{\infty} \mathbf{x}^T(i)\, (\mathbf{Q} + \mathbf{K}^T\mathbf{R}\mathbf{K})\, \mathbf{x}(i)$$

The value of the performance index for the system trajectory starting at $\mathbf{x}(0)$ is $V(\mathbf{x}(0))$. The difference

$$\begin{aligned} \Delta V(\mathbf{x}(k)) &= V(\mathbf{x}(k+1)) - V(\mathbf{x}(k)) \\ &= -\tfrac{1}{2}\mathbf{x}^T(k)\, (\mathbf{Q} + \mathbf{K}^T\mathbf{R}\mathbf{K})\, \mathbf{x}(k) \end{aligned} \tag{11.142}$$

Since $\Delta V(\mathbf{x}(k))$ is quadratic in $\mathbf{x}(k)$ and the plant equation is linear, let us assume that $V(\mathbf{x}(k))$ is also given by the quadratic form

$$V(\mathbf{x}(k)) = \tfrac{1}{2}\mathbf{x}^T(k)\, \mathbf{P}\mathbf{x}(k)$$

where $\mathbf{P}$ is a positive definite real symmetric constant matrix. Therefore

$$V(\mathbf{x}(k+1)) - V(\mathbf{x}(k)) = \Delta V(\mathbf{x}(k))$$

$$= \tfrac{1}{2}\,[(\mathbf{F}+\mathbf{GK}]^T\,\mathbf{P}(\mathbf{F}+\mathbf{GK})-\mathbf{P}]$$

Comparing this result with (11.142), we get

$$(\mathbf{F}+\mathbf{GK})^T\,\mathbf{P}(\mathbf{F}+\mathbf{GK})-\mathbf{P}+\mathbf{K}^T\mathbf{RK}+\mathbf{Q}=\mathbf{0} \tag{11.143}$$

This equation is of the form of Lyapunov equation defined in Section 8.9. Since $V(\mathbf{x}(0))$ is the value of the performance index, we have

$$J = \tfrac{1}{2}\mathbf{x}^T(0)\,\mathbf{P}\mathbf{x}(0)$$

As done in the earlier subsection, we shall consider the performance index

$$\hat{J} = \tfrac{1}{2}\,\mathrm{tr}\,[\mathbf{P}] \tag{11.144}$$

If the feedback matrix $\mathbf{K}$ is unconstrained, we shall better solve the problem by recursive relations developed in Section 11.4 rather than minimizing $\hat{J}$ of (11.144) by the gradient technique. In the following, we discuss the direct minimization procedure for the case wherein the feedback matrix $\mathbf{K}$ is constrained.

As done in the continuous-time case, we shall first derive an analytical expression for the gradient $\partial\hat{J}/\partial k_{ij}$ using perturbation analysis.

We introduce the outer product

$$\mathbf{X} = \mathbf{x}\mathbf{x}^T \tag{11.145}$$

The closed-loop system difference equation (11.140) in terms of $\mathbf{X}$ then becomes

$$\mathbf{X}(k+1) = \mathbf{D}\mathbf{X}(k)\,\mathbf{D}^T;\quad \mathbf{X}(0)=\mathbf{x}^0\mathbf{x}^{0T} \tag{11.146}$$

where

$$\mathbf{D} = \mathbf{F}+\mathbf{GK}$$

Using (11.145), the performance index J in (11.141) becomes

$$J = \tfrac{1}{2}\sum_{k=0}^{\infty} \mathbf{x}^T(k)\,\mathbf{S}\mathbf{x}(k)$$

$$= \tfrac{1}{2}\sum_{k=0}^{\infty} \mathrm{tr}\,(\mathbf{S}\mathbf{X}(k)) \tag{11.147}$$

where

$$\mathbf{S} = \mathbf{Q}+\mathbf{K}^T\mathbf{RK}$$

The perturbation of J gives

$$\delta J = \tfrac{1}{2}\sum_{k=0}^{\infty} [\mathrm{tr}\,(\mathbf{S}\,\delta\mathbf{X}(k))+\mathrm{tr}\,(\delta\mathbf{S}\mathbf{X}(k))]$$

$$= \tfrac{1}{2}\sum_{k=0}^{\infty} [\mathrm{tr}\,(\mathbf{S}\,\delta\mathbf{X}(k))+2\,\mathrm{tr}\,(\delta\mathbf{K}^T\mathbf{RKX}(k))] \tag{11.148}$$

The perturbation of (11.146) gives

$$\begin{aligned}\delta\mathbf{X}(k+1) &= \mathbf{D}\,\delta\mathbf{X}(k)\,\mathbf{D}^T + \delta\mathbf{D}\mathbf{X}(k)\,\mathbf{D}^T + \mathbf{D}\mathbf{X}(k)\,\delta\mathbf{D}^T \\ &= \mathbf{D}\,\delta\mathbf{X}(k)\,\mathbf{D}^T + \mathbf{G}\,\delta\mathbf{K}\mathbf{X}(k)\,\mathbf{D}^T + \mathbf{D}\mathbf{X}(k)\,\delta\mathbf{K}^T\mathbf{G}^T\end{aligned}$$

Multiplying both sides by $\mathbf{P}$ we get

$$\mathbf{P}\,\delta\mathbf{X}(k+1) = \mathbf{P}\mathbf{D}\,\delta\mathbf{X}(k)\mathbf{D}^T + \mathbf{P}\mathbf{G}\,\delta\mathbf{K}\mathbf{X}(k)\mathbf{D}^T + \mathbf{P}\mathbf{D}\mathbf{X}(k)\,\delta\mathbf{K}^T\mathbf{G}^T$$

or

$$\begin{aligned}\operatorname{tr}(\mathbf{P}\,\delta\mathbf{X}(k+1)) = \operatorname{tr}(\mathbf{P}\mathbf{D}\,\delta\mathbf{X}(k)\mathbf{D}^T) &+ \operatorname{tr}(\mathbf{P}\mathbf{G}\,\delta\mathbf{K}\mathbf{X}(k)\mathbf{D}^T) \\ &+ \operatorname{tr}(\mathbf{P}\mathbf{D}\mathbf{X}(k)\,\delta\mathbf{K}^T\mathbf{G}^T)\end{aligned} \tag{11.149}$$

From equation (11.143), we have

$$\mathbf{D}^T\mathbf{P}\mathbf{D}\,\delta\mathbf{X}(k) - \mathbf{P}\,\delta\mathbf{X}(k) + \mathbf{S}\,\delta\mathbf{X}(k) = \mathbf{0}$$

or

$$\operatorname{tr}(\mathbf{D}^T\mathbf{P}\mathbf{D}\,\delta\mathbf{X}(k)) = \operatorname{tr}(\mathbf{P}\,\delta\mathbf{X}(k)) - \operatorname{tr}(\mathbf{S}\,\delta\mathbf{X}(k))$$

Substituting this result in (11.149), we get

$$\begin{aligned}\operatorname{tr}(\mathbf{P}\,\delta\mathbf{X}(k+1)) = \operatorname{tr}(\mathbf{P}\,\delta\mathbf{X}(k)) - \operatorname{tr}(\mathbf{S}\,\delta\mathbf{X}(k)) &+ \operatorname{tr}(\mathbf{P}\mathbf{G}\,\delta\mathbf{K}\mathbf{X}(k)\mathbf{D}^T) \\ &+ \operatorname{tr}(\mathbf{P}\mathbf{D}\mathbf{X}(k)\,\delta\mathbf{K}^T\mathbf{G}^T)\end{aligned}$$

Using the trace identities given in Appendix III, we obtain the following result:

$$\operatorname{tr}(\mathbf{P}\,\delta\mathbf{X}(k+1)) - \operatorname{tr}(\mathbf{P}\,\delta\mathbf{X}(k)) = 2\operatorname{tr}(\delta\mathbf{K}^T\mathbf{G}^T\mathbf{P}\mathbf{D}\mathbf{X}(k)) - \operatorname{tr}(\mathbf{S}\,\delta\mathbf{X}(k))$$

Since $\delta\mathbf{X}(0) = \delta\mathbf{X}(\infty) = \mathbf{0}$, it follows that

$$\sum_{k=0}^{\infty} \operatorname{tr}(\mathbf{S}\,\delta\mathbf{X}(k)) = 2\sum_{k=0}^{\infty} \operatorname{tr}(\delta\mathbf{K}^T\mathbf{G}^T\mathbf{P}\mathbf{D}\mathbf{X}(k))$$

Substituting this result in (11.148), we get

$$\delta J = \sum_{k=0}^{\infty} [\operatorname{tr}(\delta\mathbf{K}^T\mathbf{G}^T\mathbf{P}\mathbf{D}\mathbf{X}(k)) + \operatorname{tr}(\delta\mathbf{K}^T\mathbf{R}\mathbf{K}\mathbf{X}(k))]$$

Hence

$$\begin{aligned}\mathbf{H} = \frac{\partial J}{\partial \mathbf{K}} &= [\mathbf{G}^T\mathbf{P}\mathbf{D} + \mathbf{R}\mathbf{K}]\sum_{k=0}^{\infty} \mathbf{X}(k) \\ &= (\mathbf{G}^T\mathbf{P}\mathbf{D} + \mathbf{R}\mathbf{K})\mathbf{W}\end{aligned}$$

where

$$\mathbf{W} = \sum_{k=0}^{\infty} \mathbf{X}(k)$$

From (11.146), we may write

$$\sum_{k=0}^{\infty} \mathbf{X}(k+1) = \sum_{k=0}^{\infty} \mathbf{D}\mathbf{X}(k)\,\mathbf{D}^T$$

or

$$-\mathbf{X}(0) + \mathbf{W} = \mathbf{D}\mathbf{W}\mathbf{D}^T$$

performance index to be minimized is $\hat{J} = \frac{1}{2}\operatorname{tr}[\mathbf{P}]$, then we have

$$\mathbf{H} = [\mathbf{G}^T\mathbf{P}(\mathbf{F} + \mathbf{GK}) + \mathbf{RK}]\,\mathbf{W} \tag{11.150a}$$

where

$$(\mathbf{F} + \mathbf{GK})\mathbf{W}\,(\mathbf{F} + \mathbf{GK})^T + \mathbf{I} = \mathbf{W} \tag{11.150b}$$

The solution algorithm for minimization of $\hat{J}$ is identical with that of the continuous-time case discussed in the earlier subsection.

11.8 LINEAR REGULATORS WITH LOW SENSITIVITY

Control system analysis and synthesis begin with modelling of the system under consideration. One of the requirements of modelling is that the model be simple enough to be analytically tractable and complex enough to reflect all the essential features of the system behaviour. Models may be derived by the application of physical laws with idealizations or may be obtained by identification experiment of measurement of the inputs and outputs of the system. Such model building invariably leads to a model whose parameters are known only with a certain degree of accuracy. These slight inaccuracies in modelling could lead to results with no resemblance to the behaviour of the actual system. The parameters of the physical system itself may not remain constant over the period of operation. They are subject to variations caused by several factors such as ageing, wear and tear of components, changes in environmental and operating conditions, etc.

To accomplish successful system study, it is necessary for us to consider the deviation of a system from its nominal behaviour caused by the deviation of its parameters from their normal performance characteristics. This is the essence of sensitivity analysis which we shall study in this section. We shall deal with the case of continuous-time systems. The results for discrete-time systems are analogous and can easily be obtained.

Three types of sensitivity have been studied by researchers in recent years. These are eigenvalue sensitivity, performance index sensitivity and trajectory sensitivity. The eigenvalue sensitivity can be easily computed; however, it is a less direct measure of system performance. A drawback in using performance index sensitivity is that important variations in system behaviour caused by plant parameter changes may not be reflected in the performance index (Rynaski 1968). Trajectory sensitivity is a very important measure of sensitivity. In this section, only the concept of trajectory sensitivity is considered (Frank 1978).

Trajectory Sensitivity

Let the parameters of a system be represented by a vector $\boldsymbol{\alpha}=[\alpha_1\ \ \alpha_2 \ldots \alpha_r]^T$. The mathematical model of an nth-order system relating the parameter vector $\boldsymbol{\alpha}$ to the state of the system is, say, of the form

$$\dot{\mathbf{x}} = \mathbf{f}(\mathbf{x}, \boldsymbol{\alpha}, t, \mathbf{u})\ ;\quad \mathbf{x}(t_0) \triangleq \mathbf{x}^0 \tag{11.151}$$

The nominal parameter vector of the model (11.151) will be denoted by $\boldsymbol{\alpha}_0$ in the sequel, whereas the parameter vector of the actual system is $\boldsymbol{\alpha}=\boldsymbol{\alpha}_0+\Delta\boldsymbol{\alpha}$.

We shall assume that parameter variations $\Delta\boldsymbol{\alpha}$ from the nominal value do not affect the order of the system.

Assume that (11.151) has a solution $\mathbf{x} = \mathbf{x}(t, \boldsymbol{\alpha})$. $\mathbf{x}$ is, of course, a function of $\mathbf{u}$, $\mathbf{x}(t_0)$ and t_0 as well; however, this dependence is not needed for the following considerations and will therefore be dropped for ease of notation. If the parameter takes on its nominal value $\boldsymbol{\alpha}_0$, the nominal solution $\mathbf{x}(t, \boldsymbol{\alpha}_0)$ is obtained. Therefore the parameter induced change of state vector is

$$\Delta\mathbf{x}(t, \boldsymbol{\alpha}) \stackrel{\Delta}{=} \mathbf{x}(t, \boldsymbol{\alpha}) - \mathbf{x}(t, \boldsymbol{\alpha}_0) \tag{11.152}$$

A first-order approximation of $\Delta\mathbf{x}$ can be written by the use of a Taylor series expansion in the form

$$\delta\mathbf{x}(t, \boldsymbol{\alpha}) \simeq \sum_{j=1}^{r} \left.\frac{\partial \mathbf{x}}{\partial \alpha_j}\right|_{\boldsymbol{\alpha}_0} \delta\alpha_j \tag{11.153}$$

This equation can be viewed as a definition of the parameter-induced trajectory deviation. The partial derivative

$$\boldsymbol{\sigma}_j(t, \boldsymbol{\alpha}_0) = \left.\frac{\partial \mathbf{x}(t, \boldsymbol{\alpha})}{\partial \alpha_j}\right|_{\boldsymbol{\alpha}_0} ; \; j = 1, 2, \ldots, r \tag{11.154}$$

is called the *trajectory sensitivity vector* with respect to jth parameter. Note that $\boldsymbol{\sigma}_j$ is $n \times 1$ vector. A norm of $\delta\mathbf{x}$ or $\boldsymbol{\sigma}_j$ may serve as a measure of sensitivity. Although we may be interested in obtaining sensitivity to large parameter variations (global sensitivity), differential sensitivity defined by (11.154) is of fundamental importance. As we shall see, equations characterizing differential sensitivity are linear; this often enables one to obtain analytical results not possible for finite parameter variations. Differential sensitivity is qualitatively and often approximately quantitatively indicative of global sensitivity.

Consider now a general continuous-time system described by (11.151). Taking partial derivatives of $\dot{\mathbf{x}}$ with respect to α_j, we obtain

$$\frac{\partial \dot{\mathbf{x}}}{\partial \alpha_j} = \frac{\partial \mathbf{f}}{\partial \mathbf{x}} \frac{\partial \mathbf{x}}{\partial \alpha_j} + \frac{\partial \mathbf{f}}{\partial \alpha_j} ; \frac{\partial \mathbf{x}^0}{\partial \alpha_j} = \mathbf{0} ; \; j = 1, 2, \ldots, r \tag{11.155}$$

If we now interchange the sequence of taking the derivative with respect to time t and α_j and then let $\boldsymbol{\alpha}$ approach $\boldsymbol{\alpha}_0$ we obtain

$$\dot{\boldsymbol{\sigma}}_j = \left.\frac{\partial \mathbf{f}}{\partial \mathbf{x}}\right|_{\boldsymbol{\alpha}_0} \boldsymbol{\sigma}_j + \left.\frac{\partial \mathbf{f}}{\partial \alpha_j}\right|_{\boldsymbol{\alpha}_0} ; \; \boldsymbol{\sigma}_j(0) = \mathbf{0} ; \; j = 1, 2, \ldots, r \tag{11.156}$$

This equation is called *trajectory sensitivity equation*. Note that the $n \times n$ matrix $\dfrac{\partial \mathbf{f}}{\partial \mathbf{x}}$ is the Jacobian matrix.

Solving (11.156), we obtain,

$$\boldsymbol{\sigma}(t, \alpha_0) = [\boldsymbol{\sigma}_1 \;\; \boldsymbol{\sigma}_2 \;\; \cdots \;\; \boldsymbol{\sigma}_r], \tag{11.157a}$$

the *trajectory sensitivity matrix.* The parameter-induced change of trajectory, for small $\Delta\alpha_j$'s is then (refer eqns. (11.153)-(11.154))

$$\Delta\mathbf{x}(t, \boldsymbol{\alpha}) = \boldsymbol{\sigma}(t, \boldsymbol{\alpha}_0)\, \Delta\boldsymbol{\alpha} \tag{11.157b}$$

Sensitivity functions for variables other than state variables (such as outputs) may be obtained readily from the equations relating these variables to the state:

$$\mathbf{y} = \mathbf{g}(\mathbf{x}, t, \mathbf{u}, \boldsymbol{\alpha}) \tag{11.158a}$$

The algebraic sensitivity equation is

$$\boldsymbol{\mu}_j = \left.\frac{\partial \mathbf{y}}{\partial \alpha_j}\right|_{\boldsymbol{\alpha}_0} = \left.\frac{\partial \mathbf{g}}{\partial \mathbf{x}}\right|_{\boldsymbol{\alpha}_0} \boldsymbol{\sigma}_j + \left.\frac{\partial \mathbf{g}}{\partial \alpha_j}\right|_{\boldsymbol{\alpha}_0} \tag{11.158b}$$

Consider now the linear time-invariant system

$$\dot{\mathbf{x}} = \mathbf{A}\mathbf{x} + \mathbf{B}\mathbf{u}; \mathbf{x}(t_0) \triangleq \mathbf{x}^0$$

$$\mathbf{y} = \mathbf{C}\mathbf{x} + \mathbf{D}\mathbf{u}$$

where $\mathbf{A} = \mathbf{A}(\boldsymbol{\alpha})$ is $n \times n$ matrix, $\mathbf{B} = \mathbf{B}(\boldsymbol{\alpha})$ is $n \times p$ matrix, $\mathbf{C} = \mathbf{C}(\boldsymbol{\alpha})$ is $q \times n$ matrix and $\mathbf{D} = \mathbf{D}(\boldsymbol{\alpha})$ is $q \times q$ matrix, $\mathbf{x} = \mathbf{x}(t, \boldsymbol{\alpha})$, $\mathbf{y} = \mathbf{y}(t, \boldsymbol{\alpha})$.

Following the steps of eqns. (11.151)–(11.156), we obtain the following result:

$$\dot{\boldsymbol{\sigma}}_j = \mathbf{A}(\boldsymbol{\alpha}_0) + \left.\frac{\partial \mathbf{A}}{\partial \alpha_j}\right|_{\boldsymbol{\alpha}_0} \mathbf{x}(t, \boldsymbol{\alpha}_0) + \left.\frac{\partial \mathbf{B}}{\partial \alpha_j}\right|_{\boldsymbol{\alpha}_0} \mathbf{u}(t)$$

$$\boldsymbol{\sigma}_j(t_0) = \mathbf{0}; \quad j = 1, 2, \ldots, r \tag{11.159a}$$

$$\boldsymbol{\mu}_j = \mathbf{C}(\boldsymbol{\alpha}_0)\, \boldsymbol{\sigma}_j + \left.\frac{\partial \mathbf{C}}{\partial \alpha_j}\right|_{\boldsymbol{\alpha}_0} \mathbf{x}(t, \boldsymbol{\alpha}_0) + \left.\frac{\partial \mathbf{D}}{\partial \alpha_j}\right|_{\boldsymbol{\alpha}_0} \mathbf{u}(t) \tag{11.159b}$$

It is observed that in the case of linear systems, the vector sensitivity equations have the same **A** matrix as the nominal state equations and hence the same characteristic polynomial $|\lambda\mathbf{I} - \mathbf{A}|$. They differ from the nominal original state equation only in the driving function and the initial conditions. The driving function can be obtained by solving the nominal state equation. A graphical interpretation of (11.159) is given in Fig. 11.13. If the original input **u** is applied to this structure, the sensitivity vectors $\boldsymbol{\sigma}_j$ anb $\boldsymbol{\mu}_j$ are obtained.

It may be noted that if *r* parameters are varying, *r* equations of the form (11.159) describe the situation; thus *r* sensitivity models may be used to obtain all the sensitivity functions.

Example 11.10: Consider a linear system described by the equations

$$\dot{\mathbf{x}} = \mathbf{A}\mathbf{x} + \mathbf{b}u; \quad \mathbf{x}(0) = \mathbf{0}$$

$$y = \mathbf{c}\mathbf{x}$$

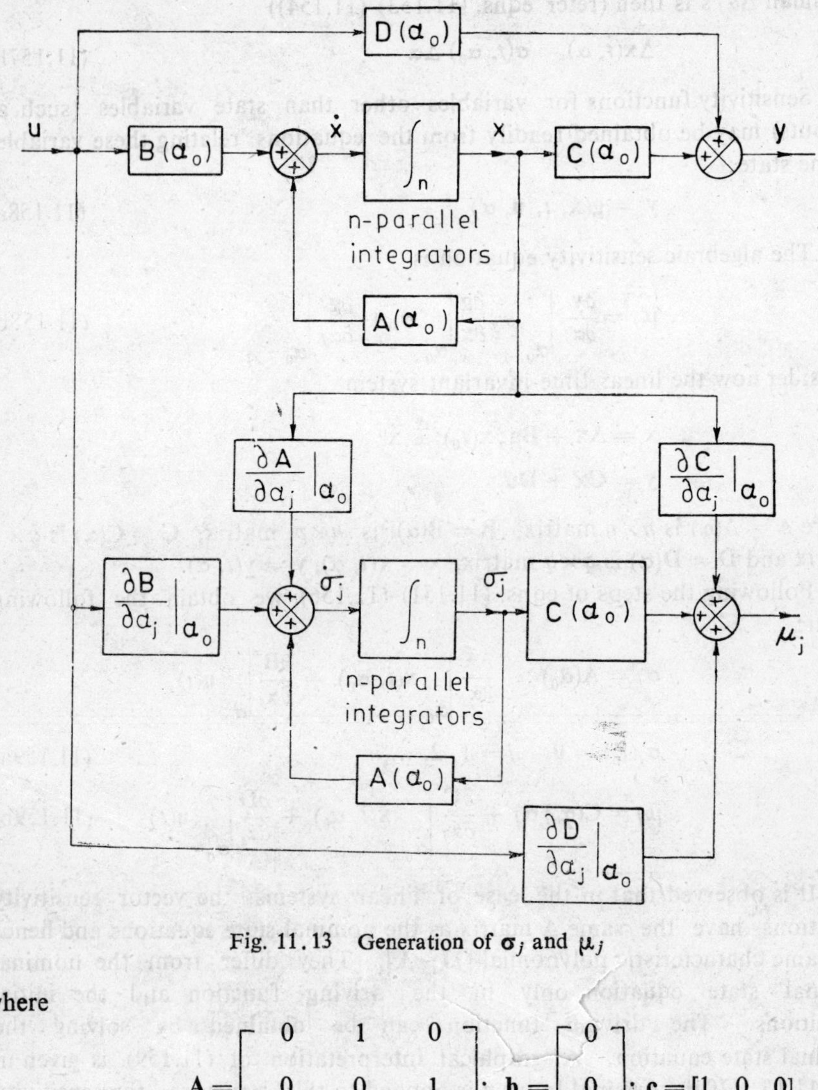

Fig. 11.13 Generation of σ_j and μ_j

where

$$\mathbf{A} = \begin{bmatrix} 0 & 1 & 0 \\ 0 & 0 & 1 \\ -\alpha_1 & -\alpha_2 & -\alpha_3 \end{bmatrix}; \ \mathbf{b} = \begin{bmatrix} 0 \\ 0 \\ 1 \end{bmatrix}; \ \mathbf{c} = [1 \ 0 \ 0]$$

The nominal values of parameters α_1, α_2, α_3 are α_{10}, α_{20}, α_{30}.
For this system

$$\frac{\partial \mathbf{A}}{\partial \alpha_1} = \begin{bmatrix} 0 & 0 & 0 \\ 0 & 0 & 0 \\ -1 & 0 & 0 \end{bmatrix}; \ \frac{\partial \mathbf{A}}{\partial \alpha_2} = \begin{bmatrix} 0 & 0 & 0 \\ 0 & 0 & 0 \\ 0 & -1 & 0 \end{bmatrix}; \ \frac{\partial \mathbf{A}}{\partial \alpha_3} = \begin{bmatrix} 0 & 0 & 0 \\ 0 & 0 & 0 \\ 0 & 0 & -1 \end{bmatrix}$$

$$\frac{\partial \mathbf{b}}{\partial \alpha_j} = \begin{bmatrix} 0 \\ 0 \\ 0 \end{bmatrix}; \quad \frac{\partial \mathbf{c}}{\partial \alpha_j} = [0 \quad 0 \quad 0] \text{ for } j = 1, 2, 3.$$

The sensitivity model for the system is given below:

$$\dot{\boldsymbol{\sigma}}_j = \begin{bmatrix} \dot{\sigma}_{j1} \\ \dot{\sigma}_{j2} \\ \dot{\sigma}_{j3} \end{bmatrix} = \begin{bmatrix} 0 & 1 & 0 \\ 0 & 0 & 1 \\ -\alpha_{10} & -\alpha_{20} & -\alpha_{30} \end{bmatrix} \boldsymbol{\sigma}_j + \left.\frac{\partial \mathbf{A}}{\partial \alpha_j}\right|_{\boldsymbol{\alpha}_0} \mathbf{x}(t, \boldsymbol{\alpha}_0); \; \boldsymbol{\sigma}_j(0) = \mathbf{0}$$

$$\mu_j = [1 \quad 0 \quad 0]\, \boldsymbol{\sigma}_j; \quad j = 1, 2, 3$$

Fig, 11.14 gives the signal flow graph of the sensitivity model.

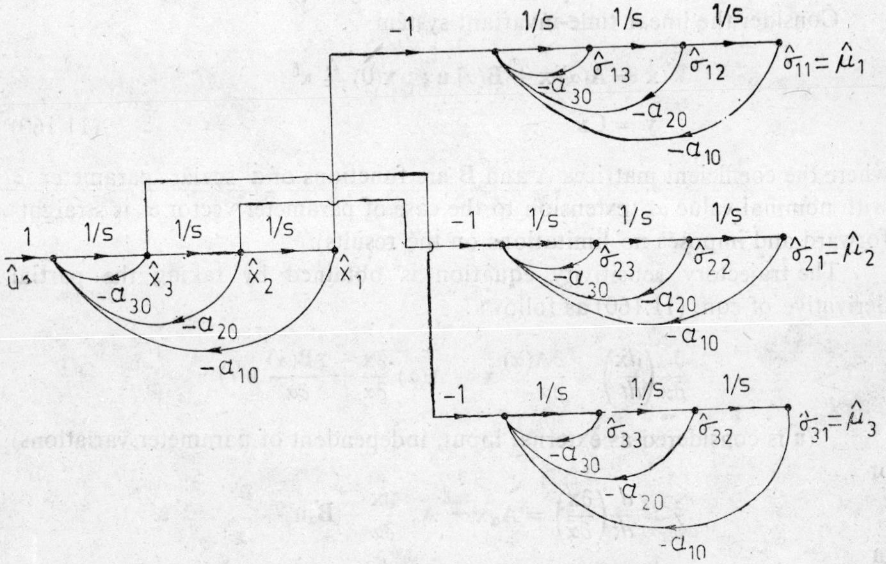

Fig. 11.14 Sensitivity model for system of Example 11.10

□

It is seen that in addition to the nominal state model, other r models of the same kind have to be drawn. Thus, in general, if a system has r parameters, r sensitivity models are needed in order to measure all trajectory sensitivity functions simultaneously according to the procedure above. Refer Wilkie and Perkins (1969) for the question of minimal structure necessary for the simultaneous measurement of all trajectory sensitivity functions.

Optimal Regulator with Low Sensitivity

The optimal linear regulator theory, discussed earlier in this chapter, provides a convenient approach to the design of control systems. Such a design procedure requires a knowledge of the values of the plant parameters. In practice, these parameters may deviate from their nominal values and the system designed to be optimal for one set of parameter values no longer remains optimal for a different set of values. Hence the trajectories may deviate considerably from the trajectories corresponding to optima control due to changes in parameter values, unless care is taken to design the control taking into account considerations of sensitivity.

We know that state variable feedback itself provides a reduction of the sensitivity of the trajectories to variations of plant parameters in comparison with an equivalent open-loop control. However, when this inherent sensitivity reduction is not sufficient, several authors have attempted to reduce the trajectory sensitivity to parameter variations by using design algorithms based on the idea of including a quadratic trajectory sensitivity term in the integrand of the cost functional and specifying a linear feedback control law comprising state and trajectory sensitivity terms. The design algorithm is outlined below:

Consider the linear time-invariant system

$$\dot{\mathbf{x}} = \mathbf{A}(\alpha)\,\mathbf{x} + \mathbf{B}(\alpha)\,\mathbf{u}\,; \quad \mathbf{x}(0) \stackrel{\Delta}{=} \mathbf{x}^0$$

$$\mathbf{y} = \mathbf{C}\mathbf{x} \tag{11.160}$$

where the coefficient matrices **A** and **B** are functions of a scalar parameter α with nominal value α_0 (extension to the case of parameter vector $\boldsymbol{\alpha}$ is straightforward and imposes no limitations on the results).

The trajectory sensitivity equation is obtained by taking the partial derivative of eqn. (11.160) as follows:

$$\frac{\partial}{\partial\alpha}\left(\frac{d\mathbf{x}}{dt}\right) = \frac{\partial \mathbf{A}(\alpha)}{\partial\alpha}\,\mathbf{x} + \mathbf{A}(\alpha)\,\frac{\partial \mathbf{x}}{\partial\alpha} + \frac{\partial \mathbf{B}(\alpha)}{\partial\alpha}\,\mathbf{u}(t)$$

(**u** is considered as external input, independent of parameter variations)

or

$$\frac{d}{dt}\left(\frac{\partial \mathbf{x}}{\partial\alpha}\right) = \mathbf{A}_\alpha\mathbf{x} + \mathbf{A}_0\,\frac{\partial \mathbf{x}}{\partial\alpha} + \mathbf{B}_\alpha\mathbf{u}$$

or

$$\dot{\boldsymbol{\sigma}} = \mathbf{A}_\alpha\mathbf{x} + \mathbf{A}_0\boldsymbol{\sigma} + \mathbf{B}_\alpha\mathbf{u}\,; \quad \boldsymbol{\sigma}(0) = \mathbf{0} \tag{11.161}$$

where

$$\boldsymbol{\sigma} = \frac{\partial \mathbf{x}}{\partial\alpha} = \text{trajectory sensitivity vector}$$

$$\mathbf{A}_\alpha = \frac{\partial \mathbf{A}(\alpha)}{\partial\alpha}$$

$$\mathbf{B}_\alpha = \frac{\partial \mathbf{B}(\alpha)}{\partial\alpha}$$

$$\mathbf{A}_0 = \mathbf{A}(\alpha_0)$$

$$\mathbf{B}_0 = \mathbf{B}(\alpha_0)$$

Equation (11.161) represents the trajectory sensitivity model of system (11.160).

The original system (11.160) when augmented with its sensitivity model (11.161) yields

$$\dot{\mathbf{z}} = \begin{bmatrix} \dot{\mathbf{x}} \\ \dot{\boldsymbol{\sigma}} \end{bmatrix} = \begin{bmatrix} \mathbf{A}_0 & \mathbf{0} \\ \mathbf{A}_\alpha & \mathbf{A}_0 \end{bmatrix} \begin{bmatrix} \mathbf{x} \\ \boldsymbol{\sigma} \end{bmatrix} + \begin{bmatrix} \mathbf{B}_0 \\ \mathbf{B}_\alpha \end{bmatrix} \mathbf{u}$$

$$= \hat{\mathbf{A}}\mathbf{z} + \hat{\mathbf{B}}\mathbf{u}\,; \quad \mathbf{z}(0) = \begin{bmatrix} \mathbf{x}^0 \\ \mathbf{0} \end{bmatrix} \tag{11.162}$$

In order to make the system less sensitive to parameter variations, the sensitivity vector is weighed in the cost function as follows:

$$J = \tfrac{1}{2}\int_0^\infty (\mathbf{x}^T\mathbf{Q}\mathbf{x} + \mathbf{u}^T\mathbf{R}\mathbf{u} + \boldsymbol{\sigma}^T\mathbf{S}\boldsymbol{\sigma})\, dt$$

$$= \tfrac{1}{2}\int_0^\infty (\mathbf{z}^T\hat{\mathbf{Q}}\mathbf{z} + \mathbf{u}^T\mathbf{R}\mathbf{u})\, dt \tag{11.163}$$

where

$$\hat{\mathbf{Q}} = \begin{bmatrix} \mathbf{Q} & \mathbf{0} \\ \mathbf{0} & \mathbf{S} \end{bmatrix}$$

Equations (11.162)–(11.163) are now in the standard form and an optimal feedback control law

$$\mathbf{u} = \mathbf{K}\mathbf{z}$$

$$= \mathbf{K}_1\mathbf{x} + \mathbf{K}_2\boldsymbol{\sigma} \tag{11.164}$$

can now be determined.

The following observations are made from the design procedure outlined above.

(i) For just one parameter variation, n differential equations are added to the nth-order system. Thus the numerical solution of a modified regulator problem creates considerable difficulties for plants of higher order.

(ii) The implementation of the optimal control law requires generation and feedback of trajectory sensitivity vectors. Thus instead of proportional state feedback, dynamic elements have to be provided in the feedback path.

A method which overcomes these problems and is also applicable when only the outputs are available for feedback, is discussed below (Subbayyan and Vaithilingam 1979; O'Reilly 1979).

Suboptimal Regulator with Low Sensitivity

Consider the linear time-invariant system described by (11.160) and the performance index J given by (11.163). We have seen that optimal control law is of the form (11.164). The main difficulties in using this law are that $\sigma(t)$ cannot be measured directly and that in practice complete $\mathbf{x}$ is unavailable for feedback control purposes.

Let us restrict the control input $\mathbf{u}(t)$ to a linear constant feedback of measurements $\mathbf{y}(t)$:

$$\mathbf{u}(t) = \mathbf{K}\,\mathbf{y}(t) \tag{11.165}$$

Using this control law, the nominal parameter closed-loop system becomes (refer (11.160))

$$\dot{\mathbf{x}} = (\mathbf{A}_0 + \mathbf{B}_0\mathbf{KC})\,\mathbf{x} = \mathbf{Dx}\,; \quad \mathbf{x}(0) = \mathbf{x}^0 \tag{11.166}$$

and the trajectory sensitivity vector is governed by

$$\begin{aligned}\dot{\boldsymbol{\sigma}} &= (\mathbf{A}_0 + \mathbf{BKC})\,\boldsymbol{\sigma} + (\mathbf{A}_\alpha + \mathbf{B}_\alpha\mathbf{KC})\,\mathbf{x} \\ &= \mathbf{D}\boldsymbol{\sigma} + (\mathbf{A}_\alpha + \mathbf{B}_\alpha\mathbf{KC})\,\mathbf{x};\ \boldsymbol{\sigma}(0) = \mathbf{0}\end{aligned} \tag{11.167}$$

Using (11.165), the performance index J given by (11.163) is expressed as

$$J = \tfrac{1}{2}\int_0^\infty [\mathbf{x}^T(\mathbf{Q} + \mathbf{C}^T\mathbf{K}^T\mathbf{RKC})\mathbf{x} + \boldsymbol{\sigma}^T\mathbf{S}\boldsymbol{\sigma}]\,dt \tag{11.168}$$

The closed-loop system (11.166) when augmented with trajectory sensitivity equation (11.167) is written as

$$\begin{bmatrix}\dot{\mathbf{x}}\\ \dot{\boldsymbol{\sigma}}\end{bmatrix} = \begin{bmatrix}\mathbf{D} & \mathbf{0}\\ \mathbf{A}_\alpha + \mathbf{B}_\alpha\mathbf{KC} & \mathbf{D}\end{bmatrix}\begin{bmatrix}\mathbf{x}\\ \boldsymbol{\sigma}\end{bmatrix};\ \begin{bmatrix}\mathbf{x}(0)\\ \boldsymbol{\sigma}(0)\end{bmatrix} = \begin{bmatrix}\mathbf{x}^0\\ \mathbf{0}\end{bmatrix} \tag{11.169}$$

The sensitivity measure J in (11.168) can be expressed as

$$J = \tfrac{1}{2}\int_0^\infty [\mathbf{x}^T \quad \boldsymbol{\sigma}^T]\begin{bmatrix}\mathbf{Q} + \mathbf{C}^T\mathbf{K}^T\mathbf{RKC} & \mathbf{0}\\ \mathbf{0} & \mathbf{S}\end{bmatrix}\begin{bmatrix}\mathbf{x}\\ \boldsymbol{\sigma}\end{bmatrix} dt \tag{11.170}$$

For the augmented system (11.169), J as in (11.170) can be expressed as

$$\begin{aligned}J &= \tfrac{1}{2}\,[\mathbf{x}^{0T} \quad \boldsymbol{\sigma}^{0T}]\,\mathbf{P}\begin{bmatrix}\mathbf{x}^0\\ \boldsymbol{\sigma}^0\end{bmatrix} \\ &= \tfrac{1}{2}\,[\mathbf{x}^{0T} \quad \boldsymbol{\sigma}^{0T}]\begin{bmatrix}\mathbf{P}_{11} & \mathbf{P}_{12}\\ \mathbf{P}_{12}^T & \mathbf{P}_{22}\end{bmatrix}\begin{bmatrix}\mathbf{x}^0\\ \boldsymbol{\sigma}^0\end{bmatrix} && (11.171\text{a}) \\ &= \tfrac{1}{2}\mathbf{x}^{0T}\ \mathbf{P}_{11}\mathbf{x}^0 && (11.171\text{b})\end{aligned}$$

where **P** is the solution of

$$\begin{bmatrix} \mathbf{D} & \mathbf{0} \\ \mathbf{A}_\alpha+\mathbf{B}_\alpha\mathbf{KC} & \mathbf{D} \end{bmatrix}^T \mathbf{P}+\mathbf{P}\begin{bmatrix} \mathbf{D} & \mathbf{0} \\ \mathbf{A}_\alpha+\mathbf{B}_\alpha\mathbf{KC} & \mathbf{D} \end{bmatrix}+\begin{bmatrix} \mathbf{Q}+\mathbf{C}^T\mathbf{K}^T\mathbf{RKC} & \mathbf{0} \\ \mathbf{0} & \mathbf{S} \end{bmatrix}=\mathbf{0} \tag{11.172}$$

This equation enfolds the following four Lyapunov-type matrix equations, each of order n:

$$\mathbf{D}^T\mathbf{P}_{22}+\mathbf{P}_{22}\mathbf{D}+\mathbf{S}=\mathbf{0} \tag{11.173a}$$

$$\mathbf{D}^T\mathbf{P}_{12}+\mathbf{P}_{12}\mathbf{D}+(\mathbf{A}_\alpha+\mathbf{B}_\alpha\mathbf{KC})^T\mathbf{P}_{22}=\mathbf{0} \tag{11.173b}$$

$$\mathbf{D}^T\mathbf{P}_{12}^T+\mathbf{P}_{12}^T\mathbf{D}+\mathbf{P}_{22}(\mathbf{A}_\alpha+\mathbf{B}_\alpha\mathbf{KC})=\mathbf{0} \tag{11.173c}$$

$$\mathbf{D}^T\mathbf{P}_{11}+\mathbf{P}_{11}\mathbf{D}+(\mathbf{A}_\alpha+\mathbf{B}_\alpha\mathbf{KC})^T\mathbf{P}_{12}^T+\mathbf{P}_{12}(\mathbf{A}_\alpha+\mathbf{B}_\alpha\mathbf{KC}) + \mathbf{Q}+\mathbf{C}^T\mathbf{K}^T\mathbf{RKC}=\mathbf{0} \tag{11.173d}$$

Note that eqn. (11.173b) is the transpose of eqn. (11.173c)

From eqn. (11.171b), we obtain

$$J=\tfrac{1}{2}\,\mathrm{tr}\,(\mathbf{P}_{11}\mathbf{x}^0\mathbf{x}^{0T})$$

Therefore the expected value of J is given as

$$\hat{J}=\tfrac{1}{2}\,\mathrm{tr}\,(\mathbf{P}_{11}) \tag{11.174}$$

with the assumption that

$$E\{\mathbf{x}^0\mathbf{x}^{0T}\}=\mathbf{I}$$

$\hat{J}$ in (11.174) is a function of **K** and the problem is to determine the optimal **K** which minimizes $\hat{J}$.

For higher-order systems, mathematical programming techniques have to be employed for minimization of $\hat{J}$ given by (11.174). Subbayyan and Vaithilingam (1979) have suggested the Hamiltonian approach to solve this problem.

11.9 MINIMUM-TIME CONTROL OF LINEAR TIME-INVARIANT SYSTEMS

So far in this chapter we considered the minimization of a quadratic performance criterion. Another equally useful performance criterion for which general results are available is time.

Consider that the linear, controllable time-invariant system

$$\dot{\mathbf{x}}=\mathbf{Ax}+\mathbf{Bu};\quad \mathbf{x}\in\mathcal{R}^n,\ \mathbf{u}\in\mathcal{R}^p \tag{11.175}$$

is initially in the state $\mathbf{x}(t_0)=\mathbf{x}^0$. We wish to apply a control $\mathbf{u}(t)$ which will drive the state to $\mathbf{x}=\mathbf{0}$ in minimum time where the control is constrained according to the following inequality:

$$\alpha_i\leqslant u_i(t)\leqslant\beta_i;\quad i=1,2,\ldots,p \tag{11.176}$$

This minimum-time problem is equivalent to the problem of minimization of the performance criterion

$$J = \int_0^{t_1} dt \tag{11.177}$$

under the constraint of (11.176). For simplicity (and without loss of generality) we shall assume the constraint on $\mathbf{u}(t)$ to be

$$|u_i(t)| \leqslant 1; \quad i = 1, 2, \ldots, p \tag{11.178}$$

We have assumed the plant (11.175) to be controllable; therefore the transition from initial state $\mathbf{x}^0$ to the origin is possible in finite time. We have shown in Section 6.4 that if $\mathbf{u}$ is unconstrained, then transition from $\mathbf{x}^0$ to the origin can be affected in any arbitrarily specified time by a suitable control that stays finite. So, for a controllable system with unconstrained control, the time-optimal (minimum-time) control problem has no solution. Would the problem have a solution if the control is constrained? The answer is yes, provided the initial and final states are such that the desired transition can be made at all with the constrained control. As we shall see in this section, mere controllability does not guarantee that a transition can be made under constrained control.

We shall use Pontryagin's minimum principle, summarized in Table 10.4, to solve the minimum-time problem. For the system under consideration, the Pontryagin function

$$\mathscr{H}(\mathbf{x}, \mathbf{u}, \boldsymbol{\lambda}) = 1 + \boldsymbol{\lambda}^T (\mathbf{A}\mathbf{x} + \mathbf{B}\mathbf{u}) \tag{11.179}$$

The initial and final conditions on $\mathbf{x}$ are specified as $\mathbf{x}^0$ and $\mathbf{0}$, respectively. Therefore, the initial and final conditions on $\boldsymbol{\lambda}$ are free (Table 10.3). The co-state equations are

$$\dot{\boldsymbol{\lambda}} = -\frac{\partial \mathscr{H}}{\partial \mathbf{x}} = -\mathbf{A}^T \boldsymbol{\lambda} \tag{11.180a}$$

The solution of this equation is given by

$$\boldsymbol{\lambda}(t) = e^{-\mathbf{A}^T t} \boldsymbol{\lambda}(0) \tag{11.180b}$$

Therefore, from (11.179) we obtain

$$\mathscr{H}(\mathbf{x}, \mathbf{u}, \boldsymbol{\lambda}) = 1 + \boldsymbol{\lambda}^T(0)\, e^{-\mathbf{A}t}\mathbf{A}\mathbf{x} + \boldsymbol{\lambda}^T(0)\, e^{-\mathbf{A}t}\mathbf{B}\mathbf{u}$$

According to the minimum principle, the time-optimal control must minimize $\mathscr{H}$ subject to the constraint (11.178). The control $\mathbf{u}$ affects $\mathscr{H}$ only through the term $\boldsymbol{\lambda}^T(0)\, e^{-\mathbf{A}t}\mathbf{B}\mathbf{u}$. Thus at every instant, any element of $\mathbf{u}$ say u_i, must be made equal to $+1$ or -1 depending upon whether at that instant the coefficient $[\boldsymbol{\lambda}^T(0)\, e^{-\mathbf{A}t}\mathbf{B}]_i$ of u_i is negative or positive:

$$u_i^*(t) = \begin{cases} +1 & \text{when } [\boldsymbol{\lambda}^T(0)\, e^{-\mathbf{A}t}\mathbf{B}]_i < 0 \\ -1 & \text{when } [\boldsymbol{\lambda}^T(0)\, e^{-\mathbf{A}t}\mathbf{B}]_i > 0 \end{cases} \tag{11.181}$$

Hence the 'control strategy' must necessarily be of what is termed the *bang-bang* type; each element $u_i(t)$ of the input vector being instantaneously switched from its maximum permissible value to its minimum permissible value (or vice-versa) whenever due to the time variation of $\boldsymbol{\lambda}(t)$, $\boldsymbol{\lambda}^T(0)\, e^{-\mathbf{A}t}\mathbf{B}$ goes through zero from negative to positive (or vice-versa).

Normality

Note that there is no indication in (11.181) of what control action to take for $u_i(t)$ when $[\boldsymbol{\lambda}^T(0)\, e^{-\mathbf{A}t}\mathbf{B}]_i = 0$ over a finite interval of time, since then $\mathscr{H}$ is independent of $u_i(t)$. This case is called a *singular control.*

The ith entry of the row vector $[\boldsymbol{\lambda}^T(0)\, e^{-\mathbf{A}t}\mathbf{B}]$ is $\boldsymbol{\lambda}^T(0)\, e^{-\mathbf{A}t}\mathbf{b}_i$ where $\mathbf{b}_i$ is the ith column of matrix $\mathbf{B}$. Denote this ith entry by $p_i(t)$, i.e.,

$$p_i(t) = \boldsymbol{\lambda}^T(0)\, e^{-\mathbf{A}t}\mathbf{b}_i$$

If there is a finite interval $[t_1, t_2]$ for which $p_i(t) = 0$, it implies that all derivatives of $p_i(t)$ with respect to time are zero on this interval. Therefore, we have for $t \in [t_1, t_2]$

$$p_i(t) = \boldsymbol{\lambda}^T(0)\, e^{-\mathbf{A}t}\mathbf{b}_i = 0$$

$$-\frac{dp_i(t)}{dt} = \boldsymbol{\lambda}^T(0)\, e^{-\mathbf{A}t}\mathbf{A}\mathbf{b}_i = 0$$

$$\frac{d^2p_i(t)}{dt^2} = \boldsymbol{\lambda}^T(0)\, e^{-\mathbf{A}t}\mathbf{A}^2\mathbf{b}_i = 0$$

$$\vdots$$

$$(-1)^{n-1}\frac{d^{n-1}p_i(t)}{dt^{n-1}} = \boldsymbol{\lambda}^T(0)\, e^{-\mathbf{A}t}\mathbf{A}^{n-1}\mathbf{b}_i = 0$$

These relations may be expressed as

$$[\,\mathbf{b}_i \quad \mathbf{A}\mathbf{b}_i \,\dots\, \mathbf{A}^{n-1}\mathbf{b}_i\,]^T\, e^{-\mathbf{A}^T t}\boldsymbol{\lambda}(0) = \mathbf{0}$$

or

$$\mathbf{U}_i^T\, e^{-\mathbf{A}^T t}\boldsymbol{\lambda}(0) = \mathbf{0} \tag{11.182}$$

If $\boldsymbol{\lambda}(0) = \mathbf{0}$, this implies that (eqn. (11.179))

$$\mathscr{H} = 1 \quad \text{at} \quad t = 0$$

However, from Table 10.3, we observe that for the case of specified $\mathbf{x}(0)$ and $\mathbf{x}(t_1)$, $\mathscr{H}^* = 0$ at $t = t_1$. Since $\mathscr{H}$ in (11.179) does not contain t explicitly, $\dfrac{\partial \mathscr{H}}{\partial t} = 0$. Now

$$\frac{d\mathcal{H}}{dt} = \left(\frac{\partial \mathcal{H}}{\partial \mathbf{x}}\right)^T \dot{\mathbf{x}} + \left(\frac{\partial \mathcal{H}}{\partial \mathbf{u}}\right)^T \dot{\mathbf{u}} + \left(\frac{\partial \mathcal{H}}{\partial \boldsymbol{\lambda}}\right)^T \dot{\boldsymbol{\lambda}} + \frac{\partial \mathcal{H}}{\partial t}$$

$$= -\dot{\boldsymbol{\lambda}}^T\dot{\mathbf{x}} + 0 + \dot{\mathbf{x}}^T\dot{\boldsymbol{\lambda}} + 0$$

$$= 0$$

Hence if $\mathcal{H}$ is not explicitly a function of t, so that its partial time derivative vanishes, then also its total time derivative vanishes, i.e., $\mathcal{H}$ is then constant throughout the control interval provided a trajectory corresponding to optimal control is being followed. Thus, $\boldsymbol{\lambda}(0) = \mathbf{0}$ contradicts the requirement that $\mathcal{H} = 0$ everywhere on the optimal trajectory. To satisfy eqn. (11.182), we must therefore have a singular matrix $\mathbf{U}_i$, i.e.,

$$\rho\,[\mathbf{U}_i] = \rho\,[\mathbf{b}_i \quad \mathbf{A}\mathbf{b}_i \ldots \mathbf{A}^{n-1}\mathbf{b}_i] < n$$

If we find that $\mathbf{U}_i$ is not singular for any i, $i = 1, 2, \ldots, p$, then we can be assured that there are no finite intervals on which $p_i(t) = 0$. A system for which $\mathbf{U}_i$ is not singular for any i is called *normal*; a controllable system which is not normal is called *singular*.

From this definition we observe that a normal system is controllable with respect to each and every component of the control vector. In this section, we shall deal with normal systems only. (The singular control problem has been treated in detail by Bell and Jacobson (1975).)

Existence and Uniqueness of Control

In the following we state and illustrate the key features of some important theorems due to Pontryagin et al. (1962).

1. *If the linear time-invariant system* (11.175) *is controllable and if all the eigenvalues of* $\mathbf{A}$ *have nonpositive real parts, then an optimal control exists that transfers any initial state* $\mathbf{x}^0$ *to the origin.*

The key feature of the proof is indicated below by restricting to single input systems with distinct real eigenvalues $\lambda_1, \lambda_2, \ldots, \lambda_n$. Such a system may be expressed in the form

$$\dot{x}_i = \lambda_i x_i + b_i u\ ; \quad i = 1, \ldots, n$$

Assume that $\lambda_i > 0.$

The solution

$$x_i(t) = e^{\lambda_i t}\left[x_i^0 + \int_0^t e^{-\lambda_i \tau}\, b_i u(\tau)\, d\tau\right]$$

A given initial condition x_i^0 can be transferred to origin in time t_1 if

$$x_i^0 = -\int_0^{t_1} e^{-\lambda_i t}\, b_i u(t)\, dt$$

From this equation we may write

$$| x_i^0 | = \left| \int_0^{t_1} e^{-\lambda_i t} \, b_i u(t) \, dt \right|$$

Since $| u(t) | \leqslant 1$, we obtain

$$| x_i^0 | \leqslant \frac{|b_i|}{\lambda_i} (1 - e^{-\lambda_i t_1})$$

Therefore,

$$(1 - e^{-\lambda_i t_1}) \geqslant \frac{\lambda_i | x_i^0 |}{| b_i |}$$

Clearly, this equation cannot be satisfied if

$$| x_i^0 | \geqslant \frac{| b_i |}{\lambda_i}$$

and therefore we cannot force x_i^0 to the origin.

This result indicates that the constraint on $u(t)$ may prevent us from driving the unstable system to the origin, despite the fact that the system is controllable.

2. *If the linear time-invariant system* (11.175) *is normal and if time-optimal control exists, then it is unique.*

We prove this result by contradiction. Assume that two different time-optimal controls u^* and $\hat{u}^*$ will drive the state from $\mathbf{x}^0$ to origin in minimum time t_1. Then

$$\mathbf{x}(t) = e^{\mathbf{A}t} \mathbf{x}^0 + \int_0^t e^{\mathbf{A}(t-\tau)} \mathbf{B}\mathbf{u}^*(\tau) \, d\tau$$

$$\hat{\mathbf{x}}(t) = e^{\mathbf{A}t} \mathbf{x}^0 + \int_0^t e^{\mathbf{A}(t-\tau)} \mathbf{B} \hat{\mathbf{u}}^*(\tau) \, d\tau$$

Since $\mathbf{x}(t_1) = \hat{\mathbf{x}}(t_1) = \mathbf{0}$, we obtain

$$\int_0^{t_1} e^{-\mathbf{A}t} \mathbf{B}\mathbf{u}^*(t) \, dt = \int_0^{t_1} e^{-\mathbf{A}t} \mathbf{B} \hat{\mathbf{u}}^*(t) \, dt \tag{11.183}$$

Assume that $\boldsymbol{\lambda}^0$ and $\hat{\boldsymbol{\lambda}}^0$ are the initial conditions on co-state vectors for the time-optimal controls $\mathbf{u}^*$ and $\hat{\mathbf{u}}^*$ respectively. Since the system (11.175) is assumed to be normal, the time-optimal controls are uniquely defined by the initial co-state vectors (refer (11.181)). Further, because each of the optimal controls must minimize the Pontryagin function in (11.179), we must have

$$\boldsymbol{\lambda}^T \mathbf{B}\mathbf{u}^*(t) \leqslant \boldsymbol{\lambda}^T \mathbf{B}\hat{\mathbf{u}}^*(t)$$

or

$$\boldsymbol{\lambda}^{0T} e^{-\mathbf{A}t} \mathbf{B}\mathbf{u}^*(t) \leqslant \boldsymbol{\lambda}^{0T} e^{-\mathbf{A}t} \mathbf{B}\, \hat{\mathbf{u}}^*(t) \tag{11.184}$$

Strict inequality must hold whenever $\mathbf{u}^*(t) \neq \hat{\mathbf{u}}^*(t)$.

If we multiply both sides of equation (11.183) by $\boldsymbol{\lambda}^{0T}$, we obtain

$$\int_0^{t_1} \boldsymbol{\lambda}^{0T} e^{-\mathbf{A}t} \mathbf{B}\mathbf{u}^*(t)\, dt = \int_0^{t_1} \boldsymbol{\lambda}^{0T} e^{-\mathbf{A}t}\, \hat{\mathbf{u}}^*(t)\, dt$$

which is impossible in view of inequality (11.184). This proves the uniqueness of the control.

3. *If the eigenvalues of matrix* **A** *of system* (11.175) *are real and a unique time-optimal control exists, then each control component can switch at most* $(n-1)$ *times.*

The key feature of this result is indicated below by restricting to the case of distinct eigenvalues of matrix **A**.

From (11.181), we observe that $u_i(t)$ can switch only at times which satisfy

$$[\boldsymbol{\lambda}^T(0)\, e^{-\mathbf{A}t}\, \mathbf{B}]_i = 0$$

or

$$\boldsymbol{\lambda}^T(0)\, e^{-\mathbf{A}t}\, \mathbf{b}_i = 0 \tag{11.185}$$

From eqn. (5.57), it is observed that there exists a nonsingular matrix **M** such that

$$e^{-\mathbf{A}t} = \mathbf{M}e^{\mathbf{J}t}\mathbf{M}^{-1}$$

$$= \mathbf{M} \begin{bmatrix} e^{-\lambda_1 t} & 0 & \dots & 0 \\ 0 & e^{-\lambda_2 t} & \dots & 0 \\ \vdots & \vdots & & \vdots \\ 0 & 0 & \dots & e^{-\lambda_n t} \end{bmatrix} \mathbf{M}^{-1}$$

Therefore, (11.185) may be expressed as

$$\boldsymbol{\lambda}^T(0)\, \mathbf{M}e^{\mathbf{J}t}\mathbf{M}^{-1}\, \mathbf{b}_i = 0$$

which can be reduced to the following form:

$$\sum_{i=1}^{n} \alpha_i\, e^{-\lambda_i t} = 0 \tag{11.186}$$

To prove the result, we have to show that eqn. (11.186) has at most $(n-1)$ roots, i.e., there are at most $(n-1)$ values of t that satisfy this equation. The result is obviously true for $n = 1$. We shall prove the general result by induction. Assume the result is true for $n = m$, i.e.,

$$\sum_{i=1}^{m} \alpha_i\, e^{-\lambda_i t} = 0 \tag{11.187a}$$

has at most $(m-1)$ roots. We show that this implies that

$$\sum_{i=1}^{m+1} \alpha_i e^{-\lambda_i t} = 0 \tag{11.187b}$$

has at most m roots. Let us show it by contradiction. Assume that (11.187b) has at least $(m+1)$ roots. Multiplying this equation by $e^{-\lambda_1 t}$ will not change the number of roots. Therefore

$$\alpha_1 + \sum_{i=2}^{m+1} \alpha_i e^{(\lambda_1 - \lambda_i)t} = 0$$

has at least $(m+1)$ roots. Differentiating this equation we get

$$\sum_{i=2}^{m+1} \alpha_i(\lambda_1 - \lambda_i) e^{-\lambda_i t} = 0$$

which must have at least m roots. But this equation is precisely of the form of eqn. (11.187a) and therefore by induction hypothesis, it can have at most $(m-1)$ roots. This contradiction proves the result.

This result is very useful for defining the structure of time-optimal control. Note that if some of the eigenvalues of $\mathbf{A}$ are complex, eqn. (11.186) will contain sine terms and therefore can have an arbitrary number of roots.

Example 11.11: For the system described by the equation

$$\dot{\mathbf{x}} = \mathbf{A}\mathbf{x} + \mathbf{b}u \tag{11.188}$$

where

$$\mathbf{A} = \begin{bmatrix} 0 & 1 \\ 0 & 0 \end{bmatrix}, \quad \mathbf{b} = \begin{bmatrix} 0 \\ 1 \end{bmatrix}$$

let us find optimal control $u^*(t)$ satisfying

$$|u(t)| \leqslant 1$$

which transfers the system from any initial state $\mathbf{x}^0$ to the origin in minimum time.

It can be verified that the given second-order system is controllable, normal and the eigenvalues of $\mathbf{A}$ are both zero. Therefore, an optimal control exists, is unique and has at most one switching.

The Pontryagin function is

$$\begin{aligned} \mathcal{H} &= 1 + \boldsymbol{\lambda}^T \mathbf{A}\mathbf{x} + \boldsymbol{\lambda}^T \mathbf{b}u \\ &= 1 + \lambda_1 x_2 + \lambda_2 u \end{aligned}$$

To minimize $\mathcal{H}$ with respect to u requires $u = -1$ if λ_2 is positive, $u = +1$ if λ_2 is negative, the switching instants being when $\lambda_2 = 0$.

$$u^*(t) = \begin{cases} -1 & \text{for } \lambda_2(t) > 0 \\ +1 & \text{for } \lambda_2(t) < 0 \end{cases}$$

$$\stackrel{\Delta}{=} - \text{sgn}\,(\lambda_2(t))$$

Now, $\lambda_2(t)$ is controlled by the co-state equation

$$\frac{\partial \mathcal{H}}{\partial \mathbf{x}} = -\dot{\boldsymbol{\lambda}}$$

or in this case

$$\dot{\lambda}_1 = 0$$

$$\dot{\lambda}_2 = -\lambda_1$$

The solutions to these equations are

$$\lambda_1(t) = \lambda_1^0, \ \lambda_2(t) = \lambda_2^0 - \lambda_1^0 t$$

where λ_i^0 are the initial conditions on $\lambda_i(t)$. The optimal control law is therefore

$$u^* = -\text{sgn}\,\lambda_2(t)$$

$$= -\text{sgn}\,(\lambda_2^0 - \lambda_1^0 t) \qquad (11.189)$$

The control can have only two values, $+1$ and -1. When $u = +1$, the solution to the system equation (11.188) is

$$x_2 = t + x_2^0$$

$$x_1 = \frac{t^2}{2} + x_2^0\, t + x_1^0$$

If t is eliminated from these equations, we obtain

$$x_1 = \frac{x_2^2}{2} + x_1^0 - \frac{{x_2^0}^2}{2} \qquad (11.190a)$$

When $u = -1$, the solution to the system equation (11.188) is

$$x_2 = -t + x_2^0$$

$$x_1 = -\frac{t^2}{2} + x_2^0\, t + x_1^0$$

Again, eliminating t we obtain

$$x_1 = -\frac{x_2^2}{2} + x_1^0 + \frac{{x_2^0}^2}{2} \qquad (11.190b)$$

Equations (11.190a) and (11.190b) each define a family of parabolas shown in Fig. 11.15. The arrows indicate the direction of increasing time. Thick lines correspond to $u = +1$ and dotted lines correspond to $u = -1$.

We know that optimal control for a second-order system can switch at

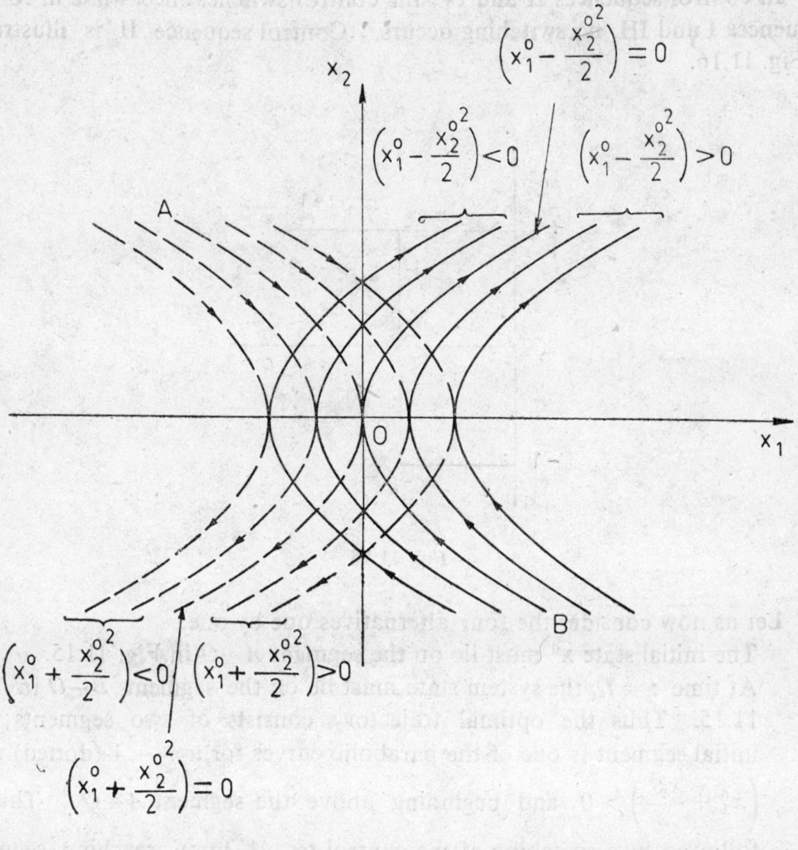

Fig. 11.15 Trajectories for Example 11.11

most once. This fact can be established from eqn. (11.189) also. This equation shows that $\lambda_2(t)$ can change sign once at most, at a time t_s given by

$$t_s = \frac{\lambda_2^0}{\lambda_1^0} \triangleq s$$

provided $s > 0$. Otherwise $\lambda_2(t)$ will not change sign. We can therefore narrow the search for the optimal control to the following four conditions.

$$\begin{aligned}
&\text{I:} && u = -1 \quad ; && t \in [0, t_1] \\
&\text{II:} && \begin{cases} u = -1 \quad ; & t \in [0, t_s] \\ u = +1 \quad ; & t \in [t_s, t_1] \end{cases} \\
&\text{III:} && u = +1 \quad ; && t \in [0, t_1] \\
&\text{IV:} && \begin{cases} u = +1 \quad ; & t \in [0, t_s] \\ u = -1 \quad ; & t \in [t_s, t_1] \end{cases}
\end{aligned}$$

In control sequences II and IV, the control switches once, while in control sequences I and III, no switching occurs. Control sequence II is illustrated in Fig. 11.16.

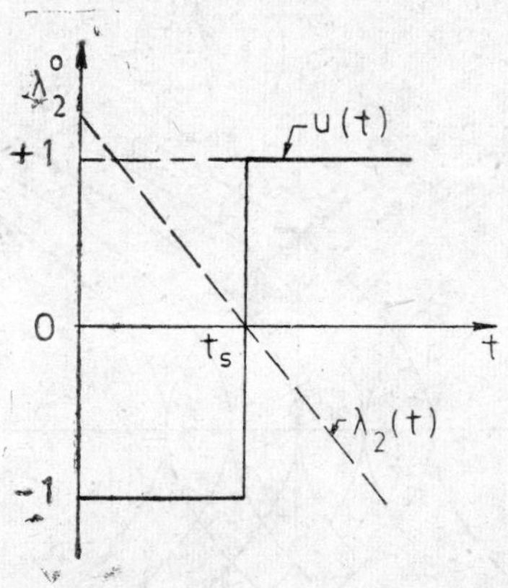

Fig. 11.16

Let us now consider the four alternatives one by one.

I : The initial state $\mathbf{x}^0$ must lie on the segment $A-O$ in Fig. 11.15.

II : At time $t = t_s$, the system state must lie on the segment $B-O$ in Fig. 11.15. Thus the optimal trajectory consists of two segments; the initial segment is one of the parabolic curves for $u = -1$ (dotted) with $\left(x_1^0 + \frac{x_2^{0\,2}}{2}\right) > 0$ and beginning above the segment $A-O$. This is followed by a switching of the control to $+1$ upon reaching segment $B-O$ and then on to the origin along $B-O$ with $u = +1$.

III : The initial state must lie on the segment $B-O$.

IV : At time $t = t_s$, the system state must lie on the segment $A-O$. Therefore the initial segment of the optimal trajectory must be a parabolic curve for $u = +1$ with $\left(x_1^0 - \frac{x_2^{0\,2}}{2}\right) < 0$ and beginning below the segment $B-O$. This is followed by a switching of the control to -1 upon reaching segment $A-O$ and then on to the origin along $A-O$ with $u = -1$.

The segments $A-O$ and $B-O$ together compose the switching curve $A-O-B$ shown in Fig. 11.17. The equation of this curve can be obtained from eqns. (11.190) by putting

$$\left(x_1^0 - \frac{x_2^{0\,2}}{2}\right) = \left(x_1^0 + \frac{x_2^{0\,2}}{2}\right) = 0$$

This gives

$$x_1(t) = -\tfrac{1}{2}x_2(t)\,|\,x_2(t)\,| \tag{11.191}$$

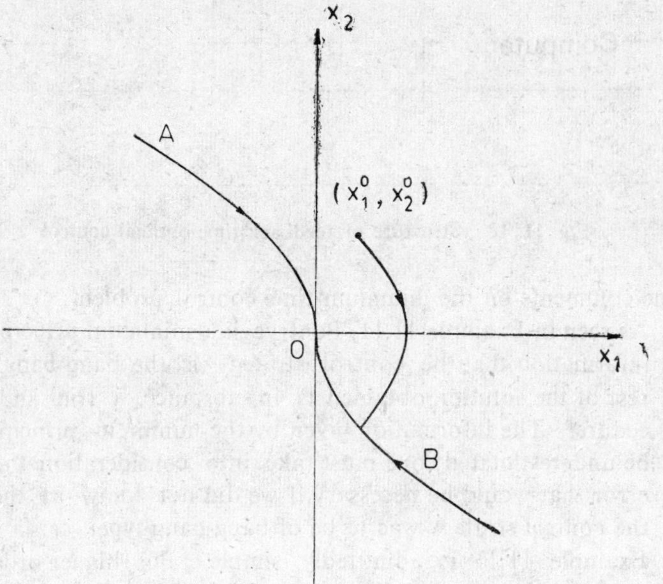

Fig. 11.17

For the control sequence of Fig. 11.16, the optimal trajectory is shown in Fig. 11.17.

$$S(\mathbf{x}(t)) \triangleq x_1(t) + \tfrac{1}{2}x_2(t)\,|\,x_2(t)\,|$$

is the switching function.

$S(\mathbf{x}(t)) > 0$ implies $\mathbf{x}(t)$ lies above the curve $A-O-B$

$S(\mathbf{x}(t)) < 0$ implies $\mathbf{x}(t)$ lies below the curve $A-O-B$

$S(\mathbf{x}(t)) = 0$ implies $\mathbf{x}(t)$ lies on the curve $A-O-B$.

In terms of the switching function, the optimal control law is

$$u^*(t) = \begin{cases} -1 \text{ when } S(\mathbf{x}(t)) > 0 \\ +1 \text{ when } S(\mathbf{x}(t)) = 0 \text{ and } x_2(t) < 0 \\ +1 \text{ when } S(\mathbf{x}(t)) < 0 \\ -1 \text{ when } S(\mathbf{x}(t)) = 0 \text{ and } x_2(t) > 0 \end{cases} \tag{11.192}$$

The structure of feedback time-optimal control is sketched in Fig. 11.18. The computer accepts the state vector and computes the switching function. It then controls the ideal relay to produce the optimal control components according to (11.192).

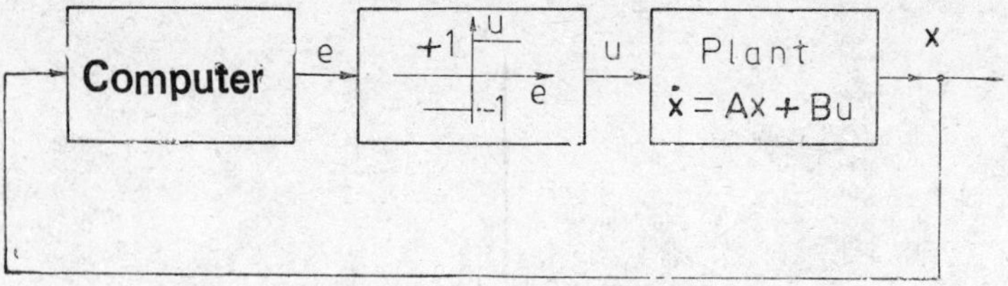

Fig. 11.18 Structure of feedback time-optimal control

□

Some comments on the minimum-time control problem:

(i) As seen in Example 11.11, Pontryagin's minimum principle gives the information that the control strategy is the bang-bang type. The rest of the solution obtained is, in substance, a trial and error procedure. The information given by the minimum principle must not be underestimated; one must take into consideration the trial and error that would be necessary if we did not know at the start that the control strategy was to be of bang-bang type.

(ii) Example 11.11 is admittedly simple. For higher-order systems ($n \geqslant 3$) it is generally difficult to obtain an analytical expression for the switching hypersurface; we may have to go for suboptimal methods of control.

The reader may refer Athans and Falb (1966) for a detailed account of time-optimal control problem.

11.10 CONCLUDING COMMENTS

This chapter has dealt with feedback control systems. In Sections 11.2–11.6 and 11.8 we have discussed quite extensively how linear state feedback control systems can be designed that are optimal in the sense of a quadratic performance index. Such systems possess many useful properties: they can be made (i) to exhibit a satisfactory transient response to nonzero initial conditions, to a change in set point etc., (ii) to have good stability characteristics, (iii) insensitive to disturbances and parameter variations. These properties can be achieved in a desired manner by an appropriate choice of the weighting matrices in the performance index.

The linear state feedback control systems require state reconstruction for the implementation of control law if all the states are not available for feedback. In Section 11.7, we discussed an alternative method of design wherein the feedback from available states only is required.

Another important class of optimal feedback control systems—the time-optimal control systems—was discussed in Section 11.9.

PROBLEMS

11.1 Consider the system

$$x(k+1) = 0.368\, x(k) + 0.632\, u(k)$$

Find the control sequence so that the following performance index is minimized:

$$J = x^2(N) + \sum_{k=1}^{3} [x^2(k) + u^2(k)]$$

Also find the control sequence when $N \to \infty$.

11.2 The linear discrete system

$$x_1(k+1) = x_1(k) + x_2(k)$$

$$x_2(k+1) = x_2(k) + u(k)$$

is to be controlled to minimize the performance index

$$J = \sum_{k=0}^{2} [4x_1^2(k) + u^2(k)]$$

Obtain the optimal control sequence $[u(0), u(1), u(2)]$; the initial state is $\mathbf{x}(0) = [1 \quad 0]^T$.

11.3 For the plant described by the equations

$$x_1(k+1) = 0.8x_1(k) + x_2(k) + u(k)$$

$$x_2(k+1) = 0.5x_2(k) + 0.5u(k)$$

find the optimal control law that minimizes the performance index

$$J = \tfrac{1}{2} \sum_{k=0}^{\infty} [x_1^2(k) + x_2^2(k) + u^2(k)]$$

11.4 It is desired to determine the control law that causes the plant

$$\dot{x}_1 = x_2$$

$$\dot{x}_2 = -x_1 - 2x_2 + u$$

to minimize the performance measure

$$J = 10x_1^2(1) + \tfrac{1}{2} \int_0^1 (x_1^2 + 2x_2^2 + u^2)\, dt$$

(i) Determine the discrete approximation for the system. Use sampling interval $T = 0.2$ sec.

(ii) Determine the optimal control law from the discrete formulation.

11.5 A first-order system

$$\dot{x} = -x + u$$

is to be controlled to minimize

$$J = \tfrac{1}{2} \int_0^1 (x^2 + u^2)\, dt$$

Find the optimal control law.

11.6 Consider the second-order system

$$\dot{x}_1 = x_2$$

$$\dot{x}_2 = u$$

and the performance index

$$J = \tfrac{1}{2}[x_1^2(5) + 2x_2^2(5)] + \int_0^5 [x_1^2(t) + 2x_2^2(t) + x_1(t)\,x_2(t) + \tfrac{1}{4}u^2(t)]\,dt$$

Set up differential equations whose solution will yield the optimal control law.

11.7 Consider the plant

$$\begin{bmatrix} \dot{x}_1 \\ \dot{x}_2 \end{bmatrix} = \begin{bmatrix} 1 & 0 \\ -1 & 2 \end{bmatrix} \begin{bmatrix} x_1 \\ x_2 \end{bmatrix} + \begin{bmatrix} 1 \\ 0 \end{bmatrix} u$$

(i) Prove that the system is unstable.
(ii) Prove that the system is controllable.
(iii) Select any values for matrices **Q** and **R** with the constraint that they are positive definite and design a controller for the plant so as to minimize

$$J = \tfrac{1}{2}\int_0^\infty (\mathbf{x}^T\mathbf{Q}\mathbf{x} + \mathbf{u}^T\mathbf{R}\mathbf{u})\,dt$$

Check that the resulting closed-loop system is stable.

11.8 Find the optimal control law for the system

$$\dot{\mathbf{x}} = \begin{bmatrix} 0 & 1 \\ 1 & 1 \end{bmatrix} \mathbf{x} + \begin{bmatrix} 1 & 1 \\ 0 & 1 \end{bmatrix} \mathbf{u}$$

with the performance index

$$J = \int_0^\infty (x_1^2 + u_1^2 + u_2^2)\,dt$$

11.9 Test whether the optimal closed-loop system for the plant

$$\dot{\mathbf{x}} = \begin{bmatrix} 0 & 0 \\ 0 & 1 \end{bmatrix} \mathbf{x} + \begin{bmatrix} 1 \\ 1 \end{bmatrix} u$$

with the performance index

$$J = \int_0^\infty (x_1^2 + u^2)\,dt$$

is stable.

11.10 Consider the Position Servo system discussed in Section 4.6. The system is described by the state equation

$$\dot{\mathbf{x}} = \begin{bmatrix} 0 & 1 \\ 0 & -5 \end{bmatrix} \mathbf{x} + \begin{bmatrix} 0 \\ 1 \end{bmatrix} u$$

where $\mathbf{x}$ has as components the angular position θ and the angular velocity $\dot{\theta}$. It is desired to regulate the angular position to a constant value θ^0. For $\rho = \frac{1}{100}, \frac{1}{10}, \frac{1}{1000}$ find optimal control law that minimizes

$$J = \int_0^\infty [(x_1 - \theta^0)^2 + \rho u^2]\, dt$$

Find closed-loop poles for various values of ρ and comment upon your result.

11.11 Discretize the state equation and performance index of Problem 11.10. Obtain the feedback control. Compare the response of the discrete-time feedback system with that of the continuous-time feedback system of Problem 11.10 for $\mathbf{x}(0) = [2 \quad 1]^T$. Take $T = 0.1$ sec.

11.12 Consider a plant consisting of a d.c. motor, the shaft of which has the angular velocity $\omega(t)$ and which is driven by the input voltage $u(t)$. The describing equation is

$$\dot{\omega}(t) = -0.5\omega + 100u(t)$$

It is desired to regulate the angular velocity at the desired value ω^0. Find the optimal feedback control that minimizes the performance index

$$J = \int_0^{t_1} [(\omega - \omega^0)^2 + 100u^2]\, dt, \quad t_1 = 1 \text{ sec}$$

Also find time-invariant feedback constant when $t_1 \to \infty$.

11.13 Reconsider the Mixing Tank system discussed in Example 11.1 (eqns. (11.12)). Find the optimal control law that minimizes the performance index

$$J = \int_0^\infty (\mathbf{y}^T \mathbf{Q} \mathbf{y} + \mathbf{u}^T \mathbf{R} \mathbf{u})\, dt;$$

$$\mathbf{Q} = \begin{bmatrix} 25 & 0 \\ 0 & 0.01 \end{bmatrix}, \mathbf{R} = \begin{bmatrix} 2 & 0 \\ 0 & 0.5 \end{bmatrix}$$

11.14 The system

$$\dot{x} = -x + u$$

is to be controlled to minimize the performance index

$$J = \frac{1}{2} \int_0^{10} [(x(t) - r(t))^2 + u^2(t)]\, dt$$

where $r(t) = \alpha e^{-t}$ (α is any real constant).

Reformulate this problem into the form of a linear regulator problem by a suitable transformation of the state variable. Find the optimal control law.

11.15 Reconsider the Position Servo system of Problem 11.10. The set point θ^0 is now variable, i.e., we assume that θ^0 is constant over long periods of time but from time to time it is shifted. Formulate the regulator problem accommodating set point changes and obtain the solution to achieve the desired objective.

11.16 Consider now that in the Position Servo of Problem 11.10, a constant disturbance of unknown magnitude can enter in the form of a constant torque τ_0 on the shaft of the motor. The system equations now become

$$\dot{\mathbf{x}} = \begin{bmatrix} 0 & 1 \\ 0 & -5 \end{bmatrix} \mathbf{x} + \begin{bmatrix} 0 \\ 1 \end{bmatrix} u + \begin{bmatrix} 0 \\ 1/J \end{bmatrix} \tau_0$$

where J = moment of inertia of rotating parts

$= 10 \text{ kg} \cdot \text{m}^2$

Find the proportional plus integral controller for this system and compare it with purely proportional scheme of Problem 11.10.

11.17 Determine the optimal control law for the system

$$\dot{\mathbf{x}} = \begin{bmatrix} 0 & 1 \\ 0 & 0 \end{bmatrix} \mathbf{x} + \begin{bmatrix} 0 \\ 1 \end{bmatrix} u$$

$$\mathbf{y} = \begin{bmatrix} 1 & 0 \\ 0 & 2 \end{bmatrix} \mathbf{x}$$

such that the following performance index is minimized.

$$J = \int_0^{\infty} (y_1^2 + y_2^2 + u^2)\, dt$$

11.18 Prove that the matrix $\mathbf{P}(t)$ as given by eqn. (11.32b) satisfies the Riccati equation (11.22) with the boundary condition $\mathbf{P}(t_1) = \mathbf{H}$.

(Hint: Differentiate $\mathbf{P}(t)$ in (11.32b) using the rule

$$\frac{d}{dt}\mathbf{M}^{-1}(t) = -\mathbf{M}^{-1}(t)\,\dot{\mathbf{M}}(t)\,\mathbf{M}^{-1}(t)$$

and substitute $\dot{\Phi}_{ij}(t, t_1)$ in terms of $\Phi_{ij}(t, t_1)$ from eqn. (11.28).)

11.19 The linearized equations of motion governing the equatorial motion of a satellite in a circular orbit are given in Problem 6.9.

(a) Find the optimal control that minimizes the performance index

$$J = \int_0^\infty (\mathbf{x}^T\mathbf{x} + \mathbf{u}^T\mathbf{u})\,dt$$

(b) Assume that the radial thruster becomes inoperable ($u_1 = 0$). Find the optimal control law that minimizes the performance index

$$J = \int_0^\infty (\mathbf{x}^T\mathbf{x} + u_2^2)\,dt$$

Compare the closed-loop response of this system with that of part (a).

11.20 Consider the Nuclear Reactor system discussed in Section 4.6. The linearized state equation is given by

$$\dot{\mathbf{x}} = \begin{bmatrix} -6.4 & 0.1 \\ 6.4 & -0.1 \end{bmatrix}\mathbf{x} + \begin{bmatrix} 10^4 \\ 0 \end{bmatrix} u$$

It is desired to maintain the neutron level close to its steady-state value of 10 KW. Suggest a suitable performance index and find the optimal control law.

11.21 Consider the plant

$$\dot{\mathbf{x}} = \begin{bmatrix} 0 & 1 \\ 0 & 0 \end{bmatrix}\mathbf{x} + \begin{bmatrix} 0 \\ 2 \end{bmatrix} u$$

with the performance index

$$J = \int_0^\infty (x_1^2 + u^2)\,dt$$

Find the optimal control law which minimizes the performance index J and in addition ensures that the largest time constant of the characteristic equation of the closed-loop system is less than or equal to 1 sec.

11.22 Consider the plant

$$\dot{\mathbf{x}} = \begin{bmatrix} 0 & 1 \\ 0 & -1 \end{bmatrix} \mathbf{x} + \begin{bmatrix} 0 \\ 1 \end{bmatrix} u$$

$$y = [1 \quad 0]\,\mathbf{x}$$

with the performance index

$$J = \int_0^\infty (x_1^2 + x_2^2 + u^2)\, dt$$

Choose a control law that minimizes J. Design a state observer for implementation of the control law, both the poles of the state observer are required to lie at $s = -3$.

11.23 For the completely controllable system

$$\dot{\mathbf{x}} = \mathbf{Ax} + \mathbf{Bu}$$

with performance index

$$J = \tfrac{1}{2} \int_0^\infty (\mathbf{x}^T \mathbf{Q}\, \mathbf{x} + \ddot{\mathbf{u}}^T \mathbf{R}\, \ddot{\mathbf{u}})\, dt;$$

find the optimal control law that minimizes J. Show how the control law can be realized.

11.24 Consider the first-order system

$$\dot{x} = -x + u;\ x(0) = x^0$$

Discuss in detail the nature of the optimal control which minimizes

$$J = \tfrac{1}{2} \int_0^{t_1} x^2(t)\, dt;\ |u(t)| \leqslant 1$$

11.25 Consider a system described by the equations

$$\begin{bmatrix} \dot{x}_1 \\ \dot{x}_2 \end{bmatrix} = \begin{bmatrix} 0 & 1 \\ 0 & 0 \end{bmatrix} \begin{bmatrix} x_1 \\ x_2 \end{bmatrix} + \begin{bmatrix} 0 \\ 1 \end{bmatrix} u$$

Choose the feedback law

$$u = -x_1 - kx_2$$

and find the value of k so that

$$J = \tfrac{1}{2} \int_0^\infty (x_1^2 + x_2^2)\, dt$$

is minimized. Find the minimum value of $E\{J\}$.

11.26 A plant is described by the equations

$$\begin{bmatrix} \dot{x}_1 \\ \dot{x}_2 \end{bmatrix} = \begin{bmatrix} 0 & 1 \\ 0 & 0 \end{bmatrix} \begin{bmatrix} x_1 \\ x_2 \end{bmatrix} + \begin{bmatrix} 0 \\ 1 \end{bmatrix} u$$

Choose the feedback law

$$u = -k[x_1 + x_2]$$

and find the value of k so that

$$J = \frac{1}{2}\int_0^\infty (x_1^2 + x_2^2 + u^2)\, dt$$

is minimized. Compare this regulator with an optimal regulator by comparing the minimum values of $E\{J\}$ in the two cases.

11.27 Is the system

$$\dot{\mathbf{x}} = \begin{bmatrix} -3 & 1 \\ -2 & 0 \end{bmatrix} \mathbf{x} + \begin{bmatrix} 1 & 0 \\ 1 & 1 \end{bmatrix} \mathbf{u}$$

(i) normal
(ii) controllable
(iii) observable ?

11.28 Use the Minimum Principle to show that it is necessary to use a Bang-Bang controller to drive the system

$$\dot{\mathbf{x}} = \begin{bmatrix} 0 & 1 \\ 0 & 0 \end{bmatrix} \mathbf{x} + \begin{bmatrix} 0 \\ 1 \end{bmatrix} u$$

from an initial state $x_1 = 1$, $x_2 = 0$ to a zero final state if the input variable is constrained. Find such a control if $-1 \leqslant u(t) \leqslant 2$.

11.29 Find the optimal control law for transferring the system

$$\dot{x}_1 = -x_1 - u$$

$$\dot{x}_2 = -2x_2 - 2u$$

from an arbitrary initial state to the origin in minimum time. The admissible controls are constrained by $|u(t)| \leqslant 1$.

REFERENCES

1. Anderson, B.D.O., and J.B. Moore, "Linear system optimization with prescribed degree of stability", *Pro. IEE*, vol. 116, pp. 2083–87, 1969.

2. Anderson, B.D.O., and J.B. Moore, *Linear Optimal Control*, Englewood Cliffs, N.J.: Prentice-Hall, 1971.

3. Armstrong, E.S., "Series representation for the weighting matrices in the sampled-data optimal regulatory problem", *IEEE Trans. Automat. Cont.*, vol. AC–23, pp. 478–479, 1978.
4. Athans, M., "The matrix minimum principle", *Information and Control*, vol. 11, pp. 592–606, 1968.
5. Athans, M., and P.L. Falb, *Optimal Control*, New York: McGraw-Hill, 1966.
6. Bell, D.J., and D.H. Jacobson, *Singular Optimal Control Problems*, London: Academic Press, 1975.
7. Bucy, R.S., and P.D. Joseph, *Filtering for Stochastic Processes with Applications to Guidance*, New York: Interscience, 1969.
8. Caines, P.E., and D.Q. Mayne, "On the discrete-time matrix Riccati equation of optimal Control", *Intern. J. Control*, vol. 12, pp. 785–794, 1970.
9. Caines, P.E., and D.Q. Mayne, "On the discrete-time matrix Riccati equation— A correction", *Intern. J. Control*, vol. 14, pp. 205–207, 1971.
10. Davis, H.T., *Introduction to Nonlinear Differential and Integral Equations*, New York: Dover, 1962.
11. Dorato, P., and H. Levis Alexander, "Optimal linear regulators: discrete-time case", *IEEE Trans. Automat. Contr.*, vol. AC-16, pp. 613-620, 1971.
12. Fath, A.F., "Computational aspects of the linear optimal regulator problem", *IEEE Trans. Automat. Contr.*, vol. AC-14, pp. 547-550, 1969.
13. Fortmann, T.E., "Stabilization of multivariable systems with constant-gain output feedback", *Proc. JACC*, 1973.
14. Frank, P.M., *Introduction to System Sensitivity Theory*, New York: Academic Press, 1978.
15. Horisberger, H.P., and P.R. Belanger, "Solution of the optimal constant output feedback problem by conjugate gradients". *IEEE Trans. Automat. Contr.*, vol. AC-19, pp. 434-35, 1974.
16. Jameson, A., "Optimization of linear systems of constrained configuration", *Intern. J. Control*, vol. 11, pp. 409-421, 1970.
17. Johnson, C.D., "Optimal control of the linear regulator with constant disturbances", *IEEE Trans. Automat. Contr.*, vol. AC-13, pp. 416-421, 1968.
18. Johnson, C.D., "Accommodation of external disturbances in linear regulator and servomechanism problems", *IEEE Trans. Automat. Contr.*, vol. AC-16, pp. 635-644, 1971.
19. Johnson, C.D., and J. E. Gibson, "Optimal control with quadratic performance index and fixed terminal time", *IEEE Trans. Automat. Contr.*, pp. 355-360, 1964.
20. Kalman, R.E., "Contributions to the theory of optimal control", Bol. Soc. Mat., Maxicana, vol. 5, pp. 102-119, 1960.
21. Kalman, R.E., "When is a linear control system optimal?" *J. Basic Eng., Trans. ASME, Ser.* D., vol. 86, pp. 51-60, 1964.
22. Klienman, D.L., "On an iterative technique for Riccati equation computation", *IEEE Trans. Automat. Contr.*, vol. 13, pp. 114-115, 1968.
23. Levine, W.S., and M. Athans, "On the determination of the optimal constant output feedback gains for linear multivariable systems", *IEEE Trans. Automat. Contr.*, vol. AC-15, pp. 44-48, 1970.
24. Levine, W.S., T.L. Johnson, and M. Athans, "Optimal limited state variable feedback controllers for linear systems", *IEEE Trans. Automat. Contr.*, vol. AC-16, pp. 785-793, 1971.
25. Levis Alexander, H., R.A. Schlueter, and M. Athans, "On the behaviour of optimal linear sampled-data regulators", *Int. J. Control*, vol. 13, pp. 343-361, 1971.
26. MacFarlane, A.G.J., "An eigenvector solution of the optimal linear regulator", *J. Electron. Control*, vol. 14, pp. 643-654, 1963.
27. Man, F.T., and H.W. Smith, "Design of linear regulators optimal for time-multiplied performance indices", *IEEE Trans. Automat. Control*, vol. AC-14, pp. 527-529, 1969.

28. Markland, C.A., "Optimal model-following control system synthesis techniques", *Proc. IEEE*, vol. 117, pp. 623-627, 1970.
29. Martensson, K., "On the matrix Riccati equation", *Information Science*, vol. 3, pp. 17-49, 1971.
30. McClamroch, N.H., "Duality and bounds for the matrix Riccati equation", *J. Math. Anal. Appl.*, vol. 25, pp. 622-627, 1969.
31. Newell, R.B., and D.G. Fisher, "Experimental evaluation of optimal, multivariable regulating controllers with model-following capabilities", *Automatica*, vol. 8, pp. 247-262, 1972.
32. Newell, R.B., D.G. Fisher, and D.E. Soberg, "Computer control using optimal multivariable feedforward-feedback algorithm", *AIChE J.*, vol. 18, pp. 976-984, 1972.
33. O'Donnell, J.J., "Asymptotic solution of the matrix Riccati equation of optimal control", *Proc. Fourth Ann. Allerton Conf. Circuit System Theory*, Urbana, Ill., pp. 577-586, 1966.
34. O'Reilly, J., "Low-sensitivity feedback controllers for linear systems with incomplete state information", *Int. J. Control*, vol. 29, pp. 1047-1058, 1979.
35. Pontryagin, L.S., V. G. Boltyanskii, R. V. Gamkrelidze, and E.F. Mishchenko, *The Mathematical Theory of Optimal Processes*, New York: Interscience, 1962.
36. Potter, J. E., "Matrix quadratic solutions", *SIAM J. Applied Math.*, vol. 14, pp. 496-501, 1966.
37. Rynaski, E., *Jt. Autom. Control Conference Workshop*, University of Michigan, p. 193, 1968.
38. Shih, Y-P, "Integral action in the optimal control of linear systems with quadratic performance index", *Ind. Engg. Chem. Fund.*, vol, 9, pp. 35–37, 1970.
39. Smith, C.L., and P.W. Murrill, "An optimal controller for multivariable systems subject to disturbance inputs", *10th Jt. Auto. Contr. Conference*, 1969.
40. Subbayyan, R., and M.C. Vaithilingam, "Sensitivity-reduced design of linear regulators", *Int. J. Control*, vol. 29, pp. 435–440, 1979.
41. Vaughan, D.R., "A negative exponential solution for the matrix Riccati equation", *IEEE Trans. Automat. Contr.*, vol. 14, pp. 72–75, 1969.
42. VanLoan, C.F., "Computing integrals involving the matrix exponential", *IEEE Trans. Automat. Contr.*, vol. AC–23, pp. 395–404, 1978.
43. VanNess, J.E., "Inverse iteration method for finding eigenvectors", *IEEE Trans. Automat. Contr.*, vol. 14, pp. 63–66, 1969.
44. West, H.H., and M.L. McGurie, "Optimal feedforward-feedback control and dead time systems", *Ind. Eng. Chem. Funda.*, vol. 8, pp. 253–257, 1969.
45. Wilkie, D.F., and W.R. Perkins, "Generation of sensitivity functions for linear systems using low-order models", *IEEE Trans. Automat. Contr.*, vol. AC–14, pp. 125–129, 1969.
46. Wonham, W.M., "On a matrix Riccati equation of stochastic control", *SIAM J. Control*, vol. 6, pp. 681–697, 1968.
47. Wonham, W.M., and W.F. Cashman, "A computational approach to optimal control of stochastic stationary systems", *Preprints, Ninth Jt. Automatic Control Conference*, University of Michigan, pp. 13–33, 1968.
48. Yore, E.E., "Optimal decoupling control", *Preprints, Ninth Jt. Automatic Control Conference*, University of Michigan, 1968.
49. Young, P.C., and J.C. Willems, "An approach to the multivariable servomechanism problem", *Int. J. Control*, vol. 15, pp. 961–979, 1972.
50. Zinober, A., "Discrete-time control using the quadratic performance index implicit model following and the specific optimal policy", *Elect. Letter*, vol. 6, pp. 841–842, 1970.

12. STOCHASTIC OPTIMAL LINEAR ESTIMATION AND CONTROL

12.1 INTRODUCTION

In Chapter 9 we found that if the entire state vector of an observable linear system is not directly available for measurement, then it can be estimated using a state observer which operates on output and input measurements. We assumed that all inputs can be specified exactly and all outputs can be measured with unlimited precision. In practice, of course, these assumptions are never really satisfied. Input and output transducers are subject to unpredictable fluctuations and disturbances. These uncertainties, usually referred to by the generic term *noise*, are inconsequential in many cases and the Luenberger observer gives a faithful estimate of the system state. In some cases, however, the effect of the noise is too great to be ignored and it must be modelled explicitly. It is with this class of systems that this chapter it concerned.

The enormous body of knowledge that we touch upon in this chapter is perhaps best described by the term 'filtering theory'. Much of this theory is summarized in the books by Meditch (1969), Astrom (1970), Jazwinski (1970), Kushner (1971), Sage and Melsa (1971) and Anderson and Moore (1979). The particular material covered here is discussed by Kalman and Bucy (1961), Anderson and Moore (1971) and Kwakernaak and Sivan (1972).

It may be clearly pointed out at this initial stage itself that the treatment we shall give will be quite incomplete and a great deal more rigorous detail is needed to put the entire subject on a sound mathematical footing. We shall first describe very briefly and heuristically a few key elements of the probability theory. The optimal estimation problem in continuous-time systems is then described. By the introduction of new variables, we shall convert the estimation problem into a deterministic optimal regulator problem of the sort discussed in Chapter 11. The solution of the regulator problem will then yield a solution for the estimation problem. Similar results for discrete-time systems are then derived.

12.2 STOCHASTIC PROCESSES AND LINEAR SYSTEMS

Stochastic Process Characterization

We begin with the characterization of a *random variable*. We shall denote a random variable (the abstract concept) and a sample of the random variable (i.e., outcome of a single trial) by the same letter x, though many books on the probability theory adopt a notation which distinguishes the two.

A random variable is completely characterized by its associated *probability density function.* However, the aggregate behaviour of many samples of a random variable can be determined from its descriptive measures. The first of these is the *mean* or *average* or *expected value.* The mean of a random variable x is denoted as $\bar{x}$ or $E\{x\}$, where the linear operator E is known as the expectation operator.

Just as important as the mean value of a random variable x, is its expected variation, i.e., the 'spread' about the mean. The most convenient measure used for this purpose is the *variance.* The variance of x is the average value of the squared difference between x and its mean, denoted as

$$V_x = \text{var}(x) \triangleq E\{(x - \bar{x})^2\}$$

The square root of variance is called the *standard deviation.*

The level of dependence between two random variables x and y is measured principally by their covariance which is defined as

$$V_{x,y} = \text{cov}(x, y) \triangleq E\{(x - \bar{x})(y - \bar{y})\}$$

(Note that $V_{x,x} = V_x$). The covariance is bounded by

$$(V_{x,y})^2 \leqslant V_x V_y$$

Therefore the correlation coefficient of x and y,

$$\rho_{x,y} \triangleq \frac{V_{x,y}}{\sqrt{V_x V_y}}$$

may take on values between -1 and $+1$. If $|\rho_{x,y}|$ is close to 1, then x and y are said to be highly correlated or dependent. When $\rho_{x,y} = +1$, the values of x and y tend to be both large or small relative to their respective means. When $\rho_{x,y} = -1$, the values of x tend to be large when values of y are small and vice-versa. When $\rho_{x,y} = 0$, then x and y are uncorrelated.

If $x_1, x_2, \ldots, x_n$ are all random variables, it is convenient to collect them into a *random vector*

$$\mathbf{x} = [x_1 \quad x_2 \ldots x_n]^T$$

with the *mean-value vector*

$$\bar{\mathbf{x}} = E\{\mathbf{x}\} = [\bar{x}_1 \quad \bar{x}_2 \ldots \bar{x}_n]^T$$

and covariance matrix

$$\mathbf{V_x} = E\{(\mathbf{x} - \bar{\mathbf{x}})(\mathbf{x} - \bar{\mathbf{x}})^T\}$$

whose *ij*-th element is $\text{cov}(x_i, x_j)$ and *i*-th diagonal element is $\text{var}(x_i)$. The matrix $\mathbf{V_x}$ is symmetric and nonnegative definite.

A *random vector process* (or *stochastic vector process*) is a collection of random vectors $\{\mathbf{x}(t): t_0 \leqslant t \leqslant t_1\}$ defined on each point on a time interval $[t_0, t_1]$. A sample of a random process consists of a single vector function $\mathbf{x}(t)$; $t_0 \leqslant t \leqslant t_1$.

The *mean-value vector* of a random process is defined as

$$\overline{\mathbf{x}}(t) = E\{\mathbf{x}(t)\} \; ; \quad t_0 \leqslant t \leqslant t_1$$

The *covariance matrix* of a random process is defined as

$$\begin{aligned} \mathbf{V}_{\mathbf{x},\mathbf{x}}(t, \tau) &= \operatorname{cov}(\mathbf{x}(t), \mathbf{x}(\tau)) \\ &= E\{(\mathbf{x}(t) - \overline{\mathbf{x}}(t))\,(\mathbf{x}(\tau) - \overline{\mathbf{x}}(\tau))^T\} \end{aligned}$$

which measures the correlation of the process with itself at different instants of time. $\mathbf{V}_{\mathbf{x},\mathbf{x}}(t, t) = \mathbf{Q}(t)$ is termed the *variance matrix*. $\mathbf{Q}(t)$ is a square, symmetric, real non-negative definite matrix.

Random processes are divided into two broad classes: *stationary* and *nonstationary*. In a *strict sense*, a stationary process is one in which the joint density function of the process is invariant with time. We are usually content with showing the *wide-sense stationarity* which only requires that its mean $\overline{\mathbf{x}}(t)$ is constant, its second-order moment $E\{\mathbf{x}(t)\,\mathbf{x}^T(t)\}$ is finite for all t and its covariance matrix $\mathbf{V}_{\mathbf{x},\mathbf{x}}(t, \tau)$ depends on $(t - \tau)$ only.

A Gaussian stochastic process ($\mathbf{x}$ is a Gaussian stochastic process if for each set of instants of time $t_1, t_2, \ldots, t_m \geqslant t_0$, the n-dimensional vector stochastic variables $\mathbf{x}(t_1), \mathbf{x}(t_2), \ldots, \mathbf{x}(t_m)$ have a Gaussian joint probability distribution) is completely characterized by its mean and covariance matrix; thus a Gaussian process is stationary if and only if it is wide-sense stationary.

A *wide-sense white noise process* $\mathbf{w}$ is one which is completely uncorrelated from one instant to the next, i.e., $\mathbf{w}$ has a property that $\mathbf{w}(t)$ and $\mathbf{w}(\tau)$ are uncorrelated even for values of $|\tau - t|$ that are quite small:

$$\mathbf{V}_{\mathbf{w},\mathbf{w}}(t, \tau) \simeq \mathbf{0} \quad \text{for } |\tau - t| > \varepsilon$$

where ε is a 'small' number. The covariance matrix of such stochastic processes can be idealized as follows:

$$\mathbf{V}_{\mathbf{w},\mathbf{w}}(t, \tau) = \mathbf{W}(t)\,\delta(t - \tau)$$

Here $\delta(t - \tau)$ is a delta function. The non-negative definite matrix $\mathbf{W}(t)$ is referred to as the intensity of the process at time t. White noise, like the delta function, is a physically unrealizable but extremely useful mathematical artifice. Also, from a strict mathematical view point, white noise processes are not really well-defined. However, once the white noise has passed at least one integration, we are on a firm mathematical ground.

In the case in which the intensity of the white noise process is constant, the process is wide-sense stationary. Taking the Fourier transform of $\mathbf{W}\,\delta(\tau)$, we see that wide-sense stationary white noise has a constant power spectral density matrix. This is why, in analogy with white light, such processes are called white noise processes.

The correlation of the process $\mathbf{x}(t)$ and another process $\mathbf{y}(t)$ is measured in terms of *cross-covariance matrix*

$$\mathbf{V}_{\mathbf{x},\mathbf{y}}(t, \tau) = \text{cov}\,(\mathbf{x}(t), \mathbf{y}(\tau))$$

$$= E\{\mathbf{x}(t) - \bar{\mathbf{x}}(t))\,(\mathbf{y}(\tau) - \bar{\mathbf{y}}(\tau))^T\}$$

The definitions for discrete-time random processes parallel closely the definitions for continuous-time random processes presented above.

A discrete-time vector stochastic process $\mathbf{x}(i)$; $i = \ldots, -1, 0, 1, 2, \ldots$ has *mean*

$$\bar{\mathbf{x}}(i) = E\{\mathbf{x}(i)\},$$

covariance matrix

$$\mathbf{V}_{\mathbf{x},\mathbf{x}}(i, j) = \text{cov}\,(\mathbf{x}(i), \mathbf{x}(j))$$

$$= E\{(\mathbf{x}(i) - \bar{\mathbf{x}}(i))\,(\mathbf{x}(j) - \bar{\mathbf{x}}(j))^T\},$$

variance matrix

$$\mathbf{V}_{\mathbf{x},\mathbf{x}}(i, i) = \mathbf{Q}(i) = E\{(\mathbf{x}(i) - \bar{\mathbf{x}}(i))\,(\mathbf{x}(i) - \bar{\mathbf{x}}(i))^T\}$$

If the process $\mathbf{x}(i)$ is *wide-sense stationary*, then its mean is constant, its second-order moment matrix $E\{\mathbf{x}(i)\,\mathbf{x}^T(i)\}$ is finite for all i and its covariance matrix $\mathbf{V}_{\mathbf{x},\mathbf{x}}(i, j)$ depends on $(i-j)$ only.

Suppose that the stochastic process $\mathbf{x}(i)$; $i = \ldots, -1, 0, 1, 2, \ldots$ consists of a sequence of mutually uncorrelated vector-valued stochastic variables. The covariance matrix of such a process is given by

$$\mathbf{V}_{\mathbf{x},\mathbf{x}}(i-j) = \begin{cases} \mathbf{W}(i) & \text{for} \quad i = j \\ \mathbf{0} & \text{for} \quad i \neq j \end{cases}$$

$$= \mathbf{W}(i)\,\mu(i-j)$$

where $\mu(i-j)$ is the unit pulse signal defined in eqn. (7.24). This process is the discrete-time equivalent of white noise and will be referred to as *discrete white noise*. When $\mathbf{W}(i)$ does not depend on i, the discrete white noise process is wide-sense stationary.

Response of Linear Continuous-Time Systems to White Noise

In this subsection, we study some of the statistical properties of the state of a linear continuous-time system with a zero-mean white noise process as input. In particular, we compute the mean and variance of the state $\mathbf{x}(t)$, governed by the equation

$$\dot{\mathbf{x}}(t) = \mathbf{A}(t)\,\mathbf{x}(t) + \mathbf{B}(t)\,\mathbf{w}(t) \tag{12.1}$$

where

$\mathbf{x}(t)$ is $n \times 1$ state vector

$\mathbf{A}(t)$ is $n \times n$ characteristic matrix

$\mathbf{B}(t)$ is $n \times p$ control distribution matrix

$\mathbf{w}(t)$ is $p \times 1$ vector stochastic process.

The vector stochastic process $\mathbf{w}(t)$ is a zero-mean white noise with the intensity $\mathbf{Q}(t)$:

$$E\{\mathbf{w}(t)\} = \mathbf{0} \tag{12.2a}$$

$$E\{\mathbf{w}(t)\,\mathbf{w}^T(\tau)\} = \mathbf{Q}(t)\,\delta(t-\tau) \tag{12.2b}$$

$\mathbf{x}(t_0)$ is assumed to be a vector stochastic process, independent of $\mathbf{w}(t)$ with mean $\bar{\mathbf{x}}^0$ and variance $\mathbf{P}_0$:

$$E\{\mathbf{x}(t_0)\,\mathbf{w}^T(t)\} = \mathbf{0} \quad \text{for all } t \tag{12.3a}$$

$$E\{\mathbf{x}(t_0)\} = \bar{\mathbf{x}}^0 \tag{12.3b}$$

$$E\{(\mathbf{x}(t_0) - \bar{\mathbf{x}}^0)(\mathbf{x}(t_0) - \bar{\mathbf{x}}^0)^T\} = \mathbf{P}_0 \tag{12.3c}$$

The solution of eqn. (12.1) is given by (eqn. (5.56a))

$$\mathbf{x}(t) = \Phi(t, t_0)\,\mathbf{x}(t_0) + \int_{t_0}^{t} \Phi(t, \tau)\,\mathbf{B}(\tau)\,\mathbf{w}(\tau)\,d\tau \tag{12.4}$$

where $\Phi(t, t_0)$ is the state transition matrix which must satisfy the relations

$$\frac{d\Phi(t, t_0)}{dt} = \mathbf{A}(t)\,\Phi(t, t_0) \tag{12.5a}$$

$$\Phi(t_0, t_0) = \mathbf{I} \tag{12.5b}$$

From eqn. (12.4) we have

$$\begin{aligned} E\{\mathbf{x}(t)\} &= E\{\Phi(t, t_0)\,\mathbf{x}(t_0)\} + E\left\{\int_{t_0}^{t} \Phi(t, \tau),\, \mathbf{B}(\tau)\,\mathbf{w}(\tau)\,d\tau\right\} \\ &= \Phi(t, t_0)\,\bar{\mathbf{x}}^0;\ t \geqslant t_0 \end{aligned} \tag{12.6}$$

We have got this result by placing the expectation operator inside the integral and using assumption (12.2a).

To find the variance matrix of the vector stochastic process $\mathbf{x}(t)$, consider the expression

$$\mathbf{P}'(t) = E\{\mathbf{x}(t)\,\mathbf{x}^T(t)\}$$

Upon differentiating this expression we obtain

$$\dot{\mathbf{P}}'(t) = E\{\dot{\mathbf{x}}(t)\,\mathbf{x}^T(t) + \mathbf{x}(t)\,\dot{\mathbf{x}}^T(t)\}$$

Substituting $\dot{\mathbf{x}}(t)$ from eqn. (12.1) and its transpose, we see that

$$\begin{aligned}\dot{\mathbf{P}}'(t) &= E\{\mathbf{A}(t)\,\mathbf{x}(t)\,\mathbf{x}^T(t) + \mathbf{B}(t)\,\mathbf{w}(t)\,\mathbf{x}^T(t) + \mathbf{x}(t)\,\mathbf{x}^T(t)\,\mathbf{A}^T(t)\\ &\qquad + \mathbf{x}(t)\,\mathbf{w}^T(t)\,\mathbf{B}^T(t)\}\\ &= \mathbf{A}(t)\,\mathbf{P}'(t) + \mathbf{P}'(t)\,\mathbf{A}^T(t) + E\{\mathbf{B}(t)\,\mathbf{w}(t)\,\mathbf{x}^T(t)\\ &\qquad + \mathbf{x}(t)\,\mathbf{w}^T(t)\,\mathbf{B}^T(t)\}\end{aligned}$$

Substituting $\mathbf{x}(t)$ from (12.4) we get

$$\begin{aligned}\dot{\mathbf{P}}'(t) &= \mathbf{A}(t)\,\mathbf{P}'(t) + \mathbf{P}'(t)\,\mathbf{A}^T(t)\\ &\quad + E\{\mathbf{B}(t)\,\mathbf{w}(t)\,[\mathbf{x}^T(t_0)\,\Phi^T(t, t_0) + \int_{t_0}^{t} \mathbf{w}^T(\tau)\,\mathbf{B}^T(\tau)\,\Phi^T(t, \tau)\,d\tau]\}\\ &\quad + E\{[\Phi(t, t_0)\,\mathbf{x}(t_0) + \int_{t_0}^{t} \Phi(t, \tau)\,\mathbf{B}(\tau)\,\mathbf{w}(\tau)\,d\tau]\,\mathbf{w}^T(t)\,\mathbf{B}^T(t)\}\end{aligned}$$

If we place the expectation operator inside the integrals we get

$$\begin{aligned}\dot{\mathbf{P}}'(t) &= \mathbf{A}(t)\mathbf{P}'(t) + \mathbf{P}'(t)\,\mathbf{A}^T(t) + \mathbf{B}(t)\,E\{\mathbf{w}(t)\,\mathbf{x}^T(t_0)\}\,\Phi^T(t, t_0)\\ &\quad + \int_{t_0}^{t} \mathbf{B}(t)\,E\{\mathbf{w}(t)\,\mathbf{w}^T(\tau)\}\,\mathbf{B}^T(\tau)\,\Phi^T(t, \tau)\,d\tau\\ &\quad + \Phi(t, t_0)\,E\{\mathbf{x}(t_0)\,\mathbf{w}^T(t)\}\,\mathbf{B}^T(t)\\ &\quad + \int_{t_0}^{t} \Phi(t, \tau)\,\mathbf{B}(\tau)\,E\{\mathbf{w}(\tau)\,\mathbf{w}^T(t)\}\,\mathbf{B}^T(t)\,d\tau \qquad (12.7)\end{aligned}$$

By virtue of assumption (12.3a), the third and fifth terms of the right-hand side of (12.7) are zero. Substituting (12.2b) in (12.7) we get

$$\begin{aligned}\dot{\mathbf{P}}'(t) &= \mathbf{A}(t)\,\mathbf{P}'(t) + \mathbf{P}'(t)\,\mathbf{A}^T(t)\\ &\quad + \int_{t_0}^{t} \mathbf{B}(t)\,\mathbf{Q}(t)\,\delta(t-\tau)\,\mathbf{B}^T(\tau)\,\Phi^T(t-\tau)\,d\tau\\ &\quad + \int_{t_0}^{t} \Phi(t, \tau)\,\mathbf{B}(\tau)\,\mathbf{Q}(\tau)\,\delta(\tau-t)\,\mathbf{B}^T(t)\,d\tau \qquad (12.8a)\\ &= \mathbf{A}(t)\,\mathbf{P}'(t) + \mathbf{P}'(t)\,\mathbf{A}^T(t) + \mathbf{B}(t)\,\mathbf{Q}(t)\,\mathbf{B}^T(t) \qquad (12.8b)\end{aligned}$$

which is solved with the initial condition

$$\mathbf{P}'(t_0) = E\{\mathbf{x}(t_0)\,\mathbf{x}^T(t_0)\}$$

The transition from (12.8a) to (12.8b) follows from the properties of a delta function[1]

It is now a simple matter to show that the variance matrix

$$\mathbf{P}(t) = E\{(\mathbf{x}(t) - \bar{\mathbf{x}}(t))(\mathbf{x}(t) - \bar{\mathbf{x}}(t))^T\} \tag{12.9a}$$

satisfies the matrix differential equation

$$\dot{\mathbf{P}}(t) = \mathbf{A}(t)\,\mathbf{P}(t) + \mathbf{P}(t)\,\mathbf{A}^T(t) + \mathbf{B}(t)\,\mathbf{Q}(t)\,\mathbf{B}^T(t) \tag{12.9b}$$

$$\mathbf{P}(t_0) = \mathbf{P}_0 \quad \text{(Given by eqn. (12.3c))} \tag{12.9c}$$

Note that the matrix equation (12.9) is of the form of a Lyapunov equation (8.35).

Response of Linear Discrete-time Systems to White Noise

Consider now the discrete-time linear system

$$\mathbf{x}(k+1) = \mathbf{F}(k)\,\mathbf{x}(k) + \mathbf{G}(k)\,\mathbf{w}(k) \tag{12.10}$$

The process $\{\mathbf{w}(k); k = \ldots, -1, 0, 1, 2, \ldots\}$ is a discrete white process for which

$$E\{\mathbf{w}(k)\} = \mathbf{0} \tag{12.11a}$$

$$E\{\mathbf{w}(k)\,\mathbf{w}^T(j)\} = \mathbf{Q}(k)\,\mu(k-j) \tag{12.11b}$$

$\mathbf{x}(k_0)$ is a vector stochastic process, independent of $\mathbf{w}(k)$ with mean $\bar{\mathbf{x}}^0$ and variance $\mathbf{P}_0$:

$$E\{\mathbf{x}(k_0)\,\mathbf{w}^T(k)\} = \mathbf{0} \quad \text{for all } k \tag{12.12a}$$

$$E\{\mathbf{x}(k_0)\} = \bar{\mathbf{x}}^0 \tag{12.12b}$$

$$E\{(\mathbf{x}(k_0) - \bar{\mathbf{x}}^0)(\mathbf{x}(k_0) - \bar{\mathbf{x}}^0)^T\} = \mathbf{P}_0 \tag{12.12c}$$

On the lines of treatment in the earlier subsection, the following results can easily be proved:

The mean of $\mathbf{x}(k)$ is

$$\bar{\mathbf{x}}(k) = \Phi(k, k_0)\,\bar{\mathbf{x}}^0;\ k \geqslant k_0 \tag{12.13}$$

where $\Phi(k, k_0)$ is the transition matrix of difference equation (12.10).

---

[1]A symmetric delta function has the following properties:

(i) $\delta(t-\tau) = \delta(\tau-t)$

(ii) $$\int_a^b f(\tau)\,\delta(\tau-t)\,d\tau = \begin{cases} 0 & \text{if } t < a \text{ or } t > b \\ f(t) & \text{if } a < t < b \end{cases}$$

If the impulse occurs at the end of the integration interval, i.e., $t = b$, the value of the integral depends on the type of delta function used. Using a symmetrical delta function with one-half of its unit area to the right of $\tau = t$ and other half to the left, we get

$$\int^b f(\tau)\,\delta(\tau-b)\,d\tau = \frac{f(b)}{2}$$

The variance matrix of $\mathbf{x}(k)$ is $\mathbf{P}(k)$ which is the solution of the matrix difference equation

$$\mathbf{P}(k+1) = \mathbf{F}(k)\,\mathbf{P}(k)\,\mathbf{F}^T(k) + \mathbf{G}(k)\,\mathbf{Q}(k)\,\mathbf{G}^T(k)\,; \qquad k = k_0,\ k_0+1, \ldots \tag{12.14a}$$

$$\mathbf{P}(k_0) = \mathbf{P}_0 \tag{12.14b}$$

The matrix equation (12.14) is of the form of a Lyapunov equation (8.78).

12.3 OPTIMAL ESTIMATION FOR LINEAR CONTINUOUS-TIME SYSTEMS

Consider the problem of state estimation for the following dynamic system.

$$\dot{\mathbf{x}}(t) = \mathbf{A}(t)\,\mathbf{x}(t) + \mathbf{B}(t)\,\mathbf{u}(t) + \mathbf{w}(t) \tag{12.15a}$$

$$\mathbf{y}(t) = \mathbf{C}(t)\,\mathbf{x}(t) + \mathbf{v}(t) \tag{12.15b}$$

where

$\mathbf{x}(t) = n \times 1$ state vector

$\mathbf{u}(t) = p \times 1$ control vector

$\mathbf{A}(t) = n \times n$ characterizing matrix

$\mathbf{B}(t) = n \times p$ control matrix

$\mathbf{y}(t) = q \times 1$ measurement vector

$\mathbf{C}(t) = q \times n$ measurement matrix

$\mathbf{w}(t) = n \times 1$ state excitation noise vector

$\mathbf{v}(t) = q \times 1$ observation noise vector

The process $\mathbf{w}(t)$ is a white-noise process for which

$$E\{\mathbf{w}(t)\} = \mathbf{0} \tag{12.16a}$$

$$E\{\mathbf{w}(t)\,\mathbf{w}^T(\tau)\} = \mathbf{Q}(t)\,\delta(t-\tau) \tag{12.16b}$$

where $\mathbf{Q}(t)$ is a positive semidefinite $n \times n$ matrix. The process $\mathbf{v}(t)$ is white-noise process for which

$$E\{\mathbf{v}(t)\} = \mathbf{0} \tag{12.17a}$$

$$E\{\mathbf{v}(t)\,\mathbf{v}^T(\tau)\} = \mathbf{R}(t)\,\delta(t-\tau) \tag{12.17b}$$

where $\mathbf{R}(t)$ is positive definite $q \times q$ matrix.

The two stochastic processes $\mathbf{w}$ and $\mathbf{v}$, as is often the case physically, are assumed to be independent of each other:

$$E\{\mathbf{v}(t)\,\mathbf{w}^T(\tau)\} = \mathbf{0} \tag{12.18}$$

Final assumptions concern the initial state of (12.15a). State estimation is assumed to commence at some time t_0 which may be minus infinity or may be finite.

The initial state $\mathbf{x}(t_0)$ is assumed to be random vector with mean

$$E\{\mathbf{x}(t_0)\} = \overline{\mathbf{x}}^0 \tag{12.19a}$$

and $n \times n$ positive semidefinite variance matrix

$$E\{(\mathbf{x}(t_0) - \overline{\mathbf{x}}^0)(\mathbf{x}(t_0) - \overline{\mathbf{x}}^0)^T\} = \mathbf{P}_0 \tag{12.19b}$$

It is further assumed that $\mathbf{x}(t_0)$ is independent of $\mathbf{w}(t)$ and $\mathbf{v}(t)$:

$$E\{\mathbf{x}(t_0)\,\mathbf{w}^T(t)\} = E\{\mathbf{x}(t_0)\,\mathbf{v}^T(t)\} = \mathbf{0} \tag{12.19c}$$

Of the requirements listed above for $\mathbf{v}$ and $\mathbf{w}$, the white noise requirement is perhaps the most difficult one to justify on a practical basis. No actual signal can ever satisfy the white-noise assumption, but if its frequency spectrum is appreciably flat over a frequency range one or more decades beyond the cross-over frequency of the system, it may be approximated by white noise with no practical loss of accuracy. From this viewpoint, it is particularly easy to justify the white noise assumption for the typically low frequency control systems.

Note that the model (12.15) assumes that the noise is injected at only two points as shown in Fig. 12.1. This restriction is not severe as might at

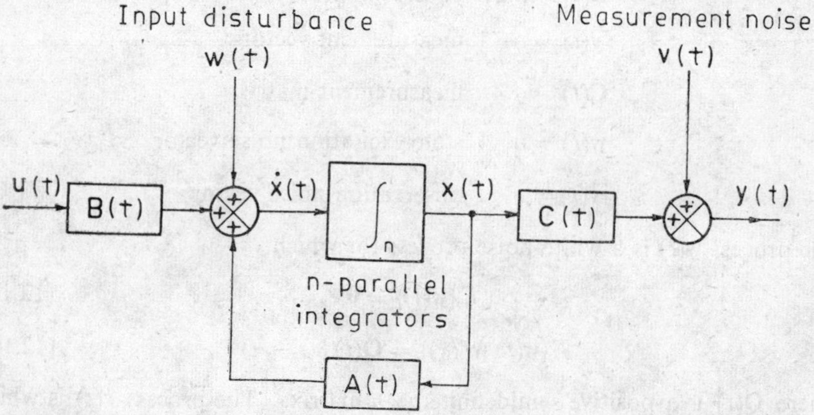

Fig. 12.1 System representation including input disturbance and measurement noise

first appear, e.g., any noise entering with $\mathbf{u}(t)$ and passing through the $\mathbf{B}(t)$ block is equivalent to some other noise entering at the same point as $\mathbf{w}(t)$.

We have made a number of assumptions regarding the estimation problem. Although these assumptions have quite often physical validity, there will undoubtedly be many cases when this is not so. Many associated extensions of the problem formulated above have been solved. However, we shall not be able to accommodate these extensions in this book.

For the problem under consideration, we postulate the existence of an estimator of the form

$$\dot{\hat{\mathbf{x}}}(t) = \mathbf{A}(t)\,\hat{\mathbf{x}}(t) + \mathbf{B}(t)\,\mathbf{u}(t) + \mathbf{K}(t)\,[\mathbf{C}(t)\,\hat{\mathbf{x}}(t) - \mathbf{y}(t)] \tag{12.20}$$

The estimator structure is shown in Fig. 12.2 which is of the same form as that considered in Section 9.5 except that noise is present at certain points in the plant.

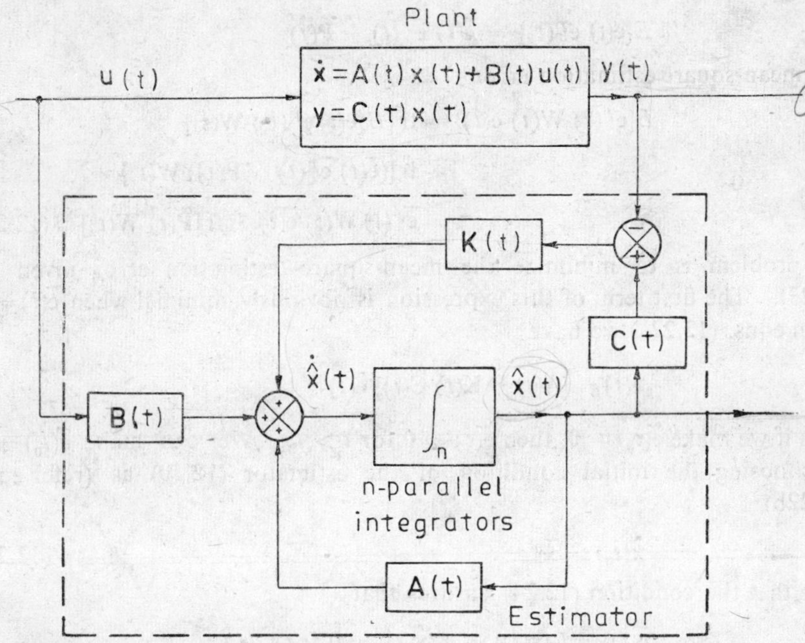

Fig. 12.2 Structure of optimal estimator

The estimation error is given by

$$\mathbf{e}(t) = \mathbf{x}(t) - \hat{\mathbf{x}}(t) \tag{12.21}$$

The mean square estimation error

$$E\{\mathbf{e}^T(t)\,\mathbf{W}(t)\,\mathbf{e}(t)\}$$

with $\mathbf{W}(t)$ a given positive definite symmetric weighing matrix, is a measure of how well the estimator reconstructs the state of the system at time t. The mean square estimation error is determined by the choice of $\hat{\mathbf{x}}(t_0)$ and of $\mathbf{K}(\tau)$, $t_0 \leqslant \tau \leqslant t$. The problem before us is how to choose $\hat{\mathbf{x}}(t_0)$ and $\mathbf{K}(\tau)$ so that mean-square estimation error is minimized.

From eqns. (12.15) and (12.20) we have

$$\dot{\mathbf{x}}(t) - \dot{\hat{\mathbf{x}}}(t) = \dot{\mathbf{e}}(t) = [\mathbf{A}(t) + \mathbf{K}(t)\,\mathbf{C}(t)]\,\mathbf{e}(t) + [\mathbf{I} + \mathbf{K}(t)]\begin{bmatrix}\mathbf{w}(t)\\ \mathbf{v}(t)\end{bmatrix} \tag{12.22a}$$

$$\mathbf{e}(t_0)=\mathbf{x}(t_0)-\hat{\mathbf{x}}(t_0) \qquad (12.22b)$$

Let us denote by $\widetilde{\mathbf{P}}(t)$ the variance matrix of $\mathbf{e}(t)$ and by $\overline{\mathbf{e}}(t)$ the mean of $\mathbf{e}(t)$:

$$E\{\mathbf{e}(t)\}=\overline{\mathbf{e}}(t)$$

$$E\{(\mathbf{e}(t)-\overline{\mathbf{e}}(t))\,(\mathbf{e}(t)-\overline{\mathbf{e}}(t))^T\}=\widetilde{\mathbf{P}}(t)$$

This gives

$$E\{\mathbf{e}(t)\,\mathbf{e}^T(t)\}=\overline{\mathbf{e}}(t)\,\overline{\mathbf{e}}^T(t)+\widetilde{\mathbf{P}}(t)$$

The mean-square estimation error

$$\begin{aligned} E\{\mathbf{e}^T(t)\,\mathbf{W}(t)\,\mathbf{e}(t)\} &= \mathrm{tr}\,[E\{\mathbf{e}(t)\,\mathbf{e}^T(t)\}\,\mathbf{W}(t)] \\ &= \mathrm{tr}\,[(\overline{\mathbf{e}}(t)\,\overline{\mathbf{e}}^T(t)+\widetilde{\mathbf{P}}(t))\,\mathbf{W}(t)] \\ &= \overline{\mathbf{e}}^T(t)\,\mathbf{W}(t)\,\overline{\mathbf{e}}(t)+\mathrm{tr}\,[\widetilde{\mathbf{P}}(t)\,\mathbf{W}(t)] \end{aligned} \qquad (12.23)$$

Our problem is to minimize the mean-square estimation error given by (12.23). The first term of this expression is obviously minimal when $\overline{\mathbf{e}}(t)=\mathbf{0}$. From eqns. (12.22a) we have

$$\dot{\overline{\mathbf{e}}}(t)=[\mathbf{A}(t)+\mathbf{K}(t)\,\mathbf{C}(t)]\,\overline{\mathbf{e}}(t)$$

Thus if we make $\overline{\mathbf{e}}(t_0)=\mathbf{0}$, then $\overline{\mathbf{e}}(t)=\mathbf{0}$ for $t>t_0$. We can make $\overline{\mathbf{e}}(t_0)=\mathbf{0}$ by choosing the initial condition of the estimator (12.20) as (refer eqn. (12.22b)

$$\hat{\mathbf{x}}(t_0)=\overline{\mathbf{x}}^0 \qquad (12.24)$$

Note that the condition (12.24) ensures that

$$\overline{\mathbf{e}}(t)=E\{\mathbf{x}(t)\}-E\{\hat{\mathbf{x}}(t)\}=\mathbf{0} \text{ for } t\geqslant t_0$$

or

$$E\{\hat{\mathbf{x}}(t)\}=E\{\mathbf{x}(t)\} \qquad \text{for } t\geqslant t_0$$

An estimator whose expected value is equal to the expected value of the quantity being estimated is called *unbiased estimator.*

Since the second term of (12.23) does not depend on $\overline{\mathbf{e}}(t)$, it can be minimized independently.

Using eqns. (12.1), (12.9) and (12.22), we get the following result:

$$\begin{aligned} \dot{\widetilde{\mathbf{P}}}(t)=[\mathbf{A}(t)+\mathbf{K}(t)\,\mathbf{C}(t)]\,\widetilde{\mathbf{P}}(t)+\widetilde{\mathbf{P}}(t)[\mathbf{A}(t)+\mathbf{K}(t)\,\mathbf{C}(t)]^T \\ +\mathbf{Q}(t)+\mathbf{K}(t)\,\mathbf{R}(t)\,\mathbf{K}^T(t) \end{aligned} \qquad (12.25a)$$

$$\begin{aligned} \widetilde{\mathbf{P}}(t_0) &= E\{(\mathbf{x}(t_0)-\hat{\mathbf{x}}(t_0))\,(\mathbf{x}(t_0)-\hat{\mathbf{x}}(t_0))^T\} \\ &= E\{(\mathbf{x}(t_0)-\overline{\mathbf{x}}^0)(\mathbf{x}(t_0)-\overline{\mathbf{x}}^0)^T\} \\ &= \mathbf{P}_0 \end{aligned} \qquad (12.25b)$$

To minimize tr $[\widetilde{\mathbf{P}}(t)\,\mathbf{W}(t)]$ optimally, we use the well known result of optimal linear regulator (Athans (1967) uses the maximum principle directly). Table 11.3 gives a summary of these results. As per these results, for any nonnegative definite matrix, say $\bar{\mathbf{P}}(t)$, other than the one satisfying the matrix Riccati equation

$$-\dot{\mathbf{P}}(t) = \mathbf{Q}(t) - \mathbf{P}(t)\mathbf{B}(t)\mathbf{R}^{-1}(t)\mathbf{B}^T(t)\mathbf{P}(t) + \mathbf{P}(t)\mathbf{A}(t) + \mathbf{A}^T(t)\mathbf{P}(t) \tag{12.26a}$$

$$\mathbf{P}(t_1) = \mathbf{H} \tag{12.26b}$$

we have

$$\mathbf{x}^T(t)\bar{\mathbf{P}}(t)\mathbf{x}(t) \geqslant \mathbf{x}^T(t)\mathbf{P}(t)\mathbf{x}(t) \quad \text{for all } \mathbf{x}(t)$$

Consider now the matrix differential equation (Wonham 1968a)

$$-\dot{\bar{\mathbf{P}}}(t) = \mathbf{Q}(t) + \mathbf{K}^T(t)\mathbf{R}(t)\mathbf{K}(t) + \bar{\mathbf{P}}(t)[\mathbf{A}(t) + \mathbf{B}(t)\mathbf{K}(t)] + [\mathbf{A}(t) + \mathbf{B}(t)\mathbf{K}(t)]^T\bar{\mathbf{P}}(t) \tag{12.27a}$$

$$\bar{\mathbf{P}}(t_1) = \mathbf{H} \tag{12.27b}$$

where $\mathbf{Q}(t)$, $\mathbf{R}(t)$, $\mathbf{A}(t)$ and $\mathbf{B}(t)$ are given time-varying matrices of appropriate dimensions with $\mathbf{Q}(t)$ nonnegative definite and $\mathbf{R}(t)$ positive definite for $t_0 \leqslant t \leqslant t_1$ and $\mathbf{H}$ nonnegative definite constant matrix. Let $\mathbf{K}(t)$ be an arbitrary continuous matrix function for $t_0 \leqslant t \leqslant t_1$. Then for $t_0 \leqslant t \leqslant t_1$,

$$\mathbf{x}^T(t)\bar{\mathbf{P}}(t)\mathbf{x}(t) \geqslant \mathbf{x}^T(t)\mathbf{P}(t)\mathbf{x}(t) \text{ for all } \mathbf{x}(t) \tag{12.28}$$

where $\mathbf{P}(t)$ is the solution of the matrix Riccati equation (12.26).

If we substitute

$$\mathbf{K}(t) = -\mathbf{R}^{-1}(t)\mathbf{B}^T(t)\bar{\mathbf{P}}(t)$$

in eqn. (12.27), we get an equation in $\bar{\mathbf{P}}(t)$ identical with matrix Riccati equation (12.26) in $\mathbf{P}(t)$; $\bar{\mathbf{P}}(t) = \mathbf{P}(t)$ and inequality (12.28) converts into an equality. Thus $\bar{\mathbf{P}}(t)$ is 'minimized' in the sense defined by (12.28) by choosing $\mathbf{K}(\tau)$ as

$$\mathbf{K}(\tau) = -\mathbf{R}^{-1}(\tau)\mathbf{B}^T(\tau)\mathbf{P}(\tau) \text{ for } t \leqslant \tau \leqslant t_1 \tag{12.29}$$

where $\mathbf{P}(t)$ satisfies the matrix Riccati equation (12.26).

Let us now go back to eqn. (12.25) and try to minimize $\tilde{\mathbf{P}}(t)$ in the sense defined above. We will have to first reverse time in (12.25) to bring it to the form of (12.27).

Introduce[2] a change in variable from t to $\hat{t} - t$ ($\hat{t} = t_0 + t_1$ with $t_1 > t_0$). This gives

---

[2]Consider the differential equations

$$\frac{dx(t)}{dt} = f(t, x(t));\ x(t_0) = x^0,\ t \geqslant t_0$$

and $$-\frac{dy(t)}{dt} = f(-t, y(t));\ y(t_1) = y^1,\ t \leqslant t_1.$$

If $x^0 = y^1$ and $\hat{t} = t_0 + t_1$ with $t_0 < t_1$, we have

$$x(t) = y(\hat{t} - t),\ t \geqslant t_0$$

$$y(t) = x(\hat{t} - t),\ t \leqslant t_1$$

$$-\dot{\widetilde{\mathbf{M}}}(t)=[\mathbf{A}^T(\hat{t}-t)+\mathbf{C}^T(\hat{t}-t)\,\mathbf{K}^T(\hat{t}-t)]^T\,\widetilde{\mathbf{M}}(t)$$
$$+\widetilde{\mathbf{M}}(t)\,[\mathbf{A}^T(\hat{t}-t)+\mathbf{C}^T(\hat{t}-t)\,\mathbf{K}^T(\hat{t}-t)]+\mathbf{Q}(\hat{t}-t)$$
$$+\mathbf{K}(\hat{t}-t)\,\mathbf{R}(\hat{t}-t)\,\mathbf{K}^T(\hat{t}-t);\ t\leqslant t_1 \tag{12.30a}$$

If

$$\widetilde{\mathbf{M}}(t_1)=\mathbf{P}_0, \tag{12.30b}$$

then

$$\widetilde{\mathbf{M}}(t)=\widetilde{\mathbf{P}}(\hat{t}-t);\ t\leqslant t_1 \tag{12.30c}$$

As per the result established in (12.27)–(12.29), the matrix $\widetilde{\mathbf{M}}(t)$ is 'minimized' in the sense defined earlier in (12.28) if $\mathbf{K}(\hat{t}-\tau);\ t\leqslant\tau\leqslant t_1$ is chosen as

$$\mathbf{K}^T(\hat{t}-\tau)=-\mathbf{R}^{-1}(\hat{t}-\tau)\,\mathbf{C}(\hat{t}-\tau)\,\mathbf{M}(\tau) \tag{12.31}$$

where $\mathbf{M}(\tau)$ is the solution of (12.30) with $\mathbf{K}$ given by (12.31), i.e.,

$$-\dot{\mathbf{M}}(t)=\mathbf{Q}(\hat{t}-t)-\mathbf{M}(t)\,\mathbf{C}^T(\hat{t}-t)\,\mathbf{R}^{-1}(\hat{t}-t)\,\mathbf{C}(\hat{t}-t)\,\mathbf{M}(t)$$
$$+\mathbf{M}(t)\,\mathbf{A}^T(\hat{t}-t)+\mathbf{A}(\hat{t}-t)\,\mathbf{M}(t);\ t\leqslant t_1$$
$$\mathbf{M}(t_1)=\mathbf{P}_0$$

The 'minimal value' of $\widetilde{\mathbf{M}}(t)$ is $\mathbf{M}(t)$. From (12.30c) and (12.31) we see that, reversing time back again, the variance matrix $\widetilde{\mathbf{P}}(t)$ of $\mathbf{e}(t)$ is minimized in the sense defined earlier by choosing

$$\mathbf{K}(\tau)=-\mathbf{P}(\tau)\,\mathbf{C}^T(\tau)\,\mathbf{R}^{-1}(\tau);\ \tau\geqslant t_0 \tag{12.32}$$

where the matrix $\mathbf{P}(t)$ satisfies the matrix Riccati equation

$$\dot{\mathbf{P}}(t)=\mathbf{Q}(t)-\mathbf{P}(t)\,\mathbf{C}^T(t)\,\mathbf{R}^{-1}(t)\,\mathbf{C}(t)\,\mathbf{P}(t)+\mathbf{P}(t)\,\mathbf{A}^T(t)$$
$$+\mathbf{A}(t)\,\mathbf{P}(t);\ t\geqslant t_0 \tag{12.33a}$$

with the initial condition

$$\mathbf{P}(t_0)=\mathbf{P}_0 \tag{12.33b}$$

Since $\mathbf{x}^T(t)\,\mathbf{P}(t)\,\mathbf{x}(t)\leqslant\mathbf{x}^T(t)\widetilde{\mathbf{P}}(t)\,\mathbf{x}(t)$ implies that

$$\mathrm{tr}\,[\mathbf{P}(t)\,\mathbf{W}(t)]\leqslant\mathrm{tr}\,[\widetilde{\mathbf{P}}(t)\,\mathbf{W}(t)]$$

for any positive definite symmetric matrix $\mathbf{W}(t)$, we conclude that the gain matrix (12.32) optimizes the estimator. We find that the solution is independent of the weighing matrix $\mathbf{W}(t)$ and hence for simplicity we assume $\mathbf{W}(t)=\mathbf{I}$.

The optimal estimator described above is known as the *Kalman filter*. (The gain $\mathbf{K}(\tau)$ given by (12.32) is referred to as the *Kalman gain*). We have derived this filter by first assuming that it has the form of an observer. In the original derivation (Kalman and Bucy 1961) it is proved that this filter is the *minimum mean square linear estimator*. It can also be proved that (Jazwinski

1970) if the initial state $\mathbf{x}(t_0)$ is Gaussian and the state excitation noise $\mathbf{w}(t)$ and observation noise $\mathbf{v}(t)$ are Gaussian white-noise processes, the Kalman filter produces an estimate $\hat{\mathbf{x}}(t)$ of $\mathbf{x}(t)$ that has the minimal mean square error among all linear and nonlinear estimates.

We have also noted in our derivation that the Kalman filter is a *minimum error variance unbiased* estimator. These are very desirable properties of an estimator. Table 12.1 provides a summary of the results derived in this section. Figure 12.3 shows the structure of a plant with the optimal estimator.

Table 12.1: Summary of Continuous-time Kalman-filter Algorithm

Message model	$\dot{\mathbf{x}}(t)=\mathbf{A}(t)\,\mathbf{x}(t)+\mathbf{B}(t)\,\mathbf{u}(t)+\mathbf{w}(t)$
Observation model	$\mathbf{y}(t)=\mathbf{C}(t)\,\mathbf{x}(t)+\mathbf{v}(t)$
Prior statistics	$E\{\mathbf{w}(t)\}=E\{\mathbf{v}(t)\}=\mathbf{0};\ E\{\mathbf{x}(t_0)\}=\bar{\mathbf{x}}^0$ $E\{\mathbf{w}(t)\mathbf{w}^T(\tau)\}=\mathbf{Q}(t)\,\delta(t-\tau)$ $E\{\mathbf{v}(t)\,\mathbf{v}^T(\tau)\}=\mathbf{R}(t)\,\delta(t-\tau)$ $E\{\mathbf{w}(t)\,\mathbf{v}^T(\tau)\}=E\{\mathbf{x}(t_0)\mathbf{v}^T(t)\}=E\{\mathbf{x}(t_0)\mathbf{w}^T(t)\}=\mathbf{0}$ $E\{(\mathbf{x}(t_0)-\bar{\mathbf{x}}^0)\,(\mathbf{x}(t_0)-\bar{\mathbf{x}}^0)^T\}=\mathbf{P}_0$
Filter algorithm	$\dot{\hat{\mathbf{x}}}(t)=\mathbf{A}(t)\,\hat{\mathbf{x}}(t)+\mathbf{B}(t)\,\mathbf{u}(t)+\mathbf{K}(t)\,[\mathbf{C}(t)\hat{\mathbf{x}}(t)-\mathbf{y}(t)];$ $\hat{\mathbf{x}}(0)=\bar{\mathbf{x}}^0$
Kalman gain algorithm	$\mathbf{K}(t)=-\mathbf{P}(t)\,\mathbf{C}^T(t)\,\mathbf{R}^{-1}(t)$
Error-variance algorithm	$\dot{\mathbf{P}}(t)=\mathbf{Q}(t)-\mathbf{P}(t)\mathbf{C}^T(t)\,\mathbf{R}^{-1}(t)\,\mathbf{C}(t)\,\mathbf{P}(t)+\mathbf{P}(t)\,\mathbf{A}^T(t)+\mathbf{A}(t)\,\mathbf{P}(t)$ $\mathbf{P}(t_0)=\mathbf{P}_0$
Mean square reconstruction error algorithm	Mean square Reconstruction Error $=$ tr [P]

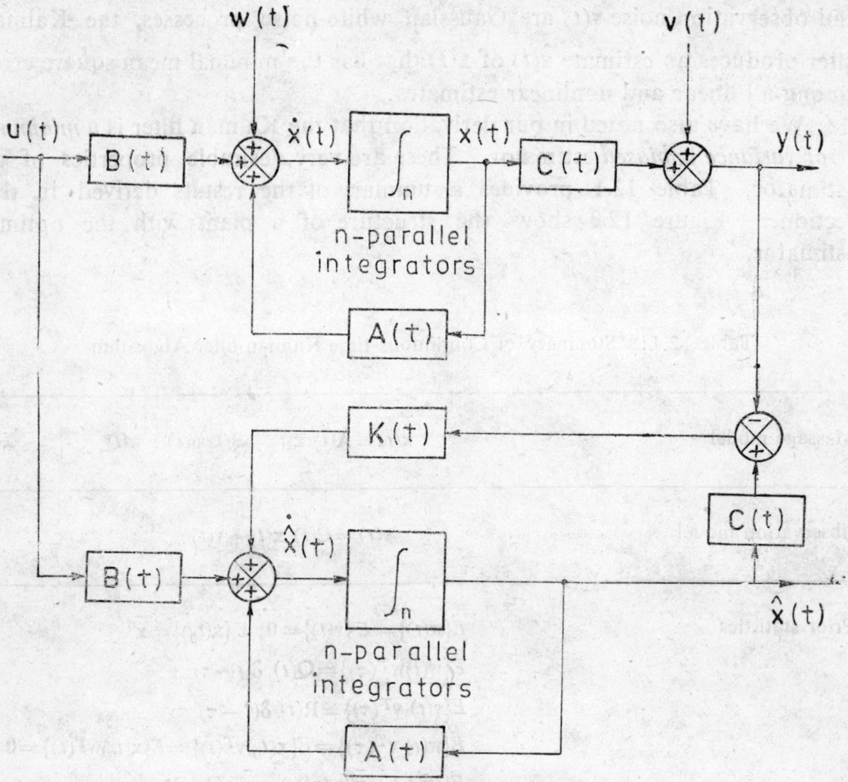

Fig. 12.3 Structure of plant with optimal estimator

Duality with Optimal Linear Regulator

Kalman and Bucy (1961) have shown that the dual of the optimal estimation problem is the optimal regulator problem.

Consider the estimation problem summarized in Table 12.1. Using time reversal (Refer Footnote 2 of this Chapter), the error-variance Riccati equation is transformed to the following:

$$-\dot{\mathbf{M}}(t) = \mathbf{Q}(\hat{t}-t) - \mathbf{M}(t)\,\mathbf{C}^T(\hat{t}-t)\,\mathbf{R}^{-1}(\hat{t}-t)\,\mathbf{C}(\hat{t}-t)\,\mathbf{M}(t) + \mathbf{M}(t)\,\mathbf{A}^T(\hat{t}-t) + \mathbf{A}(\hat{t}-t)\,\mathbf{M}(t)$$

Here $\hat{t} = t_0 + t_1$ with $t_1 > t_0$

If $\mathbf{M}(t_1) = \mathbf{P}_0$

then

$$\mathbf{M}(t) = \mathbf{P}(\hat{t}-t); \quad t \leqslant t_1$$

Let us compare this result with that of the optimal linear regulator summarized in Table 11.3. It is observed that various matrices occurring in the definitions of the optimal estimation problem and optimal regulator problem are related as follows:

Optimal Estimation Problem		*Optimal Regulator Problem*
$\mathbf{A}^T(\hat{t}-t)$	$\longrightarrow$	$\mathbf{A}(t)$
$\mathbf{C}^T(\hat{t}-t)$	$\longrightarrow$	$\mathbf{B}(t)$
$\mathbf{R}(\hat{t}-t)$	$\longrightarrow$	$\mathbf{R}(t)$
$\mathbf{Q}(\hat{t}-t)$	$\longrightarrow$	$\mathbf{Q}(t)$
$\mathbf{P}_0$	$\longrightarrow$	$\mathbf{H}$

Under these conditions the solutions of the two problems are related as follows:

Optimal Estimation Problem		*Optimal Regulator Problem*
$\mathbf{P}(\hat{t}-t)$	$\longrightarrow$	$\mathbf{P}(t)$
$\mathbf{K}^T(\hat{t}-t)$	$\longrightarrow$	$\mathbf{K}(t)$

The duality enables us to use computer programmes designed for optimal regulator problems, for optimal estimation problems and vice-versa. In the next subsection, we shall use this duality to obtain the steady-state properties of an optimal estimator from those of an optimal regulator.

Example 12.1: In Problem 11.16 we considered the Position Servo described by the state equation

$$\dot{\mathbf{x}}(t) = \mathbf{A}\mathbf{x}(t) + \mathbf{b}u(t) + \mathbf{\Gamma}\tau$$

where

$$\mathbf{A} = \begin{bmatrix} 0 & 1 \\ 0 & -5 \end{bmatrix}; \ \mathbf{b} = \begin{bmatrix} 0 \\ 1 \end{bmatrix}; \ \mathbf{\Gamma} = \begin{bmatrix} 0 \\ 0.1 \end{bmatrix}$$

We assumed that $\tau = \tau_0$ is a constant disturbing torque acting upon the shaft of the motor.

We now pose a different problem here. Assume that τ_0 is a zero-mean white noise with a constant scalar intensity 100. Further, let the measured variable be given by

$$y(t) = [1 \quad 0]\,\mathbf{x}(t) + v(t)$$

where $v(t)$ is a zero-mean white noise with constant scalar intensity which shall be taken as equal to 1 for this example. We will also assume that

$$\mathbf{P}(t_0) = \mathbf{P}_0 = \begin{bmatrix} 1 & 0 \\ 0 & 0 \end{bmatrix}$$

For this problem, the variance Riccati equation takes the form (Table 12.1)

$$\dot{\mathbf{P}}(t) = \mathbf{F}\mathbf{Q}\,\mathbf{F}^T - \mathbf{P}(t)\,\mathbf{c}^T\mathbf{R}^{-1}\,\mathbf{c}\mathbf{P}(t) + \mathbf{P}(t)\,\mathbf{A}^T + \mathbf{A}\mathbf{P}(t)$$

$$= \begin{bmatrix} 0 \\ 0.1 \end{bmatrix} 100\,[0 \quad 0.1] - \mathbf{P}(t) \begin{bmatrix} 1 \\ 0 \end{bmatrix} [1 \quad 0]\,\mathbf{P}(t)$$

$$+ \mathbf{P}(t) \begin{bmatrix} 0 & 0 \\ 1 & -5 \end{bmatrix} + \begin{bmatrix} 0 & 1 \\ 0 & -5 \end{bmatrix} \mathbf{P}(t)$$

$$\mathbf{P}(t_0) = \mathbf{P}_0$$

Kalman gain vector

$$\mathbf{k}(t) = -\,\mathbf{P}(t)\,\mathbf{c}^T(t)\,\mathbf{R}^{-1}(t)$$

$$= -\,\mathbf{P}(t) \begin{bmatrix} 1 \\ 0 \end{bmatrix}$$

Figure 12.4 illustrates the block diagram of the optimal estimator and gives the Kalman gains resulting from the computer solution of the Riccati equation. It may be noted that if observation time is long compared to the dominant time constants of the process, $\mathbf{k}(t)$ converges to a constant value $\mathbf{k}$.

Time-Invariant Linear State Estimator

In the time-invariant version of the general state estimation problem, the following three assumptions must be satisfied (refer Table 12.1):

1. The message and observation models are time-invariant.

$$\dot{\mathbf{x}}(t) = \mathbf{A}\mathbf{x}(t) + \mathbf{B}\mathbf{u}(t) + \mathbf{w}(t) \tag{12.34a}$$

$$\mathbf{y}(t) = \mathbf{C}\mathbf{x}(t) + \mathbf{v}(t) \tag{12.34b}$$

   where **A**, **B** and **C** are constant matrices of appropriate dimensions.
2. The state excitation noise $\mathbf{w}(t)$ and the measurement noise $\mathbf{v}(t)$ are at least wide-sense stationary; the matrices **Q** and **R** are therefore constant matrices.
3. The observation begins at $t_0 = -\infty$. Obviously this condition can never be true in practice; however, as long as the observation time is long compared to the dominant time constants of the process, the assumption is valid.

Under these conditions, the stochastic process $\hat{\mathbf{x}}(t)$ and hence

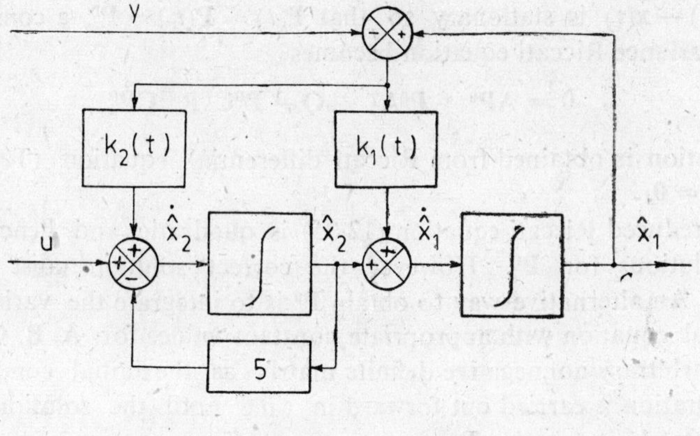

(a) Kalman filter for Example 12.1

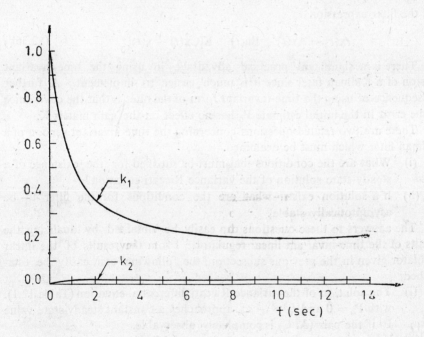

(b) Kalman gains for Example 12.1

Fig. 12.4

$\mathbf{e}(t) = \mathbf{x}(t) - \hat{\mathbf{x}}(t)$ is stationary so that $\mathbf{P}(t) = \mathbf{P}(t_0) = \mathbf{P}^0$, a constant matrix and the variance Riccati equation becomes

$$\mathbf{0} = \mathbf{A}\mathbf{P}^0 + \mathbf{P}^0\mathbf{A}^T + \mathbf{Q} - \mathbf{P}^0\mathbf{C}^T\mathbf{R}^{-1}\mathbf{C}\mathbf{P}^0 \tag{12.35}$$

The equation is obtained from Riccati differential equation (Table 12.1) by setting $\dot{\mathbf{P}} = \mathbf{0}$.

The reduced Riccati equation (12.35) is quadratic and hence there are many solutions for $\mathbf{P}^0$. However, the correct solution must be positive definite. An alternative way to obtain $\mathbf{P}^0$ is to integrate the variance Riccati differential equation with appropriate constant values for **A, B, C, Q** and **R** and any arbitrary nonnegative definite matrix as the initial condition for **P**. The integration is carried out forward in time until the solution reaches a constant steady-state value $\mathbf{P}^0$.

Corresponding to the constant matrix $\mathbf{P}^0$, we obtain the Kalman gain matrix

$$\mathbf{K} = -\mathbf{P}^0\mathbf{C}^T\mathbf{R}^{-1} \tag{12.36a}$$

and the filter expression

$$\dot{\hat{\mathbf{x}}}(t) = \mathbf{A}\hat{\mathbf{x}}(t) + \mathbf{B}\mathbf{u}(t) + \mathbf{K}[\mathbf{C}\hat{\mathbf{x}}(t) - \mathbf{y}(t)] \tag{12.36b}$$

There is a significant practical advantage in using the time-invariant version of a Kalman filter since it is much easier to implement. A further consequence of using the time-invariant form of the filter is that the covariance of the error in the initial estimate $\mathbf{P}_0$ has no effect on the gain matrix **K**.

There are two related questions concerning the time-invariant version of a Kalman filter which must be examined.

(i) What are the conditions that must be satisfied for the existence of a steady-state solution of the variance Riccati equation?

(ii) If a solution exists, what are the conditions for the filter to be asymptotically stable?

The answers to these questions can easily be obtained by dualizing the results of the time-invar ant linear regulator. From the results of the linear regulator given in the prev ous subsection, the following can easily be established.

(i) The solution of the variance Riccati differential equation (Table 12.1), with $\mathbf{P}_0 = \mathbf{0}$ and $t_0 \to -\infty$, approaches a constant steady-state value $\mathbf{P}^0$ if the pair (**A, C**) is completely observable.

(ii) If $\mathbf{P}^0$ exists, the time-invariant filter (12.36b) is asymptotically stable if the pair $(\mathbf{A}, \mathbf{\Gamma})$ is completely controllable where $\mathbf{\Gamma}$ is any matrix such that $\mathbf{\Gamma}\mathbf{\Gamma}^T = \mathbf{Q}$.

These restrictions are sufficient for the existence of asymptotically stable filter. We can dualize the results of Section 11.4 for milder restrictions.

12.4 OPTIMAL ESTIMATION FOR LINEAR DISCRETE-TIME SYSTEMS

Analogous to the continuous-time estimation problem discussed in the previous section, the discrete-time estimation problem can be formulated as follows:

Consider the system

$$\mathbf{x}(k+1) = \mathbf{F}(k)\,\mathbf{x}(k) + \mathbf{G}(k)\,\mathbf{u}(k) + \mathbf{w}(k) \tag{12.37a}$$

$$\mathbf{y}(k) = \mathbf{C}(k)\,\mathbf{x}(k) + \mathbf{v}(k) \tag{12.37b}$$

$$k = 0, 1, 2, \ldots$$

The process $\mathbf{w}(k)$ is a white-noise process for which

$$E\{\mathbf{w}(k)\} = \mathbf{0} \tag{12.38a}$$

$$E\{\mathbf{w}(k)\,\mathbf{w}^T(j)\} = \mathbf{Q}(k)\,\mu(k-j) \tag{12.38b}$$

where $\mathbf{Q}(k)$ is a positive semidefinite matrix of proper dimensions.

The process $\mathbf{v}(k)$ is a white-noise process for which

$$E\{\mathbf{v}(k)\} = \mathbf{0} \tag{12.39a}$$

$$E\{\mathbf{v}(k)\,\mathbf{v}^T(j)\} = \mathbf{R}(k)\,\mu(k-j) \tag{12.39b}$$

where $\mathbf{R}(k)$ is a positive definite matrix of proper dimensions. The two stochastic processes $\mathbf{w}$ and $\mathbf{v}$ are assumed to be independent of each other:

$$E\{\mathbf{v}(j)\,\mathbf{w}^T(k)\} = \mathbf{0} \tag{12.40}$$

The initial state $\mathbf{x}(0)$ is assumed to be random vector with mean

$$E\{\mathbf{x}(0)\} = \bar{\mathbf{x}}^0 \tag{12.41a}$$

and variance matrix

$$E\{\mathbf{x}(0) - \bar{\mathbf{x}}^0)(\mathbf{x}(0) - \bar{\mathbf{x}}^0)^T\} = \mathbf{P}_0 \tag{12.41b}$$

It is further assumed that $\mathbf{x}(0)$ is independent of $\mathbf{w}$ and $\mathbf{v}$.

$$E(\mathbf{x}(0)\,\mathbf{v}^T(k)\} = E\{\mathbf{x}(0)\,\mathbf{w}^T(k)\} = \mathbf{0} \tag{12.42}$$

Figure 12.5 shows system representation including the input disturbance and measurement noise.

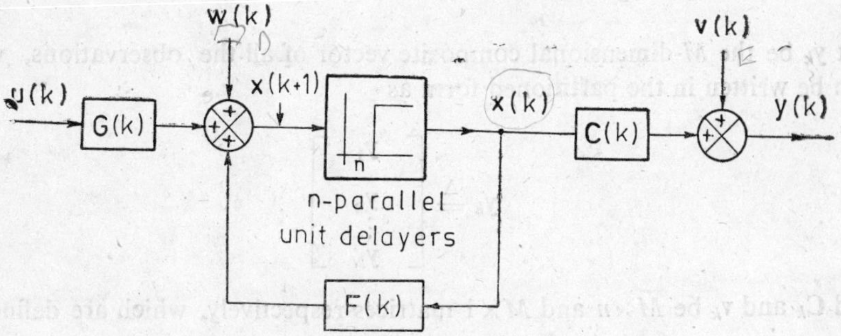

Fig. 12.5 System representation including input disturbance and measurement noise

We denote the estimate of $\mathbf{x}(k)$ by $\hat{\mathbf{x}}(k)$ and the estimation error by $\mathbf{e}(k)$ where

$$\mathbf{e}(k) = \mathbf{x}(k) - \hat{\mathbf{x}}(k) \tag{12.43}$$

In the previous section we observed that the desirable properties of the estimator are:

(1) the estimator is unbiased, i.e.,

$$E\{\hat{\mathbf{x}}\} = E\{\mathbf{x}\} \text{ for all } \mathbf{x}; \tag{12.44}$$

(2) the var $(\mathbf{e}(k)) = E\{\mathbf{e}(k)\,\mathbf{e}^T(k)\}$ is less than the variance of any other unbiased estimator.

Before deriving the filter algorithm for the multistage process given by eqns. (12.37), we consider the problem of single stage estimation of a constant parameter. The problem is to estimate a constant parameter $\mathbf{x}$ imbedded in noise $\mathbf{v}$. The linear observation equation is

$$\mathbf{y} = \mathbf{C}\mathbf{x} + \mathbf{v} \tag{12.45}$$

where $\mathbf{y}$ and $\mathbf{v}$ are vectors representing the observation and measurement noise respectively; $\mathbf{x}$ is n-dimensional (constant) parameter to be estimated and $\mathbf{C}$ is a measurement matrix. Measurement noise $\mathbf{v}$ is independent of $\mathbf{x}$ (Fig. 12.6). Assume that there are N measurements of the form (12.45), which we will symbolically represent by

$$\mathbf{y}_j = \mathbf{C}_j\mathbf{x} + \mathbf{v}_j;\; j = 1, 2, \ldots, N \tag{12.46a}$$

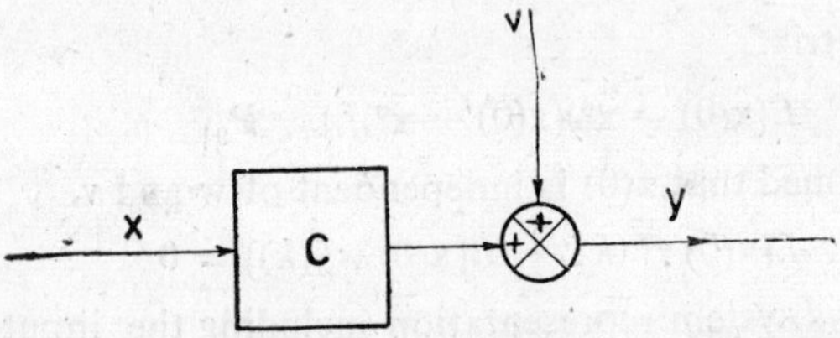

Fig. 12.6 Estimation of a constant parameter

Let $\mathbf{y}_k$ be the M-dimensional composite vector of all the observations, which can be written in the partitioned form as

$$\mathbf{y}_k \triangleq \begin{bmatrix} \mathbf{y}_1 \\ \mathbf{y}_2 \\ \vdots \\ \mathbf{y}_N \end{bmatrix};$$

and $\mathbf{C}_k$ and $\mathbf{v}_k$ be $M \times n$ and $M \times 1$ matrices respectively, which are defined as

$$\mathbf{C}_k \triangleq \begin{bmatrix} \mathbf{C}_1 \\ \mathbf{C}_2 \\ \vdots \\ \mathbf{C}_N \end{bmatrix}; \quad \mathbf{v}_k \triangleq \begin{bmatrix} \mathbf{v}_1 \\ \mathbf{v}_2 \\ \vdots \\ \mathbf{v}_N \end{bmatrix}$$

Then

$$\mathbf{y}_k = \mathbf{C}_k\mathbf{x} + \mathbf{v}_k \tag{12.46b}$$

Note that the dimension M of $\mathbf{y}_k$ may be greater than the number of observations N, since each observation may itself be a vector. We require that $M \geqslant n$, although we would normally have M much greater than n in order to 'average out' the noise in the measurements.

We shall assume that $\mathbf{v}_k$ is a zero-mean white-noise process:

$$E\{\mathbf{v}_k\} = \mathbf{0} \tag{12.47a}$$

$$E\{\mathbf{v}_k \mathbf{v}_j^T\} = \mathbf{R}_k\, \mu(k-j) \tag{12.47b}$$

where

$$\mathbf{R}_k = \left[\begin{array}{c|c|c|c} \mathbf{R}_1 & \mathbf{0} & \cdots & \mathbf{0} \\ \hline \mathbf{0} & \mathbf{R}_2 & \cdots & \mathbf{0} \\ \hline \vdots & \vdots & & \vdots \\ \hline \mathbf{0} & \mathbf{0} & \cdots & \mathbf{R}_N \end{array}\right];$$

$\mathbf{R}_1, \mathbf{R}_2, \ldots, \mathbf{R}_N$ are positive definite matrices.

Let $\hat{\mathbf{x}}_N$, the estimate of $\mathbf{x}$ based on observations $\mathbf{y}_1, \mathbf{y}_2, \ldots, \mathbf{y}_N$, be related to $\mathbf{y}_k$ by means of a linear transformation $\mathbf{L}$, i.e.,

$$\underset{(n\times 1)}{\hat{\mathbf{x}}_N} = \underset{(n\times M)}{\mathbf{L}} \quad \underset{(M\times 1)}{\mathbf{y}_k} \tag{12.48}$$

We shall determine $\mathbf{L}$ by minimizing the error variance of each parameter x_j; $j = 1, 2, \ldots, n$, subject to the constraint that the estimate must be unbiased.

From eqns. (12.46b) and (12.48) we have

$$\hat{\mathbf{x}}_N = \mathbf{L}\mathbf{C}_k\mathbf{x} + \mathbf{L}\mathbf{v}_k \tag{12.49}$$

$$E\{\hat{\mathbf{x}}_N\} = \mathbf{L}\mathbf{C}_k E\{\mathbf{x}\}$$

Therefore for the estimate to be unbiased, we must have

$$\underset{(n\times M)}{\mathbf{L}} \quad \underset{(M\times n)}{\mathbf{C}_k} = \underset{(n\times n)}{\mathbf{I}} \tag{12.50}$$

This equation may be written as

$$\underset{(n\times M)}{\mathbf{C}_k^T} \quad \underset{(M\times n)}{\mathbf{L}^T} = \underset{(n\times n)}{\mathbf{I}}$$

or

$$\underset{(n\times M)}{\mathbf{C}_k^T} \begin{bmatrix} l_{11} & l_{12} \cdots l_{1M} \\ l_{21} & l_{22} \cdots l_{2M} \\ \vdots & \vdots \qquad \vdots \\ l_{n1} & l_{n2} \cdots l_{nM} \end{bmatrix} = \underset{(n\times n)}{\mathbf{I}}$$

or

$$\mathbf{C}_k^T[\mathbf{l}_1 \mid \mathbf{l}_2 \mid \dots \mid \mathbf{l}_n] = [\mathbf{i}_1 \mid \mathbf{i}_2 \mid \dots \mid \mathbf{i}_n]$$

where

$$\mathbf{l}_j^T = [\,l_{j1} \quad l_{j2} \dots \, l_{jM}\,]$$

and

$$\mathbf{i}_j = [0 \quad 0 \dots 0 \quad \underset{\substack{\uparrow \\ j\text{th position}}}{1} \quad 0 \dots 0]^T$$

This gives

$$\mathbf{C}_k^T \mathbf{l}_j = \mathbf{i}_j \tag{12.51}$$

The error variance of the jth parameter is given by

$$\begin{aligned}
E\{(x_j - \hat{x}_{jN})^2\} &= E\{(x_j - \mathbf{l}_j^T \mathbf{y}_k)^2\} \\
&= E\{[x_j^2 - 2x_j \mathbf{y}_k^T \mathbf{l}_j + (\mathbf{y}_k^T \mathbf{l}_j)^2]\} \\
&= E\{x_j^2 - 2x_j \mathbf{x}^T \mathbf{i}_j - 2x_j \mathbf{v}_k^T \mathbf{l}_j + (\mathbf{x}^T \mathbf{i}_j + \mathbf{v}_k^T \mathbf{l}_j)^2\} \\
&= E\{x_j^2 - 2x_j^2 - 2x_j \mathbf{v}_k^T \mathbf{l}_j + x_j^2 + 2x_j \mathbf{v}_k^T \mathbf{l}_j + (\mathbf{v}_k^T \mathbf{l}_j)^2\} \\
&= E\{(\mathbf{v}_k^T \mathbf{l}_j)^2\} \\
&= E\{(\mathbf{l}_j^T \mathbf{v}_k)(\mathbf{v}_k^T \mathbf{l}_j)\} \\
&= \mathbf{l}_j^T E\{\mathbf{v}_k \mathbf{v}_k^T\} \mathbf{l}_j \\
&= \mathbf{l}_j^T \mathbf{R}_k \mathbf{l}_j
\end{aligned} \tag{12.52}$$

We shall determine $\mathbf{l}_j$ such that the performance function

$$J_j = \mathbf{l}_j^T \mathbf{R}_k \mathbf{l}_j + \boldsymbol{\lambda}_j^T [\mathbf{C}_k^T \mathbf{l}_j - \mathbf{i}_j] \tag{12.53}$$

is minimized for $j = 1, 2, \dots, n$. $\boldsymbol{\lambda}_j$ is an $n \times 1$ vector of Lagrange multipliers associated with the jth unbiasedness constraint.

Letting

$$\frac{\partial J_j}{\partial \mathbf{l}_j} = \mathbf{0} \quad \text{gives}$$

$$2\mathbf{R}_k \mathbf{l}_j + \mathbf{C}_k \boldsymbol{\lambda}_j = \mathbf{0}$$

or

$$\mathbf{l}_j = -\tfrac{1}{2}\mathbf{R}_k^{-1}\mathbf{C}_k \boldsymbol{\lambda}_j \tag{12.54}$$

Letting

$$\frac{\partial J_j}{\partial \boldsymbol{\lambda}_j} = \mathbf{0} \quad \text{gives}$$

$$\mathbf{C}_k^T \mathbf{l}_j = \mathbf{i}_j \tag{12.55}$$

From eqns. (12.54) and (12.55) we have

$$\boldsymbol{\lambda}_j = -2(\mathbf{C}_k^T \mathbf{R}_k^{-1} \mathbf{C}_k)^{-1} \mathbf{i}_j$$

$$\mathbf{l}_j = \mathbf{R}_k^{-1} \mathbf{C}_k (\mathbf{C}_k^T \mathbf{R}_k^{-1} \mathbf{C}_k)^{-1} \mathbf{i}_j$$

Therefore,

$$\begin{aligned}\mathbf{L}^T &= [\mathbf{l}_1 \vdots \mathbf{l}_2 \vdots \cdots \vdots \mathbf{l}_n] \\ &= \mathbf{R}_k^{-1}\mathbf{C}_k(\mathbf{C}_k^T\mathbf{R}_k^{-1}\mathbf{C}_k)^{-1}[\mathbf{i}_1 \vdots \mathbf{i}_2 \vdots \cdots \vdots \mathbf{i}_n] \\ &= \mathbf{R}_k^{-1}\mathbf{C}_k(\mathbf{C}_k^T\mathbf{R}_k^{-1}\mathbf{C}_k)^{-1}\end{aligned}$$

or

$$\mathbf{L} = (\mathbf{C}_k^T\mathbf{R}_k^{-1}\mathbf{C}_k)^{-1}\mathbf{C}_k^T\mathbf{R}_k^{-1} \tag{12.56}$$

Therefore

$$\hat{\mathbf{x}}_N = (\mathbf{C}_k^T\mathbf{R}_k^{-1}\mathbf{C}_k)^{-1}\mathbf{C}_k^T\mathbf{R}_k^{-1}\mathbf{y}_k \tag{12.57}$$

Note that the solution requires the inversion of the $n \times n$ matrix $(\mathbf{C}_k^T\mathbf{R}_k^{-1}\mathbf{C}_k)$; the inverse is nothing but the error-variance as seen below:

$$E\{\mathbf{e}_N\mathbf{e}_N^T\} = E\{(\mathbf{x} - \hat{\mathbf{x}}_N)(\mathbf{x} - \hat{\mathbf{x}}_N)^T\}$$

Now

$$\begin{aligned}\hat{\mathbf{x}}_N &= \mathbf{L}\,\mathbf{y}_k \\ &= \mathbf{L}(\mathbf{C}_k\mathbf{x} + \mathbf{v}_k) \\ &= \mathbf{x} + \mathbf{L}\mathbf{v}_k\end{aligned}$$

Therefore

$$\begin{aligned}E\{\mathbf{e}_N\mathbf{e}_N^T\} &= E\{\mathbf{L}\,\mathbf{v}_k\mathbf{v}_k^T\mathbf{L}^T\} \\ &= \mathbf{L}\mathbf{R}_k\mathbf{L}^T \\ &= (\mathbf{C}_k^T\mathbf{R}_k^{-1}\mathbf{C}_k)^{-1}\mathbf{C}_k^T\mathbf{R}_k^{-1}\mathbf{R}_k\mathbf{R}_k^{-1}\mathbf{C}_k(\mathbf{C}_k^T\mathbf{R}_k^{-1}\mathbf{C}_k)^{-1} \\ &= (\mathbf{C}_k^T\mathbf{R}_k^{-1}\mathbf{C}_k)^{-1}\mathbf{C}_k^T\mathbf{R}_k^{-1}\mathbf{C}_k(\mathbf{C}_k^T\mathbf{R}_k^{-1}\mathbf{C}_k)^{-1} \\ &= (\mathbf{C}_k^T\mathbf{R}_k^{-1}\mathbf{C}_k)^{-1} \\ &\triangleq \mathbf{P}_N\end{aligned} \tag{12.58}$$

Equation (12.57) may now be written as

$$\hat{\mathbf{x}}_N = \mathbf{P}_N\mathbf{C}_k^T\mathbf{R}_k^{-1}\mathbf{y}_k \tag{12.59}$$

An interesting case occurs when we have computed $\hat{\mathbf{x}}$ for N measurements and someone gives us an additional measurement. A great deal of effort would be involved in multiplying and inverting $(\mathbf{C}_k^T\mathbf{R}_k^{-1}\mathbf{C}_k)$. To repeat this procedure for $(N+1)$ measurements would probably be prohibitive with regard to computer time, particularly if 'on-line' computation is required. We therefore look for a recursive or sequential estimation scheme which allows us to add a new measurement without repeating the entire calculation. Suppose that we obtain an additional measurement of $\mathbf{x}$ given by

$$\mathbf{y}_{N+1} = \mathbf{C}_{N+1}\mathbf{x} + \mathbf{v}_{N+1} \tag{12.60}$$

We can adjoin this new observation to the previous observations to obtain

$$\mathbf{y}_{k+1} = \mathbf{C}_{k+1}\mathbf{x} + \mathbf{v}_{k+1} \tag{12.61}$$

where

$$\mathbf{y}_{k+1} = \begin{bmatrix} \mathbf{y}_k \\ \hline \mathbf{y}_{N+1} \end{bmatrix}; \quad \mathbf{C}_{k+1} = \begin{bmatrix} \mathbf{C}_k \\ \hline \mathbf{C}_{N+1} \end{bmatrix}; \quad \mathbf{v}_{k+1} = \begin{bmatrix} \mathbf{v}_k \\ \hline \mathbf{v}_{N+1} \end{bmatrix}$$

In addition, $\mathbf{R}_{k+1}$ can be written as

$$\mathbf{R}_{k+1} = \left[\begin{array}{c|c} \mathbf{R}_k & \mathbf{0} \\ \hline \mathbf{0} & \mathbf{R}_{N+1} \end{array}\right]$$

From eqns. (12.58)–(12.59) we may write

$$\hat{\mathbf{x}}_{N+1} = \mathbf{P}_{N+1}\, \mathbf{C}_{k+1}^T\, \mathbf{R}_{k+1}^{-1}\, \mathbf{y}_{k+1}$$

$$= \left\{ \begin{bmatrix} \mathbf{C}_k \\ \hline \mathbf{C}_{N+1} \end{bmatrix}^T \left[\begin{array}{c|c} \mathbf{R}_k & \mathbf{0} \\ \hline \mathbf{0} & \mathbf{R}_{N+1} \end{array}\right]^{-1} \begin{bmatrix} \mathbf{C}_k \\ \hline \mathbf{C}_{N+1} \end{bmatrix} \right\}^{-1}$$

$$\left\{ \begin{bmatrix} \mathbf{C}_k \\ \hline \mathbf{C}_{N+1} \end{bmatrix}^T \left[\begin{array}{c|c} \mathbf{R}_k & \mathbf{0} \\ \hline \mathbf{0} & \mathbf{R}_{N+1} \end{array}\right]^{-1} \begin{bmatrix} \mathbf{y}_k \\ \hline \mathbf{y}_{N+1} \end{bmatrix} \right\}$$

$$= (\mathbf{C}_k^T\, \mathbf{R}_k^{-1}\, \mathbf{C}_k + \mathbf{C}_{N+1}^T\, \mathbf{R}_{N+1}^{-1}\, \mathbf{C}_{N+1})^{-1} (\mathbf{C}_k^T\, \mathbf{R}_k^{-1}\, \mathbf{y}_k + \mathbf{C}_{N+1}^T\, \mathbf{R}_{N+1}^{-1}\, \mathbf{y}_{N+1}) \tag{12.62}$$

Therefore

$$\mathbf{P}_{N+1}^{-1} = \mathbf{P}_N^{-1} + \mathbf{C}_{N+1}^T\, \mathbf{R}_{N+1}^{-1}\, \mathbf{C}_{N+1} \tag{12.63}$$

which is a sequential equation for computing $\mathbf{P}_{N+1}^{-1}$ from $\mathbf{P}_N^{-1}$.

From eqns. (12.58)–(12.63) we have

$$\begin{aligned}
\hat{\mathbf{x}}_{N+1} &= \mathbf{P}_{N+1}\, [\mathbf{C}_k^T\, \mathbf{R}_k^{-1}\, \mathbf{y}_k + \mathbf{C}_{N+1}^T\, \mathbf{R}_{N+1}^{-1}\, \mathbf{y}_{N+1}] \\
&= \mathbf{P}_{N+1}\, [\mathbf{C}_{N+1}^T\, \mathbf{R}_{N+1}\, \mathbf{y}_{N+1} + \mathbf{P}_N^{-1}\, \hat{\mathbf{x}}_N] \\
&= \mathbf{P}_{N+1}\, [\mathbf{C}_{N+1}^T\, \mathbf{R}_{N+1}^{-1}\, \mathbf{y}_{N+1} + (\mathbf{P}_{N+1}^{-1} - \mathbf{C}_{N+1}^T\, \mathbf{R}_{N+1}^{-1}\, \mathbf{C}_{N+1})\, \hat{\mathbf{x}}_N] \\
&= \hat{\mathbf{x}}_N + \mathbf{K}_{N+1}\, [\mathbf{C}_{N+1}\, \hat{\mathbf{x}}_N - \mathbf{y}_{N+1}]
\end{aligned} \tag{12.64a}$$

where

$$\mathbf{K}_{N+1} = -\, \mathbf{P}_{N+1}\, \mathbf{C}_{N+1}^T\, \mathbf{R}_{N+1}^{-1} \tag{12.64b}$$

Equations (12.64) give a sequential algorithm for computing $\hat{\mathbf{x}}_{N+1}$ from the preceding estimate $\hat{\mathbf{x}}_N$.

We can write the expression (12.63) as

$$\mathbf{P}_{N+1} = \mathbf{P}_N - \mathbf{P}_N \mathbf{C}_{N+1}^T\, (\mathbf{R}_{N+1} + \mathbf{C}_{N+1} \mathbf{P}_N \mathbf{C}_{N+1}^T)^{-1} \mathbf{C}_{N+1} \mathbf{P}_N \tag{12.65}$$

Alternative expression for $\mathbf{K}_{N+1}$ given by eqn. (12.64b) is

$$\begin{aligned}
\mathbf{K}_{N+1} &= -\, [\mathbf{P}_N - \mathbf{P}_N \mathbf{C}_{N+1}^T\, (\mathbf{R}_{N+1} + \mathbf{C}_{N+1} \mathbf{P}_N \mathbf{C}_{N+1}^T)^{-1} \mathbf{C}_{N+1} \mathbf{P}_N]\, \mathbf{C}_{N+1}^T\, \mathbf{R}_{N+1}^{-1} \\
&= -\, \mathbf{P}_N \mathbf{C}_{N+1}^T\, (\mathbf{R}_{N+1} + \mathbf{C}_{N+1} \mathbf{P}_N \mathbf{C}_{N+1}^T)^{-1}
\end{aligned} \tag{12.66}$$

Equation (12.65) may now be written as

$$\mathbf{P}_{N+1} = [\mathbf{I} + \mathbf{K}_{N+1}\mathbf{C}_{N+1}]\mathbf{P}_N \tag{12.67}$$

We have used *matrix inversion lemma*[3] to obtain this result.
The sequential unbiased minimum variance algorithm may be written down from eqns. (12.64)–(12.67):

$$\hat{\mathbf{x}}_{N+1} = \hat{\mathbf{x}}_N + \mathbf{K}_{N+1}[\mathbf{C}_{N+1}\hat{\mathbf{x}}_N - \mathbf{y}_{N+1}] \tag{12.68a}$$

$$\mathbf{K}_{N+1} = -\mathbf{P}_N\mathbf{C}_{N+1}^T(\mathbf{R}_{N+1} + \mathbf{C}_{N+1}\mathbf{P}_N\mathbf{C}_{N+1}^T)^{-1} \tag{12.68b}$$

$$\mathbf{P}_{N+1} = [\mathbf{I} + \mathbf{K}_{N+1}\mathbf{C}_{N+1}]\mathbf{P}_N \tag{12.68c}$$

Consider now the multistage process described by eqns. (12.37). Based on a set of sequential observations

$$\mathbf{Y}(k) = \{\mathbf{y}(1), \mathbf{y}(2), \ldots, \mathbf{y}(k)\},$$

we wish to determine an estimate $\mathbf{x}(j)$, which we shall symbolically represent by $\hat{\mathbf{x}}(j \mid k)$. The estimation error will be denoted by

$$\mathbf{e}(j \mid k) = \hat{\mathbf{x}}(j) - \mathbf{x}(j \mid k) \tag{12.69}$$

---

[3] $$\mathbf{P}_{N+1}^{-1} = \mathbf{P}_N^{-1} + \mathbf{C}_{N+1}^T\mathbf{R}_{N+1}^{-1}\mathbf{C}_{N+1} \tag{i}$$

Premultiplying (i) by $\mathbf{P}_{N+1}$ gives

$$\mathbf{I} = \mathbf{P}_{N+1}\mathbf{P}_N^{-1} + \mathbf{P}_{N+1}\mathbf{C}_{N+1}^T\mathbf{R}_{N+1}^{-1}\mathbf{C}_{N+1} \tag{ii}$$

Postmultiplying (ii) by $\mathbf{P}_N$ gives

$$\mathbf{P}_N = \mathbf{P}_{N+1} + \mathbf{P}_{N+1}\mathbf{C}_{N+1}^T\mathbf{R}_{N+1}^{-1}\mathbf{C}_{N+1}\mathbf{P}_N \tag{iii}$$

Postmultiplying (iii) by $\mathbf{C}_{N+1}^T$ results in

$$\begin{aligned}\mathbf{P}_N\mathbf{C}_{N+1}^T &= \mathbf{P}_{N+1}\mathbf{C}_{N+1}^T + \mathbf{P}_{N+1}\mathbf{C}_{N+1}^T\mathbf{R}_{N+1}^{-1}\mathbf{C}_{N+1}\mathbf{P}_N\mathbf{C}_{N+1}^T \\ &= \mathbf{P}_{N+1}\mathbf{C}_{N+1}^T\mathbf{R}_{N+1}^{-1}[\mathbf{R}_{N+1} + \mathbf{C}_{N+1}\mathbf{P}_N\mathbf{C}_{N+1}^T]\end{aligned} \tag{iv}$$

Postmultiplying (iv) by $[\mathbf{R}_{N+1} + \mathbf{C}_{N+1}\mathbf{P}_N\mathbf{C}_{N+1}^T]^{-1}$ we obtain

$$\mathbf{P}_{N+1}\mathbf{C}_{N+1}^T\mathbf{R}_{N+1}^{-1} = \mathbf{P}_N\mathbf{C}_{N+1}^T[\mathbf{C}_{N+1}\mathbf{P}_N\mathbf{C}_{N+1}^T + \mathbf{R}_{N+1}]^{-1} \tag{v}$$

Postmultiplying (v) by $\mathbf{C}_{N+1}\mathbf{P}_N$ gives

$$\mathbf{P}_{N+1}\mathbf{C}_{N+1}^T\mathbf{R}_{N+1}^{-1}\mathbf{C}_{N+1}\mathbf{P}_N = \mathbf{P}_N\mathbf{C}_{N+1}^T[\mathbf{C}_{N+1}\mathbf{P}_N\mathbf{C}_{N+1}^T + \mathbf{R}_{N+1}]^{-1}\mathbf{C}_{N+1}\mathbf{P}_N \tag{vi}$$

Subtracting (vi) from $\mathbf{P}_N$ results in

$$\begin{aligned}&\mathbf{P}_N - \mathbf{P}_{N+1}\mathbf{C}_{N+1}^T\mathbf{R}_{N+1}^{-1}\mathbf{C}_{N+1}\mathbf{P}_N \\ &\qquad = \mathbf{P}_N - \mathbf{P}_N\mathbf{C}_{N+1}^T[\mathbf{C}_{N+1}\mathbf{P}_N\mathbf{C}_{N+1}^T + \mathbf{R}_{N+1}]^{-1}\mathbf{C}_{N+1}\mathbf{P}_N\end{aligned} \tag{vii}$$

From eqns. (iii) and (vii) we obtain

$$\mathbf{P}_{N+1} = \mathbf{P}_N - \mathbf{P}_N\mathbf{C}_{N+1}^T[\mathbf{C}_{N+1}\mathbf{P}_N\mathbf{C}_{N+1}^T + \mathbf{R}_{N+1}]^{-1}\mathbf{C}_{N+1}\mathbf{P}_N$$

Depending upon the relative values of j and k, the estimation is referred to as *prediction* ($j > k$), *filtering* ($j = k$) or *smoothing* ($j < k$). In the following, we shall develop a one-stage prediction solution as a step in the derivation of filter algorithm for the multistage process (12.37).

If the filtered estimate of $\mathbf{x}(k)$, namely $\hat{\mathbf{x}}(k \mid k)$ were known, then $\hat{\mathbf{x}}(k+1 \mid k)$ could he obtained as[4] (eqn. (12.13))

$$\hat{\mathbf{x}}(k+1 \mid k) = \Phi(k+1, k)\,\hat{\mathbf{x}}(k \mid k) + \mathbf{G}(k)\,\mathbf{u}(k) \qquad (12.70)$$

where $\Phi(k, k_0)$ is the state transition matrix of difference equation (12.37).

Equation (12.70) may be written as

$$\hat{\mathbf{x}}(k+1 \mid k) = \mathbf{F}(k)\,\hat{\mathbf{x}}(k \mid k) + \mathbf{G}(k)\,\mathbf{u}(k) \qquad (12.71)$$

Let us now find the variance of the filtering error $\mathbf{e}(j \mid k)$ given by eqn. (12.69). Let us denote *a priori variance* by $\mathbf{P}(k+1 \mid k)$ since it is the variance of the estimate of $\mathbf{x}(k+1)$ before the observation $\mathbf{y}(k+1)$ i received. Accordingly *a posteriori variance* will be denoted as $\mathbf{P}(k+1 \mid k+1)$.

Using eqn. (12.14a), we may write the relation between the filtered variance $\mathbf{P}(k \mid k)$ and the one-stage prediction variance $\mathbf{P}(k+1 \mid k)$ as

$$\mathbf{P}(k+1 \mid k) = \mathbf{F}(k)\,\mathbf{P}(k \mid k)\,\mathbf{F}^T(k) + \mathbf{Q}(k) \qquad (12.72)$$

The one-stage prediction given by equations (12.71) and (12.72) is now corrected with the current data $\mathbf{y}(k+1)$. Using eqns. (12.68) we may write

$$\hat{\mathbf{x}}(k+1 \mid k+1) = \hat{\mathbf{x}}(k+1 \mid k) + \mathbf{K}(k+1)\,[\mathbf{C}(k+1)\,\hat{\mathbf{x}}(k+1 \mid k) - \mathbf{y}(k+1)] \qquad (12.73a)$$

$$\mathbf{K}(k+1) = -\,\mathbf{P}(k+1 \mid k)\,\mathbf{C}^T(k+1)\,[\mathbf{R}(k+1) + \mathbf{C}(k+1)\,\mathbf{P}(k+1 \mid k)\,\mathbf{C}^T(k+1)]^{-1} \qquad (12.73b)$$

$$\mathbf{P}(k+1 \mid k+1) = [\mathbf{I} + \mathbf{K}(k+1)\,\mathbf{C}(k+1)]\,\mathbf{P}(k+1 \mid k) \qquad (12.73c)$$

Equations (12.71)–(12.73) represent the final form of the optimal linear estimator (discrete-time Kalman filter) for discrete-time systems. These equations are summarized in Table 12.2 for easy reference. Figure 12.7 gives block diagram of discrete-time Kalman filter.

Note that because of numerical errors, the symmetry of $\mathbf{P}$ may get destroyed resulting in serious degradation of the performance of the Kalman filter and even instability of the estimation procedure. We must symmetrize $\mathbf{P}$ after each step.

Further, the reader may note that the form of equations given in Table 12.2 is not the only form in which the discrete-time Kalman filter may be

---

[4] We have assumed the distributions to be Gaussian so that

$$\hat{\mathbf{x}}\,(j \mid k) = E\,\{\mathbf{x}\,(j) \mid \mathbf{Y}\,(k)\}$$

Table 12.2: Summary of Discrete-time Kalman filter Algorithm

Message model	$\mathbf{x}(k+1)=\mathbf{F}(k)\,\mathbf{x}(k)+\mathbf{G}(k)\,\mathbf{u}(k)+\mathbf{w}(k)$
Observation model	$\mathbf{y}(k)=\mathbf{C}(k)\,\mathbf{x}(k)+\mathbf{v}(k)$
Prior Statistics	$E\{\mathbf{w}(k)\}=E\{\mathbf{v}(k)\}=\mathbf{0};\ E\{\mathbf{x}(0)\}=\bar{\mathbf{x}}^0$ $E(\mathbf{w}(k)\,\mathbf{w}^T(j)\}=\mathbf{Q}(k)\,\mu(k-j)$ $E\{\mathbf{v}(k)\,\mathbf{v}^T(j)\}=\mathbf{R}(k)\,\mu(k-j)$ $E\{\mathbf{w}(k)\,\mathbf{v}^T(j)\}=E\{\mathbf{x}(0)\,\mathbf{v}^T(j)\}=E\{\mathbf{x}(0)\,\mathbf{w}^T(j)\}=\mathbf{0}$ $E\{(\mathbf{x}(0)-\bar{\mathbf{x}}^0)\,(\mathbf{x}(0)-\bar{\mathbf{x}}^0)^T\}=\mathbf{P}_0$
One-stage prediction algorithm	$\hat{\mathbf{x}}(k+1\mid k)=\mathbf{F}(k)\,\hat{\mathbf{x}}(k\mid k)+\mathbf{G}(k)\,\mathbf{u}(k);\ \hat{\mathbf{x}}(0\mid 0)=\bar{\mathbf{x}}^0$
A priori variance algorithm	$\mathbf{P}(k+1\mid k)=\mathbf{F}(k)\,\mathbf{P}(k\mid k)\,\mathbf{F}^T(k)+\mathbf{Q}(k);\ \mathbf{P}(0\mid 0)=\mathbf{P}_0$
Filter algorithm	$\hat{\mathbf{x}}(k+1\mid k+1)=\hat{\mathbf{x}}(k+1\mid k)+\mathbf{K}(k+1)\,[\mathbf{C}(k+1)\,\hat{\mathbf{x}}(k+1\mid k)$ $-\mathbf{y}(k+1)]$
Kalman gain algorithm	$\mathbf{K}(k+1)=-\mathbf{P}(k+1\mid k)\,\mathbf{C}^T(k+1)\,[\mathbf{R}(k+1)$ $+\mathbf{C}(k+1)\mathbf{P}(k+1\mid k)\,\mathbf{C}^T(k+1)]^{-1}$
A posteriori variance algorithm	$\mathbf{P}(k+1\mid k+1)=[\mathbf{I}+\mathbf{K}(k+1)\,\mathbf{C}(k+1)]\,\mathbf{P}(k+1\mid k)$

represented. However, this form is quite common. Results of the previous subsection on duality with linear regulator and on time-invariant Kalman filter apply to the discrete-time case also (Kalman 1960).

Example 12.2: In the following, we represent the continuous-time problem given in Example 12.1 by a discrete-time problem to which the algorithm derived in this section may be applied. For a sampling interval of $T = 0.1$, the discrete-time model of the Position Servo is given by (eqn. (6.60))

$$\mathbf{x}(k+1) = \mathbf{F}\mathbf{x}(k) + \mathbf{g}u(k) + \boldsymbol{\gamma}\mathbf{w}(k)$$

$$\mathbf{y}(k) = \mathbf{c}\mathbf{x}(k) + \mathbf{v}(k)$$

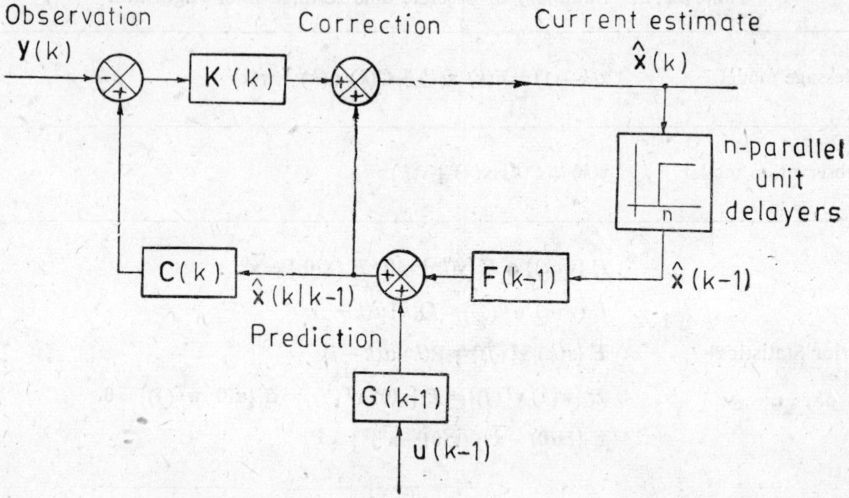

Fig. 12.7 Block diagram of discrete-time Kalman filter

where

$$\mathbf{F} = \begin{bmatrix} 1 & 0.0787 \\ 0 & 0.6065 \end{bmatrix}; \mathbf{g} = \begin{bmatrix} 0.0043 \\ 0.0717 \end{bmatrix}; \boldsymbol{\gamma} = \begin{bmatrix} 0.00043 \\ 0.00787 \end{bmatrix}$$

$$\mathbf{c} = [1 \quad 0]$$

The stochastic process $\mathbf{w}(t)$ $(= \tau_0(t))$ is assumed to be a zero-mean white noise with intensity $\mathbf{Q}(t)$:

$$E\{\mathbf{w}(t)\, \mathbf{w}^T(\tau)\} = \mathbf{Q}(t)\, \delta(t-\tau)$$

The zero-mean discrete white noise $\mathbf{w}(k)$ has the variance

$$E\{\mathbf{w}(k)\, \mathbf{w}^T(j)\} \simeq \frac{1}{T}\mathbf{Q}(kT)\, \mu(k-j)$$

Similarly, the zero-mean discrete white noise $\mathbf{v}(k)$ has the variance

$$E\{\mathbf{v}(k)\, \mathbf{v}^T(j)\} \simeq \frac{1}{T}\mathbf{R}(kT)\, \mu(k-j)$$

In Example 12.1, Q and R were taken as 100 and 1 respectively. In the present example we shall take Q and R as 1000 and 10 respectively. We shall take (from Example 12.1)

$$\mathbf{P}_0 = \begin{bmatrix} 1 & 0 \\ 0 & 0 \end{bmatrix}$$

A priori variance algorithm:

$$\mathbf{P}(k+1\,|\,k) = \mathbf{F}\mathbf{P}(k\,|\,k)\,\mathbf{F}^T + \boldsymbol{\gamma} Q \boldsymbol{\gamma}^T\,;\ \mathbf{P}(0\,|\,0) = \mathbf{P}_0$$

Kalman gain algorithm:

$$\mathbf{K}(k+1) = -\,\mathbf{P}(k+1\,|\,k)\,\mathbf{c}^T[R + \mathbf{c}\mathbf{P}(k+1\,|\,k)\,\mathbf{c}^T]^{-1}$$

A posteriori variance algorithm:

$$\mathbf{P}(k+1\,|\,k+1) = (\mathbf{I} + \mathbf{K}(k+1)\mathbf{c}]\,\mathbf{P}(k+1\,|\,k)$$

Figure 12.8 gives the Kalman gains resulting from these equations.

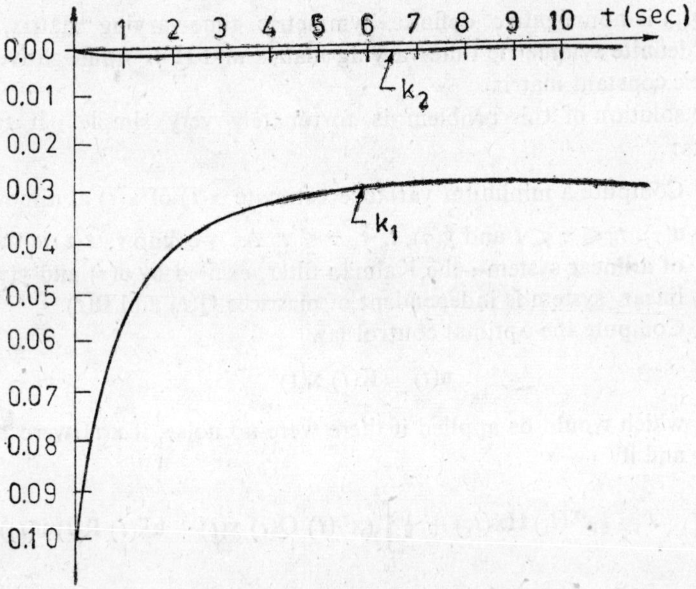

Fig. 12.8 Kalman gains for Example 12.2

12.5 STOCHASTIC OPTIMAL LINEAR REGULATOR

In Chapter 11 we discussed an optimal linear regulator problem wherein we assumed that we could make noiseless measurements of a system state vector and use these measurements in system mechanization. In the present chapter, so far we have considered estimation problems; we have not considered control uses of the estimated state variables. In this section, we now consider problems containing both control and state estimation requirements.

We assume that we are given a linear system with the additive input noise:

$$\dot{\mathbf{x}}(t) = \mathbf{A}(t)\,\mathbf{x}(t) + \mathbf{B}(t)\,\mathbf{u}(t) + \mathbf{w}(t)$$

The input noise $\mathbf{w}(t)$ is white, Gaussian, of zero mean and has intensity $\hat{\mathbf{Q}})t)$ where $\hat{\mathbf{Q}}$ is nonnegative definite symmetric matrix for all t.

The output equation is

$$\mathbf{y}(t) = \mathbf{C}(t)\,\mathbf{x}(t) + \mathbf{v}(t)$$

where $\mathbf{v}(t)$ is a white, Gaussian noise of zero mean and has intensity $\hat{\mathbf{R}}(t)$ where $\hat{\mathbf{R}}$ is positive definite symmetric matrix for all t.

The initial state $\mathbf{x}(t_0)$ at time t_0 is a Gaussian random vector of mean $\bar{\mathbf{x}}^0$ and variance $\mathbf{P}_0$. The processes $\mathbf{v}$, $\mathbf{w}$ and $\mathbf{x}(t_0)$ are independent.

Let us consider the problem of minimizing the performance index

$$J = E\{\tfrac{1}{2}\mathbf{x}^T(t_1)\,\mathbf{H}\mathbf{x}(t_1) + \tfrac{1}{2}\int_{t_0}^{t_1} (\mathbf{x}^T(t)\,\mathbf{Q}(t)\,\mathbf{x}(t) + \mathbf{u}^T(t)\,\mathbf{R}(t)\,\mathbf{u}(t))\,dt\}$$

where $\mathbf{Q}(t)$ is nonnegative definite symmetric time-varying matrix, $\mathbf{R}(t)$ is positive definite symmetric time-varying matrix and $\mathbf{H}$ is nonnegative definite symmetric constant matrix.

The solution of this problem is fortunately very simple. It falls into two parts:

1. Compute a minimum variance estimate $\hat{\mathbf{x}}(t)$ of $\mathbf{x}(t)$ at time t, using $\mathbf{u}(\tau)$, $t_0 \leqslant \tau \leqslant t$ and $\mathbf{y}(\tau)$, $t_0 \leqslant \tau \leqslant t$. As we know, $\hat{\mathbf{x}}(t)$ is the output of a linear system—the Kalman filter, excited by $\mathbf{u}(\tau)$ and $\mathbf{y}(\tau)$. This linear system is independent of matrices $\mathbf{Q}(t)$ and $\mathbf{R}(t)$.
2. Compute the optimal control law

$$\mathbf{u}(t) = \mathbf{K}(t)\,\mathbf{x}(t)$$

which would be applied if there were no noise, if $\mathbf{x}(t)$ were available and if

$$J = \tfrac{1}{2}\mathbf{x}^T(t_1)\,\mathbf{H}\mathbf{x}(t_1) + \tfrac{1}{2}\int_{t_0}^{t_1} (\mathbf{x}^T(t)\,\mathbf{Q}(t)\,\mathbf{x}(t) + \mathbf{u}^T(t)\,\mathbf{R}(t)\,\mathbf{u}(t))\,dt$$

were the performance index.

The control law

$$\mathbf{u}(t) = \mathbf{K}(t)\,\hat{\mathbf{x}}(t)$$

is optimal for the problem with noise.

The calculation of $\hat{\mathbf{x}}(t)$ and of the control law gain matrix $\mathbf{K}(t)$ are separate problems which can be tackled independently. Because of this, the result above is known as the *separation principle*. Figure 12.9 shows optimal stochastic controller.

The separation principle as stated above applies equally well to discrete-time systems.

This result can be proved using the maximum principle/dynamic programming. However, some modification in these mathematical procedures is required because we shall be taking the expected value of a deterministic performance index. For proof, refer Wonham (1968b), Edison (1971).

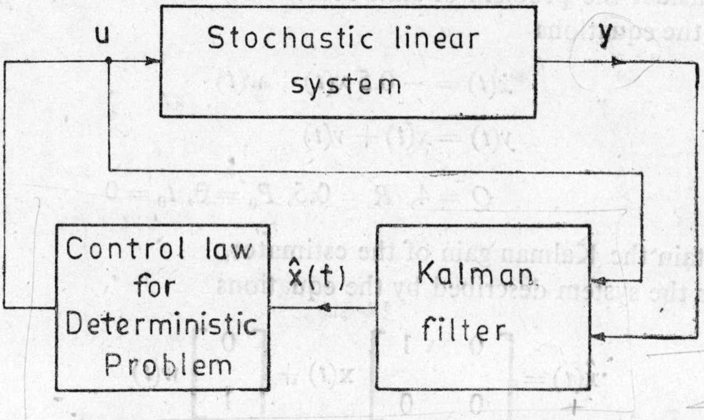

Fig. 12.9 Illustration of separation principle

12.6 CONCLUDING COMMENTS

In this chapter, we touched upon the enormous body of knowledge called the filtering theory. Wong (1973) and Kailath (1974) have surveyed the progress made in this field. We have attempted to present important results in a simple way omitting some details of mathematical rigour. This may disappoint some of the readers; however, this was necessary in the interest of confining the discussion to a reasonable length.

We have solved the problem of reconstructing the state of a linear system from incomplete and inaccurate measurements. The state reconstruction was achieved using a Kalman filter. The properties of time-invariant filters were reviewed by dualizing the results of linear state regulator.

The results of the state estimation were combined with the solution of the deterministic quadratic optimal control problem to solve the combined problem of estimation and control. It was observed that for linear systems with quadratic cost functions, subjected to additive white Gaussian inputs, the optimum stochastic controller is realized by cascading an optimal estimator with a deterministic optimum controller. When any of these restrictions do not hold, the problems become much more complex, exact solutions being exceptionally difficult to obtain and mechanize. Wonham (1963), and Jacobs and Patchell (1972) have done some probing in stochastic control. A detailed treatment of the Linear-Quadratic-Gaussian problem may be found in the 1971 Special Issue of IEEE Transactions Automatic Control (Athans 1971, Mendel and Gieseking 1971, Rhodes 1971, Tse 1971).

PROBLEMS

(Note: In all the problems given below, assume that the state excitation noise $\mathbf{w}(t)$ and observation noise $\mathbf{v}(t)$ are zero mean white-noise processes with intensities $\mathbf{Q}(t)$ and $\mathbf{R}(t)$ respectively, the initial state $\mathbf{x}(t_0)$ is a stochastic process with mean $\bar{\mathbf{x}}^0$ and variance matrix $\mathbf{P}_0$. Further, the three processes are statistically independent).

12.1 Consider the problem of state reconstruction for the system described by the equations

$$\dot{x}(t) = -0.5\,x(t) + w(t)$$
$$y(t) = x(t) + v(t)$$
$$Q = 4,\ R = 0.5,\ P_0 = 0,\ t_0 = 0$$

Obtain the Kalman gain of the estimator.

12.2 For the system described by the equations

$$\hat{\mathbf{x}}(t) = \begin{bmatrix} 0 & 1 \\ 0 & 0 \end{bmatrix} \mathbf{x}(t) + \begin{bmatrix} 0 \\ 1 \end{bmatrix} w(t)$$
$$y(t) = [1 \quad 0]\,\mathbf{x}(t) + v(t)$$
$$Q = 0.5,\ R = 8,\ \mathbf{P}_0 = \begin{bmatrix} 0.5 & 0 \\ 0 & 0 \end{bmatrix},\ t_0 = 0;$$

find the equations for the optimal estimator.

12.3 Find the time-invariant Kalman filter for the system

$$\dot{x}_1(t) = x_2(t)$$
$$\dot{x}_2(t) = u(t) + w(t)$$
$$\mathbf{y}(t) = \mathbf{x}(t) + \mathbf{v}(t)$$
$$Q = 0.5,\ \mathbf{R} = \begin{bmatrix} 8 & 0 \\ 0 & 0.5 \end{bmatrix}$$

12.4 Repeat Problem 12.3 assuming that only the y_1 output is available.

12.5 For the time-invariant system

$$\dot{x}(t) = x(t) + w(t)$$
$$y(t) = x(t) + v(t)$$
$$Q = 4\alpha,\ R = \alpha,\ \alpha = \text{constant};$$

find the time-invariant filter. Show that the answer is independent of α.

12.6 Show that the Kalman gain $\mathbf{K}(k)$ is not changed if $\mathbf{Q}(k)$, $\mathbf{R}(k)$ and $\mathbf{P}_0$ are all multiplied by a constant scalar α.

12.7 Consider the system

$$x(k+1) = 2x(k) + w(k)$$
$$y(k) = x(k) + v(k)$$
$$Q(k) = 1,\ R(k) = 0.5,\ \bar{x}^0 = 1,\ P_0 = 2$$

Find $x(k+1 \mid k)$ for $k = 1, 2, 3, 4$.

Given $y(1) = 1,\ y(2) = 2,\ y(3) = 1.5,\ y(4) = 1$.

12.8 Determine the Kalman gains $\mathbf{K}(k)$ for $k = 1$ to 5 for the following estimation problem:

$$\mathbf{x}(k+1) = \begin{bmatrix} 1 & 1 \\ 0 & 1 \end{bmatrix} \mathbf{x}(k) + \mathbf{w}(k)$$

$$y(k) = x_1(k) + v(k)$$

$$\mathbf{Q}(k) = \begin{bmatrix} 0 & 0 \\ 0 & 0.5 \end{bmatrix}, \quad R(k) = 1, \quad \mathbf{P}_0 = \begin{bmatrix} 5 & 0 \\ 0 & 5 \end{bmatrix}$$

12.9 Consider the combined estimation and control problem for the system

$$\dot{x}(t) = -0.5\, x(t) + u(t) + w(t)$$

$$y(t) = x(t) + v(t)$$

$$Q = 4, \quad R = 0.5, \quad P_0 = 0, \quad t_0 = 0.$$

It is desired to find an optimal control law that minimizes the performance index

$$J = E\{\tfrac{1}{2}x^2(2) + \tfrac{1}{2}\int_0^2 (2x^2(t) + u^2(t))\, dt\}$$

Find such a control and give a suitable scheme for the implementation of the control.

REFERENCES

1. Anderson, B.D.O., and J.B. Moore, *Linear Optimal Control*, Englewood Cliffs, N.J.: Prentice-Hall, 1971.
2. Anderson, B.D.O., and J.B. Moore, *Optimal Filtering*, Englewood Cliffs, N.J.: Prentice-Hall, 1979.
3. Astrom, K.J., *Introduction to Stochastic Control Theory*, New York: Academic Press, 1970.
4. Athans, M., "A direct derivation of the optimal linear filter using the maximum principle", *IEEE Trans. Automat. Contr.*, vol. AC–12, pp. 690–698, 1967.
5. Athans, M., "The role and use of the stochastic Linear-Quadratic-Gaussian-Problem in control system design", *IEEE Trans. Automat. Contr.*, vol. AC–16, pp. 529-551, 1971.
6. Edison, T., "On the optimal control of stochastic linear systems", *IEEE Trans. Automat. Contr.*, vol. AC–16, pp. 776–785, 1971.
7. Jacobs, O.L.R., and J.W. Patchell, "Caution and probing in stochastic control", *Int. J. Control*, vol. 16, pp. 189–199, 1972.
8. Jazwinski, A.H., *Stochastic Processes and Filtering Theory*, New York: Academic Press, 1970.

9. Kailath, T., "A view of three decades of linear filtering theory", *IEEE Trans. Infor. Theory*, vol. IT–20, pp. 146–181, **1974.**
10. Kalman, R.E., "A new approach to linear filtering and prediction problems", *J. Basic Engg., Trans. ASME,* Ser. D., vol. 82, pp. 35–45, **1960.**
11. Kalman, R.E., and R.S. Bucy, "New results in linear filtering and prediction theory", *J. Basic Engg., Trans. ASME,* Ser. D., Vol. 83, pp. 95–108, **1961**.
12. Kushner, H., *Introduction to Stochastic Control*, New York: Holt, **1971.**
13. Kwakernaak, H., and R. Sivan, *Linear Optimal Control Systems*, New York: Wiley-Interscience, **1972.**
14. Meditch, J.S., *Stochastic Optimal Linear Estimation and Control*, New York: McGraw-Hill, **1969.**
15. Mendel, J.M., and D.L. Gieseking, "Bibliography on the Linear-Quadratic-Gaussian-Problem", *IEEE Trans. Automat. Contr.*, vol. AC–16, **1971.**
16. Rhodes, I.B., "A tutorial introduction to estimation and filtering", *IEEE Trans. Automat. Contr.*, vol. AC–16, pp. 688–706, **1971.**
17. Sage, A.P., and J.L. Melsa, *Estimation Theory with Applications to Communications and Control*, New York: McGraw-Hill, **1971.**
18. Tse, E., "On the optimal control of stochastic linear systems", *IEEE Trans. Automat. Contr.*, vol. AC–16, pp. 776–784, **1971.**
19. Wong, E., "Recent progress in stochastic processes—a survey", *IEEE Trans. Infor. Theory*, vol. IT–19, pp. 262–275, **1973.**
20. Wonham, W.M., "Stochastic problems in optimal control, *1963 IEEE Convention Record*, Part 2, pp. 114–124, **1963.**
21. Wonham, W.M., "On a matrix Riccati equation of stochastic control", *SIAM J. Control*, vol. 6, pp. 681–697, **1968a.**
22. Wonham, W.M., "On the separation theorem of stochastic control", *SIAM J. Control*, vol. 6, pp. 312–326, **1968b.**

APPENDIX I

LAPLACE TRANSFORM: THEOREMS AND PAIRS

We assume that the reader has had a previous introduction to the theory of Laplace transforms. This appendix provides notational conventions for Laplace transforms of scalar and vector-valued functions and lists the main theorems and transform pairs needed to carry out certain developments in the main body of this text.

Time function	Laplace transform
1. $x(t);\ x(t) = 0$ for $t < 0$	$\mathscr{L}[x(t)] \triangleq \hat{x}(s) \triangleq \int_0^\infty x(t)e^{-st}\,dt$
2. $\mathscr{L}^{-1}[\hat{x}(s)] \triangleq \frac{1}{2\pi j}\int_{\sigma-j\infty}^{\sigma+j\infty} \hat{x}(s)\,e^{st}ds \triangleq x(t)$	$\hat{x}(s)$
3. $\mathbf{x}(t) = \begin{bmatrix} x_1(t) \\ x_2(t) \\ \vdots \\ x_n(t) \end{bmatrix}$	$\hat{\mathbf{x}}(s) = \begin{bmatrix} \hat{x}_1(s) \\ \hat{x}_2(s) \\ \vdots \\ \hat{x}_n(s) \end{bmatrix}$
4. $\mathbf{F}(t) = \begin{bmatrix} f_{11}(t) & f_{12}(t) & \cdots & f_{1n}(t) \\ f_{21}(t) & f_{22}(t) & \cdots & f_{2n}(t) \\ \vdots & \vdots & & \vdots \\ f_{m1}(t) & f_{m2}(t) & \cdots & f_{mn}(t) \end{bmatrix}$	$\hat{\mathbf{F}}(s) = \begin{bmatrix} \hat{f}_{11}(s) & \hat{f}_{12}(s) & \cdots & \hat{f}_{1n}(s) \\ \hat{f}_{21}(s) & \hat{f}_{22}(s) & \cdots & \hat{f}_{2n}(s) \\ \vdots & \vdots & & \vdots \\ \hat{f}_{m1}(s) & \hat{f}_{m2}(s) & \cdots & \hat{f}_{mn}(s) \end{bmatrix}$
5. $a\,x(t) + b\,y(t)$	$a\,\hat{x}(s) + b\,\hat{y}(s)$
6. (i) $\frac{dx(t)}{dt} \triangleq \dot{x}(t)$	$s\,\hat{x}(s) - x(0)$
(ii) $\frac{d^n x(t)}{dt^n}$	$s^n\hat{x}(s) - s^{n-1}x(0) - s^{n-2}\frac{dx}{dt}(0) - \ldots - s\frac{d^{n-2}x}{dt^{n-2}}(0) - \frac{d^{n-1}x}{dt^{n-1}}(0)$

Time function	Laplace transform
7. $\int_0^t x(\tau)\,d\tau$	$\dfrac{\hat{x}(s)}{s}$
8. $x(t-\Delta)$	$\hat{x}(s)\,e^{-s\Delta}$
9. $\int_0^t x(\tau)\,y(t-\tau)\,d\tau$	$\hat{x}(s)\,\hat{y}(s)$
10. $\delta(t)$ = unit impulse at $t=0$	1
11. $x(t)=1$	$\hat{x}(s)=\dfrac{1}{s}$
12. $\dfrac{t^n}{n!}$	$\dfrac{1}{s^{n+1}}$
13. e^{-at}	$\dfrac{1}{s+a}$
14. $\dfrac{t^n}{n!}\,e^{-at}$	$\dfrac{1}{(s+a)^{n+1}}$
15. $1-e^{-at}$	$\dfrac{a}{s(s+a)}$
16. $\sin \omega t$	$\dfrac{\omega}{s^2+\omega^2}$
17. $\cos \omega t$	$\dfrac{s}{s^2+\omega^2}$
18. $e^{-at}\sin \omega t$	$\dfrac{\omega}{(s+a)^2+\omega^2}$
19. $e^{-at}\cos \omega t$	$\dfrac{s+a}{(s+a)^2+\omega^2}$

APPENDIX II

z-TRANSFORM: THEOREMS AND PAIRS

In this appendix, we summarize some of the useful theorems and pairs which should serve the purpose of a working reference.

Time function	z-transform
1. $x(k);\ x(k) = 0$ for $k < 0$	$\mathscr{L}[x(k)] \triangleq \hat{x}(z) \triangleq \sum_{k=0}^{\infty} x(k)\, z^{-k}$
2. $\mathscr{L}^{-1}[\hat{x}(z)] \triangleq \frac{1}{2\pi j}\oint_C \hat{x}(z)\, z^{k-1}\, dz \triangleq x(k)$	$\hat{x}(z)$
3. $\mathbf{x}(k) = \begin{bmatrix} x_1(k) \\ x_2(k) \\ \vdots \\ x_n(k) \end{bmatrix}$	$\hat{\mathbf{x}}(z) = \begin{bmatrix} \hat{x}_1(z) \\ \hat{x}_2(z) \\ \vdots \\ \hat{x}_n(z) \end{bmatrix}$
4. $\mathbf{F}(k) = \begin{bmatrix} f_{11}(k) & f_{12}(k) & \cdots & f_{1n}(k) \\ f_{21}(k) & f_{22}(k) & \cdots & f_{2n}(k) \\ \vdots & \vdots & & \vdots \\ f_{m1}(k) & f_{m2}(k) & \cdots & f_{mn}(k) \end{bmatrix}$	$\hat{\mathbf{F}}(z) = \begin{bmatrix} \hat{f}_{11}(z) & \hat{f}_{12}(z) & \cdots & \hat{f}_{1n}(z) \\ \hat{f}_{21}(z) & \hat{f}_{22}(z) & \cdots & \hat{f}_{2n}(z) \\ \vdots & \vdots & & \vdots \\ \hat{f}_{m1}(z) & \hat{f}_{m2}(z) & \cdots & \hat{f}_{mn}(z) \end{bmatrix}$
5. $ax(k) + by(k)$	$a\hat{x}(z) + b\hat{y}(z)$
6. (i) $x(k+1)$	$z\hat{x}(z) - zx(0)$
(ii) $x(k+n)$	$z^n\hat{x}(z) - \sum_{i=0}^{n-1} x(i)\, z^{n-i}$
7. $x(k-n)$	$z^{-n}\hat{x}(z)$
8. $\sum_{j=0}^{k} x(k-j)\, y(j)$	$\hat{x}(z)\, \hat{y}(z)$
9. $\mu(k) =$ unit pulse at $k = 0$	1

Time function	z-transform
10. $x(k) = 1$	$\hat{x}(z) = \dfrac{z}{z-1}$
11. k	$\dfrac{z}{(z-1)^2}$
12. k^2	$\dfrac{z(z+1)}{(z-1)^3}$
13. a^k	$\dfrac{z}{z-a}$
14. a^{k-1}	$\dfrac{1}{z-a}$
15. ka^{k-1}	$\dfrac{z}{(z-a)^2}$
16. k^2a^{k-1}	$\dfrac{z(z+a)}{(z-a)^3}$
17. $\sin \omega k$	$\dfrac{z \sin \omega}{z^2 - 2z \cos \omega + 1}$
18. $\cos \omega k$	$\dfrac{z(z - \cos \omega)}{z^2 - 2z \cos \omega + 1}$

APPENDIX III

SUMMARY OF FACTS FROM MATRIX THEORY

A working knowledge of matrix manipulations has been assumed on the part of the reader. However, some useful identities for common forms in matrix algebra and matrix calculus are summarized in this appendix. This will help the reader refresh his/her memory.

To make the reader conversant with the notations we use, a few definitions and algebraic operations associated with matrices are given below.

1. $\mathbf{A} = [a_{ij}] = \begin{bmatrix} a_{11} & a_{12} & \cdots & a_{1n} \\ a_{21} & a_{22} & \cdots & a_{2n} \\ \vdots & \vdots & & \vdots \\ a_{m1} & a_{m2} & \cdots & a_{mn} \end{bmatrix}$; an $m \times n$ *rectangular matrix.*

2. If $m = n$, **A** is a *square matrix.*

3. $\mathbf{A}^T$ = *transpose* of $\mathbf{A} = [a_{ji}]$; interchange rows and columns.

4. A square matrix **A** is *symmetric* if $\mathbf{A} = \mathbf{A}^T$.

5. A square matrix **A** is *skew-symmetric* if $\mathbf{A} = -\mathbf{A}^T$.

6. A square matrix **A** is *diagonal* if $a_{ij} = 0$; $i \neq j$ and the elements a_{ii} form the principal diagonal. If all $a_{ii} = 0$, the diagonal matrix is the *null matrix* **0**. If all $a_{ii} = 1$, the diagonal matrix is *unit* or *identity* matrix **I**.

7. The *co-factor* of any a_{ij} of the square matrix **A** is $C_{ij} = (-1)^{i+j} \,|\, \mathbf{M}_{ij} \,|$ where *i*th row and *j*th column are deleted to form the reduced matrix $\mathbf{M}_{ij}$. The symbol $|\, \mathbf{M}_{ij} \,|$ (or det $\mathbf{M}_{ij}$) is used for *determinant of* $\mathbf{M}_{ij}$.

8. The *adjoint* of square matrix **A** is

$$\text{adj } \mathbf{A} = \mathbf{A}^+ = [C_{ij}]^T = [C_{ji}]$$

9. The *inverse* of the square matrix **A** (*nonsingular,* $|\, \mathbf{A} \,| \neq 0$) is

$$\mathbf{A}^{-1} = \frac{\mathbf{A}^+}{|\, \mathbf{A} \,|}$$

Special case of 2×2 *matrix*

Suppose that

$$\mathbf{A} = \begin{bmatrix} a_{11} & a_{12} \\ a_{21} & a_{22} \end{bmatrix}, \quad \det \mathbf{A} \neq 0$$

Then

$$\mathbf{A}^{-1} = \begin{bmatrix} \dfrac{a_{22}}{\det \mathbf{A}} & -\dfrac{a_{12}}{\det \mathbf{A}} \\ -\dfrac{a_{21}}{\det \mathbf{A}} & \dfrac{a_{11}}{\det \mathbf{A}} \end{bmatrix}$$

10. The *rank* $\rho(\mathbf{A})$ of a matrix **A** is the dimension of the largest array in **A** with a nonzero determinant.
11. The *trace* of square matrix **A** is

$$\operatorname{tr} \mathbf{A} = \sum_i a_{ii}$$

□

Some useful identities for common forms in matrix algebra are given below. The inverses are assumed to exist and the dimensions are assumed consistent for the required calculations.

1. $(\mathbf{A}^T)^T = \mathbf{A}$
2. $(\mathbf{A}+\mathbf{B})^T = \mathbf{A}^T + \mathbf{B}^T$
3. $(\mathbf{AB})^T = \mathbf{B}^T\mathbf{A}^T$
4. $(\mathbf{A}^{-1})^{-1} = \mathbf{A}$
5. $(\mathbf{A}^T)^{-1} = (\mathbf{A}^{-1})^T$
6. $(\mathbf{AB})^{-1} = \mathbf{B}^{-1}\mathbf{A}^{-1}$
7. $\det \mathbf{AB} = (\det \mathbf{A})(\det \mathbf{B})$
8. $\det \mathbf{B} = k^n \det \mathbf{A}$ if $\mathbf{B} = k\mathbf{A}$; **A** is $n \times n$ matrix, k is scalar
9. $\det \mathbf{A}^T = \det \mathbf{A}$
10. $\det \mathbf{A}^{-1} = \dfrac{1}{\det \mathbf{A}}$
11. $\det \mathbf{P}^{-1}\mathbf{AP} = \det \mathbf{A}$
12. $\operatorname{tr} \mathbf{A}^T = \operatorname{tr} \mathbf{A}$
13. $\operatorname{tr} (\mathbf{A}+\mathbf{B}) = \operatorname{tr} \mathbf{A} + \operatorname{tr} \mathbf{B}$
14. $\operatorname{tr} \mathbf{AB} = \operatorname{tr} \mathbf{BA}$ ($\operatorname{tr} \mathbf{AB} \neq \operatorname{tr} \mathbf{A} \operatorname{tr} \mathbf{B}$)
15. $\operatorname{tr} \mathbf{P}^{-1}\mathbf{AP} = \operatorname{tr} \mathbf{A}$
16. $\rho(\mathbf{A}^T) = \rho(\mathbf{A})$
17. $\rho(\mathbf{A}) \leqslant \min(m, n)$; **A** is $m \times n$ matrix
18. $\rho(\mathbf{AB}) \leqslant \min(\rho(\mathbf{A}), \rho(\mathbf{B}))$
19. $\rho(\mathbf{A}) = n$ if and only if $\det \mathbf{A}^T\mathbf{A} \neq 0$ (for real $m \times n$ matrix **A**)

20. $\rho(\mathbf{A}) = m$ if and only if det $\mathbf{A}\mathbf{A}^T \neq 0$ (for real $m \times n$ matrix $\mathbf{A}$)

21. A square real matrix **A** can be expressed as the sum of a symmetric matrix $\mathbf{A}_1$ and a skew-symmetric matrix $\mathbf{A}_2$.

$$\mathbf{A} = \mathbf{A}_1 + \mathbf{A}_2$$

$$\mathbf{A}_1 = \tfrac{1}{2}(\mathbf{A} + \mathbf{A}^T)$$

$$\mathbf{A}_2 = \tfrac{1}{2}(\mathbf{A} - \mathbf{A}^T)$$

□

The matrices and vectors of the book frequently are functions of real scalar argument t. The operations of differentiation and integration with respect to t are required. Furthermore, scalar and vector functions of vectors occur in system optimization problems where derivatives with respect to components of the vectors are required. The commonly used results of matrix calculus are given below.

1. For $\mathbf{x}(t) = \begin{bmatrix} x_1(t) \\ x_2(t) \\ \vdots \\ x_n(t) \end{bmatrix}$

$$\frac{d\mathbf{x}(t)}{dt} \triangleq \dot{\mathbf{x}}(t) \triangleq \begin{bmatrix} \dot{x}_1(t) \\ \dot{x}_2(t) \\ \vdots \\ \dot{x}_n(t) \end{bmatrix}$$

and

$$\int_0^t \mathbf{x}(\tau)\, d\tau \triangleq \begin{bmatrix} \int_0^t x_1(\tau)\, d\tau \\ \int_0^t x_2(\tau)\, d\tau \\ \vdots \\ \int_0^t x_n(\tau)\, d\tau \end{bmatrix}$$

2. For $\mathbf{A}(t) = \begin{bmatrix} a_{11}(t) & a_{12}(t) & \dots & a_{1n}(t) \\ a_{21}(t) & a_{22}(t) & \dots & a_{2n}(t) \\ \vdots & & & \vdots \\ a_{m1}(t) & a_{m2}(t) & \dots & a_{mn}(t) \end{bmatrix}$

$$\frac{d\mathbf{A}(t)}{dt} \triangleq \dot{\mathbf{A}}(t) \triangleq \begin{bmatrix} \dot{a}_{11}(t) & \dot{a}_{12}(t) & \dots & \dot{a}_{1n}(t) \\ \dot{a}_{21}(t) & \dot{a}_{22}(t) & \dots & \dot{a}_{2n}(t) \\ \vdots & \vdots & & \vdots \\ \dot{a}_{m1}(t) & \dot{a}_{m2}(t) & \dots & \dot{a}_{mn}(t) \end{bmatrix}$$

and

$$\int_0^t \mathbf{A}(\tau)\,d\tau \triangleq \begin{bmatrix} \int_0^t a_{11}(\tau) & \cdots & \int_0^t a_{1n}(\tau)\,d\tau \\ \vdots & & \vdots \\ \int_0^t a_{m1}(\tau)d\tau & \cdots & \int_0^t a_{mn}(\tau)\,d\tau \end{bmatrix}$$

3. For scalar-valued function

$f(\mathbf{x}) = f(x_1, x_2, \ldots, x_n)$,

$$\frac{\partial f(\mathbf{x})}{\partial \mathbf{x}} \triangleq \mathbf{g}(\mathbf{x}) \triangleq \begin{bmatrix} \partial f/\partial x_1 \\ \partial f/\partial x_2 \\ \vdots \\ \partial f/\partial x_n \end{bmatrix}$$

$\mathbf{g}(\mathbf{x})$ is called the *gradient* of $f(\mathbf{x})$.

4. For vector-valued function

$$\mathbf{f}(\mathbf{x}) = \begin{bmatrix} f_1(x_1, x_2, \ldots, x_n) \\ f_2(x_1, x_2, \ldots, x_n) \\ \vdots \\ f_m(x_1, x_2, \ldots, x_n) \end{bmatrix}$$

$$\frac{\partial \mathbf{f}(\mathbf{x})}{\partial \mathbf{x}} \triangleq \mathbf{J}(\mathbf{x}) \triangleq \begin{bmatrix} \frac{\partial f_1}{\partial x_1} & \frac{\partial f_1}{\partial x_2} & \cdots & \frac{\partial f_1}{\partial x_n} \\ \frac{\partial f_2}{\partial x_1} & \frac{\partial f_2}{\partial x_2} & \cdots & \frac{\partial f_2}{\partial x_n} \\ \vdots & \vdots & & \vdots \\ \frac{\partial f_m}{\partial x_1} & \frac{\partial f_m}{\partial x_2} & \cdots & \frac{\partial f_m}{\partial x_n} \end{bmatrix}$$

$\mathbf{J}(\mathbf{x})$ is called the *Jacobian matrix* of $\mathbf{f}$ with respect to $\mathbf{x}$.

5. The second derivative of scalar-valued function

$f(\mathbf{x}) = f(x_1, x_2, \ldots, x_n)$ is

$$\frac{\partial^2 f}{\partial \mathbf{x}^2} \triangleq \mathbf{H}(\mathbf{x}) \triangleq \begin{bmatrix} \frac{\partial^2 f}{\partial x_1^2} & \frac{\partial^2 f}{\partial x_1 \partial x_2} & \cdots & \frac{\partial^2 f}{\partial x_1 \partial x_n} \\ \frac{\partial^2 f}{\partial x_2 \partial x_1} & \frac{\partial^2 f}{\partial x_2^2} & \cdots & \frac{\partial^2 f}{\partial x_2 \partial x_n} \\ \vdots & \vdots & & \vdots \\ \frac{\partial^2 f}{\partial x_n \partial x_1} & \frac{\partial^2 f}{\partial x_n \partial x_2} & \cdots & \frac{\partial^2 f}{\partial x_n^2} \end{bmatrix}$$

$\mathbf{H}(\mathbf{x})$ is called the *Hessian matrix*.

6. For the function $f(x)$; $x \in \mathcal{R}$, $f: \mathcal{R} \to \mathcal{R}$, the Taylor series expansion about x^* is the infinite series

$$f(x) = f(x^*) + \left.\frac{\partial f}{\partial x}\right|_{x=x^*} (x - x^*) + \frac{1}{2\,!} \left.\frac{\partial^2 f}{\partial x^2}\right|_{x=x^*} (x - x^*)^2 + \dots$$
$$+ \frac{1}{k\,!} \left.\frac{\partial^k f}{\partial x^k}\right|_{x=x^*} (x - x^*)^k + \dots$$

Of particular interest are the linear approximation

$$f(x) \simeq f(x^*) + \left.\frac{\partial f}{\partial x}\right|_{x=x^*} (x - x^*)$$

and the quadratic approximation

$$f(x) \simeq f(x^*) + \left.\frac{\partial f}{\partial x}\right|_{x=x^*} (x - x^*) + \tfrac{1}{2} \left.\frac{\partial^2 f}{\partial x^2}\right|_{x=x^*} (x - x^*)^2$$

7. For the function $f(\mathbf{x})$; $\mathbf{x} \in \mathcal{R}^n$, $f: \mathcal{R}^n \to \mathcal{R}$, the Taylor series expansion about $\mathbf{x}^*$ takes the form

$$f(\mathbf{x}) = f(\mathbf{x}^*) + \mathbf{g}^T(\mathbf{x}^*)\,(\mathbf{x} - \mathbf{x}^*) + \tfrac{1}{2}\,(\mathbf{x} - \mathbf{x}^*)^T \mathbf{H}(\mathbf{x}^*)\,(\mathbf{x} - \mathbf{x}^*) + \dots$$

where

$$\mathbf{g}(\mathbf{x}^*) \overset{\Delta}{=} \left.\frac{\partial f}{\partial \mathbf{x}}\right|_{\mathbf{x}=\mathbf{x}^*}$$

and

$$\mathbf{H}(\mathbf{x}^*) \overset{\Delta}{=} \left.\frac{\partial^2 f}{\partial \mathbf{x}^2}\right|_{\mathbf{x}=\mathbf{x}^*}$$

8. For the function $\mathbf{f}(\mathbf{x})$; $\mathbf{x} \in \mathcal{R}^n$, $\mathbf{f}: \mathcal{R}^n \to \mathcal{R}^m$, the Taylor series expansion about $\mathbf{x}^*$ takes the form

$$\mathbf{f}(\mathbf{x}) = \mathbf{f}(\mathbf{x}^*) + \mathbf{J}(\mathbf{x}^*)\,(\mathbf{x} - \mathbf{x}^*)$$

$$+ \tfrac{1}{2} \begin{bmatrix} 1 \\ 0 \\ \vdots \\ 0 \end{bmatrix} (\mathbf{x} - \mathbf{x}^*)^T \mathbf{H}_1(\mathbf{x}^*)\,(\mathbf{x} - \mathbf{x}^*)$$

$$+ \tfrac{1}{2} \begin{bmatrix} 0 \\ 1 \\ 0 \\ \vdots \\ 0 \end{bmatrix} (\mathbf{x} - \mathbf{x}^*)^T \mathbf{H}_2(\mathbf{x}^*)\,(\mathbf{x} - \mathbf{x}^*)$$

$$+ \dots$$

$$+ \tfrac{1}{2} \begin{bmatrix} 0 \\ 0 \\ \vdots \\ 1 \end{bmatrix} (\mathbf{x} - \mathbf{x}^*)^T \mathbf{H}_m(\mathbf{x}^*)\,(\mathbf{x} - \mathbf{x}^*) + \dots$$

where

$$\mathbf{H}_i(\mathbf{x}) = \frac{\partial^2 f_i}{\partial \mathbf{x}^2}$$

9. $\dfrac{d}{dt}(\mathbf{A}(t) + \mathbf{B}(t)) = \dot{\mathbf{A}}(t) + \dot{\mathbf{B}}(t)$

10. $\dfrac{d}{dt}(\mathbf{A}(t)\,\mathbf{B}(t)) = \dot{\mathbf{A}}(t)\,\mathbf{B}(t) + \mathbf{A}(t)\,\dot{\mathbf{B}}(t)$

11. For $\mathbf{f} = \mathbf{f}(\mathbf{x}, t)$ and $\mathbf{x} = \mathbf{x}(t)$

$$\frac{d\mathbf{f}}{dt} = \frac{\partial \mathbf{f}}{\partial t} + \left(\frac{\partial \mathbf{f}}{\partial \mathbf{x}}\right)^T \frac{d\mathbf{x}}{dt}$$

12. For $\mathbf{x} = \mathbf{x}(t)$,

$$\frac{d(\mathbf{x}^T\mathbf{Q}\mathbf{x})}{dt} = \mathbf{x}^T(\mathbf{Q}^T + \mathbf{Q})\dot{\mathbf{x}}$$

13. $\dfrac{d(\mathbf{b}^T\mathbf{x})}{d\mathbf{x}} = \mathbf{b} = \dfrac{d(\mathbf{x}^T\mathbf{b})}{d\mathbf{x}}$

14. $\dfrac{d(\mathbf{A}\mathbf{x})}{d\mathbf{x}} = \mathbf{A}^T, \quad \dfrac{d(\mathbf{x}^T\mathbf{A}^T)}{d\mathbf{x}} = \mathbf{A}$

15. For $\mathbf{f} = \mathbf{f}(\mathbf{x})$ and $\mathbf{g} = \mathbf{g}(\mathbf{x})$,

$$\frac{d(\mathbf{f}^T\mathbf{g})}{d\mathbf{x}} = \frac{\partial \mathbf{f}}{\partial \mathbf{x}}\mathbf{g} + \frac{\partial \mathbf{g}}{\partial \mathbf{x}}\mathbf{f}$$

16. $\dfrac{d(\mathbf{x}^T\mathbf{Q}\mathbf{x})}{d\mathbf{x}} = \mathbf{Q}\mathbf{x} + \mathbf{Q}^T\mathbf{x}$

INDEX